Edited by
Mario Caironi and
Yong-Young Noh

Large Area and Flexible Electronics

Related Titles

Someya, T. (ed.)

Stretchable Electronics

2013
Print ISBN: 978-3-527-32978-6,
also available in digital formats

Brütting, W., Adachi, C. (eds.)

Physics of Organic Semiconductors
2nd Edition

2012
Print ISBN: 978-3-527-41053-8,
also available in digital formats

Klauk, H. (ed.)

Organic Electronics II
More Materials and Applications

2012
Print ISBN: 978-3-527-32647-1,
also available in digital formats

Barquinha, P.P., Martins, R.R., Pereira, L.M., Fortunato, E.E.

Transparent Oxide Electronics - From Materials to Devices

2012
Print ISBN: 978-0-470-68373-6,
also available in digital formats

Chujo, Y. (ed.)

Conjugated Polymer Synthesis
Methods and Reactions

2011
Print ISBN: 978-3-527-32267-1,
also available in digital formats

Marks, T.T., Facchetti, A.A. (eds.)

Transparent Electronics - From Synthesis to Applications

2010
Print ISBN: 978-0-470-99077-3,
also available in digital formats

Leclerc, M., Morin, J. (eds.)

Design and Synthesis of Conjugated Polymers

2010
Print ISBN: 978-3-527-32474-3,
also available in digital formats

Edited by Mario Caironi and Yong-Young Noh

Large Area and Flexible Electronics

WILEY-VCH
Verlag GmbH & Co. KGaA

The Editors

Dr. Mario Caironi
Center for Nano Science and
Technology @PoliMi
Istituto Italiano di Tecnologia
Via Pascoli, 70/3
20133 Milano
Italy

Prof. Yong-Young Noh
Dongguk University
Dept of Energy and Materials Engineering
26, Pil-dong, 3-ga, Jung-gu
100-715 Seoul
Republic of Korea

Library of Congress Card No.: applied for

British Library Cataloguing-in-Publication Data
A catalogue record for this book is available from the British Library.

Bibliographic information published by the Deutsche Nationalbibliothek
The Deutsche Nationalbibliothek lists this publication in the Deutsche Nationalbibliografie; detailed bibliographic data are available on the Internet at <http://dnb.d-nb.de>.

Print ISBN: 978-3-527-33639-5
ePDF ISBN: 978-3-527-68000-9
ePub ISBN: 978-3-527-67999-7
Mobi ISBN: 978-3-527-67998-0
oBook ISBN: 978-3-527-67997-3

Cover Design Adam-Design, Weinheim, Germany
Typesetting Laserwords Private Limited, Chennai, India
Printing and Binding Markono Print Media Pte Ltd., Singapore

Printed on acid-free paper

Contents

List of Contributors

Jong-Hyun Ahn
Yonsei University
School of Electrical and
Electronic Engineering
50 Yonsei-ro
Seodaemun-gu
Seoul, 120-749
Republic of Korea

Kang-Jun Baeg
Nano Carbon Materials Research
Group
Korea Electrotechnology
Research Institute (KERI)
12 Bulmosan-ro 10beon-gil
Seongsan-gu, Changwon
Gyeongsangnam-do, 642-120
Republic of Korea

Alejandro L. Briseno
University of Massachusetts
Department of Polymer Science
and Engineering
Conte Research Center
120 Governors Drive
Amherst
MA 01003
USA

Mario Caironi
Istituto Italiano di Tecnologia
Center for Nano Science and
Technology @PoliMi
Via Pascoli 70/3
20133 Milano
Italy

Zhuoying Chen
ESPCI/CNRS/Université Pierre
et Marie Curie
Laboratoire de Physique et
d'Etude des Matériaux
10 Rue Vauquelin
75005 Paris
France

Jeong Ho Cho
Sungkyunkwan University
Advanced Institute
Nanotechnology (SAINT)
Suwon 440746
Republic of Korea

Nicholas S. Colella
University of Massachusetts
Department of Polymer Science
and Engineering
Conte Research Center
120 Governors Drive
Amherst
MA 01003
USA

Pasquale D'Angelo
IMEM-CNR Institute of
Materials for Electronics and
Magnetism – National Research
Council
Parco Area delle Scienze 37/A
43124 Parma
Italy

Antonio Facchetti
Polyera Corporation
8045 Lamon Avenue
STE 140
Skokie
IL 60077-5318
USA

and

Northwestern University
Department of Chemistry and
the Materials Research Center
2145 Sheridan Road
Evanston
IL 60208-3113
USA

Donggeon Han
Korea Advanced Institute of
Science and Technology (KAIST)
Department of Electrical
Engineering
373-1 Guseong-dong
Daejeon 305-701
Republic of Korea

Steve D. Hoath
University of Cambridge
Department of Engineering
17 Charles Babbage Road
Cambridge, CB3 0FS
UK

Ian M. Hutchings
University of Cambridge
Department of Engineering
17 Charles Babbage Road
Cambridge, CB3 0FS
UK

Salvatore Iannotta
IMEM-CNR Institute of
Materials for Electronics and
Magnetism – National Research
Council
Parco Area delle Scienze 37/A
43124 Parma
Italy

Houk Jang
Yonsei University
School of Electrical and
Electronic Engineering
50 Yonsei-ro
Seodaemun-gu
Seoul, 120-749
Republic of Korea

Sungjune Jung
Pohang University of Science and Technology
Department of Creative IT Engineering
77 Cheongam-Ro
Pohang, 790-784
Republic of Korea

Moon Sung Kang
Soongsil University
Department of Chemical Engineering
Seoul 159-743
Republic of Korea

Seok-Ju Kang
Dongguk University
Department of Energy and Materials Engineering
26 Pil-dong
3 ga, Jung-gu
Seoul 100-715
Republic of Korea

Hoyeon Kim
Korea Advanced Institute of Science and Technology (KAIST)
Department of Electrical Engineering
373-1 Guseong-dong
Daejeon 305-701
Republic of Korea

Min-Soo Kim
Yonsei University
School of Electrical and Electronic Engineering
50 Yonsei-ro
Seodaemun-gu
Seoul, 120-749
Republic of Korea

Se Hyun Kim
Sungkyunkwan University
Advanced Institute
Nanotechnology (SAINT)
Suwon 440746
Republic of Korea

Yong-Hoon Kim
Sungkyunkwan University
School of Advanced Materials Science and Engineering
2066 Seobu-ro
Suwon 440-746
Republic of Korea

and

Sungkyunkwan University
SKKU Advanced Institute of Nanotechnology (SAINT)
2066 Seobu-ro
Suwon 440-746
Republic of Korea

Jongjin Lee
Korea Advanced Institute of Science and Technology (KAIST)
Department of Electrical Engineering
373-1 Guseong-dong
Daejeon 305-701
Republic of Korea

Seung-Hoon Lee
Dongguk University
Department of Energy and
Materials Engineering
26 Pil-dong
3-ga, Jung-gu
Seoul 100-715
Republic of Korea

and

Gwangju Institute of Science and
Technology (GIST)
Department of Nanobio
Materials and Electronics
261 Cheomdan-gwagiro
(Oryong-dong)
Buk-gu
Gwangju 500-712
Republic of Korea

Won-June Lee
Gwangju Institute of Materials
Science and Engineering (GIST)
School of Materials Science and
Engineering
123 Cheomdan-gwagiro
Buk-gu
Gwangju 500-712
Republic of Korea

Wonho Lee
Yonsei University
School of Electrical and
Electronic Engineering
50 Yonsei-ro
Seodaemun-gu
Seoul, 120-749
Republic of Korea

Vladimir Lesnyak
Istituto Italiano di Tecnologia
Department of Nanochemistry
Via Morego 30
16163 Genova
Italy

Hongbo Li
Istituto Italiano di Tecnologia
Department of Nanochemistry
Via Morego 30
16163 Genova
Italy

Chuan Liu
Sun Yat-Sen University
State Key Laboratory of
Optoelectronic Materials and
Technologies
School of Physics and
Engineering
Guangzhou 510274
China

William A. MacDonald
Dupont Teijin Films Ltd.
PO Box 2002
Wilton
Middlesbrough TS10 4RF
UK

Liberato Manna
Istituto Italiano di Tecnologia
Department of Nanochemistry
Via Morego 30
16163 Genova
Italy

Graham D. Martin
University of Cambridge
Department of Engineering
17 Charles Babbage Road
Cambridge, CB3 0FS
UK

Yong-Young Noh
Dongguk University
Department of Energy and Materials Engineering
26 Pil-dong
3-ga, Jung-gu
Seoul 100-715
Republic of Korea

Sung Kyu Park
Chung-Ang University
School of Electrical and Electronics Engineering
84 Heukseok-ro
Seoul 156-756
Republic of Korea

Sungjun Park
Gwangju Institute of Materials Science and Engineering (GIST)
School of Materials Science and Engineering
123 Cheomdan-gwagiro
Buk-gu
Gwangju 500-712
Republic of Korea

Marcos A. Reyes-Martinez
University of Massachusetts
Department of Polymer Science and Engineering
Conte Research Center
Room 120
120 Governors Drive
Amherst
MA 01003
USA

Agostino Romeo
IMEM-CNR Institute of Materials for Electronics and Magnetism – National Research Council
Parco Area delle Scienze 37/A
43124 Parma
Italy

Tsuyoshi Sekitani
Osaka University
The Institute of Scientific and Industrial Research
8-1 Mihogaoka
Ibaraki
Osaka 567-0047
Japan

and

The University of Tokyo
Department of Electrical and Electronic Engineering and Information Systems
School of Engineering
7-3-1 Hongo
Bunkyo-ku
Tokyo 113-8656
Japan

and

Exploratory Research for Advanced Technology (ERATO)
Japan Science and Technology Agency (JST)
2-11-16, Yayoi
Bunkyo-ku
Tokyo 133-0032
Japan

Takao Someya
The University of Tokyo
Department of Electrical and
Electronic Engineering and
Information Systems
School of Engineering
7-3-1 Hongo
Bunkyo-ku
Tokyo 113-8656
Japan

and

Exploratory Research for
Advanced Technology (ERATO)
Japan Science and Technology
Agency (JST)
2-11-16, Yayoi
Bunkyo-ku
Tokyo 133-0032
Japan

Chung-kun Song
Dong-A University
Department of Electronics
Engineering
College of Engineering
Building II, 840 Hadan2-dong
Saha-gu
Busan 604-714
Republic of Korea

Gi-Seong Ryu
Dongguk University
Department of Energy and
Materials Engineering
26 Pil-dong
3-ga, Jung-gu
Seoul 100-715
Republic of Korea

Sujin Sung
Gwangju Institute of Materials
Science and Engineering (GIST)
School of Materials Science and
Engineering
123 Cheomdan-gwagiro
Buk-gu
Gwangju 500-712
Republic of Korea

Taishi Takenobu
Waseda University
Department of Applied Physics
Graduate School of Advanced
Science and Engineering
3-4-1 Ohkubo
Shinjuku-ku
Tokyo 169-8555
Japan

Giuseppe Tarabella
IMEM-CNR Institute of
Materials for Electronics and
Magnetism – National Research
Council
Parco Area delle Scienze 37/A
43124 Parma
Italy

Hakan Usta
Abdullah Gül University
Department of Materials Science
and Nanotechnology
Engineering
Melikgazi
Kayseri 38039
Turkey

Yong Xu
Dongguk University
Department of Energy and
Materials Engineering
26 Pil-dong
3-ga, Jung-gu
Seoul 100-715
Republic of Korea

and

Gwangju Institute of Science and
Technology (GIST)
Department of Nanobio
Materials and Electronics
261 Cheomdan-gwagiro
(Oryong-dong)
Buk-gu
Gwangju 500-712
Republic of Korea

Tomoyuki Yokota
The University of Tokyo
Department of Electrical and
Electronic Engineering and
Information Systems
School of Engineering
7-3-1 Hongo
Bunkyo-ku
Tokyo 113-8656
Japan

and

Exploratory Research for
Advanced Technology (ERATO)
Japan Science and Technology
Agency (JST)
2-11-16, Yayoi
Bunkyo-ku
Tokyo 133-0032
Japan

Seunghyup Yoo
Korea Advanced Institute of
Science and Technology (KAIST)
Department of Electrical
Engineering
373-1 Guseong-dong
Daejeon 305-701
Republic of Korea

Myung-Han Yoon
Gwangju Institute of Materials
Science and Engineering (GIST)
School of Materials Science and
Engineering
123 Cheomdan-gwagiro
Buk-gu
Gwangju 500-712
Republic of Korea

Overview

Mario Caironi and Yong-Young Noh

Flexible and large-area electronics has the potential to mark a further technology revolution in electronics – much like the way the transition from circuits based on discrete components to integrated circuits did – by enabling the pervasive integration of electronic functionalities in all sorts of appliances, their portability, and wearability. Applications are countless: from personal devices (e.g., wearable health monitoring devices) to large-area sensors (e.g., electronic skin, biomedical devices), and smart tagging of products with radio-frequency identification tags. A huge driving force comes from the display industry, with the goal to develop flexible and/or rollable displays deployable on demand to be integrated with portable devices (e.g., smartphones or tablet PC). Conformable large-area sensor arrays (e.g., digital imagers) or chemical and biological sensor arrays also constitute another great opportunity, especially in the case of digital X-ray devices for biomedical and security applications and human healthcare systems. Lightweight, flexible, and easy-to-integrate solar cells are also one of the interesting perspectives opened by this technology, with a view to developing distributed microgeneration sources of different types to serve an information society, necessitating portability, and to provide embedded power supply to low-power, energy-independent, wireless sensor networks.

While this scenario has been considered as a mere vision until only a few years ago, the first examples of nonflat displays have made their appearance, alongside first prototypes of flexible photovoltaic modules and plastic imagers. However, most of these very promising applications are still at the research or development level, mainly because of the immaturity of technologies that would enable their ubiquitous spreading. Given the huge interest, different technologies are being developed to fulfill the requirements of the aforementioned applications. These technologies comprise the assembling of small, discrete components on large, flexible substrates, interconnected by flexible, printed, conductive tracks (Figure 1a). While this may be a suitable approach where area is not a constraint and for small production volumes (e.g., real-time monitoring of vibrations in large constructions such as bridges) [1], *integration* obviously offers a technological advantage for mass applications. Integrated flexible electronics is being developed mainly through two different

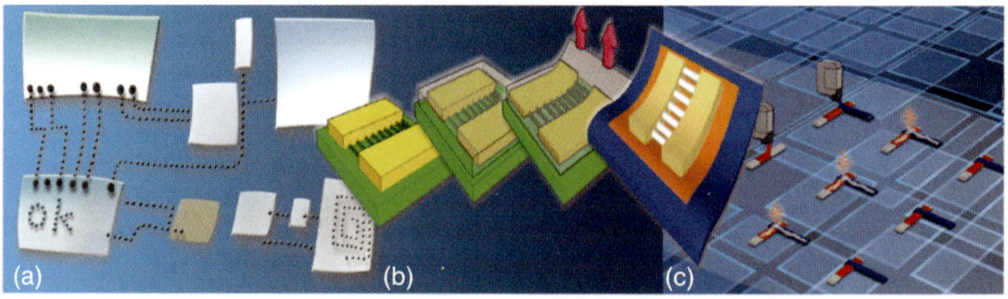

Figure 1 Illustrative representation of the main approaches toward flexible, large-area electronics. (a) Patterning and interconnections of discrete components on foils. (b) Growth of highly ordered structures and their successive transfer onto a flexible substrate. (c) Direct printing of functional inks on flexible substrates.

approaches: (i) the growth followed by sawing or exfoliation of highly ordered crystalline semiconductors and their subsequent controlled transfer onto a flexible substrate (Figure 1b), and (ii) the direct patterning on the substrates of functional materials composing the electronic systems, either via physical vapor deposition methods or by solution-based graphic art printing techniques (Figure 1c).

A widely investigated approach toward large-area flexible electronics is based on organic semiconductors within the framework of π-conjugated molecules [2], a class of materials offering a cost-effective alternative and highly tunable properties through chemical synthesis. Organic semiconductors are very suitable for the above-mentioned applications because they allow easy processing directly on the target substrate with a high degree of freedom of shape, without requiring transfer processes, with different deposition techniques: small molecules can be patterned through vacuum thermal evaporation, a process compatible with roll-to-roll fabrication lines, or, alternatively, small molecules and also polymers can be made soluble by suitable solubilizing side chains. The possibility to treat organic semiconductors in the form of functional inks, enabling a layer-by-layer approach to the realization of complex electronic devices, adds the clear advantage of resorting to low-temperature and high-throughput printing processes. These aspects, together with the superior mechanical properties offered especially by polymers, have made this class of semiconductors among the ones most commonly associated and studied toward the development of flexible electronic applications [3]. This has been a promise of organic electronics for long time. Indeed, the development toward real applications has been progressing much more slowly than predictions. The reasons for this are many, among which we can clearly indicate a technological limitations in the available printing techniques, which were not developed to pattern high-resolution electronic circuits. Another is related to the limited charge mobility of organic semiconductors. Both these problems are now finding very solid answers in the development of printing tools and methodologies enabling finer patterning with high reliability ($<10\,\mu m$)

and in the advanced synthesis of stable and high-performance materials (record mobility values exceeding 10 $cm^2 V^{-1} s^{-1}$).

Organic semiconductors are not the only possible candidates that can be directly processed onto flexible substrates. Many other very promising functional materials are being developed, among which metal oxides represent an important class (10–50 $cm^2 V^{-1} s^{-1}$), being especially applied in the field of high-resolution displays. Indeed, vacuum deposition, through sputtering, is enabling the fabrication of backplanes based on indium gallium zinc oxide (IGZO), which has overtaken hydrogenated amorphous silicon in terms of mobility and stability. Other metal oxides are being developed as well, with intense effort also dedicated to enabling solution processability, where the challenge is reducing decomposition temperatures of soluble precursors to be compatible with processing on plastic substrates [4–6].

Colloidal nanocrystals [7, 8] are also very intriguing because, in principle, they offer both solution processability and highly controlled structures, down to atomic levels. Because of their small size, with the associated large surface-to-volume ratio, their properties are highly dependent on surface states, thereby enabling the tuning of their electronic properties. Investigations on quantum dots have stimulated one of the most important fields in nanoscience and nanotechnology. After decades of continuous efforts by researchers in this field, tremendous progress has been made in the synthesis of quantum dots and their potential applications to optoelectronic devices. The implementation of quantum dots in devices relies not only on their intrinsic properties but also on the collective behavior of the dots in the assemblies, with approaches that make the preparation of quantum dots reproducible, facile, and up-scalable, and on the strategies to tune their surface states and chemical composition with the aim to impart novel properties by applying various post-synthetic treatment procedures.

It is important to note that hybrid solution-processable materials have recently received a lot of interest, with the aim to combine the best features of different materials, such as mechanical strength and processability of polymers, with high performances of 1D or 2D structures. Notably, a new emerging class of solution-processable crystals, namely hybrid metal–organic perovskites [9, 10], which are not treated in this book, has opened up vast opportunities for low-temperature fabrication of efficient solar cells and solution-processable optoelectronic devices.

Another important approach to large-area and flexible electronics is the one based on "nanomembranes," [11, 12] where circuits are developed at the wafer level and subsequently sawed or etched with specific processes in the form of thin films to be laminated on plastic carrier foils. A similar approach, which is usually considered under the same category, is the one based on the growth of high-performance 1D or 2D nanostructures (e.g., 1D crystalline nanowires, nanoribbons, or carbon nanotubes; 2D graphene, transition-metal dichalcogenides, and many others) lately patterned on plastics through different techniques such as transfer printing. These approaches allow achieving very high performances, especially in terms of carrier mobility (>100 $cm^2 V^{-1} s^{-1}$)

in transistors, and are therefore suitable for applications where high switching speeds are required (e.g., wireless electronics).

Besides semiconductors, also relevant materials such as conductors (printable, flexible, and/or stretchable) and insulators (dielectrics, substrates) are required. Among conductors, polymeric materials, such as poly(3,4-ethylenedioxythiophene):polystyrene sulfonate (PEDOT:PSS) and highly conductive metallic inks that can be sintered below 150 °C, are largely available, with the necessity to reduce costs with the challenging development of copper-based inks. Carbon-based conductors, such as those based on graphene, are an emerging opportunity. Regarding dielectrics, the challenge is to develop high-capacitance dielectrics with low leakage, compatible with large-area, flexible applications. Polymeric insulators are good candidates, but they cannot fulfill all these requirements. An interesting alternative is the use of electrolytes, enabling very high specific capacitance through the formation of electric double layers. Above all, high-performance flexible barriers are required to enable stability of applications, especially in the case of organics which are very sensitive to ambient atmosphere.

Book Structure and Aim

The book is divided in two parts. Part A covers materials, which are the key aspect of the field, their physical–chemical properties to provide the fundamentals necessary for a suitable insight, and a critical view point for applications, along with first exemplary applications in electronic devices and circuits. The first six chapters cover semiconductors, Chapters 7 and 8 deal with dielectrics, and Chapters 9 and 10 describe barriers and substrates, respectively. Different approaches for conductors are described case by case in different chapters throughout the book. Part B covers processes and applications, across flexible circuits (Chapters 11–13) and displays (Chapter 15), memories (Chapter 14), solar cells (Chapters 16 and 17), and biosensors (Chapter 18).

A peculiarity of this book is that it puts together areas that are traditionally separated, especially in terms of scientific communities, but all of which tend to the same large-area "macroelectronics" applications from different ends. As we inevitably had to make choices in selecting the topics within this very broad field, our main intention was to select the most exciting examples of approaches and technologies that can really mark the way for new products which will be able to affect in the near future our daily life, hopefully improving it.

The readers will benefit from having collected in a single source valuable information on a very expanding field, as it is the one of large area electronics. They will, first, benefit from receiving an introduction to very different approaches and have the possibility of an in-depth comparison of fundamental properties of different materials, achievable performances, and viable manufacturing methodologies. They will find both fundamental information about materials and an extensive update and description of recently developed devices, with details on the processing techniques. We hope that this book will be a useful reference source for the community of researchers in one of the fields involved in the development of large-area electronics, and also serve researchers in other fields who want to have an introduction to this exciting topic.

Acknowledgments

We are indebted *in primis* to the many authors of the chapters comprised in this book, who accepted to contribute their knowledge and expertise. Many students and young researchers in the Printed and Molecular Electronics (PME) group of the Center for Nano Science and Technology@PoliMi of the Istituto Italiano di Tecnologia, and in the Organic Printed Electronics Laboratory (OPEL) of the Department of Energy and Materials Engineering, Dongguk University, Republic of Korea, eagerly assisted throughout the revision of the book. We acknowledge Martin Preuss and Bernadette Gmeiner from Wiley for their valuable guidance throughout the project. We are also grateful to our respective institutions, Istituto Italiano di Tecnologia (MC) and Dongguk University (YYN) for supporting our scientific and dissemination activity. We sincerely thank Dr. Eun-Young Choi and Dr. Annamaria Petrozza, our respective wives, for their priceless support and understanding.

References

1. Yingzhe, H., Rieutort-Louis, W.S.A., Sanz-Robinson, J., Liechao, H., Glisic, B., Sturm, J.C., Wagner, S., and Verma, N. (2014) *IEEE J. Solid-State Circuits*, **49**, 513.
2. Sirringhaus, H. (2014) *Adv. Mater.*, **26**, 1319.
3. Baeg, K.-J., Caironi, M., and Noh, Y.-Y. (2013) *Adv. Mater.*, **25**, 4210.
4. Kim, Y.-H., Heo, J.-S., Kim, T.-H., Park, S., Yoon, M.-H., Kim, J., Oh, M.S., Yi, G.-R., Noh, Y.-Y., and Park, S.K. (2012) *Nature*, **489**, 128.
5. Banger, K.K., Yamashita, Y., Mori, K., Peterson, R.L., Leedham, T., Rickard, J., and Sirringhaus, H. (2011) *Nat. Mater.*, **10**, 45.
6. Kim, M.-G., Kanatzidis, M.G., Facchetti, A., and Marks, T.J. (2011) *Nat. Mater.*, **10**, 382.
7. Bisri, S.Z., Piliego, C., Yarema, M., Heiss, W., and Loi, M.A. (2013) *Adv. Mater.*, **25**, 4309.
8. Cheng, X., Lowe, S.B., Reece, P.J., and Gooding, J.J. (2014) *Chem. Soc. Rev.*, **43**, 2680.
9. Burschka, J., Pellet, N., Moon, S.-J., Humphry-Baker, R., Gao, P., Nazeeruddin, M.K., and Gratzel, M. (2013) *Nature*, **499**, 316.
10. Ball, J.M., Lee, M.M., Hey, A., and Snaith, H.J. (2013) *Energy Environ. Sci.*, **6**, 1739.
11. Cavallo, F. and Lagally, M.G. (2010) *Soft Matter*, **6**, 439.
12. Kim, D.-H., Xiao, J., Song, J., Huang, Y., and Rogers, J.A. (2010) *Adv. Mater.*, **22**, 2108.

Part I
Materials

1
Polymeric and Small-Molecule Semiconductors for Organic Field-Effect Transistors

Hakan Usta and Antonio Facchetti

1.1
Introduction

π-Conjugated small-molecule and polymer semiconductors are of interest because of their unique optical and electrical properties which will enable the fabrication of new optoelectronic devices having unique functionalities [1–3]. Besides the discovery of new materials, the development of organic-semiconductor-based optoelectronics requires a much better understanding of the electronic structure, charge-transport properties, as well as light–molecule/polymer and charge–charge interactions in the corresponding thin films [4]. Although these aspects are fundamental for the optimization of these materials, the goal in this chapter is to review the very recent achievements in the development of molecular and polymeric semiconductors for charge transport in thin-film transistors (TFTs). In particular, we will first introduce the basic concepts of organic semiconductor structure and organic thin-film transistor (OTFT) operation and then focus on initial studies and very recent works. Excellent review articles are available in the literature for the intermediate period [5, 6].

1.2
Organic Semiconductor Structural Design

Organic semiconductors for OTFTs must possess two essential structural features for their successful implementation in printed electronics (Figure 1.1) [7]. The first is a π-conjugated core/chain composed of linked unsaturated units. The extended π orbitals enable achieving the characteristic charge-transport and optical properties [8]. The second is core functionalization with solubilizing substituents, which is essential for inexpensive manufacture by solution methods as well as for enhancing solid-state core interactions [9]. This latter feature was not met in the

Large Area and Flexible Electronics, First Edition. Edited by Mario Caironi and Yong-Young Noh.

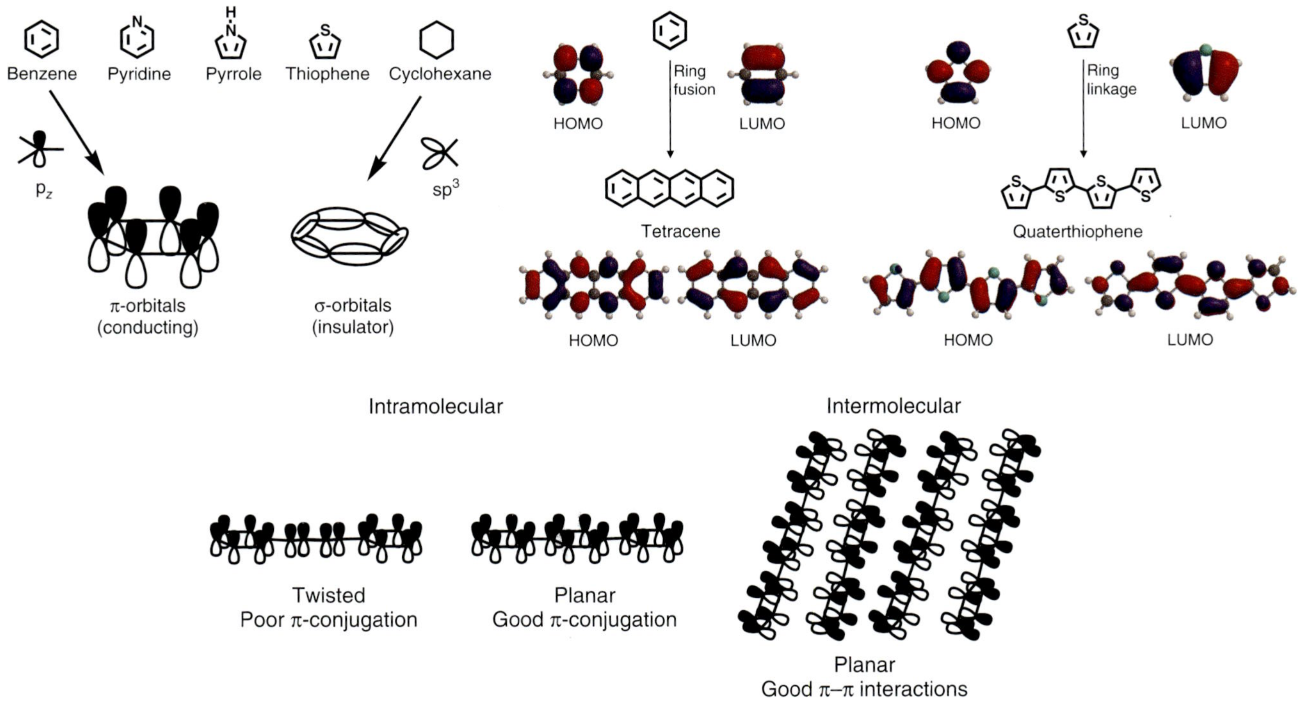

Figure 1.1 Schematic representation of a π-conjugated unit, molecular topologies of the Frontier molecular orbital (FMO), and structural connectivity affording conducting/motifs, twisted/planar connections, and intermolecular interactions.

initial studies, as most of the OTFTs were fabricated with the semiconductor film deposited by vacuum sublimation. Among the most common unsaturated units used for core construction, there are mono(poly)cyclic aromatic hydrocarbons, heterocycles, benzofused systems, and simple olefinic and acetylenic groups. The extent of conjugation/interaction between these units determines the semiconductor solution/solid state electronic structure, which in turn controls the key molecular/polymeric properties such as optical absorption/emission, redox characteristics, and frontier molecular orbital energy levels. Other important architecture parameters relevant to polymers are the molecular weight (M_w) and the polydispersity (PD) index because they influence the solubility, solution aggregation, formulation rheology, and thin-film formation and morphology for both pristine and blended materials. Because when going from low (oligomers) to high (polymer) molecular weights the electronic structure, thermal properties, and microstructure of polymers generally vary considerably, it is important to achieve an M_w/PD regime where a certain property stabilizes so that greater reproducibility of that polymer property can be achieved from batch to batch. This value is likely to be strongly dependent on the polymer structure, but for most soluble thiophene-based polymers, a number-averaged molecular weight value of about 20–30 kDa and a PD of 1.2–1.8 are reasonable for these threshold values [10].

There are several advantages in using polymeric versus molecular π-conjugated semiconductors. Thin films of polymeric materials are generally very smooth and uniform, enabling a great control over a large range of the film structural and morphological characteristics. Printing requires great control of the solution rheological properties, which can be tuned efficiently for polymeric materials. Polymer crystalline domains are typically much smaller than the length scale of several optoelectronic devices, resulting in isotropic transport characteristics. This results in low device-to-device performance variability, which is particularly important for TFT integration into circuits. Furthermore, the fabrication of multilayers from solution deposition processes requires that each stacked layer is inert to the solvents and processing temperatures that it is subsequently exposed to during device manufacture. The reduced solubility parameter window of polymers and their large bulk viscosity typically increase the options to find orthogonal solvents for solution deposition on top of polymer layers, thus expanding the choice of materials that can be used in devices. Finally, because polymers do not vaporize before decomposition and thus have negligible vapor pressure, they are not susceptible to interlayer diffusion during the typical thermal cycles during device fabrication, and typically exhibit robust mechanical properties, making nanometer-thick semiconductor films potentially compatible with roll-to-roll fabrication on flexible substrates. However, during recent years, several new approaches have been developed to improve small-molecule processability from solutions, including the use of spin coating, slot dye coating, and blade coating.

1.3 Thin-Film Transistor Applications

OTFTs are a low-cost technology alternative to amorphous hydrogenated silicon transistors for applications in large-area OTFT-based arrays, for example, backplane/driver circuits for active matrix displays, where high transistor density and switching speeds are not necessary. They may also be attractive for applications in low-end microelectronics (e.g., radio frequency identification tags and sensors), where the high cost of packaging conventional Si circuits is prohibitive for everyday items [11]. The advantages of OTFTs stems from the potential lower manufacturing costs and reduced capital investments thanks to device fabrication using common solution-based deposition and patterning techniques such as offset, gravure, screen/stencil, and inkjet printing, to cite just a few. Furthermore, OTFT-based circuits based on conjugated polymers are compatible with plastic substrates so that compact, lightweight, and structurally robust and flexible electronic devices can be fabricated.

Figure 1.2 shows a schematic structure of a bottom-gate top-contact OTFT. Note that several other device architectures can be fabricated depending on the relative position of the contacts and the dielectric/semiconductor layers (not shown). An OTFT is composed of three electrodes (source, drain, and gate), a gate dielectric layer, and an organic or polymer semiconductor layer. In this device, negligible source–drain current ($I_{SD} = 0$ A) flows when the gate voltage

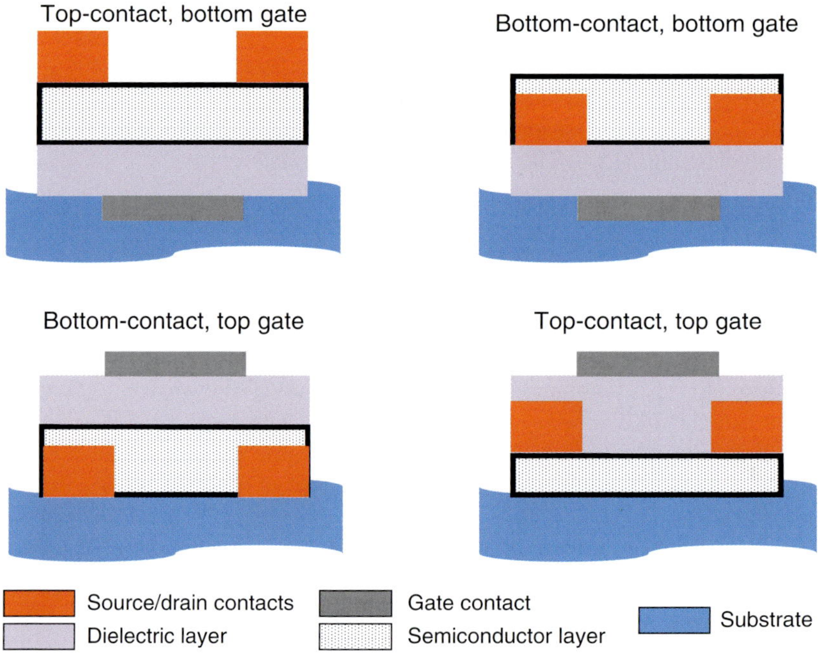

Figure 1.2 Structure and materials of four thin-film transistor architectures.

is zero ($V_G = 0$ V) independent of the bias applied between the source and the drain contacts (V_{SD}). The device turns on ($I_{SD} \neq 0$ A) when a gate field is applied ($V_G \neq 0$ V), which induces charge carriers in the semiconductor at the interface with the dielectric layer. The transistor's performance is evaluated from the output and transfer current–voltage plots, from which critical parameters such as the field-effect mobility (μ), current on/off ratio (I_{on}/I_{off}), threshold voltage (V_T), and subthreshold swing (S) are measured (Figure 1.3).

Within the metal–oxide–semiconductor field-effect transistor (MOSFET) gradual channel model approximation, the carrier mobility in the linear and saturation regimes can be extracted from the following standard MOSFET equations:

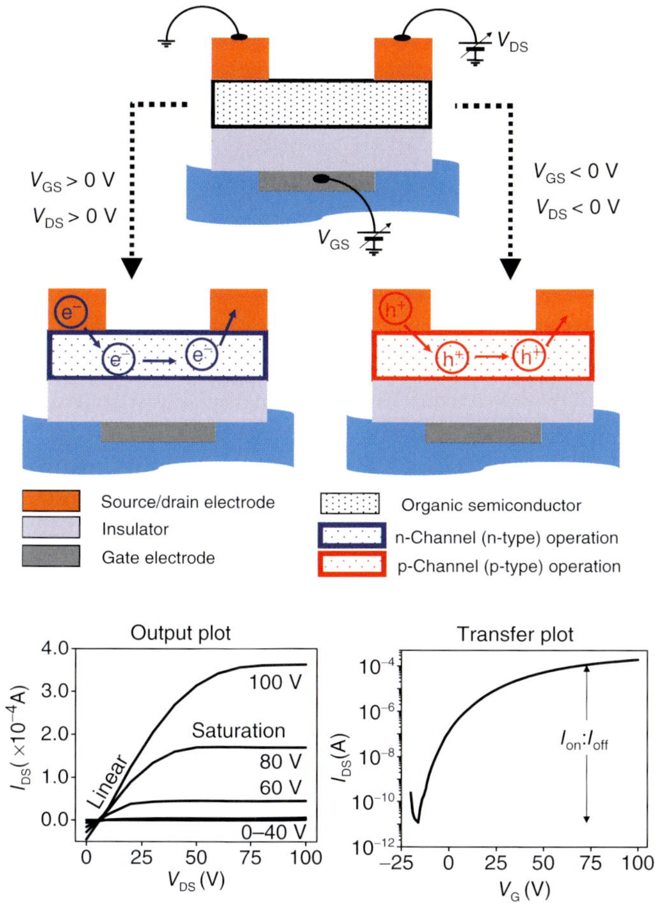

Figure 1.3 (a) Top: Schematic representation of electron and hole transport in bottom-gate top-contact TFTs. Bottom: Output plot of the source–drain current versus the source–drain voltage at a given V_G and (b) transfer plot of the source–drain current versus the gate voltage at different V_{SD}'s.

$$(I_{SD})_{lin} = \left(\frac{W}{L}\right) \mu_{FET} C_i \left(\frac{V_{SG} - V_T - V_{SD}}{2}\right) V_{SD} \tag{1.1}$$

$$(I_{SD})_{sat} = \left(\frac{W}{2L}\right) \mu_{FET} C_i (V_{SG} - V_T)^2 \tag{1.2}$$

where V_{SD} is the drain voltage with the source electrode grounded. W and L are the transistor channel width and length, respectively, and C_i is the capacitance per unit area of the dielectric layer. Polymeric semiconductors for TFTs can be divided into three classes depending on whether the majority charge carriers are holes (p-channel), electrons (n-channel), or both (ambipolar) under different gate bias conditions. Until recently, polymeric semiconductors for TFTs were particularly challenging because most of them were difficult to synthesize/reproduce in optimal batches, poorly soluble, or very sensitive to ambient conditions, and/or exhibited poor charge-carrier mobility. Insoluble polymers preclude their use with solution deposition techniques, while air sensitivity requires manufacturing in cost-prohibitive inert atmosphere. However, during the last 3 years, a few polymeric semiconductors have been found to exhibit the combination of all required properties, enabling the realization of OTFT meeting first-generation market products.

1.4 p-Channel Semiconductors

The most performing semiconductors for OFETs (organic field-effect transistors) are p-channel devices. Besides the basic structural requirements discussed previously, the common strategy to obtain good hole-transporting semiconductors for TFTs is via a delicate balance of the HOMO (highest occupied molecular orbital) energy level which should lie somewhere around −5 to −5.5 eV. When E_{HOMO} is too high (low ionization potential (IP)), easy oxidation by air and acceptor sites dramatically compromises the TFT's ambient stability and current on–off ratio. For $E_{HOMO} \ll -5.5$ eV (high IP), very large I_{on}:I_{off} and good mobilities can be achieved, but these devices typically exhibit unacceptably large threshold voltages. In this section, we will first summarize very briefly the key historical p-channel semiconductors (Figure 1.4), and then move to more recent studies regarding the development of polymers and small molecules.

1.4.1 Polymers

Poly(3-substitutedthiophene)s are one of the most studied polymer families for (semi)conductor/optical applications [8b, 12]. These systems have the advantage that the presence of the 3-substituent strongly enhances solution processability and thin-film-forming capacities. The performance of poly(3-alkylthiophene)-based OFETs, particularly Poly(3-hexylthiophene) (P3HT) (**P1**), has been

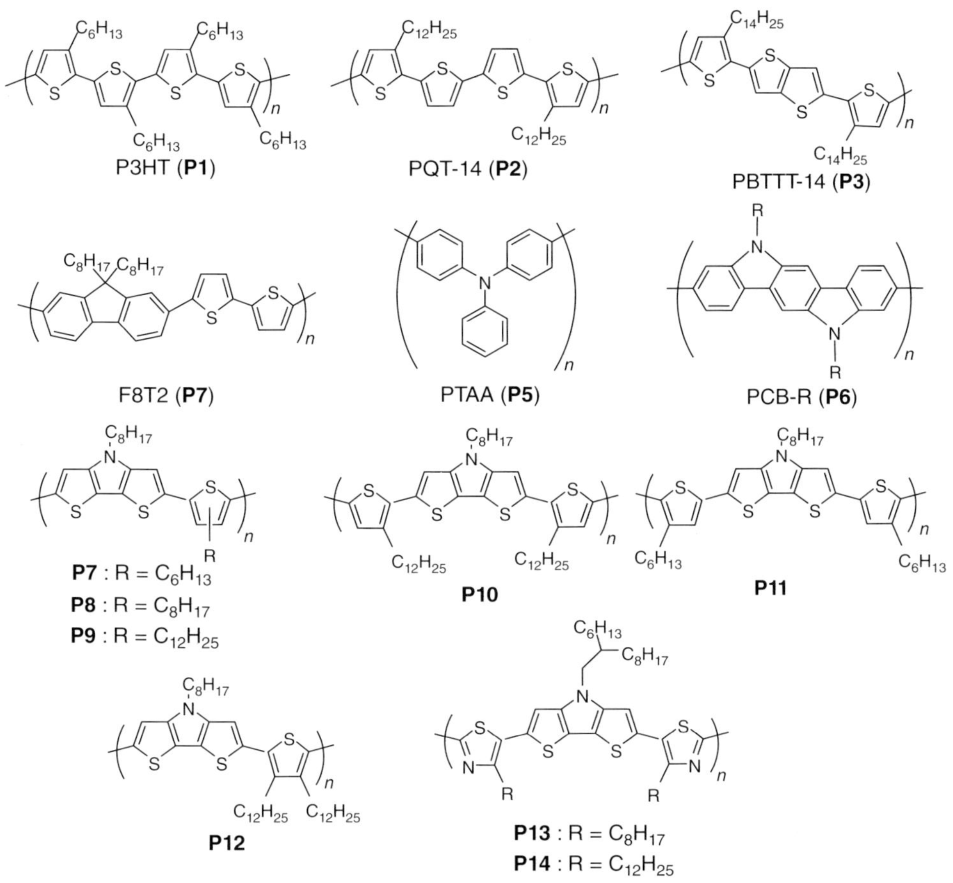

Figure 1.4 Chemical structure of the p-channel semiconductors **P1–P14**.

investigated by various groups considering the effects of the molecular weight of Poly(3-alkylthiophene) (P3AT) [13–15], film deposition solvent [16], film morphology [17, 18], film thickness [19], and fabrication process [20], as well as humidity [21] and length of the core substituent (alkyl chain) [22]. These studies have deepened our understandings of the charge-transport properties of polymeric semiconductors as a whole. However, it was challenging to achieve good control over the regioregularity in several poly(3-alkylthiophene) synthesis, and exposure of P3AT films to air usually caused an increase in the carrier density, thus degrading the transistor's I_{on}/I_{off} ratio. Therefore, high I_{on}:I_{off} ratios for P3AT are consistently achieved by preparing and testing the devices in dry N_2 [23]. To address these problems, new polythiophenes have been developed. Ong *et al.* reported a class of solution-processable regioregular polyquaterthiophenes (PQTs) that affords excellent FET performance under ambient conditions. This class of polythiophenes was designed on the basis of the following structural considerations: (i) presence long alkyl side chains for solution processability; (ii) structural regularity to induce and facilitate molecular self-assembly; and (iii) control of the π-conjugation extension to achieve a delicate balance between transistor functionality and oxidative doping stability. These semiconductors exhibit unique self-assembly ability and form highly structured thin films when deposited from solution under appropriate conditions. FETs fabricated in air with PQT-12 (**P2**, Figure 1.4) channel layers have provided high field-effect mobility of up to 0.14 $cm^2\ V^{-1}\ s^{-1}$ and high I_{on}/I_{off} ratios of $>10^7$ [24]. Chabinyc *et al.* investigated the effects of humidity on unencapsulated **P2**-based OFETs. The field-effect mobility of **P2** TFTs decreases and the rate of trapping of charge carriers increases under increasing humidity [25], pointing out that not only n-channel semiconductors are H_2O sensitive.

Instead of increasing the IP of polythiophene by sterically twisting the repeat units in the backbone, McCulloch *et al.* [26, 27] synthesized polymers incorporating thieno[3,2-*b*]thiophene (**P3**) as co-monomers. The delocalization of electrons from the fused aromatic units into the backbone is less favorable than from a single thiophene ring, which is due to the larger resonance stabilization energy of the fused ring over the single thiophene ring. The decreased delocalization along the backbone lowers the polymer HOMO level. The charge-carrier mobilities of **P3** are high, with values of 0.2–0.6 $cm^2\ V^{-1}\ s^{-1}$ obtained on annealed devices in a nitrogen atmosphere and up to 0.7 $cm^2\ V^{-1}\ s^{-1}$ for devices of 5 μm channel length. DeLongchamp *et al.* [28] investigated in detail the origin of the enhanced crystallinity of this polymer. From first-principles energy minimization using density functional theory (DFT), the authors predicted that the conjugated polymeric planes of **P3** are substantially tilted within their crystalline lamellae. This prediction was confirmed experimentally using a combination of polarized photon spectroscopies including near edge X-ray absorption fine structure (NEXAFS) and IR. Other important first-generation polymeric semiconductor families are those based on 9,9-dialkylfluorene-*alt*-bithiophene (e.g., F8T2, **P4**) [29, 30], triarylamine (**P5**) [31], and carbazole (e.g., Polycarbazole (PCB), **P6**) [32, 33] units (Figure 1.4).

During the last 3 years, several new structures, particularly based on fused units, have been developed, resulting in even greater p-channel TFT performances (Figure 1.5). Rasmussen and coworkers [34] pioneered the use of *N*-alkyldithieno[3,2-*b*:2′,3′-*d*]pyrrole (DTP) as a very promising fused aromatic building block for electronic materials. This core exhibits a completely flat crystal structure, indicating good π conjugation across the fused rings. Upon polymerization, poly(*N*-alkyl dithieno[3,2-*b*:2′,3′-*d*]pyrrole)s exhibit excellent stability in their oxidized state, have low band gaps, and show efficient red fluorescence in solution [35]. However, some Poly(dithieno[3,2-b:2′,3′-d]pyrrole) (PDTPs) have low solubilities and low molecular weights, which greatly limit their use in devices. To improve the solubility of PDTPs and to create a number of DTP-based copolymers, McCullough *et al.* [36] described the synthesis, characterization, electrical conductivity, and field-effect mobility of a series of novel soluble *N*-alkyl DTP–thiophene copolymers (**P7–P12**, Figure 1.5) and

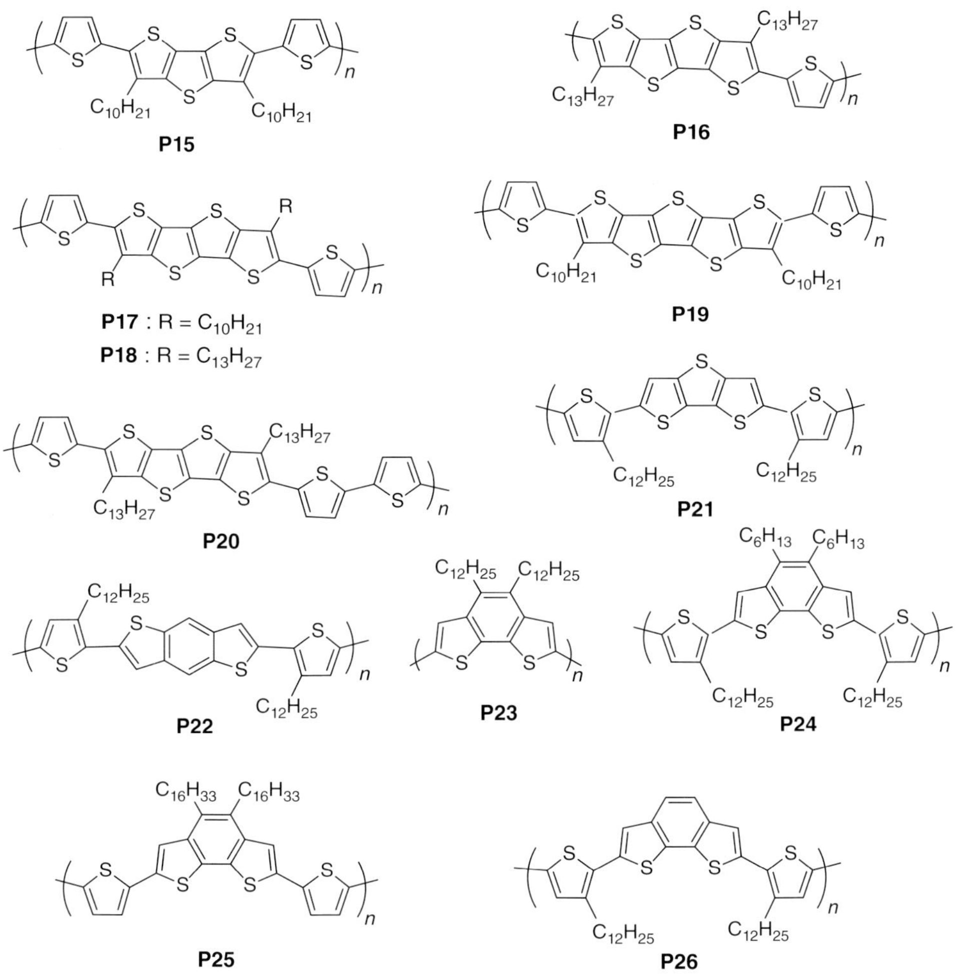

Figure 1.5 Chemical structure of fused thiophene-based p-channel semiconductors **P15–P26**.

compared them with the well-studied regioregular P3HT. These polymers were synthesized using a Stille coupling reaction, and exhibited molecular weights of 10–50 kDa. The incorporation of planar DTP units extends the π conjugation, and the introduction of thiophene subunits imparts good solubility, high conductivity, and high charge-carrier mobility. Optical characterization revealed that the bandgaps of **P7–P12** were between 1.74 and 2.00 eV, which is lower than the value for regioregular poly(3-alkylthiophenes), and the HOMO energy levels were between −4.68 and −4.96 eV. When doped, these polymers exhibited high conductivities up to 230 S cm^{-1} with excellent stability. The microstructure and surface morphologies of, for instance, poly(2-(4,4′-didodecyl-2,2′-bithiophen-5-yl)-4-octyl-4*H*-bisthieno[3,2-*b*:2′,3′-*d*]pyrrole) (**P10**) thin films were studied by X-ray diffraction and atomic force microscopy (AFM). The as-cast **P10** thin films exhibited poorly defined, randomly ordered lamellar structures, which improved significantly after thermal annealing (Figure 1.6). FET devices showed typical p-channel transistor behavior. Interestingly, the mobilities of the as-cast, less ordered samples were much higher than those observed after annealing. The highest values of the maximum and average mobilities were observed for as-cast **P10** (0.21 and 0.13 $cm^2 V^{-1} s^{-1}$, respectively). The authors' goals were to test the idea that high mobility and excellent electrical and structural reproducibility could be achieved in amorphous π-conjugated materials that could possess long-range π connectivity on the microscopic scale.

To improve the air stability of these materials, the authors included electron-deficient units into the polymer backbone to increase the IP [37]. To this end, the same group used the electron-deficient bithiazole (Tz) unit, which is known to increase IP [38]. The new polymers **P13** and **P14** (Figure 1.4), defined as "transistor paints," achieved excellent FET performance with hole mobilities as high as 0.14 and 0.10 $cm^2 V^{-1} s^{-1}$ and current on/off ratios up to 10^6 without post-deposition thermal annealing. Furthermore, these devices exhibited excellent air stability, showing no significant degradation over 60 days.

He, Malliaras, and coworkers [39] recently reported a family of fused-ring thiophene copolymers (**P15–P20**) as materials of interest for TFT applications. In an initial study, a comparison of the properties of **P15** to **P17** showed that the polymer with the even-numbered fused-thiophene core exhibited a much smaller lamellar spacing than the polymers featuring odd-numbered fused-thiophene cores. As a result, transistors fabricated from the polymer with the even-numbered fused-thiophene core (**P16**) yielded a much higher field-effect mobility than the other two (**P15** and **P17**). To obtain further insight into the structure–property relationships in these polymeric semiconductors, more recently they included additional polymers, which were used to elucidate the role of symmetry of the polymer repeat unit on structure and device performance (Figure 1.7). Devices made from the polymer with four fused rings (**P17**) showed a hole mobility of 0.087 $cm^2 V^{-1} s^{-1}$, whereas devices made from the polymers with three (**P15**) and five fused rings (**P19**) showed hole mobilities of 0.0017 and 0.0023 $cm^2 V^{-1} s^{-1}$, respectively. Devices made from **P16** and **P20** showed a hole mobility of 0.042 and 0.022 $cm^2 V^{-1} s^{-1}$, respectively, which is ~10 times lower than that achieved with **P18** (0.33 $cm^2 V^{-1} s^{-1}$) [40].

These results strongly suggest a correlation between a repeat unit that possesses a C_2 axis perpendicular to the conjugation plane, a minimum attainable lamellar spacing, and high field-effect mobility. Ong and coworkers [41] reported a new dithienothiophene-based copolymer, **P21**, which is structurally similar to **P15** and shows mobilities of ~0.01 $cm^2 V^{-1} s^{-1}$, corroborating these symmetry-driven trends.

Mullen *et al.* developed benzo[2,1-*b*;3,4-*b*0]dithiophene-containing homo- and copolymers (**P23–P26**) having solubilizing alkyl chains attached to the benzo unit [42] as an alternative to the classic Benzo[1,2-b:4,5-b']dithiophene (TBT)-based copolymer **P22** [43]. The homopolymer **P23** is very soluble in dichlorobenzene,

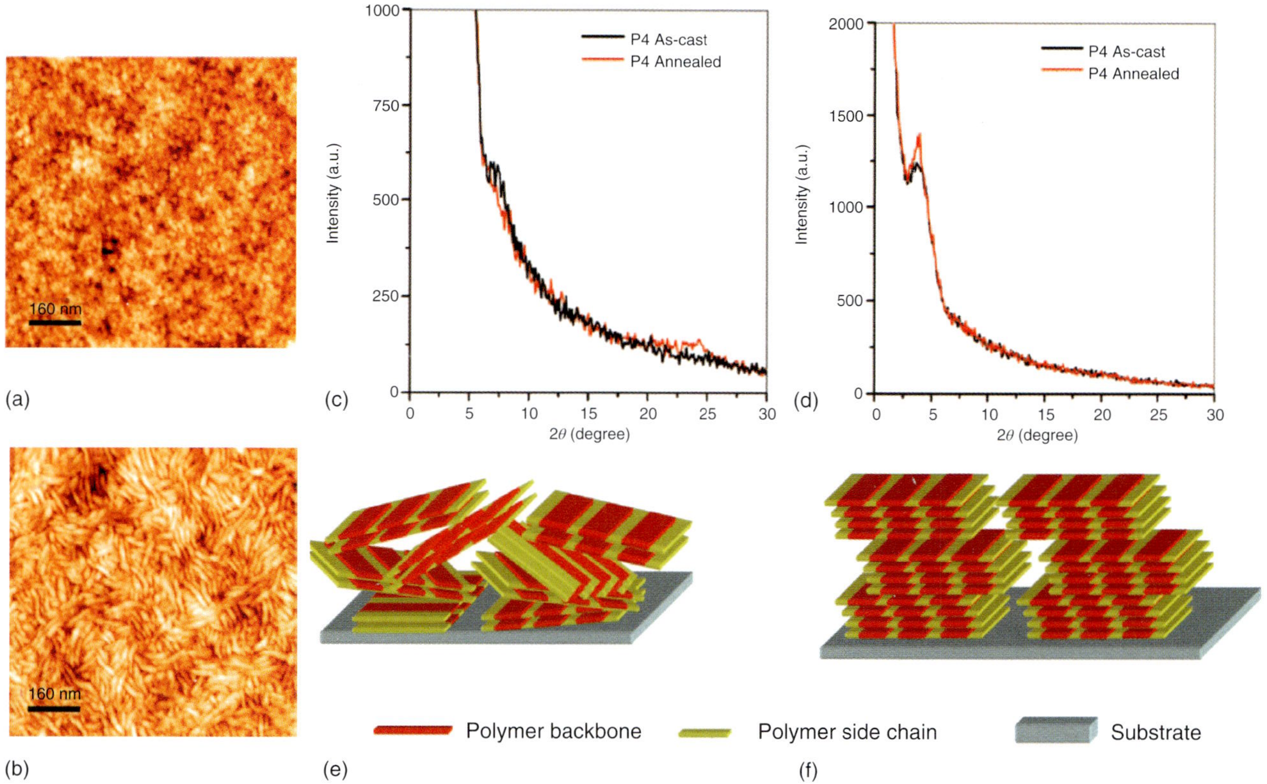

Figure 1.6 Tapping-mode AFM height images of **P10** drop-cast on octadecyltrichlorosilane (OTS)-treated SiO_2/Si surfaces (a) as-cast and (b) annealed at 120 °C for 30 min and then cooled to room temperature in a vacuum. X-ray Diffraction (XRD) profile of **P10** thin films on an OTS-treated SiO_2/Si substrate as cast (black trace) and after annealing at 120 °C for 30 min (red trace). (c) Out-of-plane XRD profile. (d) In-plane XRD profile. Scheme of molecular packing of **P10** (e) as cast and (f) after annealing at 120 °C for 30 min [36]. (Reproduced with permission. Copyright © 2008 American Chemical Society.)

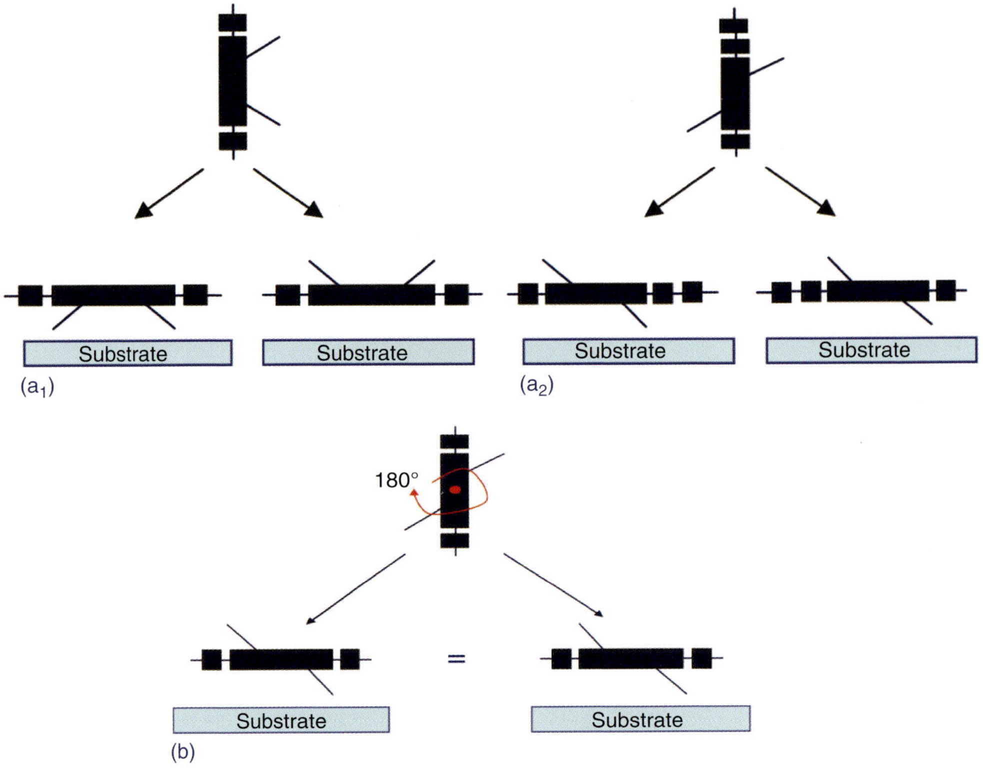

Figure 1.7 Possible configurations of the repeat unit with respect to the substrate, for repeat units (a_1 and a_2) without and (b) with C_2 symmetry. Judging from the family of closely related conjugated polymers that we have investigated, C_2 symmetry seems to enable a very small lamellar period and promote high mobility [40]. (Reproduced with permission. Copyright © 2010 American Chemical Society.)

and top-contact TFTs exhibit low charge-carrier mobility of $\sim 10^{-4}$ cm^2 V^{-1} s^{-1}, probably because the polymer backbone is too stiff. The other copolymers exhibit greater performance ranging from 0.001 to >0.1 cm^2 V^{-1} s^{-1} for bottom-gate top-contact transistors. Top-gate devices based on **P26** on a poly(ethylene terephthalate) (PET) film were also fabricated, which exhibited exceptionally large carrier mobility of >0.5 cm^2 V^{-1} s^{-1}.

Among the many approaches to increasing air stability, the incorporation of unsubstituted conjugated moieties in poly(alkylthiophene) backbones resulted in much better oxidative stability and hole mobility, as demonstrated in PQT (**P2**) and Poly(2,5p-bis(thiophenp-2yl)thieno(3,2p-b)thiophene) (PBTTT) (**P3**). These unsubstituted conjugated moieties possess rotational freedom, which curtails the effective conjugation length, lowers the HOMO level, and consequently increases oxidative stability. Unsubstituted thiophene moieties, along with lengthened alkyl side chains, play also another important role by promoting favorable interdigitation of the side chains. This leads to well-organized intermolecular

3D ordering and large crystalline domains, and consequently high mobility. Thienylenevinylene derivatives, a combination of thiophene and vinyl groups, are known to have an extended conjugated system, which is a crucial component for building organic electronic devices [44]. In addition, the incorporation of vinylene bonds in an aromatic polymer backbone leads to an increase in the degree of coplanarity of the polymer backbone, as the vinylene bond reduces steric hindrance on successive aromatic rings [45]. Recently Kim *et al.* reported high-performance OTFTs with a new thienylenevinylene polymer **P27** (Figure 1.8). An unsubstituted dithienylethylene unit is symmetrically inserted between dodecylthiophenes, thereby allowing some rotational freedom between the alkyl thiophene and unsubstituted dithienylethylene units, which lowered the HOMO level. **P27**-based OTFTs showed unoptimized charge-carrier mobility of $0.15\ cm^2\ V^{-1}\ s^{-1}$ with a relatively high oxidative stability.

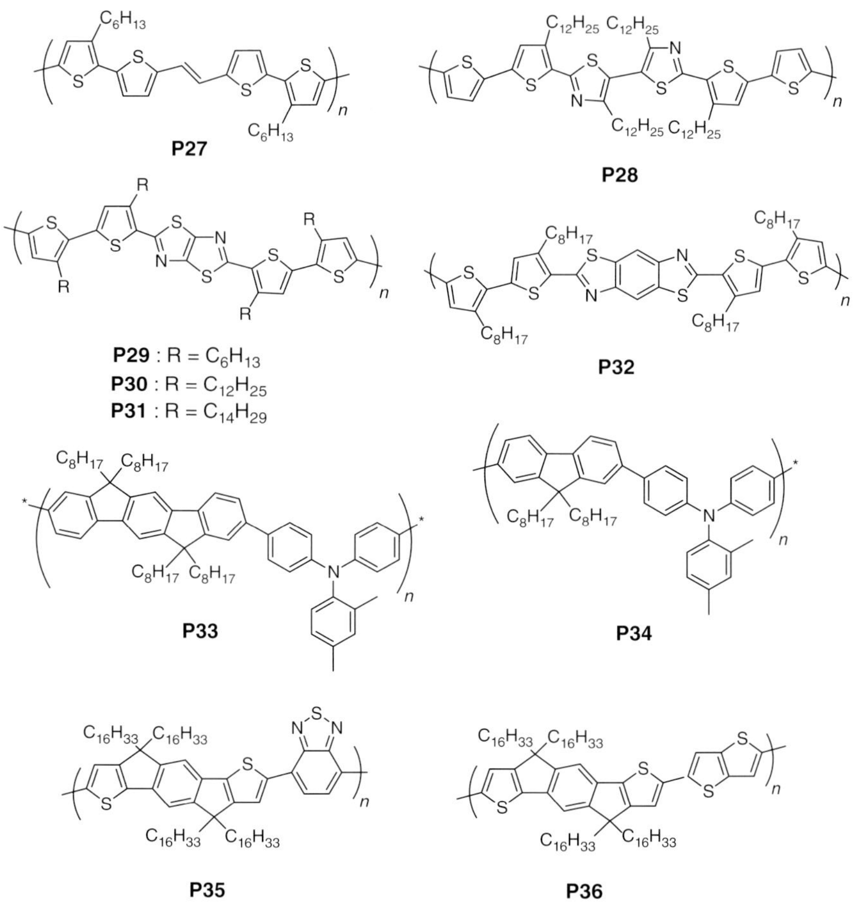

Figure 1.8 Chemical structure of p-channel semiconductors **P27–P36**.

Some interesting structures incorporating the thiazole ring have been reported, mainly with the goal to increase the oxidative stability. Another common problem with OFETs is the electrical instability under external bias stress, likely due to charge traps created through partial disorder in the structure of the thin films and the chemical characteristics of the semiconductor/insulator interface. Bias-stress instability and environmental instability can be significant challenges for semiconducting polymers [46]. To be usable in conventional electronics, OTFTs should exhibit similar characteristics with respect to electrical bias stress [47]. Although there have been a few studies aimed at enhancing the electrical stability of π-conjugated polymers under external bias stress, an adequate understanding of the relationship between the crystalline nanostructure and the bias-stress-driven electrical instability on the microscopic scale is still needed. Lee and coworkers [48] have initiated an in-depth and systematic study addressing the bias stress in polymers using highly ordered π-conjugated copolymer thin films with minimal concentration of charge traps. They used a new thiazole polymer, **P28** (Figure 1.8), having alkyl chain-substituted thiophene/thiazole blocks along the polymer backbone to increase the IP because of the enhanced rotational freedom along the backbone [27] and the electron-accepting nature of the 5,5′-bithiazole units [37b, 49, 49a,b]. In particular, **P28** presents a liquid-crystalline nature with a clear mesophase region, resulting in highly crystalline thin films forming spontaneously through the self-assembly of individual chains after thermal annealing. Most importantly, this novel polymer exhibits high field-effect mobilities of 0.33 $cm^2 V^{-1} s^{-1}$, good environmental stability, and unprecedented bias-stress stability comparable to that of amorphous silicon.

McCullough *et al.* [50] reported copolymers (**P29**–**P31**, Figure 1.8) incorporating a fused thiazolothiazole ring in the backbone. Despite the low molecular weights of this family (M_n = 4–9 kDa), they exhibit field-effect mobilities from 0.02 to 0.3 $cm^2 V^{-1} s^{-1}$ with high current on/off ratios of ~10^6. The use of thiazolothiazole-fused ring ensures a very rigid and coplanar backbone and thereby highly extended π-electron conjugation and strong π stacking. The electron-deficient nature of the thiazolothiazole affords high oxidative stability. However, the uneven placement of the alkyl side chains along the backbone reduced interdigitation and promoted amorphous-like π stacking and π connectivity along the chain while enhancing solubility. With regard to the side-chain arrangement, the key difference between the **P29**–**P31** family and PBTTT/PQT polymers is that, while the side chains in the former polymers are still arranged regiosymmetrically, they are not equally spaced along the backbone. Thus, as a consequence, the side chains are apparently disordered and do not interdigitate as seen in PBTTT or PQT (Figure 1.9). Most interestingly, despite all these factors, X-ray diffraction patterns of **P29** indicate very strong lamellar ordering, and FETs fabricated from these materials show high field-effect mobilities. All these results appear quite surprising in view of the previously mentioned findings on the role of side-chain interdigitation/ordering and indicate that strong interdigitation and formation of extended regular terrace-like structures are not necessary for high carrier mobility in polythiophene-like materials.

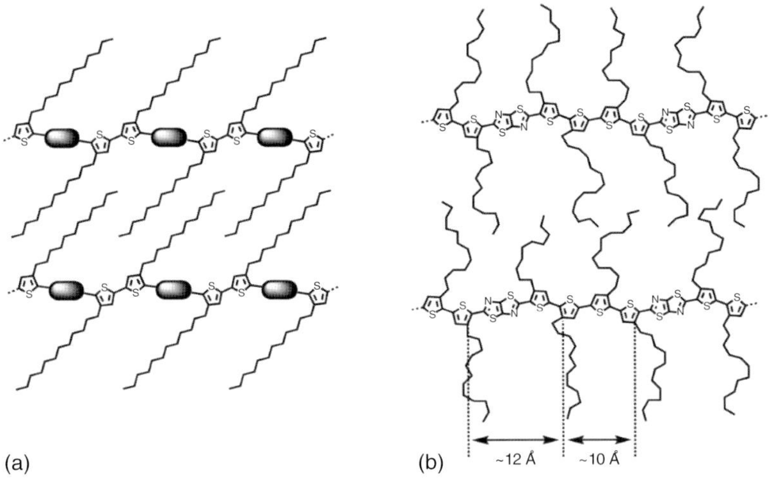

Figure 1.9 Schematic illustrations for (a) the packing structure of PQT (**P2**) and PBTTT (**P3**) with dodecyl side chain with uniform side-chain interdigitation and (b) proposed packing structure for **P30** with disordered side chains [50]. (Reproduced with permission. Copyright © 2009 American Chemical Society.)

Jenekhe and coworkers have reported an interesting polymer, **P32** (Figure 1.8), designed with the knowledge that benzobisthiazole and benzobisoxazole polymers and small molecules exhibit efficient π stacking and strong intermolecular interactions in the solid state [51] leading to high-temperature resistance with glass-transition temperatures that can exceed 300–400 °C and relatively high electron affinity [52]. Earlier studies of a benzobisthiazole polymer as an n-channel semiconductor in FETs had found a low mobility of electrons, requiring a high-electron-affinity polymer in a blend to achieve electron injection [53]. Recently, TFTs based on benzobisthiazole small molecules were shown to exhibit high field-effect mobilities for both holes and electrons (see below) [54]. The authors reported a new soluble benzobisthiazole-thiophene copolymer based on alternating benzobisthiazole and oligo-3-octylthiophene units in the backbone (**P32**), resulting in improved oxidative stability, thermal stability, interchain interactions, and thus enhanced charge-transport properties of the polymers. The highly crystalline **P32** thin films exhibit a field-effect carrier mobility of up to $0.01\,cm^2\,V^{-1}\,s^{-1}$.

Zhang and coworkers reported two classes of polymers based on fluorene and indenofluorene copolymerized with triarylamine (**P33** and **P34**, Figure 1.8) [55] and indacenodithiophene copolymerized with benzothiadiazole and thienothiophene (**P35** and **P36**) [56]. Bottom-contact, top-gate (BC–TG) (and bottom-gate) architecture FET devices based on **P33** and **P34** were fabricated with the polymer semiconductors deposited from solution. Compared to the best triarylamine homopolymers [46c] (mobility of $\sim 4 \times 10^{-3}\,cm^2\,V^{-1}\,s^{-1}$) the new polymers exhibited improved mobility by a factor of 5–$0.02\,cm^2\,V^{-1}\,s^{-1}$ by the introduction of the fluorene unit, which further increased to $0.04\,cm^2\,V^{-1}\,s^{-1}$ for the indenofluorene copolymer and current on/off ratios of $>10^6$. It is speculated that the increase in the polymer backbone planarity and persistence length in the copolymers improved the intramolecular π-orbital overlap and enhanced the local structural organization, resulting in the large measured mobilities. No evidence of thin-film crystallinity could be observed for **P33** and **P34** polymer semiconductors. For polymers **P35** and **P36** (Figure 1.8), the strategy for further improvement in charge-carrier mobility in comparison to **P33** and **P34** was to enhance the planarity of the backbone and further reduce the energetic disorder of the polymer. The aryl amine unit was replaced with more planar Benzothiadiazole (BTD) and Thienothiophene (TT) units. Grazing incidence X-ray scattering (GIXS) experiments were carried out on annealed thin films of both polymers on Si substrates to explore the microstructure, and **P35** polymer was observed to be semicrystalline while **P36** thin film was amorphous. BC–TG) architecture FET devices were fabricated with the polymer semiconductors spin-cast from a $10\,mg\,ml^{-1}$ chlorobenzene solution at 2000 rpm followed by an annealing step at 100 °C for 5 min in nitrogen. **P35** transistors yield maximum hole mobilities in the range $0.8-1.2\,cm^2\,V^{-1}\,s^{-1}$, with a current on/off ratio of $\sim 10^4$ and a threshold voltage of ~ -30 V. Despite the very high mobility values, FET operation is heavily injection-limited. The **P36** copolymer exhibits a lower mobility ($\sim 0.2\,cm^2\,V^{-1}\,s^{-1}$), which is attributed to the more amorphous nature of the thin-film microstructure.

Inspired by the work of Marks *et al.* [57] on silole-based copolymers for TFTs, Reynolds reported copolymers of dithienosilole with BTD, following the donor–acceptor concept to broaden the optical absorption (Figure 1.10) [58]. Dithienosilole (DTS)-BTD copolymers **P37–P40**, differing by the concentration of electron-donating and electron-withdrawing substituents along the backbone, were synthesized and characterized by 2D WAXS and in bottom-contact TFTs. While all copolymers self-assembled into lamellar morphologies, only **P38** and **P40** showed a propensity to form π stacks. The highest hole mobility of $0.02\,cm^2\,V^{-1}\,s^{-1}$ was observed for **P40**, in agreement with the close π stacking and lamellar distances found by structural analysis (0.36 and 1.84 nm, respectively). Following a similar strategy but by using different building blocks, Bao and coworkers [59] synthesized dithiophene and fluorene copolymers (**P41** and **P42**, Figure 1.10) containing fused aromatic thieno[3,4-*b*]pyrazine moieties. Suzuki and Stille polycondensation reactions were used for the polymerization. The bandgap (E_g) of the polymers was tuned in the range 1.15–1.6 eV. These polymers showed a field-effect mobility as high as $0.2\,cm^2\,V^{-1}\,s^{-1}$ with current on/off ratios as high as 10^6 in OTFT devices.

Takimiya and coworkers reported several advanced molecular and polymeric semiconductors for TFTs with heteroarenes (Figure 1.10). The design rationale is that fusion of thiophene rings is necessary to avoid the twist between the adjacent thiophene rings, which would reduce the π stacking. A preliminarily synthesized [1]Benzothieno[3,2-b][1]benzothiophene (BTBT)-thiophene copolymer (PBTBT, **P43**) resulted in a highly twisted backbone ($\lambda_{max} < 400$ nm), and TFTs based on this polymer did not function despite the fact that BTBT had been successful in

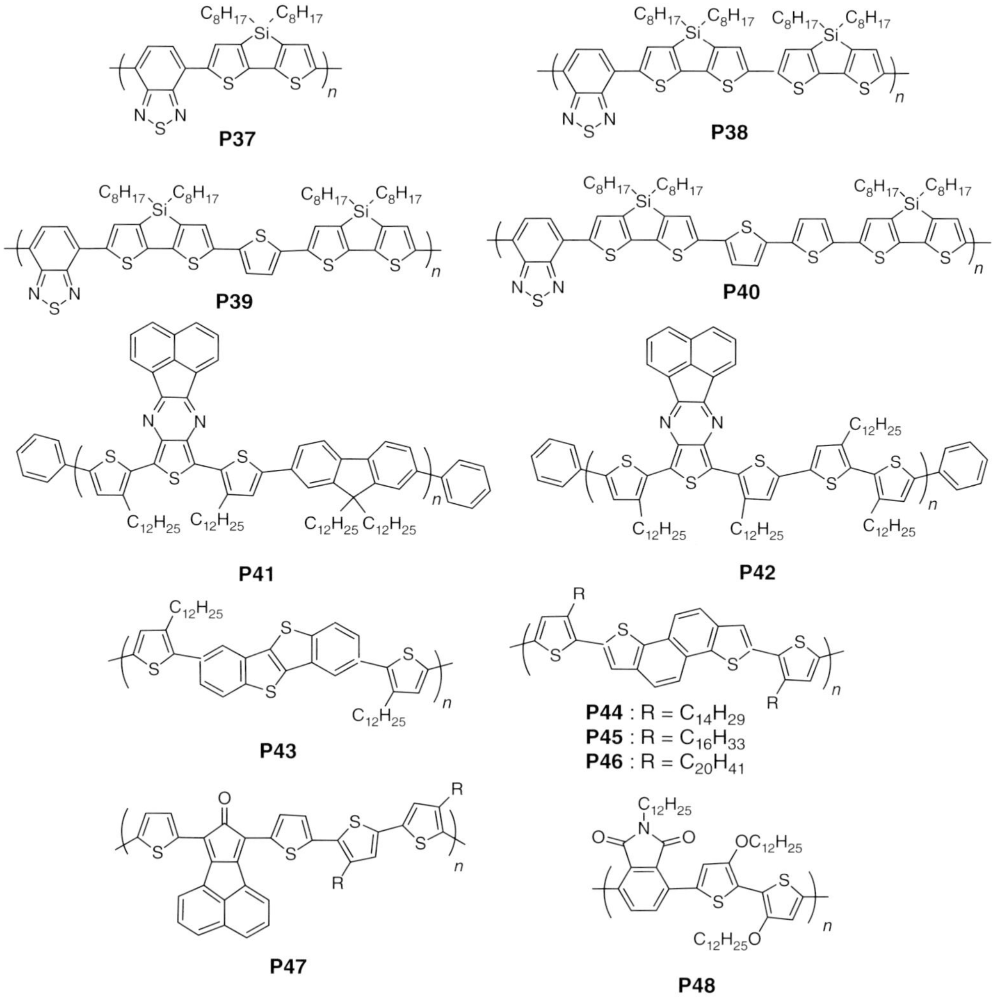

Figure 1.10 Chemical structure of p-channel semiconductors **P37–P48**.

small-molecule systems [60]. Thus, this group recently presented a new design strategy based on **P44–P46**, which incorporates naphthodithiophene (NDT) into a regiosymmetric polythiophene system [61, 62]. These polymers showed high M_n of 24–33 kDa and sufficient solubility in warm chlorinated solvents. TFTs based on **P44–P46** exhibited among the highest field-effect mobility values observed to date for semiconducting polymers (>0.3–0.5 $cm^2\,V^{-1}\,s^{-1}$).

Wudl *et al.* [63] reported **P47** (Figure 1.10) consisting of an electron donor (thiophene) and acceptor (cyclopentadienone) alternating copolymer. The results illustrated that, despite the low molecular weight of this polymer (~6 kDa), a substantial field-effect mobility (~0.02 $cm^2\,V^{-1}\,s^{-1}$) could be obtained. Watson *et al.* [64] reported a copolymer of phthalimide and

3,3′-dialkoxy-2,2′-bithiophene (**P48**, Figure 1.10). In this polymer, backbone planarity is enforced by attractive intramolecular interactions between the pendant oxygens and thienyl sulfur atoms [65], as shown by the crystal structures of some building blocks, intermolecular donor–acceptor interactions, and possibly increased quinoidal backbone character due to alternating donor and acceptor units [66]. The maximum field-effect mobilities were ~0.28 $cm^2 V^{-1} s^{-1}$ in the saturation region and current on/off current ratios of ~$10^4 - 10^5$.

Isoindigo has recently become a popular conjugated moiety in polymer semiconductor design because of its strong electron-withdrawing character. Its acceptor nature results from two lactam rings, and it can be synthesized from various natural sources. This makes isoindigo an attractive building block for sustainable synthetic sources [67]. Reynolds and coworkers reported the synthesis of the first set of isoindigo-based donor–acceptor copolymers with various electron-donating moieties (**P49**–**P54**, Figure 1.11). These polymers, in which isoindigo serves as the acceptor unit, were found to be attractive for OFET and Organic photovoltaics (OPV) applications because of their deep HOMO and LUMO (lowest unoccupied molecular orbital) levels, and their tunable absorption extended to 800 nm in the solid state [68]. Following this initial report, several new alternating copolymers bearing isoindigo units were reported, and one of these copolymers with terthiophene unit (**P55**, Figure 1.11) reached a power conversion efficiency of 6.3% in OPV devices [69]. This efficiency is a record for a polymer solar cell based on a polymer with an optical bandgap of <1.5 eV. The first air-stable and high-performance isoindigo-based polymer (**P56**, Figure 1.11) in OFETs was reported by Pei *et al.* [70] for a copolymer of isoindigo unit with bithiophene, which showed hole mobilities of up to 0.79 $cm^2 V^{-1} s^{-1}$ after thermal annealing at 150 °C. These devices showed very good ambient stability including in high humidity conditions, owing to their low-lying HOMO levels. Bao and coworkers [71] developed a novel high molecular weight ($M_n > 130$ kDa, DP ≈ 30) isoindigo-based polymer (**P57**, Figure 1.11) with siloxane-terminated solubilizing groups. This new polymer with linear alkyl chains was found to have sufficient solubility for solution processing because of its terminal siloxane groups. Record field-effect mobilities of up to 2.48 $cm^2 V^{-1} s^{-1}$ were obtained for **P57**. The same research group prepared a reference polymer **P56** bearing an identical π backbone with branched alkyl side chains. This polymer showed a lower hole mobility of 0.57 $cm^2 V^{-1} s^{-1}$. The enhanced mobility of **P57** was attributed to the stronger intermolecular interactions of the polymer chains in the solid state. Thin-film XRD characterizations indicated a smaller π–π stacking distance (3.58 vs 3.76 Å) for **P57** compared to the reference polymer **P56** (Figure 1.12). This was attributed to the fact that the branching point of **P57** solubilizing groups was placed further away from the polymer π backbone, which allowed the polymer chains to pack more closely together [72]. To further explore this phenomenon, four isoindigo-based copolymers with branching points at various positions were synthesized, achieving an exceptionally high hole mobility of 3.62 $cm^2 V^{-1} s^{-1}$ for **P58** (Figure 1.11) [73]. This study by Pei *et al.* reveals the importance of molecular engineering of alkyl side chains, not only to manipulate polymer solubility but also to enhance charge-carrier mobilities.

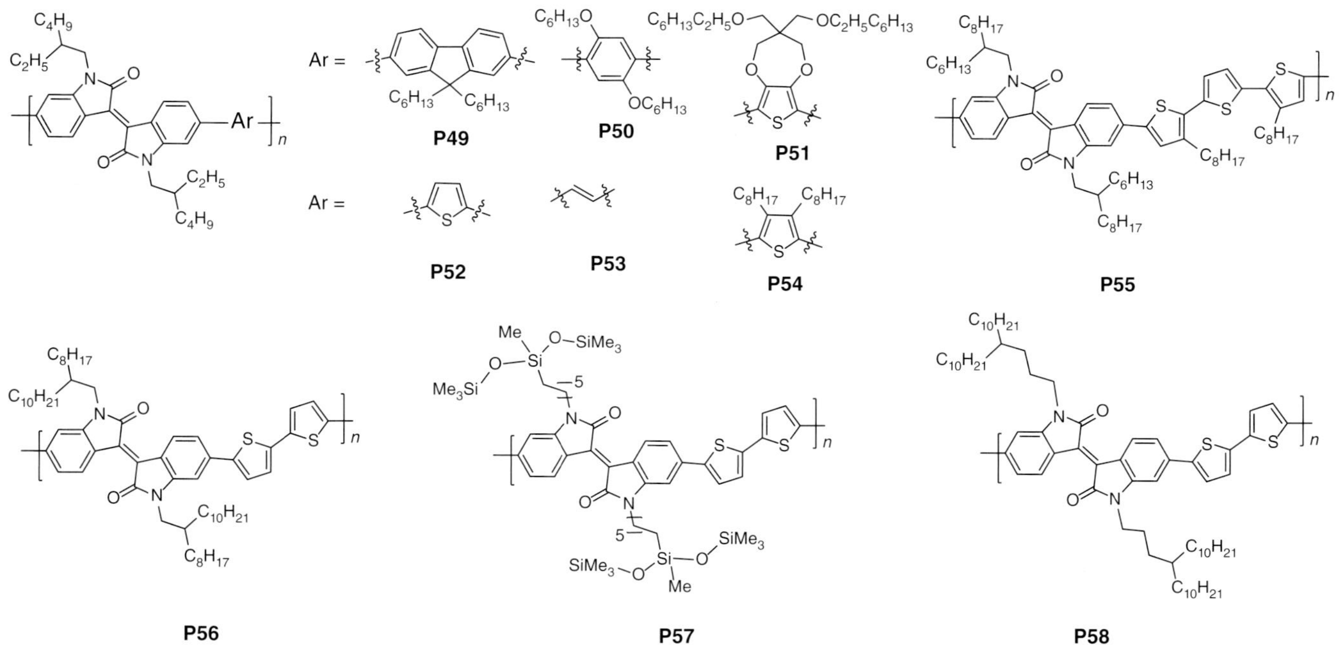

Figure 1.11 Chemical structure of p-channel semiconductors **P49–P58**.

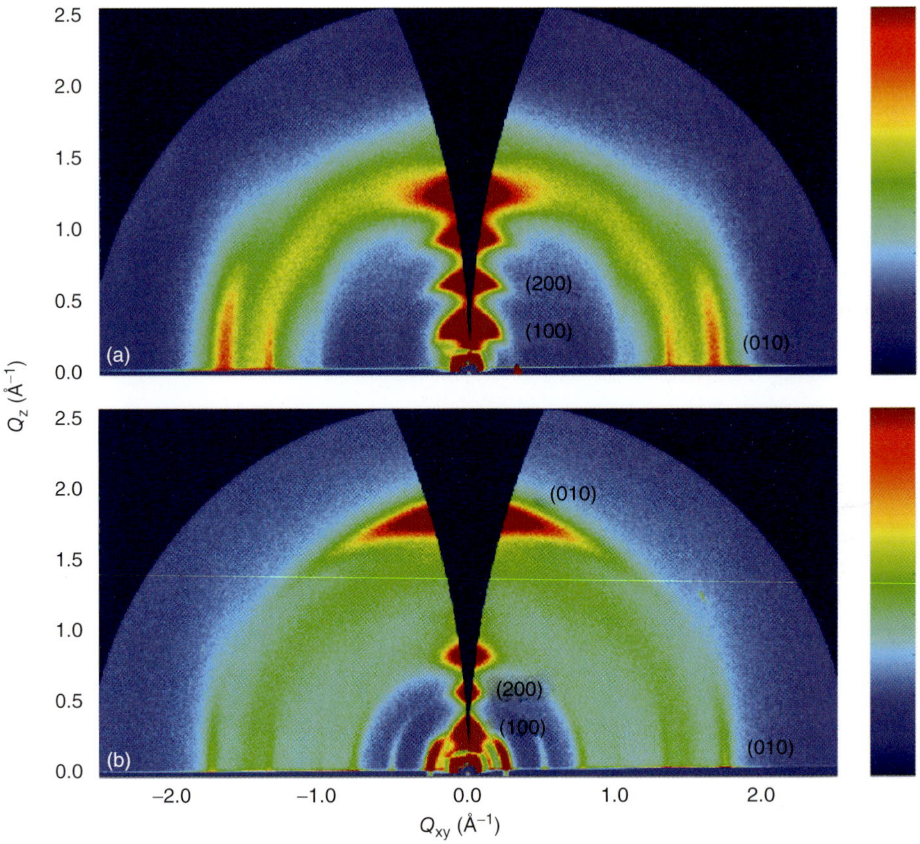

Figure 1.12 GIXD images of **P56** (a) and **P57** (b) annealed at 130 °C. **P56** displays lamellar packing common to many conjugated polymer films, with the π-stacking Bragg planes parallel to the substrate plane. In contrast, **P57** crystallites contain two kinds of textures, where the π-stacking planes are both normal and parallel to the substrate [71]. (Reproduced with permission. Copyright © 2011 American Chemical Society.)

Diketopyrrolopyrrole (DPP)-based polymers were among the first semiconducting polymers to show high charge-carrier mobilities ($\geq$1 cm^2 V^{-1} s^{-1}). As the first report of a DPP-based semiconducting polymer in 2008 [74], there have been many reports on DPP-based donor–acceptor polymer semiconductors with mobilities ranging from 0.1 to 2 cm^2 V^{-1} s^{-1}, most being around 1 cm^2 V^{-1} s^{-1} [75]. DPP-based donor–acceptor polymers exhibited extended HOMO and LUMO distributions along the polymer π backbone and optimized molecular orbital energy levels for both hole and electron injection. Therefore, most DPP-based polymers exhibited ambipolar characteristics with high hole and electron mobilities, and they will be discussed in later sections in more detail. Among these, there are some polymers that are predominantly p-channel semiconductors. Ong and coworkers [76] recently reported an extremely high hole mobility of 10.5 cm^2 V^{-1} s^{-1} for **P59** (Figure 1.13) along with long-term ambient stability. **P59** includes an unsubstituted planar thienothiophene moiety, which reduces conformational energy disorder, facilitates intermolecular packing, and promotes molecular orbital delocalization along the π backbone. High molecular weight (M_n = 110 kDa) was found to be the key to the high performance of this polymer, which is consistent with the earlier studies on P3HT semiconductors [13]. In another recent study, Kim *et al.* [77] developed a DPP-based polymer **P60** (Figure 1.13) with side chains having extended branching positions. Photophysical and structural studies indicated that, when the branching position of the side chain moved away from the π backbone, short π–π stacking distances (3.58–3.62 Å) were obtained, thereby enhancing intermolecular interactions. Additionally, in the polymer π backbone, incorporation of vinyl groups between the aryl units was believed to enhance the area available for intermolecular π-orbital overlaps, which leads to a long-range-ordered, edge-on lamellar crystalline microstructure. As a result, these polymer thin films showed unprecedented record-high hole mobilities of up to 5 cm^2 V^{-1} s^{-1} with room-temperature processing and of up to 12 cm^2 V^{-1} s^{-1} after thermal annealing. This mobility is currently the world record for a polymer semiconductor being in the mobility range of polycrystalline silicon and some other inorganic semiconductors [78]. In another study to modify aryl moieties adjacent to the DPP unit, Li and coworkers [79] synthesized a furan-containing DPP-based polymer **P61** (Figure 1.13) which showed blue shifts in UV–vis absorption characteristics and reduced HOMO energy levels compared to the thiophene analog. In furan-DPP systems, steric hindrance between lactam carbonyl and the heterocycle unit was claimed to be reduced because of the replacement of the sulfur atom with a smaller oxygen atom, which leads to enhanced backbone π coplanarity [80]. Although the polymer thin films of **P61** are found to be rather disordered in the polymer chain orientation, they showed high hole mobilities of 1.02–1.54 cm^2 V^{-1} s^{-1} after thermal annealing at 200 °C. The observed good FET performance is considered to be a result of strong intermolecular interactions and the highly interconnected morphology.

In a recent study by Liu *et al.* [81], two new DPP-based copolymers **P62** and **P63** were synthesized with sextetthiophene (6T) and heptetthiophene (7T) units (Figure 1.13). The introduction of longer β-unsubstituted oligothiophene units in DPP-based copolymers was found to have a significant effect on the molecular weight, solubility, and thin-film morphology of the DPP-oligothiophene copolymers. Although enhancing the backbone rigidity via long β-unsubstituted oliogothiophenes results in lower molecular weight polymers and poor solubility, it was found that this co-monomer strategy reduced the steric hindrance of the bulky solubilizing alkyl chains on DPP units, leading to the well-ordered lamellar structure of π–π stacking. As a result, **P62** and **P63** exhibited remarkably high hole mobilities of 3.94 and 2.82 cm^2 V^{-1} s^{-1}, respectively, through thermal annealing at 180–200 °C.

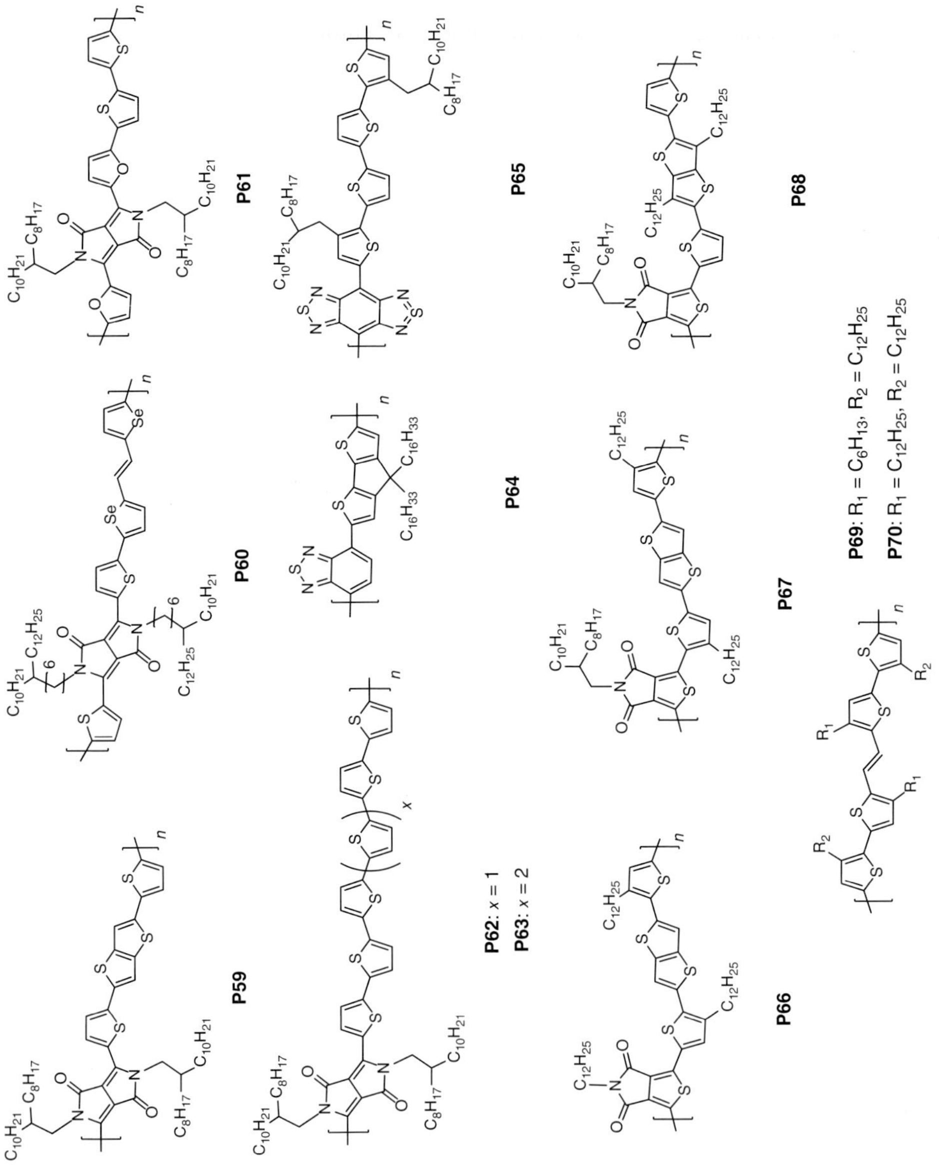

Figure 1.13 Chemical structure of p-channel semiconductors **P59–P70**.

Müllen and coworkers developed a copolymer **P64** (Figure 1.13) with cyclopentadithiophene and benzothiadiazole building blocks as donor acceptor units, respectively. Although earlier report on this polymer indicated a low hole mobility of 0.17 $cm^2\ V^{-1}\ s^{-1}$ [82], latest molecular weight optimizations ($M_n = 10-51$ kDa) by the same group resulted in much higher hole mobilities of up to 3.3 $cm^2\ V^{-1}\ s^{-1}$ [83]. It was shown that the molecular weight played a key role in improving hole mobilities due to the enhanced thin-film crystallinity, even though the film morphology did not change. Surprisingly, solid-state NMR analysis of **P64** did not show any overlaps between cyclopentadithiophene and benzothiadiazole building blocks, indicating that noncovalent donor–acceptor interactions are not the driving forces for the observed improvement in thin-film crystallinity with increasing M_n (Figure 1.14). Another new polymer, **P65**, with planar

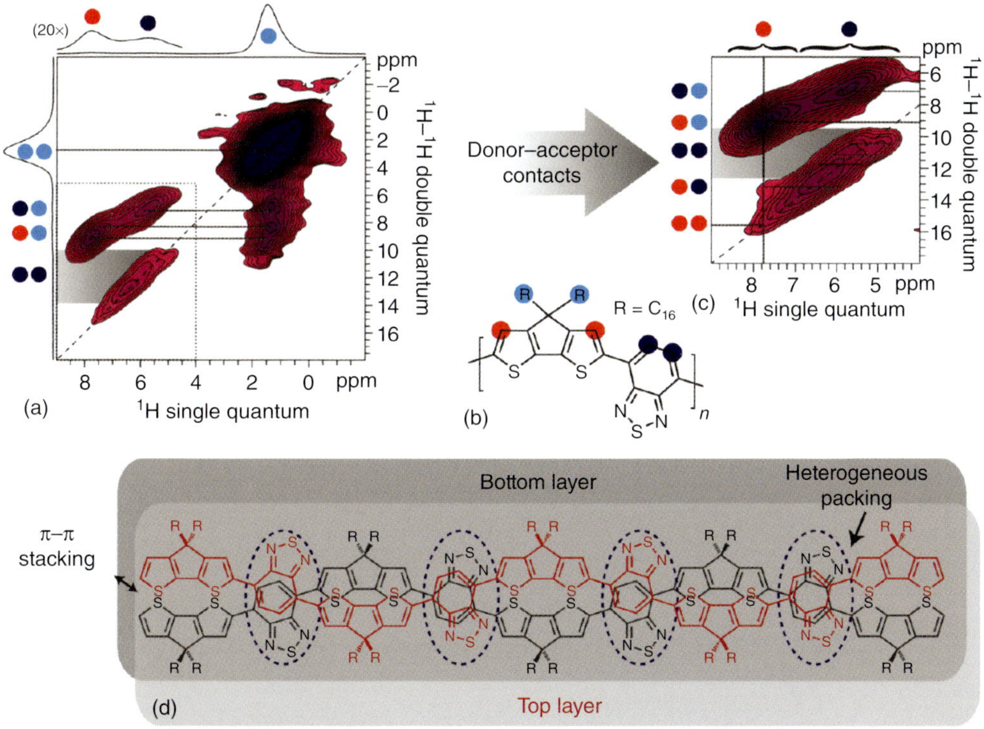

Figure 1.14 Local packing and organization of donor–acceptor groups in **P64** with $M_n = 35$ kDa. (a) 2D contour plot of the $^1H-^1H$ DQ–SQ correlation spectrum recorded at 20.0 T using a back-to-back recoupling/reconversion time of two rotor periods and a spinning frequency of 30.0 kHz. (b) Color scheme used for assignments. (c) Expansion of the backbone region (dashed box in panel a) showing the contacts between donor and acceptor groups. (d) Schematic drawing illustrating the local packing of donor–acceptor groups in two neighboring **P64** chains (here denoted as "layers"). The dashed circles (blue) mark the regions where the acceptor groups are heterogeneously packed on top of one another [83]. (Reproduced with permission. Copyright © 2011 American Chemical Society.)

triple-fused benzobisthiadiazole acceptor and β-alkylated quaterthiophene donor was reported by Wudl *et al.* (Figure 1.13) [84]. Through thermal annealing, the polymer chains adopt an "edge-on" orientation forming π-aligned polymer sheets which were stacked parallel to the substrate. Short π–π stacking distances of 3.5 Å were measured for these thin films. OFETs with **P65** exhibited a high carrier mobility of up to 2.5 $cm^2 V^{-1} s^{-1}$, which was attributed to high-order orientation in the crystalline state, extended π conjugation, and very tight packing in morphological pattern.

Copolymers **P66–P68** (Figure 1.13) containing thieno[3,4-*c*]pyrrole-4,6-dione (TPD) were strategically designed and synthesized by Li *et al.* [85] to study the effect of orientation of thw alkyl chain substituent on OFET device performance. While polymers **P66** and **P67** showed moderate hole mobilities (0.01–0.15 $cm^2 V^{-1} s^{-1}$), a maximum hole mobility of 1.29 $cm^2 V^{-1} s^{-1}$ was observed for **P68**-based devices. This mobility value is the highest reported to date for a TPD-containing polymer, indicating the favorable effects of the alkyl chain position on this particular polymer. This study showed that the orientation of the alkyl substituent on the π backbone dictated the polymer chain packing pattern, film morphology, and OFET device performance, which could lead to an enhancement in the charge-carrier mobilities of as high as two orders of magnitude.

Despite all the advances in donor–acceptor type polymer semiconductors in OFETs, p-channel polymers based on an electron-neutral thiophene backbone still show high mobilities. In a recent report by Kim *et al.* [86], two highly soluble polymers containing an alkyl-substituted thienylenevinylene unit **P69** and **P70** (Figure 1.13) were synthesized and characterized in top-gate/bottom-contact (TG–BC) OFET devices. These polymer thin films were found to be highly crystalline after a thermal annealing process ($T_{an.}$ = 200 °C), showing highly ordered lamellar microstructures and nanoribbon morphologies. Charge-carrier mobilities as high as 1.05 $cm^2 V^{-1} s^{-1}$ were achieved. The high solubility of **P70** enabled the fabrication of inkjet-printed complementary inverter circuits with Polyera's N2200 (see below), which exhibited a gain of ~16.

1.4.2 Small Molecules

Oligothiophenes are one of the most investigated organic semiconductor families for optoelectronic devices (Figure 1.15). The synthesis of these systems has been realized following different methodologies, and currently many α*n*Ts are commercially available. Polycrystalline films of the oligomers with four (α4T, **M1**) [87], five (α5T, **M2**) [87], six (α6T, **M3**) [88], and eight (α8T, **M4**) [89] thiophene rings have been widely examined as semiconductors for OFETs [90]. The crystal structures of α*n*Ts ($n = 2-8$) have been obtained, and they have been shown to exhibit similar features with very planar molecules packing in the characteristic herringbone (HB) motif [91].

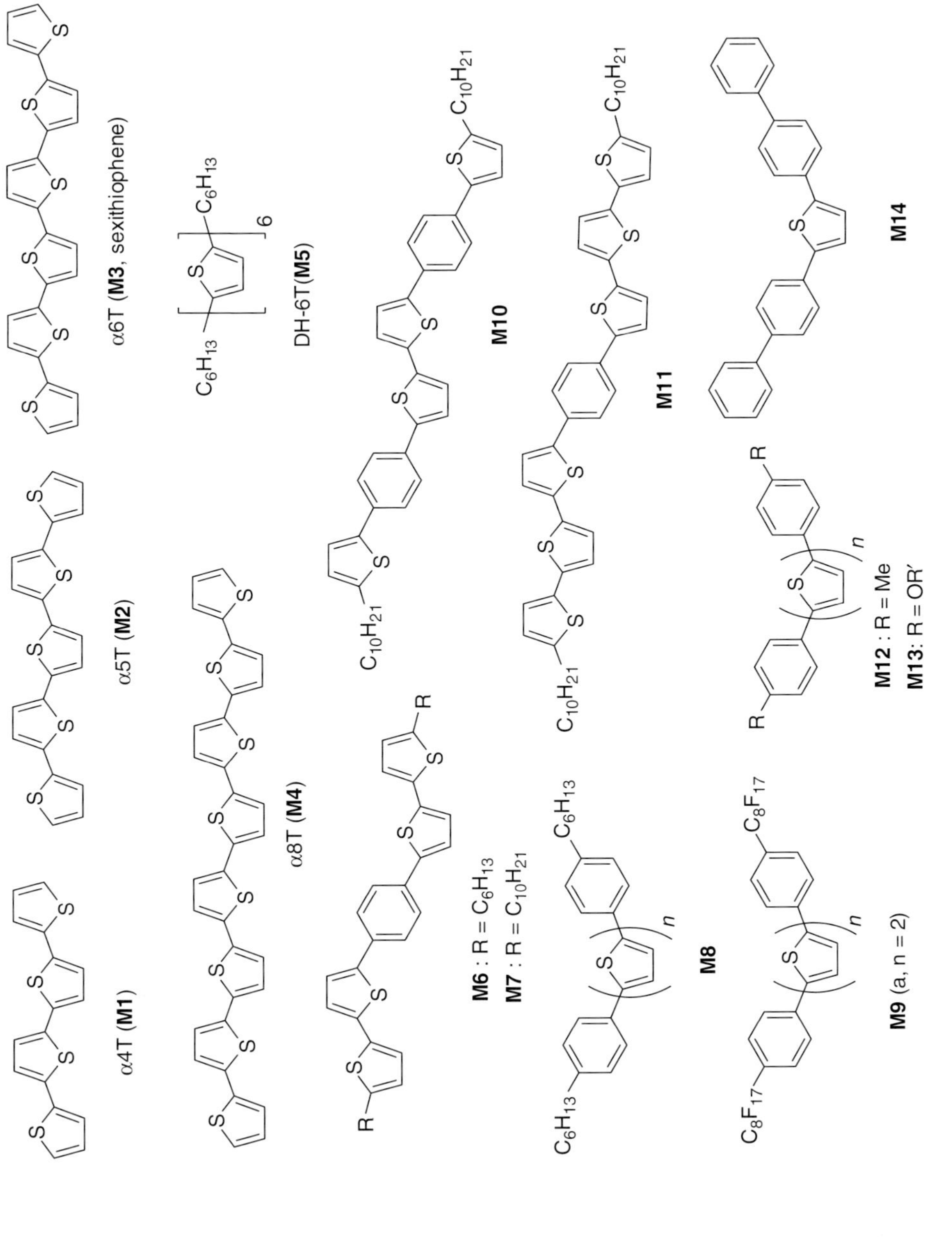

Figure 1.15 Chemical structure of p-channel small-molecule semiconductors **M1**–**M14**.

Polycrystalline films of α*n*Ts, when deposited at a relatively high temperature on SiO_2, have most of the molecules oriented with their long axis perpendicular to the substrate surface. In OFET devices, all these materials exhibit p-channel transport, and the hole mobilities are strongly dependent on the material purity, film morphology, deposition conditions, and the degree of film texturing [92]. Usually, larger cores afford greater mobilities when the devices are fabricated and measured under identical conditions [93]. The oligomers α4T and α5T exhibit mobilities of up to 0.006 and 0.08 $cm^2 V^{-1} s^{-1}$, respectively [94]. The low field-effect mobility for α4T was attributed to the poor charge injection from the contact and not to the inefficient charge transport through the material. In fact, the mobility doubles when an interfacial layer is used at the α4T source/drain contact to enhance charge injection. The highest reported mobility for α8T is 0.33 $cm^2 V^{-1} s^{-1}$ for film deposition at 120 °C [95]. This mobility value is an order of magnitude greater than previously reported for α8T and is attributed to the more favorable film morphology.

The α6T derivative is the most investigated oligomer in this series and exhibits a field-effect mobility of ~0.03 $cm^2 V^{-1} s^{-1}$ in bottom-contact Si/SiO_2 FETs [88], while single-crystalline grains of this oligomer exhibit mobilities of up to 0.075 $cm^2 V^{-1} s^{-1}$ [92c]. The film morphology and microstructure have been studied in detail. As the charge transport in OFETs occurs near the interface with the gate dielectric, Loi *et al.* have investigated the supramolecular organization of α6T ultrathin films on SiO_2 dielectric by confocal spectroscopy and microscopy. The authors demonstrated that α6T submonolayer films are composed of mixed regions where the molecules stand on and lie flat on the substrate surface. Importantly, they discovered that only after the first monolayer is completed do all the α6T molecules stand on the substrate. For films thicker than two monolayers, the photoluminescence spectra of standing molecules show a molecular H-like aggregation as in the single crystal [96].

Dinelli *et al.* have studied how the field-effect mobility of ultrathin films of α6T varies as a function of the dielectric surface coverage. FET test devices were fabricated on heavily doped n-Si wafers, with 200-nm-thick SiO_2 as dielectric which was passivated with hexamethyldisilazane. Interdigitated source and drain Ti/Au electrodes were fabricated using photolithography. Thin films of α6T were deposited by sublimation in an organic molecular beam apparatus (10^{-10} mbar base pressure). The deposition rate was monitored with a quartz oscillator. The charge-carrier mobility was found to rapidly increase with increasing surface coverage and saturated at a surface coverage of about two monolayers. For high-rate film deposition, the mobility reached a plateau with a mean value of 0.015 $cm^2 V^{-1} s^{-1}$, whereas for low-rate deposition the trend was similar but with a higher average mobility of 0.043 $cm^2 V^{-1} s^{-1}$. Therefore, the authors concluded that, for α6T films, only the two monolayers next to the dielectric interface dominated the charge transport. A quantitative analysis of spatial correlations showed that the second layer was crucial in enhancing the field-effect mobility, as it provided efficient percolation pathways for carriers generated in both the first and second layers. The upper layers do not actively contribute either because their domains are smaller than the ones in the second layer or because the carrier density is negligible [97].

A number of synthetic strategies have been developed to functionalize the thiophene ring. In the case of oligothiophenes, core functionalization enhances the solubility and may alter the molecular packing within the crystal and, ultimately, the optoelectronic properties. Functionalization at the α,ω-positions of the oligomer typically does not affect the planarity of the π-conjugated core but enhances the solubility. Furthermore, in few cases, α,ω-functionalization with alkyl chains imparts liquid-crystalline-like properties, which dramatically modifies the solid-state ordering and enhances the field-effect mobility of the resulting films. Functionalization at the oligothiophene β,β′ positions significantly enhances compound solubility but usually disrupts the *n*T core planarity and molecular staking, leading to materials of low field-effect mobility. Among the alkyl-substituted oligothiophenes, α,ω-dimethyl [98] and α,ω-dihexyl oligomers with four [99], five [100], six (e.g., DH-6T, **M5**) [101], and eight [89] thiophene rings were among the first to be synthesized and characterized. Here again, the purity of these insoluble compounds was found to be a key factor in achieving high OFET performance [92a]. The field-effect mobilities of these oligomers are as high as 0.1 $cm^2\ V^{-1}\ s^{-1}$ [100]. The field-effect mobility of DH-6T (**M5**) is typically reported to be between 0.02 and 0.07 $cm^2\ V^{-1}\ s^{-1}$ depending on the deposition conditions and the dielectric material used for device fabrication [102]. The high mobility of DH-6T, obtained almost two decade ago, was explained by the high degree of DH-6T film self-organization thanks to the presence of the alkyl chains at the α,ω positions. More recently, the highest mobility for DH-6T was reported by Dimitrakopoulos *et al.* using molecular beam deposition of the oligomer. Values of up to 0.13 $cm^2\ V^{-1}\ s^{-1}$ were measured when parylene was used as the insulating layer [103].

Phenylene-thiophene oligomers (**M6–M14**, Figure 1.15), developed separately by Hotta and Samulski [104], attracted much attention because of their potential use as the active component in organic light-emitting diodes (OLEDs) and organic laser diodes, as well as p-type semiconductors for OTFTs [105]. Compared to simple oligo/poly phenylenes and thiophenes, molecular mixing of these two aromatic rings produces structural motifs characterized by different molecular shapes and physical properties, which in turn alter the material behavior profoundly [106]. Examples of these property changes include stronger light emission and low-threshold amplified spontaneous emission (ASE) from a liquid-crystalline polymer blend [107].

Ponomarenko *et al.* investigated the end-capped bithiophene-phenylene-bithiophene core with variable alkyl chain lengths from $-C_6H_{13}$ (**M6**) to $-C_{10}H_{21}$ (**M7**). These molecules exhibit field-effect hole mobilities of ~0.3 $cm^2\ V^{-1}\ s^{-1}$ [108], which is more than an order of magnitude greater than those obtained by Hong *et al.* [109] for the compound **M6**. The reason for this large increase in performance was not clear, but the authors attributed it to material purity and the use of cross-linked poly(hydroxystyrene) as the gate dielectric. Other phenylene-thiophene derivatives **M10** and **M11** (Figure 1.15) were also synthesized.

It was found that **M11** had properties similar to those of **M7**, whereas **M10** had an order of magnitude lower mobility due to poor molecular ordering [110]. Electrochemical measurements of the oligomers containing a phenyl ring were shown to have better oxidative stability than the all-thiophene analogs, and FET devices operated in air without degradation. Recently, tolyl end-substituted oligothiophenes with three, four, and five thiophene rings have also been synthesized by Mohapatra *et al.* [105a]. These derivatives (**M12**) exhibit hole mobilities of ~0.03 $cm^2 V^{-1} s^{-1}$ and very high I_{on}:I_{off} ratios of ~10^9. Interestingly, despite the fact that odd- and even-numbered oligomers have different crystal structures and film microstructures, the FET characteristics are very similar. A phenylene-thiophene oligomer series (**M13**) end-substituted with various alkoxy chains were synthesized using Stille and Suzuki coupling reactions by Sung *et al.* [111]. Oligomers with two and three core thiophene rings functionalized with an octyloxy chain exhibit the largest mobilities of ~0.18 $cm^2 V^{-1} s^{-1}$ (I_{on}:I_{off} > 10^7) and ~0.09 $cm^2 V^{-1} s^{-1}$ (I_{on}:I_{off} ~ 10^4), respectively. Thin-film morphologies analyzed using AFM revealed higher nucleation densities for films of branched alkoxy side-chain molecules compared to their straight alkyl side-chain counterparts. However, field-effect mobilities were found to strongly decrease for the branched-substituted oligomers. Finally, Ichikawa *et al.* [105d] showed that the epitaxially grown thiophene-phenylene system BP2T (**M14**) on KCl single-crystal substrates exhibits a hole mobility (0.29–0.66 $cm^2 V^{-1} s^{-1}$) which is much larger than that of BP2T vacuum-deposited directly on SiO_2 at room temperature and close to that of high-quality oligothiophene single crystals. A number of studies have suggested that forcing oligomers to adopt a face-to-face arrangement should enhance the π–π orbital overlap, thus facilitating charge transport by maximizing the electronic coupling between adjacent molecules. This concept has been applied to a large variety of molecules including thiophene-based semiconductors. Laquindanum *et al.* explored fused thiophenes by synthesizing bis(benzodithiophene) (BDT, **M15**, Figure 1.16). FETs based on BDT exhibit field-effect mobilities of up to 0.04 $cm^2 V^{-1} s^{-1}$ for films vapor-deposited at 100 °C [112].

Heteroacenes are another promising class of fused π-conjugated p-channel semiconductors developed for OFETs over the past decade. These structures can be considered as ladder-type acenes in which one or more benzene rings in acene backbones are replaced with heterocyclic aromatic units. To this end, there have been two different design strategies followed to date: either the benzene rings are placed in the center of the semiconductor core and heterocycles are used as end-capping rings, or benzene rings are used as end-capping units while the heterocycles are in the center. Compared to their acene anologs, both of these heteroacene families have a lower degree of aromaticity, and thus they have more stabilized HOMO energy levels (<−5.2 vs −5.0 eV for pentacene) [113], which leads to good ambient stability in OFET devices. With the former strategy, the observed hole mobilities are typically in the range of those of amorphous silicon (α-Si) (0.1 – 1.0 $cm^2 V^{-1} s^{-1}$) [114].

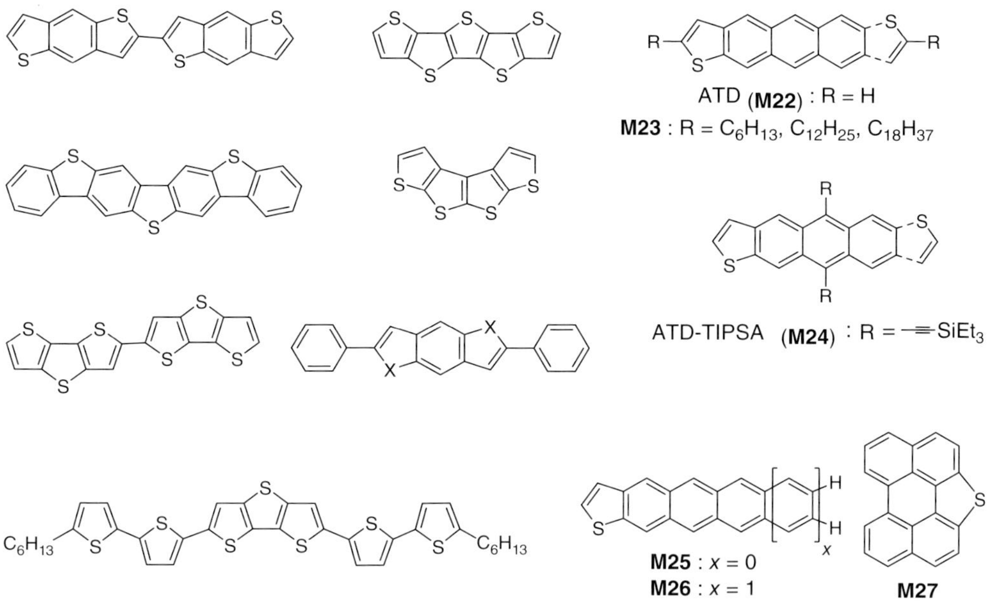

Figure 1.16 Chemical structure of p-channel small-molecule semiconductors **M15–M27**.

The fused system dibenzothienobenzothiophene (**M16**, Figure 1.16) was investigated by Sirringhaus *et al.* and coworkers [115]. As the synthesis of this compound is not regioselective, the presence of regioisomers was found to result in poor FET performance with mobilities ~0.03 $cm^2\,V^{-1}\,s^{-1}$. However, the use of a shutter during vacuum sublimation allowed selective deposition of different fractions. The greatest field-effect mobilities were ~0.15 $cm^2\,V^{-1}\,s^{-1}$ and I_{on}:I_{off} > 10^6. The same group reported the fused thiophene derivative bis(dithieno[3,2-*b*:2′,3′-*d*]thiophene) (**M17**, Figure 1.16). Crystal structures revealed that this semiconductor packs in a face-to-face stacked structure. Hole mobilities of up to ~0.05 $cm^2\,V^{-1}\,s^{-1}$ and an I_{on}:I_{off} of >10^8 were measured using this oligomer for both top- and bottom-contact FET structures [116]. Iosip *et al.* [117] have synthesized dithienothiophene in co-oligomers with thiophene (**M18**, Figure 1.16) and reported unoptimized FET hole mobilities of 0.02 $cm^2\,V^{-1}\,s^{-1}$ and I_{on}:I_{off} of ~10^6.

The dithienothiophene motif was extended up to seven fused rings by Zhang *et al.* [118], and it was found that these oligomers also pack in a face-to-face stacking motif. OFET devices were fabricated with the pentathienoacene (**M19**, Figure 1.16) and exhibited a mobility of 0.045 $cm^2\,V^{-1}\,s^{-1}$ and I_{on}:I_{off} up to ~10^3. These devices are very stable in air thanks to their large bandgap of 3.29 eV [119]. Finally, Nenajdenko *et al.* [120] have synthesized an annulated oligothiophene consisting of four fused thiophene rings (**M20**, Figure 1.16). X-ray crystallographic studies on single crystals of this compound confirmed that the molecule is planar, making it attractive for semiconductor applications. Very recently, extension to other benzo[1,2-*b*:4,5-*b*′]dichalcogenophenes (**M21**, X = S, Se, Te, Figure 1.16) [121] and the utilization of larger π cores resulted in derivatives exhibiting FET mobilities of 0.17–2.0 cm^2 V^{-1} s^{-1} [60a].

An interesting chemical approach has been to combine the molecular shape of pentacene, which leads to a favorable crystal packing geometry and orientation, with thiophene end groups, which should increase stability and also provide points of attachment for solubilizing substituents. Thus, anthradithiophene derivatives (**M22** and **M23**, Figure 1.16) were prepared and characterized for the first time by Katz *et al.* [122]. These products were obtained as a mixture of syn and anti isomers, whose separation was not possible. The field-effect mobility of Anthradithiophene (ATD) is an order of magnitude lower than that of pentacene, about 0.1 cm^2 V^{-1} s^{-1}, but the I_{on}:I_{off} is higher when I_{off} is recorded at zero gate bias. A highly ordered thin-film morphology is observed, which is consistent with the electrical characteristics. Hexyl-, dodecyl-, and octadecyl-disubstituted derivatives (**M23**) were also found to demonstrate modulation typical of FET devices, with the first two exhibiting higher mobilities (~0.15 cm^2 V^{-1} s^{-1}) than the parent compound, and with increased solubility. The third material still has significant activity (mobility ~0.06 cm^2 V^{-1} s^{-1}), even though it consists mostly of nonconjugated carbon atoms. Anthony *et al.* [46a] fabricated FETs based on solution-deposited films of functionalized anthradithiophene **M24** (Figure 1.16). This compound forms uniform films of excellent quality, yielding hole mobility of 1.0 cm^2 V^{-1} s^{-1} with an excellent I_{on}:I_{off} of ~10^7. The high OFET performance of this material was attributed to the close π-stacked interactions in the crystal. Loo *et al.* showed that the performance of bottom-contact **M24**-based OFETs can be improved dramatically by a simple and straightforward solvent-vapor annealing process after device fabrication. Exposing the transistors to dichloroethane vapor for 2 min induces structural rearrangement and crystallization of the semiconducting film. This procedure resulted in drastic increases in I_{on} currents, and mobility increased by two orders of magnitude [123].

Very recently, thiophene-fused acenes anthra[2,3-*b*]thiophene (**M25**) and tetraceno[2,3-*b*]thiophene (**M26**) were synthesized by Valiyev *et al.* [124] (Figure 1.16). The crystal structures of **M25** and **M26** were described. The addion of the thiophene ring to the acene unit resulted in a weak dipole, and the crystal packing changes from the triclinic structure of acenes to orthorhombic. TFT devices were fabricated from these materials. Compound **M25** exhibited a field-effect mobility of ~0.1 cm^2 V^{-1} s^{-1} and I_{on}:I_{off} of ~10^8 (T_D = 25 °C), whereas the compound **M26**a showed a mobility of ~0.25 cm^2 V^{-1} s^{-1} and I_{on}:I_{off} of ~10^6 (T_D = 80 °C). Finally, Zhu *et al.* investigated the OFET behavior of perylo[1,12-*b*,*c*,*d*]thiophene (**M27**, Figure 1.16). Vacuum-deposited films showed moderate hole mobilities of ~0.05 cm^2 V^{-1} s^{-1}, I_{on}:I_{off} of ~10^5, and a low threshold voltage of −6.3 V. Interestingly, single-crystal micrometer wires were grown from solution, and OFET based on the single wire exhibited mobilities of up to ~0.8 cm^2 V^{-1} s^{-1}. The authors believe that the solid-state packing arrangement with the likelihood of a double-channel arrangement induced by marked S···S

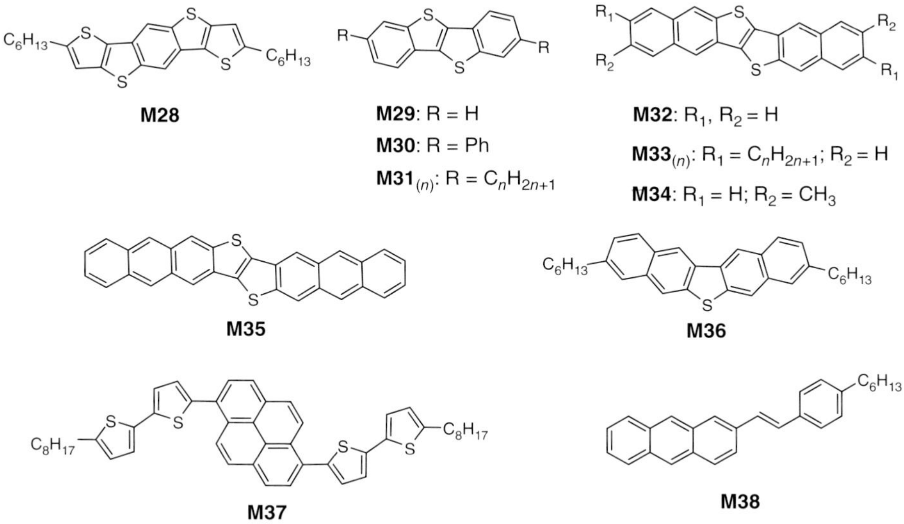

Figure 1.17 Chemical structure of p-channel small-molecule semiconductors **M28–M38**.

interactions may be the reason for the high performance. Among heteroacenes, one of the highest performances was achieved with **M28** (Figure 1.17), which was developed by Müllen *et al.* [125]. This semiconductor exhibited hole mobilities of up to 1.7 $cm^2\,V^{-1}\,s^{-1}$ and I_{on}/I_{off} ratios of 10^7 in solution-processed OFET devices. This study was one of the early reports indicating that small heteroacene semiconductors with an extended aromatic core and solubilizing alkyl chains could be very promising for solution-processed optoelectronic devices.

A different design strategy was adopted by Takimiya and coworkers, in which the heterocycles were placed in the center of the semiconductor core and benzenes were used as end-capping rings. This approach was found to result in much higher hole mobilities compared to the earlier heteroacene-based semiconductors. In one class of such heteroacene semiconductors, a thienothiophene unit was incorporated in the center of an acene core while having benzene (BTBT, **M29–31**), naphthalene (Dinaphtho[2,3-b:2′,3′-f]thieno[3,2-b]thiophene (DNTT), **M32–34**), and anthracene (Dianthra[2,3-b:2′,3′-*f*]thieno[3,2-b]thiophene (DATT), **M35**) as end-capping moieties (Figure 1.17). The OFET devices fabricated with these semiconductors were found to be ambient-stable because of their low-lying HOMO energy levels. Among these, the BTBT (**M29**) core was developed earlier than the others via a simple thermal reaction of benzylidene dichloride with elemental sulfur or from commercially available *o*-chlorobenzaldehyde. Although the yield of **M29** synthesis is moderate (30–40%), its simplicity and fewer steps make it quite attractive for large-scale preparations. The earliest version of BTBT, **M30**, was prepared with phenyl substituents, and it showed excellent FET characteristics under ambient conditions with hole mobilities of 2.0 $cm^2\,V^{-1}\,s^{-1}$ and I_{on}/I_{off} ratios of 10^7 [60a]. Thereafter, alkyl-substituted

BTBTs (**M31**$_{(n=5-14)}$) with various alkyl chain lengths ($n=5-14$) were developed, which were found to be highly soluble in common organic solvents (up to ≈90 mg ml^{-1} in chloroform at room temperature). However, it is noteworthy that BTBT derivatives with alkyl chains longer than $C_{12}H_{25}$ showed reduced solubilities compared to those with shorter alkyl chains. Solution-processed OFETs fabricated with **M31**$_{(n=5-14)}$ showed typical p-channel behavior, with field-effect mobilities as high as 2.75 cm^2 V^{-1} s^{-1} and I_{on}/I_{off} ratios of 10^7 [60b]. Inspired by this initial report, several BTBT-based OFET devices were fabricated with various device configurations and solution-processing techniques, which recently yielded mobilities as high as 5 cm^2 V^{-1} s^{-1} [126]. In one of these studies, Hasegawa and coworkers [127] showed that single-crystalline thin films of **M31**$_{(n=8)}$ could be grown at the liquid–air interfaces by mixing fine droplets of an antisolvent (*N*,*N*-dimethylformamide) and semiconductor in 1,2-dichlorobenzene solution (Figure 1.18). This approach yielded inkjet-printed, patterned semiconductors with average hole mobilities of 16.4 cm^2 V^{-1} s^{-1} and a maximum value as high as 31.3cm^2 V^{-1} s^{-1}. Nonetheless, further improvements of this technique are needed to address the issues of the broad distribution of crystal orientations and mobilities and moderate single-crystal formation yields (50%).

DNTT semiconductors (**M32–34**) have lower solubilities in common organic solvents compared to their BTBT counterparts owing to their further π-extended core structures. Therefore, they were typically characterized in vapor-deposited OFET devices. Although the DNTT core itself (**M32**) shows hole mobilities of 2.9 cm^2 V^{-1} s^{-1} in vapor-deposited films [128], 2–4 times higher mobilities were achieved after attaching alkyl chains. Alkylated DNTT semiconductors (**M33**$_{(n=6-12)}$) showed very high hole mobilities close to 8 cm^2 V^{-1} s^{-1}, which is among the highest for OFETs based on polycrystalline thin films [129]. The authors claimed that the long alkyl chains attached to the DNTT core enhanced molecular ordering and resulted in tighter packing in the thin-film phase. This is attributed to the van der Waals intermolecular interactions between the alkyl groups, which is the so-called molecular fastener effect [130]. Very recently, highly crystalline OFET devices showing band-like charge transport were fabricated with decyl ($C_{10}H_{21}$)-substituted DNTT (**M33**$_{(n=10)}$) semiconductor by using a new hot-solution processing technique [131]. This new solution processing method was developed by Takeya and coworkers, and yielded devices with excellent hole mobilities of up to 11 cm^2 V^{-1} s^{-1}. The authors also showed that their method was applicable to flexible plastic substrates by fabricating top-gate devices on a flexible polyethylene naphthalate (PEN substrate with hole mobilities as high as 9 cm^2 V^{-1} s^{-1}. Another interesting DNTT derivative developed by Takimiya *et al.* [132] is **M34** with two methyl groups attached at 3 and 10 positions. Although this semiconductor did not show as high a mobility as earlier derivatives ($\mu_h = 0.8$ cm^2 V^{-1} s^{-1}), it was unique in that, different from the other DNTT-based semiconductors, it showed three-dimensional (3D) HB arrangements both in single-crystal and thin-film phases.

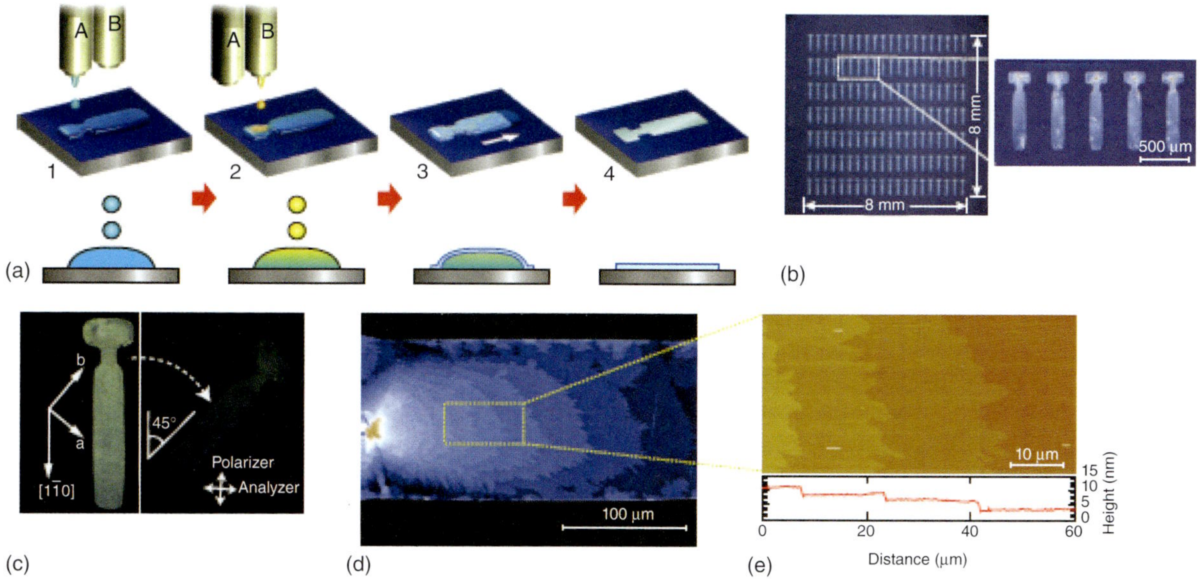

Figure 1.18 Inkjet printing of organic single-crystal thin films. (a) Schematic of the process. Antisolvent ink (A) is first inkjet-printed (step 1), and then solution ink (B) is overprinted sequentially to form intermixed droplets confined to a predefined area (step 2). Semiconducting thin films grow at liquid–air interfaces of the droplet (step 3), before the solvent fully evaporates (step 4). (b) Micrographs of a 20×37 array of inkjet-printed **M31**$_{(n=8)}$ single-crystal thin films. (c) Crossed-Nicols polarized micrographs of the film. (d) Expanded micrograph of the film, showing stripes caused by molecular-layer steps. (e) Atomic force microscopy image and the height profile (below) showing the step-and-terrace structure on the film surfaces [127]. (Reproduced with permission. Copyright © 2011, Nature Publishing Group.)

The longest π-extended ladder-type heteroacene compound developed by Takimiya *et al.* [133] is **M35** with eight fused aromatic rings. This semiconductor has a fairly stable device operation although it has a highly extended acene-like π system with an IP of 5.1 eV. A remarkable field-effect mobility of $3.0\,cm^2\,V^{-1}\,s^{-1}$ was observed in vapor-deposited films, even though the thin-film microstructure was found to include both face-on and edge-on molecular orientations.

In another study, Okamoto and coworkers [134] developed a V-shaped extended heteroacene semiconductor, **M36** (Figure 1.17), which exhibited sufficient solubility in common solvents and high-mobility charge-transport characteristics (μ_h up to $9.5\,cm^2\,V^{-1}\,s^{-1}$) in solution-crystallized OFET devices. The observed high mobility for **M36** was attributed to an almost optimized arrangement of semiconductor π backbones, which led to the maximum bandwidth and high transfer integrals (39–68 meV). It is noteworthy that these solution-crystallized films exhibit high thermal stability up to 150 °C, which is important for practical applications. It was claimed by the authors that the synthetic route followed during the synthesis of **M36** could enable mass production.

Compared to heteroacene and oligothiophene compounds, pyrene-based organic semiconductors have rarely been studied in OFETs. Only recently, a high-mobility pyrene-based liquid-crystalline semiconductor **M37** (Figure 1.17) was reported by Lee *et al.* [135]. In this semiconductor, pyrene core was disubstituted at 1 and 6 positions with 5′-octyl-2,2′-bithiophene-5-yl aromatic units. A high-performance transistor behavior with a field-effect mobility of $2.1\,cm^2\,V^{-1}\,s^{-1}$ and I_{on}/I_{off} ratio of $\sim 10^7$ was achieved along with an improved long-term stability compared to the pentacene devices. The promising results obtained in this study indicate that pyrene-based semiconductors are good candidates for high-performance optoelectronic applications.

Finally, styrylacenes have emerged as a new promising class of p-channel semiconductors, in which the stability of an acene core is enhanced via extending the π conjugation through a styrene group. Recently, Perepichka and coworkers [136] developed a structurally simple styrylacene p-channel semiconductor **M38** (Figure 1.17), which exhibited a high field-effect hole mobility of $1.5\,cm^2\,V^{-1}\,s^{-1}$ in thin films and $2.6\,cm^2\,V^{-1}\,s^{-1}$ in single crystals. AFM characterizations showed that the high hole mobility of **M38**-based thin-films benefited from the perfect layer-by-layer growth at room temperature with an upright molecular orientation, which led to continuous semiconductor films at a low nominal thickness (Figure 1.19). Additionally, this semiconductor showed strong bright-blue solid-state emission with 70% photoluminescence quantum yield. This study indicates that strong π–π interactions, required for efficient charge transport in the solid-state, do not necessarily lead to quenching of the luminescence when the singlet-fission process is turned off.

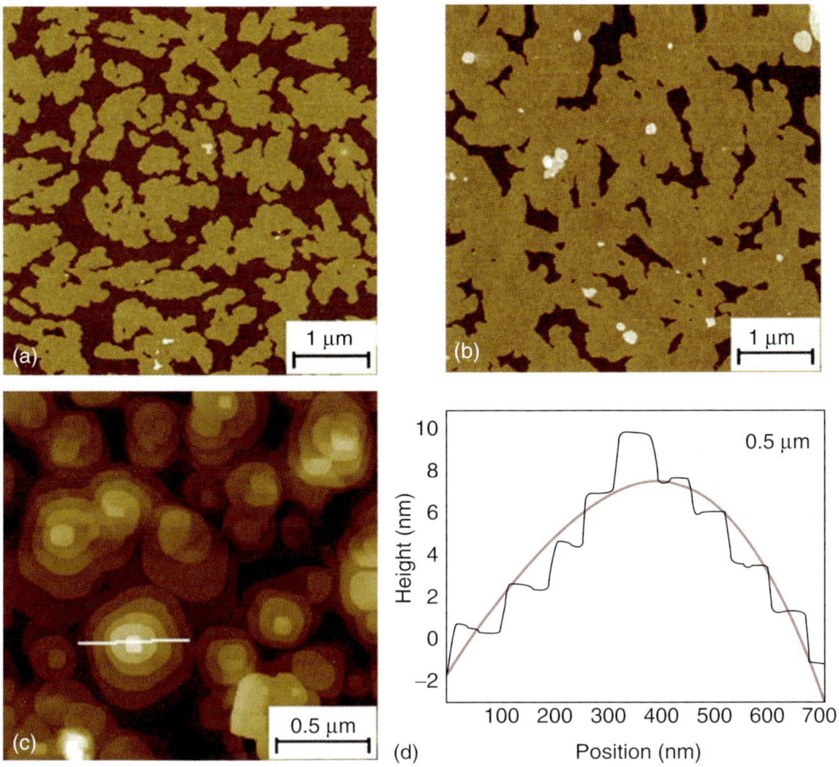

Figure 1.19 AFM images of (a) submonolayer, (b) monolayer, and (c) multilayer films of **M38** grown on Si/SiO_2. (d) Height profile of image (c) [136]. (Reproduced with permission. Copyright © 2012, John Wiley & Sons, Inc.)

1.5 n-Channel Semiconductors

1.5.1 Polymers

n-Channel polymers for TFTs, like n-channel molecular systems, have been traditionally underdeveloped compared to p-channel semiconductors. The reason of this deficiency in materials development was surprising considering that several experimental and theoretical studies suggested that organic semiconductors should transport electrons as efficiently as or even more efficiently than holes. More recently, our understanding on how to enable efficient electron transport in TFTs has improved after discovering the key role played by other factors, besides the semiconductor electronic structure, affecting electron transport. These include the following: (i) Gate dielectric surface chemistry. Most of the first-generation dielectric materials for TFTs were oxides, such as SiO_x, which efficiently trap electrons on the surface. (ii) Metal contacts [137]. This high workfunction results in a large energy barrier for electron injection, and therefore, from an energetic perspective, low-workfunction metals such as Al or Ca should be preferable. However, air-induced oxidation of these conductor surfaces results in an insulating layer, which nullifies the benefit of having a formal lower energy barrier. (iii) Ambient atmosphere composition. At the beginning of the TFT field, very few research groups carrying out material development had the capabilities to perform measurements in an inert atmosphere or vacuum. It is known that O_2 and H_2O can efficiently inhibit electron transport, preventing several semiconductor classes to be screened for n-channel transport.

Today, several of these issues can be controlled, and, particularly, the use of dielectric surface passivation strategies and polymeric dielectric materials has resulted in impressive progress in performance. From an energetic perspective, it is now believed that a very narrow energetic window for the LUMO level, located at about −4.0 to −4.3 eV, must be achieved to enable polymeric semiconductors with good charge transport in ambient atmosphere. For higher LUMO energies (low electron affinity), the polymer performance rapidly degrades after exposure to ambient atmosphere even when using electron-trap-free dielectrics. On the other hand, for very low-lying LUMOs systems (this has been well-established for molecular semiconductors), the corresponding devices are very difficult to turn off. In this section, we will describe a few first-generation n-channel polymers, followed by more recent developments.

The first report of an n-channel TFT-active polymer was poly(benzobisimidazobenzophenanthroline) (BBL, **P71**) and the corresponding Semiladder poly(benzobisimidazobenzophenanthroline) (BBB) (**P72**) (Figure 1.20). This ladder-type polymer exhibits a high electron mobility of 0.1 $cm^2\,V^{-1}\,s^{-1}$ as spin-coated polycrystalline film [53b] and ~0.01 $cm^2\,V^{-1}\,s^{-1}$ for nanobelts [138]. In a series of papers, the Northwestern University (NU) group reported polymers based on indenofluorene and bisindenofluorene cores having C≡O and $C{\equiv}C(CN)_2$ substituents (**P73** family, Figure 1.20) [57b]. A novel design approach was employed using computational modeling to identify favorable monomer properties, such as core planarity, solubilizing substituent tailorability, and appropriate electron affinity, with gratifying results. Monomeric model compounds were also synthesized to confirm these properties, and crystal structures of several cores revealed short (<3.5 Å) π–π stacking distances with favorable solubilizing substituent orientations. A family of 10 homopolymers and bithiophene **P73** copolymers was synthesized via Yamamoto and Stille polymerizations, which exhibited n-channel mobilities approaching $10^{-3}\,cm^2\,V^{-1}\,s^{-1}$ under inert conditions.

The Northwestern group have also designed new homopolymers and copolymers based on the dithenodiimide core (**P74** and **P75**, Figure 1.20) [139]. Two of these polymers are processable in common organic solvents: the homopolymer poly(*N*-(2-octyldodecyl)-2,2′-bithiophene-3,3′-dicarboximide) (P(DTI), **P74**) exhibits n-channel FET activity, and the copolymer poly(*N*-(2-octyldodecyl)-2,2′:5′,2″:5″,2‴-quaterthiophene-3,3′-dicarboximide) (P(DTI-T2), **P75**) exhibits air-stable p-channel FET operation. After annealing, **P74** films exhibit a very high degree of crystallinity and an electron mobility >0.01 $cm^2\,V^{-1}\,s^{-1}$ with a current

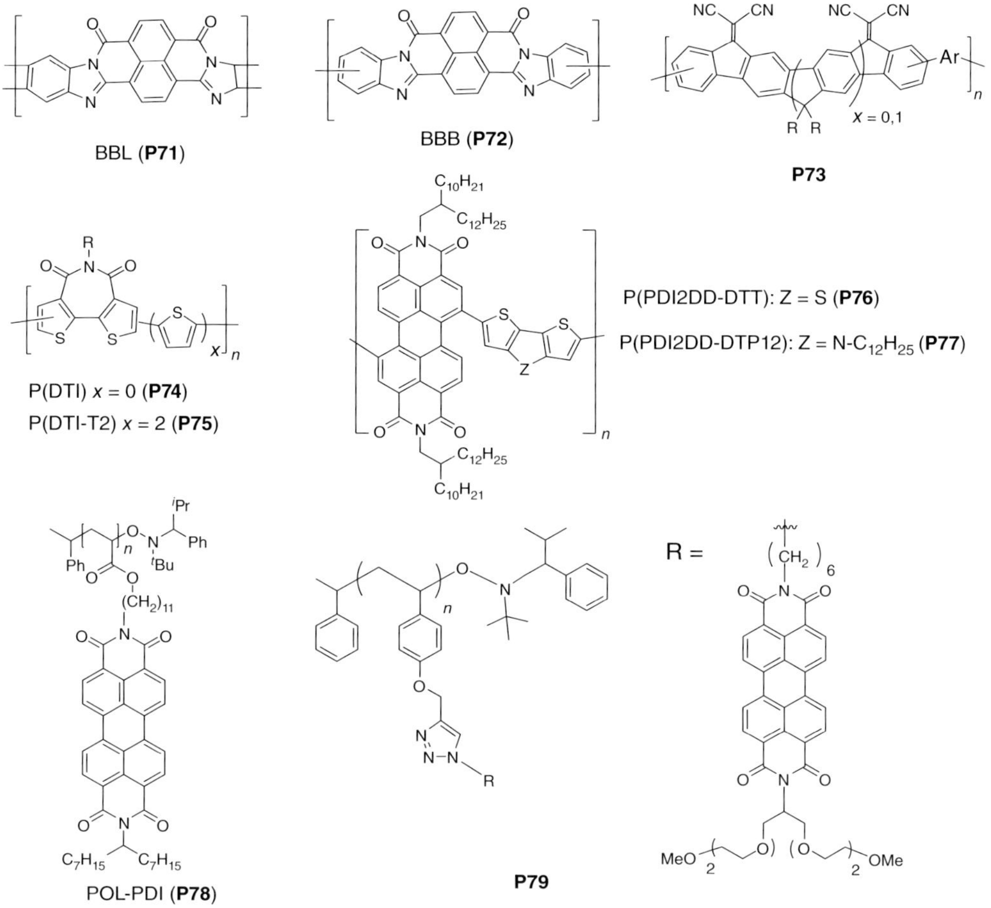

Figure 1.20 Chemical structure of n-channel polymer semiconductors **P71–P79**.

on–off ratio of 10^7, which is remarkably independent of the film deposition conditions. Extraordinarily, P(DTI) films also exhibit terracing in AFM images with a step height matching the X-ray diffraction *d* spacing, a rare phenomenon for polymeric organic semiconductors (Figure 1.21).

Relevant n-channel polymers are those based on rylene dicarboximide cores, particularly perylene and naphthalene (Figures 1.20 and 1.22). The first perylene-based polymer (P(PDI2DD-DTT), **P76**, Figure 1.20) was synthesized by Stille coupling of *N*,*N*′-dialkyl-1,7-dibromo-3,4,9,10-perylene diimide (PDI) with a distannyl derivative of dithienothiophene [140]. The M_w of P(PDI2DD-DTT) was not very high (~15 kDa using gel permeation chromatography (GPC)) and it was soluble in several solvents and could readily be processed from solution. Differential scanning calorimetry (DSC) showed a glass-transition temperature of 215 °C, while thermal gravimetric analysis (TGA) suggested excellent thermal stability with an onset decomposition temperature under nitrogen of 410 °C.

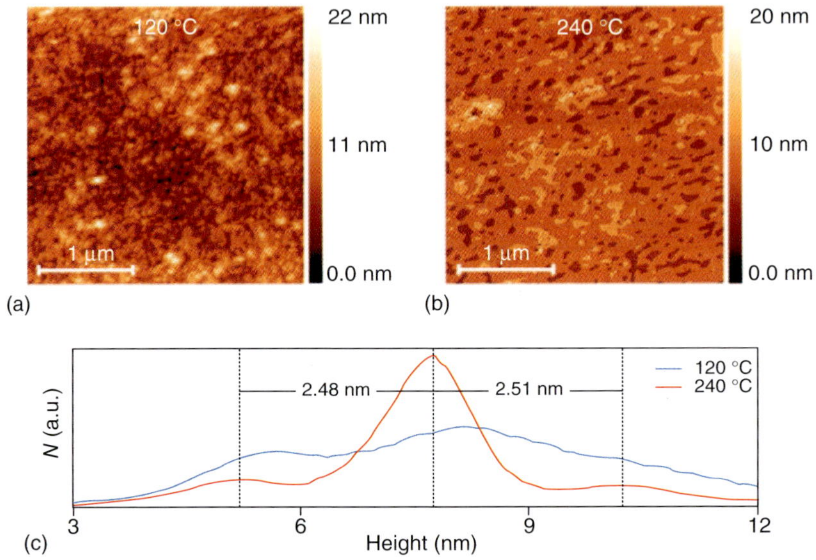

Figure 1.21 AFM images of **P74** films spin-cast from 1,2,3-TRICHLOROBENZENE (TCB) and annealed at 120 °C for 30 min (a) and 240 °C for 2 h (b), and a height histogram (c) for both images (120 °C in blue and 240 °C in red) revealing the formation of a terraced surface having a 2.50-nm step height [139]. (Reproduced with permission. Copyright © 2008 American Chemical Society.)

Polymer **P76**-based OFETs (Al source/drain electrodes, top-contact/bottom-gate geometry) were measured under nitrogen and exhibited electron mobilities as high as ~0.01 $cm^2 V^{-1} s^{-1}$ and I_{on}:$I_{off} > 10^4$. Very recently, a dithienopyrrole analog, **P77** (Figure 1.20), was reported to show an electron mobility of 7.4×10^{-4} $cm^2 V^{-1} s^{-1}$, which increased to 1.2×10^{-3} $cm^2 V^{-1} s^{-1}$ on annealing at 100 °C for 60 min under inert atmosphere [141]. The lower mobility observed for **P77** may be related to the dilution of the electron-transporting unit by the presence of the additional N-substituents of the dithienopyrrole donors. Thelakkat *et al.* [142] have reported OFETs based on polymers containing perylene unit as pendant groups, such as **P78** (Figure 1.20). For this polymer, after thermal annealing at 210 °C for 60 min, the threshold voltage dropped significantly to 7 V, while the current and charge-carrier mobility both increased by 100 times, approaching 1.2×10^{-3} $cm^2 V^{-1} s^{-1}$. Unfortunately, OFETs based on these polymers are unstable under ambient conditions. In a recent work by Thelakkat and coworkers [143], electron-deficient semiconducting side-chain polymers carrying perylenediimide pendant moieties were synthesized with a combination of "click" chemistry and nitroxide-mediated radical polymerization. Space-charge-limited current (SCLC) measurements indicated appreciable electron mobilities of up to 0.01 $cm^2 V^{-1} s^{-1}$ for **P79** (Figure 1.20), which carries oligoethyleneglycol substituents. This study demonstrated that, by varying the

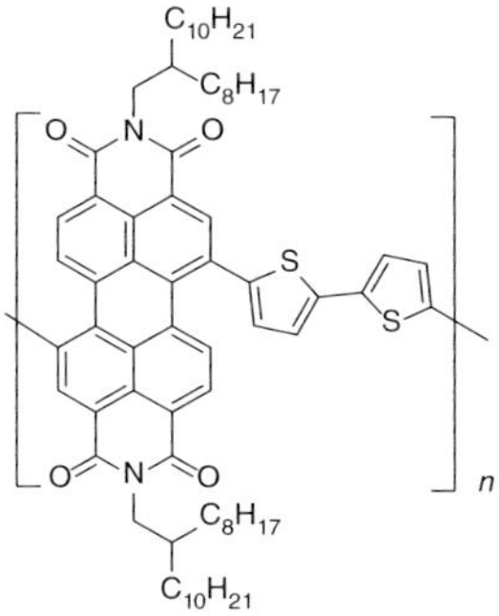

P(PDI2OD-T2) (**P80**)

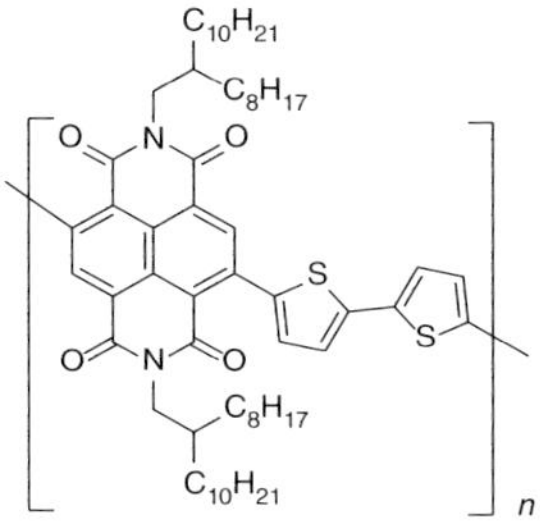

P(NDI2OD-T2) (**P81**)

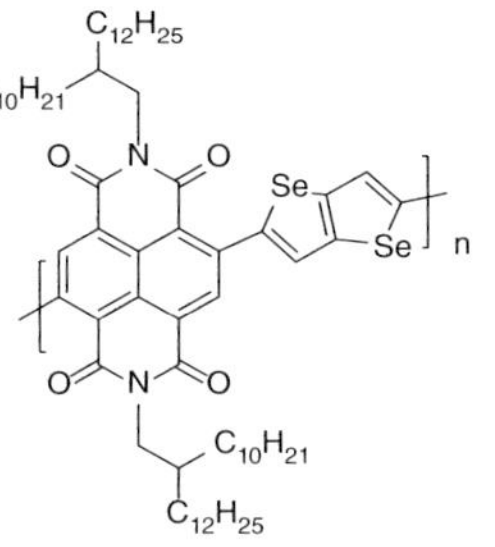

P82

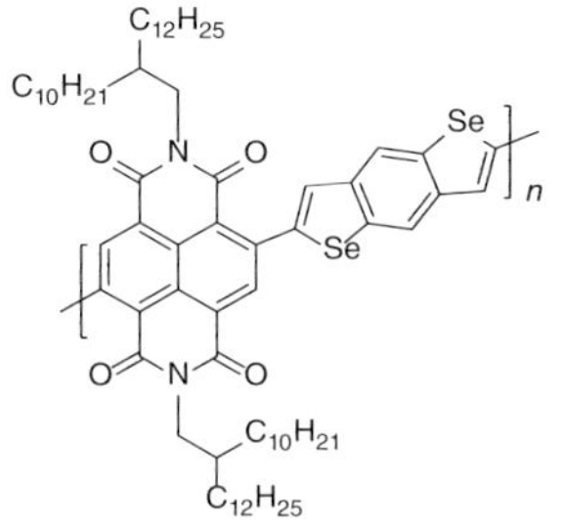

P83

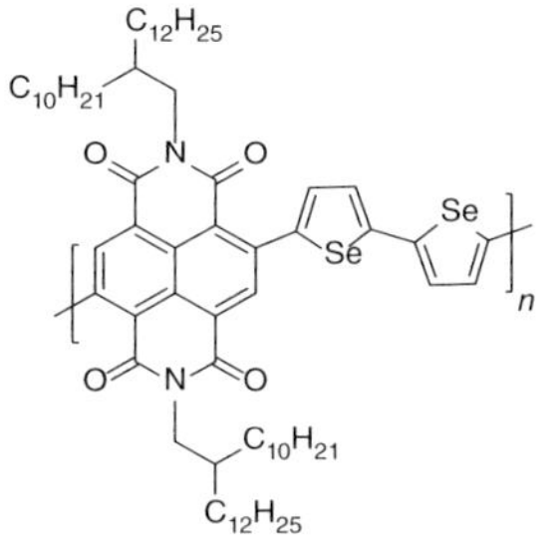

P84

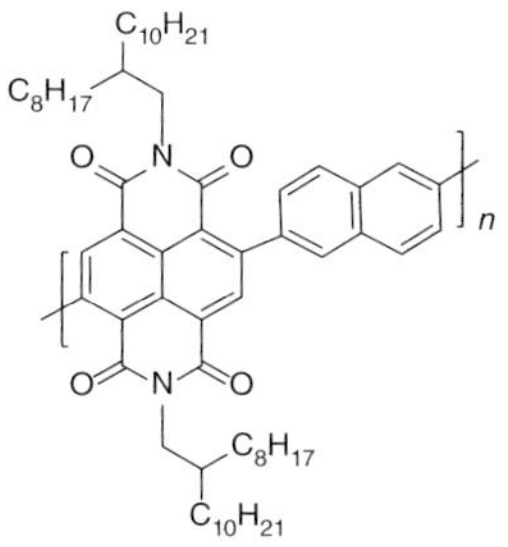

P85

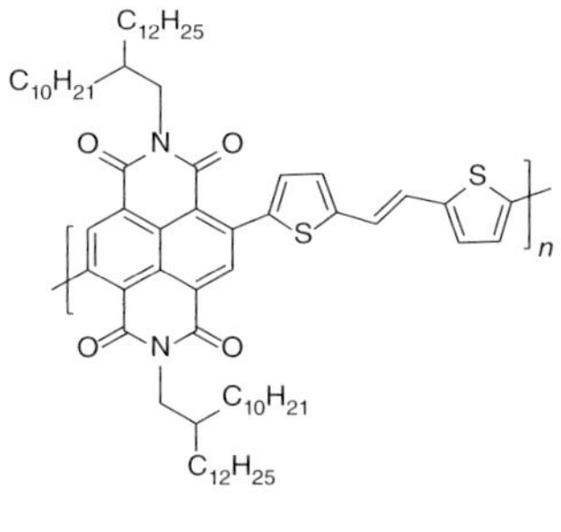

P86

Figure 1.22 Chemical structure of n-channel polymer semiconductors **P80–P86**.

substituents of side-chain polymers, optical and charge-transport properties can be fine-tuned.

Recently, the Polyera Corporation team has reported the synthesis, characterization, and comparative properties of *N,N′*-dialkylperylenedicarboximide-dithiophene (PDIR-T2, **P80**, Figure 1.22) and *N,N′*-dialkylnaphthalenedicarboximide-dithiophene (NDIR-T2, **P81**, Figure 1.22) copolymers and the fabrication of the corresponding bottom-gate TFTs on Si/SiO_2 substrates [144]. Their results demonstrate that the choice of the NDIR versus PDIR co-monomer is strategic to achieving both high-performance bottom-gate n-channel TFTs and device functioning under ambient conditions. The rylene building block and the polymer structural design rationale were as follows: (i) The electron-poor NDIR co-monomer was selected because of the large electron affinity of this core, comparable to that of the far more π-extended PDIR systems. (ii) Equally important, NDIR-Br_2 can be easily isolated as pure 2,6-diastereoisomers, enabling the synthesis of a regioregular polymeric backbone. Note that the isolation of PDIR-Br_2 regioisomers, although demonstrated, is tedious. Therefore, compared to PDIR-based polymers, it should lead to a more π-conjugated structure and, consequently, better charge-transport efficiencies. (iii) Proper alkyl (R) functionalization at the rylene nitrogen atoms, in that study 2-octyldodecyl (2OD), should result in highly soluble and processable, yet charge-transport-efficient, polymers. (iv) Finally, the dithiophene (T2) unit is utilized because of the commercial availability, stability, and known electronic structure and geometric characteristics of this core, likely providing highly conjugated, planar, and rod-like polymers. The new NDIR- and PDIR-based polymers were synthesized in high yields via a Pd-catalyzed Stille polymerization. Using the reported synthetic procedure, the polymer M_ws are larger for P(NDI2OD-T2) (**P81**) (~250 K, PD ~ 5) than for P(PDI2OD-T2) (**P80**) (~32 K PD ~ 3). The optical and electrochemical properties of these new systems reveal important aspects of the polymer electronic structures and NDIR versus PDIR co-monomer effects. Bottom-gate top-contact OTFTs were fabricated on n^{++}-Si/SiO_2/OTS substrates on which the semiconducting polymer solutions (~3–10 mg ml^{-1} in 1,2-Dichlorobenzene (DCB)-$CHCl_3$) were spin-coated to afford ~100-nm-thick films. The films were annealed at 110 °C for 4 h before the TFT structure was completed by Au source/drain vapor deposition. Electrical measurements were performed both under high vacuum and in ambient atmosphere. Electron mobilities of ~0.08–0.06 $cm^2 V^{-1} s^{-1}$ for P(NDI2OD-T2) and ~0.003–0.001 $cm^2 V^{-1} s^{-1}$ for P(PDI2OD-T2) were measured in vacuum. However, when the same TFT array was measured in ambient atmosphere, the P(NDI2OD-T2)-based devices continued to function also after 16 weeks from fabrication ($\mu_e \sim 0.01$ $cm^2 V^{-1} s^{-1}$), while P(PDI2OD-T2) mobility dropped to ~2×10^{-4} $cm^2 V^{-1} s^{-1}$ within 1 week, in agreement with previous studies on PDI-based polymers (Figure 1.23). In a very recent study, we addressed the effect of the regioregularity of these systems by synthesizing the regioregular (only 1,7-linked) P(PDI2OD-T2) as well as the regioirregular (2,6- + 2,7-linked) P(NDI2OD-T2) polymers. The TFT results showed that the regioirregular P(NDI2OD-T2)-based devices exhibited 10 times lower electron mobilities, corroborating the key role of the polymer architecture regiochemistry [145].

The same team also developed high-performance polymeric TG–BC TFTs and the first all-polymeric complementary metal–oxide–semiconductor (CMOS) circuit functioning in ambient based on P(NDI2OD-T2) (**P81**) [146]. These TG–BC TFTs were fabricated on glass or PET and had the structure substrate/Au(source–drain contacts)/P(NDI2OD-T2)/polymeric dielectric/Au(gate contact). This structure was selected because of the superior injection characteristics of typical staggered (top-gate) architectures and considering the facile channel miniaturization for bottom-contact TFTs which

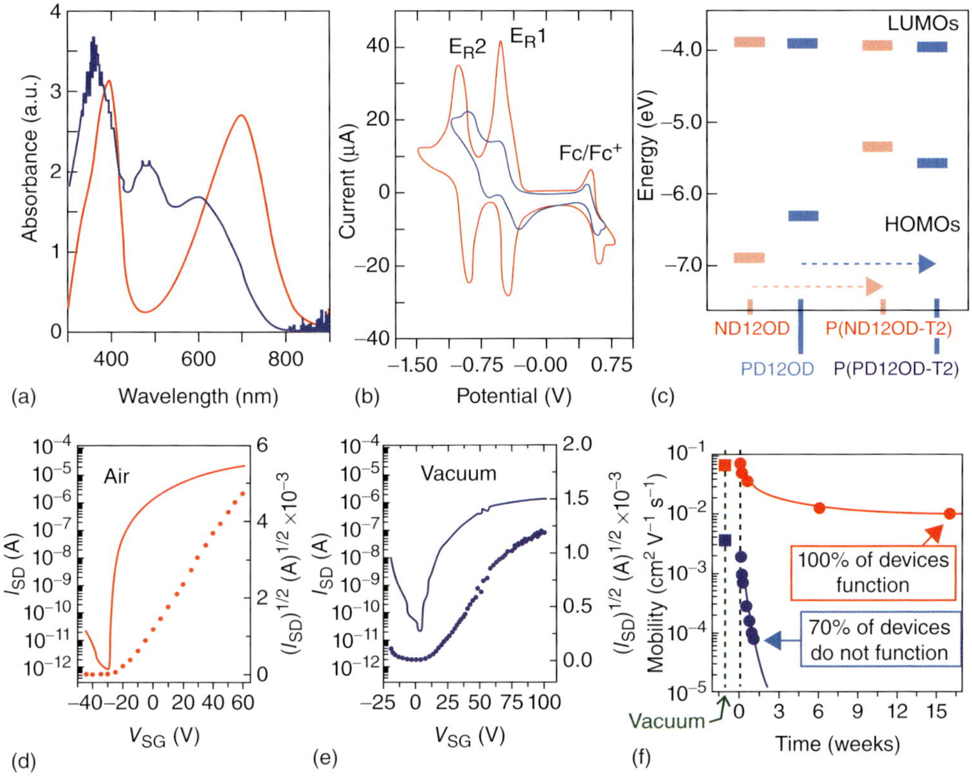

Figure 1.23 (a) Optical absorption spectra of spin-coated P(NDI2OD-T2) (red line) and P(PDI2OD-T2) (blue line) films (~30 nm thick) on glass. (b) Thin-film cyclic voltammetries (Fc (+0.54 V vs SCE (saturated calomel electrode)) internal standard) of P(NDI2OD-T2) (red line) and P(PDI2OD-T2) (blue line) thin films on a Pt electrode. The ER1 values of NDI2OD and PDI2OD (not shown) are −0.49 and 0.46 V versus SCE, respectively. (c) Energy diagram for the specified rylene monomers and polymers. (d) I–V transfer plots for P(NDI2OD-T2) TFT in air for 1 h and (e) P(PDI2OD-T2) TFT in vacuum. (f) Polymer TFT electron mobility plots in vacuum and ambient (RH) 20–40%, ($T \approx 25\,°C$) versus time [144]. (Reproduced with permission. Copyright © 2009 American Chemical Society.)

could lead to high-frequency circuits. These devices were fabricated with the P(NDI2OD-T2) film deposited by spin-coating as well as gravure, flexographic, and inkjet printing and with the dielectric layer deposited by spin-coating. Furthermore, TFTs in which both the semiconductor and dielectric layers were gravure-printed were demonstrated. All device fabrication processes were performed under ambient conditions with the exception of the Au contact vapor-deposition and the film-drying steps (110 °C). The TG–BC TFTs based on this polymer exhibited excellent n-channel OTFT characteristics in ambient, with electron mobilities of up to ~0.45–0.85 $cm^2\,V^{-1}\,s^{-1}$, I_{on}:$I_{off} > 10^5$, and $V_{on} \sim 0$–5 V. Importantly, the carrier mobility of P(NDI2OD-T2)-based TFTs

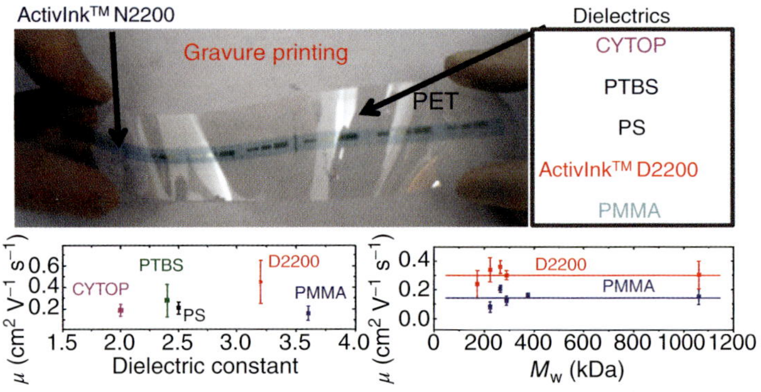

Figure 1.24 Top: Optical image of the first gravure-printed n-channel polymeric TFTs on a PET web fabricated at Polyera. Bottom: Mobility versus dielectric constant (k) of the polymeric gate dielectric (left) and molecular weight (M_w) (right) [146]. (Reproduced with permission. Copyright © 2009, Nature Publishing Group.)

was insensitive to the dielectric constant (k) of the gate dielectric material (Figure 1.24). This is of great importance to broaden the compatibility of this n-channel semiconductor family with several p-channel materials *using the same gate dielectric*. Furthermore, this polymer's TFT properties are independent of the polymer molecular weight (M_w) over a large range of values ($M_w \sim 200$ kDa to >1 MDa, Figure 1.20). The insensitivity of the device performance on the polymer chain-length extension is of extreme importance for large-scale synthesis and batch-to-batch reproducibility of the TFT characteristics. P(NDI2OD-T2)-based TFTs are also exceptionally stable in ambient up to ~70% relative humidity. Because of the stability of this n-channel polymer family, excellent TFT performance with high-work-function metal contacts, and compatibility with Polyera UV-curable dielectrics with both p- and n-channel semiconductors, they also enabled the first polymeric complementary logic. We have also fabricated printed inverters with P(NDI2OD-T2) (**P81**, n-channel), P3HT (p-channel), and ActivInk™ D2200 (gate dielectric). These inverters show remarkably small hysteresis, reflecting the transistor's threshold voltage stability. The voltage gains for the gravure-printed devices are very large ($dV_{OUT}/dV_{IN}(\max) > 25-60$). More recently, Noh *et al.* fabricated monolithically integrated polymeric complementary circuits using P(NDI2OD-T2) and two p-type polymers P3HT and a new dithiophene-based polymer (Polyera ActivInk P2100). Inkjet-printed TG–BC FETs exhibited very high hole and electron mobilities (μ_{FET}) of $0.2-0.5\ cm^2\ V^{-1}\ s^{-1}$. The active regions of the FET were patterned and via holes were defined by direct inkjet printing of the conjugated polymer solutions and the polymer gate dielectric solvent, respectively, enabling high-performance CMOS inverters (gain >30) and ring oscillators (f_{osc} up to ~50 kHz) [147].

In a recent paper, Salleo *et al.* investigated the molecular packing and structure of P(NDI2OD-T2) (**P81**). The importance of molecular packing and microstructure on performance of p-type thiophene-based semiconductors is widely appreciated and has been extensively studied [28, 148]. It is generally believed that the best transport properties are attained when there is a high degree of in-plane π stacking of the thiophene rings, because this allows two-dimensional (2D) transport along the chain backbone and along the π-stacking direction in the plane of the substrate [148e, 149]. To date, similar structure–property studies are not as well established in n-type polymers because of the lack of high-performing materials. X-ray scattering experiments on P(NDI2OD-T2) films revealed that this polymer exhibits an exceptional degree of in-plane ordering and adopts a largely face-on packing (π-stacking direction normal to the substrate), which is an uncommon crystallographic texture for semiconducting polymers with high field-effect mobility (Figure 1.25). Furthermore, we employed X-ray scattering from aligned films as well as AFM characterization of the top and bottom interface, and, to support these findings, discussed the implications of this unexpected crystalline texture for charge transport [150].

Finally, recent bulk electron transport studies using both time-of-flight and electron-only current measurements suggest a bulk mobility of $\sim 5 \times 10^{-3}\ cm^2\ V^{-1}\ s^{-1}$ for **P81**, which is the highest value reported for time-of-flight (TOF) electron transport in conjugated polymers (Figure 1.26) [151].

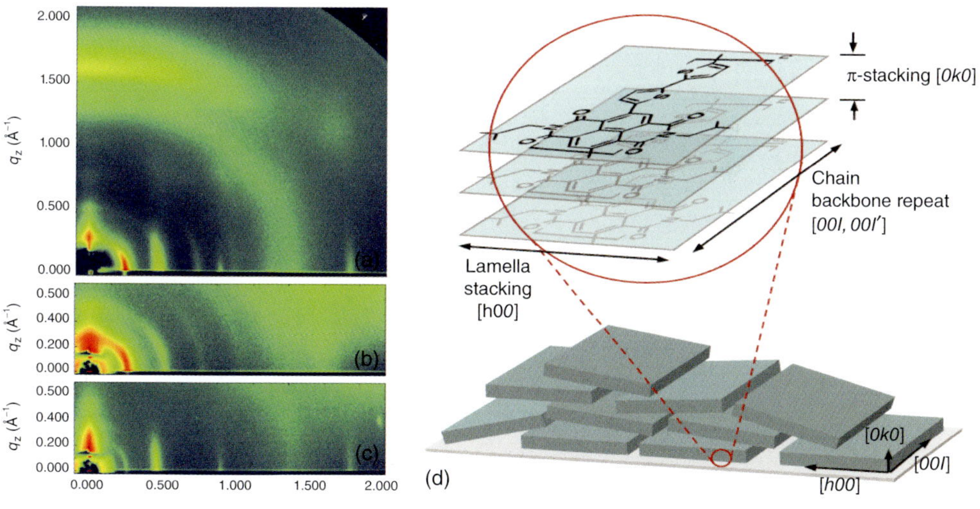

Figure 1.25 X-ray characterization of P(NDI2OD-T2) structure. (a–c) 2D grazing incidence diffraction pattern from spun-cast, isotropic film (a), dip-coated, aligned film with scattering vector *q* nominally perpendicular to the fiber direction (b), and *q* nominally parallel to the fiber direction (c). (d) Schematic of face-on molecular packing of P(NDI2OD-T2) inferred from X-ray data (top) indicating the repeat directions referenced. Proposed microstructural arrangement of the crystallites (bottom) indicating slight disorder in the π-stacking and lamella stacking directions of the flat, platelet-like crystallites [150]. (Reproduced with permission. Copyright © 2010, John Wiley & Sons, Inc.)

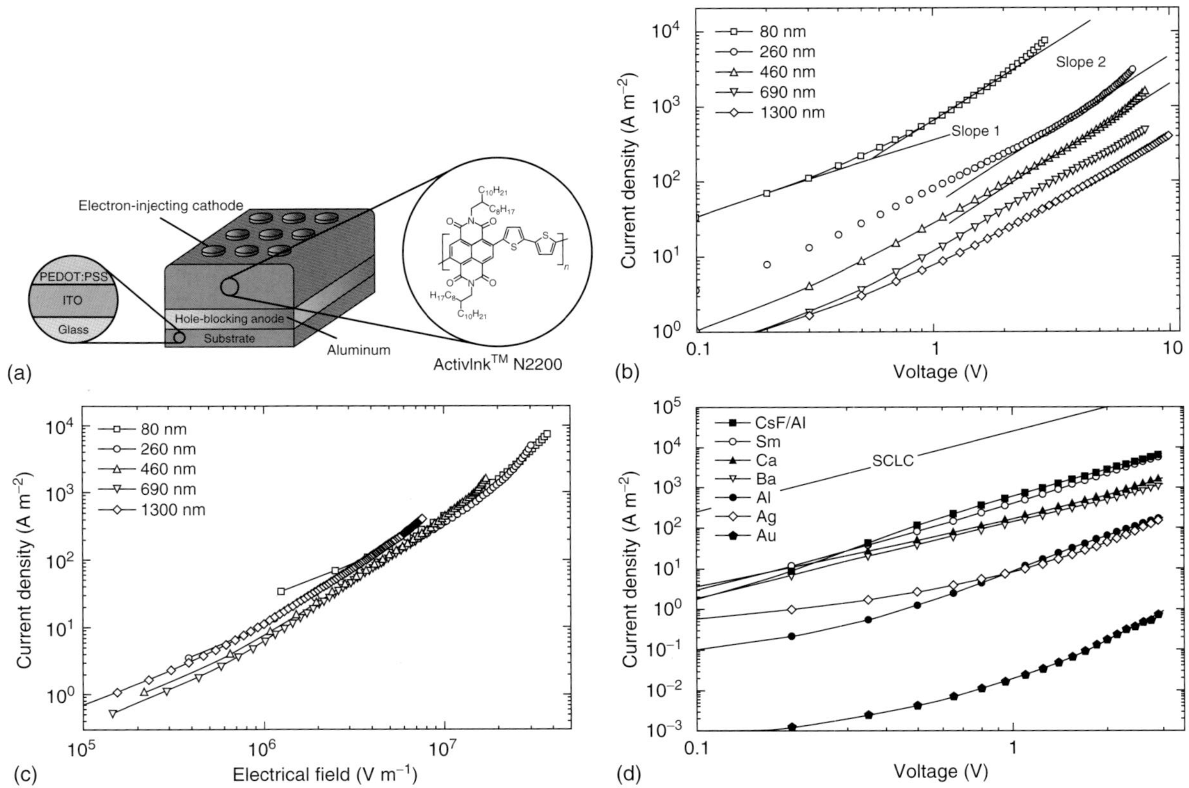

Figure 1.26 (a) Schematic illustration of an electron-only device used to investigate the charge transport of the conjugated n-type polymer **P81** used in this work. PEDOT:PSS is used as a smoothing layer for the hole-blocking bottom aluminum anode to avoid negative differential resistance effects. (b) *J–V* characteristics of **P81** electron-only devices with a barium cathode for several N2200 layer thicknesses. (c) *J–E* characteristics of **P81** electron-only devices with barium cathode for several layer thicknesses. (d) *J–V* characteristics of **P81** electron-only devices ($d = 85$ nm) with different electron-injecting top electrodes. The straight line shows the expected current according to the Mott–Gurney law using the average TOF mobility ($\mu_{TOF} = 5 \times 10^{-3}$ cm^2 V^{-1} s^{-1}) [151]. (Reproduced with permission. Copyright © 2010, John Wiley & Sons, Inc.)

Importantly, the electron-only device currents were found to be injection-limited for a wide range of electrode work functions and semiconductor layer thicknesses, despite the rather high electron affinity of this polymer. Contact-limited currents were observed even when low-work-function metals such as Sm, Ca, Ba, or Cs were employed, which are known to enable ohmic contacts with other n-type polymers [152]. However, the previously investigated polymers typically exhibit rather low bulk-transport-limited currents owing to, for example, severe electron trapping in an exponential density-of-states distribution. Thus, we believe that evaporation of reactive metals onto layers of conjugated polymers may generally cause injection barriers (through the formation of oxides and chemical defects), but that these barriers are masked by the low bulk currents in the majority of n-type polymers that have been investigated to date.

Recently, Jenekhe and coworkers reported the synthesis and physicochemical properties of new NDI (naphthalene diimide)-based polymers **P82–P84** (Figure 1.22) with selenophene-based donor units instead of the bithiophene unit used earlier [153]. Bottom-gate/top-contact OFETs gave average electron mobilities as high as 0.24 $cm^2\ V^{-1}\ s^{-1}$ with I_{on}/I_{off} ratios of 10^4-10^6. The authors also mentioned that in the phenyl end-capped high molecular weight **P84**, large enhancement of the field-effect mobilities were observed compared to the uncapped polymers. In a parallel study by Yang *et al.* [154], an acene-based centrosymmetric copolymer of NDI, **P85** (Figure 1.22), was synthesized and characterized in bottom-gate/top-contact OFET devices. Electron mobilities as high as 0.056 $cm^2\ V^{-1}\ s^{-1}$ were obtained for **P85**-based devices, which indicates that, in addition to the donor strength, the geometric feature of the donor moiety also strongly affects the charge-transport characteristics. Noh and coworkers [155] recently reported another high-performance NDI-based n-channel polymer, **P86** (Figure 1.22), which is systematically designed and synthesized with alternating long alkyl (2-decyltetradecyl) substituted NDI (acceptor) and thienylene–vinylene–thienylene (donor) building blocks. Compared to the famous P(NDI2OD-T2), this new polymer has an extended π-conjugated backbone with a comparable LUMO energy level (−4.00 eV). TG–BC devices fabricated with polymer **P86** demonstrated an extremely high electron mobility of 1.8 $cm^2\ V^{-1}\ s^{-1}$ ($I_{on}/I_{off} = 10^6$) with high air and bias-stress stability. In this device architecture, a typical polymer dielectric, poly(methyl methacrylate) (PMMA), was used. This OFET device is among the best performing polymer-based n-channel device reported to date. The high performance of **P86** was attributed to the π-extended conjugation of its polymer backbone, enhanced π–π intermolecular interactions, improved crystallinity in thin-film phase with a highly interdigitated lamellar structure, and mixed face-on/edge-on orientations.

In a recent report by Pei and coworkers [156], a novel highly electron-deficient poly(*p*-phenylenevinylene) (PPV) derivative, benzodifurandione-based PPV **P87** (Figure 1.27), was developed. This new polymer includes strong electron-withdrawing carbonyl groups on the double bonds, which extends the backbone π conjugation and forms intramolecular hydrogen bonds to afford a "locked" aromatic plane. Solubilizing alkyl groups (4-octadecyldocosyl) with distant

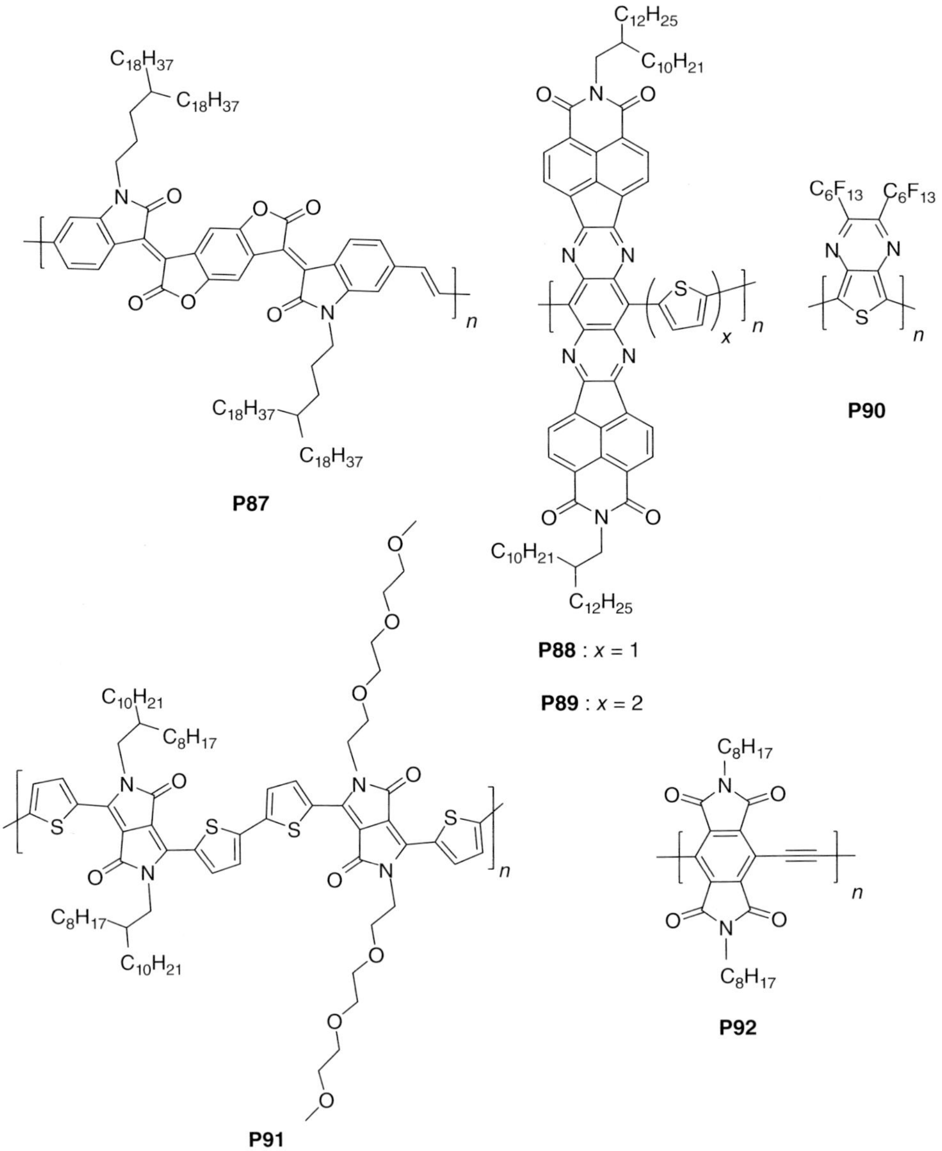

Figure 1.27 Chemical structure of n-channel polymer semiconductors **P87–P92**.

branching points are used as side chains to ensure both good solubility and interchain π–π stacking. The authors claim that this new polymer overcomes the common defects in PPVs, such as conformational disorder, weak interchain interaction, and a high LUMO level, all of which should lead to high n-channel mobilities. Grazing-incident X-ray diffraction (GIXD) and AFM characterizations indicated the formation of crystalline fiber-like intercalating networks with

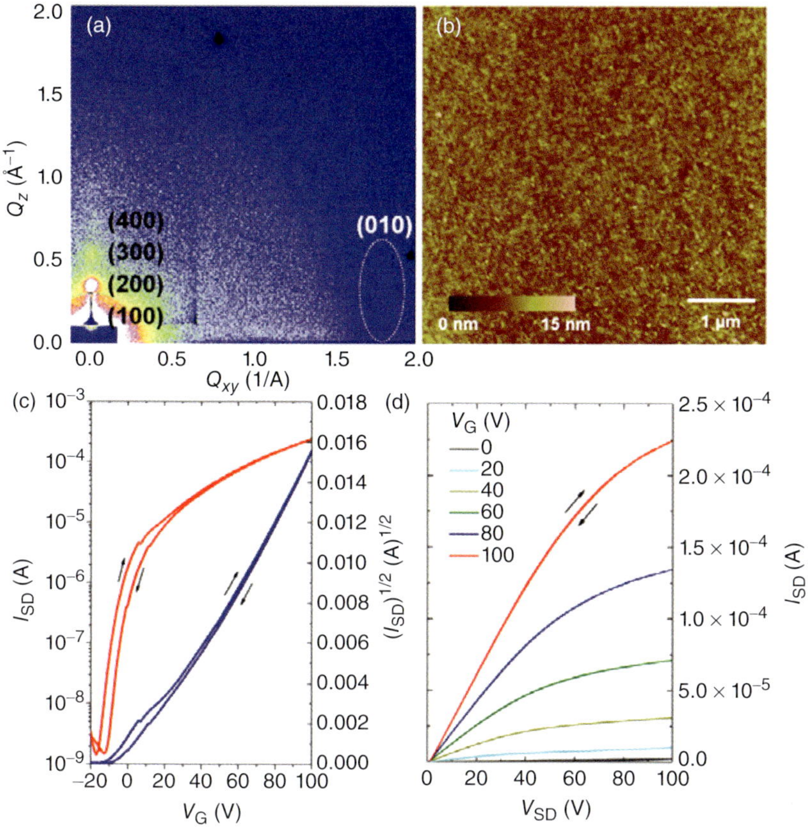

Figure 1.28 (a) 2D-GIXD pattern and (b) AFM height images of polymer **P87** film prepared by spin-coating its DCB solution (3 mg ml^{-1}). (c) Transfer and (d) output characteristics of a **P87** device under ambient conditions ($L = 10$ µm, $W = 200$ µm, $C_i = 3.7$ nF cm^{-2}), measured $V_T = 5$ V [156]. (Reproduced with permission. Copyright © 2013, American Chemical Society.)

a good lamellar edge-on packing and a short interchain π–π stacking distance of ~3.55 Å (Figure 1.28a,b). Unlike other PPV derivatives, **P87** clearly exhibited a much stronger aggregation tendency and more ordered packing. Using **P87** as the semiconductor layer in a TG–BC OFET device, electron mobilities as high as 1.1 cm^2 V^{-1} s^{-1} were achieved under ambient conditions (Figure 1.28c,d). This is one of the highest mobilities achieved to date in n-channel polymer OFETs.

The first example of a heterocyclic diimide building block, tetraazabenzodifluoranthenediimide (BFI), was recently synthesized by Jenekhe *et al.* [157] via fusion of a synthetically tunable tetraazaanthracene core and two naphthalene imide units. This new core forms a large, rigid, ladder-type macromolecular structure with 11 aromatic rings (2.0-nm-long π-conjugated framework) and shows high electron affinity (3.6–4.3 eV) and tunable optoelectronic properties. In a later report by the same group, two copolymers, **P88** and **P89** (Figure 1.27), were

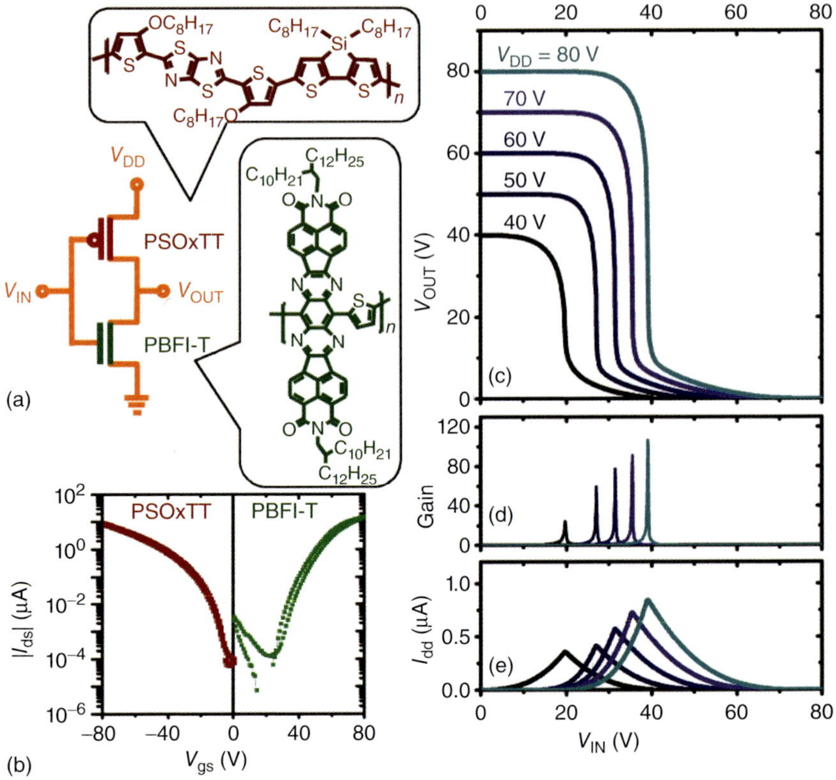

Figure 1.29 (a) Schematic of a complementary p/n polymer inverter and molecular structures of p-channel polymer PSOxTT and n-channel polymer Poly(tetraazabenzodifluoranthenediimide-thiophene) (PBFI-T) (**P88**). (b) Transfer curves of the p-channel and n-channel polymer transistors. Static switching characteristics of the inverter. (c) Output voltage, (d) gain, and (e) current as a function of input voltage (V_{IN}) with various power supply voltages (V_{DD}) [158]. (Reproduced with permission. Copyright © 2013, American Chemical Society.)

prepared by incorporating BFI as the key building block along with thiophene and bithiophene monomers, respectively [158]. These polymers exhibit a novel two-dimensional (2D) π conjugation both along the main chain and in the lateral direction, leading to efficient n-channel transport in OFETs with electron mobilities of $0.3\,cm^2\,V^{-1}\,s^{-1}$. Complementary inverters with **P88**-based transistors showed nearly ideal switching characteristics with a high gain of 107 (Figure 1.29).

Although the fluorination of π-conjugated small molecules has shown great promise in designing high-performance n-channel semiconductors, there have been only limited reports on fluorinated n-channel polymers. Swager and coworkers [159] recently reported the synthesis of a highly fluorinated, low-bandgap ($E_g^{opt} = 0.87\,eV$), π-conjugated polymer, poly-(2,3-bis(perfluorohexyl)thieno[3,4-*b*]pyrazine) **P90** (Figure 1.27). This new polymer exhibits exclusive solubility in perfluorinated solvents such as perfluorooctane and perfluoro(methylcyclohexane). Additionally, it has high air- and thermal stability, high electron affinity (4.12 eV), and good electrochemical n-doping characteristics. TG–BC OFET devices with **P90** showed low electron mobilities of $\sim 10^{-5}\,cm^2\,V^{-1}\,s^{-1}$, which was attributed to the unfavorable thin-film morphology. Although the carrier mobilities obtained for this polymer is low, the authors believe that their design approach is important for the realization of orthogonally processed organic photovoltaic devices.

DPP is an attractive building block for n-channel polymer semiconductors owing to its several favorable characteristics including high electron-deficiency, enhanced backbone coplanarity, and good intermolecular $\pi-\pi$ interactions. However, to date, DPP-based polymers have been used only in p-channel and ambipolar OFET devices. In a recent study by Patil *et al.* [160], a novel DPP–DPP-based conjugated copolymer **P91** was reported with alternating alkyl and triethylene glycol side chains (Figure 1.27). It was found that this side-chain architecture provides good solubility and high molecular weight, and induces spontaneous chain crystallization. After a careful device engineering, strictly unipolar n-channel OFET devices (TG–BC) were fabricated with electron mobilities of up to $3\,cm^2\,V^{-1}\,s^{-1}$. The authors mentioned that this unipolar behavior of **P91** with high electron mobility is crucial to printed, low-power, large-scale complementary logic circuitry.

A new class of n-channel semiconducting polymer based on the smallest rylenediimide building block pyromellitic diimide (PyDI) was recently developed by Katz and coworkers [161]. Prior to this report, the same group also developed small-molecule n-channel semiconductors based on PyDI core [162]. **P92** is an example of this new semiconductor family with ethynyl linkages and octyl (C_8H_{17}) side chains (Figure 1.27). Unipolar n-channel transfer characteristics with moderate electron mobility of $2\times10^{-4}\,cm^2\,V^{-1}\,s^{-1}$ were observed after ambient processing. The authors claimed that, to the best of their knowledge, **P92** was the first example of a PyDI semionducting polymer, and it was the simplest n-channel polymer reported up to that time. Further chemical structure and device optimizations with this polymer are in progress.

1.5.2 Small Molecules

To the best of our knowledge, the first n-channel OFET was fabricated by Guillaud *et al.* [163] using vapor-deposited films of lutetiumphthalocyanine ($LuPc_2$, **M39**, Figure 1.30). These devices exhibit electron mobilities of $10^{-4}\,cm^2\,V^{-1}\,s^{-1}$ in vacuum. Bao *et al.* [164] demonstrated the first air-stable n-channel OTFTs using a semiconductor layer vacuum-deposited by perfluorinated phthalocyanines ($F_{16}CuPc$, **M40**, Figure 1.30). These FETs exhibited maximum electron mobilities of $\sim 0.03\,cm^2\,V^{-1}\,s^{-1}$ in air for sublimed films deposited at a gate-dielectric substrate temperature of 130 °C. This is one of the first contributions where

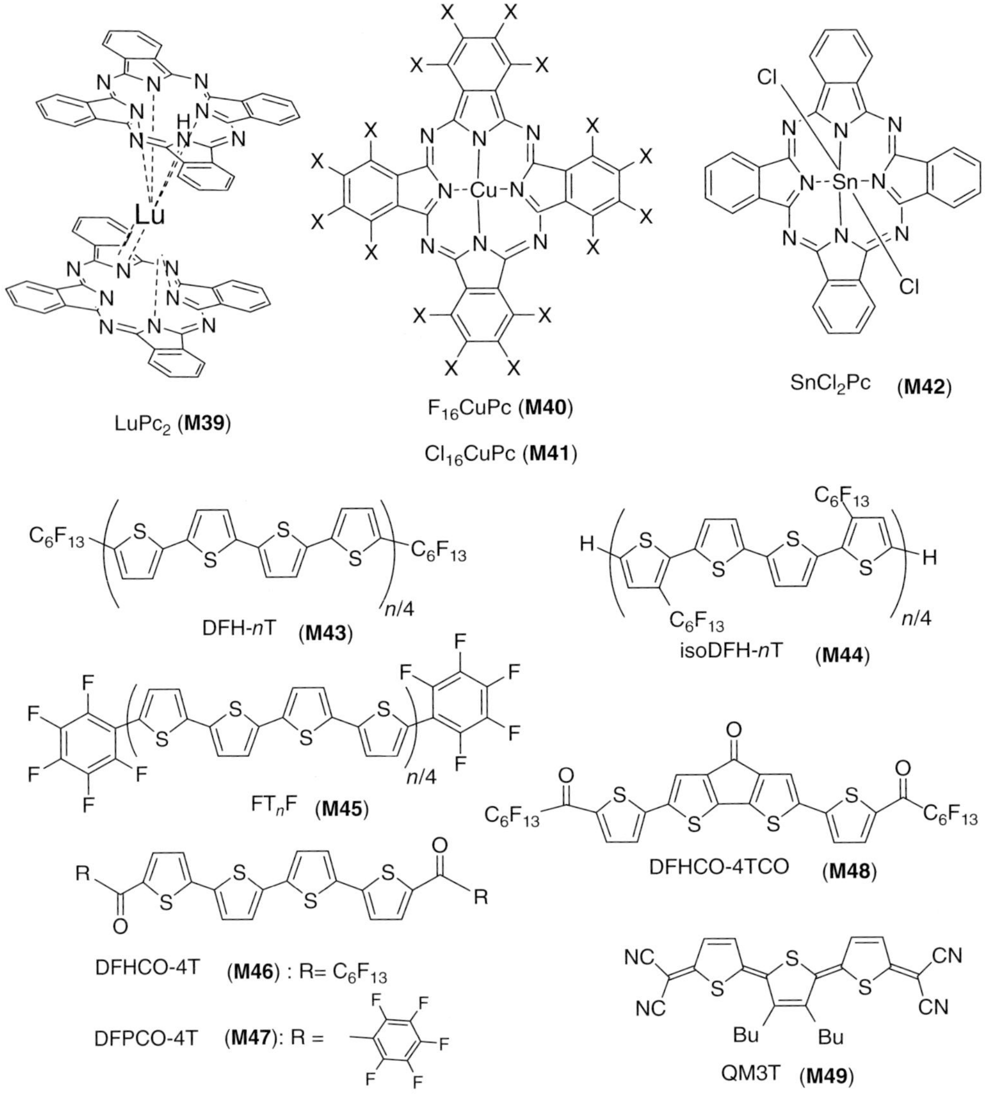

Figure 1.30 Chemical structure of n-channel small-molecule semiconductors **M39–M49**.

core functionalization of a p-channel material (CuPc) with strong electron-withdrawing F groups resulted in ambient-stable n-channel OTFTs. This material is also the most used for the fabrication of single-crystal devices [165]. A seed-induced growth by vapor processing was recently developed for *in situ* patterning of organic single-crystalline nanoribbons of $F_{16}CuPc$ on a Si/SiO_2 surface [166]. With this method, devices based on individual $F_{16}CuPc$ submicro/nanometer-sized ribbons were fabricated with Ag/Au asymmetrical source/drain electrodes, showing that the asymmetrical devices possessed much higher mobilities (0.2 $cm^2 V^{-1} s^{-1}$) than the devices with Au/Au symmetrical electrodes (0.01 $cm^2 V^{-1} s^{-1}$). This is due to the asymmetrical source/drain electrode configuration which established a stepwise energy level between the electrodes and the LUMO of F_{16}CuPc, promoting electron injection and transport. Recently, Ling *et al.* [167] have used copper hexachlorophthalocyanine (Cl_{16}CuPc, **M41**, Figure 1.30) as a semiconductor for n-channel OTFTs, reporting electron mobilities of ~0.01 $cm^2 V^{-1} s^{-1}$. Remarkably, these devices show no performance degradation after storage in air for more than 50 days. Wang *et al.* [168] have reported n-channel FETs where the F_{16}CuPc film was epitaxy grown on *para*-hexaphenyl (*p*-6P); these devices exhibited electron mobilities as high as 0.11 $cm^2 V^{-1} s^{-1}$, which approach those of the single-crystal-based devices [168]. Very recently, impressive electron mobilities of 0.30 $cm^2 V^{-1} s^{-1}$ in air were reported by Song *et al.* [169] using phthalocyanato tin(IV) dichloride ($SnCl_2$Pc, **M42**, Figure 1.30) as a semiconductor.

The first n-channel oligothiophene-based transistor was fabricated at NU in 2000. In a series of papers, our group has described the synthesis, comparative physicochemical properties, and solid-state structures of several oligothiophenes substituted with perfluorohexyl chains [170]. These series include the n-channel α,ω-diperfluorohexyl-*n*Ts (DFH-*n*Ts, **M43n**, Figure 1.30) and β,β′-diperfluorohexyl-*n*Ts (isoDFH-*n*Ts, **M44n**, Figure 1.30), which were compared with the corresponding p-channel hexyl-substituted and unsubstituted oligothiophenes (α-*n*Ts, $n=2-6$). The crystal structures of key fluorocarbon-substituted oligomers were also analyzed, which showed close π–π intermolecular interactions between the aromatic cores, whereas the fluorocarbon chains segregated into lamellar structures. X-ray structural analysis was performed for DFH-3T ($n=3$) and DFH-4T ($n=5$) which exhibit an all-anti, fully planar geometry, with dihedral angles between the mean plane of the rings of <2°. This value compares well with that reported for α4T but is much smaller than the 6–9° reported for α3T. The perfluorohexyl substituents exhibit a zigzag helical conformation that is characteristic of fluorocarbon chains [171] and are positioned at ~140° with respect to the oligothiophene backbone axes in both cases. The molecular packing of DFH-4T shares the familiar "HB" motif with an angle of 50° between the mean planes of adjacent molecules. Typical HB angles for oligothiophene α*n*Ts ($n=4-6$, 8) range between 55° and 70°. The minimum interplanar distances between neighboring molecules in DFH-4T is 3.52 Å, which is comparable to 3.5–3.9 Å in analogous oligothiophenes. The crystal structures and packing characteristics of β,β′-disubstituted systems isoDFH-5T are considerably different from those of the end-capped compounds. The effect of α,ω and β,β′ regiochemical substitution on the π core structure is dramatic, with a large torsional angle (up to 64°) forced between adjacent thiophene rings compared to the DFH-*n*Ts series. To a lesser degree, this has been observed as well for some β-alkylsubstituted oligothiophenes [172] and is due to the steric repulsion between the fluorohexyl chain CF_2 group and either the sulfur atom or the C–H moiety on the adjacent thiophene ring. It was also observed that, regardless of the differences in molecular structure and packing characteristics, all fluorocarbon-substituted oligothiophenes exhibit similar intermolecular separations between the oligothiophene core and fluorocarbon regions.

From the electrochemical and optical data, the energy positions of the HOMOs and LUMOs of these oligothiophenes were estimated and, as expected, fluorocarbon functionalization of the oligothiophene core reduces both HOMO and LUMO energies (~0.1 – 0.5 eV depending on the core length). The shift of HOMO and LUMO energies of DFH/isoDFH-*n*Ts and the corresponding alkyl-substituted oligothiophenes is in quantitative agreement with the expected effects of fluorocarbon and hydrocarbon substituent-derived Hammett parameters. Overall, the molecular orbital energy trends within each series are determined by a balance between the nature of the substituent and density and the extension of the aromatic core. The net results from this interplay are frontier MOs where the LUMO and HOMO energies are practically independent of the conjugation length for fluorocarbon- and alkyl-substituted *n*Ts, respectively. The experimental results were confirmed by DFT computational modeling, the results of which closely paralleled the experimental trends. OFET devices were fabricated and characterized on Si/SiO_2 substrates treated with Hexamethyldisilazane (HMDS). The results clearly demonstrated the crucial importance of oligothiophene core substitution in determining the p- or n-channel activity of the semiconductor. Fluorocarbon-substituted systems are majority-carrier electron-transporting materials, while oligothiophenes (with or without $-C_6H_{13}$ chains) are majority-carrier hole-transporting semiconductors. Trends in field-effect mobilities and I_{on}/I_{off} ratios between and within these series were explained by the interplay of molecular structure and film microstructure and morphology. The films that exhibited the greatest carrier mobilities possessed the appropriate combination of large grain size and smooth, interconnected morphology, exhibit molecular orientation directed along the substrate normal, and had sufficient core conjugation/length. The second distinctive material is DFH-4T, which was shown to exhibit one of the highest n-type carrier mobility (0.22 $cm^2\,V^{-1}\,s^{-1}$) and I_{on}/I_{off} ratio reported for an n-type transporting material. More recently, an extensive IR/electrochemical characterization of these fluorocarbon-substituted oligothiophenes has been performed by Casado and López Navarrete [170e].

The NU group also reported the comparative properties of arene-thiophene and fluoroarene-thiophene (FT_nF (**M45**), $n = 2-4$, Figure 1.30) semiconductors [173] with respect to regiochemical modifications of the core and oligothiophene core-shortening. These extensive studies included thermal analysis, optical spectroscopy, cyclic voltammetry, single-crystal XRD structural data for the majority of these compounds, film microstructure and morphology, and OTFT fabrication and characterization. For the fluoroarene-thiophene series, the majority charge-transport type and mobilities (0.00001 – 0.5 $cm^2\,V^{-1}\,s^{-1}$) and the I_{on}/I_{off} current ratios ($10^1 - 10^8$) change dramatically with molecular regiochemistry as well as with the substrate deposition temperature. The large electron mobility of FTTTTF was attributed to the favorable interaction between the electron-rich and electron-deficient groups, thereby achieving optimum molecular $\pi-\pi$ overlap and excellent film texture. The shorter fluoroarene-thiophene oligomers ($n = 2, 3$)

are also n-type semiconductors, demonstrating that fluoroarene end-substitution promotes majority charge flip independently of the thiophene core extension. DFT band structure calculations are shedding light on the majority charge-carrier flip in this family [174].

An effective strategy to tune oligothiophene core energy levels, optical characteristics, and solid-state packing is to functionalize the oligothiophene core with strong inductive/mesomeric carbonyl (C≡O) groups. In the first systems, the investigated carbonyl chemical functionalities included perfluorohexylcarbonyl groups (e.g., 5,5″-diperfluorohexylcarbonyl-2,2′:5′,2″:5″,2‴ (DFHCO-4T), **M46**, Figure 1.30) installed at the oligothiophene α,ω positions as well as bridged carbonyl groups positioned at the center of the molecular core (e.g., DFHCO-4TCO, **M48**, Figure 1.30) [175]. OFETs were fabricated on HMDS-treated Si/SiO_2 substrates, and DFHCO-4T showed monopolar n-type activity (in vacuum) with an exceptionally high mobility of $0.32\,cm^2\,V^{-1}\,s^{-1}$ for semiconducting films deposited at a substrate temperature of 25 °C. Note that similar carbonyl group effect on n-type (electron) transport was previously demonstrated in electroactive aromatic polyketones and polyesters [176]. Upon additional carbonyl group introduction into the quaterthiophene core, DFHCO-4TCO exhibited stable n-type activity even in air, although the ambient electron mobility ($0.01\,cm^2\,V^{-1}\,s^{-1}$) was somewhat lower than that recorded under vacuum ($0.08\,cm^2\,V^{-1}\,s^{-1}$). After proper dielectric surface modification, n-type mobilities were substantially improved to $1.7\,cm^2\,V^{-1}\,s^{-1}$ for DFHCO-4T. More recently, Schols *et al.* [177] have improved the electron mobilities of DFHCO-4T Au top-contact transistors to $4.6\,cm^2\,V^{-1}\,s^{-1}$; the drastically reduced performance using Al/LiF as top contact was also explained as a consequence of an electron-transfer reaction occurring at the metal/DFHCO-4T interface. Another promising carbonyl-containing oligothiophene is DFPCO-4T (**M47**, Figure 1.30) [178]. DFPCO-4T, which crystallizes in an HB motif, with the shortest intercore distance being 3.50 Å and the average dihedral angle between the phenyl substituent and the adjacent thiophene subunit ~53°. FETs of DFPCO-4T were fabricated with Au top-contact electrodes. Semiconductor films (50 nm) were deposited onto temperature-controlled HMDS-treated SiO_2/p^+-Si substrates by vapor deposition and drop-casting. A 50-nm layer of Au was then deposited through a shadow mask to define the source and drain electrodes. OFET characterization was performed under Argon. High electron mobilities of $\sim 0.5\,cm^2\,V^{-1}\,s^{-1}$ were observed for vapor-deposited DFPCO-4T films ($T_D = 80\,°C$) with a threshold voltage of ~30 V ($I_{on}:I_{off} > 10^8$). In solution-cast devices, electron mobilities were exceptionally high with a maximum of $\sim 0.25\,cm^2\,V^{-1}\,s^{-1}$ ($I_{on}:I_{off} = 10^5$; $V_T = 50-70\,V$). Recently, the NU group has analyzed the temperature dependence of FET mobility for a series of n-channel oligothiophenes along with those of other p-channel and ambipolar organic semiconductors [179].

Frisbie *et al.* reported that quinomethane terthiophene (QM3T, **M49**, Figure 1.30) exhibits $\mu_s \sim 0.002-0.5\,cm^2\,V^{-1}\,s^{-1}$ [180]. An extended series showing even greater performance and ambipolar transport has been recently reported

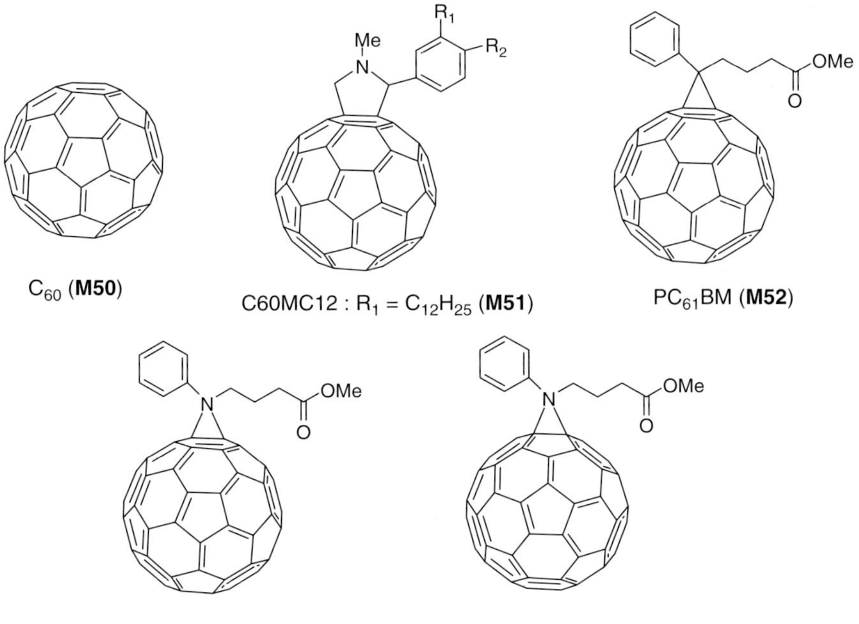

C_{60} (**M50**)

C60MC12 : $R_1 = C_{12}H_{25}$ (**M51**)

$PC_{61}BM$ (**M52**)

[6,6]-closed APCBM (**M53**)

[6,6]-open APCBM (**M54**)

Figure 1.31 Chemical structure of n-channel small-molecule semiconductors **M50–M54**.

[181, 182]. Handa *et al.* [183] have shown a new hybrid-type dicyanomethylene-substituted terthienoquinoid compound for use as a solution-processable n-channel semiconductor; electron mobilities up to $0.16\,cm^2\,V^{-1}\,s^{-1}$ were obtained upon annealing the spin-coated films.

Figure 1.31 collects the chemical structure of several fullerene derivatives. The first report of C_{60} (**M50**)-based TFT evidenced low electron mobilities of $\sim10^{-4}\,cm^2\,V^{-1}\,s^{-1}$ [184]. Following that report, C_{60} thin films were grown and studied in ultrahigh vacuum (UHV) [185]. TFTs were fabricated without air exposure and consisted of heavily doped n-type silicon wafers which were oxidized to leave a 3000 Å thick layer of silicon dioxide dielectric and bottom-contact Cr/Au electrodes. The authors reported electron mobilities of $0.56-0.2\,cm^2\,V^{-1}\,s^{-1}$, comparable to that obtained with time-of-flight measurements on C_{60} single crystals. n-Channel FETs based on C_{70} were first reported a few years later [186] and exhibited field effect mobilities up to $2\times10^{-3}\,cm^2\,V^{-1}\,s^{-1}$ in UHV and I_{on}/I_{off} ratios as high as 10^5. Unfortunately, the performance of C_{60}- and C_{70}-based FETs degrades quickly upon exposure to air. More recently, Itaka *et al.* [187] have enhanced the crystallinity of vapor-deposited C_{60} films by coating the dielectric with a thin layer of pentacene. These FETs exhibit electron mobilities of $\sim2.0-5.0\,cm^2\,V^{-1}\,s^{-1}$. Jang *et al.* [188] have recently reported air-stable C_{60} TFTs with electron mobilities of $\sim0.05\,cm^2\,V^{-1}\,s^{-1}$ by using a perfluoropolymer as the gate dielectric. Very recently, Anthopoulos *et al.* [189] have reported the greatest field-effect electron mobilities for C_{60} transistors, which approached $\mu \sim 6\,cm^2\,V^{-1}\,s^{-1}$. They also demonstrated ring oscillators based on C_{60} films grown by hot-wall epitaxy. Low-voltage C_{60} n-channel OTFTs with high electron mobilities of $2.3\,cm^2\,V^{-1}\,s^{-1}$ were fabricated by engineering the electrode/semiconductor and dielectric/semiconductor interfaces [190].

Functionalized fullerenes are promising candidates for solution-processed n-channel OTFTs. Chikamatsu *et al.* [191] have synthesized soluble C_{60}-fused *N*-methylpyrrolidine-*meta*-C12 phenyl (C_{60}MC12, **M51**, Figure 1.31) and fabricated FETs exhibiting high electron mobilities of $\sim 0.07\,cm^2\,V^{-1}\,s^{-1}$. More recently, Wöbkenberg *et al.* developed several fluorine-containing C_{60} derivatives. Solution-processed OTFTs based on these compounds exhibit electron mobilities up to $0.15\,cm^2\,V^{-1}\,s^{-1}$ [192]. These devices show enhanced stability in ambient conditions as compared to standard methanofullerene OTFTs. Far more ambient-stable C_{60}-based FETs were developed by Chikamatsu *et al.* [193] who synthesized new soluble perfluoroalkyl-substituted C_{60} derivatives exhibiting electron mobilities as high as $0.25\,cm^2\,V^{-1}\,s^{-1}$. TFTs based on [6,6]-phenyl-C61-butyric acid methyl ester ([60]PCBM) (**M52**, Figure 1.31) and [6,6]-phenyl-C71-butyric acid methyl ester ([70]PCBM), materials widely used for the fabrication of organic photovoltaic devices, have also been recently reported [194]. Despite the fact that both derivatives form glassy films when processed from solution, their electron mobilities are high at ~ 0.21 and $\sim 0.1\,cm^2\,V^{-1}\,s^{-1}$ for [60]PCBM and [70]PCBM, respectively. Although the derived mobility of [60]PCBM is comparable to the best values reported in the literature, the electron mobility of [70]PCBM is the highest value reported to date for any C_{70}-based derivative. Wudl reported the one-pot preparation of two isomeric imino-PCBMs, that is, [5,6]-open azafulleroid (**M53**, Figure 1.31) and [6,6]-closed aziridinofullerene (**M54**, Figure 1.31). OFETs were fabricated on heavily doped Si wafers with a 200-nm-thick SiO_2 layer with top-contact geometry, yielding electron mobilities of ~ 0.04, ~ 0.02, and $0.03\,cm^2\,V^{-1}\,s^{-1}$ for [5,6]-open APCBM, [6,6]-closed APCBM, and PCBM, respectively. The higher electron mobility in the [5,6]-open APCBM OFET can be attributed to its 60 π-electron nature, which affords a stronger electron-acceptor strength than the [6,6] junctions [195]. Using C_{60} and pentacene CMOS NOT logic circuits were also fabricated by Kanbara *et al.* [196]. The voltage gain demonstrated was ~4. Low-voltage CMOS inverters with C_{60} (and pentacene) were reported by Kitamura and Arakawa on glass substrates [197]. The inverter operated at low voltages (1–5 V) and the C_{60} and pentacene TFTs had high field-effect mobilities of 0.68 and $0.59\,cm^2\,V^{-1}\,s^{-1}$ and threshold voltages of 0.80 and −0.84 V, respectively. As expected from the threshold voltages of the individual transistors, the inverter operates at supply voltages of only $V_{DD} > 1.64$ V. The signal gain calculated is in the range −50 to −150. More recently, Kippelen *et al.* demonstrated complementary inverters on PEN substrates [198].

The chemical structures of several perylenes and quaterylene derivatives are shown in Figure 1.32. Horowitz *et al.* [199] first demonstrated electron mobilities of $\sim 10^{-5}\,cm^2\,V^{-1}\,s^{-1}$ with an *N*,*N*′-diphenyl-substituted perylene, **M55-A**. In 2001, Malenfant *et al.* [200] reported n-channel OFETs based on *N*,*N*′-dioctyl PDI

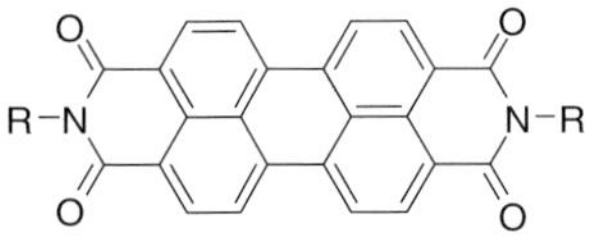

M55-A: R = C_6H_5

M55-B (PDI8): R = C_8H_{17}

M55-C: R = $C_{13}H_{27}$

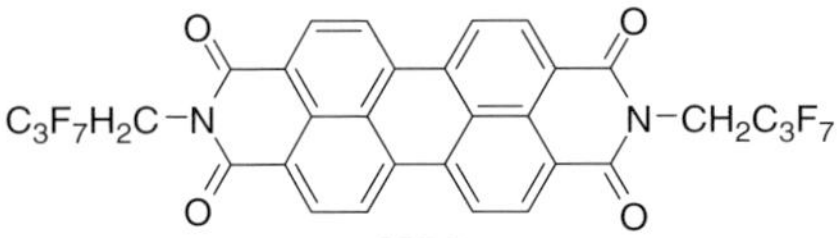

M56

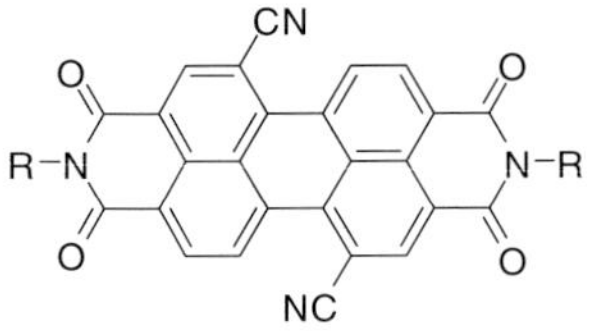

PDICy-CN_2 (**M57**-A): R = 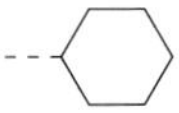

PDIF-CN_2 (**M57**-B): R = $CH_2C_3F_7$

PDI-D (**M57**-C): R = C_8H_{17}

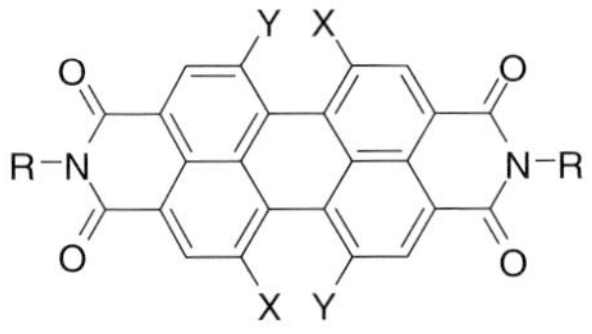

M58-A : X = H, Y = F, R = $CH_2C_3F_7$

M58-B : X = F, Y = F, R = $CH_2C_3F_7$

M58-C : X = Cl, Y = Cl, R = C_6H_5

M58-D : X = Cl, Y = Cl, R = $CH_2C_3F_7$

M58-E : X = H, Y = Br, R = $CH_2C_3F_7$

M58-F : X = Br, Y = Br, R = $CH_2C_3F_7$

M58-G : X = Cl, Y = Cl, R = H

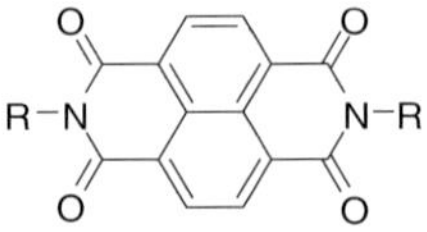

M61-A: R = H

M61-B(NDI8) : R = C_8H_{17}

M61-C: R = $CH_2C_7H_{15}$

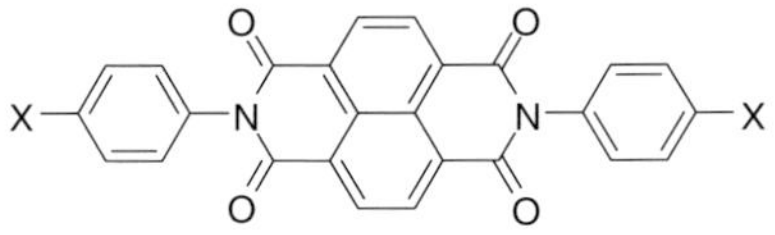

M62-A: X = CH_3

M62-B: R = CF_3

M62-C: R = $CH_2CH_2C_8F_{17}$

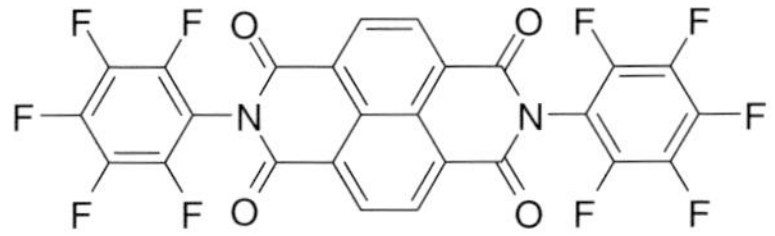

M63

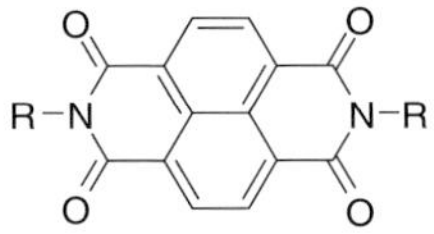

M64-A : R = $CH_2CH_2C_6F_5$

M64-B : R = 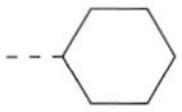

M64-C : R = C_6H_{13}

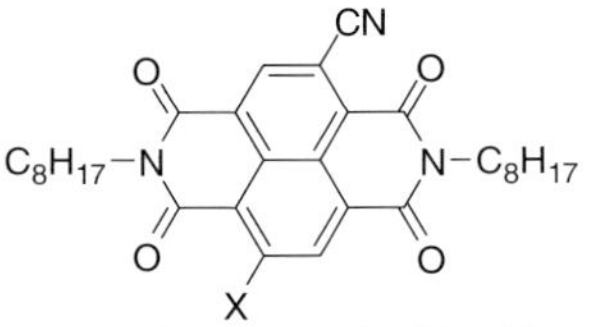

NDI8-CN (**M65**-A): X = H

NDI8-CN_2 (**M65**-B): X = CN

Figure 1.32 Chemical structure of n-channel small-molecule semiconductors **M55–M66**.

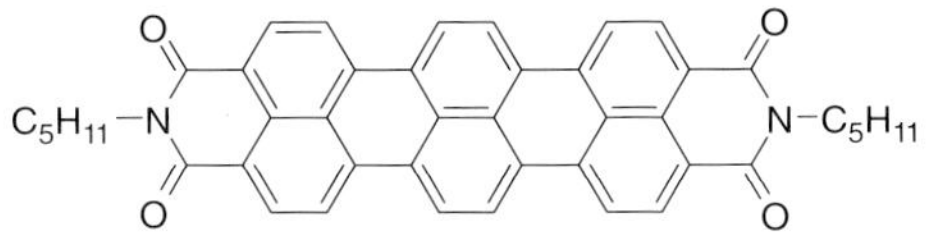

M59

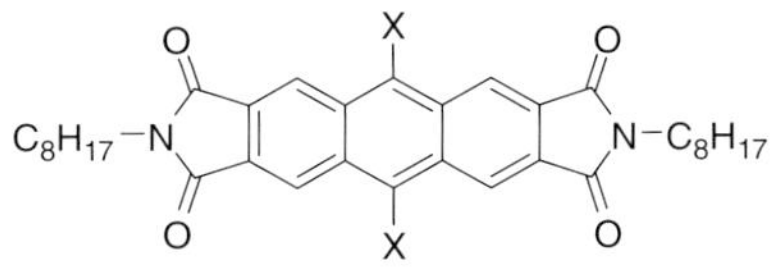

ADI8 (**M66**-A): X = H

ADI8-CN_2 (**M66**-B): X = CN

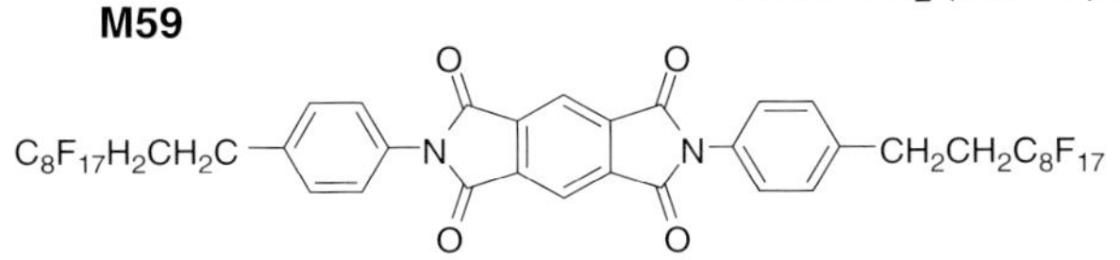

PMDIF (**M60**)

Figure 1.32 (*Continued*).

M55-B (or PDI8) with electron mobility of 0.6 $cm^2\,V^{-1}\,s^{-1}$ under nitrogen, but with a high threshold voltage of +75 V, which was attributed to a large trap density. Later, Chesterfield *et al.* [201] demonstrated that devices with a maximum mobility of 1.7 $cm^2\,V^{-1}\,s^{-1}$, an on/off current ratio of 10^7, and threshold voltages of 10–15 V can be obtained using **M55-B** by coating the SiO_2 gate dielectric with poly(α-methylstyrene). Coating of the dielectric with polymers also considerably improves the air stability of device operation for **M55-B**, presumably by passifying the acidic silanol groups on the SiO_2 surface which can act as electron traps [202]. Ichikawa *et al.* [203] demonstrated that the mobility of OFETs based on *N,N′*-bis-tridecyl **M55-C** can be increased 10^3 times to 2.1 $cm^2\,V^{-1}\,s^{-1}$ by thermal annealing; the thermal treatments improve both the thin-film crystallinity and morphology. Bao, Würthner, and coworkers [204] reported that OFETs based on PDI **M56** exhibit mobilities as high as 0.72 $cm^2\,V^{-1}\,s^{-1}$, which decrease only slightly after air exposure and remain stable for more than 50 days. As the partial fluorination has only a small effect on the redox potential (LUMO energy) relative to *N,N′*-di(alkyl) analogs, the stability was attributed to the hindrance of O_2 and H_2O diffusion by the dense packing of the cores and by the fluoroalkyl chains.

Core-cyanated PDIs (**M57**, Figure 1.32) were first synthesized by Wasielewski *et al.* [205]. These systems are significantly more readily reduced than their unsubstituted analogs (by about 0.36 V); the associated high electron affinity is believed to be a factor contributing to the high electron mobility (0.10 $cm^2\,V^{-1}\,s^{-1}$) achieved in air for OFETs based on PDICy-CN_2. Combining partial fluorination of the N,N′ substituents and 1,7-dicyano substitution in PDIF-CN_2 affords a still higher electron mobility (0.64 $cm^2\,V^{-1}\,s^{-1}$). The effects of PDIF-CN_2 film-growth conditions on n-channel OFET performance have also been investigated [43]; dramatic enhancements of the on/off current ratio and mobility are obtained with increased substrate temperature (T_s) during film growth, the increased mobility being correlated with higher levels of molecular ordering and with minimization of film-surface irregularities [206]. In addition, the effects modifying the SiO_2 surface of the gate dielectric with octadecyltrichlorosilane- or hexamethyldisilazane-derived monolayers, as well as with polystyrene, were investigated for PDIF-CN_2 films deposited at $T_s = 130\,°C$; the SiO_2 surface treatments substantially modulated the mobility and growth morphology of PDIF-CN_2 films. Recently, Morpurgo *et al.* [207] fabricated OFETs based on PDIF-CN_2 single crystals with PMMA as the gate dielectric, which exhibited electron mobilities approaching $6\,cm^2\,V^{-1}\,s^{-1}$, which is 10 times greater than those of the corresponding thin-film devices, both in air and in vacuum. Furthermore, these devices exhibit near-zero threshold voltage and subthreshold slopes and current on/off ratios ($10^3 - 10^4$) comparable to the very best p-channel single-crystal devices when the same gate dielectric is employed.

Weitz *et al.* [208] reported air-stable n-channel OFETs based on five dicyano PDIs with fluorinated linear and cyclic N,N′-substituents (mobilities up to $0.1\,cm^2\,V^{-1}\,s^{-1}$) and investigated the relationships between molecular structure, thin-film morphology, substrate temperature, device performance, and air stability. Interestingly, the mobility degradation rate in air was found to be similar for all compounds and at all substrate temperatures, raising the question of whether air stability can always be explained on the basis of kinetic barriers to O_2/H_2O diffusion formed by densely packed fluorine substituents. In addition to core cyanation, core halogenation is an effective way to functionalize the perylene "bay" positions. Würthner *et al.* studied a series of core-halogenated *N,N′*-bis(heptafluorobutyl) PDIs. While the introduction of halogens in the "bay" positions facilitates reduction, with 1,6,7,12-tetrahalo derivatives being slightly more readily reduced than their 1,7-dihalo analogs, the more highly substituted examples tend to exhibit lower mobilities; this can be attributed to the disruption of core planarity and, therefore, of effective $\pi-\pi$ overlap due to steric interactions [209]. Thus, both the difluoro compound **M58-A** (Figure 1.32) and the parent **M56** (Figure 1.32) exhibit densely π-stacked, more-or-less planar PDI units (torsion angles of 3.0° and 1.5°, respectively, between the two constituent naphthalene units according to X-ray crystal structures), with the FET electron mobility of **M58-A** being around half that of **M56**. On the other hand, the perylene core of **M58-B** (Figure 1.32) is distinctly nonplanar because of F–F steric interactions (torsion angles 20° – 25°), leading to less dense and less regular packing and to mobility that is an order of magnitude lower than that of **M58-A**. The tetrabromo derivative **M58-F** (Figure 1.32) also exhibits about 10 times lower mobility than its dibromo analog, **M58-E** (Figure 1.32). Furthermore, 1,6,7,12-tetrachloro and bromo derivatives **M58-C** (Figure 1.32) and **M58-F** (Figure 1.32) exhibit mobilities 10^3 times lower than that of their tetrafluoro counterpart, **M58-B**, presumably due to the increased bulk of the substituents leading to significantly increased torsion angles and reduced intermolecular $\pi-\pi$ interactions. Interestingly, replacing the *N,N′*-fluoroalkyl substituents of the 1,6,7,12-tetrachloro **M58-C** (Figure 1.32) by an *N,N′*-pentafluorophenyl groups in **M58-D** (Figure 1.32) led to $\sim 10^4$ times increase in the mobility. Müllen *et al.* [210] have pioneered the synthesis of higher rylene diimide dyes and other species based on extended PDI cores, investigating in detail their thermotropic behavior and optical properties as well as the details of their microstructure. However, exploitation of higher rylene diimides in organic electronics is limited to a report by Petit *et al.* [211] on the FET properties of vapor-deposited films of *N,N′*-dipentylterrylene-3,4:11,12-tetracarboxylic diimide derivative **M59** (Figure 1.32) on Si/SiO_2 substrates; a maximum electron mobility of ~0.07 $cm^2\ V^{-1}\ s^{-1}$ and an on/off current ratio in excess of 10^4 were obtained.

Katz *et al.* [212] reported the only example of an OFET based on a PyDI (**M60**, Figure 1.32). The synthesis of PyDI derivatives involves one simple conventional reaction between amines and pyromellitic dianhydride in dimethylformamide (DMF) at 110 °C. Single crystals of **M60** were obtained by slowly cooling hot, saturated DMF solutions. The unit cell of the single crystal is monoclinic with $a = 10.24$ Å, $b = 11.53$ Å, and $c = 9.28$ Å, with the crystal packing exhibiting a close $\pi-\pi$ packing between the side-chain benzene ring and the PyDI core. **M60** exhibited a mobility of 0.054 $cm^2\ V^{-1}\ s^{-1}$ in air and very high on/off current ratios.

The first OFET based on a rylene diimide was fabricated using the unsubstituted NDI, **M61-A** (Figure 1.32), and yielded electron mobilities of ~10^{-4} $cm^2\ V^{-1}\ s^{-1}$ [213]. Later, Katz *et al.* demonstrated that the OFET performance could be greatly improved by functionalizing the NDI core at the nitrogen positions with *n*-octyl groups. The compound **M61-B** (or NDI-8, Figure 1.32) exhibits electron mobilities of ~0.16 $cm^2\ V^{-1}\ s^{-1}$ in vacuum, although almost no FET activity was measurable in air [214]. It was found that N,N′ substitution with *n*-$CH_2C_7F_{15}$ groups in **M61-C** (Figure 1.32) significantly improved the device air stability, with mobilities of 0.05–0.1 $cm^2\ V^{-1}\ s^{-1}$ in air [215]. Replacing the methyl substituents on the *N,N′* benzyl groups of **M62-A** (Figure 1.32) with CF_3 groups in NDI-E led to 10^5 times enhancement of the mobility in air. Elongating CF_3 to *n*-$CH_2CH_2C_8F_{17}$ further enhances the mobility from 0.12 $cm^2\ V^{-1}\ s^{-1}$ (**M62-B**, Figure 1.32) to 0.57 $cm^2\ V^{-1}\ s^{-1}$ (**M62-C**, Figure 1.32) [25]. Strikingly, inserting an ethylene bridge between the nitrogen atoms and the perfluorophenyl substituents of **M63** (Figure 1.32) to give **M64-A** (Figure 1.32) led to a crystalline rather than an amorphous material and increased the mobility from $<10^{-6}$ to 0.31 $cm^2\ V^{-1}\ s^{-1}$ [216]. Shukla *et al.* [217] reported that, compared to linear *n*-hexyl chains, cyclohexyl substituents assist in directing intermolecular $\pi-\pi$ stacking, affording a dramatic increase in mobility from 0.70 (**M64-C**, Figure 1.32) to 6.2 $cm^2\ V^{-1}\ s^{-1}$ (**M64-B**, Figure 1.32) in vacuum.

The studies described above show the great potential of core-unsubstituted NDIs. Wasielewski *et al.* [218] reported two new core-cyanated NDI semiconductors, NDI8-CN (**M65-A**, Figure 1.32) and NDI8-CN_2 (**M65-B**, Figure 1.32), which are air-stable, high-mobility, transparent organic n-type semiconductors. The syntheses of NDI8-CN and NDI8-CN_2 were achieved via a new NDI core bromination–cyanation sequence. The electronic structures of NDI8-CN and NDI8-CN_2 were examined by cyclic voltammetry, optical spectroscopy, and photoluminescence. Electrochemical reduction potentials in dichloromethane versus SCE were −0.22 V for NDI8-CN and +0.08 V for NDI8-CN_2, consistent with systematic LUMO energy depression with increasing cyanation. Importantly, NDI8-CN_2 has a reduction potential similar to that of *N,N*-dialkylsubstituted core-cyanated perylenes (PDI, see above) (−0.07 V vs SCE);

therefore, the LUMO/charge-carrier energies in the NDI and PDI materials should be similar. Optical and photoluminescence spectroscopy of these NDI derivatives reveals a band gap of >3 eV, reflecting the smaller conjugated core dimensions. Thus, thin films of these NDIs are transparent in the visible region. OFET measurements performed in vacuum ($<10^{-6}$ Torr) reveal optimal average electron mobilities for NDI8-CN and NDI8-CN_2 films of 4.7×10^{-3} and $0.15\ cm^2\ V^{-1}\ s^{-1}$, for films vapor-deposited at 130 and 110 °C, respectively. Interestingly, OFET operation in ambient atmosphere reveals that NDI8-CN devices undergo severe $I-V$ curve degradation, whereas the NDI-8CN2 devices exhibit stable operation with only a slightly lower maximum average mobility of $0.11\ cm^2\ V^{-1}\ s^{-1}$. The current on/off ratios (I_{on}:I_{off}) can be as high as $\sim10^5$ for NDI-8CN and $\sim10^3$ for NDI8-CN_2 thin films. The lower I_{on}/I_{off} ratio of NDI8-CN_2 was ascribed to a high I_{off}, which is likely due to dopants in the NDI8-CN_2 thin films from contacts or donor sites in the dielectric. Top-contact, bottom-gate, transparent, channel-flexible n-type OFETs were fabricated with NDI8-CN_2 to demonstrate its unique material properties. Thin NDI8-CN_2 films (50 nm) were vapor-deposited onto overhead transparency film coated with a spin-cast poly(3,4-ethylenedioxythiophene):polystyrene sulfonate (PEDOT:PSS) polymeric gate, a P-UV-013 (Polyera Corporation) polymer dielectric, and Au contacts. This air-stable, flexible, transparent OFET exhibited a mobility of $0.03\ cm^2\ V^{-1}\ s^{-1}$ in ambient atmosphere. An analogous rigid device fabricated on an ultrasmooth ITO (indium tin oxide)/glass substrate as gate gave a mobility value of $\sim0.08\ cm^2\ V^{-1}\ s^{-1}$ in ambient atmosphere.

Marks *et al.* [219] reported the only examples of OFETs based on anthracene-2,3:6,7-tetracarboxylic diimides. Devices based on ADI8 (**M66-A**, Figure 1.32) exhibited a mobility of $0.02\ cm^2\ V^{-1}\ s^{-1}$ in vacuum, which is 10 times lower than its NDI counterpart. However, OFETs based on ADI8 cannot be operated in air because of the material's relatively low electron affinity. Also, in this case, the introduction of cyano groups in the 9,10 positions of the anthracene ring significantly increases the electron affinity and, therefore, affords improved air stability; devices based on ADI8-CN_2 (**M66-B**, Figure 1.32) exhibit a mobility of $0.02\ cm^2\ V^{-1}\ s^{-1}$ but very high current on/off ratios ($>10^7$).

In the last few years, several unconventional structures have been designed and synthesized as n-channel semiconductor candidates. Figure 1.33 collects the chemical structures of these derivatives. Yamashita *et al.* [220] have developed several molecular materials, such as functionalized anthracenes exhibiting electron mobilities up to $\sim5 \times 10^{-3}\ cm^2\ V^{-1}\ s^{-1}$, trifluoromethylphenyl-subtituted thiazole (**M67**) and thiazolothiazole (**M73**) oligomers enabling electron mobilities up to $\sim2\ cm^2\ V^{-1}\ s^{-1}$, and tetrathiafulvalene (**M68** and **M69**) derivatives with electron mobilities up to $\sim0.1\ cm^2\ V^{-1}\ s^{-1}$. Interesting are the performances of indenofluorenedione (**M74**)- and diindenopyrazinedione-based TFTs, which exhibit electron mobilities approaching $0.2\ cm^2\ V^{-1}\ s^{-1}$. Tetracene (**M72**) and perylene BF_2 complexes are also very interesting new electron-deficient arene semiconductors. Perfluoropentacene (**M70**) was developed by Sakamoto *et al.* [221] and exhibits the highest electron mobility of $0.11\ cm^2\ V^{-1}\ s^{-1}$.

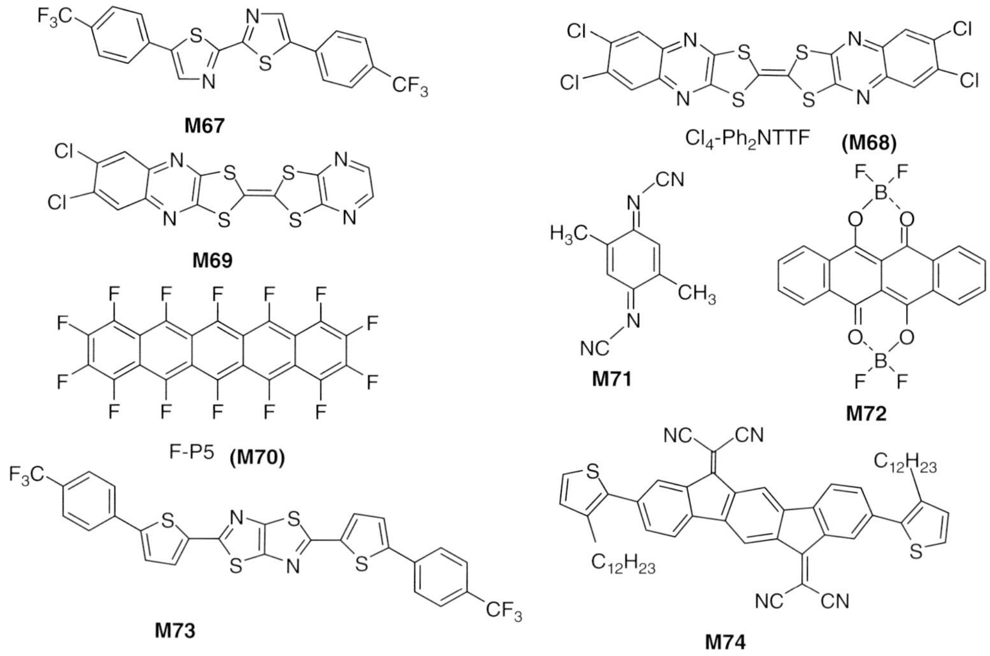

Figure 1.33 Chemical structure of n-channel small-molecule semiconductors **M67–M74**.

The same group also demonstrated the first CMOS circuit using pentacene as the p-channel counterpart. Particularly interesting are two new trifluoromethyltriphenodioxazine derivatives, the devices of which exhibit electron mobilities approaching $0.1\,cm^2\,V^{-1}\,s^{-1}$ in ambient conditions [222]. Recently, dimethyldicyanoquinonediimine (**M71**, Figure 1.32) was developed as an n-channel semiconductor to achieve ambient-stable TFTs with electron mobilities of $\sim 0.01\,cm^2\,V^{-1}\,s^{-1}$ [223].

Dicyanomethylene-substituted quinoidal π-conjugated small molecules are excellent electron acceptors with high structural coplanarity and low-lying LUMO energy levels. Recently, they have been extensively studied as n-channel semiconductors for applications in OFETs, and some of them exhibited high charge-carrier mobilities of $\geq 0.5\,cm^2\,V^{-1}\,s^{-1}$. In one of their papers, Zhu and coworkers [224] reported the design, synthesis, and characterization of a novel series of DPP-containing quinoidal small-molecule semiconductors with branched alkyl substituents. **M75** and **M76** were the first demonstration of DPP-based small molecules offering unipolar electron transport in OFET devices (Figure 1.34). Ambient-stable electron mobilities as high as 0.55 and $0.35\,cm^2\,V^{-1}\,s^{-1}$ ($I_{on}/I_{off} = 10^5 - 10^6$) were obtained for vapor-deposited films of **M75** and solution-processed films of **M76**, respectively. These results showed that incorporation of a DPP unit to construct a quinoidal semiconductor architecture was an effective design strategy to enhance the charge-transport characteristics.

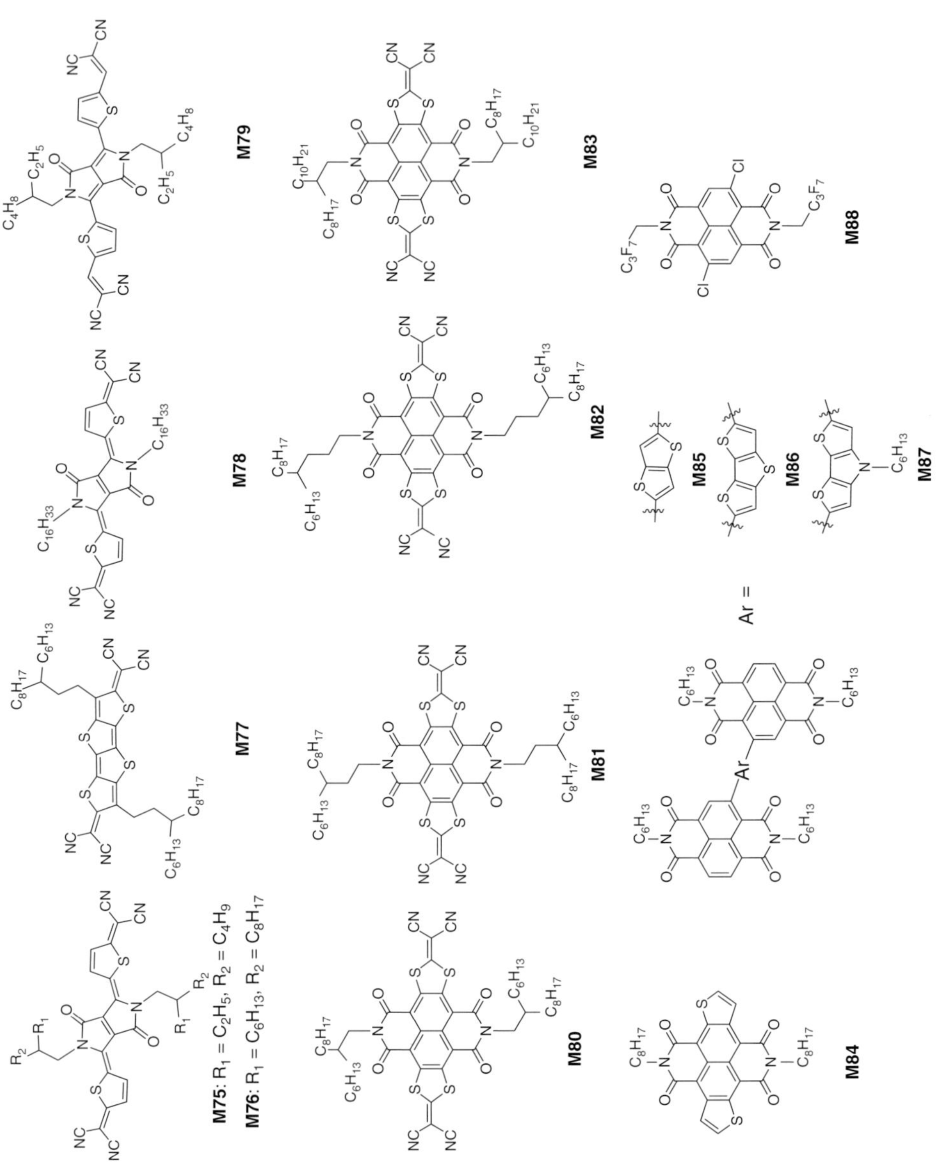

Figure 1.34 Chemical structure of n-channel small-molecule semiconductors **M75–M88**.

The same group recently reported another quinoidal semiconductor **M77** (Figure 1.34), which is a dicyanomethylene-substituted fused tetrathienoquinoid molecule with branched alkyl substituents [225]. OFET devices fabricated with solution-cast films of **M77**, without any post treatment, displayed good ambient stability and exceptionally high n-channel performance with an electron mobility of up to 0.9 $cm^2 V^{-1} s^{-1}$. In a parallel study, Heeney and coworkers [226] developed another DPP-based semiconductor **M78** (Figure 1.34), which has the same π backbone as **M75** and **M76** but with a linear alkyl chain ($C_{16}H_{33}$). Solution-processed thin films of **M78** were prepared with an insulating polymer binder (PαMS), and the resulting top-gate OFET devices yielded maximum electron mobilities of 0.5 $cm^2 V^{-1} s^{-1}$ and I_{on}/I_{off} ratios of 10^2-10^3. **M78**-based devices showed excellent operating stability when exposed to ambient air for a prolonged period owing to their low-lying LUMO levels (−4.2 eV). Detailed thin-film analysis indicated that blended films of **M78** and the polymer binder exhibited strong vertical phase separation, with the DPP molecules diffusing and crystallizing on the surface of the composite film.

In another study, a new dicyanovinyl-substituted DPP-based molecule **M79** (Figure 1.34) was developed as a potential n-channel semiconductor by Park *et al.* [227]. The strong electron-withdrawing dicyanovinyl functionalities stabilize the LUMO level while conserving the aromaticity of the DPP-thiophene core. The authors claimed that the solubility and thin-film crystallinity of **M79** were slightly higher than those of dicyanomethylene-substituted quinoidal molecules **M75** and **M76**. Exceptionally high electron mobilities of up to 0.96 $cm^2 V^{-1} s^{-1}$ were obtained with solution-processed single-crystal OFET devices, which is one of the highest achieved to date for DPP-based small molecules. Polycrystalline OFETs prepared via vapor deposition of **M79** also showed great promise with electron mobilities up to 0.64 $cm^2 V^{-1} s^{-1}$. The remarkable crystallinity of **M79** in the solid state displaying uniform terraces along with the well-defined lamellar microstructure contributed to this high OFET performance.

New promising classes of rylene diimide-based small molecule n-channel semiconductors have recently been reported by several research groups. One example of this family is the core-expanded NDI derivatives bearing two 2-(1,3-dithiol-2-ylidene)malonitrile moieties at the central naphthalene core (**M80**–**M83**, Figure 1.34). These molecules were recently reported by Zhu *et al.* [228, 229] with varied *N*-alkyl chain lengths and different branching points. This study indicated that, although the *N*-alkyl substituent length has a moderate influence on thin-film microstructure and OFET performances, the position of branching point results in significant improvements in molecular packing and charge-transport characteristics. With this new core-expanded NDI structure, ambient-stable electron mobilities ranging from ~0.001 to 3.5 $cm^2 V^{-1} s^{-1}$ were observed depending on the nature of side-chain substituent (Figure 1.35). The highest electron mobility was achieved with the semiconductor **M81** which includes three-branched *N*-alkyl chains. Thin films of **M81** exhibit large grain sizes with efficient in-plane packing, which enables achieving large electron

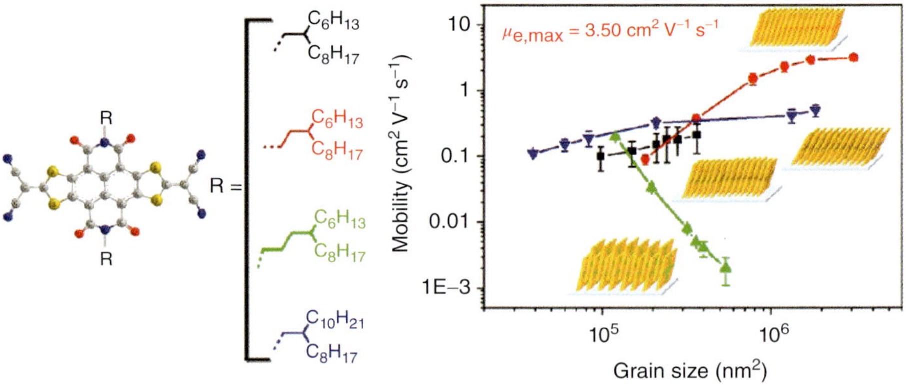

Figure 1.35 Charge-carrier mobilities as a function of grain size for OTFTs based on semiconductors **M80–M83** with different branching alkyl chains. The inset images are the schematic molecular packing demonstrating different grain sizes [229]. (Reproduced with permission. Copyright © 2013, American Chemical Society.)

mobilities of up to 3.5 $cm^2 V^{-1} s^{-1}$. This value of the mobility is currently a world record for ambient-stable, solution-processed, small-molecule n-channel OTFTs. Furthermore, an n-channel transistor array was fabricated by inkjet-printing **M83** on flexible substrates, which showed high mobility of up to 0.45 $cm^2 V^{-1} s^{-1}$ along with good saturation and sharp turn-on characteristics (Figure 1.36) [230].

Another promising family of core-expanded NDI small molecules was developed by fusing heterocyclic rings such as thiophene, indole, pyrrole, pyrazine, and thiazole to the NDI core [231]. Among such semiconductors, the thiophene-fused NDI derivative **M84** (Figure 1.34) was recently designed and synthesized by Takimiya and coworkers [232] as an attractive structure for the development of functional π-conjugated materials. Single-crystal X-ray analysis showed that **M84** had a highly planar core with relatively close π–π stacking (~3.43 Å). Electrochemical and optical studies demonstrated that the compound had a low-lying LUMO energy level (−4.0 eV) and a small HOMO–LUMO gap (2.1 eV). Electron mobility as high as 0.05 $cm^2 V^{-1} s^{-1}$ was achieved under vacuum, which is comparable to those of related core-expanded NDIs such as indole-fused NDI (~0.03 $cm^2 V^{-1} s^{-1}$) and thiazole-fused NDIs (~0.15 $cm^2 V^{-1} s^{-1}$). Another rylene diimide-based semiconductor family, which was recently reported by Marder *et al.* [233], includes bis(NDI) derivatives with conjugated bridging groups based on heterocyclic ring systems (**M85–M87**, Figure 1.34). Solution-processed thin films of **M87** exhibited electron mobility values of up to 1.5 $cm^2 V^{-1} s^{-1}$. It is noteworthy that, because of the presence of a strong donor unit in **M87**, these devices show slight ambipolar behavior with low hole mobilities ($\mu_h = 9.8 \times 10^{-3}$ $cm^2 V^{-1} s^{-1}$). On the other hand, OFETs based on **M85** and **M86** showed unipolar n-channel charge transports with

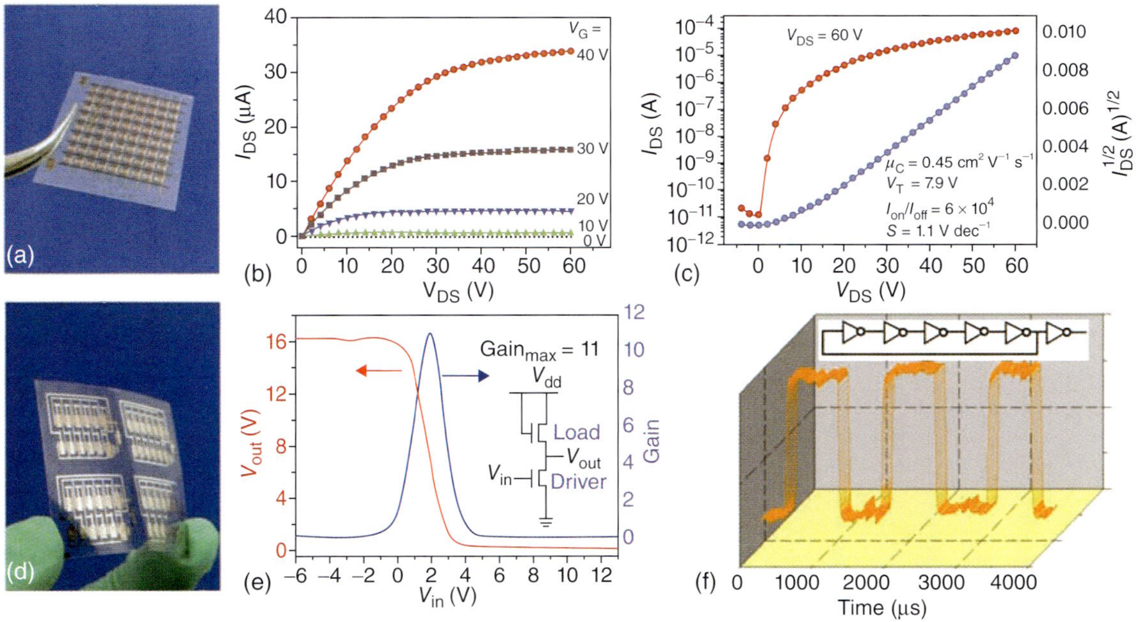

Figure 1.36 (a) Optical images, (b) output characteristics, and (c) transfer characteristics of all-solution-processed flexible OTFTs based on **M83**. (d) Optical images of a flexible five-stage organic oscillator. Electrical characteristics of (e) an organic inverter and (f) an organic oscillator [230]. (Reproduced with permission. Copyright © 2011, John Wiley & Sons, Inc.)

appreciable electron mobilities of $0.14\,cm^2\,V^{-1}\,s^{-1}$ as a result of the decreased donor strength of the bridging unit. Finally, more recently, Würthner *et al.* [234] reported the preparation of high-quality thin films based on the dichlorinated NDI derivative **M88** (Figure 1.34) on bare SiO_2 via a solution shearing method. The OFET devices with these active layers exhibited electron mobilities of up to $0.95\,cm^2\,V^{-1}\,s^{-1}$ in air. In this study, the authors optimized the device's processing parameters to enable the directional growth of large crystalline domains with high anisotropy. Remarkably, a reversible up to fourfold increase in charge-carrier mobility ($\mu_e = 4.26\,cm^2\,V^{-1}\,s^{-1}$) was observed under positive bias stress.

1.6 Ambipolar Semiconductors

The vast majority of known organic semiconductors are either hole (p-channel) or electron (n-channel) transporting materials. However, very recently, ambipolar organic semiconductors have been attracting attention for their potential use in numerous technologically relevant applications [235]. The discovery of ambipolarity as a general characteristic of several semiconducting polymers was made possible by the understanding of the crucial role played by traps of electrons on the surface of several dielectrics, such as hydroxyl, silanol, and carbonyl groups. Representative technological examples include the area of organic microelectronics where patterning of p- and n-channel semiconductors is one of the major hurdles for the implementation of organic complementary logic. In this context, the use of ambipolar materials should enable the fabrication of complementary-like circuits through the use of a single semiconductor that functions both as p- and/or n-channel, thereby significantly reducing fabrication complexity [236]. The latest application of ambipolar organic semiconductors is in bifunctional TFTs such as light-sensing organic organic thin-film transistors (LS-OTFTs) and organic light-emitting transistors (OLETs) [237]. These types of OTFTs can combine electrical switching with additional functionalities such as light sensing or light emission in a single device, making them attractive for various optoelectronic applications including nanoscale light sources and image-sensor arrays. Ambipolar OFETs based on a number of different materials have been reported. These include thermally evaporated small molecules, spin-coated poly(9,9-dioctylfluorene-*alt*-benzothiadiazole) (F8BT), single crystals of copper and iron phthalocyanines, solution-processed nickel dithiolenes, and spin-cast squaraines. One of the common features in several ambipolar polymers is the lower bandgap (<2 eV) compared to the corresponding unipolar semiconductors. This is the result of the typical (but not essential) donor–acceptor-like structure characterized by copolymerization of electron-rich and electron-poor heteroaromatic units. However, despite intensive research on ambipolar organic semiconductors and OFETs, the key material and device properties that enable ambipolar charge transport have been investigated to a much less extent than for unipolar charge transport.

1.6.1 Polymers

Sirringhaus *et al.* [238] reported the general observation of ambipolar charge transport in a series of regioregular polyselenophene-based polymers. The HOMO of polythiophenes has little contribution from the sulfur heteroatom, whereas the LUMO has significant electron density on the heteroatom [239]. Polyselenophenes were initially developed as promising alternatives to polythiophenes for OPV cell applications, mainly because of their reduced optical bandgaps. The regioregular polyselenophenes investigated in that work were poly(3-octyl)selenophene (**P93**, Figure 1.37) and poly(3,3″-di-*n*-alkylterselenophene) (**P94**, Figure 1.37). Two-dimensional, grazing-incidence wide-angle X-ray scattering (GIWAXS) indicated that the films were polycrystalline with the side-chain stacking mainly along the out-of-plane direction (perpendicular to the substrate) and π–π stacking in the in-plane direction. TG–BC TFT configurations with gold source–drain electrodes were employed for all polymers. Polymer **P94**

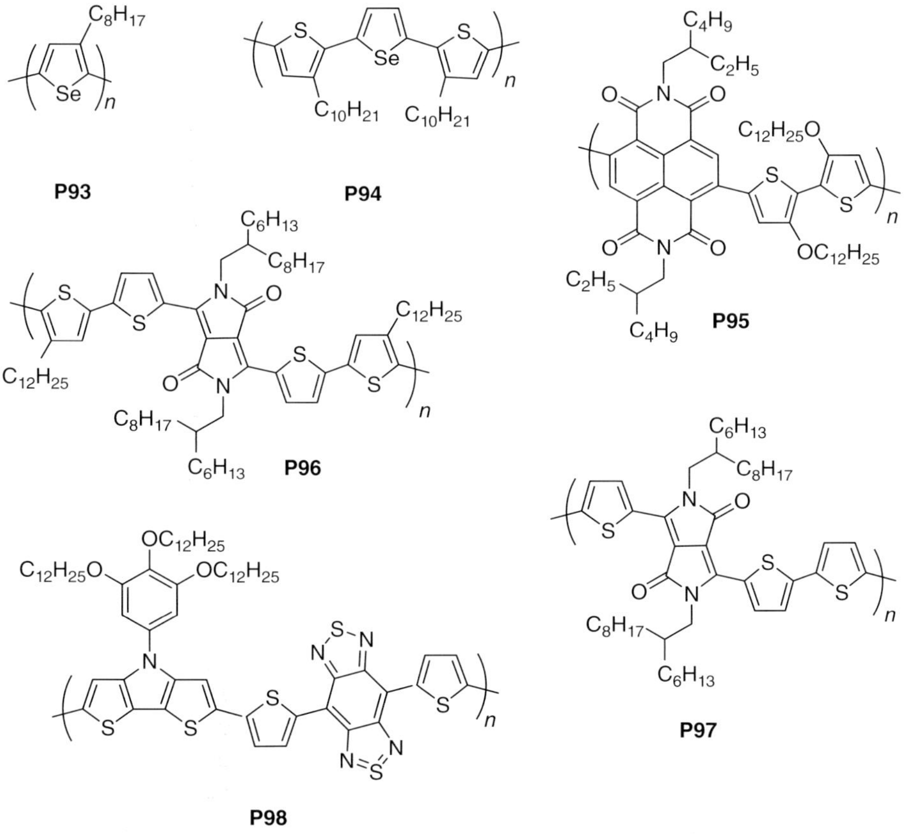

Figure 1.37 Chemical structure of ambipolar polymer semiconductors **P93–P98**.

showed clean ambipolar transport characteristics with similar hole and electron saturation and linear mobilities of >0.01 $cm^2 V^{-1} s^{-1}$ (PMMA as gate dielectric). While the saturation mobility values for holes and electrons were similar, some reversible hysteresis was systematically observed in the transfer characteristics in the electron transport regime but not in the hole transport regime, thus indicating the presence of a larger number of shallow traps for electrons than for holes. Using the same device configuration, TFTs based on **P93** exhibited ambipolar properties in as-spun films, with hole mobilities of ~0.02–0.09 $cm^2 V^{-1} s^{-1}$ and electron mobilities of ~0.004–0.009 $cm^2 V^{-1} s^{-1}$. Using **P94**, complementary-like inverters based on two identical TG–BC ambipolar transistors with a common gate as input and a common drain as output were fabricated, eliminating the need for semiconductor patterning. Despite the general fact that none of these TFTs cannot be fully switched off in such an inverter, the authors obtained very high gain in switching (absolute value as high as 86), much higher than the previously reported gain values in inverters composed of ambipolar OFETs [240].

Watson and Jenekhe *et al.* [241] reported a new naphthalenebiscarboximide-bithiophene copolymer semiconductor, **P95** (Figure 1.37), with an alternating donor–acceptor architecture consisting of electron-donating dialkoxybithiophene and electron-accepting naphthalene bisimide. High-mobility ambipolar transistors and high-gain complementary-like inverters were fabricated, which exhibited electron and hole mobilities as high as 0.04 and 0.003 $cm^2 V^{-1} s^{-1}$, respectively, and output voltage gains as high as 30. **P95**-based devices showed typical ambipolar features, such as a diode-like current increase with current saturation at high gate voltage in output curves and V-shaped transfer curves with a narrow off-state. However, the film annealing temperature (T_a = 100–250 °C) had notable effects on the ambipolar **P95** transistors. Figure 1.38a shows the

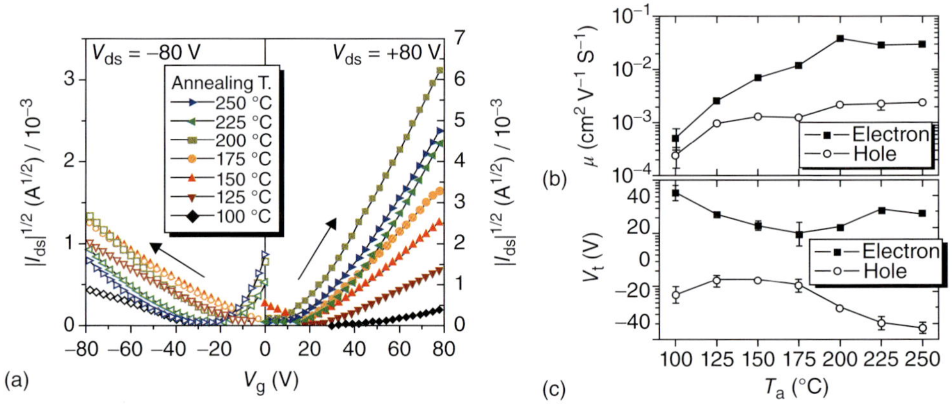

Figure 1.38 Ambipolar characteristics of **P95**. (a) Square-root of source–drain current versus gate voltage curves at various annealing temperatures (T_a). (b) Saturation mobility and (c) threshold voltage as a function of the annealing temperature [241]. (Reproduced with permission. Copyright © 2010, John Wiley & Sons, Inc.)

transfer curves of the ambipolar TFTs, $I_{SD}^{1/2}$ versus V_G, as a function of the annealing temperature T_a. The saturation electron and hole mobilities extracted from these transfer curves are shown in Figure 1.38b as a function of T_a. As T_a was increased from 100 to 200 °C, the mobility increased monotonically from $\sim 5 \times 10^{-4}$ to $\sim 0.03-0.04\,cm^2\,V^{-1}\,s^{-1}$ for electrons and from $\sim 2 \times 10^{-4}$ to $\sim 0.002-0.003\,cm^2\,V^{-1}\,s^{-1}$ for holes. About two orders of magnitude improvement in electron mobility was observed, while the hole mobility improved by 10 times. No significant change in the charge-carrier mobilities was observed for $T_a > 200$ °C. Unlike the carrier mobility and threshold voltage (Figure 1.38c), the current on/off ratios for the p- and n-channel operation were not affected by the annealing temperature.

Winnewisser *et al.* [74] reported on a new low-bandgap DPP-based polymer semiconductor (**P96**, Figure 1.37) with marked ambipolar charge-transport properties. Solution-processed devices using polymeric insulators (PVP (poly(vinyl pyrrolidone)) or PMMA) and an inorganic gate dielectric (octyltrichlorosilane-treated SiO_2) showed ambipolar behavior. The latter insulator resulted in the highest field-effect mobilities, reaching ~0.1 and up to $\sim 0.09\,cm^2\,V^{-1}\,s^{-1}$ for holes and electrons, respectively. These values are larger by an order of magnitude than the highest ones previously reported for solution-processed ambipolar transistors. Ambipolarity in this material is not limited to one particular transistor architecture but has been observed in five different configurations including transistors with solution-processed gate dielectrics in bottom-gate as well as top-gate structures. When driven under appropriate bias conditions, the ambipolar **P96** transistors emit near-infrared light. This was the first time that NIR light emission was reported for organic transistors with polymer gate dielectrics as well as for top-gate transistors (Figure 1.39). A similar structure, **P97** (Figure 1.37), was reported by Janssen *et al.* [242], which exhibited ambipolar transport with balanced electron and hole mobilities in the range of $\sim 0.01\,cm^2\,V^{-1}\,s^{-1}$, making it an interesting candidate for CMOS-like circuits.

To enable ambipolarity and simultaneously produce soluble low-bandgap polymers, Reynolds *et al.* [243] utilized the strong donor dithieno[3,2-*b*:2′,3′-*d*]pyrrole (DTP) functionalized with a trialkoxyphenyl group, combined with a strong acceptor based on benzo[1,2-*c*;4,5-*c*′]bis[1,2,5]thiadiazole (BBT) to produce the interesting polymer **P98** (Figure 1.37). This strategy provides a high-lying HOMO, planarity for π stacking, and solubility in the polymers due to the long-chain alkoxy substituents. This polymer is spray-processable and shows an optical bandgap of only 0.5–0.6 eV, which is the lowest value reported for a soluble polymer. In electrochemical cells, four differently colored redox states of the polymer can be accessed at moderate potentials and have good stability. This polymer also shows potential for use in ambipolar OFETs, with respectable mobilities of 1.2×10^{-3} and $5.8 \times 10^{-4}\,cm^2\,V^{-1}\,s^{-1}$ measured for p-channel and n-channel operation, respectively.

In a recent report by Pei *et al.* [244], ambipolar charge-transport behavior was observed for the first time in isoindigo-based donor–acceptor π-conjugated

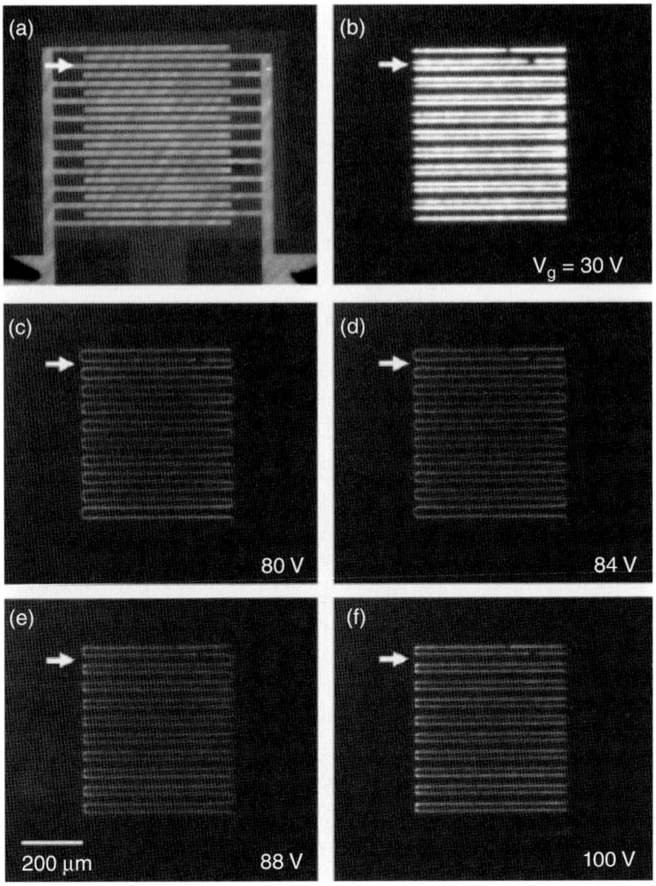

Figure 1.39 Microscopy images of an operating light-emitting bottom-gate FET. (a) Transistor with interdigitated source–drain electrodes under external illumination. The electrode fingers are 20 mm wide and the distance between them (i.e., the channel length L) is 10 mm. (b–f) NIR light emitted by the transistor when driven with $V_d = 140\,V$ (the drain electrode is the one contacted from the right side) and $V_g = 30$, 80, 84, 88, and 100 V, respectively. The camera settings were not changed during this series. The white arrow marks the position of the second source finger from the top [74]. (Reproduced with permission. Copyright © 2008, John Wiley & Sons, Inc.)

polymers. The new polymer **P99** (Figure 1.40) was synthesized with alternating fluorinated isoindigo and bithiophene units. As a result of fluorination, **P99** had an effectively lowered LUMO level (ΔLUMO = −0.2 eV), and the electron mobility increased from 10^{-2} to $0.43\,cm^2\,V^{-1}\,s^{-1}$ in ambient compared to the non-fluorinated isoindigo-based polymers. The most impressive feature of **P99** is that it still maintains a high hole mobility of up to $1.85\,cm^2\,V^{-1}\,s^{-1}$ in ambient (Figure 1.41). This polymer represents the first ambipolar polymer based on an isoindigo building block. GIXD and AFM studies indicated that the introduction

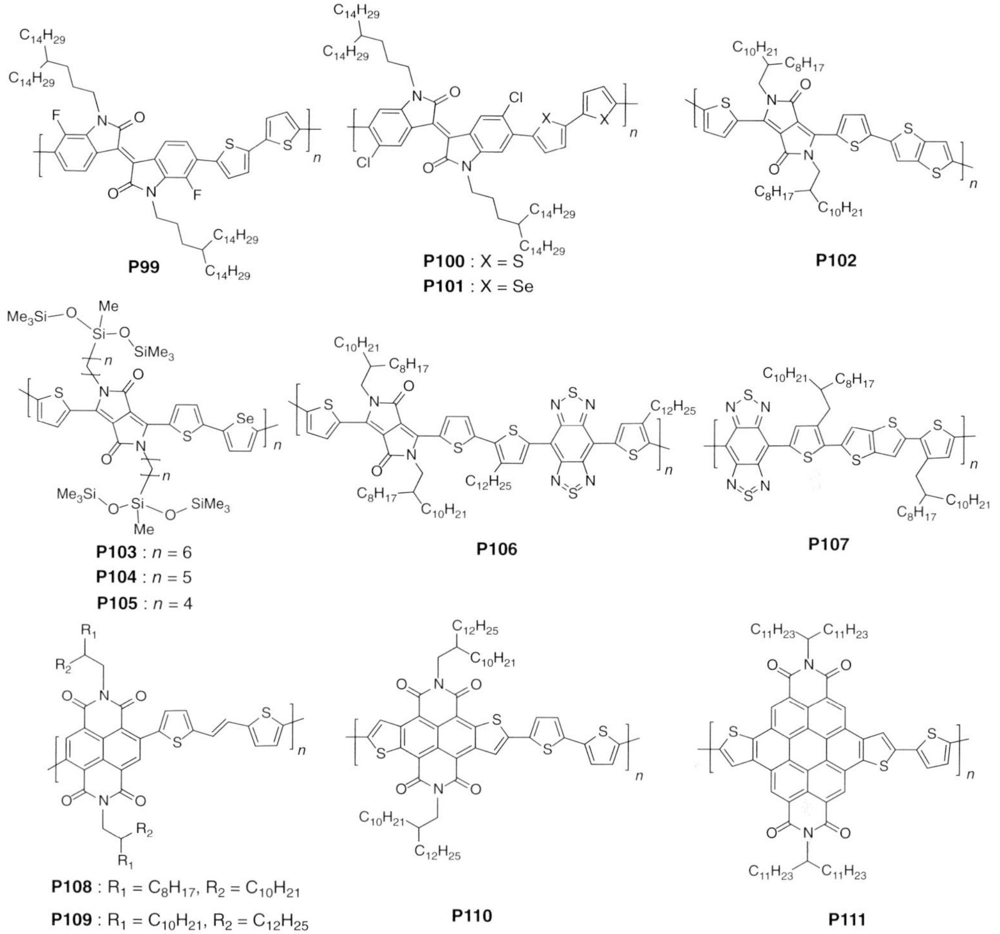

Figure 1.40 Chemical structure of ambipolar polymer semiconductors **P99–P111**.

of fluorine in an electron-deficient isoindigo core led to enhanced crystallinity with a different molecular packing. The structural influence of fluorine is based on the planarizing interaction of the fluorine on the isoindigo with the β-hydrogen on the bithiophene. The authors also believe that the strong crystallinity and dense, ordered packing of **P99** contributes to its improved ambient stability. The same group recently extended the halogenation strategy to chlorine atoms as well by developing an efficient chlorination reaction to engineer the isoindigo-based polymers [245]. Two new copolymers **P100** and **P101** containing a dichlorinated isoindigo unit as acceptor and bithiophene and biselenophene building blocks as donor were realized (Figure 1.40). Polymer FETs fabricated with the new ambipolar polymers in a TG–BC configuration showed high hole mobilities of up to 1.05 $cm^2\ V^{-1}\ s^{-1}$ and balanced electron mobilities of up to 0.72 $cm^2\ V^{-1}\ s^{-1}$. In addition, complementary-like inverters based on **P100** and **P101** showed

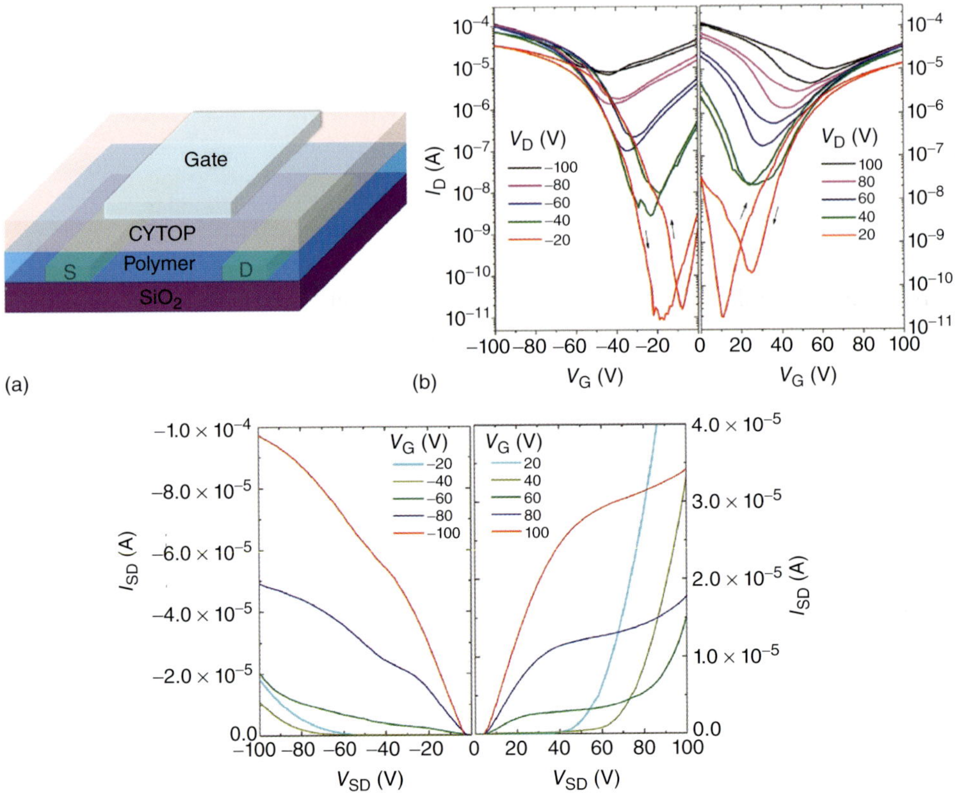

Figure 1.41 (a) Schematic diagram of a TG–BC OFET device structure based on **P99**. (b) The transfer and (c) output characteristics of **P99**-based devices fabricated in ambient. OFET devices ($L = 50\,\mu m$, $W = 1000\,\mu m$) were fabricated with CYTOP (500 nm thickness, $C_i = 3.7\,nF\,cm^{-2}$) [244]. (Reproduced with permission. Copyright © 2012, American Chemical Society.)

sharp signal-switching with high gains of up to 48. This report is the first study demonstrating that chlorination is an effective design strategy to tune the optoelectronic properties of semiconducting polymers and improve their device performances.

DPP-based donor–acceptor polymers have recently become very promising to demonstrate balanced and high ambipolar charge-transport characteristics. Ambipolarity of DPP-containing copolymers is attributed to their optimized molecular orbital architectures and energy levels, which enables donor and acceptor orbital hybridization and extends HOMO and LUMO distributions along the π backbone. Although the first DPP-based ambipolar polymer was reported with mobilities of 0.1 and $0.09\,cm^2\,V^{-1}\,s^{-1}$ for holes and electrons, respectively, much higher mobilities ($\geq 1\,cm^2\,V^{-1}\,s^{-1}$) were recently achieved with modified chemical structures [246]. In one of these reports, a DPP-based conjugated polymer, **P102** (Figure 1.40), was synthesized by Sirringhaus and

McCulloch *et al.* [247] with thiophene and thienothiophene comonomers. This polymer possesses proper HOMO (−5.33 eV) and LUMO (−4.07 eV) energy levels with a low bandgap of 1.26 eV, which enables efficient and balanced ambipolar charge injection and transport. As a result, ambipolar OFETs with high, balanced hole and electron mobilities of 1.4–1.6 $cm^2 V^{-1} s^{-1}$ were achieved upon careful optimization of the device architecture, charge injection, and polymer processing. In another recent study, Yang and coworkers developed a new polymer **P103** based on dithienyl-diketopyrrolopyrrole (TDPP) and selenophene building blocks with siloxane-terminated solubilizing groups (Figure 1.40) [248]. Extraordinarily high hole and electron mobilities of 3.97 and 2.20 $cm^2 V^{-1} s^{-1}$ with relatively well-balanced polarities were achieved for solution-sheared OFET devices made from **P103**. This ambipolar behavior is attributed to the synergistic combination of rational polymer backbone design, side-chain dynamics, and favorable solution processing. Additionally, in this study, CMOS-like inverters based on two identical ambipolar transistors of **P103** were fabricated, which yielded a gain of 18.0. Following this initial report, the same group demonstrated further improvement of charge-carrier mobilities based on **P103** π backbone via hybrid side-chain engineering [249]. In this study, two additional polymers (**P104** and **P105**) were synthesized with different alkyl spacer lengths to systematically study its effect on the ambipolar performance (Figure 1.40). The main goal was to induce denser molecular packing in polymer thin film and to facilitate charge transport through 3D conduction channels. Thin films of **P104** with pentyl spacers were found to exhibit the most optimized three-dimensional charge transport, yielding exceptionally high hole and electron mobilities of 8.84 and 4.34 $cm^2 V^{-1} s^{-1}$, respectively. In this study, it was found by GIXD characterizations that shorter alkyl spacer groups induced smaller lamellar spacing with a close π–π stacking distance compared to the longer chains, which led to enhanced charge transport in the as-prepared films (Figure 1.42). Additionally, the solution-shearing method resulted in dramatically improved performances compared to drop-casting technique as a result of improved alignment of the polymer grains. Charge-carrier mobility of **P104** currently represents the best ambipolar OFET performance reported to date with relatively well-balanced polarities.

Another promising acceptor building block, which was only recently incorporated into donor–acceptor polymers, is benzobisthiadiazole (BBT). BBT is a 14π electron system with a tetravalent sulfur atom. This core has a high electron affinity asbecause it generates a more stable Kekulé-type thiadiazole moiety upon accepting an electron. In the past few years, Wudl and Heeger *et al.* [250, 251] reported a family of BBT-based polymers with various donor spacers. The general observation was that ambipolar behavior is universal for these polymers owing to their very low band gaps and proper HOMO/LUMO energy levels. One of these polymers, **P106** (Figure 1.40), consists of two coupled acceptor units, that is, BBT and DPP, along with a thiophene-based donor moiety. This polymer exhibited excellent ambipolar transistor performance with mobilities exceeding 1 $cm^2 V^{-1} s^{-1}$ for both electrons and holes. In this polymer, the BBT moiety was

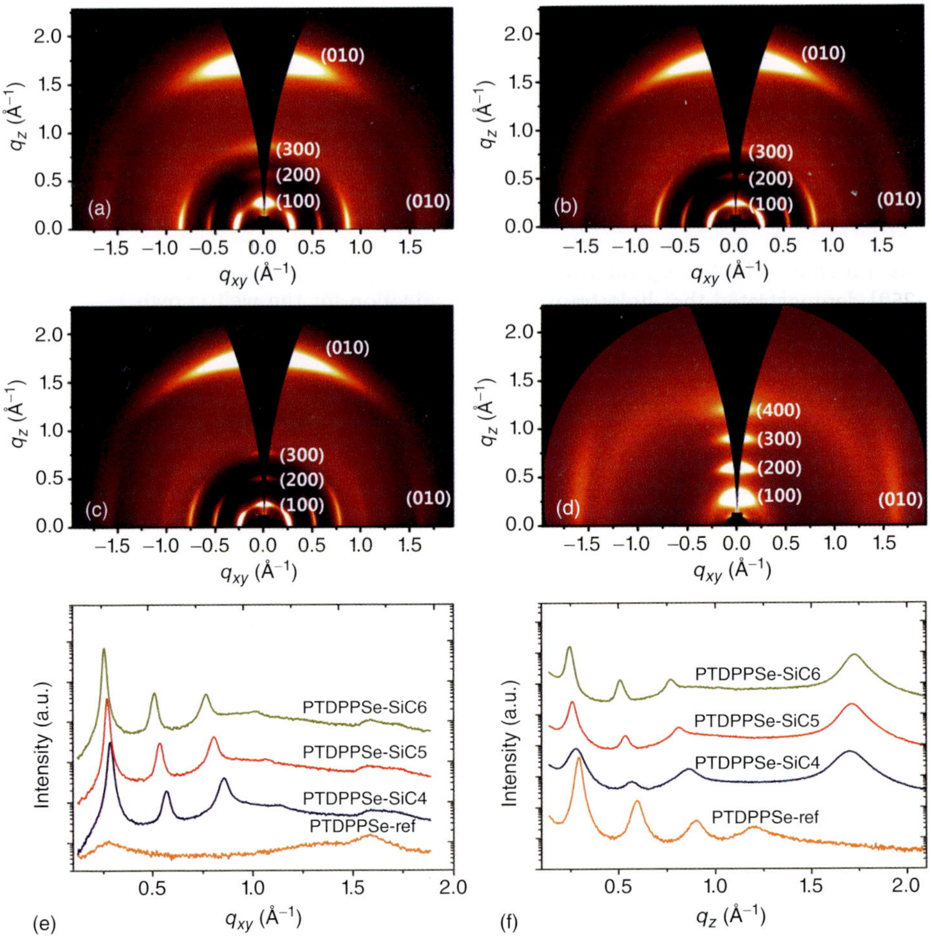

Figure 1.42 GIXD images of drop-cast TDPPSe-based copolymer films annealed at 220 °C. (a) **P105** (PTDPPSe-SiC4), (b) **P104** (PTDPPSe-SiC5), (c) **P103** (PTDPPSe-SiC6), and (d) TDPPSe-based copolymer containing 2-octyldodecyl chains (PTDPPSe-ref). The corresponding GIXD diffractogram profiles. (e) In-plane and (f) out-of-plane GIXD patterns [249]. (Reproduced with permission. Copyright © 2013, American Chemical Society.)

found to strengthen the interchain interactions providing higher thermal stability and higher OFET performance, compared to the reference polymers with no BBT unit. **P107** (Figure 1.40) is another recently synthesized ambipolar polymer containing a bi(thiophen-2-yl)-thieno[3,2-*b*]thiophene donor alternating with a BBT acceptor in the polymer backbone. This polymer exhibited an ultralow bandgap of 0.56 eV with HOMO and LUMO energy levels located at −4.36 and −3.80 eV, respectively, as a result of enhanced interactions between strong acceptor and strong donor units. This semiconducting polymer exhibited nearly balanced electron and hole mobilities as high as 0.7 and 1.0 $cm^2 V^{-1} s^{-1}$, respectively, under an inert atmosphere. Inverters fabricated with only **P107**-based transistors exhibited a high gain of 35, which is much higher than values usually obtained for unipolar logic.

NDI is a well-known electron-deficient core with favorable optoelectronic properties, which has been successfully applied to high-performance n-channel polymer semiconductors. Nonetheless, there have been only a limited number of reports on its ambipolar behavior, especially in polymers with both electron and hole mobilities exceeding 0.1 $cm^2 V^{-1} s^{-1}$. Recently, Noh and Facchetti *et al.* [252] demonstrated that hole transport and injection for the well-known n-channel polymer P(NDI2OD-T2) could be significantly enhanced by using a fluorinated high-*k* dielectric (Figure 1.43). Balanced hole and electron mobilities of ~0.1 $cm^2 V^{-1} s^{-1}$ were obtained in the TG–BC device configuration. The observed hole mobility increase was attributed to the dielectric surface C–F dipoles and not to the bulk effects. In a parallel study, Liu and coworkers [253] reported the synthesis of two novel donor–acceptor copolymers **P108** and **P109** based on (NDI) acceptor and (*E*)-2-(2-(thiophen-2-yl)vinyl)thiophene (TVT) donor (Figure 1.40). The incorporation of a TVT unit into polymer backbone effectively promoted intermolecular π–π stacking and enhanced charge transport. Additionally, the HOMO level increased from −5.82 to −5.61 eV while the energy level of LUMO was maintained at −3.90 eV, thus facilitating both hole and electron injection/transport. TG–BC FETs fabricated with these polymers yielded high hole and electron mobilities of 0.30 and 1.57 $cm^2 V^{-1} s^{-1}$, respectively. This is among the best ambipolar performance reported to date in NDI-based polymers. Additionally, ambipolar inverters fabricated with **P108** exhibited a sharp signal switching with a high gain of 155 in ambient.

Recently, two new thiophene-fused, core-extended, diimide-based building blocks were reported in ambipolar semiconducting polymers [232, 254]. In these polymers **P110** and P111 (Figure 1.40), the acceptor units are thiophene-fused NDI and coronenediimide cores, respectively, while the donor units are based on electron-rich thiophene. These polymers showed small optical bandgaps (1.9–2.1 eV) with favorable HOMO and LUMO energy levels located at −5.6 and −3.7 to −4.4 eV, respectively. Solution-processed ambipolar OFETs have been demonstrated, yielding charge-carrier mobilities up to 0.30 $cm^2 V^{-1} s^{-1}$ (electron) and 0.1 $cm^2 V^{-1} s^{-1}$ (hole) for **P110**, and 0.30 $cm^2 V^{-1} s^{-1}$ (electron) and 0.04 $cm^2 V^{-1} s^{-1}$ (hole) for **P111**. Although these results are not among the best, it indicates that through rational design and synthesis, core-extended, π-conjugated, diimide-based building blocks are promising for ambipolar OFETs.

1.6.2 Small Molecules

Bao *et al.* reported the acene-based semiconductors **M89–M91** (Figure 1.44). Compound **M89** exhibited balanced ambipolar performance with a hole mobility of ~0.1 $cm^2 V^{-1} s^{-1}$ and electron mobility of 0.133 $cm^2 V^{-1} s^{-1}$ in nitrogen [255].

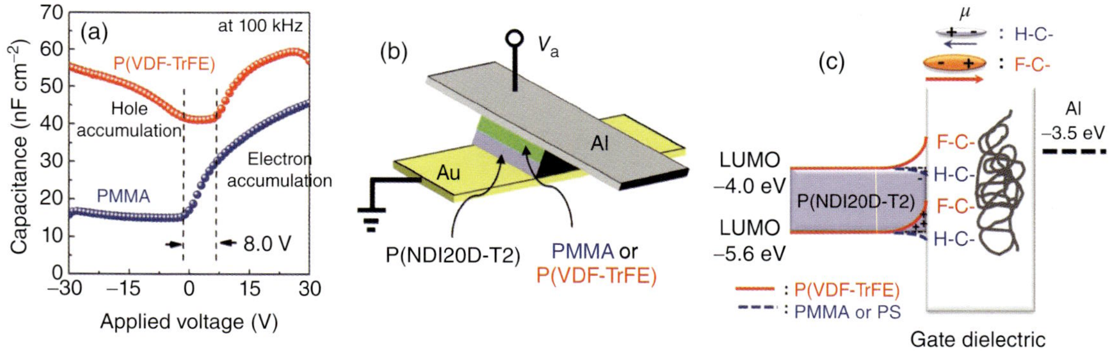

Figure 1.43 Capacitance versus voltage (*C–V*) characteristics at 100 kHz frequency for P(NDI2OD-T2)/PMMA and P(NDI2OD-T2)/P(VDF-TrFE) capacitors of the structure reported in (b). (c) Schematic description of the energy band diagram in the gate-to-channel direction from the source to the drain electrodes [252]. (Reproduced with permission. Copyright © 2012, John Wiley & Sons, Inc.)

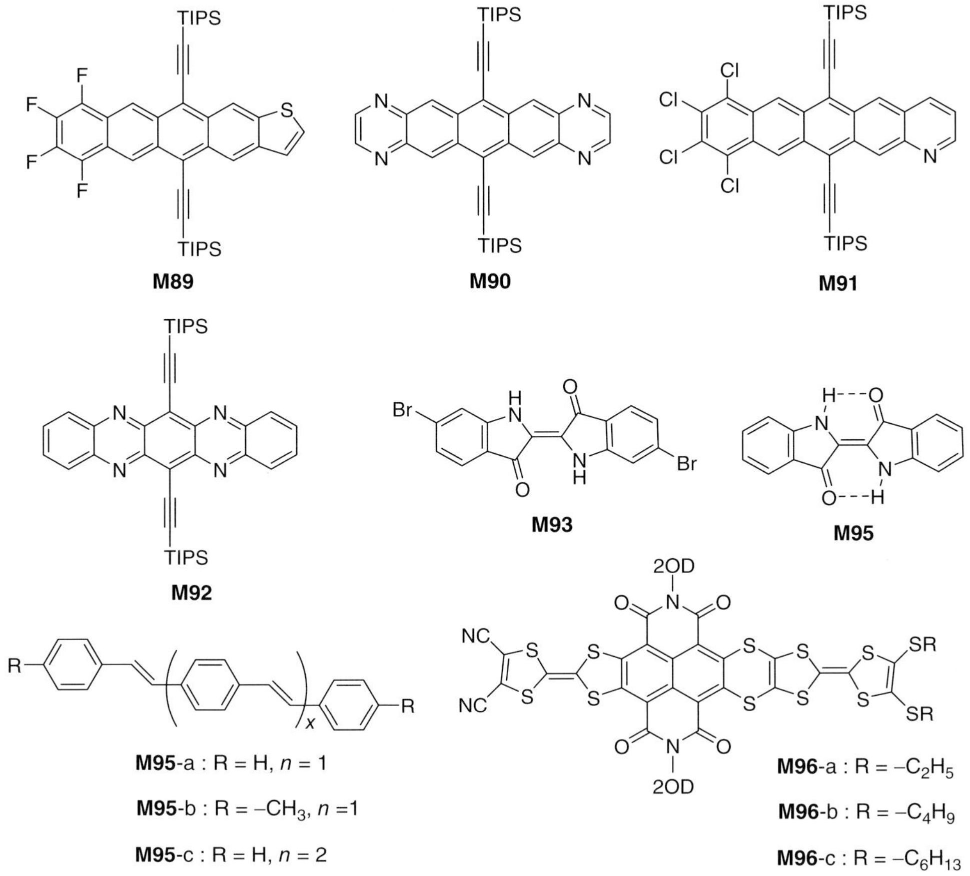

Figure 1.44 Chemical structure of ambipolar small-molecule semiconductors **M89–M96**.

M90 was reported as the constitutional isomer of **M92** with only a different position of the N atom [256]. **M92** (Figure 1.44) showed only high electron mobility while **M90** exhibited ambipolar performance with a hole mobility of $0.22\,cm^2\,V^{-1}\,s^{-1}$ and an electron mobility of $1.1\,cm^2\,V^{-1}\,s^{-1}$. Note that the high electron performance could only be obtained in vacuum, and when tested in ambient air, the electron mobility decreased to the range of $10^{-3}\,cm^2\,V^{-1}\,s^{-1}$ because of the electron trapping of oxygen or water. **M91** OFETs showed performance similar to that of **M89** [257]; An **M90**-based CMOS-like inverter with high gain up to 180 was also reported [258]. All these three acene-based semiconductors showed ambipolar performance only in vacuum or inert atmosphere because of their relatively high LUMO level. **M93** and **M94** were reported with similar chemical structures by the same group (Figure 1.44). They are a kind of natural material for ambipolar OFETs and circuits. **M93**- and **M94**-based OFET devices were fabricated on the natural resin shellac substrates, and Al was employed as gate electrode and AlO_x/tetratetracontane for dielectric layer.

Ambipolar performance was observed in vacuum, but after extensive exposure to air, n-channel performance showed deterioration in both **M93** and **M94** OFETs [259]. One interesting thing is that, apart from tetratetracontane, only evaporated polyethylene films allowed the observation of ambipolar transport in **M94**-based OFETs. **M94** did not show any semiconductor behavior on other investigated dielectrics, that is, poly(vinyl alcohol), shellac, melamine, adenine, and guanine, as well as plain aluminum oxide. Complementary-like inverter circuits were fabricated with both **M93** and **M94**, and the highest gain of 255 was obtained, which is one of the highest reported for organic ambipolar devices.

Single-crystal OFETs based on highly luminescent oligo(*p*-phenylenevinylene) derivatives **M95a–M95c** (Figure 1.44) were reported by Nakanotani *et al.* [260]. All the OFETs based on **M95a–M95c** showed ambipolar behavior, in which **M95c**-based OFETs exhibited balanced hole and electron mobilities both higher than 0.1 $cm^2\ V^{-1}\ s^{-1}$. Intense electroluminescence was also observed in the channel of **M95c** single-crystal-based OFETs. **M96a–M96c** (Figure 1.44) are recently reported tetrathiafulvalene-fused NDI derivatives with a low LUMO energy level of −4.3 eV. Ambipolar behaviors were obtained for all the **M96a–M96c**-based OFETs in ambient air because of their low LUMO level [261]. However, the mobility was relatively low (about $10^{-4}\ cm^2\ V^{-1}\ s^{-1}$) for **M96a–M96c** OFETs. The highest mobilities of up to 0.03 and 0.003 $cm^2\ V^{-1}\ s^{-1}$ were observed for holes and electrons, respectively, for **M96c**-based devices.

Wang and coworkers developed an ambient-stable, ambipolar, DPP-based oligomer **M97** with cyano terminal functionalities (Figure 1.45). **M97** was found to exhibit a very low bandgap of 1.48 eV with HOMO and LUMO energies of −5.45 and −3.74 eV, respectively. Bottom-gate/top-contact OFETs fabricated by solution-processing of **M97** exhibited hole and electron mobilities of 0.07 and 0.03 $cm^2\ V^{-1}\ s^{-1}$, respectively, under ambient conditions [262]. This performance is currently among the highest for solution-processable, ambipolar, small-molecule semiconductors measured under ambient conditions. The authors also synthesized a reference oligomer without cyano groups, which exhibited only p-channel behavior. The ambient ambipolar characteristic of **M97** was attributed to its relatively low-lying LUMO and dense intermolecular packing.

In a recent study by Nishinaga and Kunugi *et al.* [263], a new sulfur-bridged tetra[2,3-thienylene] semiconductor (**M98**, Figure 1.45) with (triisopropylsilyl)ethynyl groups was synthesized and characterized in single-crystal FET devices. **M98** has a narrow HOMO–LUMO gap of 1.9 eV because of the antiaromatic cyclooctatetraene core. OFET devices based on **M98** single crystals exhibited ambipolar characteristics with hole and electron mobilities of up to 0.40 and 0.18 $cm^2\ V^{-1}\ s^{-1}$, respectively. Different from the previous donor–acceptor approaches, ambipolarity of a small-molecule semiconductor was achieved for the first time with antiaromaticity. This is a novel and effective strategy to realize high and balanced ambipolar charge transport, because it does not lead to HOMO and LUMO localization on separate donor and acceptor units. Another example of an antiaromatic, small-molecule, ambipolar semiconductor,

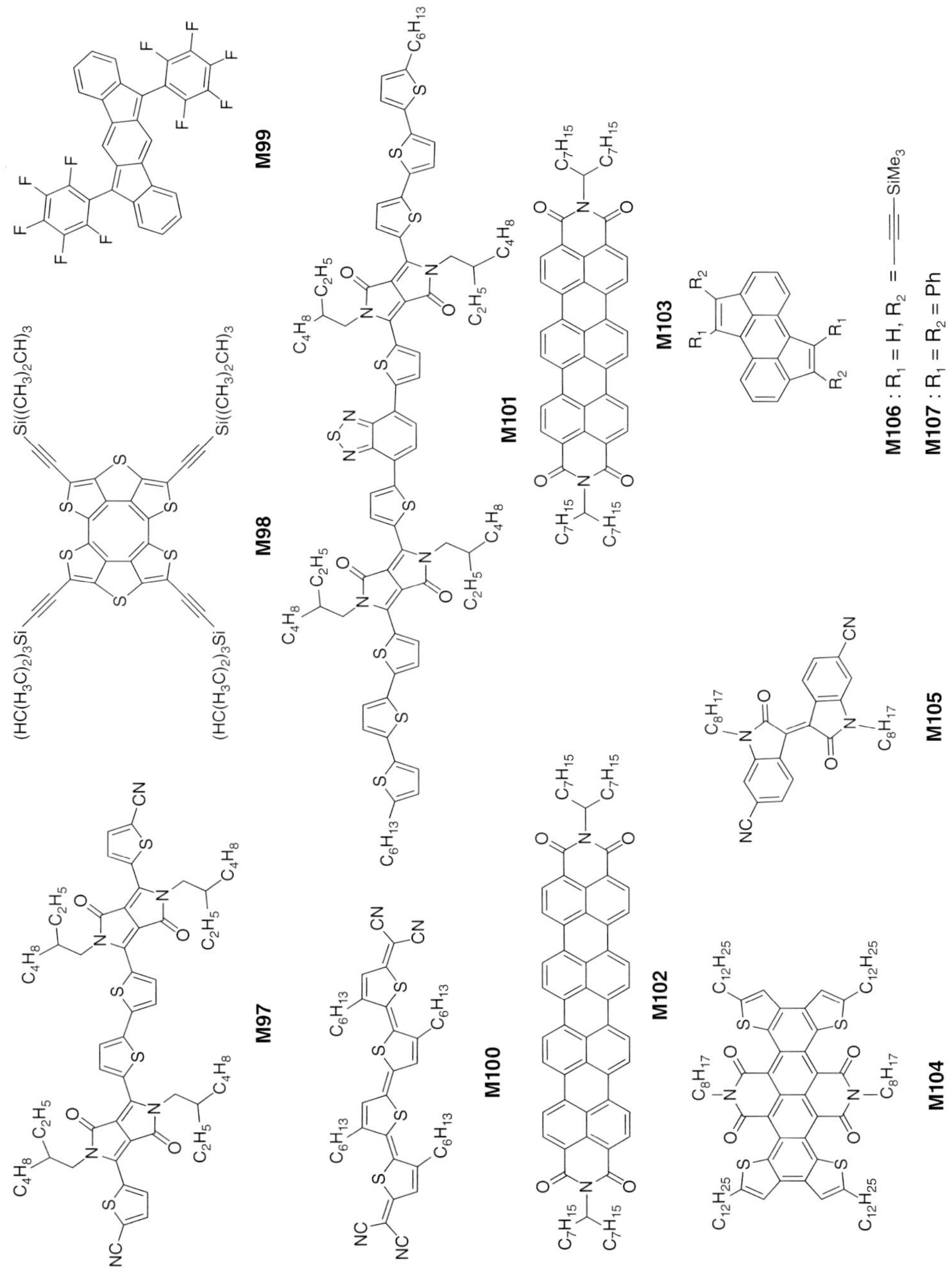

Figure 1.45 Chemical structure of ambipolar small-molecule semiconductors **M97–M107**.

which was recently reported by Haley and coworkers, is 6,12-diarylindeno[1,2-*b*]fluorene (IF). A single crystal of the pentafluorophenyl-substituted IF derivative **M99** (Figure 1.45) yielded ambipolar OFET devices with hole and electron mobilities of 7×10^{-4} and 3×10^{-3} cm^2 V^{-1} s^{-1}, respectively [264].

Quinoidal oligothiophene derivatives have shown great promise for the realization of ambipolar OFETs. Aoyama *et al.* [265] reported high-performance, air-stable, solution-processed ambipolar **M100** (Figure 1.45) with hole and electron mobilities of 0.1 and 0.006 cm^2 V^{-1} s^{-1}, respectively. This semiconductor exhibits an extremely low bandgap of ~0.9 eV in the solid state, which enables efficient injection of both charge-carrier types. The same group recently reported more optimized device structures with **M100**, which yielded more balanced hole and electron mobilities of 0.08 and 0.015 cm^2 V^{-1} s^{-1}, respectively [266]. In this study, the relationship between OFET performance and semiconductor film thickness was explained on the basis of the effects of grain boundaries and contact resistances.

In another study by Nguyen *et al.* [267], ambipolar charge transport in a solution-processed thin film of newly synthesized bis-DPP compound, **M101** (Figure 1.45), was investigated with a low-work-function (Φ_m = 2.7 eV) Ba electrode. **M101** includes two electron-accepting units, DPP and benzothiadiazole, along with thiophene-based donor units. Ambipolar OFETs fabricated with **M101** showed hole and electron mobilities of up to 0.016 and 0.015 cm^2 V^{-1} s^{-1}, respectively. Thermal annealing at 150 °C was found to be critical to enhance field-effect mobilities of **M101** due to the formation of continuous grain boundaries with low surface roughness (Figure 1.46). Larger crystalline domain formation at 240 °C leads to discontinuous grain boundaries, which deteriorates charge-transport characteristics. Additionally, in **M101**-based OFET devices, electron injection was greatly enhanced by using low-work-function Ba contacts, leading to an improvement of electron current at the saturation regime.

Müllen *et al.* [268] utilized a discotic dye **M102** (Figure 1.45) as an ambipolar semiconductor in solution-processed OFETs. **M102** includes a quaterrylenediimide π backbone with swallow tails, and the design rationale was that the conjugation of perylenediimides was extended along the long axis to effectively increase the HOMO energy level without effecting the LUMO level. Hole and electron mobilities of ~10^{-3} cm^2 V^{-1} s^{-1} were obtained in solution-processed bottom-gate/bottom-contact OFETs. For this semiconductor, ambipolarity was found to be lost after a thermal treatment, which was attributed to a change in morphology. Another solution-processable, rylenediimide-based, ambipolar semiconductor, terrylenediimide **M103** (Figure 1.45), was developed by Müllen and Sirringhaus *et al.* [269]. When applied in a TG–BC OFET structure with a polycyclohexylethylene-based gate dielectric, **M103**-based devices exhibited ambipolar transport with electron and hole mobilities of 7.2×10^{-3} and 2.2×10^{-3} cm^2 V^{-1} s^{-1}, respectively. Spin-coated, annealed **M103** films indicated a terraced crystalline microstructure with an "edge-on" molecular orientation, a favorable packing arrangement for charge transport in the plane of the film.

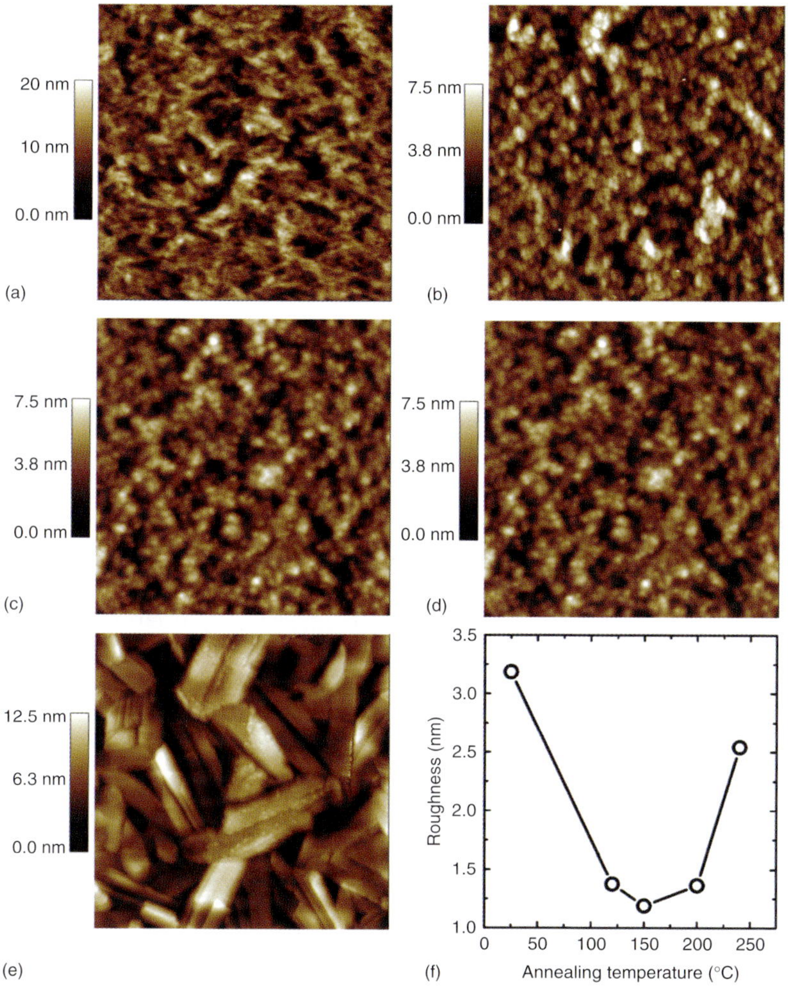

Figure 1.46 2 μm × 2 μm AFM topographic images of **M101** films prepared at different conditions. (a) As cast, and annealed at (b) 120 °C, (c) 150 °C, (d) 200 °C, and (e) 240 °C. (f) root-mean-square (RMS) film roughness versus the annealing temperature [267]. (Reproduced with permission. Copyright © 2012, John Wiley & Sons, Inc.)

The first successful synthesis of a tetrathienyl-fused tetracene diimide **M104** (Figure 1.45) was reported by Chi *et al.* [270] via a $FeCl_3$-mediated oxidative cyclodehydrogenation reaction. The design strategy was to make a core expansion from a smaller NDI building block with the aim of obtaining a low bandgap (1.52 eV) as a result of strong intramolecular donor–acceptor interactions. **M104** showed a liquid-crystalline behavior and ambipolar charge transport in thin-film FETs. The devices exhibited typical ambipolar behavior in ambient, with hole and electron mobilities approaching 10^{-3} cm^2 V^{-1} s^{-1} and I_{on}/I_{off} ratios of 10^3-10^4. It is noteworthy that the hole transport of **M104** could only be obtained when the devices were measured in ambient, which indicated that it was most likely due to doping by molecular oxygen.

Solution-processed, small-molecule, bulk heterojunction (BHJ), ambipolar OTFTs were very recently reported by Chu *et al.* [271] based on a combination of two heteroacene-based p-channel small molecules and C_{60}. An optimized blending of two p-channel semiconductors with the n-channel semiconductor C_{60} results in an efficient charge-transport network with balanced hole and electron mobilities of 0.03 and 0.02 $cm^2\,V^{-1}\,s^{-1}$, respectively. Complementary-like inverter with a large transfer gain of 115 was achieved by using these blend-based ambipolar transistors.

Würthner and coworkers [272] recently reported a new core-cyanated isoindigo derivative **M105** (Figure 1.45) with a decreased LUMO level of −3.88 eV. The synthesis was achieved via palladium-catalyzed cyanation of a core-brominated isoindigo compound. As shown in Figure 1.47, the crystal structure of **M105** reveals a highly planar core adopting a favorable packing with face-to-face π−π contacts with a minimum interplanar spacing of 3.22 Å and hydrogen bonding between cyano and CH units (Figure 1.47c). OFET devices fabricated via vacuum deposition on pentadecylfluorooctadecylphosphonic acid (FOPA)-modified substrates exhibit ambipolar behavior with very good I_{on}/I_{off} ratios of 10^5-10^6 and electron and hole mobilities of 0.11 and 0.045 $cm^2\,V^{-1}\,s^{-1}$, respectively. However, only n-channel transport was observed with other self-assembled monolayers such as *n*-tetradecylphosphonic acid, which is most likely due to the absence of polarization of the active layer by the electronegative fluorine substituents of the FOPA.

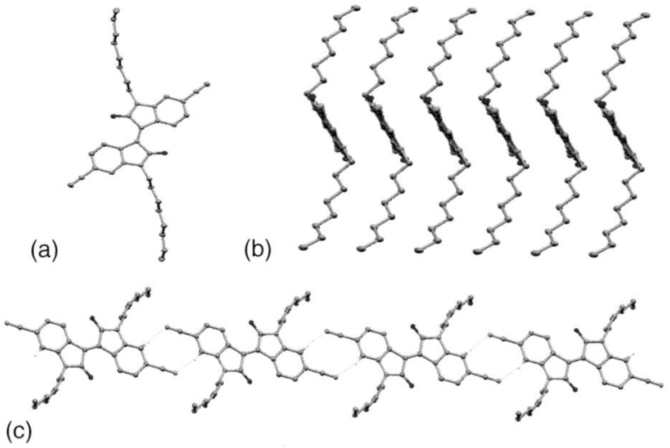

Figure 1.47 Molecular structure of **M105** (a) and solid-state packing (b,c). In (b), the one-dimensional face-to-face π−π stacking and in (c) the hydrogen bonding between cyano and CH groups are shown (the details for the C–H···N H-bonding are as follows: H···N 2.40 Å, C···N 3.3339(15) Å, and C–H···N 168°). Hydrogen atoms are partially omitted for clarity [272]. (Reproduced with permission. Copyright © 2014, The Royal Society of Chemistry.)

A new family of cyclopent[*hi*]aceanthrylene derivatives, which have a cyclopentadiene moiety to accept electrons, was synthesized by Miao *et al.* [273] as potential ambipolar semiconductors. Similar to fullerenes, cyclopenta-fused polycyclic aromatic hydrocarbons exhibit low LUMO energy levels because the five-membered ring tends to form aromatic cyclopentadienide structure (6π-electrons) by accepting one additional electron. **M106** and **M107** (Figure 1.45) are two examples of this family with low LUMO energy levels of about −3.6 eV and balanced HOMO levels of about −5.4 eV. OFET devices with thermally evaporated films of **M106** and **M107** as semiconductor layers exhibited ambipolar characteristics, and hole and electron mobilities as high as 0.21 and 0.10 $cm^2 V^{-1} s^{-1}$, respectively, were achieved. This study demonstrates that the incorporation of a cyclopentadiene unit into π-conjugated cores is an effective strategy to realize ambipolar charge transport.

1.7 Conclusions

In this chapter, we reviewed several first-generation and recently reported molecular and polymeric semiconductors for TFTs. It is now clear that during the last 2–3 years impressive performance, now approaching that of polycrystalline silicon, have been achieved. p-Channel semiconductors based on small molecules now achieve field-effect mobilities of ~30 $cm^2 V^{-1} s^{-1}$, also when processed from solution. Polymeric p-channel TFTs have reached new heights, with hole mobilities unthinkable only few years back and surpassing 10 $cm^2 V^{-1} s^{-1}$. Implemented into a commercially relevant process, these materials could be useful in fabricating unipolar backplane arrays for electrophoretic as well as LCD and possibly OLED displays. The performance of n-channel semiconductors continues to lag behind that of p-channel devices, but small molecules have now demonstrated electron mobilities as high as 7 $cm^2 V^{-1} s^{-1}$ whereas polymers as high as ~1 $cm^2 V^{-1} s^{-1}$. These results are encouraging for enabling the design of complementary circuits. Ambipolar semiconductors, where both hole and electron transport is promoted by the sign of the gate bias, have also experienced dramatic performance improvements, with balanced ambipolar mobilities surpassing 1 $cm^2 V^{-1} s^{-1}$. However, the stability and reproducibility of this performance remains an issue for practical application in, for example, CMOS and light-emitting transistors. All these results combined clearly demonstrate that electronics based on organic semiconductors will be part of a new generation of widely employed products.

References

1. (a) Farchioni, R. and Grosso, G. (eds) (2001) *Organic Electronic Materials: Conjugated Polymers and Low Molecular Weight Organic Solids*, Springer, Berlin; (b) Miller, L.S. and Mullin, J.B. (eds) (1991) *Electronic Materials: From Silicon to Organics*, Plenum Press, New York.

2. (a) Nalwa, H.S. (ed.) (1997) *Handbook of Organic Conductive Molecules and Polymers*, vol. 1-4, John Wiley & Sons, Ltd, Chichester; (b) Müllen, K. and Wegner, G. (eds) (1998) *Electronic Materials: The Oligomer Approach*, John Wiley & Sons, Inc, New York.
3. (a) Coughlin, J.E., Henson, Z.B., Welch, G.C., and Bazan, G.C. (2014) *Acc. Chem. Res.*, **47**, 257; (b) Heeger, A.J. (2014) *Adv. Mater.*, **26**, 10; (c) Dang, M.T., Hirsch, L., Wantz, G., and Wuest, J.D. (2013) *Chem. Rev.*, **113**, 3734; (d) Di, C.-A., Zhang, F., and Zhu, D. (2013) *Adv. Mater.*, **25**, 313; (e) Lee, S.S., Loth, M.A., Anthony, J.E., and Loo, Y.-L. (2012) *J. Am. Chem. Soc.*, **134**, 5436; (f) Natali, D. and Caironi, M. (2012) *Adv. Mater.*, **24**, 1357; (g) Baeg, K.-J., Khim, D., Kim, J., Yang, B.-D., Kang, M., Jung, S.-W., You, I.-K., Kim, D.-Y., and Noh, Y.-Y. (2012) *Adv. Funct. Mater.*, **22**, 2881; (h) Usta, H., Facchetti, A., and Marks, T.J. (2011) *Acc. Chem. Res.*, **44**, 501; (i) Beaujuge, P.M. and Fréchet, J.M.J. (2011) *J. Am. Chem. Soc.*, **133**, 20009; (j) Kelley, T.W., Baude, P.F., Gerlach, C., Ender, D.E., Muyres, D., Haase, M.A., Vogel, D.E., and Theiss, S.D. (2004) *Chem. Mater.*, **16**, 4413; (k) Forrest, S.R. (2004) *Nature*, **428**, 911; (l) Tsutsui, T. and Fujita, K. (2002) *Adv. Mater.*, **14**, 949; (m) Shaw, J.M. and Seidler, P.F. (2001) *IBM J. Res. Dev.*, **45**, 3.
4. (a) Cornil, J., Verlaak, S., Martinelli, N., Mityashin, A., Olivier, Y., Van Regemorter, T., D'Avino, G., Muccioli, L., Zannoni, C., Castet, F., Beljonne, D., and Heremans, P. (2013) *Acc. Chem. Res.*, **46**, 434; (b) Wang, C., Dong, H., Hu, W., Liu, Y., and Zhu, D. (2012) *Chem. Rev.*, **112**, 2208; (c) Rivnay, J., Mannsfeld, S.C.B., Miller, C.E., Salleo, A., and Toney, M.F. (2012) *Chem. Rev.*, **112**, 5488; (d) Saeki, A., Koizumi, Y., Aida, T., and Seki, S. (2012) *Acc. Chem. Res.*, **45**, 1193; (e) Figueira-Duarte, T.M. and Müllen, K. (2011) *Chem. Rev.*, **111**, 7260; (f) Mas-Torrent, M. and Rovira, C. (2011) *Chem. Rev.*, **111**, 4833; (g) Pingel, P., Zhu, L.P., Kue, S., Vogel, J.-O., Janietz, S., Kim, E.-G., Rabe, J.P., Bredas, J.-L., and Koch, N. (2010) *J. Phys. Chem. Lett.*, **1**, 2037; (h) Chang, H.-C., Ruden, P.P., Liang, Y., and Frisbie, C.D. (2010) *J. Appl. Phys.*, **107**, 104502/1; (i) Gutierrez Lezama, I. and Morpurgo, A.F. (2009) *Phys. Rev. Lett.*, **103**, 066803/1; (j) Bredas, J.-L., Norton, J.E., Cornil, J., and Coropceanu, V. (2009) *Acc. Chem. Res.*, **42**, 1691; (k) Troisi, A. (2010) *Adv. Polym. Sci.*, **223**, 259; (l) Choi, S.H., Risko, C., Ruiz Delgado, M.C., Kim, B.S., Bredas, J.-L., and Frisbie, C.D. (2010) *J. Am. Chem. Soc.*, **132**, 4358; (m) Liu, C. and Sirringhaus, H. (2010) *Org. Electron.*, **11**, 558.
5. (a) Sekitani, T. and Someya, T. (2010) *Adv. Mater.*, **22**, 2228; (b) Ma, H., Yip, H.-L., Huang, F., and Jen, A.K.-Y. (2010) *Adv. Funct. Mater.*, **20**, 1371; (c) Wu, W., Liu, Y., and Zhu, D. (2010) *Chem. Soc. Rev.*, **39**, 1489; (d) Ortiz, R.P., Yan, H., Facchetti, A., and Marks, T. (2010) *J. Mater.*, **3**, 1533; (e) Wen, Y. and Liu, Y. (2010) *Adv. Mater.*, **22**, 1331; (f) Lee, W.H., Cho, J.H., and Cho, K. (2010) *J. Mater. Chem.*, **20**, 2549; (g) Miozzo, L., Yassar, A., and Horowitz, G. (2010) *J. Mater. Chem.*, **20**, 2513; (h) Smith, J., Hamilton, R., McCulloch, I., Stingelin-Stutzmann, N., Heeney, M., Bradley, D.D.C., and Anthopoulos, T.D. (2010) *J. Mater. Chem.*, **20**, 2562; (i) Singh, M., Haverinen, H.M., Dhagat, P., and Jabbour, G.E. (2010) *Adv. Mater.*, **22**, 673; (j) Hoppe, A., Balster, T., Muck, T., and Wagner, V. (2009) in *Organic Electronics: Structural and Electronic Properties of OFETs* (ed. C., Wöll), Wiley-VCH Verlag GmbH & Co. KGaA, Weinheim, pp. 469–497, ISBN: 9783527408108; (k) Hamilton, R., Heeney, M., Anthopoulos, T., and McCulloch, I. (2010) Development of Polymer Semiconductors for Field-Effect Transistor Devices in *Organic Electronics: Materials, Processing, Devices and Applications*, (ed. So, F.) CRC Press, Boca Raton, FL, p. 393; (l) Kaake, L.G., Barbara, P.F., and Zhu, X.-Y. (2010) *J. Phys. Chem. Lett.*, **1**, 628; (m) Sirringhaus, H. (2009) *Adv. Mater.*, **21**, 3859; (n) Ortiz, R.P., Facchetti, A., and Marks, T. (2010) *J. Chem. Rev.*,

110, 205; (o) Yamashita, Y. (2009) *Chem. Lett.*, **38**, 870; (p) Ohshita, J. (2009) *Macromol. Chem. Phys.*, **210**, 1360; (q) Katz, H.E. and Huang, J. (2009) *Annu. Rev. Mater. Res.*, **39**, 71; (r) Di, C.-A., Liu, Y., Yu, G., and Zhu, D. (2009) *Acc. Chem. Res.*, **42**, 1573; (s) Roberts, M.E., Sokolov, A.N., and Bao, Z. (2009) *J. Mater. Chem.*, **19**, 3351; (t) DiBenedetto, S.A., Facchetti, A., Ratner, M.A., and Marks, T. (2009) *J. Adv. Mater.*, **21**, 1407; (u) Braga, D. and Horowitz, G. (2009) *Adv. Mater.*, **21**, 1473; (v) McCulloch, I., Heeney, M., Chabinyc, M.L., De Longchamp, D., Kline, R.J., Colle, M., Duffy, W., Fischer, D., Gundlach, D., Hamadani, B., Hamilton, R., Richter, L., Salleo, A., Shkunov, M., Sparrowe, D., Tierney, S., and Zhang, W. (2009) *Adv. Mater.*, **21**, 1091; (w) Zhan, X., Barlow, S., and Marder, S.R. (2009) *Chem. Commun.*, 1948; (x) Panzer, M.J. and Frisbie, C.D. (2008) *Adv. Mater.*, **20**, 3177; (y) de Boer, B. and Facchetti, A. (2008) *Polym. Rev.*, **48**, 423.

6. (a) Watanabe, M., Chen, K.-Y., Chang, Y.J., and Chow, T.J. (2013) *Acc. Chem. Res.*, **46**, 1606; (b) Bujak, P., Kulszewicz-Bajer, I., Zagorska, M., Maurel, V., Wielgus, I., and Pron, A. (2013) *Chem. Soc. Rev.*, **42**, 8895; (c) Jiang, W., Li, Y., and Wang, Z. (2013) *Chem. Soc. Rev.*, **42**, 6113; (d) Dong, H., Fu, X., Liu, J., Wang, Z., and Hu, W. (2013) *Adv. Mater.*, **25**, 6158; (e) Baeg, K.-J., Caironi, M., and Noh, Y.-Y. (2013) *Adv. Mater.*, **25**, 4210; (f) Zhao, Y., Guo, Y., and Liu, Y. (2013) *Adv. Mater.*, **25**, 5372; (g) Lin, Y., Fan, H., Li, Y., and Zhan, X. (2012) *Adv. Mater.*, **24**, 3087; (h) Lin, Y., Li, Y., and Zhan, X. (2012) *Chem. Soc. Rev.*, **41**, 4245; (i) Gaudiana, R. (2010) *J. Phys. Chem. Lett.*, **1**, 1288; (j) Giridharagopal, R. and Ginger, D.S. (2010) *J. Phys. Chem. Lett.*, **1**, 1160; (k) Steim, R., Kogler, F.R., and Brabec, C.J. (2010) *J. Mater. Chem.*, **20**, 2499; (l) Bredas, J.-L. and Durrant, J.R. (2009) *Acc. Chem. Res.*, **42**, 1689; (m) Nelson, J., Kwiatkowski, J.J., Kirkpatrick, J., and Frost, J.M. (2009) *Acc. Chem. Res.*, **42**, 1768; (n) Cheng, Y.-J., Yang, S.-H., and Hsu, C.-S. (2009) *Chem. Rev.*, **109**, 5868; (o) Heremans, P., Cheyns, D., and Rand, B.P. (2009) *Acc. Chem. Res.*, **42**, 1740; (p) Potscavage, W.J. Jr., Sharma, A., and Kippelen, B. (2009) *Acc. Chem. Res.*, **42**, 1758; (q) Roncali, J. (2009) *Acc. Chem. Res.*, **42**, 1719; (r) Chen, J. and Cao, Y. (2009) *Acc. Chem. Res.*, **42**, 1709; (s) Dennler, G., Scharber, M.C., and Brabec, C. (2009) *Adv. Mater.*, **21**, 1323; (t) Pagliaro, M., Ciriminna, R., and Palmisano, G. (2008) *ChemSusChem*, **1**, 880.
7. (a) *Organic Semiconductors and Devices* (Ed. G. Malliaras), John Wiley & Sons, Inc., Hoboken, NJ 2003; (b) Karl, N. (2003) *Synth. Met.*, **133–134**, 649.
8. (a) Sirringhaus, H. (2003) *Nat. Mater.*, **2**, 641; (b) Pron, A. and Rannou, P. (2002) *Prog. Polym. Sci.*, **27**, 135; (c) Cacialli, F. (2000) *Philos. Trans. R. Soc. London, Ser. A*, **358**, 173; (d) Kowalsky, W., Becker, E., Benstem, T., Johannes, H.-H., Metzdorf, D., Neuner, H., and Schobel, J. (2000) *Adv. Solid State Phys.*, **40**, 795; (e) Horowitz, G. (1990) *Adv. Mater.*, **2**, 287.
9. (a) Katz, H.E. (2004) *Chem. Mater.*, **16**, 4748; (b) Allard, S., Forster, M., Souharce, B., Thiem, H., and Scherf, U. (2008) *Angew. Chem. Int. Ed.*, **47**, 4070.
10. Hamilton, R., Bailey, C., Duffy, W., Heeney, M., Shkunov, M., Sparrowe, D., Tierney, S., McCulloch, I., Kline, R.J., DeLongchamp, D.M., and Chabinyc, M. (2006) *Proc. SPIE-Int. Soc. Opt. Eng.*, **6336**, 158.
11. (a) Klauk, H. (ed.) (2006) *Organic Electronics: Materials, Manufacturing, and Applications*, Wiley-VCH Verlag GmbH , Weinheim; (b) Crawford, G.P. (ed.) (2005) *Flexible Flat Panel Display*, John Wiley & Sons, Inc., New York; (c) Gamota, D., Brazis, P., Kalyanasumdaram, K., and Zhang, J. (eds) (2004) *Printed Organic and Molecular Electronics*, Kluwer Academic Publishers, Boston, MA.
12. Perepichka, I.F., Perepichka, D.F., Meng, H., and Wudl, F. (2005) *Adv. Mater.*, **17**, 2281.
13. Kline, R.J., McGehee, M.D., Kadnikova, E.N., Liu, J., Frechet, J.M.J., and Toney, M.F. (2005) *Macromolecules*, **38**, 3312.

14. Zen, A., Saphiannikova, M., Neher, D., Grenzer, J., Grigorian, S., Pietsch, U., Asawapirom, U., Janietz, S., Scherf, U., Lieberwirth, I., and Wegner, G. (2006) *Macromolecules*, **39**, 2162.
15. Zhang, R., Li, B., Iovu, M.C., Jeffries-EL, M., Sauve, G., Cooper, J., Jia, S., Tristram-Nagle, S., Smilgies, D.M., Lambeth, D.N., McCullough, R.D., and Kowalewski, T. (2006) *J. Am. Chem. Soc.*, **128**, 3480.
16. Chang, J.-F., Sun, B., Breiby, D.W., Nielsen, M.M., Soelling, T.I., Giles, M., McCulloch, I., and Sirringhaus, H. (2004) *Chem. Mater.*, **16**, 4772.
17. Majewski, L.A., Kingsley, J.W., Balocco, C., and Song, A.M. (2006) *Appl. Phys. Lett.*, **88**, 222108/1.
18. Surin, M., Leclere, P., Lazzaroni, R., Yuen, J.D., Wang, G., Moses, D., Heeger, A.J., Cho, S., and Lee, K. (2006) *J. Appl. Phys.*, **100**, 033712.
19. Jia, H., Gowrisanker, S., Pant, G.K., Wallace, R.M., and Gnade, B.E. (2006) *J. Vac. Sci. Technol., A*, **24**, 1228.
20. Kim, D.H., Jang, Y., Park, Y.D., and Cho, K. (2006) *J. Phys. Chem. B*, **110**, 15763.
21. Hoshino, S. *et al.* (2004) *J. Appl. Phys.*, **95**, 5088.
22. Park, Y.D., Kim, D.H., Jang, Y., Cho, J.H., Hwang, M., Lee, H.S., Lim, J.A., and Cho, K. (2006) *Org. Electron.*, **7**, 514.
23. Bao, Z., Feng, Y., Dodabalapur, A., Raju, V.R., and Lovinger, A.J. (1997) *Chem. Mater.*, **9**, 1299.
24. Ong, B.S., Wu, Y., Liu, P., and Gardner, S. (2004) *J. Am. Chem. Soc.*, **126**, 3378.
25. Chabinyc, M.L., Endicott, F., Vogt, B.D., DeLongchamp, D.M., Lin, E.K., Wu, Y., Liu, P., and Ong, B.S. (2006) *Appl. Phys. Lett.*, **88**, 113514/1.
26. Heeney, M., Bailey, C., Genevicius, K., Shkunov, M., Sparrowe, D., Tierney, S., and McCulloch, I. (2005) *J. Am. Chem. Soc.*, **127**, 1078.
27. McCulloch, I., Heeney, M., Bailey, C., Genevicius, K., MacDonald, I., Shkunov, M., Sparrowe, D., Tierney, S., Wagner, R., Zhang, W., Chabinyc, M.L., Kline, R.J., McGehee, M.D., and Toney, M.F. (2006) *Nat. Mater.*, **5**, 328.
28. DeLongchamp, D.M., Kline, R.J., Lin, E.K., Fischer, D.A., Richter, L.J., Lucas, L.A., Heeney, M., McCulloch, I., and Northrup, J.E. (2007) *Adv. Mater.*, **19**, 833.
29. Chua, L.-L., Ho, P.K.H., Sirringhaus, H., Friend, R.H. *et al.* (2004) *Adv. Mater.*, **16**, 1609.
30. Yan, H., Yoon, M.-H., Facchetti, A., and Marks, T.J. (2005) *Appl. Phys. Lett.*, **87**, 183501.
31. (a) Fielke, D., Huebler, A.C., Hahn, U., Brandt, N., Bartzsch, M., Fuegmann, U., Fischer, T., Veres, J., and Ogier, S. (2005) *Appl. Phys. Lett.*, **87**, 123508/1; (b) Veres, J., Ogier, S.D., Leeming, S.W., Cupertino, D.C., and Khaffaf, S.M. (2003) *Adv. Funct. Mater.*, **13**, 199.
32. Li, Y., Wu, Y., and Ong, B.S. (2006) *Macromolecules*, **39**, 6521.
33. Drolet, N., Morin, J.-F., Leclerc, N., Wakim, S., Tao, Y., and Leclerc, M. (2005) *Adv. Funct. Mater.*, **15**, 1671.
34. (a) Ogawa, K. and Rasmussen, S.C. (2003) *J. Org. Chem.*, **68**, 2921; (b) Ogawa, K., Stafford, J.A., Rothstein, S.D., Tallman, D.E., and Rasmussen, S.C. (2005) *Synth. Met.*, **152**, 137.
35. (a) Ogawa, K. and Rasmussen, S.C. (2006) *Macromolecules*, **39**, 1771; (b) Koeckelberghs, G. and Samyn, C. (2005) *Macromolecules*, **38**, 4545.
36. Liu, J., Zhang, R., Sauve, G., Kowalewski, T., and McCullough, R.D. (2008) *J. Am. Chem. Soc.*, **130**, 13167.
37. (a) Yamamoto, T., Kokubo, H., Kobashi, M., and Sakai, Y. (2004) *J. Mater. Chem.*, **16**, 4616; (b) Yamamoto, T., Arai, M., Kokubo, H., and Sasaki, S. (2003) *Macromolecules*, **36**, 7986; (c) Li, W., Katz, H., Lovinger, A., and Laquindanum, J. (1999) *Chem. Mater.*, **11**, 458.
38. Liu, J., Zhang, R., Osaka, I., Mishra, S., Javier, A.E., Smilgies, D., Kowalewski, T., and McCullough, R.D. (2009) *Adv. Funct. Mater.*, **19**, 3427.
39. (a) Fong, H.H., Pozdin, V.A., Amassian, A., Malliaras, G.G., Smilgies, D.-M., He, M., Gasper, S., Zhang, F., and Sorensen, M. (2008) *J. Am. Chem. Soc.*, **130**, 13202; (b) He, M., Li, J., Sorensen, M.L., Zhang, F., Hancock, R.R., Fong,

H.H., Pozdin, V.A., Smilgies, D.-M., and Malliaras, G.G. (2009) *J. Am. Chem. Soc.*, **131**, 11930.
40. He, M., Li, J., Tandia, A., Sorensen, M., Zhang, F., Fong, H.H., Pozdin, V.A., Smilgies, D., and Malliaras, G.G. (2010) *Chem. Mater.*, **22**, 2770.
41. Li, J., Qin, F., Li, C.M., Bao, Q., Chan-Park, M.B., Zhang, W., Qin, J., and Ong, B.S. (2008) *Chem. Mater.*, **20**, 2057.
42. Rieger, R., Beckmann, D., Pisula, W., Steffen, W., Kastler, M., and Mullen, K. (2010) *Adv. Mater.*, **22**, 83.
43. Prosa, T.J., Winokur, M.J., Moulton, J., Smith, P., and Heeger, A.J. (1992) *Macromolecules*, **25**, 4364.
44. Nakayama, J. and Fujimiri, T. (1991) *Heterocycles*, **32**, 991.
45. (a) Fu, Y., Cheng, H., and Elsenbaumer, R.L. (1997) *Chem. Mater.*, **9**, 1720; (b) Elandaloussi, E.H., Fre're, P., Richomme, P., Orduna, J., Garin, J., and Roncali, J. (1997) *J. Am. Chem. Soc.*, **119**, 10774.
46. (a) Payne, M., Parkin, S., Anthony, J., Kuo, C., and Jackson, T. (2005) *J. Am. Chem. Soc.*, **127**, 4986; (b) Stingelin-Stutzmann, N., Smits, E., Wondergem, H., Tanase, C., Blom, P., Smith, P., and de Leeuw, D.M. (2005) *Nat. Mater.*, **4**, 601; (c) Mathijssen, S.G.J., Cölle, M., Gomes, H., Smits, E.C.P., de Boer, B., McCulloch, I., Bobbert, P.A., and de Leeuw, D.M. (2007) *Adv. Mater.*, **19**, 2785; (d) Mathijssen, S.G.J., Kemerink, M., Sharma, A., Colle, M., Bobbert, P.A., Janssen, R.A.J., and de Leeuw, D.M. (2008) *Adv. Mater.*, **20**, 975; (e) Street, R.A., Chabinyc, M.L., Endicott, F., and Ong, B. (2006) *J. Appl. Phys.*, **100**, 114518; (f) Street, R.A., Salleo, A., and Chabinyc, M. (2003) *Phys. Rev. B*, **68**, 085316; (g) Salleo, A. and Street, R.A. (2004) *Phys. Rev. B*, **70**, 235324; (h) Salleo, A., Endicott, F., and Street, R.A. (2005) *Appl. Phys. Lett.*, **86**, 263505.
47. Tsukada, T. (2000) in *Technology and Applications of Amorphous Silicon*, vol. 37 (ed. R.A. Street), Springer-Verlag, Heidelberg, p. 49.
48. Kim, D.H., Lee, B., Moon, H., Kang, H.M., Jeong, E.J., Park, J., Han, K., Lee, S., Yoo, B.W., Koo, B.W., Kim, J.Y., Lee, W.H., Cho, K., Becerril, H.A., and Bao, Z. (2009) *J. Am. Chem. Soc.*, **131**, 6124.
49. (a) Yamamoto, T., Kokudo, H., Kobashi, M., and Sakai, Y. (2004) *Chem. Mater.*, **16**, 4616; (b) Kokudo, H., Sato, T., and Yamamoto, T. (2006) *Macromolecules*, **39**, 3959.
50. Osaka, I., Zhang, R., Sauve, G., Smilgies, D., Kowalewski, T., and McCullough, R.D. (2009) *J. Am. Chem. Soc.*, **131**, 2521.
51. (a) Jenekhe, S.A. and Osaheni, J.A. (1994) *Science*, **265**, 765; (b) Osaheni, J.A. and Jenekhe, S.A. (1995) *J. Am. Chem. Soc.*, **117**, 7389; (c) So, Y.-H., Zaleski, J.M., Murlick, C., and Ellaboudy, A. (1996) *Macromolecules*, **29**, 2783; (d) Mike, J.F., Makowski, A.J., and Jeffries-EL, M. (2008) *Org. Lett.*, **10**, 4915.
52. (a) Wolfe, J.F., Loo, B.H., and Arnold, F.E. (1981) *Macromolecules*, **14**, 915; (b) Choe, E.W. and Kim, S.N. (1981) *Macromolecules*, **14**, 920; (c) Roberts, M.F. and Jenekhe, S.A. (1993) *Chem. Mater.*, **5**, 1744; (d) Osaheni, J.A. and Jenekhe, S.A. (1995) *Chem. Mater.*, **7**, 672; (e) Osaheni, J.A. and Jenekhe, S.A. (1993) *Macromolecules*, **26**, 4726; (f) Alam, M.M. and Jenekhe, S.A. (2002) *Chem. Mater.*, **14**, 4775; (g) Jenekhe, S.A., Osaheni, J.A., Meth, J.S., and Vanherzeele, H. (1992) *Chem. Mater.*, **4**, 683; (h) Osaheni, J.A. and Jenekhe, S.A. (1992) *Chem. Mater.*, **4**, 1282; (i) Dotrong, M., Mehta, R., Balchin, G.A., Tomlinson, R.C., Sinsky, M., Lee, C.Y.C., and Evers, R.C.J. (1993) *Polym. Sci., Part A*, **31**, 723.
53. (a) Babel, A. and Jenekhe, S.A. (2002) *J. Phys. Chem. B*, **106**, 6129; (b) Babel, A. and Jenekhe, S.A. (2003) *J. Am. Chem. Soc.*, **125**, 13656.
54. (a) Pang, H., Vilela, F., Skabara, P.J., McDouall, J.J.W., Crouch, D.J., Anthopoulos, T.D., Bradley, D.D.C., de Leeuw, D., Horton, P.N., and Hursthouse, M.B. (2007) *Adv. Mater.*, **19**, 4438; (b) Mamada, M., Nishida, J.-I., Tokito, S., and Yamashita, Y. (2008) *Chem. Lett.*, **37**, 766.
55. Zhang, W., Smith, J., Hamilton, R., Heeney, M., Kirkpatrick, J., Song, K., Watkins, S.E., Anthopoulos, T., and

McCulloch, I. (2009) *J. Am. Chem. Soc.*, **131**, 10814.
56. Zhang, W., Smith, J., Watkins, S.W., Gysel, R., McGehee, M., Salleo, A., Kirkpatrick, J., Ashraf, S., Anthopoulos, T., Heeney, M., and McCulloch, I. (2010) *J. Am. Chem. Soc.*, **132**, 11437.
57. (a) Usta, H., Lu, G., Facchetti, A., and Marks, T.J. (2006) *J. Am. Chem. Soc.*, **128**, 9034–9035; (b) Usta, H., Facchetti, A., and Marks, T.J. (2008) *J. Am. Chem. Soc.*, **130**, 8580–8581.
58. Beaujuge, P.M., Tsao, H.N., Hansen, M.R., Amb, C.M., Risko, C., Subbiah, J., Choudhury, K.R., Mavrinskiy, A., Pisula, W., Brédas, J.-L., So, F., Müllen, K., and Reynolds, J.R. (2012) *J. Am. Chem. Soc.*, **134**, 8944–8957.
59. Mondal, R., Miyaki, N., Becerril, H.A., Norton, J.E., Parmer, J., Mayer, A.C., Tang, M.L., Bredas, J.-L., McGehee, M.D., and Bao, Z. (2009) *Chem. Mater.*, **21**, 3618.
60. (a) Takimiya, K., Ebata, H., Sakamoto, K., Izawa, T., Otsubo, T., and Kunugi, Y. (2006) *J. Am. Chem. Soc.*, **128**, 12604; (b) Ebata, H., Izawa, T., Miyazaki, E., Takimiya, K., Ikeda, M., Kuwabara, H., and Yui, T. (2007) *J. Am. Chem. Soc.*, **129**, 15732.
61. Osaka, I., Abe, T., Shinamura, S., Miyazaki, E., and Takimiya, K. (2010) *J. Am. Chem. Soc.*, **132**, 5000.
62. Shinamura, S., Miyazaki, E., and Takimiya, K. (2010) *J. Org. Chem.*, **75**, 1228.
63. Yang, C., Cho, S., Chiechi, R.C., Walker, W., Coates, N.E., Moses, D., Heeger, A.J., and Wudl, F. (2008) *J. Am. Chem. Soc.*, **130**, 16524.
64. Guo, X., Kim, F.S., Jenekhe, S.A., and Watson, M.D. (2009) *J. Am. Chem. Soc.*, **131**, 7206–7207.
65. (a) Bleiholder, C., Gleiter, R., Werz, D.B., and Koeppel, H. (2007) *Inorg. Chem.*, **46**, 2249; (b) Pomerantz, M. (2003) *Tetrahedron Lett.*, **44**, 1563; (c) Irvin, J.A., Schwendeman, I., Lee, Y., Abboud, K.A., and Reynolds, J.R. (2001) *J. Polym. Sci. A: Polym. Chem.*, **39**, 2164; (d) Cloutier, R. and Leclerc, M. (1991) *J. Chem. Soc., Chem. Commun.*, 1194; (e) Vangheluwe, M., Verbiest, T., and Koeckelberghs, G. (2008) *Macromolecules*, **41**, 1041.
66. (a) Thompson, B.C., Kim, Y.G., McCarley, T.D., and Reynolds, J.R. (2006) *J. Am. Chem. Soc.*, **128**, 12714; (b) Roncali, J. (1997) *Chem. Rev.*, **97**, 173.
67. FitzHugh, E.W. (ed.) (1997) *Artists' Pigments: A Handbook of Their History and Characteristics*, vol. 3, Oxford University Press, New York.
68. Stalder, R., Mei, J., and Reynolds, J.R. (2010) *Macromolecules*, **43**, 8348.
69. Wang, E., Ma, Z., Zhang, Z., Vandewal, K., Henriksson, P., Inganäs, O., Zhang, F., and Andersson, M.R. (2011) *J. Am. Chem. Soc.*, **133**, 14244.
70. Lei, T., Cao, Y., Fan, Y., Liu, C.-J., Yuan, S.-C., and Pei, J. (2011) *J. Am. Chem. Soc.*, **133**, 6099.
71. Mei, J., Kim, D.H., Ayzner, A.L., Toney, M.F., and Bao, Z. (2011) *J. Am. Chem. Soc.*, **133**, 20130.
72. Holliday, S., Donaghey, J.E., and McCulloch, I. (2014) *Chem. Mater.*, **26**, 647.
73. Lei, T., Dou, J.-H., and Pei, J. (2012) *Adv. Mater.*, **24**, 6457.
74. Burgi, L., Turbiez, M., Pfeiffer, R., Bienewald, F., Kirner, H.-J., and Winnewisser, C. (2008) *Adv. Mater.*, **20**, 2217.
75. (a) Li, Y., Sonar, P., Singh, S.P., Soh, M.S., Meurs, M.v., and Tan, J. (2011) *J. Am. Chem. Soc.*, **133**, 2198; (b) Zhang, X., Richter, L.J., DeLongchamp, D.M., Kline, R.J., Hammond, M.R., McCulloch, I., Heeney, M., Ashraf, R.S., Smith, J.N., Anthopoulos, T.D., Schroeder, B., Geerts, Y.H., Fischer, D.A., and Toney, M.F. (2011) *J. Am. Chem. Soc.*, **133**, 15073; (c) Ha, J.S., Kim, K.H., and Choi, D.H. (2011) *J. Am. Chem. Soc.*, **133**, 10364; (d) Bronstein, H., Chen, Z., Ashraf, R.S., Zhang, W., Du, J., Durrant, J.R., Tuladhar, P.S., Song, K., Watkins, S.E., Geerts, Y., Wienk, M.M., Janssen, R.A.J., Anthopoulos, T., Sirringhaus, H., Heeney, M., and McCulloch, I. (2011) *J. Am. Chem. Soc.*, **133**, 3272.
76. Li, J., Zhao, Y., Tan, H.S., Guo, Y., Di, C.-A., Yu, G., Liu, Y., Lin, M., Lim,

S.H., Zhou, Y., Su, H., and Ong, B.S. (2012) *Sci. Rep.*, **2**, 754.
77. Kang, I., Yun, H.-J., Chung, D.S., Kwon, S.-K., and Kim, Y.-H. (2013) *J. Am. Chem. Soc.*, **135**, 14896.
78. Li, Y., Sonar, P., Murphy, L., and Hong, W. (2013) *Energy Environ. Sci.*, **6**, 1684.
79. Li, Y., Sonar, P., Singh, S.P., Zeng, W., and Soh, M.S. (2011) *J. Mater. Chem.*, **21**, 10829.
80. Nielsen, C.B., Turbiez, M., and McCulloch, I. (2013) *Adv. Mater.*, **25**, 1859.
81. Yi, Z., Ma, L., Chen, B., Chen, D., Chen, X., Qin, J., Zhan, X., Liu, Y., Ong, W.J., and Li, J. (2013) *Chem. Mater.*, **25**, 4290.
82. Zhang, M., Tsao, H.N., Pisula, W., Yang, C., Mishra, A.K., and Müllen, K. (2007) *J. Am. Chem. Soc.*, **129**, 3472.
83. Tsao, H.N., Cho, D.M., Park, I., Hansen, M.R., Mavrinskiy, A., Yoon, D.Y., Graf, R., Pisula, W., Spiess, H.W., and Müllen, K. (2011) *J. Am. Chem. Soc.*, **133**, 2605.
84. Fan, J., Yuen, J.D., Cui, W., Seifter, J., Mohebbi, A.R., Wang, M., Zhou, H., Heeger, A., and Wudl, F. (2012) *Adv. Mater.*, **24**, 6164.
85. Wu, Q., Wang, M., Qiao, X., Xiong, Y., Huang, Y., Gao, X., and Li, H. (2013) *Macromolecules*, **46**, 3887.
86. Kim, J., Lim, B., Baeg, K.-J., Noh, Y.-Y., Khim, D., Jeong, H.-G., Yun, J.-M., and Kim, D.-Y. (2011) *Chem. Mater.*, **23**, 4663.
87. Hajlaoui, R., Horowitz, G., Garnier, F., Arce-Brouchet, A., Laigre, L., El Kassmi, A., Demanze, F., and Kouki, F. (1997) *Adv. Mater.*, **9**, 389.
88. Dodabalapur, A., Torsi, L., and Katz, H.E. (1995) *Science*, **268**, 270.
89. Hajlaoui, R., Fichou, D., Horowitz, G., Nessakh, B., Constant, M., and Garnier, F. (1997) *Adv. Mater.*, **9**, 557.
90. (a) Garnier, F., Horowitz, G., Fichou, D., and Yassar, A. (1996) *Synth. Met.*, **81**, 163; (b) Fichou, D., Horowitz, G., and Garnier, F. (1992) *Mol. Cryst. Liq. Cryst. Sci. Technol., Sect. A*, **217**, 193.
91. (a) Lovinger, A.J., Davis, D.D., Dodabalapur, A., and Katz, H.E. (1996) *Chem. Mater.*, **8**, 2836; (b) Lovinger, A.J., Davis, D.D., Dodabalapur, A., Katz, H.E., and Torsi, L. (1996) *Macromolecules*, **29**, 4952.
92. (a) Katz, H.E., Bao, Z., and Gilat, S.L. (2001) *Acc. Chem. Res.*, **34**, 359; (b) Fichou, D., Teulade-Fichou, M.P., Horowitz, G., and Demanze, F. (1997) *Adv. Mater.*, **9**, 75; (c) Horowitz, G., Garnier, F., Yassar, A., Hajlaoui, R., and Kouki, F. (1996) *Adv. Mater.*, **8**, 52; (d) Katz, H.E., Torsi, L., and Dodabalapur, A. (1995) *Chem. Mater.*, **7**, 2235.
93. (a) Nagamatsu, S., Kaneto, K., Azumi, R., Matsumoto, M., Yoshida, Y., and Yase, K. (2005) *J. Phys. Chem. B*, **109**, 9374; (b) Horowitz, G., Peng, X.Z., Fichou, D., and Garnier, F. (1991) *J. Mol. Electron.*, **7**, 85.
94. Ostoja, P., Maccagnani, P., Gazzano, M., Cavallini, M., Kengne, J.C., Kshirsagar, R., Biscarini, F., Melucci, M., Zambianchi, M., and Barbarella, G. (2004) *Synth. Met.*, **146**, 243.
95. Hajlaoui, M.E., Garnier, F., Hassine, L., Kouki, F., and Bouchriha, H. (2002) *Synth. Met.*, **129**, 215.
96. Loi, M.A., Da Como, E., Dinelli, F., Murgia, M., Zamboni, R., Biscarini, F., and Muccini, M. (2005) *Nat. Mater.*, **4**, 81.
97. Dinelli, F., Murgia, M., Levy, P., Cavallini, M., Biscarini, F., and de Leeuw, D.M. (2004) *Phys. Rev. Lett.*, **92**, 116802/1.
98. (a) Waragai, K., Akimichi, H., Hotta, S., Kano, H., and Sakaki, H. (1993) *Synth. Met.*, **57**, 4053; (b) Waragai, K., Akimichi, H., Hotta, S., Kano, H., and Sakaki, H. (1995) *Phys. Rev. B*, **52**, 1786.
99. (a) Garnier, F., Hajlaoui, R., El Kassmi, A., Horowitz, G., Laigre, L., Porzio, W., Armanini, M., and Provasoli, F. (1998) *Chem. Mater.*, **10**, 3334; (b) Katz, H.E., Lovinger, A.J., and Laquindanum, J.G. (1998) *Chem. Mater.*, **10**, 457.
100. Li, W., Katz, H.E., Lovinger, A.J., and Laquindanum, J.G. (1999) *Chem. Mater.*, **11**, 458.127.
101. (a) Facchetti, A., Yoon, M.-H., Hutchison, G.R., Stern, C.L., Ratner, M.A., and Marks, T.J. (2004) *J. Am. Chem. Soc.*, **126**, 13480–13501; (b) Facchetti, A., Mushrush, M., Yoon, M.-H., Hutchison, G.R., Ratner, M.A., and

Marks, T.J. (2004) *J. Am. Chem. Soc.*, **126**, 13859–13874.

102. (a) Katz, H.E., Dodabalapur, A., Torsi, L., and Elder, D. (1995) *Chem. Mater.*, **7**, 2238; (b) Horowitz, G., Deloffre, F., Garnier, F., Hajlaoui, R., Hmyene, M., and Yassar, A. (1993) *Synth. Met.*, **54**, 435; (c) Garnier, F., Hajlaoui, R., Yassar, A., and Srivastava, P. (1994) *Science*, **265**, 1684.
103. Dimitrakopoulos, C.D., Furman, B.K., Graham, T., Hegde, S., and Purushothaman, S. (1998) *Synth. Met.*, **92**, 47.
104. (a) Hotta, S. and Lee, S.A. (1999) *Synth. Met.*, **101**, 551; (b) Dingemans, T.J., Bacher, C.J., Thelakkat, M., Pederson, L.G., Samulski, E.T., and Schimidt, H.-W. (1999) *Synth. Met.*, **105**, 171.
105. (a) Mohapatra, S., Holmes, B.T., Newman, C.R., Prendergast, C.F., Frisbie, C.D., and Ward, M.D. (2004) *Adv. Funct. Mater.*, **14**, 605; (b) Yanagi, H., Araki, Y., Ohara, T., Hotta, S., Ichikawa, M., and Taniguchi, Y. (2003) *Adv. Funct. Mater.*, **13**, 767; (c) Casado, J., Ruiz Delgado, M.C., Shirota, Y., Hernandez, V., and Lopez Navarrete, J.T. (2003) *J. Phys. Chem. B*, **107**, 2637; (d) Ichikawa, M., Yanagi, H., Shimizu, Y., Hotta, S., Suganuma, N., Koyama, T., and Taniguchi, Y. (2002) *Adv. Mater.*, **14**, 1272; (e) Apperloo, J.J., Groenendaal, L.B., Verheyen, H., Jayakannan, M., Janssen, R.A.J., Dkhissi, A., Beljonne, D., Lazzaroni, R., and Bredas, J.-L. (2002) *Chemistry*, **8**, 2384; (f) Moreno Castro, C., Ruiz Delgado, M.C., Hernandez, V., Hotta, S., Casado, J., and Lopez Navarrete, J.T. (2002) *J. Phys. Chem. B*, **116**, 10419; (g) Lee, S.A., Yoshida, Y., Fukuyama, M., and Hotta, S. (1999) *Synth. Met.*, **106**, 39.
106. (a) Hotta, S. and Katagiri, T. (2003) *J. Heterocycl. Chem.*, **40**, 845; (b) Hotta, S., Kimura, H., Lee, S.A., and Tamaki, T. (2000) *J. Heterocycl. Chem.*, **37**, 281; (c) Hotta, S., Lee, S.A., and Tamaki, T. (2000) *J. Heterocycl. Chem.*, **37**, 25.
107. (a) Yanagi, H., Yoshiki, A., Hotta, S., and Kobayashi, S. (2004) *J. Appl. Phys.*, **96**, 4240; (b) Ichikawa, M., Hibino, R., Inoue, M., Haritani, T., Hotta, S., Koyama, T., and Taniguchi, Y. (2003) *Adv. Mater.*, **15**, 213; (c) Nagawa, M., Hibino, R., Hotta, S., Yanagi, H., Ichikawa, M., Koyama, T., and Taniguchi, Y. (2002) *Appl. Phys. Lett.*, **80**, 544; (d) Kim, Y.C., Lee, T.-W., Park, O.O., Kim, C.Y., and Cho, H.N. (2001) *Adv. Mater.*, **13**, 646.
108. Ponomarenko, S.A., Kirchmeyer, S., Halik, M., Klauk, H., Zschieschang, U., Schmid, G., Karbach, A., Drechsler, D., and Alpatova, N.M. (2005) *Synth. Met.*, **149**, 231.
109. Hong, X.M., Katz, H.E., Lovinger, A.J., Wang, B.C., and Raghavachari, K. (2001) *Chem. Mater.*, **13**, 4686.
110. Ponomarenko, S.A., Kirchmeyer, S., Elschner, A., Alpatova, N.M., Halik, M., Klauk, H., Zschieschang, U., and Schmid, G. (2006) *Chem. Mater.*, **18**, 579.
111. Sung, A., Ling, M.M., Tang, M.L., Bao, Z., and Locklin, J. (2007) *Chem. Mater.*, **19**, 2342.
112. Laquindanum, J.G., Katz, H.E., Lovinger, A.J., and Dodabalapur, A. (1997) *Adv. Mater.*, **9**, 36.
113. Takimiya, K., Shinamura, S., Osaka, I., and Miyazaki, E. (2011) *Adv. Mater.*, **23**, 4347.
114. (a) Wang, C., Dong, H., Hu, W., Liu, Y., and Zhu, D. (2011) *Chem. Rev.*, **112**, 2208; (b) Dong, H., Wang, C., and Hu, W. (2010) *Chem. Commun.*, **46**, 5211; (c) Takimiya, K., Kunugi, Y., and Otsubo, T. (2007) *Chem. Lett.*, **36**, 578; (d) Shinamura, S., Osaka, I., Miyazaki, E., Nakao, A., Yamagishi, M., Takeya, J., and Takimiya, K. (2011) *J. Am. Chem. Soc.*, **133**, 5024.
115. Sirringhaus, H., Friend, R.H., Wang, C., Leuninger, J., and Mullen, K. (1999) *J. Mater. Chem.*, **9**, 2095.
116. (a) Sirringhaus, H., Friend, R.H., Li, X.C., Moratti, S.C., Holmes, A.B., and Feeder, N. (1997) *Appl. Phys. Lett.*, **71**, 3871; (b) Li, X.C., Sirringhaus, H., Garnier, F., Holmes, A.B., Moratti, S.C., Feeder, N., Clegg, W., Teat, S.J., and Friend, R.H. (1998) *J. Am. Chem. Soc.*, **120**, 2206.
117. Iosip, M.D., Destri, S., Pasini, M., Porzio, W., Pernstich, K.P., and Batlogg, B. (2004) *Synth. Met.*, **146**, 251.

118. (a) Zhang, X., Cote, A.P., and Matzger, A.J. (2005) *J. Am. Chem. Soc.*, **127**, 10502; (b) Zhang, X. and Matzger, A.J. (2003) *J. Org. Chem.*, **68**, 9813.
119. Xiao, K., Liu, Y., Qi, T., Zhang, W., Wang, F., Gao, J., Qiu, W., Ma, Y., Cui, G., Chen, S., Zhan, X., Yu, G., Qin, J., Hu, W., and Zhu, D. (2005) *J. Am. Chem. Soc.*, **127**, 13281.
120. Nenajdenko, V.G., Sumerin, V.V., Chernichenko, K.Y., and Balenkova, E.S. (2004) *Org. Lett.*, **6**, 3437.
121. Takimiya, K., Kunugi, Y., Konda, Y., Niihara, N., and Otsubo, T. (2004) *J. Am. Chem. Soc.*, **126**, 5084.
122. Laquindanum, J.G., Katz, H.E., and Lovinger, A.J. (1998) *J. Am. Chem. Soc.*, **120**, 664.
123. Dickey, K.C., Anthony, J.E., and Loo, Y.-L. (2006) *Adv. Mater.*, **18**, 1721.
124. Valiyev, F., Hu, W.-S., Chen, H.-Y., Kuo, M.-Y., Chao, I., and Tao, Y.-T. (2007) *Chem. Mater.*, **19**, 3018.
125. Gao, P., Beckmann, D., Tsao, H.N., Feng, X., Enkelmann, V., Baumgarten, M., Pisula, W., and Müllen, K. (2009) *Adv. Mater.*, **21**, 213.
126. Uemura, T., Hirose, Y., Uno, M., Takimiya, K., and Takeya, J. (2009) *Appl. Phys. Express*, **2**, 111501.
127. Minemawari, H., Yamada, T., Matsui, H., Tsutsumi, J., Haas, S., Chiba, R., Kumai, R., and Hasegawa, T. (2011) *Nature*, **475**, 364.
128. Yamamoto, T. and Takimiya, K. (2007) *J. Am. Chem. Soc.*, **129**, 2224.
129. Kang, M.J., Doi, I., Mori, H., Miyazaki, E., Takimiya, K., Ikeda, M., and Kuwabara, H. (2011) *Adv. Mater.*, **23**, 1222.
130. (a) Inokuchi, H., Imaeda, K., Enoki, T., Mori, T., Maruyama, Y., Saito, G., Okada, N., Yamochi, H., Seki, K., Higuchi, Y., and Yasuoka, N. (1987) *Nature*, **329**, 39; (b) Inokuchi, H., Saito, G., Wu, P., Seki, K., Tang, T.B., Mori, T., Imaeda, K., Enoki, T., Higuchi, Y., Inaka, K., and Yasuoka, N. (1986) *Chem. Lett.*, 1263.
131. Nakayama, K., Hirose, Y., Soeda, J., Yoshizumi, M., Uemura, T., Uno, M., Li, W., Kang, M.J., Yamagishi, M., Okada, Y., Miyazaki, E., Nakazawa, Y., Nakao, A., Takimiya, K., and Takeya, J. (2011) *Adv. Mater.*, **23**, 1626.
132. Kang, M.J., Yamamoto, T., Shinamura, S., Miyazaki, E., and Takimiya, K. (2010) *Chem. Sci.*, **1**, 179.
133. Niimi, K., Shinamura, S., Osaka, I., Miyazaki, E., and Takimiya, K. (2011) *J. Am. Chem. Soc.*, **133**, 8732.
134. Okamoto, T., Mitsui, C., Yamagishi, M., Nakahara, K., Soeda, J., Hirose, Y., Miwa, K., Sato, H., Yamano, A., Matsushita, T., Uemura, T., and Takeya, J. (2013) *Adv. Mater.*, **25**, 6392.
135. Cho, H., Lee, S., Cho, N.S., Jabbour, G.E., Kwak, J., Hwang, D.-H., and Lee, C. (2013) *ACS Appl. Mater. Interfaces*, **5**, 3855.
136. Dadvand, A., Moiseev, A.G., Sawabe, K., Sun, W.-H., Djukic, B., Chung, I., Takenobu, T., Rosei, F., and Perepichka, D.F. (2012) *Angew. Chem. Int. Ed.*, **51**, 3837.
137. (a) Yoon, M.-Y., Kim, C.S., Facchetti, A., and Marks, T.J. (2006) *J. Am. Chem. Soc.*, **128**, 12851; (b) Chua, L.L., Zaumseil, J., Chang, J.F., Ou, E.C.W., Ho, P.K.H., Sirringhaus, H., and Friend, R.H. (2005) *Nature*, **434**, 194.
138. Briseno, A.L., Mannsfeld, S.C.B., Shamberger, P.J., Ohuchi, F.S., Bao, Z., Jenekhe, S.A., and Xia, Y. (2008) *Chem. Mater.*, **20** (14), 4712.
139. Letizia, J., Salata, M., Tribout, C., Facchetti, A., Ratner, M.A., and Marks, T.J. (2008) *J. Am. Chem. Soc.*, **130**, 9679.
140. Zhan, X., Tan, Z., Domercq, B., An, Z., Zhang, X., Barlow, S., Li, Y., Zhu, D., Kippelen, B., and Marder, S.R. (2007) *J. Am. Chem. Soc.*, **129**, 7246.
141. Zhan, X., Tan, Z., Zhou, E., Li, Y., Misra, R., Grant, A., Domercq, B., Zhang, X., An, Z., Zhang, X., Barlow, S., Kippelen, B., and Marder, S.R. (2009) *J. Mater. Chem.*, **19**, 5794.
142. Hüttner, S., Sommer, M., and Thelakkat, M. (2008) *Appl. Phys. Lett.*, **92**, 093302.
143. Lang, A.S., Muth, M.-A., Heinrich, C.D., Carassco-Orozco, M., and Thelakkat, M. (2013) *J. Polym. Sci. Pol. Phys.*, **51**, 1480.

144. Chen, Z., Zheng, Y., Yan, H., and Facchetti, A. (2009) *J. Am. Chem. Soc.*, **131**, 8.
145. Facchetti, A. (2010) Presented at the 9th International Symposium on Functional π-Electron Systems, "*Semiconducting Polymers for Printed Transistors and Circuits*", Atlanta, GA.
146. Yan, H., Chen, Z., Zheng, Y., Newman, C.E., Quin, J., Dolz, F., Kastler, M., and Facchetti, A. (2009) *Nature*, **457**, 679.
147. Baeg, K.-J., Khim, D., Kim, D.-Y., Jung, S.-W., Koo, J.B., You, I.-K., Yan, H., Facchetti, A., and Noh, Y.-Y. (2011) *J. Poly. Sci. B: Polym. Phys.*, **49**, 62.
148. (a) Sirringhaus, H., Brown, P.J., Friend, R.H., Nielsen, M.M., Bechgaard, K., Langeveld-Voss, B.M.W., Spiering, A.J.H., Janssen, R.A.J., Meijer, E.W., Herwig, P., and de Leeuw, D.M. (1999) *Nature*, **401**, 685; (b) Kline, R.J., McGehee, M.D., and Toney, M.F. (2006) *Nat. Mater.*, **5**, 222; (c) Kim, D.H., Park, Y.D., Jang, Y., Yang, H., Kim, Y.H., Han, J.I., Moon, D.G., Park, S., Chang, T., Chang, C., Joo, M., Ryu, C.Y., and Cho, K. (2005) *Adv. Funct. Mater.*, **15**, 77; (d) Street, R.A., Northrup, J.E., and Salleo, A. (2005) *Phys. Rev. B*, **71**, 165202; (e) Tsao, H.N., Cho, D., Andreasen, J.W., Rouhanipour, A., Breiby, D.W., Pisula, W., and Müllen, K. (2009) *Adv. Mater.*, **21**, 209; (f) Yang, H., Shin, T.J., Yang, L., Cho, K., Ryu, C.Y., and Bao, Z. (2005) *Adv. Funct. Mater.*, **15**, 671.
149. (a) Kline, R.J., DeLongchamp, D.M., Fischer, D.A., Lin, E.K., Heeney, M., McCulloch, I., and Toney, M.F. (2007) *Appl. Phys. Lett.*, **90**, 062117; (b) Salleo, A. (2007) *Mater. Today*, **10**, 38.
150. Rivnay, J., Toney, M.F., Zhang, Y., Kauvar, I.V., Wagner, V., Facchetti, A., and Salleo, A. (2010) *Adv. Mater.*, **22**, 4359–4363.
151. Steyrleuthner, R., Schubert, M., Jaiser, F., Blakesley, J.C., Chen, Z., Facchetti, A., and Neher, D. (2010) *Adv. Mater.*, **22**, 2799.
152. (a) Mandoc, M.M., de Boer, B., and Blom, P.W.M. (2006) *Phys. Rev. B*, **73**, 155205; (b) Steyrleuthner, R., Bange, S., and Neher, D. (2009) *J. Appl. Phys.*, **105**, 8.
153. Hwang, Y.-J., Murari, N.M., and Jenekhe, S.A. (2013) *Polym. Chem.*, **4**, 3187.
154. Kim, Y., Hong, J., Oh, J.H., and Yang, C. (2013) *Chem. Mater.*, **25**, 3251.
155. Kim, R., Amegadz, P.S.K., Kang, I., Yun, H.-J., Noh, Y.-Y., Kwo, S.-K., and Kim, Y.-H. (2013) *Adv. Funct. Mater.*, **23**, 5719.
156. Lei, T., Dou, J.-H., Cao, X.-Y., Wang, J.-Y., and Pei, J. (2013) *J. Am. Chem. Soc.*, **135**, 12168.
157. Li, H., Kim, F.S., Ren, G., Hollenbeck, E.C., Subramaniyan, S., and Jenekhe, S.A. (2013) *Angew. Chem. Int. Ed.*, **52**, 5513.
158. Li, H., Kim, F.S., Ren, G., and Jenekhe, S.A. (2013) *J. Am. Chem. Soc.*, **135**, 14920.
159. Takeda, Y., Andrew, T.L., Lobez, J.M., Mork, A.J., and Swager, T.M. (2012) *Angew. Chem. Int. Ed.*, **51**, 9042.
160. Kanimozhi, C., Yaacobi-Gross, N., Chou, K.W., Amassian, A., Anthopoulos, T.D., and Patil, S. (2012) *J. Am. Chem. Soc.*, **134**, 16532.
161. Kola, S., Kim, J.H., Ireland, R., Yeh, M.-L., Smith, K., Guo, W., and Katz, H.E. (2013) *ACS Macro Lett.*, **2**, 664.
162. Zheng, Q., Huang, J., Sarjeant, A., and Katz, H.E. (2008) *J. Am. Chem. Soc.*, **130**, 14410.
163. Guillaud, G., Al Sadound, M., and Maitrot, M. (1990) *Chem. Phys. Lett.*, **67**, 503.
164. Bao, Z., Lovinger, A.J., and Brown, J. (1998) *J. Am. Chem. Soc.*, **120** (1), 207–208.
165. (a) Tang, Q., Jiang, L., Tong, Y., Li, H., Liu, Y., Wang, Z., Hu, W., Liu, Y., and Zhu, D. (2008) *Adv. Mater.*, **20**, 2497; (b) Reese, C. and Bao, Z. (2007) *Mater. Today*, **10** (3), 20; (c) de Boer, R.W.I., Gershenson, M.E., Morpurgo, A.F., and Podzorov, V. (2005) *Phys. Org. Semicond.*, 393.
166. (a) Tang, Q.X., Li, H.X., Song, Y.B., Xu, W., Hu, W.P., Jiang, L., Liu, Y., Wang, X., and Zhu, D.B. (2006) *Adv. Mater.*, **18**, 3010; (b) Tang, Q.X., Li, H.X., Liu, Y.L., and Hu, W.P. (2006) *J. Am. Chem. Soc.*, **128**, 4634; (c) Tang, Q., Jiang, L., Tong, Y., Li, H., Liu, Y., Wang, Z., Hu,

W., Liu, Y., and Zhu, D. (2008) *Adv. Mater.*, **20**, 2947–2951.
167. Ling, M.M., Bao, Z., and Erk, P. (2006) *Appl. Phys. Lett.*, **89**, 163516.
168. Wang, H., Zhu, F., Yang, J., Geng, Y., and Yan, D. (2007) *Adv. Mater.*, **19**, 2168.
169. Song, D., Wang, H., Zhu, F., Yang, J., Tian, H., Geng, Y., and Yan, D. (2008) *Adv. Mater.*, **20** (11), 2142.
170. (a) Facchetti, A., Mushrush, M., Yoon, M.-H., Hutchison, G.R., Ratner, M.A., and Marks, T.J. (2004) *J. Am. Chem. Soc.*, **126**, 13859; (b) Facchetti, A., Yoon, M.-H., Stern, C.L., Hutchison, G.R., Ratner, M.A., and Marks, T.J. (2004) *J. Am. Chem. Soc.*, **126**, 13480; (c) Facchetti, A., Mushrush, M., Katz, H.E., and Marks, T.J. (2003) *Adv. Mater.*, **15**, 33; (d) Facchetti, A., Deng, Y., Wang, A., Koide, Y., Sirringhaus, H., Marks, T.J., and Friend, R.H. (2000) *Angew. Chem. Int. Ed.*, **39**, 4547; (e) Casado, J., Ponce Ortíz, R., Hernández, V., López Navarrete, J.T., Facchetti, A., and Marks, T.J. (2005) *J. Am. Chem. Soc.*, **127**, 13364.
171. Bunn, C.W. and Howells, E.R. (1954) *Nature*, **174**, 549.
172. (a) Barbarella, G., Zambianchi, M., Bongini, A., and Antolini, L. (1992) *Adv. Mater.*, **4**, 282; (b) Herrema, J.K., Wildeman, J., van Bolhuis, F., and Hadziioannou, G. (1993) *Synth. Met.*, **60**, 239; (c) Dihexyl-α6T:Sato, T., Fujitsuka, M., Shiro, M., and Tanaka, K. (1998) *Synth. Met.*, **95**, 143.
173. (a) Yoon, M.H., Facchetti, A., Stern, C.F., and Marks, T.J. (2006) *J. Am. Chem. Soc.*, **128**, 5792; (b) Facchetti, A., Yoon, M.H., Stern, C.L., Katz, H.E., and Marks, T.J. (2003) *Angew. Chem. Int. Ed.*, **42**, 3900.
174. Koh, S.E., Delley, B., Medvedeva, J.E., Facchetti, A., Freeman, A.J., Marks, T.J., and Ratner, M.A. (2006) *J. Phys. Chem. B*, **110**, 24361.
175. (a) Yoon, M.H., DiBenedetto, S.A., Russell, M.T., Facchetti, A., and Marks, T.J. (2007) *Chem. Mater.*, **19**, 4864; (b) Yoon, M.-H., DiBenedetto, S., Facchetti, A., and Marks, T.J. (2005) *J. Am. Chem. Soc.*, **127**, 1348.
176. (a) Chiechi, R.C., Sonmez, G., and Wudl, F. (2005) *Adv. Funct. Mater.*, **15**, 427; (b) Donat-Bouillud, A., Mazerolle, L., Gagnon, P., Goldenberg, L., Petty, M.C., and Leclerc, M. (1997) *Chem. Mater.*, **9**, 2815.
177. Schols, S., Willigenburg, L.V., Müller, R., Bode, D., Debucquoy, M., Jonge, S.D., Genoe, J., Heremans, P., Lu, S., and Facchetti, A. (2008) *Appl. Phys. Lett.*, **93**, 263303.
178. Letizia, J.A., Facchetti, A., Stern, C.L., Ratner, M.A., and Marks, T.J. (2005) *J. Am. Chem. Soc.*, **127**, 13476.
179. Letizia, J.A., Rivnay, J., Facchetti, A., Ratner, M.A., and Marks, T.J. (2010) *Adv. Funct. Mater.*, **20**, 50.
180. Pappenfus, T.M., Chesterfield, R.J., Frisbie, C.D., Mann, K.R., Casado, J., Raff, J.D., and Miller, L.L. (2002) *J. Am. Chem. Soc.*, **124**, 4184.
181. Casado, J. *et al.* (2002) *J. Am. Chem. Soc.*, **124**, 12380.
182. Chesterfield, R.J. (2003) *Adv. Mater.*, **15**, 1278.
183. Handa, S., Miyazaki, E., Takimiya, K., and Kunugi, Y. (2007) *J. Am. Chem. Soc.*, **129**, 11684.
184. Kastner, J., Paloheimo, J., and Kuzmany, H. (1993) *Solid State Sciences*, Springer, New York.
185. Haddon, R.C., Perel, A.S., Morris, R.C., Palstra, T.T.M., Hebard, A.F., and Fleming, R.M. (1995) *Appl. Phys. Lett.*, **67** (1), 121.
186. Haddon, R.C. (1996) *J. Am. Chem. Soc.*, **118** (12), 3041.
187. Itaka, K., Yamashiro, M., Yamaguchi, J., Haemori, M., Yaginuma, S., Matsumoto, Y., Kondo, M., and Koinuma, H. (2006) *Adv. Mater.*, **18**, 1713.
188. Jang, J., Kim, J.W., Park, N., and Kim, J.J. (2008) *Org. Elctron.*, **9**, 481.
189. Anthopoulos, T.D., Singh, B., Marjanovic, N., Sarciftci, N.S., Montaigne, A., Sitter, H., Cölle, M., and de Leeuw, D.M. (2006) *Appl. Phys. Lett.*, **89**, 213504.
190. Zhang, X.H. and Kippelen, B. (2008) *Appl. Phys. Lett.*, **93**, 133305.
191. Chikamatsu, M., Nagamatsu, S., Yoshida, Y., Saito, K., Yase, K., and Kikuchi, K. (2005) *Appl. Phys. Lett.*, **87**, 203504.

192. Wöbkenberg, P.H., Ball, J., Bradley, D.D.C., Anthopoulos, T.D., Kooistra, F., Hummelen, J.C., and de Leeuw, D.M. (2008) *Appl. Phys. Lett.*, **92**, 143310.
193. Chikamatsu, M., Itakura, A., Yoshida, Y., Azumi, R., and Yase, K. (2008) *Chem. Mater.*, **20**, 7365.
194. Wobkenberg, P.H., Bradley, D.H., Kronholm, D., Hummelen, J.C., de Leeuw, D.M., Colle, M., and Anthopoulos, T.D. (2008) *Synth. Met.*, **158** (11), 468.
195. Yang, C., Cho, S., Heeger, A.J., and Wudl, F. (2009) *Angew. Chem. Int. Ed.*, **48** (9), 1592.
196. Kanbara, T., Shibata, K., Fujiki, S., Kubozono, Y., Kashino, S., Urisu, T., Sakai, M., Fujiwara, A., Kumashiro, R., and Tanigaki, K. (2003) *Chem. Phys. Lett.*, **379** (3,4), 223.
197. Kitamura, M. and Arakawa, Y. (2007) *Appl. Phys. Lett.*, **91** (5), 53505.
198. Zhang, X.-H., Potscavage, W.J. Jr., Choi, S., and Kippelen, B. (2009) *Appl. Phys. Lett.*, **94** (4), 43312.
199. Horowitz, G., Kouki, F., Spearman, P., Fichou, D., Nogues, C., Pan, X., and Garnier, F. (1996) *Adv. Mater.*, **8**, 242.
200. Malenfant, P.R.L., Dimitrakopoulos, C.D., Gelorme, J.D., Kosbar, L.L., Graham, T.O., Curioni, A., and Andreoni, W. (2002) *Appl. Phys. Lett.*, **80**, 2517.
201. Chesterfield, R.J., McKeen, J.C., Newman, C.R., Ewbank, P.C., da Silva Filho, D.A., Brédas, J.L., Miller, L.L., Mann, K.R., and Frisbie, C.D. (2004) *J. Phys. Chem. B*, **108**, 19281.
202. Chen, F.-C. and Liao, C.H. (2008) *Appl. Phys. Lett.*, **93**, 103310.
203. Tatemichi, S., Ichikawa, M., Koyama, T., and Taniguchi, Y. (2006) *Appl. Phys. Lett.*, **89**, 112108.
204. Oh, J.H., Liu, S., Bao, Z., Schmidt, R., and Würthner, F. (2007) *Appl. Phys. Lett.*, **91**, 212107.
205. Jones, B.A., Ahrens, M.J., Yoon, M.H., Facchetti, A., Marks, T.J., and Wasielewski, M.R. (2004) *Angew. Chem. Int. Ed.*, **43**, 6363.
206. Jones, B.A., Facchetti, A., Wasielewski, M.R., and Marks, T.J. (2008) *Adv. Funct. Mater.*, **18**, 1329.
207. Molinari, A.S., Alves, H., Chen, Z., Facchetti, A., and Morpurgo, A.F. (2009) *J. Am. Chem. Soc.*, **131**, 2462.
208. Weitz, R.T., Amsharov, K., Zschieschang, U., Villas, E.B., Goswami, D.K., Burghard, M., Dosch, H., Jansen, M., Kern, K., and Klauk, H. (2008) *J. Am. Chem. Soc.*, **130**, 4637.
209. Schmidt, R., Oh, J.H., Sun, Y.S., Deppisch, M., Krause, A.-M., Radacki, K., Braunschweig, H., Könemann, M., Erk, P., Bao, Z., and Würthner, F. (2009) *J. Am. Chem. Soc.*, **131**, 6215.
210. (a) Nolde, F., Pisula, W., Mueller, S., Kohl, C., and Müllen, K. (2006) *Chem. Mater.*, **18**, 3715; (b) Nolde, F., Qu, J., Kohl, C., Pschirer, N.G., Reuther, E., and Müllen, K. (2005) *Chemistry*, **11**, 3959.
211. Petit, M., Hayakawa, R., Shirai, Y., Wakayama, Y., Hill, J.P., Ariga, K., and Chikyow, T. (2008) *Appl. Phys. Lett.*, **92**, 163301.
212. Zheng, Q., Huang, J., Sarjeant, A., and Katz, H.E. (2008) *J. Am. Chem. Soc.*, **130**, 14410.
213. Laquindanum, J.G., Katz, H.E., Dodabalapur, A., and Lovinger, A.J. (1996) *J. Am. Chem. Soc.*, **118**, 11331.
214. Katz, H.E., Johnson, J., Lovinger, A.J., and Li, W. (2000) *J. Am. Chem. Soc.*, **122**, 7787.
215. Katz, H.E., Lovinger, A.J., Johnson, J., Kloc, C., Siegrist, T., Li, W., Lin, Y.-Y., and Dodabalapur, A. (2000) *Nature*, **404**, 478.
216. Jung, B.J., Sun, J., Lee, T., Sarjeant, A., and Katz, H.E. (2009) *Chem. Mater.*, **21**, 94.
217. Shukla, D., Nelson, S.F., Freeman, D.C., Rajeswaran, M., Ahearn, W.G., Meyer, D.M., and Carey, J.T. (2008) *Chem. Mater.*, **20**, 7486.
218. Jones, B.A., Facchetti, A., Marks, T.J., and Wasielewski, M.R. (2007) *Chem. Mater.*, **19**, 2703.
219. (a) Wang, Z., Kim, C., Facchetti, A., and Marks, T.J. (2007) *J. Am. Chem. Soc.*, **129**, 13362; (b) Usta, H., Wang, Z., Kim, C., Huang, H., Lu, S., Facchetti, A., and Marks, T.J. (2012) *J. Mater. Chem.*, **22**, 4459.
220. (a) Ando, S., Nishida, J.-I., Fujiwara, E., Tada, H., Inoue, Y., Tokito, S., and

Yamashita, Y. (2005) *Chem. Mater.*, **17**, 1261; (b) Ando, S., Nishida, J.-I., Fujiwara, E., Tada, H., Inoue, Y., Tokito, S., and Yamashita, Y. (2005) *J. Am. Chem. Soc.*, **127**, 5336; (c) Ando, S., Murakami, R., Nishida, J.-I., Fujiwara, E., Tada, H., Inoue, Y., Tokito, S., and Yamashita, Y. (2005) *J. Am. Chem. Soc.*, **127**, 14996; (d) Naraso, J.-I., Nishida, D., Kumaki, S., Tokito, S., and Yamashita, Y. (2006) *J. Am. Chem. Soc.*, **128**, 9598; (e) Nakagawa, T., Nishida, J.-I., Tokito, S., and Yamashita, Y. (2008) *Chem. Mater.*, **20**, 2015; (f) Ono, K., Yamaguchi, H., Taga, K., Saito, K., Nishida, J.-I., and Yamashita, Y. (2009) *Org. Lett.*, **11**, 149.

221. Sakamoto, Y., Suzuki, T., Kobayashi, M., Gao, Y., Fukai, Y., Inoue, Y., Sato, F., and Tokito, S. (2004) *J. Am. Chem. Soc.*, **126**, 8138.

222. Di, C.-A., Li, J., Yu, G., Xiao, Y., Guo, Y., Liu, Y., Qian, X., and Zhu, D. (2008) *Org. Lett.*, **10**, 3025.

223. Wada, H., Shibata, K., Bando, Y., and Mori, T. (2008) *J. Mater. Chem.*, **18**, 4165.

224. Qiao, Y., Guo, Y., Yu, C., Zhang, F., Xu, W., Liu, Y., and Zhu, D. (2012) *J. Am. Chem. Soc.*, **134**, 4084.

225. Wu, Q., Li, R., Hong, W., Li, H., Gao, X., and Zhu, D. (2011) *Chem. Mater.*, **23**, 3138.

226. Zhong, H., Smith, J., Rossbauer, S., White, A.J.P., Anthopoulos, T.D., and Heeney, M. (2012) *Adv. Mater.*, **24**, 3205.

227. Yoon, W.S., Park, S.K., Cho, I., Oh, J.-A., Kim, J.H., and Park, S.Y. (2013) *Adv. Funct. Mater.*, **23**, 3519.

228. Gao, X., Di, C.-A., Hu, Y., Yang, X., Fan, H., Zhang, F., Liu, Y., Li, H., and Zhu, D. (2010) *J. Am. Chem. Soc.*, **132**, 3697.

229. Zhang, F., Hu, Y., Schuettfort, T., Di, C.-A., Gao, X., McNeill, C.R., Thomsen, L., Mannsfeld, S.C.B., Yuan, W., Sirringhaus, H., and Zhu, D. (2013) *J. Am. Chem. Soc.*, **135**, 2338.

230. Zhao, Y., Di, C.-A., Gao, X., Hu, Y., Guo, Y., Zhang, L., Liu, Y., Wang, J., Hu, W., and Zhu, D. (2011) *Adv. Mater.*, **23**, 2448.

231. (a) Suraru, S.-L., Zschieschang, U., Klauk, H., and Würthner, F. (2011) *Chem. Commun.*, **47**, 11504; (b) Li, C., Xiao, C., Li, Y., and Wang, Z. (2013) *Org. Lett.*, **15**, 682; (c) Ye, Q., Chang, J., Huang, K.-W., Shi, X., Wu, J., and Chi, C. (2013) *Org. Lett.*, **15**, 1194; (d) Chen, X., Guo, Y., Tan, L., Yang, G., Li, Y., Zhang, G., Liu, Z., Xu, W., and Zhang, D. (2013) *J. Mater. Chem. C*, **1**, 1087.

232. Fukutomi, Y., Nakano, M., Hu, J.-Y., Osaka, I., and Takimiya, K. (2013) *J. Am. Chem. Soc.*, **135**, 11445.

233. Polander, L.E., Tiwari, S.P., Pandey, L., Seifried, B.M., Zhang, Q., Barlow, S., Risko, C., Bredas, J.-L., Kippelen, B., and Marder, S.R. (2011) *Chem. Mater.*, **23**, 3408.

234. Stolte, M., Gsänger, M., Hofmockel, R., Suraru, S.-L., and Würthner, F. (2012) *Phys. Chem. Chem. Phys.*, **14**, 14181.

235. (a) Shkunov, M., Simms, R., Heeney, M., Tierney, S., and McCulloch, I. (2005) *Adv. Mater.*, **17**, 2608; (b) Kang, S.-M. and Leblebici, Y. (1996) *CMOS Digital Integrated Circuits: Analysis and Design*, McGraw-Hill, New York.

236. (a) Anthopoulos, T.D., de Leeuw, D.M., Cantatore, E., Setayesh, S., Meijer, E.J., Tanase, C., Hummelen, J.C., and Blom, P.W.M. (2004) *Appl. Phys. Lett.*, **85**, 4205; (b) Anthopoulos, T.D., de Leeuw, D.M., Cantatore, E., van 't Hof, P., Alma, J., and Hummelen, J.C. (2005) *J. Appl. Phys.*, **98**, 6; (c) Anthopoulos, T.D., Setayesh, S., Smits, E., Cölle, M., Cantatore, E., de Boer, B., Blom, P.W.M., and de Leeuw, D.M. (2006) *Adv. Mater.*, **18**, 1900.

237. (a) Capelli, R., Toffanin, S., Generali, G., Usta, H., Facchetti, A., and Muccini, M. (2010) *Nat. Mater.*, **9**, 496; (b) Swensen, J.S., Soci, C., and Heeger, A.J. (2005) *Appl. Phys. Lett.*, **87**, 3; (c) Generali, G., Capelli, R., Toffanin, S., Facchetti, A., and Muccini, M. (2010) *Microelectron. Reliab.*, **50**, 1861; (d) Zaumseil, J., Friend, R.H., and Sirringhaus, H. (2006) *Nat. Mater.*, **5**, 69; (e) Muccini, M. (2006) *Nat. Mater.*, **5**, 605.

238. Chen, Z., Lemke, H., Albert-Seifried, S., Caironi, M., Nielsen, M.M., Heeney, M., Zhang, W., McCulloch, I., and

Sirringhaus, H. (2010) *Adv. Mater.*, **22**, 2371.
239. (a) Heeney, M., Zhang, W., Crouch, D.J., Chabinyc, M.L., Gordeyev, S., Hamilton, R., Higgins, S.J., McCulloch, I., Skabara, P.J., Sparrowe, D., and Tierney, S. (2007) *Chem. Commun.*, 5061; (b) Oyaizu, K., Iwasaki, T., Tsukahara, Y., and Tsuchida, E. (2004) *Macromolecules*, **37**, 1257.
240. Meijer, E.J., De Leeuw, D.M., Setayesh, S., Van Veenendaal, E., Huisman, B.H., Blom, P.W.M., Hummelen, J.C., Scherf, U., and Klapwijk, T.M. (2003) *Nat. Mater.*, **2**, 678.
241. Kim, F.S., Guo, X., Watson, M.D., and Jenekhe, S.A. (2010) *Adv. Mater.*, **22**, 478.
242. Bijleveld, J.C., Zoombelt, A.P., Mathijssen, S.G.J., Wienk, M.M., Turbiez, M., de Leeuw, D.M., and Janssen, R.A.J. (2009) *J. Am. Chem. Soc.*, **131**, 16616.
243. Steckler, T.T., Zhang, Z., Hwang, J., Honeyager, R., Ohira, S., Zhang, X., Grant, A., Ellinger, S., Odom, S.A., Sweat, D., Tanner, D.B., Rinzler, A.G., Barlow, S., Bredas, J.-L., Kippelen, B., Marder, S.R., and Reynolds, J.R. (2009) *J. Am. Chem. Soc.*, **131**, 2824.
244. Lei, T., Dou, J.-H., Ma, Z.-J., Yao, C.-H., Liu, C.-J., Wang, J.-Y., and Pei, J. (2012) *J. Am. Chem. Soc.*, **134**, 20025.
245. Lei, T., Dou, J.-H., Ma, Z.-J., Liu, C.-J., Wang, J.-Y., and Pei, J. (2013) *Chem. Sci.*, **4**, 2447.
246. Bürgi, L., Turbiez, M., Pfeiffer, R., Bienewald, F., Kirner, H.-J., and Winnewisser, C. (2008) *Adv. Mater.*, **20**, 2217.
247. Chen, Z., Lee, M.J., Ashraf, R.S., Gu, Y., Albert-Seifried, S., Nielsen, M.M., Schroeder, B., Anthopoulos, T.D., Heeney, M., McCulloch, I., and Sirringhaus, H. (2012) *Adv. Mater.*, **24**, 647.
248. Lee, J., Han, A.-R., Kim, J., Kim, Y., Oh, J.H., and Yang, C. (2012) *J. Am. Chem. Soc.*, **134**, 20713.
249. Lee, J., Han, A.-R., Yu, H., Shin, T.J., Yang, C., and Oh, J.H. (2013) *J. Am. Chem. Soc.*, **135**, 9540.
250. Fan, J., Yuen, J.D., Wang, M., Seifter, J., Seo, J.-H., Mohebbi, A.R., Zakhidov, D., Heeger, A., and Wudl, F. (2012) *Adv. Mater.*, **24**, 2186.
251. Yuen, J.D., Fan, J., Seifter, J., Lim, B., Hufschmid, R., Heeger, A.J., and Wudl, F. (2011) *J. Am. Chem. Soc.*, **133**, 20799.
252. Baeg, K.-J., Khim, D., Jung, S.-W., Kang, M., You, I.-K., Kim, D.-Y., Facchetti, A., and Noh, Y.-Y. (2012) *Adv. Mater.*, **24**, 5433.
253. Chen, H., Guo, Y., Mao, Z., Yu, G., Huang, J., Zhao, Y., and Liu, Y. (2013) *Chem. Mater.*, **25**, 3589.
254. Usta, H., Newman, C., Chen, Z., and Facchetti, A. (2012) *Adv. Mater.*, **24**, 3678.
255. Tang, M.L., Reichardt, A.D., Miyaki, N., Stoltenberg, R.M., and Bao, Z. (2008) *J. Am. Chem. Soc.*, **130**, 6064.
256. Liang, Z., Tang, Q., Mao, R., Liu, D., Xu, J., and Miao, Q. (2011) *Adv. Mater.*, **23**, 5514.
257. Song, C.L., Ma, C.B., Yang, F., Zeng, W.J., Zhang, H.L., and Gong, X. (2011) *Org. Lett.*, **13**, 2880.
258. Zeng, W.J., Zhou, X.Y., Pan, X.J., Song, C.L., and Zhang, H.L. (2013) *AIP Adv.*, **3**, 012101.
259. (a) Glowacki, E.D., Leonat, L., Voss, G., Bodea, M.A., Bozkurt, Z., Ramil, A.M., Irimia-Vladu, M., Bauer, S., and Sariciftci, N.S. (2011) *AIP Adv.*, **1**, 042132; (b) Irimia-Vladu, M., Glowacki, E.D., Troshin, P.A., Schwabegger, G., Leonat, L., Susarova, D.K., Krystal, O., Ullah, M., Kanbur, Y., Bodea, M.A., Razumov, V.F., Sitter, H., Bauer, S., and Sariciftci, N.S. (2011) *Adv. Mater.*, **24**, 375.
260. Nakanotani, H., Saito, M., Nakamura, H., and Adachi, C. (2009) *Appl. Phys. Lett.*, **95**, 033308.
261. Tan, L.X., Guo, Y.L., Yang, Y., Zhang, G.X., Zhang, D.Q., Yu, G., Xu, W., and Liu, Y.Q. (2012) *Chem. Sci.*, **3**, 2530.
262. Wang, L., Zhang, X., Tian, H., Lu, Y., Geng, Y., and Wang, F. (2013) *Chem. Commun.*, **49**, 11272.
263. Nishinaga, T., Ohmae, T., Aita, K., Takase, M., Iyoda, M., Arai, T., and Kunugi, Y. (2013) *Chem. Commun.*, **49**, 5354.
264. Chase, D.T., Fix, A.G., Kang, S.J., Rose, B.D., Weber, C.D., Zhong, Y., Zakharov,

L.N., Lonergan, M.C., Nuckolls, C., and Haley, M.M. (2012) *J. Am. Chem. Soc.*, **134**, 10349.

265. Ribierre, J.C., Watanabe, S., Matsumoto, M., Muto, T., and Aoyama, T. (2010) *Appl. Phys. Lett.*, **96**, 083303.

266. Ribierre, J.C., Watanabe, S., Matsumoto, M., Muto, T., Hashizume, D., and Aoyama, T. (2011) *J. Phys. Chem. C*, **115**, 20703.

267. Zhang, Y., Kim, C., Lin, J., and Nguyen, T.-Q. (2012) *Adv. Funct. Mater.*, **22**, 97.

268. Tsao, H.N., Pisula, W., Liu, Z.H., Osikowicz, W., Salaneck, W.R., and Müllen, K. (2008) *Adv. Mater.*, **20**, 2715.

269. Liu, C., Liu, Z., Lemke, H.T., Tsao, H.N., Naber, R.C.G., Li, Y., Banger, K., Müllen, K., Nielsen, M.M., and Sirringhaus, H. (2010) *Chem. Mater.*, **22**, 2120.

270. Ye, Q., Chang, J., Huang, K.-W., and Chi, C. (2011) *Org. Lett.*, **13**, 5960.

271. Cheng, S.-S., Huang, P.-Y., Ramesh, M., Chang, H.-C., Chen, L.-M., Yeh, C.-M., Fung, C.-L., Wu, M.C., Liu, C.-C., Kim, C., Lin, H.-C., Chen, M.-C., and Chu, C.-W. (2014) *Adv. Funct. Mater.*, **24**, 2057.

272. Yue, W., He, T., Stolte, M., Gsänger, M., and Würthner, F. (2014) *Chem. Commun.*, **50**, 545.

273. Xia, H., Liu, D., Xua, X., and Miao, Q. (2013) *Chem. Commun.*, **49**, 4301.

2
Metal-Oxide Thin-Film Transistors for Flexible Electronics

Yong-Hoon Kim and Sung Kyu Park

2.1
Introduction

From the last decade, flexible electronic devices have been extensively researched in industries and research institutions with an aim to create new conceptual electronics such as flexible and bendable devices. Generally array-type electronics can be categorized into two types by their addressing method: passive-matrix addressing and active-matrix addressing. Active-matrix addressing is superior to passive-matrix addressing in that the latter has limitations in increasing the resolution and operating speed of a device. In devices with active-matrix addressing, for example, sensors and displays, an electrical switching device or a thin-film transistor (TFT) is placed in each pixel either to write or read electrical signals individually.

In flat-panel displays such as TFT-LCDs and AMOLEDs (active-matrix organic light-emitting diodes), hydrogenated amorphous silicon thin-film transistors (a-Si:H TFTs) and low-temperature polycrystalline silicon thin-film transistors (LTPS TFTs) are the most widely used TFT devices. This is because these TFTs use a well-known semiconductor material, silicon, as a channel layer and, most importantly, their mass production technology is mature. Although LTPS TFTs outperform a-Si:H TFTs in terms of carrier mobility and electrical stability, the former's limited scalability (large-area manufacturing) and complicated fabrication process compared to a-Si:H TFT technology have prevented their further extension in flat-panel display manufacturing.

However, the performance and stability of a-Si:H TFTs do not fully satisfy the requirements for high-resolution TFT-LCD and AMOLED displays. Especially, an AMOLED backplane requires TFTs with high carrier mobility (typically over $10\,cm^2\,V^{-1}\,s^{-1}$) and better immunity to bias-illumination-stress conditions for stable operation. For these reasons, LTPS TFTs have been the only option to be used in AMOLED backplanes.

In the past years, solution-processed organic TFTs also have been widely investigated to replace silicon-based ones. But, their electrical properties and operational stability do not yet fully satisfy the requirements for high-end flat-panel displays. More recently, metal-oxide (MO) semiconductors have attracted a

Large Area and Flexible Electronics, First Edition. Edited by Mario Caironi and Yong-Young Noh.

great deal of interest because of their relatively high carrier mobility (10–50 cm^2 V^{-1} s^{-1}) and better operational stability over a-Si:H and organic TFTs [1–6]. Also, their high carrier mobility even in an amorphous phase may allow good device-to-device uniformity over a large area using current manufacturing technology. Generally, MO semiconductors such as indium gallium zinc oxide (IGZO) films are deposited by a vacuum process, that is, sputtering. However, there have been continuous attempts to replace the vacuum process by a non-vacuum process such as roll-to-roll printing so as to reduce the manufacturing cost of the components and also to allow their integration on flexible and soft materials.

2.2 Metal-Oxide TFTs

2.2.1 Advantages and Applications

The most intriguing advantages of using MO semiconductor TFTs in flat-panel displays is the large-area scalability (with a good uniformity) by relatively simple fabrication processes compared to LTPS TFT technology, not to mention their moderate electrical performance, which lies in between those of a-Si:H and LTPS TFTs. Previously, reliability problems, such as threshold voltage (V_{TH}) shift under positive-bias stress (PBS) and negative-bias-illumination stress (NBIS), have been critical issues for commercialization. However, at this moment, most of these reliability issues have been settled and the physical mechanisms behind these instabilities are being extensively studied by many research groups [7–10]. In the early stage of development, molecular adsorption at the back channel surface was suggested as one of the origins of negative and positive shift of V_{TH} during operation [11, 12]. It was found that such instability could be effectively controlled by stacking a dense oxide passivation layer on top of the oxide semiconductor channel layer. Later, charge trapping at the semiconductor/gate dielectric interface and mid-gap defect states such as oxygen vacancies were also suggested as the origins of the V_{TH} shift under PBS or NBIS conditions [13, 14]. By optimizing the device structure, materials, and fabrication processes, successful commercialization is now possible by several panel makers.

At the current stage of development, the deposition of oxide semiconductors is mostly done by sputtering, which is a vacuum process. Other types of deposition methods such as atomic layer deposition (ALD), pulsed-laser deposition, spray deposition, spin coating, ink-jet printing, and slot-die printing can be also used for depositing oxide semiconductors even at large scales [15–17].

2.2.2 Vacuum Deposition

The most prominent feature of the vacuum deposition process is compatibility with current manufacturing processes. Vacuum deposition processes are now widely used in flat-panel display manufacturing, and sputtering technology is one of the most mature technologies among many other deposition techniques. Besides sputtering, the ALD process is also another reliable method to deposit MO semiconductors, and much work is going on in this field. One of the advantages of the ALD process is that it can provide relatively high-quality and stoichiometric MO films because of its unique deposition mechanism. However, relatively slow deposition rates and scalability could be a limitation to its application in massive and large-area manufacturing.

2.2.3 Solution Processing

Solution processing of MO semiconductors may offer a significant reduction of manufacturing cost by eliminating vacuum deposition processes and replacing them with potentially cheaper printing processes. Printing processes such as slot-die coating, which is currently used in organic solar cells manufacturing, can be used in oxide semiconductor deposition. Also, direct printing methods such as ink-jet printing can be another possible way to deposit oxide semiconductors [16, 17]. The main advantage of using ink-jet printing is the removal of the channel patterning process. In addition, low consumption of precursor materials could further reduce the manufacturing cost. However, the incompatibility with present manufacturing processes is one of the obstacles delaying the adoption of ink-jet printing technology in display manufacturing. Nevertheless, solution processing still has a great potential in realizing high-performance oxide semiconductors and devices at lower cost.

There are several ways of forming oxide semiconductor thin films by solution processing. Typically, these can be categorized into three types: (i) nanoparticle-based process, (ii) sol–gel-based process, and (iii) hybrid-type approach (Figure 2.1).

2.3 Solution-Processed MO Thin Films

2.3.1 Nanoparticle-Based Process

Nanoparticle-based process utilizes presynthesized oxide nanoparticles for depositing MO thin films. This approach has been widely used in forming transparent conductive oxides. Even though the synthesis of oxide nanoparticles may involve a high-temperature process, the actual film formation requires much lower temperature just to evaporate the remaining solvent and percolate the nanoparticles. For example, in the case of indium tin oxide (ITO), a curing temperature of 230 °C can provide a thin film with a sheet resistance of ~350 Ω/□ [18]. Likewise, oxide semiconductor nanoparticles can be synthesized and

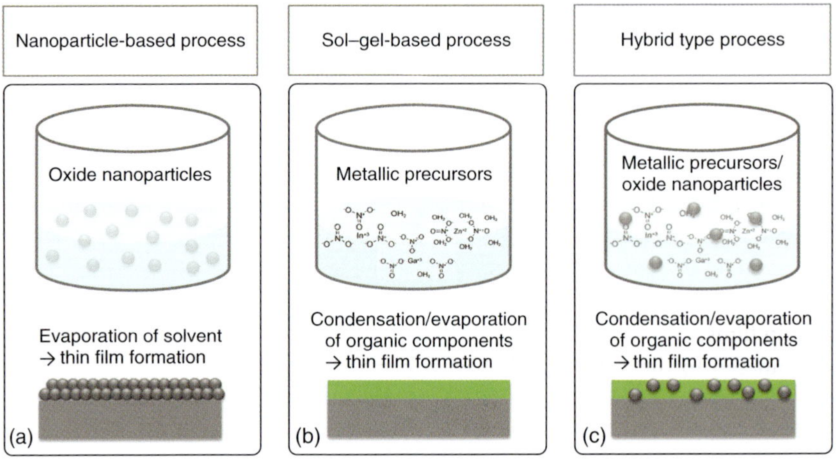

Figure 2.1 (a–c) Various approaches of forming oxide semiconductor thin films by solution processing.

Figure 2.2 Schematic illustration of nanoparticle-based oxide semiconductor films.

formulated into a semiconductor ink for a solution process [19, 20]. Since curing is needed only for the evaporation of the solvents and other remaining organic compounds such as dispersion agents, curing at relatively low temperatures is required for film formation. However, one drawback can be the build-up of an energy barrier between the nanoparticles, which may hinder efficient charge transport between the nanoparticles throughout the film as shown in Figure 2.2.

2.3.2 Sol–Gel-Based Process

Another well-known method of forming oxide thin films from solution processing is the sol–gel process. The general sol–gel process can be described as the formation of an oxide network from polycondensation of molecular precursors. Different from nanoparticle-based thin film formation, the sol–gel processed films are generally in the amorphous state (<500 °C), although the crystalline phase can be observed in In-rich or Zn-rich films at much lower annealing temperatures [21]. Also, the sol–gel approach provides a facile and simple route to study various combinations of metallic precursors during the optimization process of unknown materials, and definitely most of the well-known oxide semiconductor combinations such as ZnO, IZO (indium zinc oxide), ZTO (zinc tin oxide), and IGZO can be easily produced from the sol–gel method. Metallic precursors can be present in many different structures such as nitrate, alkoxide, hydroxides, carboxylate, and β-diketonate, and each type of the precursors may bring different chemical reactions during the sol–gel process, as well as the electrical and optical properties of the thin films.

Compared to the nanoparticle-based formation method, the sol–gel process requires a relatively high annealing temperature, especially when a more crystallized film is demanded (>500 °C). Even for amorphous films, an annealing temperature of >300 °C is generally needed to remove the unwanted solvent molecules, byproducts, and other organic compounds for condensation and densification of thin films. However, despite the high annealing temperature, sol–gel-processed oxide semiconductors and TFTs generally show better electrical properties compared to nanoparticle-based devices.

2.3.3 Hybrid Type

Several groups have demonstrated a combination of nanoparticle-based and sol–gel-based processes. In this approach, presynthesized nanoparticles are mixed with sol–gel precursor solutions for a hybrid-type film structure having crystallized nanoparticles embedded in the amorphous- or crystal-phase sol–gel matrix [22]. By inserting controlled nanoparticles in the sol–gel matrix, performance enhancement or memory effects can be obtained depending on the properties of the nanoparticles. Also, embedding stoichiometric nanoparticles can have the benefits of precisely controlling the electrical or optical properties of the oxide semiconductor films.

2.4 Low-Temperature-Processed MO TFTs for Flexible Electronics

As described earlier, sol–gel-based oxide formation requires a relatively high annealing temperature for sufficient densification and removal of organic components from the film. Therefore, this has been one of the obstacles for achieving high-performance MO TFTs on flexible substrates. Recently, several attempts to lower the annealing temperature of oxide semiconductor films from the sol–gel process have been reported, and they can be categorized into three distinct approaches: (i) modifying the annealing environment, (ii) ink formulation, and (iii) alternate annealing process.

2.4.1 Low-Temperature-Processed MO TFTs

2.4.1.1 Annealing Environment

Controlling the annealing environment has been shown to have a great influence on lowering the annealing temperature of sol–gel films. For example, Banger *et al.* [23] have reported a novel annealing process called *sol–gel on chip* to derive high-performance oxide semiconductors at temperatures as low as 230 °C. The sol–gel on chip process involves an *in situ* hydrolyzed process in which a controlled amount of water vapor is introduced upon annealing the spun oxide films. The oxide semiconductor film obtained from the sol–gel on chip process was found to retain relatively small amounts of carbon and hydrogen atoms as well as a significantly lower defect density [23]. Another interesting approach for lowering the annealing temperature is the use of an O_2/O_3 atmosphere for annealing [24]. Annealing in the presence of O_2/O_3 molecules can induce the formation of In_2O_3 films with more fully coordinated In sites with six nearest neighbor oxygen atoms compared to conventional air-annealed films. Using the O_2/O_3 atmospheric annealing process, TFTs with field-effect mobility as high as 22.1 $cm^2\,V^{-1}\,s^{-1}$ were obtained at a sufficiently low temperature. Also, high-pressure annealing (HPA) in an O_2 environment has been reported to improve the electrical properties of solution-processed ZTO TFTs by eliminating defect states which originated from oxygen vacancies. Using the HPA-processed ZTO films, TFTs with field-effect mobility of 2.35 $cm^2\,V^{-1}\,s^{-1}$ were obtained with an annealing temperature of 350 °C [25].

2.4.1.2 Ink Formulation

Another possible way to achieve low-temperature-processed oxide films from a sol–gel method is by modifying the formulation of the oxide semiconductor solution. Previously, low-temperature fabrication of MO TFTs from a combustion process had been reported [2]. In this method, a combustion synthesis is utilized by introducing a balanced pair of fuels such as acetylacetone or urea and oxidizers such as metal nitrates. Because of the exothermic nature of combustion synthesis, high self-generated energies are provided, thereby lowering the thermal energy required for the sol–gel reaction [2]. Using the combustion process, an In_2O_3 TFT with a field-effect mobility of 6.0 $cm^2\,V^{-1}\,s^{-1}$ could be fabricated on a flexible substrate with an annealing temperature of 200 °C.

In addition, impurity-free ink formulation may help in reducing the annealing temperature. As described before, the relatively high annealing temperature required for the sol–gel process is due to the removal of organic-related (mostly carbon-related) components during the annealing process and successful densification of the films. Therefore, minimizing or excluding carbon-related components from the oxide semiconductor inks may reduce the required annealing temperature. Widely used metallic precursors such as acetates and alkoxides contain a significant amount of carbon atoms in their chemical structures. Also, alcohol-based solvents such as 2-methoxyethanol possess a considerable amount of carbon atoms. These unwanted carbons as well as other possible byproducts should be removed during the film formation process in order to achieve highly dense oxide thin films. From this perspective, an impurity-free precursor solution having metal hydroxides as the metallic precursors and an aqueous-type solvent will be more favorable in achieving dense oxide films at a lower temperature. Several groups have reported the utilization of zinc hydroxide for forming ZnO semiconductor films [26, 27], and an exceptionally low annealing temperature (<200 °C) has been reported for flexible ZnO TFTs having an average field-effect mobility of 0.27 $cm^2\ V^{-1}\ s^{-1}$ on polyimide substrates.

2.4.1.3 Alternate Annealing Process

Besides the conventional thermal annealing method, other types of annealing methods can be employed in forming oxide semiconductors by solution processing. Previously, microwave-assisted annealing had been reported by Jun *et al.* [28]. Microwave irradiation was found to have an effect of reducing the annealing time required for ZnO formation and also significantly lowering the number of hydroxyl groups in oxide films even at an annealing temperature of 140 °C [28]. With microwave-assisted annealing, ZnO TFTs with field-effect mobility of 1.7 $cm^2\ V^{-1}\ s^{-1}$ could be successfully obtained at a temperature of 140 °C. Also, the hysteresis characteristics were greatly reduced by microwave-assisted annealing.

In addition, photochemical activation of sol–gel films using deep ultraviolet (DUV) irradiation is other alternate approach for forming dense oxide semiconductors at a low temperature [29]. Without any intentional heating, the DUV irradiation can induce a dense oxide network by photochemical cleavage and condensation/densification of sol–gel films.

2.4.2 Photochemical Activation of Oxide Semiconductors

DUV-induced photochemical activation of sol–gel films is schematically illustrated in Figure 2.3 [29]. The as-spun sol–gel film contains a considerable amount of carbon-related and also other components inside the film. By DUV irradiation using a low-pressure mercury lamp having main peaks at 184.9 nm (10%) and 253.7 nm (90%), photochemical cleavage of alkoxy groups and formation of an M–O–M network are induced. As an evidence of these reactions, a significant decrease in the carbon content in the initial 30 min of irradiation was observed. Further irradiation of DUV without any intentional heating leads to densification of the oxide film.

The DUV irradiation for photochemical activation is conducted in a N_2 atmosphere. It has been found that controlling the atmosphere during the photochemical activation is very important. For example, when the sol–gel film is photo-annealed in air atmosphere, the electrical performance is significantly degraded compared to when annealed in N_2 [29]. It is suggested that the molecular oxygen present in the environment absorbs the 184.9 nm light emission from

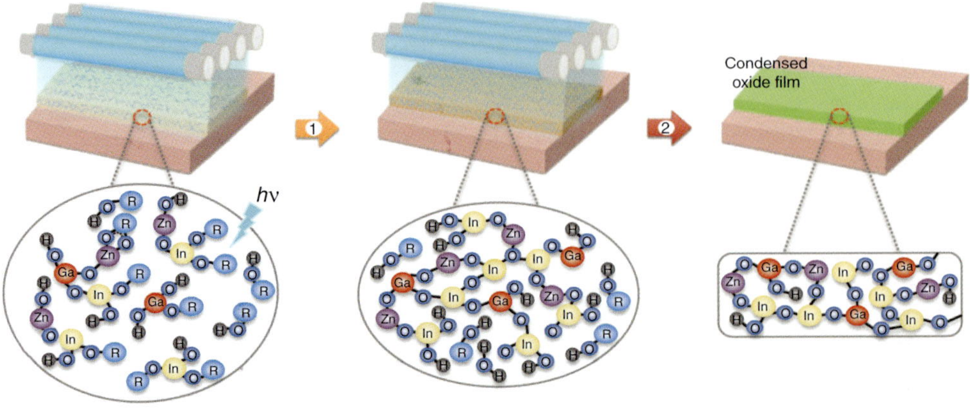

Figure 2.3 Schematic showing the photochemical activation of a sol–gel film by DUV irradiation. The DUV radiation is emitted from a low-pressure mercury lamp having main peaks at 184.9 nm (10%) and 253.7 nm (90%) [29]. (Reproduced with permission. Copyright © 2012, Nature Publishing Group.)

the mercury lamp and therefore the efficiency is significantly lowered, thereby causing insufficient photochemical cleavage and poor densification of the film.

The atomic binding states of oxide films are important factors in determining the quality of sol–gel-derived oxide films. Figure 2.4 shows the X-ray photoelectron spectroscopy results of the O 1s peaks of as-spun, photo-annealed, and thermally annealed IGZO films at 350 °C. Here, the peaks at ~530.0, ~531.0, and ~532.0 eV

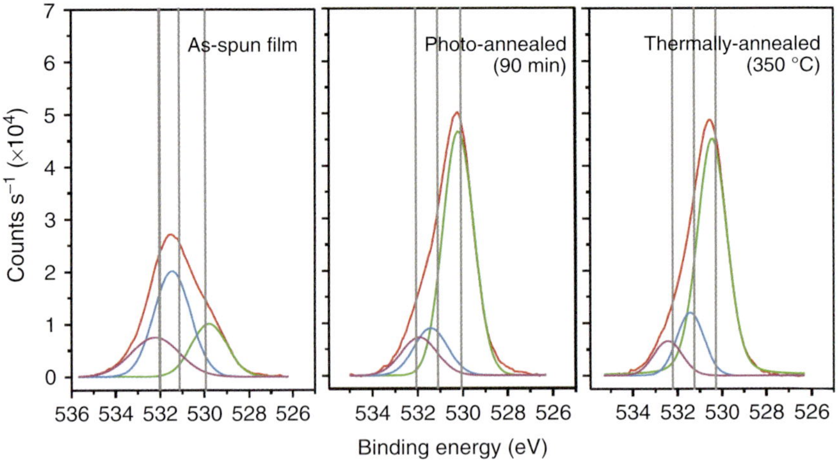

Figure 2.4 X-ray photoelectron spectroscopy data of as-spun, photo-annealed, and thermally annealed IGZO films (peaks at ~530.0, ~531.0, and ~532.0 eV correspond to oxygen atoms in M–O–M lattice, oxygen atoms near oxygen vacancies and in M–OC bonds, and oxygen atoms in M–OH, respectively) [29]. (Reproduced with permission. Copyright © 2012, Nature Publishing Group.)

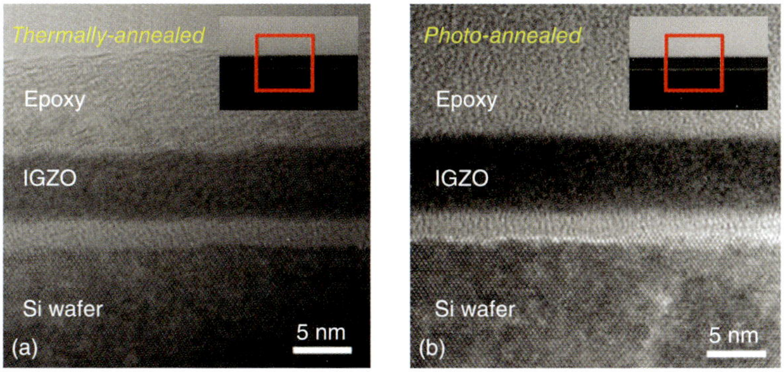

Figure 2.5 Cross-sectional high-resolution transmission electron microscopy (HRTEM) images of (a) thermally annealed (350 °C) and (b) photo-annealed (120 min) IGZO films [29]. (Reproduced with permission. Copyright © 2012, Nature Publishing Group.)

correspond to oxygen atoms in M–O–M lattice, oxygen atoms near oxygen vacancies and in M–OC bonds, and oxygen atoms in M–OH, respectively. As illustrated, the as-spun IGZO film contains large proportions of $M–O_{vac}$ (%area of 50.9) and M–OH (%area of 24.7) binding states, and only a small proportion of M–O binding states (%area of 24.4). However, by DUV irradiation for 90 min, the proportion of M–O binding states is significantly increased up to a %area of 71.9 [29]. Simultaneously, the proportions of $M–O_{vac}$ and M–OH binding states are decreased considerably. The thermally annealed IGZO film at 350 °C, for instance, shows M–O binding states with a %area of 74.3, which is similar to that of photo-annealed IGZO film. These results support the idea that photochemical activation can facilitate the M–O–M network of sol–gel films at a low temperature.

In oxide TFTs, the density of the oxide semiconductor film (or the interatomic distance) is an important factor which can determine the electrical performance of the device. The film density, which is related to the overlap of orbitals between neighboring metallic ions, may largely affect the charge-transport properties in oxide semiconductor films. In Figure 2.5, the cross-sectional high-resolution transmission electron microscopy (HRTEM) images of both thermally annealed (350 °C) and photo-annealed (120 min) IGZO films on Si wafers are shown. The thicknesses of both IGZO films are almost identical, showing that sufficient densification is brought about by the photochemical activation without a high-temperature annealing process. The densification of oxide films also can be determined by Rutherford backscattering spectrometry (RBS) and ellipsometry measurement. The areal densities obtained from RBS measurements show that the photo-annealed IGZO film has a slightly higher areal density (52.88×10^{15} atoms cm^{-2}) compared to that of an IGZO film annealed at 350 °C (52.43×10^{15} atoms cm^{-2}) [29]. Also, the refractive index data and estimated thickness from ellipsometry show similar densification of IGZO films by photochemical activation.

DUV-assisted photochemical activation can be applied to a wide range of sol–gel precursor solutions. One of the important aspects for sol–gel precursor solutions to be used in photochemical activation is the absorption in the UV region. For example, precursor solutions having $ZnCl_2$ as a metallic precursor may not be applicable to the photochemical activation process because of negligible DUV absorption, leading to inefficient photochemical activation and cleavage by the high-energy DUV photons [29]. Therefore, in preparing sol–gel precursor solutions for oxide thin film formation, the absorption properties of the precursor solutions in the UV region should be carefully considered.

The photochemical activation process can be easily integrated into MO TFT fabrication, as shown in Figure 2.6. On a glass or polymer substrate, a metallic gate electrode is deposited by sputtering or thermal evaporation. After deposition, a patterning process is carried out followed by standard photolithography and wet etching. On the gate electrode, a gate dielectric layer comprising Al_2O_3 or SiO_2 is deposited by ALD or plasma-enhanced chemical vapor deposition. For the channel layer, an oxide precursor solution is coated and photo-annealed in a N_2 atmosphere for a controlled time. After patterning the channel layer by photolithography and wet etching, via holes can be formed by etching away a portion of Al_2O_3 gate dielectric, and finally source/drain electrodes are deposited and patterned. Also, similar to vacuum-deposited MO TFTs, an etch-stop layer (or a passivation layer), typically composed of dense SiO_2 or Al_2O_3, can be deposited on top of the channel layer to improve the reliability of the TFTs. In addition to the above-described bottom-gate TFT structure, the photochemical activation process can be also applied to top-gate-structured TFTs where the channel lies beneath the gate dielectric layer.

Nevertheless, photochemical activation is a versatile method for forming oxide semiconductors at a low temperature. For instance, oxide TFTs other than IGZO TFTs, such as IZO and In_2O_3 TFTs, can be fabricated by the photochemical activation process. Figure 2.7 shows the typical transfer characteristics of various oxide TFTs fabricated by photochemical activation. The average mobilities of IGZO, IZO, and In_2O_3 TFTs are 8.76, 4.43, and 11.29 $cm^2\ V^{-1}\ s^{-1}$, respectively, showing comparable electrical performance to the TFTs thermally annealed at 350 °C [29].

It is also important to note that the best electrical performance and uniformity are achieved at a photo-annealing period of 90–120 min. However, the atomic composition ratio of the film changes only gradually after the annealing period of 30 min. These distinctive trends suggest that the photochemical activation process can be divided into two stages: first, a rapid condensation from photo-cleavage and, second, gradual structure rearrangement and further densification of the film.

One of the most important features of photochemical activation is its exceptionally low temperature process which enables application to flexible devices where low thermal budge polymers are used as substrates. Unlike glass or silicon wafers, polymers are vulnerable to high-temperature processes. The temperature required for the sol–gel process is generally higher than the glass transition temperature of most engineering polymers.

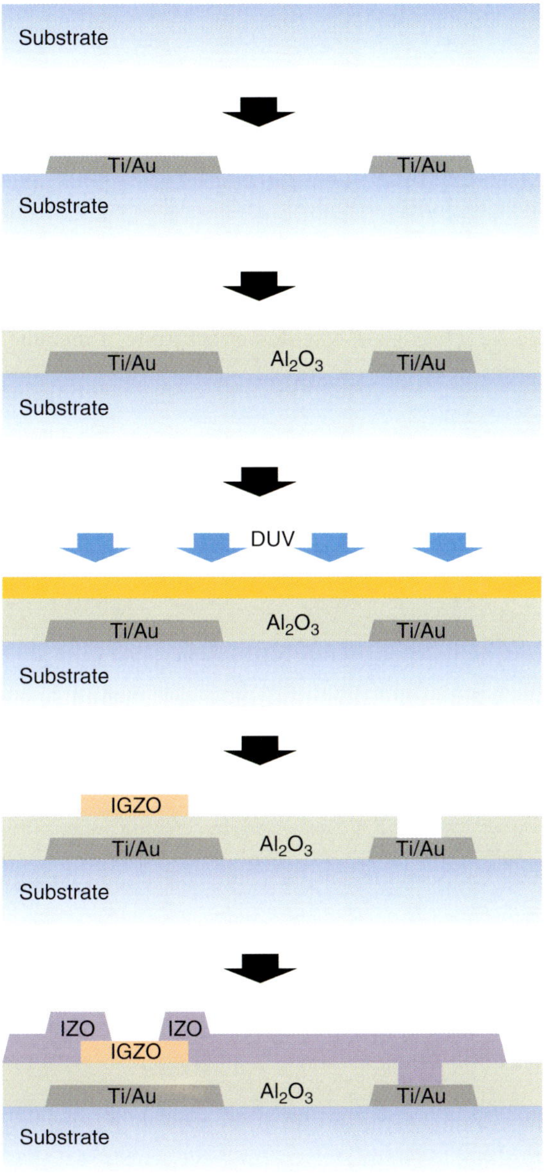

Figure 2.6 Fabrication process of bottom-gate IGZO TFTs using photochemical activation.

Photochemical activation, however, allows a successful sol–gel process while keeping the substrate temperature below 150 °C. Figure 2.8 shows the transfer and output characteristics of flexible IGZO TFTs fabricated on a polyarylate (PAR) substrate using photochemical activation. Similar to those fabricated on glass substrates, the TFTs showed an average field-effect mobility of 3.77 $cm^2\ V^{-1}\ s^{-1}$

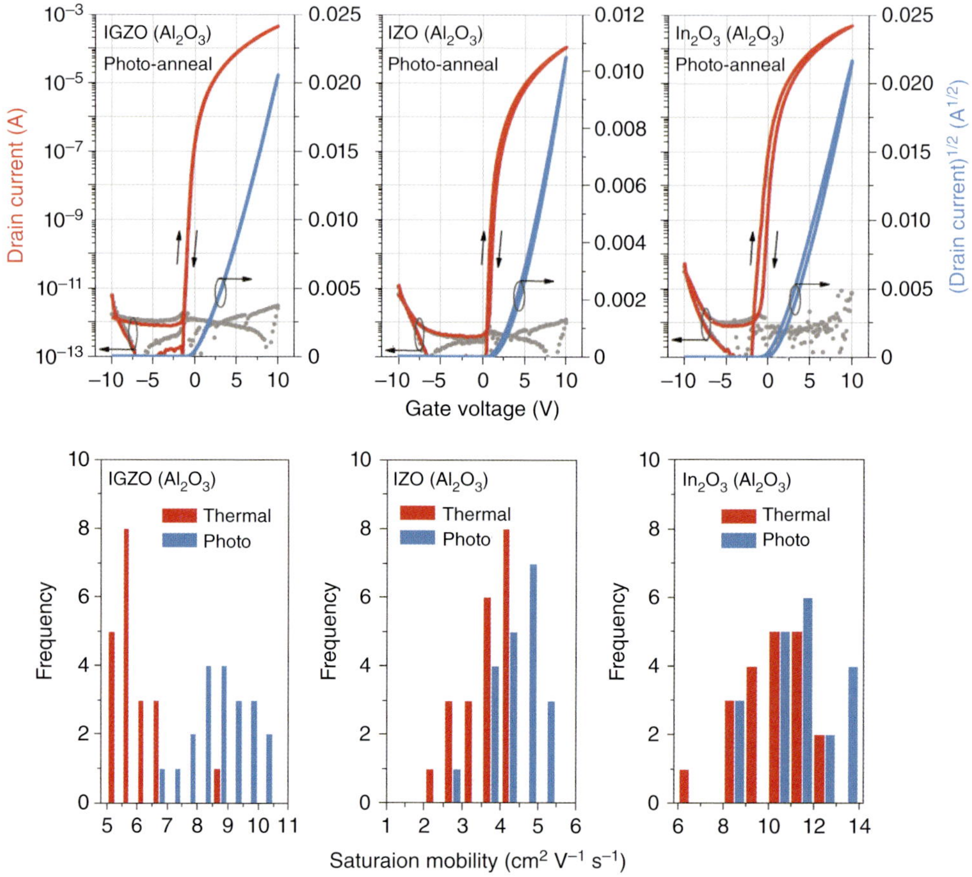

Figure 2.7 Transfer characteristics and mobility distribution of photo-annealed IGZO, IZO, and In_2O_3 TFTs fabricated on glass with Al_2O_3 gate dielectric layer ($V_{DS} = 10\,V$) [29]. (Reproduced with permission. Copyright © 2012, Nature Publishing Group.)

with a very narrow distribution of the device performance [29]. The devices also exhibit negligible hysteresis, showing the formation of a good interface with the gate dielectric layer.

Gate bias instability is especially important in MO TFTs because the current output is largely affected by threshold voltage shift. It has been widely accepted that the PBS instability is mainly concerned with interfacial states at the oxide semiconductor/gate dielectric interface. Also, bulk defects and the back-channel effect can be other possible reasons for instability. The PBS stability of photo-annealed IGZO TFTs is almost comparable to that of TFTs annealed at 350 °C, as shown in Figure 2.8. For instance, the photo-annealed IGZO TFTs show threshold voltage shift of 1.12 V after 10 ks of PBS ($V_{GS} = 5\,V$, $V_{DS} = 0.1\,V$), while the thermally annealed TFTs show a threshold voltage shift of 0.86 V under the same condition.

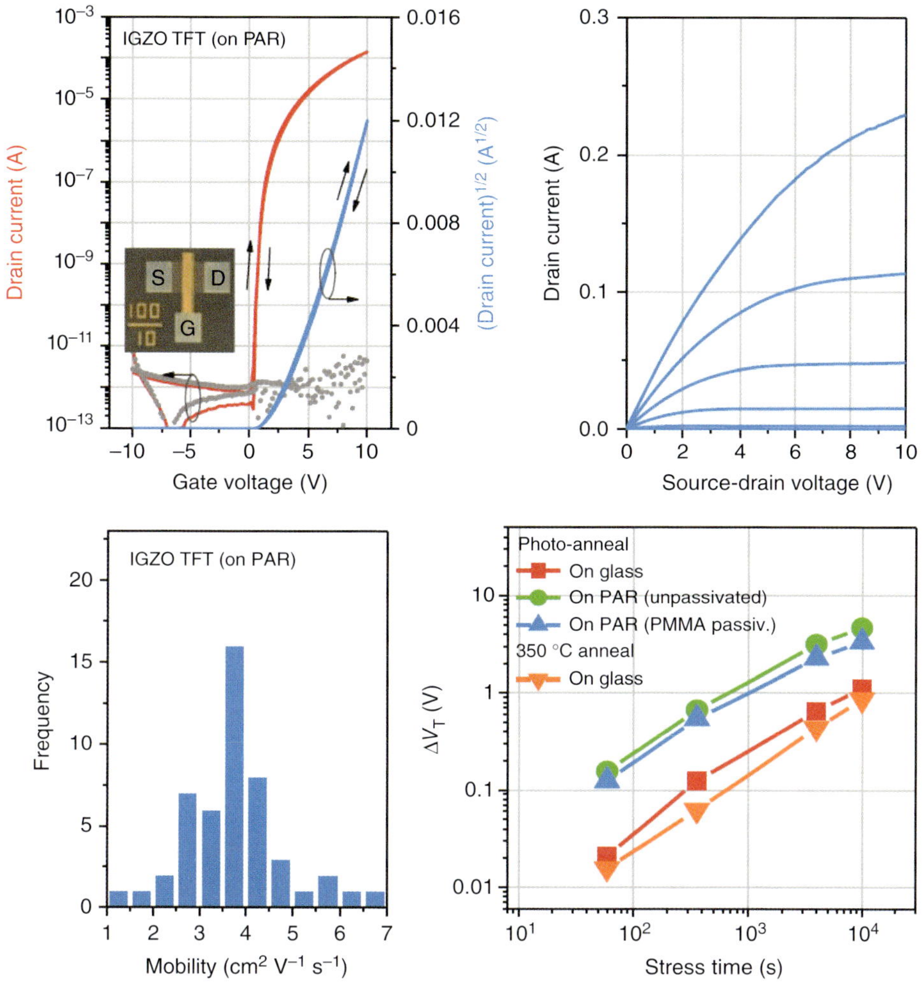

Figure 2.8 (a–d) Transfer and output characteristics, mobility distribution, and PBS stability of photo-annealed IGZO TFTs fabricated on a PAR substrate ($V_{DS} = 10$ V for transfer curve; $V_{GS} = 0–10$ V for output curves; $V_{GS} = 5$ V, $V_{DS} = 0.1$ V for PBS) [29]. (Reproduced with permission. Copyright © 2012, Nature Publishing Group.)

In the case of flexible TFTs, the threshold voltage shifts were 4.5 and 3 V for unpassivated and poly(methyl methacrylate)-passivated devices, respectively. The rather large threshold voltage shift in flexible TFTs can be traced to the high surface roughness of the gate dielectric layer.

Figure 2.9 shows a seven-stage ring oscillator fabricated on a PAR substrate using photo-annealed IGZO TFTs. The inverters in the seven-stage ring oscillator had a β-ratio of 2, with $(W/L)_{drive} = 100$ μm/7 μm, and $(W/L)_{load} = 50$ μm/7 μm. Successful oscillation was observed with an oscillation frequency up to ~340 kHz

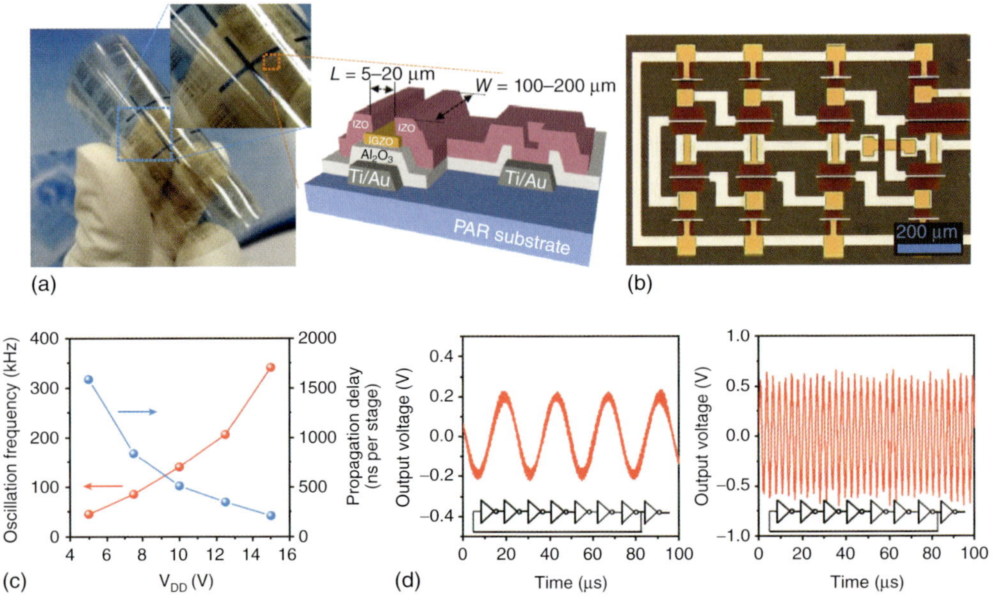

Figure 2.9 (a) Optical images and a schematic diagram of photo-annealed IGZO TFTs and circuits on a plastic substrate. (b) Optical image of a seven-stage ring oscillator. (c) Oscillation frequency and propagation delay of seven-stage ring oscillator as a function of the supply voltage V_{DD}. (d) Output waveforms of a seven-stage ring oscillator operating with V_{DD} of 5 and 15 V [29]. (Reproduced with permission. Copyright © 2012, Nature Publishing Group.)

and propagation delay of ~210 ns per stage at a V_{DD} of 15 V [29], demonstrating reasonable device-to-device uniformity from the photochemical activation process.

2.5 Summary

Solution-processed MO TFT is now becoming one of the essential technologies for future flexible electronics because of its high electrical performance with reasonable reliability. Furthermore, several methods are now available for the low-temperature processing of sol–gel oxide semiconductors, including control of the annealing environment, modification of precursor solutions, and non-thermal annealing methods such as photochemical activation. Many successful and intriguing results have been reported in the past few years, and it is now possible to integrate high-performance MO TFTs and circuits on plastics with processing temperatures not exceeding 200 °C. The next important step for the solution-processed MO TFTs, as well as for the vacuum-deposited MO TFTs, is the realization of stable p-type TFTs for complementary metal oxide semiconductor (MOS) architecture solely using MO semiconductors.

References

1. Nomura, K., Ohta, H., Takagi, A., Kamiya, T., Hirano, M., and Hosono, H. (2004) Room-temperature fabrication of transparent flexible thin-film transistors using amorphous oxide semiconductors. *Nature*, **432**, 488–492.
2. Kim, M.G., Kanatzidis, M.G., Facchetti, A., and Marks, T.J. (2011) Low-temperature fabrication of high-performance metal oxide thin-film electronics via combustion processing. *Nat. Mater.*, **10**, 382–388.
3. Facchetti, A. and Marks, T.J. (2010) *Transparent Electronics: From Synthesis to Applications*, John Wiley & Sons, Ltd, Chichester.
4. Dusastre, V. (2010) *Materials for Sustainable Energy*, World Scientific Publishing Co Pte Ltd, Singapore.
5. Kamiya, T., Nomura, K., and Hosono, H. (2010) Present status of amorphous In-Ga-Zn-O thin film transistors. *Sci. Technol. Adv. Mater.*, **11**, 044305.
6. Jeon, S., Park, S., Song, I., Hur, J.H., Park, J., Kim, H., Kim, S., Kim, S., Yin, H., Chung, H.I., Lee, E., and Kim, C. (2011) Nanometer-scale oxide thin film transistor with potential for high density image sensor applications. *ACS Appl. Mater. Interfaces*, **3**, 1–6.
7. Suresh, A. and Muth, J.F. (2008) Bias stress stability of indium gallium zinc oxide channel based transparent thin film transistors. *Appl. Phys. Lett.*, **92**, 033502.
8. Yang, S., Cho, D.-H., Ryu, M.K., Park, S.-H.K., Hwang, C.-S., and Jeong, J.K. (2010) Improvement in the photon-induced bias stability of Al-Sn-Zn-In-O thin film transistors by adopting AlO_x passivation layer. *Appl. Phys. Lett.*, **96**, 213511.
9. Oh, H., Yoon, S.-M., Ryu, M.K., Hwang, C.-S., Yang, S., and Park, S.-H.K. (2010) Photon-accelerated negative bias instability involving subgap states creation in amorphous In-Ga-Zn-O thin film transistor. *Appl. Phys. Lett.*, **97**, 183502.
10. Kwon, J.-Y., Jung, J.S., Son, K.S., Lee, K.-H., Park, J.S., Kim, T.S., Park, J.-S., Choi, R., Jeong, J.K., Koo, B., and Lee, S.Y. (2010) The impact of gate dielectric materials on the light-induced bias instability in Hf-In-Zn-O thin film transistor. *Appl. Phys. Lett.*, **97**, 183503.
11. Jeong, J.K., Yang, H.W., Jeong, J.H., Mo, Y.G., and Kim, H.D. (2008) Origin of threshold voltage instability in indium-gallium-zinc oxide thin film transistors. *Appl. Phys. Lett.*, **93**, 123508.
12. Kim, Y.H., Kim, H.S., Han, J.I., and Park, S.K. (2010) Solvent-mediated threshold voltage shift in solution-processed transparent oxide thin-film transistors. *Appl. Phys. Lett.*, **97**, 092105.
13. Ryu, B., Noh, H.K., Choi, E.A., and Chang, K.J. (2010) O-vacancy as the origin of negative bias illumination stress instability in amorphous In–Ga–Zn–O thin film transistors. *Appl. Phys. Lett.*, **97**, 022108.
14. Nomura, K., Kamiya, T., Hirano, H., and Hosono, H. (2009) Origins of threshold voltage shifts in room-temperature deposited and annealed a-In–Ga–Zn–O thin-film transistors. *Appl. Phys. Lett.*, **95**, 013502.
15. Adamopoulos, G., Bashir, A., Thomas, S., Gillin, W.P., Georgakopoulos, S., Shkunov, M., Baklar, M.A., Stingelin, N., Maher, R.C., Cohen, L.F., Bradley, D.D.C., and Anthopoulos, T.D. (2010) Spray-deposited Li-Doped ZnO transistors with electron mobility exceeding 50 cm^2/Vs. *Adv. Mater.*, **22** (42), 4764–4769.
16. Kim, Y.H., Kim, K.H., Oh, M.S., Kim, H.J., Han, J.I., Han, M.K., and Park, S.K. (2010) Ink-jet-printed zinc–yin–oxide thin-film transistors and circuits with rapid thermal annealing process. *IEEE Electron. Dev. Lett.*, **31** (8), 836–838.
17. Avis, C. and Jang, J. (2011) A high performance inkjet printed zinc tin oxide transparent thin-film transistor manufactured at the maximum process temperature of 300 °C and its stability test. *Electrochem. Solid-State Lett.*, **14** (2), J9–J11.
18. ULVAC *www.ulvac.com*; ITO nano-metal ink (accessed 16 July 2014).
19. Faber, H., Hirschmann, J., Klaumünzer, M., Braunschweig, B., Peukert, W., and Halik, M. (2012) Impact of oxygen

plasma treatment on the device performance of zinc oxide nanoparticle-based thin-film transistors. *ACS Appl. Mater. Interfaces*, **4**, 1693–1696.

20. Cho, S.Y., Kang, Y.H., Jung, J.Y., Nam, S.Y., Lim, J., Yoon, S.C., Choi, D.H., and Lee, C. (2012) Novel zinc oxide inks with zinc oxide nanoparticles for low-temperature, solution-processed thin-film transistors. *Chem. Mater.*, **24** (18), 3517–3524.
21. Kim, Y.H., Han, M.K., Han, J.I., and Park, S.K. (2010) Effect of metallic composition on electrical properties of solution-processed indium-gallium-zinc-oxide thin-film transistors. *IEEE Trans. Electron Devices*, **57** (5), 1009–1014.
22. Hu, Q., Lee, S.C., Baek, Y.J., Lee, H.H., Kang, C.J., Kim, H.M., Kim, K.B., and Yoon, T.S. (2013) Non-volatile nano-floating gate memory with Pt-Fe_2O_3 composite nanoparticles and indium gallium zinc oxide channel. *J. Nanopart. Res.*, **15**, 1435–1436.
23. Banger, K.K., Yamashita, Y., Mori, K., Peterson, R.L., Leedham, T., Rickard, J., and Sirringhaus, H. (2011) Low-temperature, high-performance solution-processed metal oxide thin-film transistors formed by a 'sol–gel on chip' process. *Nat. Mater.*, **10**, 45–50.
24. Han, S.-Y., Herman, G.S., and Chang, C.-H. (2011) Low-temperature, high-performance, solution-processed indium oxide thin-film transistors. *J. Am. Chem. Soc.*, **133**, 5166–5169.
25. Lim, H.S., Rim, Y.S., Kim, D.L., and Kim, H.J. (2012) Performance enhancement of solution-processed Zn-Sn-O TFTs using high-pressure annealing. *Proc. SID*, **43** (1), 1104–1107.
26. Meyers, S.T., Anderson, J.T., Hung, C.M., Thompson, J., Wager, J.F., and Keszler, D.A. (2008) Aqueous inorganic inks for low-temperature fabrication of ZnO TFTs. *J. Am. Chem. Soc.*, **130** (51), 17603–17609.
27. Song, K., Noh, J., Jun, T., Jung, Y., Kang, H.Y., and Moon, J. (2010) Fully flexible solution-deposited ZnO thin-film transistors. *Adv. Mater.*, **22**, 4308–4312.
28. Jun, T., Song, K., Jeong, Y., Woo, K., Kim, D., Bae, C., and Moon, J. (2011) High-performance low-temperature solution-processable ZnO thin film transistors by microwave-assisted annealing. *J. Mater. Chem.*, **21**, 1102–1108.
29. Kim, Y.H., Heo, J.S., Kim, T.H., Park, S., Yoon, M.H., Kim, J., Oh, M.S., Yi, G.R., Noh, Y.Y., and Park, S.K. (2012) Flexible metal-oxide devices made by room temperature photochemical activation of sol–gel films. *Nature*, **489**, 128–132.

3
Carbon Nanotube Thin-Film Transistors

Taishi Takenobu

3.1
Introduction

Single-walled carbon nanotubes (SWCNTs) can be considered as monolayer graphene sheets with a honeycomb structure that are rolled into seamless, hollow cylinders (Figure 3.1). Because of their small size (~1–2 nm in diameter) and their superior electronic properties without dangling surface bonds, SWCNTs hold great potential for a wide range of applications in solid-state devices and are envisioned as a promising candidate for beyond-silicon electronics. SWCNTs can be categorized by their chiral vectors, which are defined on the hexagonal crystal lattice using two integers (m and n). The chiral vectors correspond to the direction along which a graphene sheet is wrapped to result in a SWCNT. The electronic properties of SWCNTs are strongly dependent on their chiral vectors, leading to SWCNTs that are either metallic ($m = n$ or $m - n$ is a multiple of 3) or semiconducting (all other cases) [1–4]. Using this rule of thumb, one can infer from the possible (n, m) values that one-third of SWCNTs are metallic and the other two-thirds are semiconducting. For practical use as the active channel component of electronic devices, semiconducting SWCNTs are commonly used. The advantages of semiconducting SWCNTs over other conventional semiconductors are multifold. First, the charge carriers in SWCNTs have long mean free paths, on the order of a few hundred nanometers, for the acoustic phonon scattering mechanism. As a result, scattering-free ballistic transport of carriers at low electric fields can be achieved in SWCNTs with moderate channel lengths (e.g., sub-100 nm) [5]. Second, the carrier mobility of semiconducting nanotubes has been experimentally measured to be >10 000 $cm^2\ V^{-1}\ s^{-1}$ [1, 6, 7] at room temperature, which is higher than that of state-of-the-art silicon transistors. Finally, their small diameters enable excellent electrostatics with efficient gate control of the channel for highly miniaturized devices. Thereby, SWCNTs have stimulated enormous interest in both fundamental research and practical applications in nano and macroelectronics.

Large Area and Flexible Electronics, First Edition. Edited by Mario Caironi and Yong-Young Noh.

Figure 3.1 Molecular structure of a single-walled carbon nanotube.

3.2
Individual SWCNTs and SWCNT Thin Films

Researchers had previously demonstrated excellent field-effect transistors (FETs) [5, 8–12] and integrated circuits [13–17] using individual SWCNTs. Despite the tremendous progress made with individual nanotube transistors and circuits, major technological challenges remain, including the need for deterministic assembly of nanotubes on a handling substrate with nanometer-scale accuracy, minimal device-to-device performance variation, and the development of a fabrication process that is scalable and compatible with industry standards. Hence, the use of carbon nanotubes for commercial nanoelectronic applications is still far from being realized. In contrast, the use of SWCNT networks, especially those based on semiconductor-enriched samples, present a highly promising path for the realization of high-performance thin-film transistors (TFTs) for macro and flexible electronic applications. The most significant advantages obtained by using SWCNT random networks as an active layer of TFTs lie in the fact that SWCNT thin films are mechanically flexible and optically transparent and can be prepared using solution-based room-temperature processing; none of these features can be provided by amorphous or polysilicon technologies [18–20]. Compared with organic semiconductors [21–25], a competing platform for flexible TFTs, SWCNT thin films offer significantly better carrier mobility (nearly two orders of magnitude higher). Thereby, large-area TFT applications may offer an ideal niche for CNT-based electronics, because of their superb physical and chemical properties without hindrance from precise assembly limitations down to the nanometer scale.

3.3
Chemical Vapor Deposition Growth of SWCNT TFTs

Numerous research efforts have been devoted to the successful realization of large-scale chemical vapor deposition (CVD) growth of high-density, horizontally aligned SWCNTs on single-crystal quartz or sapphire substrates [26–34]. In CVD growth, a substrate with catalyst particles is placed in a furnace (generally at temperatures >800 °C) with a supply of carbon feedstock gas and hydrogen gas; the SWCNTs are then grown directly on the substrate surface. The main advantage of CVD-grown SWCNTs is that they exhibit superior electrical performance compared to SWCNTs produced by other methods. Previous reports indicate that some of the highest mobilities, of more than ~10 000 $cm^2 V^{-1} s^{-1}$ [1, 6, 7], for single SWCNT transistors have been obtained using CVD-grown SWCNTs. The high performance of CVD-grown SWCNTs arises from the fact that they can be relatively long (from tens to hundreds of micrometers) and are nearly free of bundles. These properties are advantageous for obtaining high-performance TFTs because tube–tube junctions and bundles have been shown to significantly increase the resistance of SWCNT films.

Transfer techniques have been further developed, enabling the demonstration of high-performance transistors and integrated circuits using aligned nanotubes on various types of rigid and flexible substrates [35–42]. However, considering the fact that roughly one-third of the as-grown nanotubes are metallic, techniques such as electrical breakdown [43] are necessary to remove the leakage-causing metallic paths, a process that adds complexity, is not scalable, and significantly degrades the device performance due to the high fields applied during the process. Preferential growth of aligned semiconducting SWCNTs has been reported recently [32, 44, 45], which is an important step forward; however, the purity is not yet high enough to achieve transistors with a high on/off current ratio (I_{on}/I_{off}) for digital applications. Therefore, to obtain devices with a better I_{on}/I_{off} ratio, it is preferable to have networks of SWCNTs with a higher percentage of semiconducting tubes and/or with a random orientation in which individual nanotubes do not directly bridge the source/drain electrodes, thereby minimizing the metallic pathways [46–50].

CVD-grown *random* nanotube networks have also been widely explored for TFTs [51, 52], and medium-scale flexible integrated circuits were demonstrated by Rogers *et al.* [51]. For this method, metal catalysts are typically deposited over the entire substrate using either evaporation or spin-coating methods followed by CVD growth with hydrocarbon precursors such as methane, ethylene, ethanol, and methanol. Despite the tremendous success achieved in producing flexible nanotube TFTs and circuits with promising electrical performance, the existence of metallic nanotubes still presents a drawback by degrading the current on/off ratio of the devices. Although stripe-patterning has been proposed to improve the device on/off ratio by cutting the percolative transport through metallic paths in the transistors [51], the channel length must be relatively large in this case, limiting the degree of integration for future applications. The dry filtration method has been used by Ohno and coworkers [53] to achieve high-performance flexible nanotube TFTs and D-flip-flop circuits. In this method, SWCNTs grown by plasma-enhanced CVD are captured using a filter membrane, and the density of the nanotubes can be easily controlled by the collection time. The collected nanotube networks can be subsequently transferred to fabrication substrates by dissolving the filter using acetone. Although the devices fabricated via this method show excellent performance, throughput and scalability remain as problems.

3.4
Solution-Based Methods for SWCNT TFTs

The most common methods for assembling nanotube networks from solution are evaporation assembly [54], spin coating [55–57], drop coating [58–63], and printing [64–68]. The IBM group used a novel evaporation assembly method to obtain aligned nanotube strips with high-purity semiconducting nanotubes (Figure 3.2) [54]. Although submicrometer devices with good performance have been achieved, the scalability of this assembly method may be a potential problem. SWCNT networks can also be obtained by dropping a nanotube solution onto a spinning substrate [55]. This method is also limited in its scalability because the deposited SWCNTs often align along different orientations depending on the location on the substrate, preventing wafer-scale fabrication with high uniformity. The other two solution-based SWCNT assembly methods – drop coating and printing – have been found to be more promising for large-scale applications of nanotube TFTs. In the drop-coating method, the substrates are first functionalized with amine-containing molecules, which are effective adhesives for SWCNTs. By simply immersing the substrate into a nanotube solution, highly uniform nanotube networks can be obtained throughout the wafer, enabling the fabrication of nanotube TFTs with high yield and small device-to-device variation [59, 60, 62, 63]. Printing represents another low-cost approach for fabricating large-scale nanotube TFTs and circuits. Here, the SWCNT channel, electrodes, and gate dielectric can all be printed using inkjet [64, 65] or gravure printing [66–68] processes. This approach is useful for producing cost-effective large-area nanotube circuits requiring only moderate performance, as the resolution that can be achieved using printing processes is generally lower than that of conventional photolithography. Each of the methods discussed above presents unique opportunities and challenges. In this chapter, we primarily focus on the inkjet printing of SWCNT TFTs. We will first discuss the inkjet techniques for SWCNT networks and fabrication schemes for high-performance TFTs on mechanically rigid and flexible substrates.

3.5
Inkjet Printing of Flexible SWCNT TFTs

Printable technology has the potential to drastically reduce ecological impact, energy consumption during manufacturing, and wasted materials by controlling the quantity and location of ink deposition. Inkjet technology is exceptionally promising because patterns can be generated without any material waste, leading to drastic reductions in production costs and environmental impact. Materials for printable electronics must satisfy several requirements, such as high transport

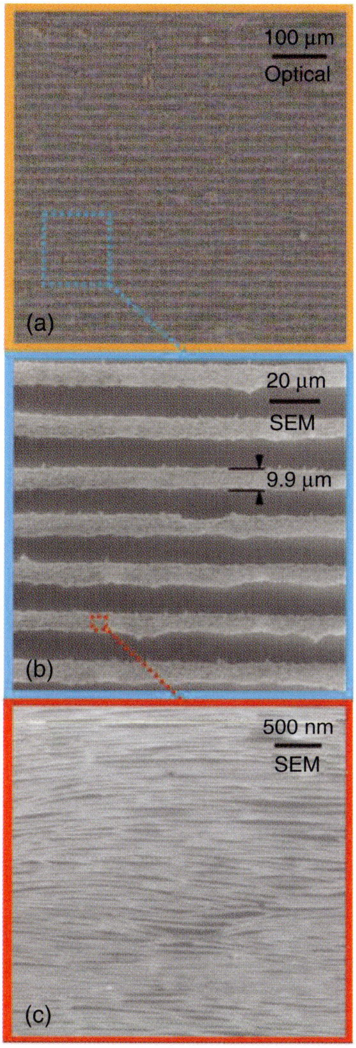

Figure 3.2 (a–c) SWCNT film fabricated by the evaporation assembly method [54]. (Reprinted with permission. Copyright © 2008, American Chemical Society.)

properties, chemical stability, and low-temperature processability. In SWCNTs, high performance (>1 cm^2 V^{-1} s^{-1}) has been reported for spaghetti-like, solution-processed random networks of SWCNTs, which can be attained from an inkjet process [69–75]. Moreover, the process temperature is typically <100 °C, which is applicable to plastic flexible substrates. Therefore, SWCNTs are one of the most promising materials for inkjet printable technology.

The first inkjet-printed flexible TFTs were demonstrated in 2009 [75]. An inkjet-printable aqueous SWCNT dispersion was prepared from 100 mg of HiPCO SWCNTs (Carbon Nanotechnologies, Inc.) and 1 g of sodium deoxycholate

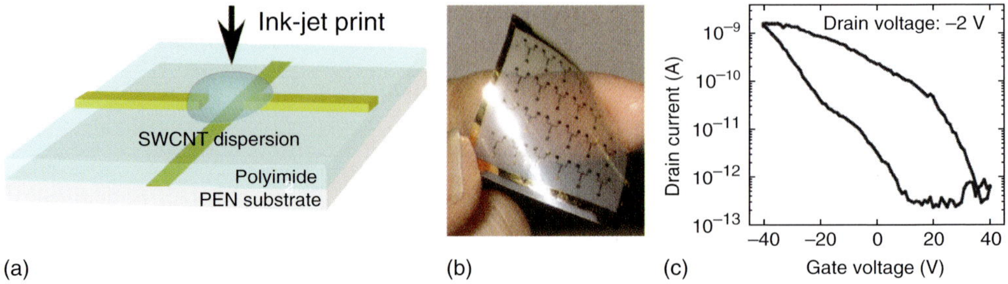

Figure 3.3 (a–c) First reported inkjet-printed SWCNT TFTs [75]. (Reproduced with permission. Copyright © 2009, The Japan Society of Applied Physics.)

(Wako Pure Chemical Industries, Ltd.) with 5 h of ultrasonic agitation and gentle stirring (270 rpm). Subsequently, the dispersion was centrifuged at $10\,000 \times g$ for 5 h and then decanted, leaving behind the sediment formed by centrifugation. The solution was then ejected from a piezo-driven dispenser head. To make the solution easier to apply, the author added 0.15% glycerin by weight to adjust the viscosity of the SWCNT dispersion.

Figure 3.3a shows a cross-sectional illustration of the resultant device. First, Au gate electrodes were thermally deposited through a shadow mask on flexible polyethylenenaphthalate (PEN) films (Teijin Dupont). A high-purity polyimide precursor (KEMITITE CT4112, Kyocera Chemical) was then spin-coated to obtain gate dielectric layers. Before the SWCNT films were printed, gold source and drain electrodes were deposited on the polyimide layer. Finally, the SWCNT dispersion was inkjet-printed between two Au electrodes, and the substrate was then washed with deionized water. Figure 3.3b shows an optical image of the flexible SWCNT TFTs, and Figure 3.3c shows the transfer characteristics of a typical flexible SWCNT TFT. The observed on/off current ratio was $\sim 10^4$ without any additional processing steps to remove the effect of metallic SWCNTs. However, the typical mobility was $10^{-3}-10^{-4}\,\text{cm}^2\,\text{V}^{-1}\,\text{s}^{-1}$, which is three or four orders of magnitude smaller than that obtained from other solution-phase methods.

3.6
Fabrication Schemes for High-Performance Inkjet-Printed SWCNT TFTs

In networks with high SWCNT coverage, especially in the form of small bundles, the metallic tubes (normally coexisting with semiconducting tubes at a 1 : 2 ratio) form a percolating network that behaves like a conducting film. With moderate SWCNT coverage, only the semiconducting tubes create a percolating network, and the film displays semiconducting properties. In the past, SWCNT TFTs have been inkjet-printed without precise density control, resulting in low transistor performance ($10^{-3}-10^{-4}\,\text{cm}^2\,\text{V}^{-1}\,\text{s}^{-1}$). To solve this problem, very dilute SWCNT dispersions have been adapted to control both the density and the electrical properties of the networks by optimizing the inkjet printing process (Figure 3.4) [76]. Figure 3.4c shows a schematic illustration of the materials,

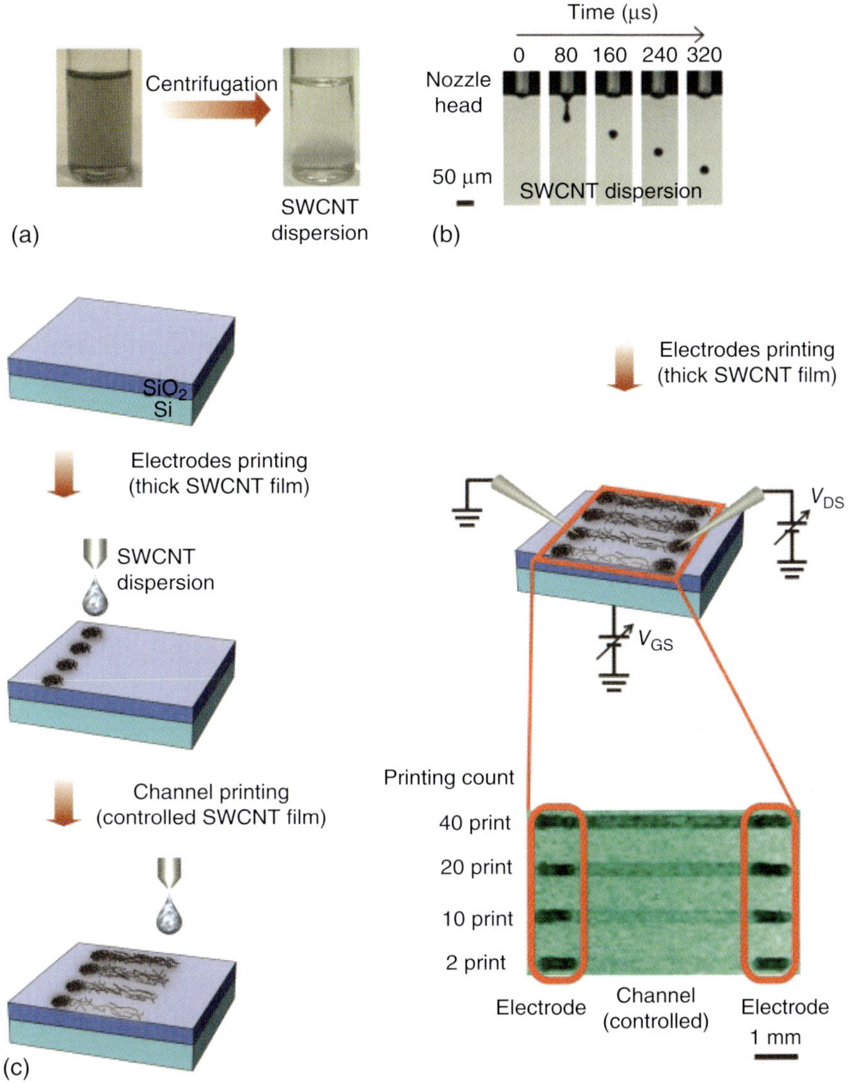

Figure 3.4 Fabrication of inkjet-printed SWCNT TFTs. (a) Optical micrograph of a dimethylformamide (DMF)-based SWCNT dispersion. (b) Time-dependent snapshots of a DMF-based SWCNT droplet during inkjet printing. The top gray stripe is the nozzle of the inkjet head and the black sphere is the injected droplet of the SWCNT dispersion. (c) Schematic illustration of the fabrication procedures for inkjet-printed SWCNT-TFTs. The black lines correspond to SWCNT bundles. An optical micrograph of inkjet-printed SWCNT-TFTs on SiO_2 substrate is also shown [76]. (Reproduced with permission. Copyright © 2010, WILEY-VCH Verlag GmbH & Co. KGaA, Weinheim.)

device layouts, and fabrication procedures. The authors inkjet-printed SWCNT networks on a SiO_2 (500 nm)/Si wafer, creating both source/drain electrodes and semiconducting active layers. In the electrode regions, the SWCNT dispersion was deposited 100 times at each position. For the active semiconducting layers, the authors printed four types of networks by depositing 40, 20, 10, or 2 droplets per position. Figure 3.4c shows an optical micrograph of printed SWCNT transistors with clear contrast between networks, strongly suggesting successful density control.

Figure 3.5 shows atomic force microscopy (AFM) images of inkjet-printed SWCNT films. The film density was well controlled and uniform. Networks with high (4.20% surface coverage, Figure 3.5a) and moderate (0.61% surface coverage, Figure 3.5d) coverage acted as metallic and semiconducting films, respectively. Importantly, the authors observed clear boundaries between metallic (0.85% surface coverage) and semiconducting (0.61% surface coverage) percolating networks. These results demonstrate that the electrical properties of SWCNT networks can be tailored via the inkjet process. The authors tuned the on/off current ratio from 1.5 to 10^4 and the carrier mobility from 4.2 to 49 $cm^2\ V^{-1}\ s^{-1}$ (1.6–4.2 $cm^2\ V^{-1}\ s^{-1}$ with an on/off ratio of 10^4–10^5).

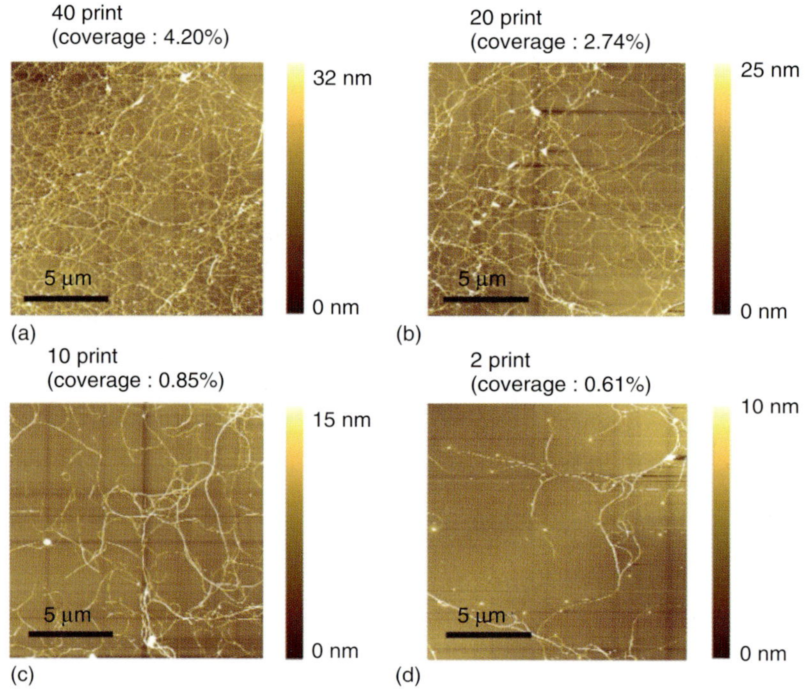

Figure 3.5 AFM images and printing count dependence of surface coverage in inkjet-printed SWCNT films. AFM images of the four types of networks created by deposition of (a) 40, (b) 20, (c) 10, or (d) 2 prints per position [76]. (Reproduced with permission. Copyright © 2010, WILEY-VCH Verlag GmbH & Co. KGaA, Weinheim.)

Inkjet-printed SWCNT TFTs can potentially be used in flexible electronics because of their high performance and low-temperature processability. However, various properties need to be improved before application to achieve low hysteresis, low-voltage operation, and full printability with printable dielectric materials. To this end, the authors changed the dielectric layer from solid-state SiO_2 to an ionic liquid (Figure 3.6a) as a high-capacitance gate dielectric compatible with TFTs and solution processing [76, 77]. Figure 3.6b shows a schematic illustration of device layouts and fabrication procedures for exclusively inkjet-printed SWCNT TFTs. The authors fabricated high-performance SWCNT TFTs using the inkjet method and then printed the SWCNT gate electrode (Figure 3.6c).

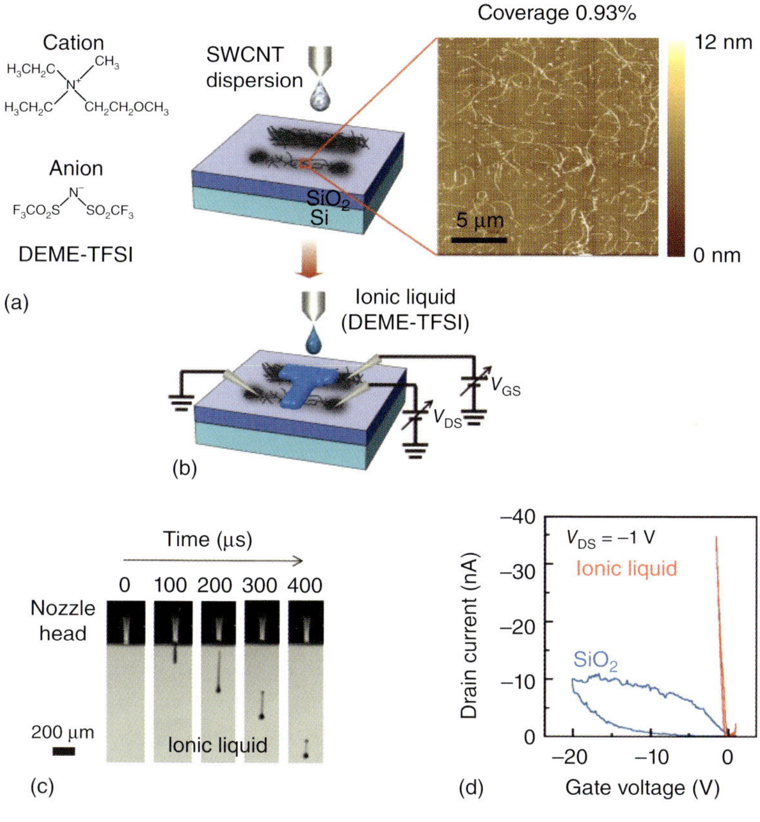

Figure 3.6 Fabrication of exclusively inkjet-printed SWCNT TFTs. (a) Schematic representation of *N*,*N*-diethyl-*N*-methyl-*N*-(2-methoxyethyl) ammonium bis(trifluoromethanesulfonyl) imide (DEME-TFSI). (b) Schematic representation of fabrication procedures for exclusively inkjet-printed SWCNT-TFTs and a corresponding AFM image. (c) Time-dependent snapshots of an ionic-liquid droplet during inkjet printing. Because of the higher viscosity of the ionic liquid, the authors heated the inkjet head to 65 °C. (d) Transfer characteristics of SiO_2 back-gated (blue) and ionic-liquid-gated (red) SWCNT TFTs [76]. (Reproduced with permission. Copyright © 2010, WILEY-VCH Verlag GmbH & Co. KGaA, Weinheim.)

The surface coverage of this film was 0.93%. In the final fabrication step, the authors inkjet-printed an ionic liquid as the gate dielectric material. Figure 3.6d compares the characteristics of back-gate-operated solid-state SiO_2 and ionic-liquid gating. With ionic-liquid gating, the operating voltage was significantly reduced, and the hysteretic response against gate voltages was improved, possibly as a result of the efficient screening of charged impurities by the ionic liquid.

3.7 Inkjet Printing of SWCNT CMOS Inverters

Currently, it is difficult for most materials to dope carriers via inkjet printing; although organic materials can provide excellent complementary logic circuits, p- and n-type semiconductors must be combined for their preparation because the doping method is poorly defined. Furthermore, doping techniques for inorganic materials generally require high-temperature and/or vacuum processes, which are incompatible with current printing technology. Among these materials, it may be possible to control the transistor polarity with SWCNTs using a printing technique.

The doping of SWCNTs differs from the doping of conventional semiconductor materials because it is difficult for SWCNTs to incorporate foreign atoms, such as nitrogen or boron, into their substitutional or interstitial sites. As a result, adsorbates such as atoms [78, 79], molecules [79–82], and polymers [83–85] with the appropriate functional groups have been introduced on the nanotube surface. The motion of the electrons between the adsorbates and the SWCNTs leads to a change in the Fermi level of the SWCNTs. For low-temperature inkjet-carrier doping, the authors selected polyethyleneimine (PEI, average molecular weight ~2000, Sigma Aldrich, Japan) as the electron dopant because it is a well-known soluble electron donor for SWCNTs [83–85]. Importantly, the PEI reaction occurs at room temperature.

A PEI solution in ethanol and SWCNT TFTs were inkjet-printed using a method described by Okimoto *et al.* [76, 77]. First, Matsuzaki and coworkers [86] printed seven droplets per dot with a dot-to-dot distance of 10 μm (Figure 3.7a). The Si substrate was used as a back gate electrode, and the evaporated Au was used as the source and drain electrodes (Figure 3.7b). Finally, the authors printed PEI inks onto the SWCNT films with selective positioning using an inkjet system, and the sample was baked at 45 °C for 5 min to remove the solvent (Figure 3.7c,d). Figure 3.7e shows the transfer characteristics before and after PEI printing on a SWCNT TFT. This device clearly displayed n-type transistor characteristics after inkjet-carrier doping.

As indicated in Figure 3.7c, the authors inkjet-printed SWCNT complementary circuits by combining n- and p-type SWCNT TFTs. The PEI ink was dropped onto specific locations on the SWCNT TFT in the designed complementary circuit by inkjet printing. Because it is difficult to tune a doping shift of threshold voltages, we prepared several p-type transistors with different threshold voltages

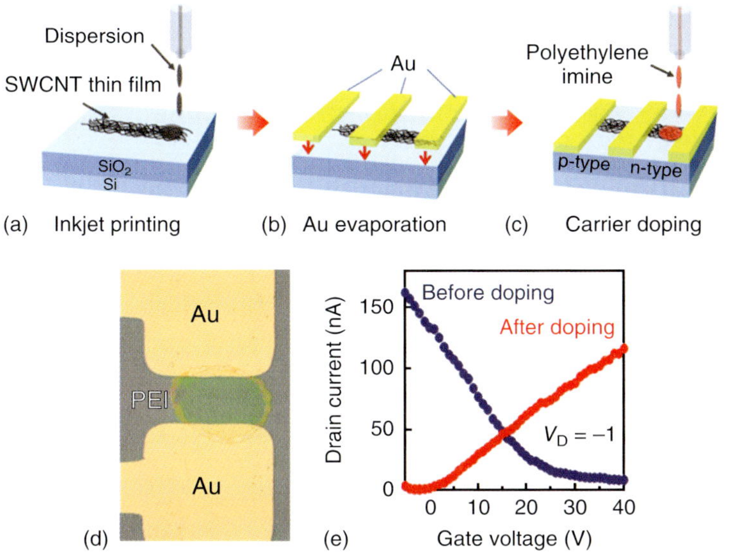

Figure 3.7 Schematic of the inkjet-doped SWNT TFT fabrication procedure. (a) Inkjet printing of SWCNT TFTs. (b) Evaporation of gold electrodes. (c) Inkjet printing of PEI for electron doping onto SWCNT TFTs. (d) Photograph of a PEI-printed SWCNT TFT. (e) Transfer characteristics of SWCNT TFTs before and after PEI printing [86]. (Reproduced with permission. Copyright © 2011, The Japan Society of Applied Physics.)

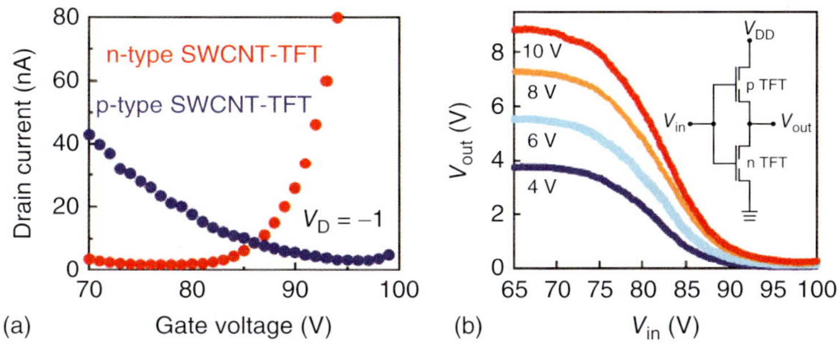

Figure 3.8 (a) Transfer characteristics of p-type and n-type SWCNT-TFTs. (b) Transfer characteristics of a SWCNT complementary inverter at different supply voltages (V_{DD}). The inset shows a schematic circuit of SWCNT complementary inverters [86]. (Reproduced with permission. Copyright © 2011, The Japan Society of Applied Physics.)

and selected one value for inverter fabrication. Figure 3.8a,b shows the characteristics of the simplest logic gate, an inverter (NOT). As shown in Figure 3.8b, for a supply voltage (V_{DD}) of 4–10 V, clear on and off states were observed, suggesting correct switching action of the inkjet-printed SWCNT complementary inverter. The inverter performance can be determined by the signal gain ($\Delta V_{out}/\Delta V_{in}$).

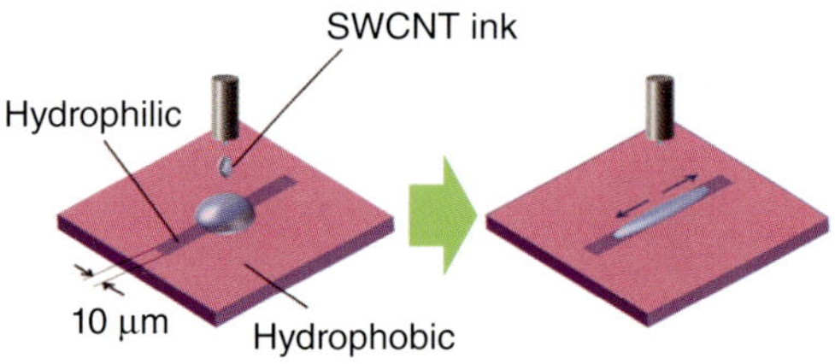

Figure 3.9 Schematic illustration of the inkjet-printed droplet movement on a HMDS-SAM-coated SiO_2 substrate, which is converted to become partially hydrophilic by UV irradiation through a 3-μm-gap photomask [87]. (Reproduced with permission. Copyright © 2013, AIP Publishing LLC.)

The transfer curves in Figure 3.9b exhibit gains of up to 1.4 in the given voltage range. Importantly, although the obtained gain is smaller than that of other SWCNT inverters, this is the first successful demonstration of inkjet-printed SWCNT complementary logic gates prepared at moderate temperatures.

3.8 Inkjet Printing of Aligned SWCNT Films

Individual SWCNTs have excellent carrier mobility; however, a random network of SWCNTs exhibits electrostatic screening at SWCNT–SWCNT junctions. Consequently, the geometry of a SWCNT network has a strong influence on the device characteristics, and reducing or eliminating the number of junctions can improve the charge transport [12–16]. Therefore, the alignment of SWCNTs is extremely important for high-performance film devices, and the achievement of efficient alignment using printing techniques is highly desirable. Recently, Takagi and coworkers [87] succeeded in the alignment of SWCNTs via inkjet printing.

SWCNT films were inkjet-printed using the method described by Okimoto *et al.* [76, 77]. The authors used self-assembled monolayers (SAMs) to control the wettability of the substrates [88, 89]. Highly doped silicon wafers with 500-nm-thick, thermally grown oxide layers were used as substrates. The authors immersed these substrates in a solution of 1,1,1,3,3,3-hexamethyldissilazane (HMDS) and then rinsed them with chloroform. These HMDS SAMs provide uniform hydrophobicity over the entire substrate surface as a result of the fabrication of SAMs. Importantly, the HMDS SAMs can be removed under UV irradiation (wavelength of 184.9–253.7 nm) to convert the hydrophobic surface to a hydrophilic one. To enable control of the drying process of the SWCNT ink, the authors adopted the SAM-based patterning method described by Nobusa *et al.* [88, 89]. Figure 3.9 schematically illustrates their method for controlling the drying process. The authors converted specific areas by applying UV irradiation through a 3-μm-gap photomask and the inkjet-printed SWCNT dispersions on these substrates to control both the movement of the SWCNT ink and the drying process.

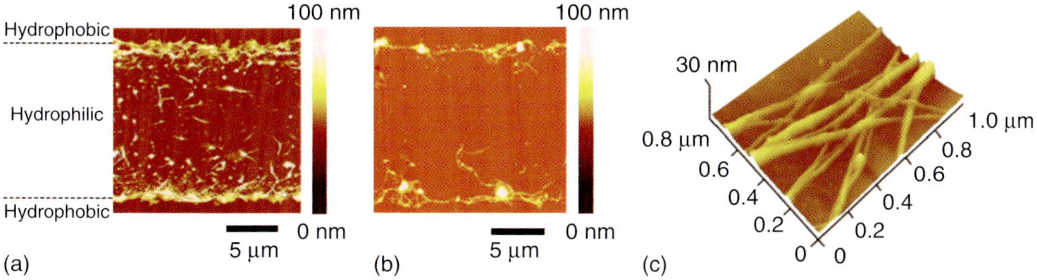

Figure 3.10 AFM images of inkjet-printed SWCNT films. (a) s-SWCNT and (b) laser-SWCNT thin films inkjet-printed onto HMDS-SAM-coated SiO_2 substrates, which are converted to become partially hydrophilic by UV irradiation through a 3-μm-gap photomask. These substrates were sonicated in deionized water before the inkjet printing of the SWCNT dispersions. (c) High-resolution AFM image of laser-SWCNTs [87]. (Reproduced with permission. Copyright © 2013, AIP Publishing LLC.)

The topography of the SWCNT films was characterized by AFM (Multimode 8, Bruker). In these films, the SWCNTs were primarily concentrated along the boundary of the wettability contrast; this tendency was observed more frequently for the longer laser-SWCNTs (Figure 3.10b) than for the shorter s-SWCNTs (Figure 3.10a). This difference between the two types of SWCNTs arises from the differences in the interactions between the nanotubes, which are simply proportional to the tube length. These results indicate that we can obtain highly oriented SWCNT films from longer SWCNTs via the inkjet printing method (Figure 3.10c).

3.9 Conclusion

Although it was very difficult to fabricate SWCNT TFTs by inkjet printing techniques prior to 2009, because of the rapid progress of this field a wide variety of inkjet-printed SWCNT films, such as density-controlled transistors, fully printed electrolyte-gated transistors, inkjet-printed SWCNT CMOS (complementary metal–oxide–semiconductor) inverters, and inkjet-printed aligned SWCNT films, have been realized. These potentially large-area, flexible SWCNT devices open routes for developing high-performance and environmentally friendly SWCNT printed electronics. The next target may be fully printed, stretchable SWCNT devices.

References

1. Dresselhaus, M.S., Dresselhaus, G., and Avouris, P. (eds) (2001) *Carbon Nanotubes – Synthesis, Structure, Properties, and Applications*, Springer-Verlag.
2. Javey, A. and Kong, J. (eds) (2009) *Carbon Nanotube Electronics*, Springer.
3. Wildoer, J., Venema, L., Rinzler, A., Smalley, R., and Dekker, C. (1998) *Nature*, **391**, 59–62.

4. Odom, T., Huang, J., Kim, P., and Lieber, C. (1998) *Nature*, **391**, 62–64.
5. Javey, A., Guo, J., Wang, Q., Lundstrom, M., and Dai, H. (2003) *Nature*, **424**, 654–657.
6. Durkop, T., Getty, S.A., Cobas, E., and Fuhrer, M.S. (2004) *Nano Lett.*, **4**, 35–39.
7. Zhou, X., Park, J.Y., Huang, S., Liu, J., and McEuen, P.L. (2005) *Phys. Rev. Lett.*, **95**, 146805.
8. Tans, S., Verschueren, A., and Dekker, C. (1998) *Nature*, **393**, 49–52.
9. Martel, R., Schmidt, T., Shea, H.R., Hertel, T., and Avouris, P. (1998) *Appl. Phys. Lett.*, **73**, 2447–2449.
10. Javey, A., Guo, J., Farmer, D., Wang, Q., Wang, D., Gordon, R., Lundstrom, M., and Dai, H. (2004) *Nano Lett.*, **4**, 447–450.
11. Javey, A., Guo, J., Farmer, D., Wang, Q., Yenilmez, E., Gordon, R., Lundstrom, M., and Dai, H. (2004) *Nano Lett.*, **4**, 1319–1322.
12. Franklin, A.D., Luisier, M., Han, S.-J., Tulevski, G., Breslin, C.M., Gignac, L., Lundstrom, M.S., and Haensch, W. (2012) *Nano Lett.*, **12**, 758–762.
13. Bachtold, A., Hadley, P., Nakanishi, T., and Dekker, C. (2001) *Science*, **294**, 1317–1320.
14. Derycke, V., Martel, R., Appenzeller, J., and Avouris, P. (2001) *Nano Lett.*, **1**, 453–456.
15. Javey, A., Wang, Q., Ural, A., Li, Y., and Dai, H. (2002) *Nano Lett.*, **2**, 929–932.
16. Chen, Z., Appenzeller, J., Lin, Y., Oakley, J.S., Rinzler, A.G., Tang, J., Wind, S.J., Solomon, P.M., and Avouris, P. (2006) *Science*, **311**, 1735.
17. Ding, L., Zhang, Z., Liang, S., Pei, T., Wang, S., Li, Y., Zhou, W., Liu, J., and Peng, L.-M. (2012) *Nat. Commun.*, **3**, 667.
18. Street, R.A. (ed.) (2000) *Technology and Applications of Amorphous Silicon*, Springer-Verlag.
19. Ucjikoga, S. (2002) *MRS Bull.*, **27**, 881–886.
20. Snell, A.J., Mackenzie, K.D., Spear, W.E., LeComber, P.G., and Hughes, A.J. (1981) *Appl. Phys. A*, **24**, 357–362.
21. Dimitrakopoulos, C.D. and Mascaro, D.J. (2001) *IBM J. Res. Dev.*, **45**, 11–27.
22. Forrest, S.R. (2004) *Nature*, **428**, 911–918.
23. Gelinck, G.H., Edzer, H., Huitema, A., Van Veenendaal, E., Cantatore, E., Schrijnemakers, L., Van Der Putten, J.B.P.H., Geuns, T.C.T., Beenhakkers, M., Giesbers, J.B., Hiusman, B.-H., Meijer, E.J., Benito, E.M., Touwslager, F.J., Marsman, A.W., Van Rens, B.J.E., and De Leeuw, D.M. (2004) *Nat. Mater.*, **3**, 106–110.
24. Sekitani, T., Zschieschang, U., Klauk, H., and Someya, T. (2010) *Nat. Mater.*, **9**, 1015–1022.
25. Sekitani, T. and Someya, T. (2010) *Adv. Mater.*, **22**, 1–19.
26. Ismach, A., Segev, L., Wachtel, E., and Joselevich, E. (2004) *Angew. Chem. Int. Ed.*, **43**, 6140–6143.
27. Ismach, A., Kantorovich, D., and Joselevich, E. (2005) *J. Am. Chem. Soc.*, **127**, 11554–11555.
28. Han, S., Liu, X., and Zhou, C. (2005) *J. Am. Chem. Soc.*, **127**, 5294–5295.
29. Kocabas, C., Hur, S., Gaur, A., Meitl, M., Shim, M., and Rogers, J.A. (2005) *Small*, **1**, 1110–1116.
30. Ding, L., Yuan, D., and Liu, J. (2008) *J. Am. Chem. Soc.*, **130**, 5428–5429.
31. Patil, N., Lin, A., Myers, E.R., Ryu, K., Badmaev, A., Zhou, C., Wong, H.-S.P., and Mitra, S. (2009) *IEEE Trans. Nanotechnol.*, **8**, 498–504.
32. Ding, L., Tselev, A., Wang, J., Yuan, D., Chu, H., McNicholas, T.P., Li, Y., and Liu, J. (2009) *Nano Lett.*, **9**, 800–805.
33. Hong, S.W., Banks, T., and Rogers, J.A. (2010) *Adv. Mater.*, **22**, 1826–1830.
34. Wang, C., Ryu, K.M., Gomez, L., Badmaev, A., Zhang, J., Lin, X., Che, Y., and Zhou, C. (2010) *Nano Res.*, **3**, 831–842.
35. Liu, X., Han, S., and Zhou, C. (2006) *Nano Lett.*, **6**, 34–39.
36. Kang, S.J., Kocabas, C., Ozel, T., Shim, M., Pimparkar, N., Alam, M.A., Rotkin, S.V., and Rogers, J.A. (2007) *Nat. Nanotechnol.*, **2**, 230–236.
37. Kang, S.J., Kocabas, C., Kim, H.S., Cao, Q., Meitl, M.A., Khang, D.Y., and Rogers, J. (2007) *Nano Lett.*, **7**, 3343–3348.
38. Jiao, L., Fan, B., Xian, X., Wu, Z., Zhang, J., and Liu, Z. (2008) *J. Am. Chem. Soc.*, **130**, 12612–12613.

39. Wang, C., Ryu, K., Badmaev, A., Patil, N., Lin, A., Mitra, S., Wong, H.-S.P., and Zhou, C. (2008) *Appl. Phys. Lett.*, **93**, 033101.
40. Ryu, K., Badmaev, A., Wang, C., Lin, A., Patil, N., Gomez, L., Kumar, A., Mitra, S., Wong, H.-S.P., and Zhou, C. (2009) *Nano Lett.*, **9**, 189–197.
41. Ishikawa, F., Chang, H., Ryu, K., Chen, P., Badmaev, A., De Arco Gomez, L., Shen, G., and Zhou, C. (2009) *ACS Nano*, **3**, 73–79.
42. Lin, A., Patil, N., Ryu, K., Badmaev, A., De Arco Gomez, L., Zhou, C., Mitra, S., and Wong, H.-S.P. (2009) *IEEE Trans. Nanotechnol.*, **8**, 4–9.
43. Collins, P.G., Arnold, M.S., and Avouris, P. (2001) *Science*, **292**, 706–709.
44. Hong, G., Zhang, B., Peng, B., Zhang, J., Choi, W.M., Choi, J.-Y., Kim, J.M., and Liu, Z. (2009) *J. Am. Chem. Soc.*, **131**, 14642–14643.
45. Che, Y., Wang, C., Liu, J., Liu, B., Lin, X., Parker, J., Beasley, C., Wong, H.-S.P., and Zhou, C. (2012) *ACS Nano*, **6**, 7454–7462.
46. Cao, Q. and Rogers, J.A. (2008) *Nano Res.*, **1**, 259–272.
47. Cao, Q. and Rogers, J.A. (2009) *Adv. Mater.*, **21**, 29–53.
48. Rouhi, N., Jain, D., and Burke, P.J. (2011) *ACS Nano*, **5**, 8471–8487.
49. Kocabas, C., Pimparkar, N., Yesilyurt, O., Kang, S.J., Alam, M.A., and Rogers, J.A. (2007) *Nano Lett.*, **7**, 1195–1202.
50. Pimparkar, N., Kocabas, C., Kang, S.J., Rogers, J., and Alam, M.A. (2007) *IEEE Electron Device Lett.*, **28**, 593–595.
51. Cao, Q., Kim, H.S., Pimparkar, N., Kulkarni, J.P., Wang, C., Shim, M., Roy, K., Alam, M.A., and Rogers, J.A. (2008) *Nature*, **454**, 495–500.
52. Snow, E.S., Novak, J.P., Campbell, P.M., and Park, D. (2003) *Appl. Phys. Lett.*, **82**, 2145–2147.
53. Sun, D., Timmermans, M.Y., Tian, Y., Nasibulin, A.G., Kauppinen, E.I., Kishimoto, S., Mizutani, T., and Ohno, Y. (2011) *Nat. Nanotechnol.*, **6**, 156–161.
54. Engel, M., Small, J.P., Steiner, M., Freitag, M., Green, A.A., Hersam, M.C., and Avouris, P. (2008) *ACS Nano*, **2**, 2445–2452.
55. Meitl, M.A., Zhou, Y.X., Gaur, A., Jeon, S., Usrey, M.L., Strano, M.S., and Rogers, J.A. (2004) *Nano Lett.*, **4**, 1643–1647.
56. LeMieux, M.C., Roberts, M., Barman, S., Jin, Y.W., Kim, J.M., and Bao, Z. (2008) *Science*, **321**, 101–104.
57. Vosgueritchain, M., LeMieux, M.C., Dodge, D., and Bao, Z. (2010) *ACS Nano*, **4**, 6137–6145.
58. Snow, E.S., Campbell, P.M., Ancona, M.G., and Novak, J.P. (2005) *Appl. Phys. Lett.*, **86**, 033105.
59. Wang, C., Zhang, J., Ryu, K., Badmaev, A., Gomez, L., and Zhou, C. (2009) *Nano Lett.*, **9**, 4285–4291.
60. Wang, C., Zhang, J., and Zhou, C. (2010) *ACS Nano*, **4**, 7123–7132.
61. Rouhi, N., Jain, D., Zand, K., and Burke, P.J. (2011) *Adv. Mater.*, **23**, 94–99.
62. Takahashi, T., Takei, K., Gillies, A.G., Fearing, R.S., and Javey, A. (2011) *Nano Lett.*, **11**, 5408–5413.
63. Wang, C., Chien, J.-C., Takei, K., Takahashi, T., Nah, J., Niknejad, A.M., and Javey, A. (2012) *Nano Lett.*, **12**, 1527–1533.
64. Ha, M., Xia, Y., Green, A.A., Zhang, W., Renn, M.J., Kim, C.H., Hersam, M.C., and Frisbie, C.D. (2010) *ACS Nano*, **4**, 4388–4395.
65. Chen, P., Fu, Y., Aminirad, R., Wang, C., Zhang, J., Wang, K., Galatsis, K., and Zhou, C. (2011) *Nano Lett.*, **11**, 5301–5308.
66. Jung, M., Kim, J., Noh, J., Lim, N., Lim, C., Lee, G., Kim, J., Kang, H., Jung, K., Leonard, A.D., Tour, J.M., and Cho, G. (2010) *IEEE Trans. Electron Devices*, **57**, 571–580.
67. Noh, J., Jung, M., Jung, K., Lee, G., Kim, J., Lim, S., Kim, D., Choi, Y., Kim, Y., Subramanian, V., and Cho, G. (2011) *IEEE Electron Device Lett.*, **32**, 638–640.
68. Noh, J., Kim, S., Jung, K., Kim, J., Cho, S., and Cho, G. (2011) *IEEE Electron Device Lett.*, **32**, 1555–1557.
69. Takenobu, T., Takahashi, T., Kanbara, T., Tsukagoshi, K., Aoyagi, Y., and Iwasa, Y. (2006) *Appl. Phys. Lett.*, **88**, 033511.
70. Fukao, T., Nakamura, S., Kataura, H., and Shiraishi, M. (2006) *Jpn. J. Appl. Phys., Part 1*, **45**, 6524.

71. Fujii, S., Tanaka, T., Nishiyama, S., and Kataura, H. (2011) *Phys. Status Solidi B*, **248**, 2692.
72. Fu, D., Okimoto, H., Lee, C.W., Takenobu, T., Iwasa, Y., Kataura, H., and Li, L.-J. (2010) *Adv. Mater.*, **22**, 4867.
73. Lee, C.W., Weng, C.H., Wei, L., Chen, Y., Chan-Park, M.B., Tsai, C.H., Leou, K.C., Poa, C.H.P., Wang, J.L., and Li, L.J. (2008) *J. Phys. Chem. C*, **112**, 12089.
74. Beecher, P., Servati, P., Rozhin, A., Colli, A., Scardaci, V., Pisana, S., Hasan, T., Flewitt, A.J., Robertson, J., Hsieh, G.W., Li, F.M., Nathan, A., Ferrari, A.C., and Milne, W.I. (2007) *J. Appl. Phys.*, **102**, 043710.
75. Takenobu, T., Miura, N., Lu, S.Y., Okimoto, H., Asano, T., Shiraishi, M., and Iwasa, Y. (2009) *Appl. Phys. Express*, **2**, 025005.
76. Okimoto, H., Takenobu, T., Yanagi, K., Miyata, Y., Shimotani, H., Kataura, H., and Iwasa, Y. (2010) *Adv. Mater.*, **22**, 3981.
77. Okimoto, H., Takenobu, T., Yanagi, K., Shimotani, H., Miyata, Y., Kataura, H., and Iwasa, Y. (2010) *Jpn. J. Appl. Phys.*, **49**, 02BD09.
78. Rao, A.M., Eklund, P.C., Bandow, S., Thess, A., and Smalley, R.E. (1997) *Nature*, **388**, 257.
79. Takenobu, T., Takano, T., Shiraishi, M., Murakami, Y., Ata, M., Kataura, H., Achiba, Y., and Iwasa, Y. (2003) *Nat. Mater.*, **2**, 683.
80. Kazaoui, S., Minami, N., Jacquimin, R., Kataura, H., and Achiba, Y. (1999) *Phys. Rev. B*, **60**, 13339.
81. Takenobu, T., Kanbara, T., Akima, N., Takahashi, T., Shiraishi, M., Tsukagoshi, K., Kataura, H., Aoyagi, Y., and Iwasa, Y. (2005) *Adv. Mater.*, **17**, 2430.
82. Takenobu, T., Takahashi, T., Akima, N., Shiraishi, M., Kataura, H., and Iwasa, Y. (2007) *J. Nanosci. Nanotechnol.*, **7**, 3533.
83. Shim, M., Javey, A., Kam, N.W.S., and Dai, H.J. (2001) *J. Am. Chem. Soc.*, **123**, 11512.
84. Zhou, Y., Gaur, A., Hur, S.H., Kocabas, C., Meitl, M.A., Shim, M., and Rogers, J.A. (2004) *Nano Lett.*, **4**, 2031.
85. Nouchi, R., Tomita, H., Ogura, A., Kataura, H., and Shiraishi, M. (2008) *Appl. Phys. Lett.*, **92**, 253507.
86. Matsuzaki, S., Nobusa, Y., Yanagi, K., Kataura, H., and Takenobu, T. (2011) *Appl. Phys. Express*, **4**, 105101.
87. Takagi, Y., Nobusa, Y., Gocho, S., Kudou, H., Yanagi, K., Kataura, H., and Takenobu, T. (2013) *Appl. Phys. Lett.*, **102**, 143107.
88. Nobusa, Y., Yomogida, Y., Matsuzaki, S., Yanagi, K., Kataura, H., and Takenobu, T. (2011) *Appl. Phys. Lett.*, **99**, 183106.
89. Nobusa, Y., Takagi, Y., Gocho, S., Matsuzaki, S., Yanagi, K., and Takenobu, T. (2012) *Jpn. J. Appl. Phys.*, **51**, 06FD15.

4
Organic Single-Crystalline Semiconductors for Flexible Electronics Applications

Marcos A. Reyes-Martinez, Nicholas S. Colella, and Alejandro L. Briseno

4.1
Introduction

There is no question that the outstanding electronic properties of organic single-crystalline semiconductors have opened opportunities to a new generation of high-performance organic electronic devices. For over 50 years, organic single crystals have been utilized for fundamental charge-transport studies and for determining the performance limitations of organic semiconductors [1–3]. Nonetheless, there is still a great deal of knowledge that needs to be acquired from these highly ordered solids.

For example, in the field of flexible electronics, a great amount of effort has been placed into the investigation of inorganic structures [4], organic polycrystalline thin films [5], graphene sheets, and carbon nanotubes [6]. The efforts to understand the effects of mechanical strain on performance and the determination of the limits of flexibility of organic single crystals are minuscule in comparison. This apparent standstill is attributed to the intrinsic fragility of the crystals, the associated handling difficulty, and the low throughput in device fabrication. Nevertheless, recent developments in the processing and integration of organic single crystals into large-area arrays of field-effect devices promise to move this field forward [7, 8].

In this chapter, we review the electronic and structural properties that give organic single crystals their remarkable characteristics when compared to organic polycrystalline films. We look into novel materials design/synthesis, suitable for different types of crystallization and various applications. We examine published work where organic single crystals have been successfully applied into flexible electronics and other advances that promise to make a significant impact on the field of organic single-crystal-based flexible electronics. Finally, we discuss some of the obstacles that need to be overcome in order to realize practical applications of organic single crystals in fundamental flexible studies and consumer electronics technology.

Large Area and Flexible Electronics, First Edition. Edited by Mario Caironi and Yong-Young Noh.

4.2
Electronic and Structural Properties of Single Crystals

Because of their relative ease of fabrication, thin films are the most commonly studied structures of organic semiconductors. However, as a result of the inherent grain boundaries and the dependence of the morphology on processing, it is exceptionally difficult to create films with reproducibly high mobilities and virtually impossible to characterize the intrinsic properties of the material under study [2]. Single crystals of organic semiconductors are nearly defect free, and thus provide an opportunity to probe the true transport properties of the material without any reduction in performance (see Table 4.1) due to grain boundaries or impurities (Figure 4.1).

Furthermore, single-crystal transistors exhibit reproducible behavior because of their molecular perfection. Immutable van der Waals forces create the same unit cell in the crystal structure, with few variations due to polymorphs, which are

Table 4.1 Notable organic single-crystal field-effect mobilities.

Carrier type	Material	Mobility ($cm^2\ V^{-1}\ s^{-1}$)	Growth technique
Holes	Rubrene [9]	20	Vapor
	C_8-BTBT [9]	16.4	Solution
	TIPS-Pentacene [10]	8.1	Solution
	Tetracene [11]	2.4	Vapor
	Pentacene [12]	2.2	Vapor
Electrons	PDIF-CN_2 [13]	3	Vapor
	TCNQ [14]	1.6	Vapor
	BPE-PTCDI [15]	1.4	Solution

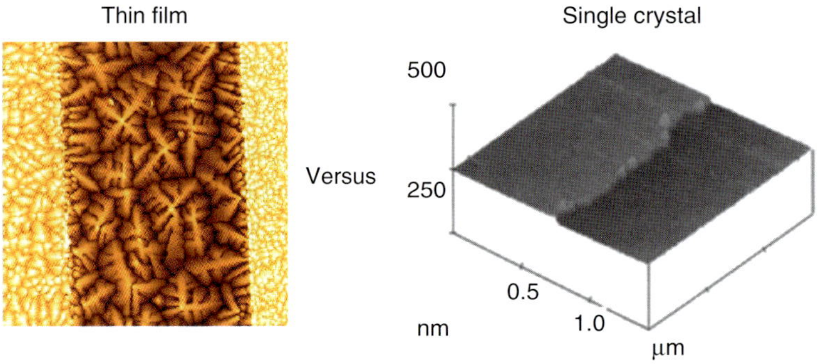

Figure 4.1 Grain boundaries present in thin films conflate intrinsic material performance with device processing. Few molecular steps are present in single crystals [25]. (Reproduced with permission. Copyright © 2006, WILEY-VCH Verlag GmbH & Co. KGaA, Weinheim.)

well-documented, consistent, and easily controlled [16, 17]. The crystal structure, particularly the intermolecular distance, π–π stacking, and orbital overlap, dictate the electronic coupling in the long-range-order crystal.

4.2.1
Intrinsic Transport Properties

There are two main mechanisms of carrier transport in organic semiconductors: hopping and band transport [3]. Hopping occurs between the aromatic π orbitals of neighboring molecules, generally the lowest unoccupied molecular orbitals (LUMOs) and highest occupied molecular orbitals (HOMOs) for electron and hole transport, respectively, and exhibits the highest transport when there is strong overlap [3]. The directionality of these orbitals necessarily produces mobility anisotropy, which has been characterized and correlated to the crystal structure of many materials. In the hopping regime, charge carriers can be considered to be localized, as polarons, on individual molecules. In contrast, the charge carriers in the band transport regime are completely delocalized; their location cannot be determined. These two regimes are analogous to the particle–wave duality of quantum mechanics. An excellent example of band transport, in which mobility is dictated solely by the electronic structure of the material, is naphthalene at low temperatures (below 150 K). Above 150 K, charge transport in naphthalene shows temperature dependence, indicating the presence of charge traps in the hopping regime [18].

While transport through delocalized orbitals does occur, it is imperative to note the differences between organic and inorganic semiconductors. Because of the strong polarizability of organic molecules and the defects inherent in even the best grown "single crystals," polarons in the hopping regime dominate transport in most organic semiconductors. The residence time of a charge on organic molecules is typically an order of magnitude longer than the time needed to form an electronic polaron (on the order 10^{-15} s) [19]. These polarons interact strongly with vibrations of the crystal lattice; mobility decreases as the temperature increases, according to an inverse power law. However, it is important to note that charge transport is observed only above the temperature at which thermal energy allows charge carriers to escape trap states caused by defects.

Time-of-flight (TOF) studies on ultrapure crystals were the first experiments to examine the intrinsic properties of semiconducting organic materials [20]. Pulses of light excite charge carriers in the bulk of the crystals, and the response is measured to characterize the mobility without loss of performance due to static disorder. In transistors, however, the conducting channel is generated by a field on surface of the semiconducting crystal, which has a much higher defect density than the bulk.

Field-effect transistors (FET) constitute the main experimental tool to probe the charge carrier mobilities of semiconducting materials. Figure 4.2 shows the general structure of a transistor. Transistors consist of three terminals: gate (G), source (S), and drain (D). The active semiconductor material connects the source

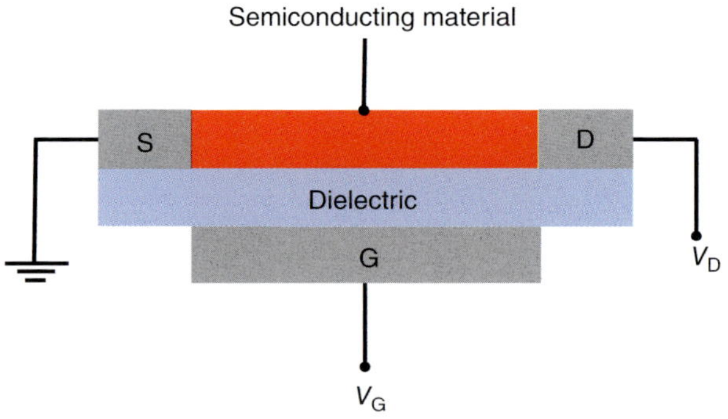

Figure 4.2 General structure of a field-effect transistor.

and drain, and a dielectric layer insulates the gate electrode from the rest of the device. When a voltage is applied to the gate, charge carriers are induced in the dielectric–semiconductor interface, creating a conductive channel. By applying a bias to the drain electrode, a current across the source and drain is generated. The distance between the source and drain is referred to as the *channel length* (L), while the width is known as the *channel width* (W). The field-effect mobility of the active layer can be extracted from the drain current at the saturation regime from the expression

$$I_{DS} = \frac{WC\mu}{2L}(V_G - V_T)^2$$

where C is the capacitance per unit area of the insulating layer, V_G is the gate voltage, V_T is the threshold voltage, and μ is the field-effect mobility.

4.2.2 Crystal Dimensionality

The orientation of single-crystal semiconductors has a profound impact on their mobilities, as the anisotropy of the crystal structure exhibits varying degrees of orbital overlap and delocalization in different directions [1]. It is important to note that the same features that promote charge transport, for example, π–π stacking, also encourage crystal growth. Molecules that pack face to face due to π–π interactions and van der Waals forces tend to form one-dimensional nanowire crystals, with their highest mobilities coincident with their longest axes [21]. Two notable examples of nanowire single crystals are perylenetetracarboxyldiimide (PTCDI) and hexathiapentacene (HTP), which have intermolecular spacings of 3.3 and 3.54 Å, respectively, producing nanowires with excellent mobilities along their long axis (over 1.4 and 0.04 $cm^2\,V^{-1}\,s^{-1}$, respectively) [15, 21]. HTP molecules stack face to face along the [100] crystallographic a-axis (Figure 4.3) [21]. In pentacene single crystals, the electrons of the aromatic rings interact with the peripheral electron-deficient hydrogen atoms, inducing an edge-to-face geometry. In HTP, the chalcogen atoms interrupt this conformation. Instead, the sulfur atoms

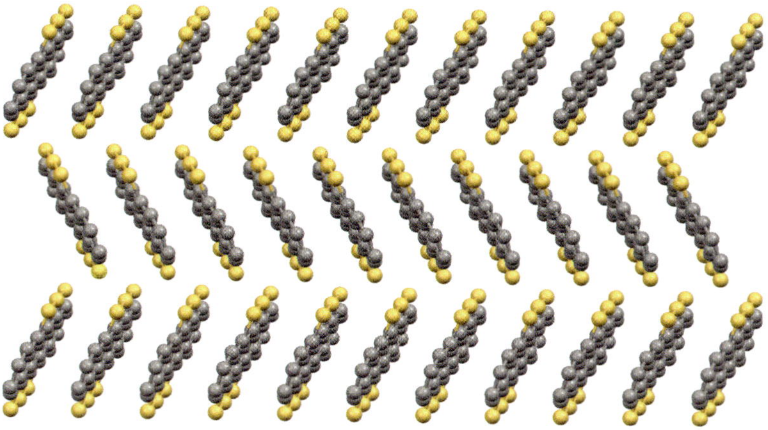

Figure 4.3 Hexathiapentacene (HTP) crystal packing. One-dimensional nanowires grow such that there is the orbital overlap in the π-stacking direction.

Figure 4.4 Pentacene packing motif. The herringbone structure of pentacene forms two-dimensional plates.

induce dipoles between each other, creating an intermolecular distance that is smaller than the van der Waals radii of the two atoms in what is commonly referred to as the *sulfur–sulfur interaction*. This small intermolecular spacing allows the nanowires to exhibit high mobilities along their axes.

Another typical packing motif is the herringbone structure exhibited by pentacene (Figure 4.4); this crystal structure occurs as a result of the interaction between the aromatic rings and the edges of the molecules. This herringbone structure creates similar carrier mobilities along different axes and results in two-dimensional plates [22]. A compilation of the field-effect mobilities of the most relevant organic single crystal is shown in Table 4.1.

4.3 Crystallization Techniques

4.3.1 Growth from Vapor Phase

In order to grow well-faceted crystals of macroscopic dimensions, the process of physical vapor transport (PVT) is often used [23]. Figure 4.5a shows a schematic of a furnace tube used to grow high-quality organic single crystals [24]. The same figure shows representative organic single crystals grown via physical transport. A temperature gradient is induced in the furnace, and then the source organic material is placed in the area with the highest temperature, where it sublimes. The vapor molecules are carried down the temperature gradient by a steady flow of an inert gas, usually argon or nitrogen. The molecules crystallize once they reach a cooler region below the material's crystallization temperature. The quality of the single crystals obtained by PVT has been confirmed using polarized microscopy and diffraction experiments (Figure 4.5b) [25].

The growth conditions in the furnaces can be easily adjusted [25]. One can control the growth dimensions of the crystals to fulfill specific needs; this will be discussed in depth in a later section of this chapter. A comprehensive analysis on crystal growth conditions can be found in Ref. [23].

4.3.2 Growth from Solution

HTP is a representative material to illustrate the growth of single crystals from solution. HTP is a stable, relatively insoluble semiconductor that can be readily synthesized by reacting pentacene with elemental sulfur (Figure 4.6a) [21]. After the synthesis of the bulk polycrystalline material, it can be recrystallized to obtain single crystals. HTP can be dispersed in an organic solvent with a high boiling point such as nitrobenzene or *o*-dichlorobenzene. Heating this solution allows the HTP to dissolve, and subsequent cooling results in the formation of single-crystal nanowires (Figure 4.6b). Additionally, the recrystallization conditions can be modified to control the nanowire widths via the addition of methanol during cooling. This method of HTP crystallization is quite robust, creating nanowires with lengths on the order of tens to hundreds of micrometers and widths of hundreds of nanometers. Additionally, the nanowires produced are oxidatively stable in most organic solvents, likely due to their high ionization potential ($\text{HOMO}_{\text{HTP}} = -5.35$ eV; $\text{HOMO}_{\text{pentacene}} = -5.0$ eV) (Figure 4.6c,d).

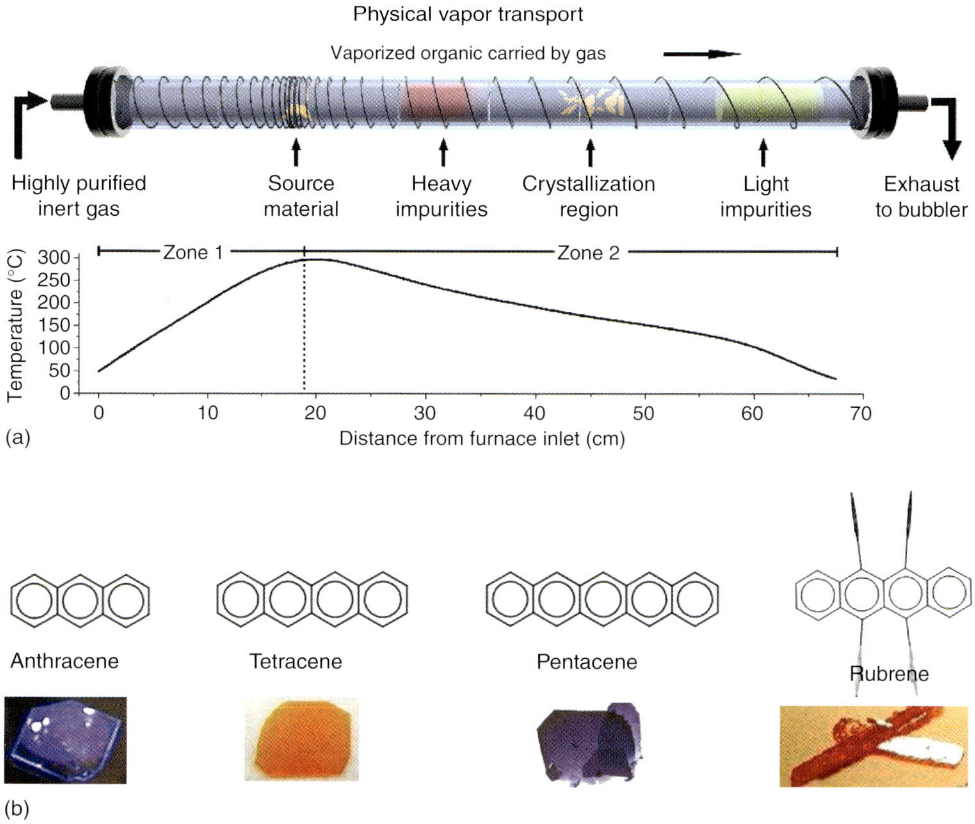

Figure 4.5 Crystal growth by physical vapor transport. (a) Physical vapor transport furnace. A temperature gradient is induced in a cylindrical glass tube by coiling a heating rope around it. The source material is placed in the area with the highest temperature, where it sublimes. The molecules are then carried down the temperature gradient with the help of a steady flow of an inert gas. Subsequently, the material crystallizes at a cooler zone. (b) Common oligoacenes crystallized by the physical vapor transport method [24]. (Reprinted with permission. Copyright © 2007, Elsevier Ltd.)

4.4 Single-Crystal Flexible Electronic Devices

4.4.1 Fundamental Mechanics for Flexible Electronics

One of the most basic requirements for the development of flexible electronics is bendability. This property is quantified by the bending stiffness M, which relates the Young's modulus E of the beam or plate material, its second moment of inertia I, and the resulting curvature κ through the expression

$$M = EI\kappa$$

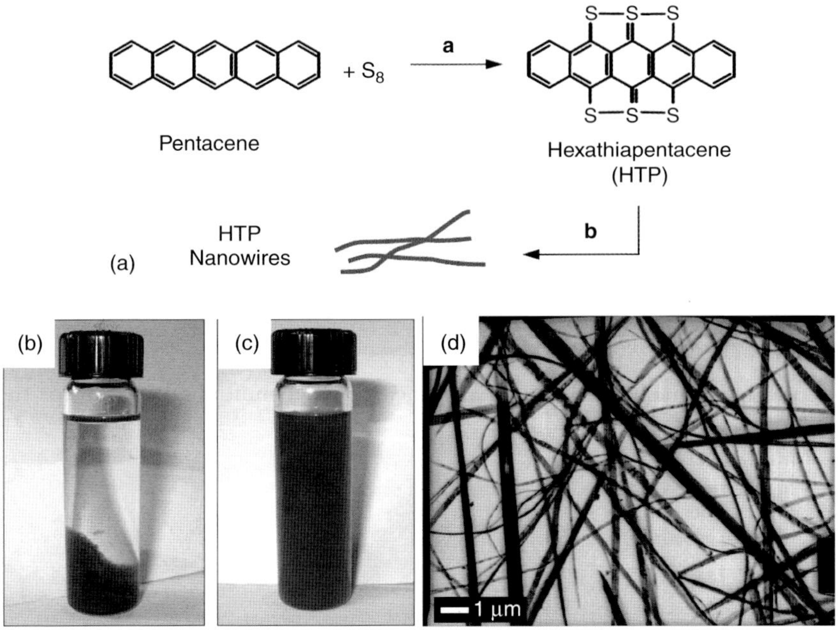

Figure 4.6 (a) Pentacene is reacted with elemental sulfur to form hexathiapentacene. (b) Hexathiapentacene wires show good oxidative stability, but settle in solution. (c) The nanowires can be easily redispersed by simple agitation. (d) TEM image of a network of hexathiapentacene nanowires [43]. (Reprinted with permission. Copyright © 2007, American Chemical Society.)

For the case of a rectangular beam

$$I = \frac{bh^3}{12}$$

where b is the cross-sectional length and h the thickness (Figure 4.7) [28]. The cubic relationship between bending stiffness and thickness indicates that it is easier to bend an object as its thickness decreases. This concept has been applied to fabricate stretchable devices using ultrathin films of traditionally stiff semiconductor materials, such as silicon. Stretchability is achieved by taking advantage of surface buckling instabilities, allowing devices to be deformed in a plane in a way analogous to an accordion [29, 30]. The intrinsic properties of the active materials in buckled stretchable devices are not impaired upon stretching.

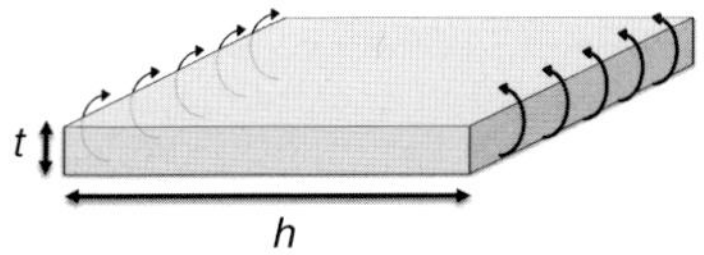

Figure 4.7 Bending of a rectangular plate.

Another strategy adopted by several researchers has been to take advantage of the mechanical "neutral plane" for fabricating functional devices. In a bending plate, the outside of the curve is stretched in tension while the interior of the curve is compressed. The neutral plane is the region between these two opposite forces where there are no net longitudinal strains. Thus, fabricating devices such that the active layer(s) are encapsulated in the neutral plane reduces the chances of degradation as a product of straining the device. Moreover, it increases the total amount of strain that the whole device can endure before breaking down.

4.4.2 Mechanical Versatility of Organic Single Crystals

As previously mentioned, modifying the growth conditions can control the dimensions of the crystals obtained by the PVT method [25]. Slow growth with a minimum number of growth-nucleation sites on the inner walls of the glass cylinder produces large, thick crystals. For the case of rubrene, growth times range from 30 min to 48 h, and sublimation temperatures are in the range of ~280–300 °C [25]. Typical gas flow rates used are 50 ml min^{-1}. The large dimensions of the crystals obtained under these conditions (hundreds of micrometers in thickness) [2] have been advantageous in the fabrication of free-standing transistors, where the crystal itself is utilized as a substrate for all the components of the transistor [2]. However, the intrinsic brittleness of thick crystals makes them unsuitable for applications where they must undergo bending and stretching; ultrathin crystals are more suitable to withstand mechanical deformations. In order to obtain ultrathin (100 nm to 1 μm) crystals, rapid growth and an increased flow rate are required. In the case of rubrene, the source material is sublimed at temperatures between 280 and 300 °C. After nucleation is observed inside the glass cylinder, the temperature of the sublimation zone is increased to 330 °C and the argon flow rate is increased from 50 to 100 cm^3 min^{-1}. Figure 4.8 shows a comparison of thin, nearly transparent rubrene single crystals and thick crystals ($\gg$3 μm). The flexibility of a thin rubrene single crystal as it bends by gravity on the tip of tweezers can also be appreciated in the same figure.

There are additional advantages to using thin crystals beyond exploiting their bendability for flexible devices. Detailed surface analysis has shown that thin single crystals have a nearly defect-free surface morphology. AFM scans of a random 2-μm square area of different samples of ultrathin rubrene crystals revealed only one surface step [25]. The surface roughness of a 300-nm-thick rubrene single crystal is 0.23 nm, indicating the smoothness of the surface. However, a scan of a crystal >3 mm thick showed more frequent and much larger surface steps with a surface roughness of at least twice that of thin crystals. Figure 4.9a shows the SEM images comparing an ultrathin rubrene single crystal with a crystal of larger dimensions, both electrostatically adhered to the same kind of test structures. A high degree of interfacial adhesion and surface conformity is observed for the thin crystal compared to the thicker crystal. The thick crystal in Figure 4.9a exhibits a less intimate contact with the underlying dielectric layer and electrodes.

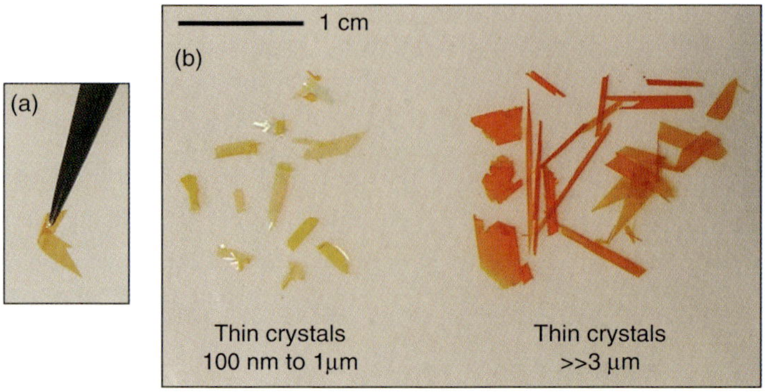

Figure 4.8 Comparison between thick and thin rubrene single crystals. (a) Digital photograph of rubrene single crystal as it bends. (b) Digital photograph of thick and thin rubrene [25]. (Reproduced with permission. Copyright © 2006, WILEY-VCH Verlag GmbH & Co. KGaA, Weinheim.)

The differences shown between both devices clearly reflect on their respective performances. Field-effect experiments were carried out to monitor the effective mobility at various crystal thicknesses. The results shown on Figure 4.9b display the gradual drop in mobility as a function of crystal thickness. Exceptional adhesion and conformability of thin crystals to the dielectric surface is likely the reason for higher carrier mobility as opposed to the poor crystal–dielectric interface with thicker crystals.

4.4.3
Importance of Mechanical Properties Knowledge

The field of organic flexible electronics, in addition to improving processing and fabrication techniques, is moving toward gaining a more fundamental understanding of the mechanics of materials. In recent literature, numerous studies of the mechanical properties of organic semiconductors can be found. This understanding of the mechanical properties of semiconducting materials is necessary for their application in flexible electronic devices; this knowledge is imperative as a tool for the selection of materials for engineering design and processing [31].

The wrinkling instability is the metrology tool of choice when studying the mechanical properties of thin films. It has been used to determine the Young's modulus of polycrystalline thin films of various semiconductor materials [32]. A summary of their results can be seen in Table 4.2.

Delongchamp and coworkers took the investigation of the mechanical properties of organic materials a step further [33]. They drew a qualitative correlation between mechanical properties and electrical properties by studying the Young's modulus, field-effect mobility, and degree of crystallinity of poly(3-hexylthiophene) (P3HT) and poly-(2,5-bis(3-alkylthiophene-2-yl)thieno[3,2-*b*]thiophene) (PBTTT) thin films. Figure 4.10a,b shows micrographs of wrinkled P3HT

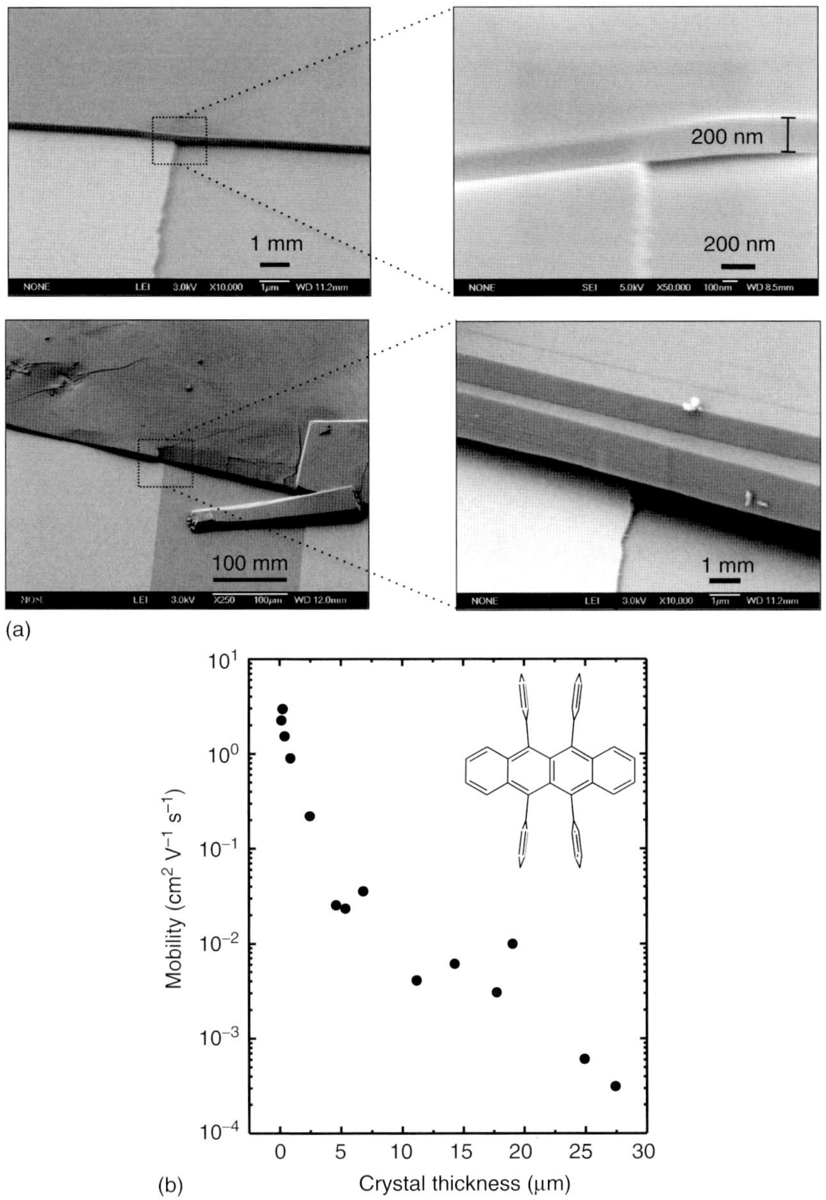

Figure 4.9 (a) An SEM image of a thin (~1 mm) rubrene single crystal electrostatically adhered to a substrate. The curved crystal shows no interfacial or surface damage even after an acute bending radius. Single crystals thicker than 3 mm tend to shatter or break when they are bent. (b) Trend of field-effect mobility as a function of crystal thickness [25]. (Reproduced with permission. Copyright © 2006, WILEY-VCH Verlag GmbH & Co. KGaA, Weinheim.)

Table 4.2 Elastic moduli of organic electronic calculated from wrinkling.

Material	Young's modulus (GPa)
rr-P3HT	1.33 (±0.01)
P3HT/PCBM	6.02 (±0.03)
PEDOT:PSS	2.26 (±0.05)
PANI	0.03
Pentacene	16.09 (±2.83)

Source: Reprinted with permission from Ref. [32]. Copyright © 2009, American Chemical Society.

and PBTTT films, respectively. They then used the wrinkling wavelengths to obtain the materials' respective Young's moduli. Figure 4.10c shows how a higher mobility correlates to a higher modulus and higher level of crystallinity. Even though this study focused on polycrystalline films of polymeric materials, their conclusions are of great interest to this field, as it sets the groundwork for future studies. The reader should not forget that the correlations established in this study are qualitative; an analytical approach is still needed to elucidate to the precise relationship between mechanical and electrical properties in organic semiconductors.

Most of the work related to the understanding of the mechanical properties of organic materials has been carried out on polycrystalline or amorphous thin films. In order to extract the intrinsic mechanical properties of a material, these measurements must be carried out on the single-crystalline form of the material.

4.4.4
The Elastic Constants of Rubrene Single Crystals

An in-depth study of the elastic constants of rubrene single crystals (Figure 4.11) was performed utilizing wrinkling instability [34]. For wrinkling experiments, thin rubrene single crystals were grown using the PVT conditions outlined in Section 4.4.2. In this study, researchers employed two wrinkling methods (Figure 4.12) to overcome the difficulty of measuring mechanical properties on crystals with limited lateral dimensions. In both techniques, crystals are manually laminated on poly(dimethylsiloxane) (PDMS) substrates (Sylgard 184). Through this process, crystals adhere to the PDMS surface by van der Waals forces. In the first wrinkling method (Figure 4.12a), the PDMS substrate is deformed by applying a point-like normal load on the elastomer near the edge of the crystal where wrinkles are to be induced. The load on the elastomeric substrate generates a radial stress σ_{rr} which deforms the substrate under the crystal; at the same time, a compressive stress $\sigma_{\theta\theta}$ develops in the azimuthal direction. Once a critical azimuthal strain is reached, wrinkles develop on the crystal. An inverted microscope captures micrographs of the wrinkled crystals (Figure 4.12b,c).

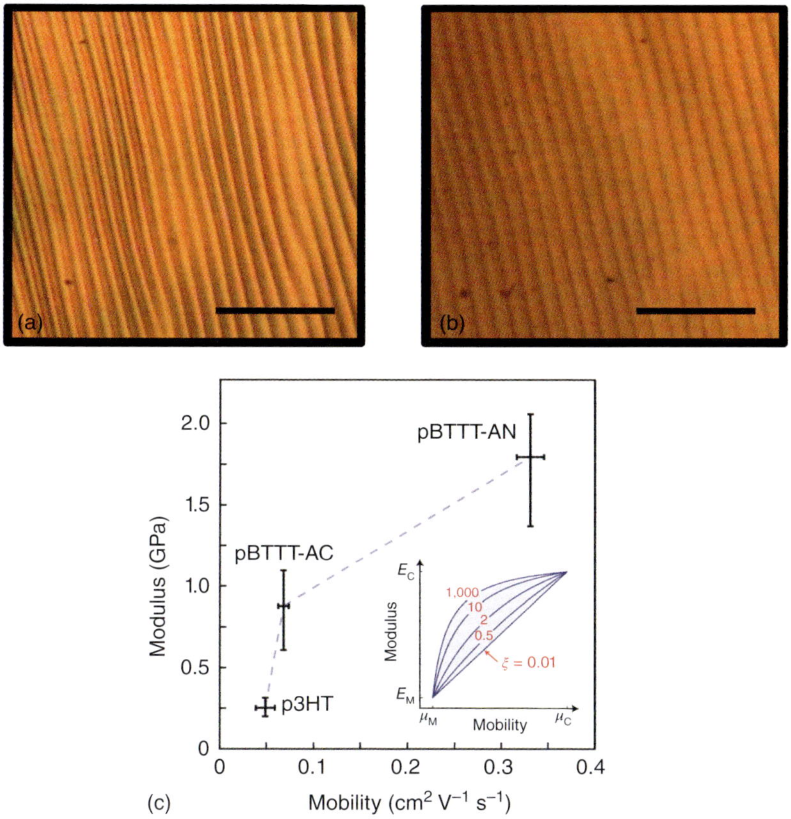

Figure 4.10 Optical micrographs of buckling pattern observed for (a) P3HT film and (b) PBTTT film on PDMS under compression. The scale bar on the images is 100 μm. (c) Comparison of the field-effect mobility and elastic modulus of P3HT, as-cast pBTTT-AC, and annealed pBTTT-AN films. Inset: the charge mobility and elastic modulus of a semicrystalline polymer with percent crystallinity [33]. (Reprinted with permission. Copyright © 2010, American Chemical Society.)

In the second wrinkling method, the crystal is subjected to an in-plane stress by compressing a cylindrical PDMS substrate (Figure 4.12d) with a custom-built strain stage. At a critical strain, wrinkles develop on the crystal, as shown in Figure 4.12e,f. The use of a cylindrical PDMS provides the opportunity to compress the crystal samples along any angle, generating wrinkles along different directions of the crystal. Wrinkles developed using both of these methods are reversible; in other words, when the load is removed, the crystal returns to an optically planar configuration.

A key finding in this investigation is the observation of different wrinkle wavelengths for wrinkles that are induced along different crystal directions. Figure 4.13a,b shows the optical micrographs portraying one crystal as it wrinkles along the [010] and [001] directions, respectively. According to wrinkling

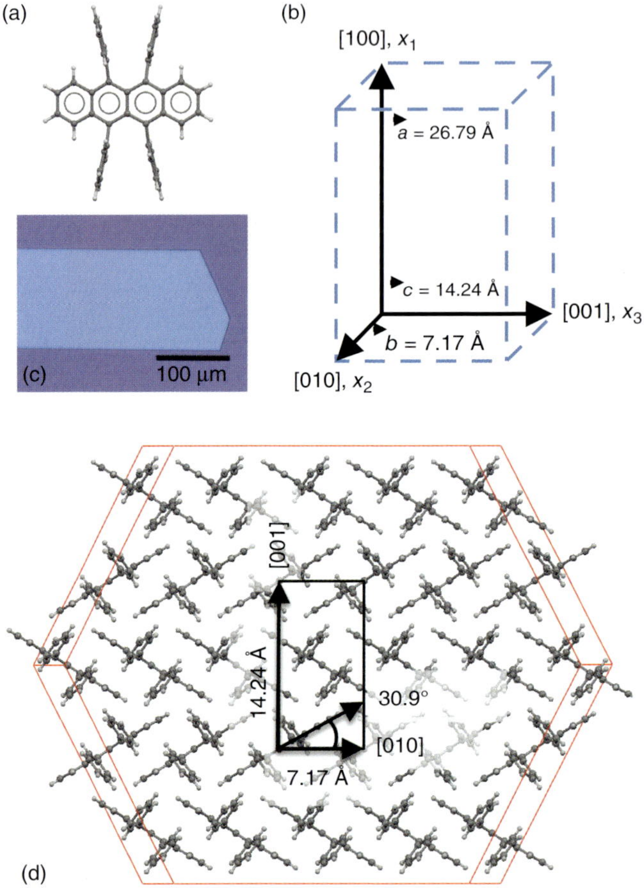

Figure 4.11 Rubrene single crystals. (a) Rubrene molecule, (b) unit cell parameters for rubrene, (c) optical micrograph of ultrathin rubrene single crystal, and (d) crystal packing for (100) face of rubrene [34]. (Reproduced with permission. Copyright © 2012, WILEY-VCH Verlag GmbH & Co. KGaA, Weinheim.)

mechanics, the observed differences in wrinkle wavelengths suggest differences in the elastic constants along distinct directions of the crystal plane [35]. Figure 4.13c shows wrinkle wavelength data obtained from both experimental wrinkling techniques along the [001] and [010] directions. The authors define a scaled wrinkle wavelength as $\lambda^* = (\lambda/2\pi)(3\overline{E}_s)^{1/3}$, which allows the integration of data obtained from crystals attached to substrates of different moduli. From these experimental data, the in-plane elastic constants for rubrene are found to be $\overline{C}_{22} = 14.89 \pm 0.73\,\text{GPa}$ and $\overline{C}_{33} = 9.89 \pm 0.60\,\text{GPa}$. Moreover, these results are in the same order of magnitude as those of polycrystalline films, as shown in Table 4.2.

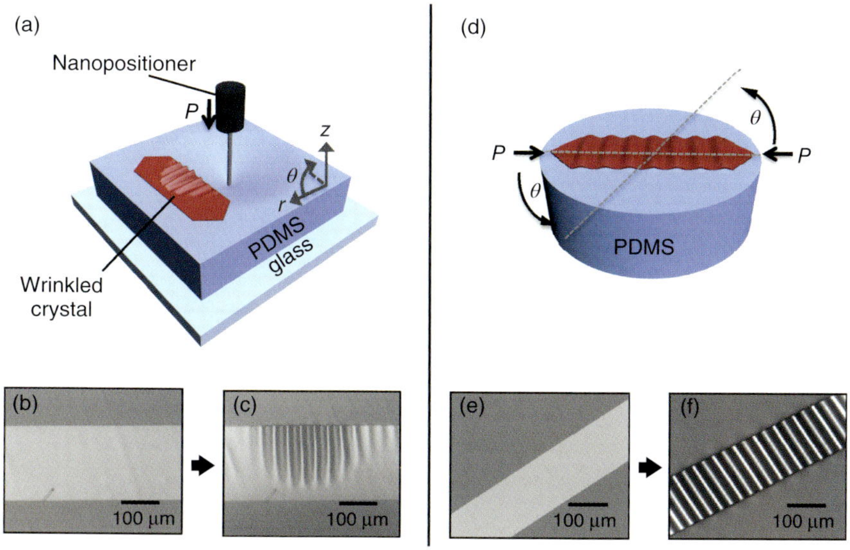

Figure 4.12 Wrinkling techniques. (a) Schematic of wrinkling by contact-point deformation. (b,c) Optical micrographs of rubrene crystal before and after wrinkling, probe is not shown. (d) Schematic of wrinkling by mechanical compression. (e,f) Optical micrographs of a rubrene crystal before and after wrinkling is induced [34]. (Reproduced with permission. Copyright © 2012, WILEY-VCH Verlag GmbH & Co. KGaA, Weinheim.)

Because of the in-plane anisotropy of rubrene, the buckling wavelengths will be different in any loading direction, as seen in Figure 4.14a. Thus, to further quantify the mechanical response of the crystal, the buckling of the rubrene film along different in-plane directions was systematically examined. If the sample is loaded along an axis at angle θ measured counterclockwise in plane from the [010] axis, the elastic constant $\overline{C'}_{22}$ in the direction of compression can be defined as

$$\overline{C'}_{22}(\theta) = \overline{C}_{22}\cos^4\theta + \overline{C}_{33}\sin^4\theta + 2(\overline{C}_{23} + 2\overline{C}_{44})\cos^2\theta\sin^2\theta$$

such that the buckling wavelength correspondingly varies with orientation as

$$\frac{\lambda(\theta)}{t} = 2\pi\left[\frac{\overline{C'}_{22}(\theta)}{\overline{E}_s}\right]^{\frac{1}{3}}$$

It is well known that the unique packing of the rubrene molecules in the crystal results in pronounced anisotropic carrier mobilities [1]. Differences in wrinkling wavelengths along different crystallographic directions provide clear evidence of anisotropic mechanical properties. Figure 4.14b indicates a buckling wavelength peak at ~30° with relation to the b-axis. Interestingly, this value is very close to the pitch angle of rubrene, which is the angle that the backbone of the molecule makes with the [010] direction (Figure 4.11d) [36–38]. The observation of the peak value for the buckling wavelength not corresponding to the high charge-carrier mobility

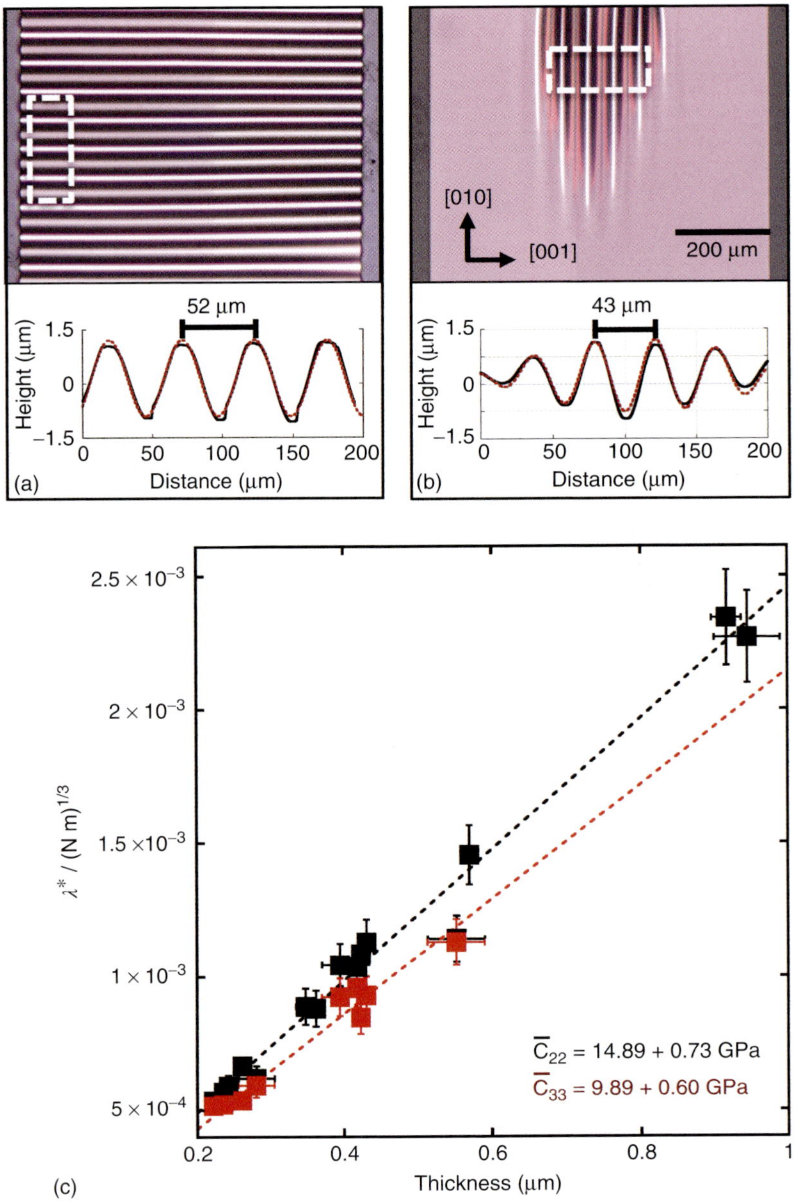

Figure 4.13 Anisotropic wrinkling in rubrene single crystals. (a) Optical micrographs and height profile of wrinkles along the [010] direction (*b*-axis). (b) Optical micrographs and height profile of wrinkles along the [001] direction (*c*-axis). (c) In-plane elastic constants determination using a scaled wrinkled wavelength [34]. (Reproduced with permission. Copyright © 2012, WILEY-VCH Verlag GmbH & Co. KGaA, Weinheim.)

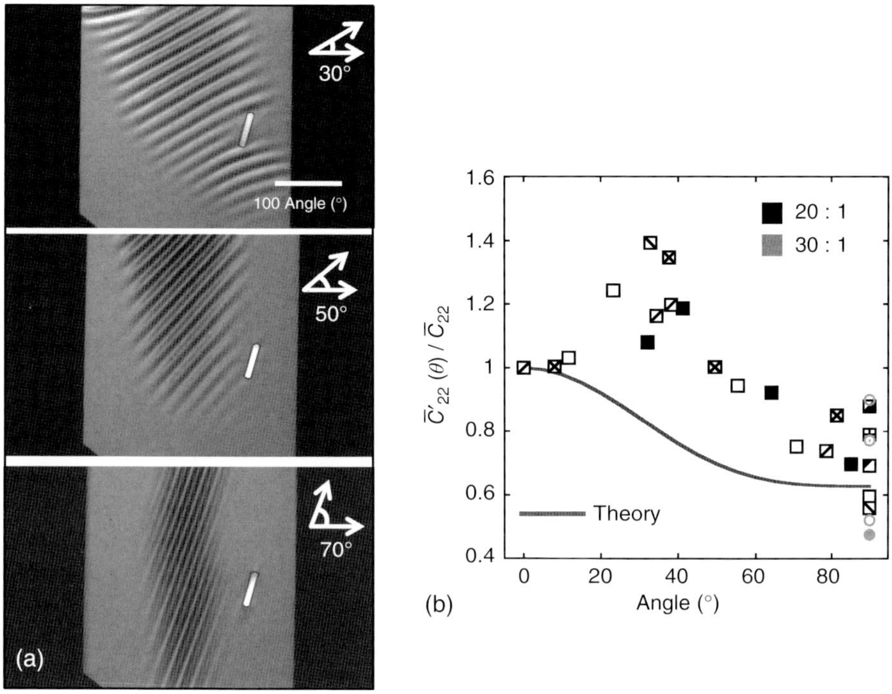

Figure 4.14 Anisotropic wrinkling wavelength. (a) Optical micrographs of a wrinkled rubrene crystal at different load angles with respect to the [010] direction. (b) Normalized angular dependence of in-plane stiffness [34]. (Reproduced with permission. Copyright © 2012, WILEY-VCH Verlag GmbH & Co. KGaA, Weinheim.)

axis ([010] direction) might suggest nonlinear correlations between mechanical and electrical properties.

The wrinkling techniques introduced in this chapter are currently being utilized to characterize the mechanical properties of other important materials in the organic electronics field.

4.5 Strategies for Flexible Organic Single-Crystal Device Fabrication

It is clear that fabricating single-crystal transistors is virtually the only means by which to measure the intrinsic mobilities of organic semiconductor surfaces, and the search for facile methods of producing these devices is an area of intense research. One of the more apparent methods is the so-called pick-and-place method, which involves growing single crystals by PVT or solution methods and electrostatically "bonding" the crystal to the other components of the device (dielectric, source, drain, and gate) fabricated separately [2]. Unfortunately, this is a time-consuming method which requires a lengthy crystallization period and manual placement of fragile crystals. Devices fabricated by this method may also exhibit lower performance because of the imperfect contact between the semiconductor and the dielectric due to surface defects of the crystal. The use of parylene as a dielectric has been used to overcome this obstacle; it can be directly deposited onto organic single crystals without damaging them, creating a conformal surface [26]. Here we present some of the successful approaches taken by some researchers in fabricating organic single-crystal devices on plastic substrates.

4.5.1 Discrete Ultrathin Single-Crystal Transistor

The 2006 study published by Briseno *et al.* [25] on flexible organic single crystal transistors was the first demonstration of the use of macroscopic organic single crystals in high-performance flexible electronics. Figure 4.15a illustrates the schematic of the flexible single-crystal devices fabricated on a polyimide substrate. The crystals are laminated onto the substrates by using the mentioned "pick-and-place" technique. Output and transfer curves can be seen on the plots in Figure 4.15b–d, respectively, indicating ideal transistor behavior. The authors measured the transfer characteristics at different bending radii, as can be seen in Figure 4.15e. For smaller bending radii, a decrease in mobility is observed (Figure 4.15f); the measured mobility after releasing the strain is restored to over 90% of the original "non-bent" position. The authors hypothesize that the decrease in mobility can be attributed to strain-induced defects on the dielectric–crystal interface. This study is a fundamental proof of concept; there is more to learn about the effects of strain on the crystal and how this affects its performance.

4.5.2 Transistor Arrays Based on Micropatterned Single Crystals

The patterning of single crystals is essential for fabricating high-throughput devices [7]. Because of the sensitivity of organic materials to damage from radiation, traditional patterning techniques used for inorganic semiconductors are unsuitable. Furthermore, the transfer of crystals after fabrication is problematic because of their fragility. One method of fabrication large-area arrays of patterned single crystals was recently reported (Figure 4.16). A PDMS stamp was used to pattern a SiO_2 or plastic substrate with "inked" octadecyltriethoxysilane (OTS), causing the crystallization of organic material to selectively nucleate on the nanoscopically rough patterned areas during the PVT growth process [7, 39, 40].

Arrays of transistors were fabricated on flexible substrates using the aforementioned OTS pattering technique [7]. Figure 4.17a shows the structure of the patterned single-crystal transistors on a polyimide substrate. In the same figure, an optical photograph of patterned rubrene single-crystal devices attached to a vial with a diameter of 11.7 mm can be seen, while the rightmost figure is a SEM image of the patterned rubrene crystals. The output and transfer characteristics

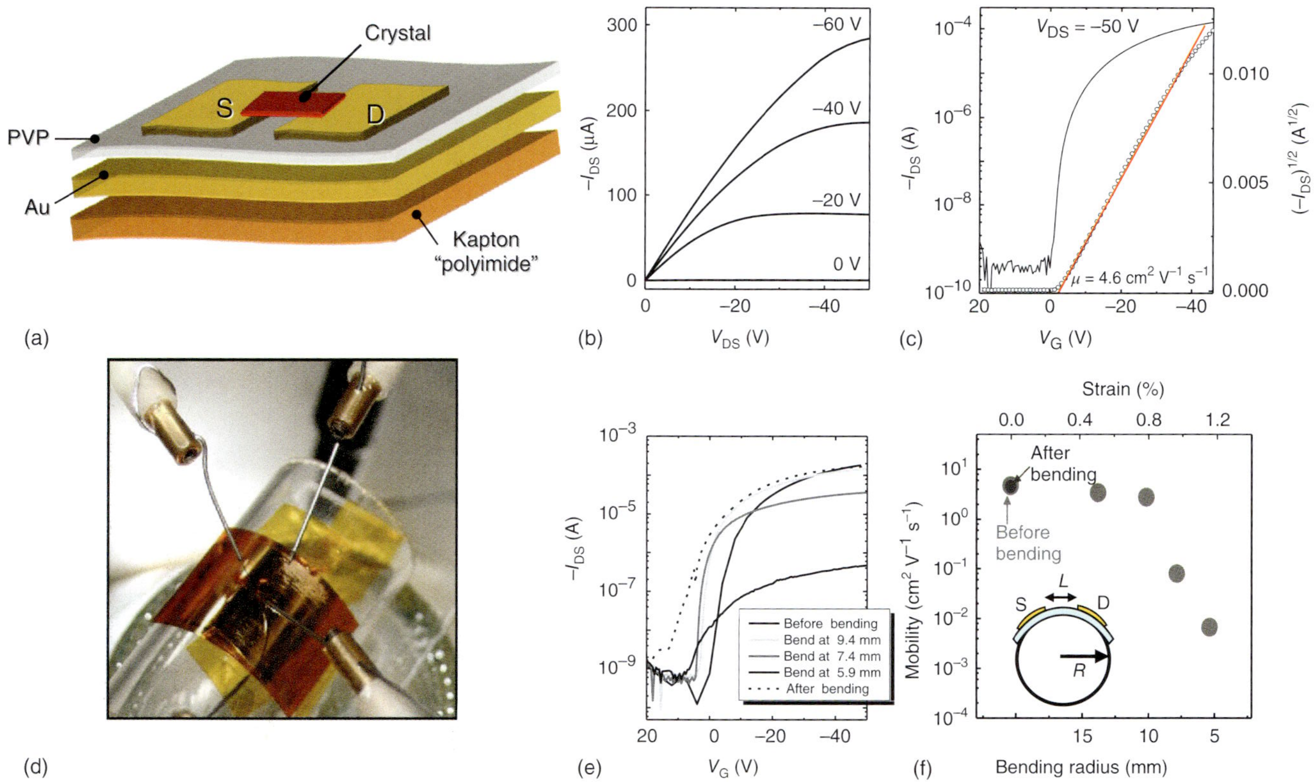

Figure 4.15 Organic single-crystal field-effect transistors on flexible substrates. (a) Bent ultra-thin rubrene crystal. (b-c) Output and transfer characteristics of the thin rubrene FET. (d) A digital photograph of a flexible polyimide single-crystal device as it is measured on a curved surface. (e) Overlay of transfer curves at different bending radii. (f) Field-effect mobility as a function of bending radius. All bending measurements were performed on substrates bent across the length (L) of the channel as shown in the inset [25]. (Reproduced with permission. Copyright © 2006, WILEY-VCH Verlag GmbH & Co. KGaA, Weinheim.)

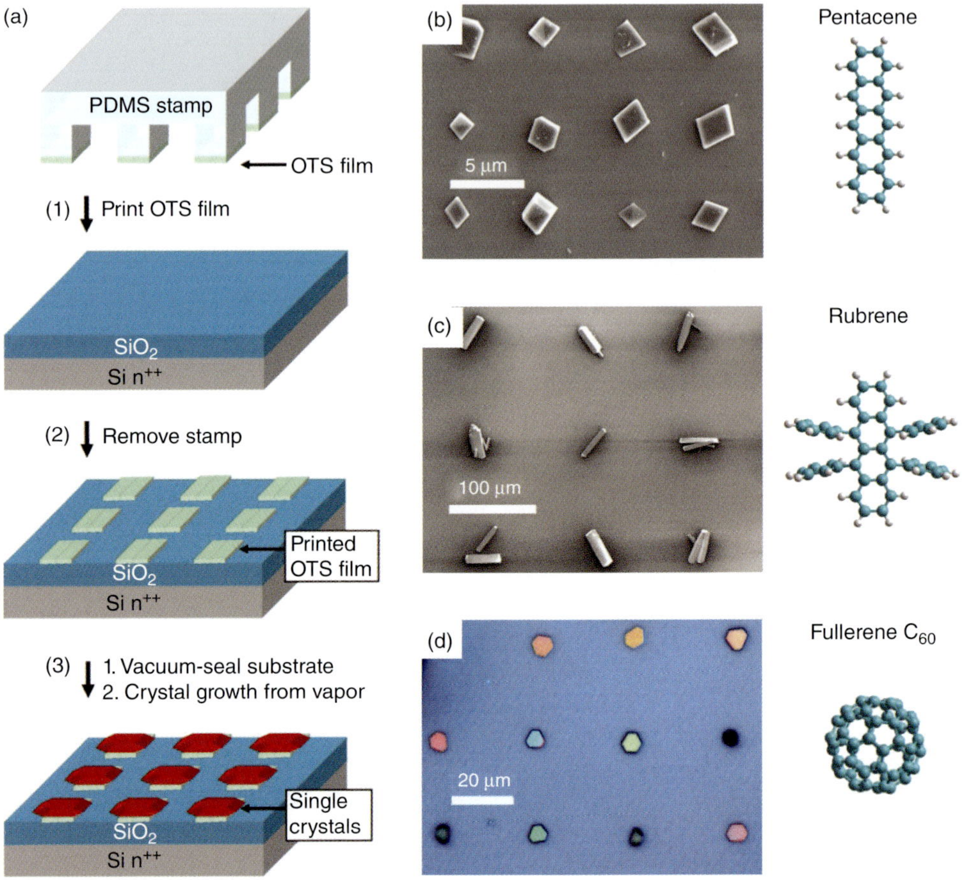

Figure 4.16 Patterning of organic single crystals. (a) Schematic of the procedure used to grow organic single crystals on substrates that have been patterned with OTS using a PDMS stamp. To grow the patterned single crystals, the patterned substrate was placed in a physical vapor transport furnace with the organic source material under vacuum (0.38 mmHg). (b–d) Patterned single-crystal arrays of different semiconductor materials [7]. (Reprinted by permission. Copyright © 2006, Macmillan Publishers Ltd.)

of a plastic rubrene transistor before and after bending (bending direction across the channel) are presented in Figure 4.17b. A mobility of 0.9 $cm^2\ V^{-1}\ s^{-1}$, current on/off ratio of 10^4, and a threshold voltage ~1.5 V were measured in this flexible rubrene transistor. Patterned pentacene transistors fabricated in a similar way exhibited mobilities of 0.1 $cm^2\ V^{-1}\ s^{-1}$ with on/off ratios of $\sim 10^3$. As can be seen in Figure 4.17b, no significant loss in performance was observed when the devices were bent to a radius of 6 mm. In addition to rubrene and pentacene, high-vapor-pressure materials such as anthracene and tetracene were successfully patterned on flexible and transparent polyethylene terephthalate (PET) substrates without degradation/decomposition of the plastic.

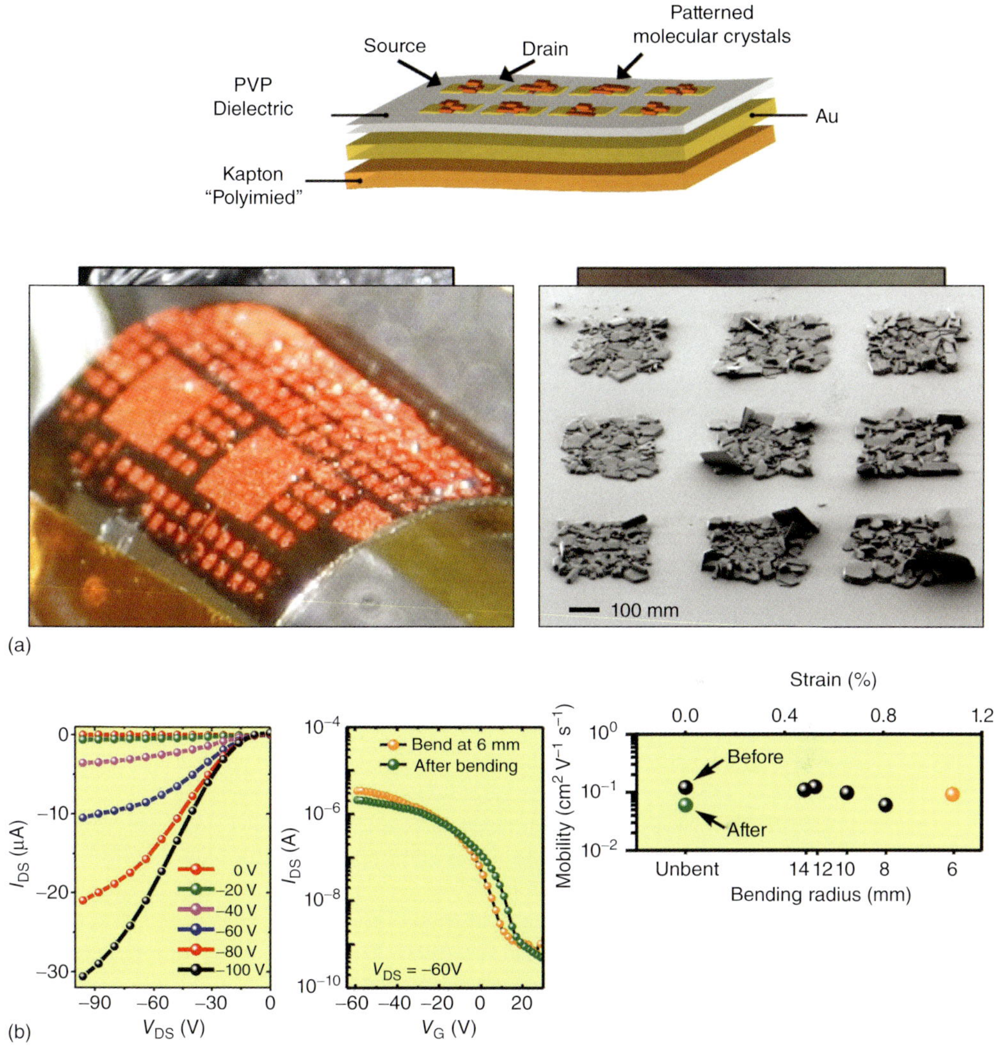

Figure 4.17 Micropaterned arrays of organic single-crystal transistors on flexible substrates. (a) Structure of micropatterned arrays of organic single-crystal transistors on flexible substrates. A digital photograph shows the transistor array bent on a cylinder. (b) Output and transfer curves. The transfer curve shows a slight difference in drain current on the saturation regime between the bent and after bending measurements. A plot of the measured mobilities as a function of bending radius shows differences in performance [7]. (Reprinted by permission. Copyright © 2006, Macmillan Publishers Ltd.)

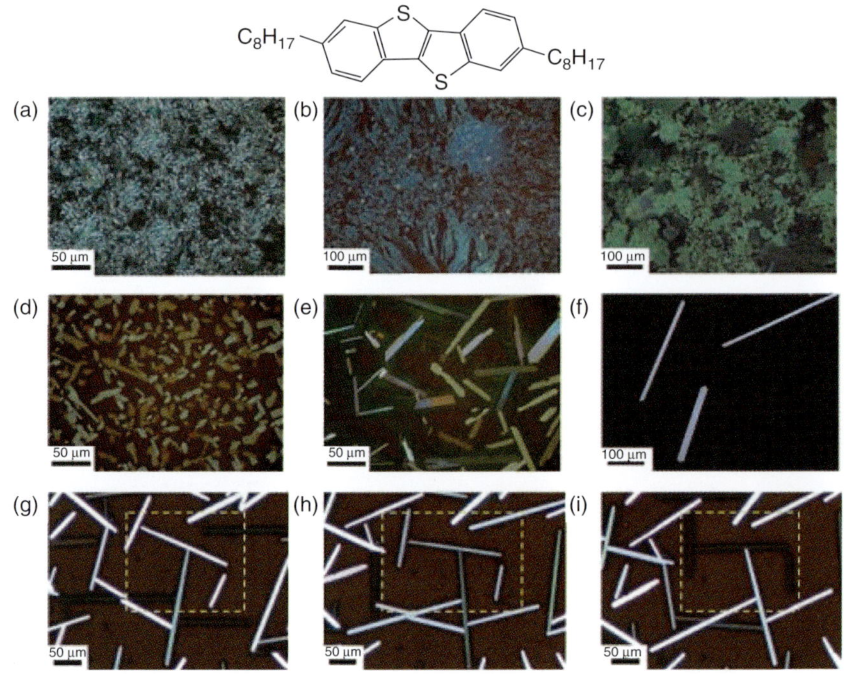

Figure 4.18 (a–c) Optical images of C8-BTBT thin films on SiO_2, poly(methyl methacrylate (PMMA), and as a blend of C8-BTBT and PMMA, respectively. (d,f) The same films after solvent vapor annealing. (g–i) The crystals as viewed through crossed polarizers, with a 15° rotation counterclockwise per image. Note that visible crystals in the dashed box of (g) are barely visible in (i) [42]. (Reproduced with permission. Copyright © 2011, WILEY-VCH Verlag GmbH & Co. KGaA, Weinheim.)

Another approach to patterning is to employ solution-processable materials such as 2,9-alkyl-dinaphtho[2,3-*b*:2′,3′-*f*]thieno[3,2-*b*]thiophene (C_{10}-DNTT). This material can be patterned from solution via the controlled evaporation of the solvent directly onto surface patterned domains [41]. Because of its structure and the resulting intermolecular interactions, this material inherently forms nanowires during solvent evaporation at relatively low temperatures, allowing it to be processed on flexible substrates that are unsuitable for direct physical vapor deposition. Another material that inherently forms nanowires via solvent evaporation is C_8-BTBT [42] (Figure 4.18).

Yet another technique to pattern single crystals, also from solution, was reported by Minemawari and coworkers [8]. Figure 4.19a shows a schematic of their method which integrates the use of inkjet printing technology and an "antisolvent" crystallization technique as a way to pattern organic single crystals over large areas. A solution of C_8-BTBT in dichlorobenzene is printed onto predeposited patterned domains of dimethylformamide, the "antisolvent." Crystallization occurs at the liquid–air interface, creating single-crystal thin films of C_8-BTBT as shown in Figure 4.19b. They used this technique to make

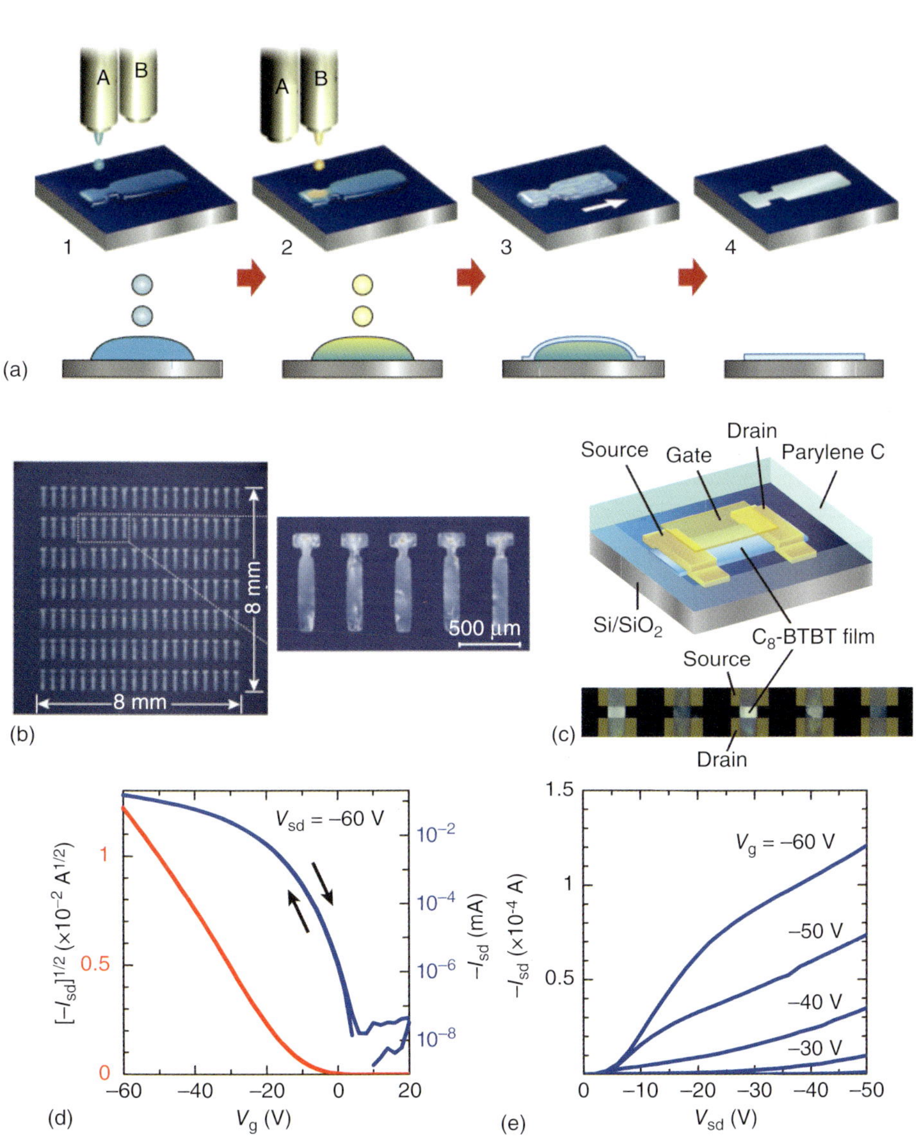

Figure 4.19 Inkjet printing of C8-BTBT single crystal. (a) Schematic of the process; ink (A) denotes the solvent while (B) indicates the solution. (b) Micrographs of an array of printed single-crystal thin films. (c) Device structure and micrograph of transistors. (d) Transfer characteristics. (e) Output characteristics [8]. (Reprinted by permission. Copyright © 2011, Macmillan Publishers Ltd.)

large arrays of single-crystal transistors (Figure 4.19c–e). This process has not yet been demonstrated on flexible substrates but can easily be adapted. The "filtration and transfer" method can also be used to align from solution [15]. This method, involving filtration through a PDMS mask on porous anodized aluminum oxide, produced aligned HTP nanowires.

4.5.3 Flexible Single-Crystal Nanowire Devices

One way of increasing single-crystal device throughput is by using single-crystal nanowires grown from solution to fabricate arrays of transistors [43, 44]. Earlier in this chapter we explored some of the properties HTP nanowires. Briseno *et al.* reported the fabrication and characterization of discrete HTP nanowire transistors on mechanically flexible substrates. Figure 4.20a illustrates the device configuration of a nanowire across source–drain electrodes on a plastic substrate. The goal of their study was to evaluate their potential for applications in flexible electronics and to determine how strain affects the performance of an individual nanowire (Figure 4.20b). Dispersions of HTP nanowires in ethanol were drop-cast onto plastic devices and allowed to dry in a vacuum chamber for several hours. Devices containing only one discrete nanowire were tested. Output and transfer characteristics exhibiting transistor behavior can be seen in Figure 4.20d,e. The highest mobility obtained was $0.19\,cm^2\,V^{-1}\,s^{-1}$, while the average mobility over six devices was $0.08 \pm 0.06\,cm^2\,V^{-1}\,s^{-1}$. In order to investigate the mechanical versatility of HTP nanowires, devices were subjected to measurements under bending (Figure 4.20c). All experiments were conducted on substrates bent across the channel length. The devices were first measured in the planar geometry, followed by compressive strain, tensile strain, and, finally, back to the original planar geometry. No significant changes in mobility were observed in any of the measured devices during or after the strains were applied (Figure 4.20h). Figure 4.20f,g shows the overlays of transfer curves of a single-wire device subjected to different bending radii.

The unaffected electrical performance under the effect of flexing can be explained by the absorption of the stress by the substrate, which reduces the stress experienced by the single crystal [45, 46]. The stress is concentrated on the polyimide substrate because it is considerably thicker than the combined dielectric and crystal thicknesses. Moreover, the effective composite Young's modulus of this multilayer device is not significantly different from most of the materials used. These results demonstrate the consistency of the performance of single-crystal nanowire transistors under mechanical stress.

A recent paper that promises to influence the field of organic flexible electronics in the near feature comes from Hu and coworkers. They demonstrated the flexibility of a single copper phthalocyanine (CuPc) nanowire [47]. In Figure 4.21a–d, a series of SEM images demonstrates the bendability of a nanowire. It has also recently been demonstrated that single-crystal p–n nanowire junctions can be fabricated by PVT [48] (Figure 4.21e–f). Bending experiments of p–n nanowire

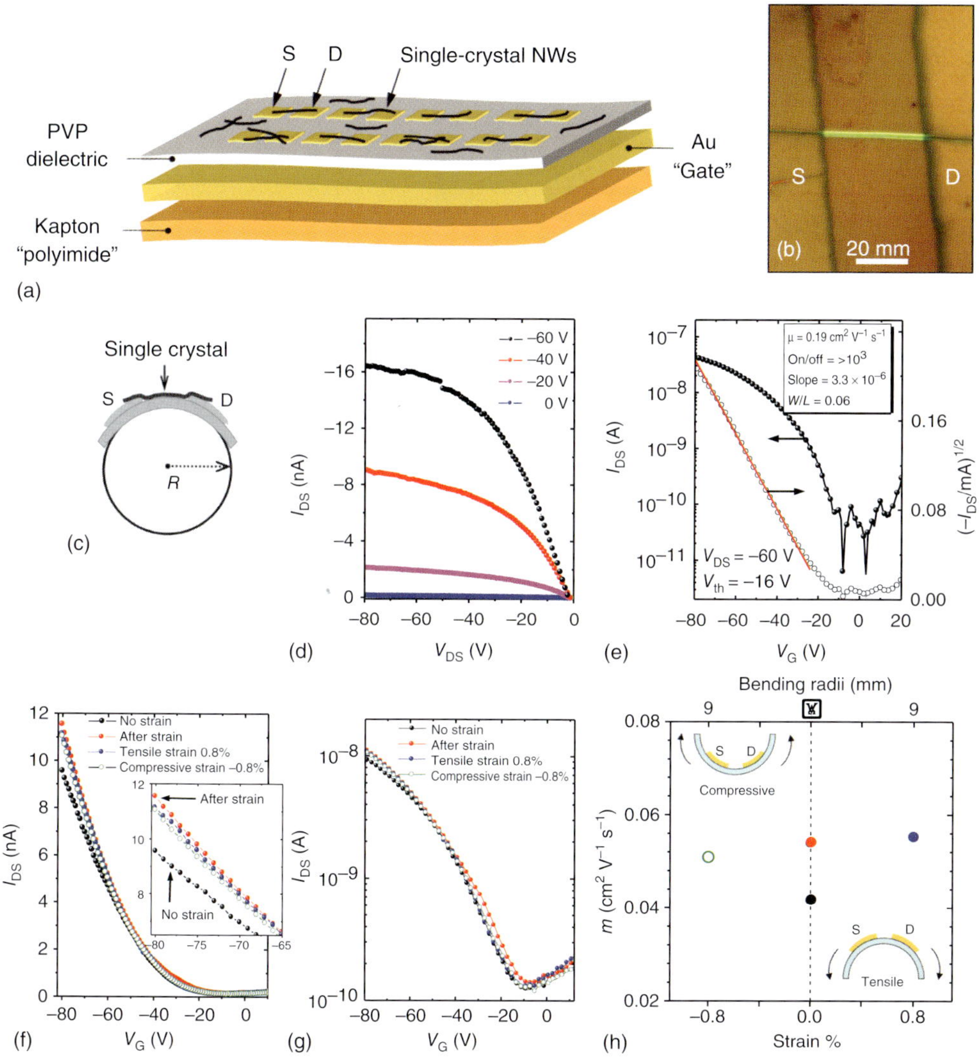

Figure 4.20 Flexible nanowire field-effect transistor. (a) Structure of flexible single-crystal nanowire field effect transistor. (b,c) Output and transfer curve for nanowire transistor on its "planar" configuration. (d) Optical micrograph on a single nanowire transistor. (e–g) Plots illustrating different device performances at different levels of strain. (h) Field-effect mobility as a function of applied strain [43]. (Reprinted with permission. Copyright © 2007, American Chemical Society.)

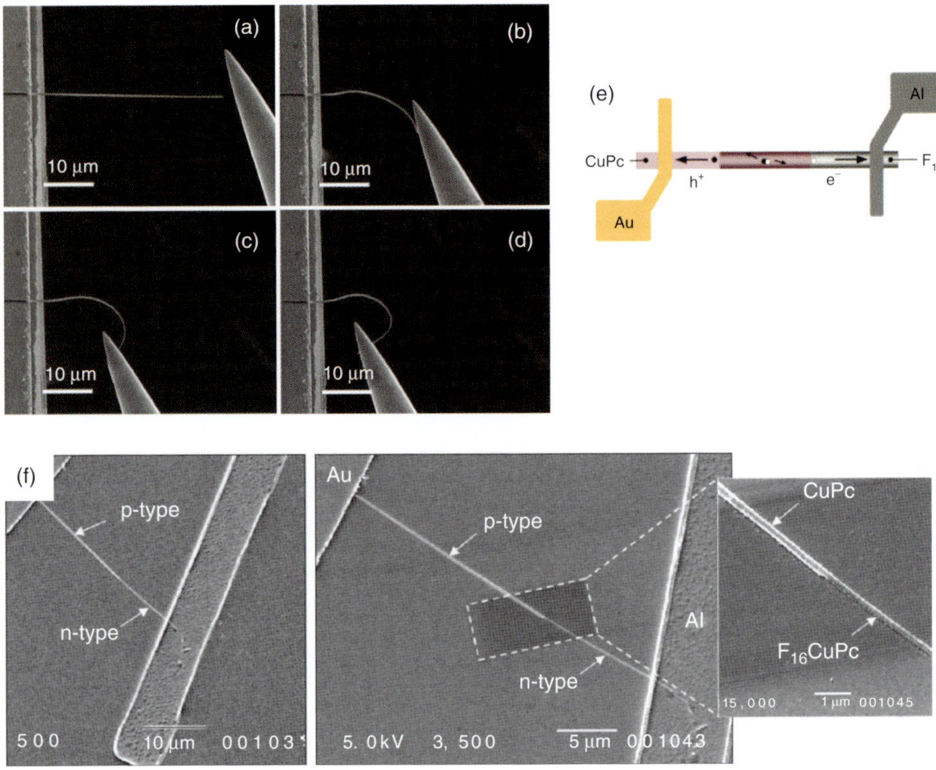

Figure 4.21 Flexible single-crystal nanowire. (a–d) SEM images by Hu, Bjornholm and coworkers, showing the bending of a single CuPc nanowire (400 nm wide) by a mechanical micromanipulator [47]. (Reprinted with permission. Copyright © 2011, WILEY-VCH Verlag GmbH & Co. KGaA, Weinheim.) (e) Schematic of F16CuPc/CuPc single-crystal p–n junctions. (f) SEM images of F16CuPc/CuPc single-crystal p–n junctions [48]. (Reprinted with permission. Copyright © 2010, American Chemical Society.)

junctions would likely expand the knowledge of photo-induced transport in semiconductors. This proposed investigation has the possibility of opening a new exciting realm of fundamental excitonic studies on the nanoscale.

4.6 Conclusions

The outstanding electrical properties of organic single crystals make them ideal candidates for consumer applications such as flexible drivers for displays, sensor arrays, and photovoltaic devices. Nevertheless, there is still a strong need to advance the understanding of the processing and patterning of organic single-crystal semiconductors in order to scale-up their production. Recent advances indicate that the processing of organic single crystals from solution is the most promising route for the large-scale production of single-crystal devices. For example, the work by Minemawari and coworkers [8] represents an important step in processability, demonstrating that high-quality single-crystal thin films can be printed over large areas. In 2013, Bao and coworkers [10] reported a modified blade coating technique, giving unique orientation control of single-crystalline domains of TIPS-pentacene. Despite these advances, we believe that the control of crystal orientation in deposited films will continue to attract significant research efforts.

In addition to processing, flexible electronics requires a fundamental understanding of the mechanical properties of organic single crystals and the effects of mechanical deformation on their electrical properties. Mechanical studies will allow scientists and engineers to take advantage of the exceptional electronic properties of organic crystals in applications where both flexibility and exceptional performance are required. The studies highlighted in this chapter demonstrate that organic single crystals will continue to pave the way for the next generation of high-performance flexible electronic technologies.

Acknowledgments

MAR acknowledges NEAGEP and the National Science Foundation for supporting this work (DMR-1112455). We acknowledge the NSF NSEC, Center for Hierarchical Manufacturing, which provided support for NSC (CMMI-0531171). ALB acknowledges support by the Office of Naval Research (N000141110636). We also thank Prof. A. J. Crosby and Prof. J. J. Watkins for fruitful conversations and feedback.

References

1. Sundar, V.C., Zaumseil, J., Podzorov, V., Menard, E., Robert, L., Someya, T., Gershenson, M.E., and Rogers, J.A. (2004) *Science*, **303**, 1644–1646.
2. De Boer, R., Gershenson, M., Morpurgo, A., and Podzorov, V. (2004) *Phys. Status Solidi A*, **201**, 1302–1331.
3. Gershenson, M.E. and Podzorov, V. (2006) *Rev. Mod. Phys.*, **78**, 973–989.
4. Sun, Y. and Rogers, J.A. (2004) *Nano Lett.*, **4**, 1953–1959.
5. Someya, T., Kato, Y., Sekitani, T., Iba, S., Noguchi, Y., Murase, Y., Kawaguchi, H., and Sakurai, T. (2005) *Proc. Natl. Acad. Sci. U.S.A.*, **102**, 12321–12325.
6. Cao, Q., Hur, S.-H., Zhu, Z.-T., Sun, Y.G., Wang, C.-J., Meitl, M.A., Shim, M., and Rogers, J.A. (2006) *Adv. Mater.*, **18**, 304–309.
7. Briseno, A.L., Mannsfeld, S.C.B., Ling, M.M., Liu, S., Tseng, R.J., Reese, C., Roberts, M.E., Yang, Y., Wudl, F., and Bao, Z. (2006) *Nature*, **444**, 913–917.
8. Minemawari, H., Yamada, T., Matsui, H., Tsutsumi, J., Haas, S., Chiba, R., Kumai, R., Hasegawa, T. *et al.* (2011) *Nature*, **475**, 364–367.
9. Podzorov, V., Menard, E., Borissov, A., Kiryukhin, V., Rogers, J.A., and Gershenson, M.E. (2004) *Phys. Rev. Lett.*, **93**, 1–4.
10. Diao, Y., Tee, B.C.-K., Giri, G., Xu, J., Kim, D.H., Becerril, H.A., Stoltenberg, R.M., Lee, T.H., Xue, G., Mannsfeld, S.C.B., and Bao, Z. (2013) *Nat. Mater.*, **12**, 665–671.

11. Reese, C., Chung, W.-J., Ling, M., Roberts, M., and Bao, Z. (2006) *Appl. Phys. Lett.*, **89**, 202108.
12. Roberson, L.B., Kowalik, J., Tolbert, L.M., Kloc, C., Zeis, R., Chi, X., Fleming, R., and Wilkins, C. (2005) *J. Am. Chem. Soc.*, **127**, 3069–3075.
13. Molinari, A.S., Alves, H., Chen, Z., Facchetti, A., and Morpurgo, A.F. (2009) *J. Am. Chem. Soc.*, **131**, 2462–2463.
14. Menard, E., Podzorov, V., Hur, S.-H., Gaur, A., Gershenson, M.E., and Rogers, J.A. (2004) *Adv. Mater.*, **16**, 2097–2101.
15. Oh, J.H., Lee, H.W., Mannsfeld, S., Stoltenberg, R.M., Jung, E., Jin, Y.W., Kim, J.M., Yoo, J.B., and Bao, Z. (2009) *Proc. Natl. Acad. Sci. U.S.A.*, **106**, 6065.
16. Moulton, B. and Zaworotko, M.J. (2001) *Chem. Rev.*, **101**, 1629–1658.
17. Kitamura, M. and Arakawa, Y. (2008) *J. Phys. Condens. Matter*, **20**, 184011.
18. Warta, W. and Karl, N. (1985) *Phys. Rev. B: Condens. Matter*, **32**, 1172.
19. Horowitz, G. (2010) *Interfaces in Orfanic Field-Effect Transistors*. In Advances in Polymer Science: Organic Electronics; (eds Meller, G., Grasser, T.), Springer-Verlag Berlin Heidelberg, **223**, 113–153.
20. Anthony, J.E. (2008) *Angew. Chem. Int. Ed.*, **47**, 452–483.
21. Briseno, A.L., Miao, Q., Ling, M.-M., Reese, C., Meng, H., Bao, Z., and Wudl, F. (2006) *J. Am. Chem. Soc.*, **128**, 15576–15577.
22. Drummy, L.F., Miska, P.K., Alberts, D., Lee, N., and Martin, D.C. (2006) *J. Phys. Chem. B*, **110**, 6066–6071.
23. Laudise, R.A., Kloc, C., Simpkins, P.G., and Siegrist, T. (1998) *J. Cryst. Growth*, **187**, 449–454.
24. Reese, C. and Bao, Z. (2007) *Mater. Today*, **10**, 20–27.
25. Briseno, A.L., Tseng, R.J., Ling, M.-M., Falcao, E.H.L., Yang, Y., Wudl, F., and Bao, Z. (2006) *Adv. Mater.*, **18**, 2320–2324.
26. Podzorov, V., Sysoev, S.E., Loginova, E., Pudalov, V.M., and Gershenson, M.E. (2003) *Appl. Phys. Lett.*, **83**, 3504.
27. Bao, Z. and Xia, Y. (2007) *Nano Lett.*, **7** (3), 668–675.
28. Timoshenko, S. (1934) *Theory of Elasticity*, 1st edn, McGraw-Hill, New York and London, pp. 216–227.
29. Song, J., Jiang, H., Huang, Y., and Rogers, J.A. (2009) *J. Vac. Sci. Technol., A*, **27**, 1107.
30. Sun, Y. and Rogers, J.A. (2007) *J. Mater. Chem.*, **17**, 832.
31. Ashby, M.F. (1998) *Proc. R. Soc. London, Ser. A*, **454**, 1301–1321.
32. Tahk, D., Lee, H.H., and Khang, D.-Y. (2009) *Macromolecules*, **42**, 7079–7083.
33. O'Connor, B., Chan, E.P., Chan, C., Conrad, B.R., Richter, L.J., Kline, R.J., Heeney, M., McCulloch, I., Soles, C.L., and DeLongchamp, D.M. (2010) *ACS Nano*, **4**, 7538–7544.
34. Reyes-Martinez, M.A., Ramasubramaniam, A., Briseno, A.L., and Crosby, A.J. (2012) *Adv. Mater.*, **24**, 5548–5552.
35. Cerda, E. and Mahadevan, L. (2003) *Phys. Rev. Lett.*, **90**, 1–5.
36. Jurchescu, O.D., Meetsma, A., and Palstra, T.T.M. (2006) *Acta Crystallogr., Sect. B: Struct. Sci.*, **62**, 330–334.
37. Matsukawa, T., Yoshimura, M., Uchiyama, M., Yamagishi, M., Nakao, A., Takahashi, Y., Takeya, J., Kitaoka, Y., Mori, Y., and Sasaki, T. (2010) *Jpn. J. Appl. Phys.*, **49**, 085502.
38. Li, A., Wen, S.-H., Song, J.-L., and Deng, W.-Q. (2009) *Org. Electron.*, **10**, 1054–1059.
39. Mannsfeld, S.C.B., Briseno, A.L., Liu, S., Reese, C., Roberts, M.E., and Bao, Z. (2007) *Adv. Funct. Mater.*, **17**, 3545–3553.
40. Liu, S., Wang, W.M., Briseno, A.L., Mannsfeld, S.C.B., and Bao, Z. (2009) *Adv. Mater.*, **21**, 1217–1232.
41. Nakayama, K., Hirose, Y., Soeda, J., Yoshizumi, M., Uemura, T., Uno, M., Li, W., Kang, M.J., Yamagishi, M., Okada, Y., Miyazaki, E., Nakazawa, Y., Nakao, A., Takimiya, K., and Takeya, J. (2011) *Adv. Mater.*, **23**, 1626–1629.
42. Liu, C., Minari, T., Lu, X., Kumatani, A., Takimiya, K., and Tsukagoshi, K. (2011) *Adv. Mater.*, **23**, 523–526.
43. Briseno, A.L., Mannsfeld, S.C.B., Lu, X., Xiong, Y., Jenekhe, S.A., Bao, Z., and Xia, Y. (2007) *Nano Lett.*, **7**, 668–675.
44. Briseno, A.L., Mannsfeld, S.C.B., Jenekhe, S.A., Bao, Z., and Xia, Y. (2008) *Mater. Today*, **11**, 38–47.

45. Suo, Z., Ma, E.Y., Gleskova, H., and Wagner, S. (1999) *Appl. Phys. Lett.*, **74**, 1177.
46. Gleskova, H., Wagner, S., and Suo, Z. (1999) *Appl. Phys. Lett.*, **75**, 3011.
47. Tang, Q., Tong, Y., Zheng, Y., He, Y., Zhang, Y., Dong, H., Hu, W., Hassenkam, T., and Bjørnholm, T. (2011) *Small*, **7**, 189–193.
48. Zhang, Y., Dong, H., Tang, Q., Ferdous, S., Liu, F., Mannsfeld, S.C.B., Hu, W., and Briseno, A.L. (2010) *J. Am. Chem. Soc.*, **132**, 11580–11584.

5
Solution-Processable Quantum Dots

Hongbo Li, Vladimir Lesnyak, and Liberato Manna

5.1
Introduction

Nanocrystals (NCs) are a class of materials having sizes typically in the range 1 – 100 nm, at the same time preserving crystal structures of their bulk counterparts [1]. Typically, they are composed of metal, metal oxide, and semiconductor materials, as well as their combination in various heterostructures [2–5]. Significant efforts have been made in the liquid colloidal synthesis of NCs because of their unique size-dependent optical, electronic, and magnetic properties. Owing to their small size, with the related large surface-to-volume ratio, their properties are highly dependent on surface states. Therefore, surface modification of NCs performed by using chemicals with different affinity to surface atoms is a versatile tool to fine-tune their properties. In addition, although solid-state reactions in bulk materials are slow and unfavorable, nanosized particles with a limited reaction zone make fast ion diffusion possible, which allows researchers to tune their chemical composition via chemical reactions such as ion exchange, oxidation, and reduction processes. NCs exhibit excellent colloidal stability in solution, which enables their processing into bulk-like arrays for various practical applications where each single nanoparticle still possesses nanosize properties. This peculiarity of NCs provides many opportunities to fabricate various electronic devices, including solar cells, detectors, light-emitting and laser diodes, field-effect transistors (FETs), and others [6]. Among the wide variety of nanoparticles, the most important for electronic applications are semiconductor NCs, also known as *quantum dots* (*QDs*).

Recent progress toward the colloidal synthesis of QDs has provided a platform that allows researchers to obtain the desired materials in their nanoparticulate form to accommodate various applications. A general description of the colloidal synthesis of NCs in liquid solution can be interpreted as a chemical reaction of molecular precursors in a mixture of high-boiling-point organic solvents [7] or in aqueous media [8] in the presence of organic molecules as stabilizing ligands to control the nucleation and growth processes, as schematically shown in Figure 5.1. Developments in the field of synthesis can be classified into several

Large Area and Flexible Electronics, First Edition. Edited by Mario Caironi and Yong-Young Noh.

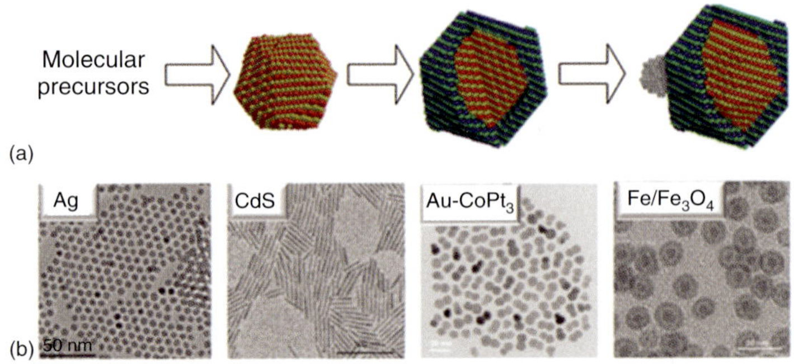

Figure 5.1 (a) Nanomaterial synthesis often starts with simple molecular precursors that undergo chemical transformation into nanoparticles which can be further transformed into core–shell and even complex nano-heterostructures. (b) Several examples of nanomaterials with different shapes and compositions [9]. (Reproduced with permission. Copyright © 2013, The Royal Society of Chemistry.)

main directions: optimization of existing synthetic protocols toward the control of the size, shape, and crystal structure, the development of new approaches, and syntheses of novel materials including multifunctional heterostructures with advanced properties.

5.2
Optimization of the Colloidal Synthesis of Quantum Dots by Selection of Suitable Solvents, Ligands, and Precursors

The synthesis of QDs relies on the chemical reaction between cation and anion precursors to form solid materials under certain conditions [7]. This process can be influenced by many factors, such as the solvent, ligands, temperature and most importantly molecular precursors [10]. Therefore, it is essential to choose appropriate compounds that can be used in the form of stable stock solutions. The first requirement for precursors is to provide sufficient chemical reactivity. By varying the types of precursors, one can control the crystal phase, size range, and, consequently, the optical and electronic properties of QDs. Taking CdSe as an example, we will illustrate the evolution of the QD synthesis, especially focusing on the recent developments in the choice of the precursors.

At the very beginning stage of the colloidal synthesis of QDs, organometallic compounds for both the cation and the anion of the target QDs were employed as precursors, because these compounds are widely used for the fabrication of the corresponding films using techniques such as metal-organic chemical vapor deposition [11]. However, this approach was very sophisticated and laborious and could be performed only in a few laboratories. The first successful synthesis of CdSe QDs with a narrow size distribution (below 5%) was based on the injection of dimethylcadmium ($CdMe_2$) as a Cd precursor along with Se powder dissolved in tri-*n*-octylphosphine (TOP) into hot tri-*n*-octylphosphine oxide (TOPO) used as

Table 5.1 Common cation and anion precursors used for the syntheses of various QDs.

Cation precursor	$CdMe_2$ [12, 14, 16]	CdO [13], ZnO [17–19, 27], PbO [22, 28, 29]	$Cd(Zn)(AcO)_2$ [13], $Cd(Zn, Hg)Cl_2$ [8], $Cd(Zn, Hg)(ClO_4)_2$ [8]		
Anion precursor	$S(Me_3Si)_2$ [29], $Se(Me_3Si)_2$ [28]	TOP-S (-Se, -Te) [13, 30] S(Se)-ODE [33, 34]	Na_2S [8, 31], NaHSe [8], NaHTe [8, 32, 36, 37]	Elemental S, Se, Te [17, 38–42]	SeO_2 [43]

a solvent [12]. It is important to stress that all manipulations involving the $CdMe_2$ precursor need to be done very carefully, since it is a highly pyrophoric substance and sensitive to moisture, in addition to being highly toxic and volatile. A landmark development in the synthesis was represented by the discovery of alternative Cd precursors which were easier to handle than $CdMe_2$ [13, 14]. This sort of evolution in the precursors has been summarized in a recent review [15]. Currently, $CdMe_2$ is not a frequently employed precursor in the synthesis of NCs. Its uses are currently limited to the growth of a CdS or CdSe shell on presynthesized NCs, at relatively low temperatures [14, 16]. To date, the most extensively employed cadmium precursor is CdO, which can be dissolved in standard surfactants such as oleic acid or *n*-octadecylphosphonic acid (ODPA) via chemical reaction. CdO reacts with these acids, and water is released and pumped off the reaction flask and the corresponding Cd-carboxylate or Cd-phosphonate complexes are formed. Other options for Cd precursors, summarized in Table 5.1, are cadmium choride ($CdCl_2$), and acetate $Cd(AcO)_2$, which can be dissolved with the assistance of long-chain alkylamines and alkylphosphines as ligands. This approach is also applicable to other materials, such as ZnO for the synthesis of zinc chalcogenide QDs [17–19]; or PbO for lead chalcogenide nanoparticles [20–22], indium chloride ($InCl_3$), and acetate ($In(AcO)_3$) for the synthesis of indium-based QDs such as InP [23, 24], and ternary copper indium selenide [25, 26].

By analogy with cations, several options for chalcogen sources are available for the synthesis of QDs (see Table 5.1): bis(trimethylsilyl) chalcogenides (Me_3Si-A-$SiMe_3$, A = S/Se/Te), complexes of the chalcogen with TOP (TOP-S/Se/Te), solutions of S and Se in 1-octadecene (ODE) or in oleylamine, Na_2S, NaHSe(Te) (produced by the reduction of corresponding chalcogens by $NaBH_4$), as well as in the form of elemental powders. Among the various precursors, bis(trimethylsilyl) chalcogenides can be employed at relatively low temperatures because of their high reactivity toward different cation precursors. However, bis(trimethylsilyl) chalcogenides are not used very often because of their high cost and unpleasant odor, even under proper operation. Therefore, their application is confined to specific purposes, for example, in the synthesis of ultrasmall PbSe QDs [28]. Among the different phosphines, TOP, a ternary phosphine, is most widely used, owing to its strong complexing ability, which allows the dispersion of the chalcogenide powder at quite high concentration (an essential prerequisite for the large-scale synthesis of NCs). The resulting stable TOP-chalcogen solution can be stored for a long period, without affecting the quality of the synthesized QDs. Recent studies have indicated that, when the chalcogen is bound to a secondary phosphine, its reactivity is higher than when bound to a tertiary phosphine such as TOP. It was found that this increased reactivity ensures higher yield of NCs, improves the synthetic reproducibility of NCs and, in addition, influences their nucleation and shape evolution [44–46].

On the other hand, the chalcogen can be used in other forms without phosphines [35, 47]. The exclusion of phosphines reduces the cost and simplifies the synthesis procedure; for example, in some cases the synthesis can be performed under air without the need of a Schlenk line (which is used to carry out reactions under inert atmosphere). It was found that sulfur and selenium powders are soluble in ODE or oleylamine at high temperatures. This kind of chalcogen precursors is proven to exhibit high reactivity [34]. Alternatively, Se power can be dissolved in a mixture of oleylamine, ODE, and 1-dodecanethiol (DDT), as a new Se precursor, which is used for the synthesis of CdSe as well as quaternary $Cu_2ZnSnSe_4$ and $Cu(InGa)Se_2$ NCs because of its high reactivity [48]. One more example from Zeng *et al.* shows that a highly reactive Se precursor could be prepared by dissolving Se powder in oleylamine in the presence of $NaBH_4$. This precursor was then used successfully for the synthesis of Mn-doped ZnSe/ZnS core/shell QDs [49]. Although it is believed that a homogenous solution is essential to guarantee the production of high-quality QDs, in many cases it has been demonstrated that one can use directly a precursor in the form of a powder for the synthesis of QDs and disks [39, 40, 50].

5.3
Large-Scale Synthesis of Quantum Dots

One critical goal in the synthetic chemistry of QDs, with regard to their applications, is the development of techniques to efficiently manufacture materials in large quantities (from grams to kilograms), at the same time preserving their high quality. From this perspective, upscaling of the hot-injection synthesis faces challenges because of the inhomogeneous nucleation in a large reaction volume as well as owing to the poor temperature regulation of the reaction mixture, where rapid termination of the growth is essential for the control of the QD sizes and consequently their optoelectronic properties. To address this issue, methods such as the heat-up approach or the synthesis in a continuous flow reactor were exploited for the preparation of NCs. These approaches provide the opportunity to realize the large-scale production of QDs and thus pave the way to their commercialization.

The *heat-up approach* involves the mixing of precursors and ligands in a solvent followed by the controlled ramp rate of temperature to induce the reaction, thereby avoiding precursor injection [29, 39, 51–54]. Although in this case the growth of QDs is slower as compared to hot-injection synthesis, this approach allows a more precise control of the reaction kinetics, thereby making the synthesis more reproducible. Thus, the nucleation and growth processes can be well controlled by adjusting the heating rate, which can be easily realized by using commercially available heaters. Recently, Yang *et al.* [39] demonstrated a one-pot non-injection synthesis of CdSe and CdTe QDs, which can be easily upscaled. In a typical procedure, Se powder and cadmium myristate were mixed with oleic acid and ODE in a reactor, and the dispersion obtained was stirred and heated to 240 °C under an inert gas flow for the nucleation and growth of CdSe particles. In this method, the reaction temperature is dependent on the reactivity of the precursors. SeO_2 is another optional Se precursor that is suitable for heat-up synthesis of various metal selenide NCs [43]. Although in this approach the NC growth mechanism is different from the hot-injection method [54] NCs of many different materials have been reported recently, including ZnSe [27], PbS [29], Cu_7S_4 [55], as well as alloyed CdS_xSe_{1-x} NCs [56]. A moderate NC growth rate can be employed for the synthesis of core/shell structures by choosing precursors that can be sequentially stimulated at different temperatures, as in the case of the reported syntheses of CdSe/CdS [57], InP/ZnS [58], and $CuInS_2$/ZnS core/shell QDs [59]. In this respect, special attention is devoted to the development of low-toxicity, air-stable precursors, allowing operation under atmospheric conditions.

The *continuous flow synthesis* is another approach to the large-scale production of QDs. The flow in the QDs synthesis can be realized in different ways, such as, for example, chemical aerosol flow and microfluidic flow. Besides the ability to scale up the synthesis, continuous flow systems facilitate rapid mass and thermal transfer, thereby providing efficient kinetic control. Chemical aerosol flow synthesis is performed as shown in Figure 5.2a [60, 61]. In this system, precursors, ligands, and solvent are mixed with toluene to form a solution, which is then burst into micro-droplets by applying the ultrasonic nebulization technique [62]. These liquid droplets containing the precursor are carried into the hot furnace by inert gas flow. In furnace 1, toluene evaporates from the droplets and leaves a concentrated solution of reactants in the high-boiling-point solvent. In furnace 2, the precursors react inside these isolated droplets and induce the formation of QDs, which are subsequently collected in bubblers filled with toluene. In this method, the growth process is easily controlled by the flow rate and by the temperature of the furnaces, allowing the large-scale synthesis of CdSe and CdTe QDs [60] as well as ternary alloyed CdTeSe and CdTeS NCs [61], and even water-soluble CdTe/CdS core/shell QDs [32, 63]. One more example of the continuous flow system for an in-line synthesis is presented in Figure 5.2 [64–66]. In this system, precursors are mixed and reacted in the channel, which can be stainless steel coils (1 mm in diameter) [67] or silica glass capillaries (150 μm in diameter) [65, 68]. Nucleation and growth of QDs are initiated when the precursors flow through the channel and, as in the case of aerosol flow system, are controlled by adjustment of the flow rate and temperature. Chan *et al.* [64] first demonstrated the application of microfabricated, chip-based reactors for the continuous synthesis of high-quality CdSe NCs. Evaluation of the size distribution of the QDs prepared by the continuous approach yielded values comparable to those of CdSe NCs synthesized by a conventional method. A dual-temperature-stage coil-reactor system was designed to produce

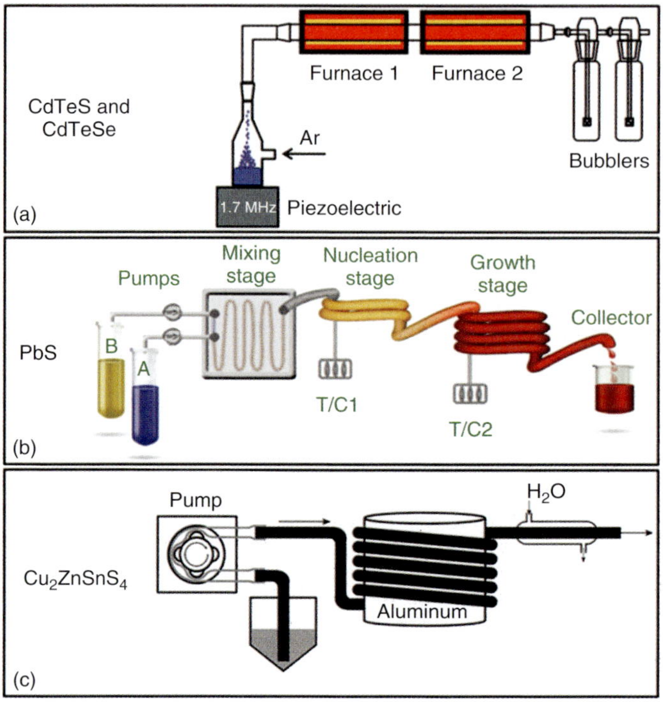

Figure 5.2 Schematics of (from (a) to (c)): chemical aerosol flow apparatus for the synthesis of alloyed CdTeS and CdTeSe QDs. The precursors are carried into the furnace in the form of droplets acting as isolated mircoreactors [61]. (Reproduced with permission. Copyright © 2008, American Chemical Society.) A dual-stage continuous-flow reactor setup for PbS QDs synthesis from precursor A (Pb-oleate in ODE) and precursor B ($S(Me_3Si)_2$ in ODE) [70]. (Reproduced with permission. Copyright © 2013, American Chemical Society.) A flow reactor setup for the production of Cu_2SnZnS_4 NCs [71]. (Reproduced with permission. Copyright © 2012, American Chemical Society.)

high-quality PbS QDs by controlling the nucleation and growth processes, which are carried out separately in different temperature zones (see Figure 5.2b). By using a single molecular precursor, alloyed CdSeS and core/shell CdSe/CdS QDs have been prepared [68]. The in-line synthesis approach has been applied successfully for the synthesis of SnTe nanorods [69] and even quaternary Cu_2SnZnS_4 NCs [71], as shown in Figure 5.2c.

The main issue of the in-line synthesis is associated with the difficulty of mixing precursors due to the high viscosity of the solvent, which is especially difficult in microchannels (or microcapillary reactors). In this respect, the use of hexane as a solvent at high pressure in a continuous microflow synthesis overcomes the inherent limitations of the conventional solvents. As demonstrated by Marre *et al.* [72], the supercritical continuous flow yields a narrower size distribution of NCs and results in better reproducibility. Recently, Toyato *et al.* designed the combinatorial synthesis system composed of several microreactors and an online spectrometer as detector. This combination makes possible the correlation of the reaction conditions in order to achieve optimal optical properties of QDs under various parameters by *in situ* reaction monitoring at as many as 125 points [73]. Such system reliability makes this synthetic approach very attractive for practical industrial applications.

5.4
Surface Chemistry of Quantum Dots

QDs are defined as a condensed crystalline inorganic core covered by a layer of organic surfactant (also known as a *ligand*) bound to the surface sites. The reader may find various aspects of the role of surfactants in several recently published reviews [74–77]. Ligand molecules are incorporated during the colloidal synthesis. Their main function is to impart colloidal stability to the QDs in solution, which facilitates their following handling, assembly, and processing into materials [75]. On the other hand, surfactants are considered as the size- and shape-controlling agents [76]. Nucleation and growth of QDs are directed by ligands that adhere to the growing nuclei, thus establishing a dynamic organic–inorganic interface that regulates the incorporation of active species into the solid NCs [77]. Ligand molecules, by covering the surface of NCs, strongly influence their physicochemical properties and, consequently, the electronic response of a final device based on such NCs. Therefore, in some cases, after the synthesis additional functionalization of the QD surface is needed to accommodate different environments and applications by the replacement of the initial surfactants. As we mentioned in the previous section, the surface capping of the QDs is mainly represented by alkylphosphonates, amines, oleates, and thiols, as schematically shown in Figure 5.3 [80]. Since mainly a mixture of different capping agents is employed in the synthesis, it is necessary to thoroughly investigate surface states of QDs in order to determine the actual species bound to the inorganic core, their density and geometry, type of binding, and so on, using various well-established analytical techniques [78, 79]. Here, we provide some recent insights gained in the surface chemistry of QDs and show the effect of the ligand exchange on their optical and electronic properties, as well as ligand-tailored device performance.

High-quality QDs should possess high values of the photoluminescence (PL) quantum yield and sufficient stability, which are strongly dependent on their surface states, considering that a large portion of constituting atoms are exposed to the environment. These atoms left unbound by ligands detaching from the surface of NCs act as traps for excitons and induce nonradiative pathways, thereby lowering the quantum yields. The surface states of QDs are dependent on the synthesis protocol, method of purification, and the post-synthetic treatment of nanoparticles. Jasieniak and Mulvaney [81] demonstrated that by using a phosphine-free selenium precursor (Se dissolved in ODE) it was possible to manipulate the overall surface stoichiometry of CdSe QDs from a cadmium-dominated surface to a selenium-dominated one. In the latter case, their emission is greatly quenched, suggesting that unpassivated Se atoms can efficiently trap the exciton. These Se

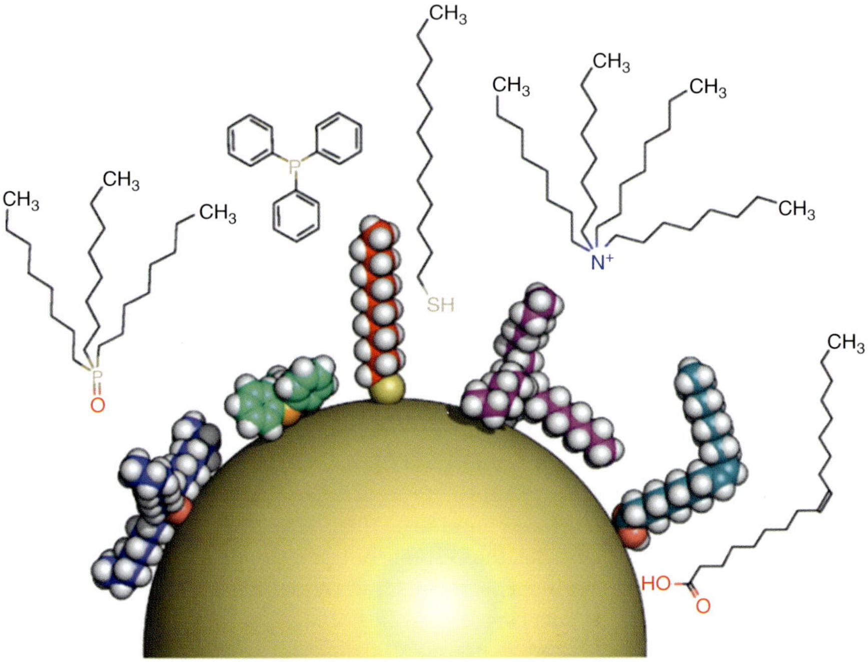

Figure 5.3 A nanoparticle of 5 nm core diameter with different hydrophobic ligand molecules drawn to scale. (Left to right): TOPO, triphenylphosphine, dodecanethiol, tetraoctylammonium bromide, and oleic acid [80]. (Reproduced with permission. Copyright © 2010, The Royal Society.)

atoms could be protected in the presence of tertiary phosphines (e.g., TOP), as proven by a marked enhancement in the PL quantum yield. However, TOP binds specifically to Se instead of Cd, because the enhancement is observed only in the case of QDs having a Se-rich surface. Another recent report has demonstrated that the quantum yield of CdS QDs is determined by their surface states [82]. Wei *et al.* developed a protocol to alternately change the surface state of CdS QDs into a terminating monolayer containing solely Cd or S atoms via a successive ion-layer adsorption and reaction (SILAR) method by using highly reactive diphenylphosphine sulfide as a sulfur precursor. Figure 5.4 displays the absorption and PL spectra evolution during the growth steps. The initial CdS QDs are characterized by completely quenched bandgap emission in the case of S termination (dark state), while their emission recovers after a sequential layer growth by adding cadmium oleate (bright state). Related calculations suggest that the sulfur-terminating layer provides pathways for effective nonradiative recombination and thus is responsible for the PL quenching. A similar tendency has recently been demonstrated by Soni and Sapra [83] for the case of core/shell QDs. They found that the cation-rich surface enhances the emission intensity while the anion-rich surface decreases the fluorescence signal of CdSe/CdS and CdSe/ZnS core/shell QDs. Purification

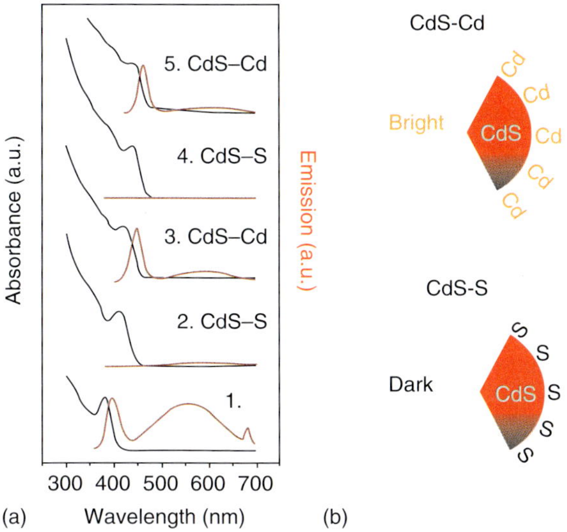

Figure 5.4 (a) Absorption (black lines) and PL (red lines) spectra of as-prepared CdS QDs (step 1) and after sequential additions of S (steps 2 and 4) and Cd monolayers (steps 3 and 5), demonstrating how terminating surface atomic layer, schematically shown on (b) as bright and dark states, affects the optical properties of QDs [82]. (Reproduced with permission. Copyright © 2012, American Chemical Society.)

of QDs from the excess of ligands, unreacted precursors, and byproducts, which is commonly performed immediately after the synthesis, may also strongly impact their optical properties. Thus, as has been shown by Hassinen *et al.* [84], precipitation of PbSe and CdSe QDs by the addition of protic short-chain alcohols as nonsolvents results in the reduction of their quantum yields, indicating that the initial surface ligands could be stripped. On the contrary, the use of aprotic acetonitrile as a nonsolvent does not quench the emission: this provides a method for purification of semiconductor NCs with preservation of their initial optical properties. Post-synthetic treatment of nanoparticles can modify their surface states and therefore impart chemical and photostability. It was found that surface modification of PbSe QDs through the reaction with molecular chlorine induced a thin $PbCl_x$ passivation layer, which could enhance their stability against oxidation [85].

Apart from influencing the optical properties of QDs, surfactant capping has a strong impact on their electronic properties, which is especially important for the fabrication of devices built of QD arrays. For example, Tagliazucchi *et al.* [86] investigated the effect of the length of the ligands on the rates of photoinduced electron transfer (PET), a vital process of intense interest in solar energy conversion, photodetection, and light-emitting technologies. First, to displace the native ligands, CdSe QDs were subjected to the ligand exchange reaction, another technique of post-synthetic treatment, by using mercaptocarboxylic acids HS–$(CH_2)_n$–COOH ($n = 1$, 2, 5, 7, 10). Ultrafast transient absorption measurements revealed that the rate of the PET process is dependent on the length of the acid chain; that is, the rate constant for PET decreases with increasing *n*. Furthermore, when *n* is larger than 10, the formation of bundles of ligands could induce a much slower decrease in the rate of PET, indicating that the molecular-level morphology of ligands plays a role in determining interfacial PET. It is also worth mentioning the emergence of approaches to coat the surface of NCs with short inorganic ligands. This gives the possibility to fabricate NC films with high conductivity owing to the relatively short distances between nanoparticles, as compared to those capped by long-chain organic molecules [87, 88]. Zanella *et al.* [89] developed a method of Cl passivation of a wide variety of NCs by using propyltrichlorosilane to remove the initial surfactants, which also reduces the interparticle separation and therefore greatly increases the film conductivity. A solid-state halide anion treatment of PbS QDs has been realized to improve the performance of photovoltaic cells [90, 91]. In parallel with the PL dependence on the terminating layer on the surface of QDs, as mentioned above, Kim *et al.* [92] demonstrated the same trend in the electronic properties of CdSe nanowires in FETs, where the cadmium-rich surface enhances electron currents while selenide- or sulfide-rich surface suppresses them. The group of Talapin [88, 93] has recently introduced a wide range of inorganic molecular metal chalcogenide surface ligands, such as $Sn_2S_6{}^{4-}$, $Sn_2Se_6{}^{4-}$, $In_2Se_4{}^{2-}$, $HgSe_2{}^{2-}$, $AsS_3{}^{3-}$, $Sb_2S_3{}^{-}$, and $Sb_2Se_3{}^{-}$, which can be converted into semiconducting phases upon moderate heat treatment, generating inorganic NC solids. A band-like electron transport in arrays of CdSe QDs capped with $In_2Se_4{}^{2-}$ has been reported, with measured electron mobilities as high as $16\,cm^2\,V^{-1}\,s^{-1}$, which is about an order of magnitude higher than in the best solution-processed organic and NC-based devices to date [94]. The structure of a FET based on $In_2Se_4{}^{2-}$-capped CdSe QDs and its characteristics are presented in Figure 5.5. By employing CdSe/CdS core/shell nanoparticles with the same capping, photodetectors with normalized detectivity of $D^* > 1 \times 10^{13}$ Jones ($1\,Jones = 1\,cm\,Hz^{1/2}\,W^{-1}$) were fabricated, which again was a record for II–VI QDs. Later on, the same group developed metal-free inorganic surface ligands for colloidal NCs, such as S^{2-}, HS^-, Se^{2-}, HSe^-, Te^{2-}, HTe^-, $TeS_3{}^{2-}$, OH^-, and $NH_2{}^-$ [95].

Applications of QDs in electronics rely on their solution-processing into bulk-like arrays, mostly in the form of films deposited onto substrates, as is schematically shown in Figure 5.6. Various approaches were developed to fabricate NC films using common solution processing techniques including drop casting, spin coating, self-assembly, layer-by-layer deposition, and ink-jet printing. For electronic applications, it is required that a film should be composed of well-packed QDs, which is critical to provide stable and reproducible device performance. In order to improve the conductivity of the resulting assemblies, one can apply chemical or/and thermal treatments to remove or decompose long-chain organic ligands from the surface of QDs [96]. However, it still remains a challenging task to develop efficient approaches capable of eradicating the ligands without changing the surface state and the structure of the inorganic cores [97]. Besides ligands, the unoccupied interparticle space after processing is another factor affecting the conductivity of the assembly. This boosted the interest in understanding how to

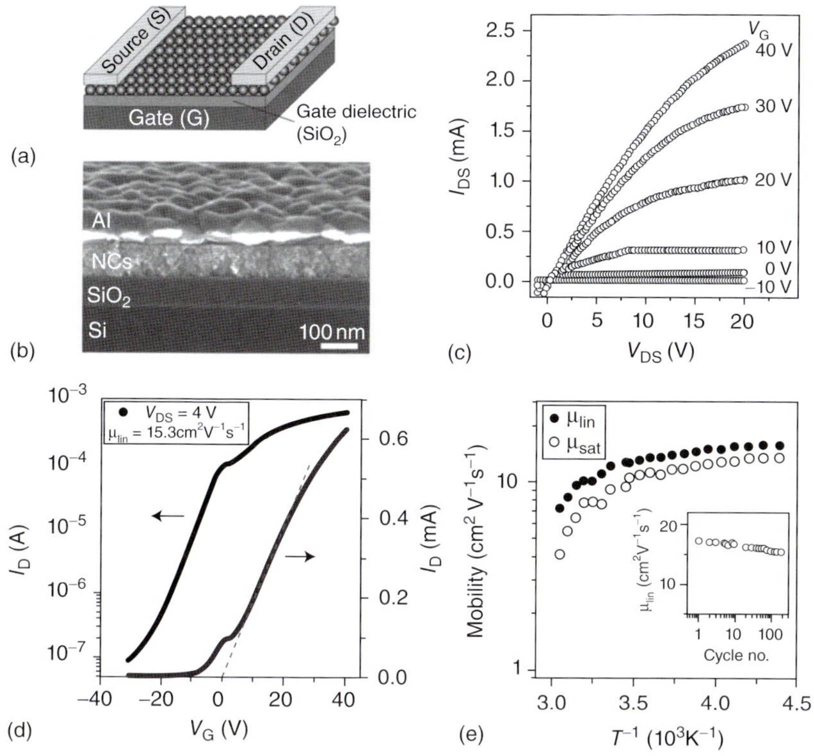

Figure 5.5 (a) Schematic of a FET with the channel assembled from $In_2Se_4^{2-}$-capped 3.9 nm CdSe QDs. (b) SEM image of FET cross section. (c) Plots of drain current I_D versus drain–source voltage V_{DS} as a function of gate voltage V_G. (d) Plots of I_D versus V_G at $V_{DS} = 4$ V used to calculate current modulation and linear-regime field-effect mobility for a n-channel FET. (e) Temperature dependence of field-effect mobility for a n-channel FET. Inset: evolution of field-effect mobility over 200 consecutive cycles of the gate voltage between −30 and +30 V at $V_{DS} = 2$ V. The QD layer was annealed at 200 °C for 30 min ($L = 72$ μm, $W = 954$ μm) [94]. (Reproduced with permission. Copyright © 2011, Nature Publishing Group.)

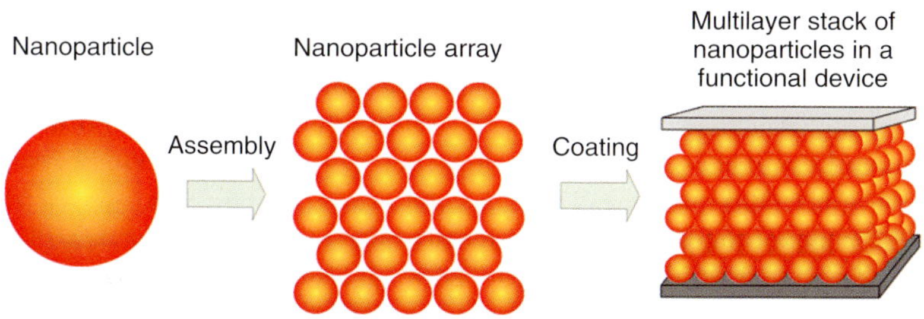

Figure 5.6 A scheme of processing NCs from solutions into 2D nanoparticle arrays on substrates and further into multilayered stacks sandwiched in between two electrodes in a functional device.

pack the NCs with nonspherical shapes and at the same time to make the assembly of NCs with higher volume fraction more attractive [98, 99]. For instance, a combination of multiple types of crystals with different sizes or crystals with unique shapes provides new opportunities to pack them into denser films, as compared to spherical particles of one type [100–104].

5.5
Post-Synthetic Chemical Modification of Nanocrystals

Beside the surface modification of nanoparticles employed to tune their properties, their whole inorganic core can be tailored by post-synthetic chemical treatments such as ion exchange and oxidation and reduction reactions. It is worth mentioning two recent reviews discussing the cation exchange reactions at the nanoscale from various aspects [105, 106], as well as a review introducing new nanomaterials with tunable plasmonic properties [107]. Recent developments show that additional chemical treatments after the synthesis provide a versatile alternative strategy for the fabrication of NCs having tunable composition and structure and exhibiting various photo/electronic characteristics that would be inaccessible by conventional methods [108–111]. Moreover, these modifications can be applied even to already fabricated NC arrays. Thus, for example, oxidation and reduction of copper chalcogenide NCs, a rapidly developing class of nanomaterials, after their deposition in the form of films, result in a controllable variation of their electronic properties, which might be used in energy-related devices and detectors [112, 113]. In this section, we illustrate how NCs can benefit from post-synthetic chemical modification and what its advantages are compared to conventional colloidal synthesis.

In an ionic compound, the sublattice of cations can be partially or totally replaced with a new sublattice of other cations while the anion sublattice remains practically intact owing to the much larger size of the anions with respect to the cations and consequently the much lower mobility of the former. This process is known as *cation exchange*. This kind of a transformation at the nanoscale was first demonstrated by Son *et al.* [114], who converted CdSe NCs to Ag_2Se NCs by exchanging Cd^{2+} with Ag^+ ions, with their subsequent recovery back to CdSe via the reverse reaction. Cation exchange at the nanoscale was found to be extremely fast, as opposed to the macroscale reaction in the bulk solid state, providing many opportunities for modification of nanoparticles [108]. In many studies of cation exchange reactions in NCs, it was found that they can preserve their sizes and the shapes, at the same time their crystal structure could even evolve to a metastable phases (at least for the corresponding bulk materials) [115–117]. Figure 5.7a shows a typical Cd-to-Cu cation exchange reaction performed on CdSe/CdS octapod-shaped NCs with complete conservation of their initial morphology [116]. As seen from the figure, this approach allows the isolation of intermediate hybrid $Cd_xCu_{2-0.5x}Se/Cd_yCu_{2-0.5y}S$ structures by dosing the amount of copper precursor added. The reaction started at the tips of the pods, yielding intermediate heterostructures, with outer sections of the pods

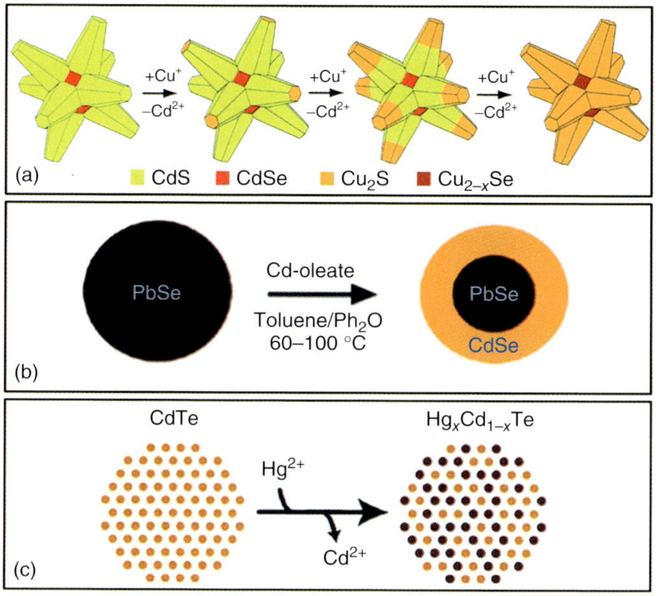

Figure 5.7 (a) Cation exchange reaction from CdSe/CdS to Cu_2Se/Cu_2S octapod NCs demonstrating preservation of the shape of initial particles [116]. (Reproduced with permission. Copyright © 2011, American Chemical Society.) (b) Surface-specific replacement of Pb^{2+} with introduced Cd^{2+} (cadmium oleate) by performing cation exchange reaction on PbSe QDs under low temperature, resulting in the formation of PbSe/CdSe core/shell NCs with enhanced infrared emission [118]. (Reproduced with permission. Copyright © 2008, American Chemical Society.) (c) Incorporation of mercury ions into CdTe QDs leading to the formation of alloyed $Hg_xCd_{1-x}Te$ NCs [119]. (Reproduced with permission. Copyright © 2011, American Chemical Society.)

composed of chalcocite Cu_2S while the inner pod sections were still wurtzite CdS. When the exchange was complete, the pods were transformed entirely to Cu_2S while the central region became cubic Cu_2Se. Depending on the structural compatibility (similarity of the crystal symmetry and lattice constants) of the new forming phase and the initial one, they can form either a heterostructure where the two phases are separated by the interface boundary, as in the cases of octapod conversion shown in Figure 5.7a [116] and $Pb^{2+} \rightarrow Cd^{2+}$ exchange shown in Figure 5.7b [118], or a continuous homogeneous alloyed structure, as demonstrated by a partial $Cd^{2+} \rightarrow Hg^{2+}$ cation exchange in Figure 5.7c [119].

Cation exchange reaction from Pb^{2+} to Cd^{2+} provides an interesting system to investigate its facet dependency, since lead chalcogenide and cadmium chalcogenide are immiscible and, in consequence, a well defined Cd/Pb interface are formed [118, 120]. Moreover, this reaction can be well controlled by regulating the temperature and the time. The incorporation of Cd ions into PbSe QDs by the addition of cadmium oleate yields a layer of CdSe which provides sufficient colloidal stability to the particles [121]. The resulting core/shell

QDs exhibit stable infrared emission with greatly enhanced quantum efficiency. Their emission can be tuned by varying the size of the initial PbSe NCs and by controlling the thickness of the newly formed CdSe layer. We note that core/shell PbSe/CdSe QDs are inaccessible from conventional colloidal synthesis because the reaction temperature required for CdSe shell growth would affect the initial PbSe particles, which would coalesce via Ostwald ripening. While it is possible to grow a PbS shell using conventional approaches, related studies have shown that such a PbS shell does not provide a substantial stability to the NCs [122]. The cation exchange method was applied also to lead chalcogenide NCs with different shapes, including rods, cubes, and star-shaped particles [123]. In the case of PbSe NCs with cubic shapes used as the host particles, exchange proceeds following the specific [111] direction, resulting in the formation of well-defined [111]PbSe/[111]CdSe interfaces. Complete cation exchange allows for the synthesis of PbS nanorods from the corresponding CdS "predecessor." [115] At the same time, a partial $Cd^{2+} \rightarrow Pb^{2+}$ exchange reaction resulted in the formation of interesting heterostructures, made of multiple PbS dots inside the CdS nanorods, which were characterized by near-infrared emission with high quantum yields (45–55%) [124].

As we mentioned earlier, cation exchange reactions can be also used to synthesize alloyed QDs if the pristine host and the incorporating guest materials have similar structures, as shown in Figure 5.7c for the example of CdTe NC hosting HgTe [119, 125, 126]. We note that $Cd_xHg_{1-x}Te$ QDs are inaccessible by direct hot-injection synthesis because of the lack of appropriate Hg precursors, although their direct synthesis in aqueous media is quite well developed [8]. Instead, the $Cd^{2+} \rightarrow Hg^{2+}$ exchange reaction ensured the formation of $Cd_xHg_{1-x}Te$ alloyed particles even at room temperature [119]. Figure 5.8b displays the typical absorption and emission spectra of these NCs, which are tunable by varying the content of mercury. As is evident from the spectra, their absorption and emission characteristic peak positions are both shifted toward the near-infrared region as the extent of mercury exchange increases, leading to a narrowing of the bandgap

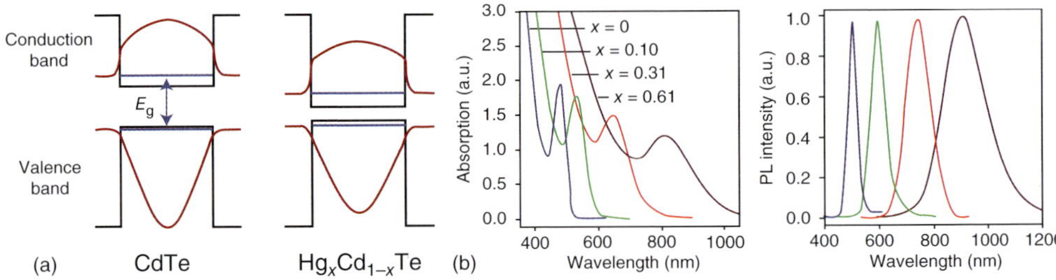

Figure 5.8 (a) Diagrams depicting the potential energy wells (black lines), quantum-confined kinetic energy levels (blue lines), and wave functions (red) of electrons and holes in CdTe and $Hg_xCd_{1-x}Te$ QDs. (b) Absorption and PL spectra of alloyed $Hg_xCd_{1-x}Te$ QDs with different contents of Hg obtained by the partial cation exchange on initial CdTe QDs with size of 2.3 nm [119]. (Reproduced with permission. Copyright © 2011, American Chemical Society.)

of the alloy (see the energy diagrams in Figure 5.8a). Another example of alloyed QDs is represented by $Cd_xZn_{1-x}Se$ NCs, which were prepared via a partial cation exchange reaction starting from ZnSe particles [125]. By controlling the amount of incorporated Cd ions, their emission could be tuned from the violet to the orange region. Recent works have demonstrated that the cation exchange strategy represents an efficient alternative in the preparation of doped NCs [127–130]. For example, De Trizio *et al.* [131] observed that the incorporation of Zn into the $Cu_{1-x}InS_2$ NCs increased the PL quantum efficiency of the starting NCs. Sahu *et al.* [132] reported that the addition of a few Ag atoms into the CdSe NCs by cation exchange also improves their quantum efficiency. For a doped quantum emitter, slight variation of the dopant concentration can induce significant changes of its optoelectronic properties [133]. In this context, the cation exchange strategy could be a facile and reliable way for the controllable doping of QDs [127]. Nowadays, cation exchange is widely exploited for the synthesis of QDs with novel structures and chemical compositions including alloyed, core/shell, and doped materials, which are unattainable by conventional methods.

Besides cation exchange, the chemical composition of nanoparticles can be tailored by oxidation and reduction reactions. This kind of post-synthetic treatment is especially important for copper chalcogenide NCs (i.e., copper sulfide, selenide, and telluride), which exhibit localized surface plasmon resonances (LSPR) owing to heavy p-doping resulting from a high level of copper vacancies. Their oxidation and reduction allows for the variation of the vacancy concentration and hence for the LSPR tuning in the following way: the greater the number of vacancies, the higher the LSPR frequency [42, 134–138]. As shown in Figure 5.9a, oxidation of an as-synthesized $Cu_{1.96}Se$ NC solution by exposure to air or by addition of an oxidizing agent (e.g., Ce(IV) complex) leads to an extraction of copper ions yielding nonstoichiometric $Cu_{1.81}Se$ NCs, accompanied by the development of a LSPR [138]. Interestingly, this transformation is reversible: addition of a reducing agent (a Cu(I) complex) reverts the composition of NCs back to stoichiometric and damps the plasmon band. This behavior is observed in all copper chalcogenides. Kriegel *et al.* [134] demonstrated that diisobutylaluminum hydride can be used as a reducing agent. Xie *et al.* [135] reported a chemical process for transforming strongly plasmonic covellite $Cu_{1.1}S$ NCs into Cu_2S NCs, which do not show the plasmon band. A recent study demonstrated that structural rearrangement of the oxidized chalcocite Cu_2S NCs to roxbyite $Cu_{1.75}S$ releases the copper vacancies [42]. This suggests that, besides the copper deficiency, crystal structure also plays an additional role in controlling the plasmonic properties of copper chalcogenide NCs.

This unique ability of controllable oxidation of copper chalcogenide NCs can be harnessed in electronic devices with tunable electronic properties. Thus, solid-state conversion of stoichiometric Cu_2Se NCs to substoichiometric $Cu_{1.8}Se$ NCs resulted in 3000-fold increase in conductivity of a thin-film chip made thereof (see Figure 5.9b,c) [113]. This straightforward approach allows the implementation of the same material as both semiconducting and conducting domain by controllable alteration of its electronic properties via a simple oxidation reaction.

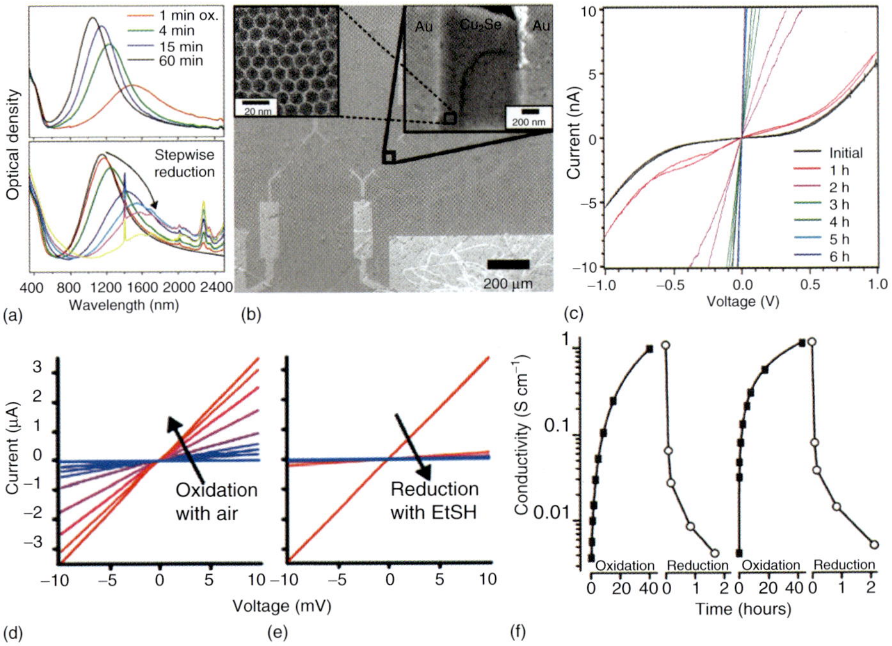

Figure 5.9 (a) Evolution of the LSPR of $Cu_{1.96}Se$ NCs over time by their gradual oxidation in air (top) followed by reduction (bottom) [138]. (Reproduced with permission. Copyright © 2011, American Chemical Society.) (b) SEM image of a Cu_2Se NC thin-film device. Insets: (left) TEM image of the NCs, (right) a high-resolution SEM image of the thin film overlying two gold electrodes. (c) Current versus voltage measurements of the device from (b) as a function of time of its air exposure [113]. (Reproduced with permission. Copyright © 2011, American Chemical Society.) (d) Current versus voltage measurements of the Cu_2S nanorod thin-film FET upon exposure to air and (e) upon ethanethiol treatment. (f) Conductivity measurements on the same FET, demonstrating reversible oxidation and reduction steps [112]. (Reproduced with permission. Copyright © 2013, WILEY VCH Verlag GmbH & Co. KGaA.)

Moreover, as in the case of NCs dispersed in solution, oxidation of solid-state NC bulks has been demonstrated to be reversible, which opens up the opportunity of using semiconductor nanoparticles for plasmonic probing of processes, such as redox reactions, electrochemical charging or discharging, ligand binding, and impurity doping, occurring in nanoscale volumes [112]. This reversible formation and filling of copper vacancies in Cu_2S nanorod films accompanied by their electrical probing has recently been realized by Jain *et al.* Like in the case of Cu_2Se NC films mentioned above, the small-field conductivity of Cu_2S nanorod thin-film FETs increases by over two orders of magnitude upon exposure to air and completely reverses upon soaking in a solution of ethanethiol (EtSH) in ethanol (see Figure 5.9d,e). This process is reversible and can be repeated several times, as revealed by conductivity measurements on a single device, results of which are displayed in Figure 5.9f [112].

5.6 Conclusions and Outlook

In this chapter, we have summarized the general aspects of semiconductor NC synthesis, with special emphasis on its optimization achieved within last years by a proper design of the precursors, ligands, and solvents employed. A further step toward commercialization of QDs has been made by the development of their continuous flow synthesis, providing large amounts of nanoparticles with reproducible properties. This approach overcomes the drawbacks of a conventional batch synthesis, especially the hot-injection approach, which is inappropriate for an efficient upscaling. Since the surface chemistry of QDs has crucial importance for optoelectronic applications, we discussed the impact of ligand capping on related properties of QDs, such as PL efficiency and electrical conductivity. Adjustment of the QD surface via, for example, ligand exchange can lead to a drastic enhancement of collective electronic characteristics of NC-based bulks. On the other hand, post-synthetic treatments of the inorganic core of nanoparticles also provide a powerful tool to control their physicochemical properties by varying their composition through cation exchange, oxidation, and reduction reactions. The latter are particularly useful for the treatment of already fabricated NC arrays, as has recently been demonstrated for Cu_2Se and Cu_2S NC-based devices, which results in a vast increase of their conductivity. The recent progress made in the field of the QD synthesis and their characterization paves the way to commercialization of these materials and their exploitation in working devices, such as photodetectors, solar cells, FETs, and light-emitting and laser diodes.

References

1. Goesmann, H. and Feldmann, C. (2010) Nanoparticulate functional materials. *Angew. Chem. Int. Ed.*, **49** (8), 1362–1395.
2. Park, J., Joo, J., Soon, G.K., Jang, Y., and Hyeon, T. (2007) Synthesis of monodisperse spherical nanocrystals. *Angew. Chem. Int. Ed.*, **46** (25), 4630–4660.
3. Cozzoli, P.D., Pellegrino, T., and Manna, L. (2006) Synthesis, properties and perspectives of hybrid nanocrystal structures. *Chem. Soc. Rev.*, **35** (11), 1195–1208.
4. Donega, C.D.M. (2011) Synthesis and properties of colloidal heteronanocrystals. *Chem. Soc. Rev.*, **40** (3), 1512–1546.
5. Kwon, S.G. and Hyeon, T. (2008) Colloidal chemical synthesis and formation kinetics of uniformly sized nanocrystals of metals, oxides, and chalcogenides. *Acc. Chem. Res.*, **41** (12), 1696–1709.
6. Talapin, D.V., Lee, J.-S., Kovalenko, M.V., and Shevchenko, E.V. (2010) Prospects of colloidal nanocrystals for electronic and optoelectronic applications. *Chem. Rev.*, **110** (1), 389–458.
7. Sowers, K.L., Swartz, B., and Krauss, T.D. (2013) Chemical mechanisms of semiconductor nanocrystal synthesis. *Chem. Mater.*, **25** (8), 1351–1362.
8. Lesnyak, V., Gaponik, N., and Eychmüller, A. (2013) Colloidal semiconductor nanocrystals: the aqueous approach. *Chem. Soc. Rev.*, **42** (7), 2905–2929.
9. Yin, Y. and Talapin, D. (2013) The chemistry of functional nanomaterials. *Chem. Soc. Rev.*, **42** (7), 2484–2487.
10. Hughes, B.K., Luther, J.M., and Beard, M.C. (2012) The subtle chemistry of

colloidal, quantum-confined semiconductor nanostructures. *ACS Nano*, **6** (6), 4573–4579.
11. Brennan, J.G., Siegrist, T., Carroll, P.J., Stuczynski, S.M., Brus, L.E., and Steigerwald, M.L. (1989) The preparation of large semiconductor clusters via the pyrolysis of a molecular precursor. *J. Am. Chem. Soc.*, **111** (11), 4141–4143.
12. Murray, C.B., Norris, D.J., and Bawendi, M.G. (1993) Synthesis and characterization of nearly monodisperse CdE (E = sulfur, selenium, tellurium) semiconductor nanocrystallites. *J. Am. Chem. Soc.*, **115** (19), 8706–8715.
13. Peng, Z.A. and Peng, X. (2000) Formation of high-quality CdTe, CdSe, and CdS nanocrystals using CdO as precursor. *J. Am. Chem. Soc.*, **123** (1), 183–184.
14. Talapin, D.V., Koeppe, R., Götzinger, S., Kornowski, A., Lupton, J.M., Rogach, A.L., Benson, O., Feldmann, J., and Weller, H. (2003) Highly emissive colloidal CdSe/CdS heterostructures of mixed dimensionality. *Nano Lett.*, **3** (12), 1677–1681.
15. Peng, X.G. (2009) An essay on synthetic chemistry of colloidal nanocrystals. *Nano Res.*, **2** (6), 425–447.
16. Peng, X., Schlamp, M.C., Kadavanich, A.V., and Alivisatos, A.P. (1997) Epitaxial growth of highly luminescent CdSe/CdS Core/Shell nanocrystals with photostability and electronic accessibility. *J. Am. Chem. Soc.*, **119** (30), 7019–7029.
17. Li, L.S., Pradhan, N., Wang, Y., and Peng, X. (2004) High quality ZnSe and ZnS nanocrystals formed by activating zinc carboxylate precursors. *Nano Lett.*, **4** (11), 2261–2264.
18. Chen, H.-S., Lo, B., Hwang, J.-Y., Chang, G.-Y., Chen, C.-M., Tasi, S.-J., and Wang, S.-J.J. (2004) Colloidal ZnSe, ZnSe/ZnS, and ZnSe/ZnSeS quantum dots synthesized from ZnO. *J. Phys. Chem. B*, **108** (44), 17119–17123.
19. Shen, H.B., Wang, H.Z., Li, X.M., Niu, J.Z., Wang, H., Chen, X., and Li, L.S. (2009) Phosphine-free synthesis of high quality ZnSe, ZnSe/ZnS, and Cu-, Mn-doped ZnSe nanocrystals. *Dalton Trans.*, (47), 10534–10540.
20. Pietryga, J.M., Schaller, R.D., Werder, D., Stewart, M.H., Klimov, V.I., and Hollingsworth, J.A. (2004) Pushing the band gap envelope: mid-infrared emitting colloidal PbSe quantum dots. *J. Am. Chem. Soc.*, **126** (38), 11752–11753.
21. Cademartiri, L., Montanari, E., Calestani, G., Migliori, A., Guagliardi, A., and Ozin, G.A. (2006) Size-dependent extinction coefficients of PbS quantum dots. *J. Am. Chem. Soc.*, **128** (31), 10337–10346.
22. Li, H., Chen, D., Li, L., Tang, F., Zhang, L., and Ren, J. (2010) Size- and shape-controlled synthesis of PbSe and PbS nanocrystals via a facile method. *CrystEngComm*, **12** (4), 1127–1133.
23. Battaglia, D. and Peng, X. (2002) Formation of high quality InP and InAs nanocrystals in a noncoordinating solvent. *Nano Lett.*, **2** (9), 1027–1030.
24. Xie, R., Battaglia, D., and Peng, X. (2007) Colloidal InP nanocrystals as efficient emitters covering blue to near-infrared. *J. Am. Chem. Soc.*, **129** (50), 15432–15433.
25. Panthani, M.G., Akhavan, V., Goodfellow, B., Schmidtke, J.P., Dunn, L., Dodabalapur, A., Barbara, P.F., and Korgel, B.A. (2008) Synthesis of $CuInS_2$, $CuInSe_2$, and $Cu(In_xGa_{1-x})Se_2$ (CIGS) nanocrystal "Inks" for printable photovoltaics. *J. Am. Chem. Soc.*, **130** (49), 16770–16777.
26. Xie, R., Rutherford, M., and Peng, X. (2009) Formation of high-quality I–III–VI semiconductor nanocrystals by tuning relative reactivity of cationic precursors. *J. Am. Chem. Soc.*, **131** (15), 5691–5697.
27. Liu, Y., Tang, Y., Ning, Y., Li, M.J., Zhang, H., and Yang, B. (2010) "One-pot" synthesis and shape control of ZnSe semiconductor nanocrystals in liquid paraffin. *J. Mater. Chem.*, **20** (21), 4451–4458.
28. Ma, W., Swisher, S.L., Ewers, T., Engel, J., Ferry, V.E., Atwater, H.A., and Alivisatos, A.P. (2011) Photovoltaic performance of ultrasmall PbSe quantum dots. *ACS Nano*, **5** (10), 8140–8147.
29. Liu, T.Y., Li, M.J., Ouyang, J.Y., Zaman, M.B., Wang, R.B., Wu, X.H., Yeh, C.S., Lin, Q., Yang, B., and Yu, K. (2009)

Non-injection and low-temperature approach to colloidal photoluminescent PbS nanocrystals with narrow bandwidth. *J. Phys. Chem. C*, **113** (6), 2301–2308.
30. Cozzoli, P.D., Manna, L., Curri, M.L., Kudera, S., Giannini, C., Striccoli, M., and Agostiano, A. (2005) Shape and phase control of colloidal ZnSe nanocrystals. *Chem. Mater.*, **17** (6), 1296–1306.
31. Levina, L., Sukhovatkin, V., Musikhin, S., Cauchi, S., Nisman, R., Bazett-Jones, D.P., and Sargent, E.H. (2005) Efficient infrared-emitting PbS quantum dots grown on DNA and stable in aqueous solution and blood plasma. *Adv. Mater.*, **17** (15), 1854–1857.
32. Yan, C., Tang, F., Li, L., Li, H., Huang, X., Chen, D., Meng, X., and Ren, J. (2010) Synthesis of aqueous CdTe/CdS/ZnS core/shell/shell quantum dots by a chemical aerosol flow method. *Nanoscale Res. Lett.*, **5** (1), 189–194.
33. Dilena, E., Dorfs, D., George, C., Miszta, K., Povia, M., Genovese, A., Casu, A., Prato, M., and Manna, L. (2012) Colloidal $Cu_{2-x}(S_ySe_{1-y})$ alloy nanocrystals with controllable crystal phase: synthesis, plasmonic properties, cation exchange and electrochemical lithiation. *J. Mater. Chem.*, **22** (26), 13023–13031.
34. Bullen, C., van Embden, J., Jasieniak, J., Cosgriff, J.E., Mulder, R.J., Rizzardo, E., Gu, M., and Raston, C.L. (2010) High activity phosphine-free selenium precursor solution for semiconductor nanocrystal growth. *Chem. Mater.*, **22** (14), 4135–4143.
35. Deng, Z.T., Cao, L., Tang, F.Q., and Zou, B.S. (2005) A new route to zinc-blende CdSe nanocrystals: mechanism and synthesis. *J. Phys. Chem. B*, **109** (35), 16671–16675.
36. Nakashima, T., Kobayashi, Y., and Kawai, T. (2009) Optical activity and chiral memory of thiol-capped CdTe nanocrystals. *J. Am. Chem. Soc.*, **131** (30), 10342–10343.
37. Kapitonov, A.M., Stupak, A.P., Gaponenko, S.V., Petrov, E.P., Rogach, A.L., and Eychmüller, A. (1999) Luminescence properties of thiol-stabilized CdTe nanocrystals. *J. Phys. Chem. B*, **103** (46), 10109–10113.
38. Cao, Y.C. and Wang, J. (2004) One-pot synthesis of high-quality zinc-blende CdS nanocrystals. *J. Am. Chem. Soc.*, **126** (44), 14336–14337.
39. Yang, Y.A., Wu, H., Williams, K.R., and Cao, Y.C. (2005) Synthesis of CdSe and CdTe nanocrystals without precursor injection. *Angew. Chem. Int. Ed.*, **44** (41), 6712–6715.
40. Li, Z. and Peng, X. (2011) Size/shape-controlled synthesis of colloidal CdSe quantum disks: ligand and temperature effects. *J. Am. Chem. Soc.*, **133** (17), 6578–6586.
41. Flamee, S., Dierick, R., Cirillo, M., Van Genechten, D., Aubert, T., and Hens, Z. (2013) Synthesis of metal selenide colloidal nanocrystals by the hot injection of selenium powder. *Dalton Trans*, **42** (35), 12654–12661.
42. Saldanha, P. L.; Brescia, R.; Prato, M.; Li, H.; Povia, M.; Manna, L.; Lesnyak, V. (2014) Generalized one-pot synthesis of copper sulfide, selenide-sulfide, and telluride-sulfide nanoparticles. *Chem. Mater.*, **26** (3), 1442–1449.
43. Chen, O., Chen, X., Yang, Y.A., Lynch, J., Wu, H.M., Zhuang, J.Q., and Cao, Y.C. (2008) Synthesis of metal-selenide nanocrystals using selenium dioxide as the selenium precursor. *Angew. Chem. Int. Ed.*, **47** (45), 8638–8641.
44. Evans, C.M., Evans, M.E., and Krauss, T.D. (2010) Mysteries of TOPSe revealed: insights into quantum dot nucleation. *J. Am. Chem. Soc.*, **132** (32), 10973–10975.
45. Ruberu, T.P.A., Albright, H.R., Callis, B., Ward, B., Cisneros, J., Fan, H.-J., and Vela, J. (2012) Molecular control of the nanoscale: effect of phosphine–chalcogenide reactivity on CdS–CdSe nanocrystal composition and morphology. *ACS Nano*, **6** (6), 5348–5359.
46. Yu, K., Liu, X., Zeng, Q., Leek, D.M., Ouyang, J., Whitmore, K.M., Ripmeester, J.A., Tao, Y., and Yang, M. (2013) Effect of tertiary and secondary

phosphines on low-temperature formation of quantum dots. *Angew. Chem. Int. Ed.*, **52** (18), 4823–4828.
47. Jasieniak, J., Bullen, C., van Embden, J., and Mulvaney, P. (2005) Phosphine-free synthesis of CdSe nanocrystals. *J. Phys. Chem. B*, **109** (44), 20665–20668.
48. Liu, Y., Yao, D., Shen, L., Zhang, H., Zhang, X., and Yang, B. (2012) Alkylthiol-enabled Se powder dissolution in oleylamine at room temperature for the phosphine-free synthesis of copper-based quaternary selenide nanocrystals. *J. Am. Chem. Soc.*, **134** (17), 7207–7210.
49. Zeng, R., Zhang, T., Dai, G., and Zou, B. (2011) Highly emissive, color-tunable, phosphine-free Mn:ZnSe/ZnS core/shell and Mn:ZnSeS shell-alloyed doped nanocrystals. *J. Phys. Chem. C*, **115** (7), 3005–3010.
50. Saldanha, P.L., Brescia, R., Prato, M., Li, H., Povia, M., Manna, L., and Lesnyak, V. (2014) Generalized one-pot synthesis of copper sulfide, selenide-sulfide, and telluride-sulfide nanoparticles. *Chem. Mater.*, **26**, 1442–1449.
51. Pradhan, N. and Efrima, S. (2003) Single-precursor, one-pot versatile synthesis under near ambient conditions of tunable, single and dual band fluorescing metal sulfide nanoparticles. *J. Am. Chem. Soc.*, **125** (8), 2050–2051.
52. Joo, J., Na, H.B., Yu, T., Yu, J.H., Kim, Y.W., Wu, F., Zhang, J.Z., and Hyeon, T. (2003) Generalized and facile synthesis of semiconducting metal sulfide nanocrystals. *J. Am. Chem. Soc.*, **125** (36), 11100–11105.
53. Yu, K., Ouyang, J., Zaman, M.B., Johnston, D., Yan, F.J., Li, G., Ratcliffe, C.I., Leek, D.M., Wu, X.H., Stupak, J., Jakubek, Z., and Whitfiel, D. (2009) Single-sized CdSe nanocrystals with bandgap photoemission via a noninjection one-pot approach. *J. Phys. Chem. C*, **113** (9), 3390–3401.
54. Kanehara, M., Arakawa, H., Honda, T., Saruyama, M., and Teranishi, T. (2012) Large-scale synthesis of high-quality metal sulfide semiconductor quantum dots with tunable surface-plasmon resonance frequencies. *Chem. Eur. J.*, **18** (30), 9230–9238.
55. Kwon, S.G. and Hyeon, T. (2011) Formation mechanisms of uniform nanocrystals via hot-injection and heat-up methods. *Small*, **7** (19), 2685–2702.
56. Zou, Y., Li, D.S., and Yang, D.R. (2010) Noninjection synthesis of CdS and alloyed CdS_xSe_{1-x} nanocrystals without nucleation initiators. *Nanoscale Res. Lett.*, **5** (6), 966–971.
57. Chen, O., Zhao, J., Chauhan, V.P., Cui, J., Wong, C., Harris, D.K., Wei, H., Han, H.S., Fukumura, D., Jain, R.K., and Bawendi, M.G. (2013) Compact high-quality CdSe-CdS core-shell nanocrystals with narrow emission linewidths and suppressed blinking. *Nat. Mater.*, **12** (5), 445–451.
58. Li, L. and Reiss, P. (2008) One-pot synthesis of highly luminescent InP/ZnS nanocrystals without precursor injection. *J. Am. Chem. Soc.*, **130** (35), 11588–11589.
59. Wang, H.Z., Li, X.M., Shen, H.B., Zhou, C.H., and Li, L.S. (2013) Non-injection synthesis of $CuInS_2$/ZnS nanocrystals and the effect of hydrophilic surface modification on photoluminescence. *Sci. Adv. Mater.*, **5** (1), 51–56.
60. Didenko, Y.T. and Suslick, K.S. (2005) Chemical aerosol flow synthesis of semiconductor nanoparticles. *J. Am. Chem. Soc.*, **127** (35), 12196–12197.
61. Bang, J.H., Suh, W.H., and Suslick, K.S. (2008) Quantum dots from chemical aerosol flow synthesis: preparation, characterization, and cellular imaging. *Chem. Mater.*, **20** (12), 4033–4038.
62. Bang, J.H. and Suslick, K.S. (2010) Applications of ultrasound to the synthesis of nanostructured materials. *Adv. Mater.*, **22** (10), 1039–1059.
63. Li, H., Tang, F., Li, L., Yan, C., Huang, X., and Chen, D. (2009) Ultrafast synthesis of water-soluble nanocrystals by the chemical aerosol flow method. *CrystEngComm*, **11** (7), 1231–1234.
64. Chan, E.M., Mathies, R.A., and Alivisatos, A.P. (2003) Size-controlled growth of CdSe nanocrystals in microfluidic reactors. *Nano Lett.*, **3** (2), 199–201.
65. Yen, B.K.H., Stott, N.E., Jensen, K.F., and Bawendi, M.G. (2003) A continuous-flow microcapillary reactor

for the preparation of a size series of CdSe nanocrystals. *Adv. Mater.*, **15** (21), 1858–1862.

66. Nightingale, A.M. and de Mello, J.C. (2010) Microscale synthesis of quantum dots. *J. Mater. Chem.*, **20** (39), 8454–8463.
67. Moghaddam, M.M., Baghbanzadeh, M., Sadeghpour, A., Glatter, O., and Kappe, C.O. (2013) Continuous-flow synthesis of CdSe quantum dots: a size-tunable and scalable approach. *Chem. Eur. J.*, **19** (35), 11629–11636.
68. Abdelhady, A.L., Afzaal, M., Malik, M.A., and O'Brien, P. (2011) Flow reactor synthesis of CdSe, CdS, CdSe/CdS and CdSeS nanoparticles from single molecular precursor(s). *J. Mater. Chem.*, **21** (46), 18768–18775.
69. Jin, H.D. and Chang, C.-H. (2011) Continuous synthesis of SnTe nanorods. *J. Mater. Chem.*, **21** (33), 12218–12220.
70. Pan, J., El-Ballouli, A.A.O., Rollny, L., Voznyy, O., Burlakov, V.M., Goriely, A., Sargent, E.H., and Bakr, O.M. (2013) Automated synthesis of photovoltaic-quality colloidal quantum dots using separate nucleation and growth stages. *ACS Nano*, **7** (11), 10158–10166.
71. Shavel, A., Cadavid, D., Ibáñez, M., Carrete, A., and Cabot, A. (2012) Continuous production of Cu_2ZnSnS_4 nanocrystals in a flow reactor. *J. Am. Chem. Soc.*, **134** (3), 1438–1441.
72. Marre, S., Park, J., Rempel, J., Guan, J., Bawendi, M.G., and Jensen, K.F. (2008) Supercritical continuous-microflow synthesis of narrow size distribution quantum dots. *Adv. Mater.*, **20** (24), 4830–4834.
73. Toyota, A., Nakamura, H., Ozono, H., Yamashita, K., Uehara, M., and Maeda, H. (2010) Combinatorial synthesis of CdSe nanoparticles using microreactors. *J. Phys. Chem. C*, **114** (17), 7527–7534.
74. Green, M. (2010) The nature of quantum dot capping ligands. *J. Mater. Chem.*, **20** (28), 5797–5809.
75. Baranov, D., Manna, L., and Kanaras, A.G. (2011) Chemically induced self-assembly of spherical and anisotropic inorganic nanocrystals. *J. Mater. Chem.*, **21** (42), 16694–16703.
76. Weiss, E.A. (2013) Organic molecules as tools to control the growth, surface structure, and redox activity of colloidal quantum dots. *Acc. Chem. Res.*, **46** (11), 2607–2615.
77. Yin, Y. and Alivisatos, A.P. (2005) Colloidal nanocrystal synthesis and the organic–inorganic interface. *Nature*, **437** (7059), 664–670.
78. Morris-Cohen, A.J., Malicki, M., Peterson, M.D., Slavin, J.W.J., and Weiss, E.A. (2013) Chemical, structural, and quantitative analysis of the ligand shells of colloidal quantum dots. *Chem. Mater.*, **25** (8), 1155–1165.
79. Hens, Z. and Martins, J.C. (2013) A solution NMR toolbox for characterizing the surface chemistry of colloidal nanocrystals. *Chem. Mater.*, **25** (8), 1211–1221.
80. Sperling, R.A. and Parak, W.J. (2010) Surface modification, functionalization and bioconjugation of colloidal inorganic nanoparticles. *Philos. Trans. R. Soc. London, Ser. A*, **368** (1915), 1333–1383.
81. Jasieniak, J. and Mulvaney, P. (2007) From Cd-rich to Se-rich - the manipulation of CdSe nanocrystal surface stoichiometry. *J. Am. Chem. Soc.*, **129** (10), 2841–2848.
82. Wei, H.H.-Y., Evans, C.M., Swartz, B.D., Neukirch, A.J., Young, J., Prezhdo, O.V., and Krauss, T.D. (2012) Colloidal semiconductor quantum dots with tunable surface composition. *Nano Lett.*, **12** (9), 4465–4471.
83. Soni, U. and Sapra, S. (2010) The importance of surface in core–shell semiconductor nanocrystals. *J. Phys. Chem. C*, **114** (51), 22514–22518.
84. Hassinen, A., Moreels, I., De Nolf, K., Smet, P.F., Martins, J.C., and Hens, Z. (2012) Short-chain alcohols strip X-type ligands and quench the luminescence of PbSe and CdSe quantum dots, acetonitrile does not. *J. Am. Chem. Soc.*, **134** (51), 20705–20712.
85. Bae, W.K., Joo, J., Padilha, L.A., Won, J., Lee, D.C., Lin, Q., Koh, W.-K., Luo, H., Klimov, V.I., and Pietryga, J.M. (2012) Highly effective surface passivation of PbSe quantum dots through reaction

with molecular chlorine. *J. Am. Chem. Soc.*, **134** (49), 20160–20168.
86. Tagliazucchi, M., Tice, D.B., Sweeney, C.M., Morris-Cohen, A.J., and Weiss, E.A. (2011) Ligand-controlled rates of photoinduced electron transfer in hybrid CdSe nanocrystal/poly(viologen) films. *ACS Nano*, **5** (12), 9907–9917.
87. Dong, A., Ye, X., Chen, J., Kang, Y., Gordon, T., Kikkawa, J.M., and Murray, C.B. (2011) A generalized ligand-exchange strategy enabling sequential surface functionalization of colloidal nanocrystals. *J. Am. Chem. Soc.*, **133** (4), 998–1006.
88. Kovalenko, M.V., Scheele, M., and Talapin, D.V. (2009) Colloidal nanocrystals with molecular metal chalcogenide surface ligands. *Science*, **324** (5933), 1417–1420.
89. Zanella, M., Maserati, L., Pernia Leal, M., Prato, M., Lavieville, R., Povia, M., Krahne, R., and Manna, L. (2013) Atomic ligand passivation of colloidal nanocrystal films via their reaction with propyltrichlorosilane. *Chem. Mater.*, **25** (8), 1423–1429.
90. Tang, J., Kemp, K.W., Hoogland, S., Jeong, K.S., Liu, H., Levina, L., Furukawa, M., Wang, X., Debnath, R., Cha, D., Chou, K.W., Fischer, A., Amassian, A., Asbury, J.B., and Sargent, E.H. (2011) Colloidal-quantum-dot photovoltaics using atomic-ligand passivation. *Nat. Mater.*, **10** (10), 765–771.
91. Ip, A.H., Thon, S.M., Hoogland, S., Voznyy, O., Zhitomirsky, D., Debnath, R., Levina, L., Rollny, L.R., Carey, G.H., Fischer, A., Kemp, K.W., Kramer, I.J., Ning, Z., Labelle, A.J., Chou, K.W., Amassian, A., and Sargent, E.H. (2012) Hybrid passivated colloidal quantum dot solids. *Nat. Nanotechnol.*, **7** (9), 577–582.
92. Kim, D.K., Fafarman, A.T., Diroll, B.T., Chan, S.H., Gordon, T.R., Murray, C.B., and Kagan, C.R. (2013) Solution-based stoichiometric control over charge transport in nanocrystalline CdSe devices. *ACS Nano*, **7** (10), 8760–8770.
93. Kovalenko, M.V., Bodnarchuk, M.I., Zaumseil, J., Lee, J.S., and Talapin, D.V. (2010) Expanding the chemical versatility of colloidal nanocrystals capped with molecular metal chalcogenide ligands. *J. Am. Chem. Soc.*, **132** (29), 10085–10092.
94. Lee, J.-S., Kovalenko, M.V., Huang, J., Chung, D.S., and Talapin, D.V. (2011) Band-like transport, high electron mobility and high photoconductivity in all-inorganic nanocrystal arrays. *Nat. Nanotechnol.*, **6** (6), 348–352.
95. Nag, A., Kovalenko, M.V., Lee, J.-S., Liu, W., Spokoyny, B., and Talapin, D.V. (2011) Metal-free inorganic ligands for colloidal nanocrystals: S^{2-}, HS^-, Se^{2-}, HSe^-, Te^{2-}, HTe^-, $TeS_3{}^{2-}$, OH^-, and $NH_2{}^-$ as surface ligands. *J. Am. Chem. Soc.*, **133** (27), 10612–10620.
96. Jasieniak, J., MacDonald, B.I., Watkins, S.E., and Mulvaney, P. (2011) Solution-processed sintered nanocrystal solar cells via layer-by-layer assembly. *Nano Lett.*, **11** (7), 2856–2864.
97. Cademartiri, L., Ghadimi, A., and Ozin, G.A. (2008) Nanocrystal plasma polymerization: from colloidal nanocrystals to inorganic architectures. *Acc. Chem. Res.*, **41** (12), 1820–1830.
98. Damasceno, P.F., Engel, M., and Glotzer, S.C. (2012) Predictive self-assembly of polyhedra into complex structures. *Science*, **337** (6093), 453–457.
99. de Graaf, J. and Manna, L. (2012) A roadmap for the assembly of polyhedral particles. *Science*, **337** (6093), 417–418.
100. Wang, T., Wang, X., LaMontagne, D., Wang, Z., Wang, Z., and Cao, Y.C. (2012) Shape-controlled synthesis of colloidal superparticles from nanocubes. *J. Am. Chem. Soc.*, **134** (44), 18225–18228.
101. Chen, J., Dong, A., Cai, J., Ye, X., Kang, Y., Kikkawa, J.M., and Murray, C.B. (2010) Collective dipolar interactions in self-assembled magnetic binary nanocrystal superlattice membranes. *Nano Lett.*, **10** (12), 5103–5108.
102. Ye, X., Millan, J.A., Engel, M., Chen, J., Diroll, B.T., Glotzer, S.C., and Murray, C.B. (2013) Shape alloys of nanorods and nanospheres from self-assembly. *Nano Lett.*, **13** (10), 4980–4988.
103. Dong, A.G., Chen, J., Vora, P.M., Kikkawa, J.M., and Murray, C.B. (2010)

Binary nanocrystal superlattice membranes self-assembled at the liquid-air interface. *Nature*, **466** (7305), 474–477.
104. Shevchenko, E.V., Talapin, D.V., Kotov, N.A., O'Brien, S., and Murray, C.B. (2006) Structural diversity in binary nanoparticle superlattices. *Nature*, **439** (7072), 55–59.
105. Rivest, J.B. and Jain, P.K. (2013) Cation exchange on the nanoscale: an emerging technique for new material synthesis, device fabrication, and chemical sensing. *Chem. Soc. Rev.*, **42** (1), 89–96.
106. Beberwyck, B.J., Surendranath, Y., and Alivisatos, A.P. (2013) Cation exchange: a versatile tool for nanomaterials synthesis. *J. Phys. Chem. C*, **117** (39), 19759–19770.
107. Comin, A. and Manna, L. (2014) New materials for tunable plasmonic colloidal nanocrystals. *Chem. Soc. Rev.* **43** (11), 3957–3975.
108. Jain, P.K., Amirav, L., Aloni, S., and Alivisatos, A.P. (2010) Nanoheterostructure cation exchange: anionic framework conservation. *J. Am. Chem. Soc.*, **132** (29), 9997–9999.
109. Li, H., Brescia, R., Krahne, R., Bertoni, G., Alcocer, M.J.P., D'Andrea, C., Scotognella, F., Tassone, F., Zanella, M., De Giorgi, M., and Manna, L. (2012) Blue-UV-emitting ZnSe(Dot)/ZnS(Rod) core/shell nanocrystals prepared from CdSe/CdS nanocrystals by sequential cation exchange. *ACS Nano*, **6** (2), 1637–1647.
110. Beberwyck, B.J. and Alivisatos, A.P. (2012) Ion exchange synthesis of III–V nanocrystals. *J. Am. Chem. Soc.*, **134** (49), 19977–19980.
111. Li, H., Brescia, R., Povia, M., Prato, M., Bertoni, G., Manna, L., and Moreels, I. (2013) Synthesis of uniform disk-shaped copper telluride nanocrystals and cation exchange to cadmium telluride quantum disks with stable red emission. *J. Am. Chem. Soc.*, **135** (33), 12270–12278.
112. Jain, P.K., Manthiram, K., Engel, J.H., White, S.L., Faucheaux, J.A., and Alivisatos, A.P. (2013) Doped nanocrystals as plasmonic probes of redox chemistry. *Angew. Chem. Int. Ed.*, **52** (51), 13671–13675.
113. Riha, S.C., Johnson, D.C., and Prieto, A.L. (2011) Cu_2Se nanoparticles with tunable electronic properties due to a controlled solid-state phase transition driven by copper oxidation and cationic conduction. *J. Am. Chem. Soc.*, **133** (5), 1383–1390.
114. Son, D.H., Hughes, S.M., Yin, Y.D., and Alivisatos, A.P. (2004) Cation exchange reactions-in ionic nanocrystals. *Science*, **306** (5698), 1009–1012.
115. Luther, J.M., Zheng, H., Sadtler, B., and Alivisatos, A.P. (2009) Synthesis of PbS nanorods and other ionic nanocrystals of complex morphology by sequential cation exchange reactions. *J. Am. Chem. Soc.*, **131** (46), 16851–16857.
116. Miszta, K., Dorfs, D., Genovese, A., Kim, M.R., and Manna, L. (2011) Cation exchange reactions in colloidal branched nanocrystals. *ACS Nano*, **5** (9), 7176–7183.
117. Li, H., Zanella, M., Genovese, A., Povia, M., Falqui, A., Giannini, C., and Manna, L. (2011) Sequential cation exchange in nanocrystals: preservation of crystal phase and formation of metastable phases. *Nano Lett.*, **11** (11), 4964–4970.
118. Pietryga, J.M., Werder, D.J., Williams, D.J., Casson, J.L., Schaller, R.D., Klimov, V.I., and Hollingsworth, J.A. (2008) Utilizing the lability of lead selenide to produce heterostructured nanocrystals with bright, stable infrared emission. *J. Am. Chem. Soc.*, **130** (14), 4879–4885.
119. Smith, A.M. and Nie, S. (2011) Bright and compact alloyed quantum dots with broadly tunable near-infrared absorption and fluorescence spectra through mercury cation exchange. *J. Am. Chem. Soc.*, **133** (1), 24–26.
120. Lambert, K., Geyter, B.D., Moreels, I., and Hens, Z. (2009) PbTe/CdTe core/shell particles by cation exchange, a HR-TEM study. *Chem. Mater.*, **21** (5), 778–780.
121. Justo, Y., Geiregat, P., Hoecke, K.V., Vanhaecke, F., De Mello Donega, C., and Hens, Z. (2013) Optical properties of PbS/CdS core/shell quantum dots. *J. Phys. Chem. C*, **117** (39), 20171–20177.
122. Stouwdam, J.W., Shan, J., van Veggel, F.C.J.M., Pattantyus-Abraham, A.G., Young, J.F., and Raudsepp, M. (2006)

Photostability of colloidal PbSe and PbSe/PbS core/shell nanocrystals in solution and in the solid state. *J. Phys. Chem. C*, **111** (3), 1086–1092.

123. Casavola, M., van Huis, M.A., Bals, S., Lambert, K., Hens, Z., and Vanmaekelbergh, D. (2011) Anisotropic cation exchange in PbSe/CdSe core/shell nanocrystals of different geometry. *Chem. Mater.*, **24** (2), 294–302.

124. Justo, Y., Goris, B., Kamal, J.S., Geiregat, P., Bals, S., and Hens, Z. (2012) Multiple dot-in-rod PbS/CdS heterostructures with high photoluminescence quantum yield in the near-infrared. *J. Am. Chem. Soc.*, **134** (12), 5484–5487.

125. Zhong, X., Feng, Y., Zhang, Y., Gu, Z., and Zou, L. (2007) A facile route to violet- to orange-emitting $Cd_xZn_{1-x}Se$ alloy nanocrystals via cation exchange reaction. *Nanotechnology*, **18** (38), 385606.

126. Taniguchi, S., Green, M., and Lim, T. (2011) The room-temperature synthesis of anisotropic CdHgTe quantum dot alloys: a "Molecular Welding" effect. *J. Am. Chem. Soc.*, **133** (10), 3328–3331.

127. Mocatta, D., Cohen, G., Schattner, J., Millo, O., Rabani, E., and Banin, U. (2011) Heavily doped semiconductor nanocrystal quantum dots. *Science*, **332** (6025), 77–81.

128. Eilers, J., Groeneveld, E., de Mello Donegá, C., and Meijerink, A. (2012) Optical properties of Mn-doped ZnTe magic size nanocrystals. *J. Phys. Chem. Lett.*, **3** (12), 1663–1667.

129. Wills, A.W., Kang, M.S., Wentz, K.M., Hayes, S.E., Sahu, A., Gladfelter, W.L., and Norris, D.J. (2012) Synthesis and characterization of Al- and In-doped CdSe nanocrystals. *J. Mater. Chem.*, **22** (13), 6335–6342.

130. Vlaskin, V.A., Barrows, C.J., Erickson, C.S., and Gamelin, D.R. (2013) Nanocrystal diffusion doping. *J. Am. Chem. Soc.*, **135** (38), 14380–14389.

131. De Trizio, L., Prato, M., Genovese, A., Casu, A., Povia, M., Simonutti, R., Alcocer, M.J.P., D'Andrea, C., Tassone, F., and Manna, L. (2012) Strongly fluorescent quaternary Cu-In-Zn-S nanocrystals prepared from $Cu_{1-x}InS_2$ nanocrystals by partial cation exchange. *Chem. Mater.*, **24** (12), 2400–2406.

132. Sahu, A., Kang, M.S., Kompch, A., Notthoff, C., Wills, A.W., Deng, D., Winterer, M., Frisbie, C.D., and Norris, D.J. (2012) Electronic impurity doping in CdSe nanocrystals. *Nano Lett.*, **12** (5), 2587–2594.

133. Jain, P.K., Beberwyck, B.J., Fong, L.-K., Polking, M.J., and Alivisatos, A.P. (2012) Highly luminescent nanocrystals from removal of impurity atoms residual from ion-exchange synthesis. *Angew. Chem. Int. Ed.*, **51** (10), 2387–2390.

134. Kriegel, I., Jiang, C., Rodríguez-Fernández, J., Schaller, R.D., Talapin, D.V., da Como, E., and Feldmann, J. (2011) Tuning the excitonic and plasmonic properties of copper chalcogenide nanocrystals. *J. Am. Chem. Soc.*, **134** (3), 1583–1590.

135. Xie, Y., Riedinger, A., Prato, M., Casu, A., Genovese, A., Guardia, P., Sottini, S., Sangregorio, C., Miszta, K., Ghosh, S., Pellegrino, T., and Manna, L. (2013) Copper sulfide nanocrystals with tunable composition by reduction of covellite nanocrystals with Cu^+ ions. *J. Am. Chem. Soc.*, **135** (46), 17630–17637.

136. Yang, H.-J., Chen, C.-Y., Yuan, F.-W., and Tuan, H.-Y. (2013) Designed synthesis of solid and hollow $Cu_{2-x}Te$ nanocrystals with tunable near-infrared localized surface plasmon resonance. *J. Phys. Chem. C*, **117** (42), 21955–21964.

137. Liu, X., Wang, X., and Swihart, M.T. (2013) $Cu_{2-x}S_{1-y}Se_y$ alloy nanocrystals with broadly tunable near-infrared localized surface plasmon resonance. *Chem. Mater.*, **25** (21), 4402–4408.

138. Dorfs, D., Härtling, T., Miszta, K., Bigall, N.C., Kim, M.R., Genovese, A., Falqui, A., Povia, M., and Manna, L. (2011) Reversible tunability of the near-infrared valence band plasmon resonance in $Cu_{2-x}Se$ nanocrystals. *J. Am. Chem. Soc.*, **133** (29), 11175–11180.

6 Inorganic Semiconductor Nanomaterials for Flexible Electronics

Houk Jang, Wonho Lee, Min-Soo Kim, and Jong-Hyun Ahn

6.1 Introduction

Flexible electronics has been receiving increasing attention because of its use in displays as well as in a variety of new application areas, including sensory skins and wearable electronic devices [1–4]. These applications require a means to fabricate thin-film transistors (TFTs) and circuits with similar (or increased) performance requirements on thin, flexible substrates with decreased cost per unit area as those based on rigid substrates such as glass and wafers.

Various semiconductors, from organic molecules to carbon-based materials, have been investigated to demonstrate flexible devices using cost-effective large-area printing techniques [5–9]. However, the relatively low mobilities and on/off ratios, as well as the instability of the materials, result in devices with poor performance, thereby restricting the range of application possibilities. As an alternative, direct printing of high-quality, single-crystal inorganic semiconductor nanomembranes (NMs) (or nanoribbons), directly created from source wafers via typical patterning and etching process, onto plastic substrates at relatively low temperatures has been proposed to yield flexible TFTs and circuits. Such materials have attracted much attention because of their excellent properties and their compatibility with current printing technologies. Although many studies have been conducted on various organic materials in different structures including zero-dimensional quantum dots or one-dimensional nanotubes and nanowires [10–14], inorganic nanomembranes (NMs) which exhibit promising electrical properties and have direct compatibility with established industry standard make them good candidates in emerging, large area flexible and stretchable electronics.

This chapter introduces the recent progress in research on inorganic semiconductor nanomaterials in the form of membranes and ribbons for large-area, flexible, and stretchable electronics. The chapter is organized into three main sections: Section 6.2 summarizes the general methods for producing semiconductor nanomaterials and their properties. Section 6.3 presents various

Large Area and Flexible Electronics, First Edition. Edited by Mario Caironi and Yong-Young Noh.

applications of semiconductor nanomaterials, such as flexible solar cells, memories, and radio frequency (RF) devices. The last section summarizes the results and provides the perspectives and trends for future work.

6.2 Characteristics and Synthesis of Inorganic Semiconducting NMs

Unusual properties of zero-dimensional (0D) and one-dimensional (1D) nanomaterials such as quantum dots and nanowires are expected to be similar to those in two-dimensional NMs [15–21]. In this section, the unique properties of inorganic NMs, including mechanical, electromechanical, and optoelectrical, are introduced, as well as the various methods to fabricate them.

6.2.1 Characteristics of Inorganic NMs

6.2.1.1 Mechanical Properties of Inorganic NMs in Bending and Stretching

The fracture strain of inorganic materials is much lower than that of organic materials, with its magnitude ranging from 0.5% to 1.0% [22, 23], which is not suitable for flexible or stretchable electronics [24]. However, their extraordinary electrical properties and well-established processes are still fascinating for large-area flexible electronics. Many researchers have found promising features for flexible electronics using inorganic materials in very thin NM form [25].

Figure 6.1a shows the thin inorganic NMs, less than few hundred nanometers in thickness. The mechanism – how the inorganic NMs exhibit flexibility even though its fracture strain is very small – is shown in Figure 6.1b and is summarized in the following equation [22, 26]:

$$\varepsilon = \frac{t}{2R}$$

where ε is the induced strain, t is the thickness of the NM, and R is the bending radius. Because of the linear relationship between the induced strain and thickness, the induced strain on NMs will be reduced as they get thinner so that even a brittle inorganic material can be integrated into flexible electronics [27]. The mechanical advantages of NMs are shown in Figure 6.1c, indicating the exponential decrease of flexural rigidity and energy release rate. The low flexural rigidity, which is defined as force required to bend a nonrigid structure to a unit curvature or the resistance offered by a structure through the bending process, suggests that the NM form can be bent with a small force, for example, the surface tension of water [28]. With the low flexural rigidity combined with good flexibility of NMs, a stretchable geometry is formed and then the induced strain becomes negligible when a compressive strain (i.e., pre-tensile strain) is applied to NMs to form a wrinkled structure [29, 30]. When a tensile strain

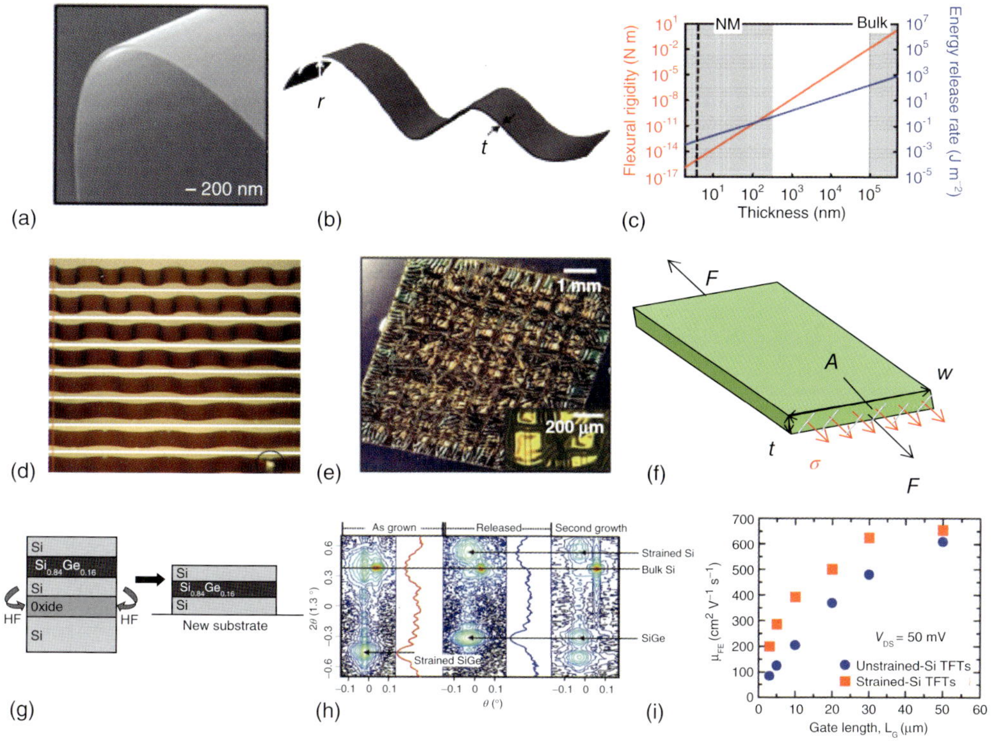

Figure 6.1 (a) SEM image of a Si NM bent at a radius of 1 µm [32]. (Reproduced with permission. Copyright © 2011, Nature Publishing Group.) (b) Schematic illustration showing how thickness provides flexibility to the material [22]. (Reproduced with permission. Copyright © 2008, WILEY-VCH Verlag GmbH & Co. KGaA, Weinheim.) (c) Flexural rigidity and energy release rate as a function of the thickness of Si [32]. (Reproduced with permission. Copyright © 2011, Nature Publishing Group.) (d) Wrinkle structure of Si NMs formed by compressive strain applied by the PDMS substrate [30]. (Reproduced with permission. Copyright © 2009, WILEY-VCH Verlag GmbH & Co. KGaA, Weinheim.) (e) Optical image of circuits on the carrier wafer and magnified view of a single complementary metal–oxide–semiconductor (CMOS) inverter (inset) in a wavy configuration on PDMS [33]. (Reproduced with permission. Copyright © 2008, American Association for the Advancement of Science.) (f) Schematic illustration showing the relationship between force and strain applied to a NM. (g) Schematic diagram of membrane formation and transfer. Etching the oxide by HF releases the membrane, which is then transferred to a new substrate [34]. (Reproduced with permission. Copyright © 2006, Nature Publishing Group.) (h) XRD reciprocal-space maps from a Si/SiGe/Si membrane on an SOI wafer (left), a membrane transferred to a newly oxidized Si substrate (middle), and a transferred membrane with the second SiGe film at a higher Ge concentration (right) [34]. (Reproduced with permission. Copyright © 2006, Nature Publishing Group.) (i) Effective mobility as a function of gate length from the unstrained Si TFTs (blue) and the strained Si TFTs (red) [35]. (Reproduced with permission. Copyright © 2006, AIP Publishing LLC.)

is applied to prestrained NMs with wrinkles, they become flat as the stored compressive strain is released. Then the NMs remain extended as long as a pretensile strain is applied, as shown in Figure 6.1d. In Figure 6.1e, wrinkles appear in both 1D and 2D structures. When an in-plane prestrain is applied, the NMs are folded to form a herringbone structure, which is a 2D wrinkled structure. With negligible bending-induced strain at the surface, the NMs can be integrated into 2D stretchable electronics. So far, the deformation of NMs based on a bending mechanism is discussed in terms of the force applied in the out-of-plane direction. In the meantime, it is also worthwhile addressing the deformation of NMs in in-plane directions. The strain (ε) can be expressed as

$$\left(\frac{F}{A}\right) \times E = \varepsilon$$

$$\varepsilon = \frac{FE}{tw}$$

where F is the applied force perpendicular to the plane, A is the area of the plane, E is the Young's modulus, and w and t are the width and thickness of NMs, respectively. Since force and thickness are linearly dependent, the strain can be applied with a small force effectively, as illustrated in Figure 6.1f. Therefore, it is possible to stretch very thin NM devices with a small mechanical energy. This is useful for nanoelectromechanical application such as nanogenerators [31].

Many attempts have been made to improve the intrinsic electrical properties, such as the mobility, of silicon (Si) by applying strain [36]. In the form of an NM, mobility can be easily enhanced by stacking films despite the existence of stress, because only a small force is needed to generate a tensile strain as discussed above.

Roberts *et al.* [34] presented a novel way to fabricate strained Si NMs by stacking and then isolating Si with a SiGe film from the substrate. Above all, it is grown on an SOI (silicon-on-insulator) wafer of a 56-nm-thick top Si layer followed by a 128-nm-thick undoped $Si_{0.84}Ge_{0.16}$ layer to apply the strain. The trilayer structure (Si/SiGe/Si) is formed by depositing undoped Si up to 48 nm on the SiGe layer as shown in Figure 6.1g. This trilayer structure exerts a compressive strain on the Si because of the lattice mismatch between Si and SiGe. Then the trilayer is floated and released by etching the bottom sacrificial layer. The release process causes tensile strain to develop on the Si layer. Throughout the etching process of the sacrificial layer, alternative flexible or stretchable substrates can be used for fabricating the strained Si layer. The applied strain can be estimated by the X-ray diffraction (XRD) reciprocal-space map, as shown in Figure 6.1h. The difference in the theta value between the released and transferred membrane peak in the $\theta/2\theta$ scan is observed to be as much as $0.091 \pm 0.008°$. These peak separations indicate that a tensile strain of $0.3 \pm 0.02\%$ and a compressive strain of $0.29 \pm 0.02\%$ are applied. The applied strain in the trilayer structure (Si/SiGe/Si) can be expressed as

$$\varepsilon_{Si} = -\varepsilon_m \frac{h_{SiGe}M_{SiGe}}{h_{Si}M_{Si} + h_{SiGe}M_{SiGe}}$$

$$\varepsilon_{SiGe} = \varepsilon_m \frac{h_{Si}M_{Si}}{h_{Si}M_{Si} + h_{SiGe}M_{SiGe}}$$

where ε_m is the mismatch strain, ε is the layer strain, M is the biaxial modulus, and h is the thickness of the layer. The strained semiconducting NM would be implemented in a TFT with enhanced mobility. Figure 6.1i shows that the enhanced mobilities of strained Si TFTs based on the trilayer structure (Si/SiGe/Si) can be as much as 38%. This mobility improvement is attributed to the splitting of the Si conduction band into two valleys and the lowering of the Schottky barrier in the metal contacts [21]. The strained NMs are also good candidates for fast RF applications, as discussed in Section 6.3.3.2.

6.2.1.2 Optoelectrical Properties

Inorganic NMs with thickness over 10 nm have not shown significant differences from the bulk in their band structure and electrical properties [37]. Therefore, the electrical properties from the bulk inorganic semiconductors will also appear in NMs with thickness over 10 nm. In contrast, ultrathin NMs (<10 nm in thickness) show unusual electrical properties as a result of quantum confinement effects [15]. As the thickness of inorganic NMs is controllable, the range of applications can be extended from flexible electronics to optoelectronics including transparent electronics [38]. The fabrication of ultrathin single crystals of Si NM is feasible by repeating the oxidation and etching processes [39]. It has been reported that the bandgap of Si NM can be modified as it becomes thin enough. As can be seen in Figure 6.2a, the energy bandgap increases from 1.1 to 1.8 eV upon decreasing its thickness below 5 nm when the density of state (DOS) in ultrathin Si NM changes as a result of the quantum confinement effect [40]. Their optical property is also closely related to the change in the bandgap structure. The normalized $\alpha^{1/2}$ $(=h\nu - E_g)$ values (α is the absorption coefficient) of Si NMs in Figure 6.2b increase with increase in the incident photon energy near the absorption edge. This indicates that a higher photon energy from the incident light is absorbed by the thinner Si NMs which exhibit larger energy bandgaps [39].

In the meantime, the absorbance spectra also change with the film thickness. InAs NM, which is another kind of 2D semiconductor, shows the quantum confinement effect in absorption [41]. Absorbance spectra of InAs NMs with various thicknesses are shown in Figure 6.2c. The absorbance as a function of photon energy shows an absorbance step of ~1.6%. This absorbance step is attributed to the quantized interband transition between the 2D subbands. This quantum confinement effect in 2D semiconductors can have broad applications for novel photonic and optoelectronic devices. Additionally, the intensity of the transmitted light within a sample is given by $I(s) = I_0 e^{-\alpha s}$, where s is the

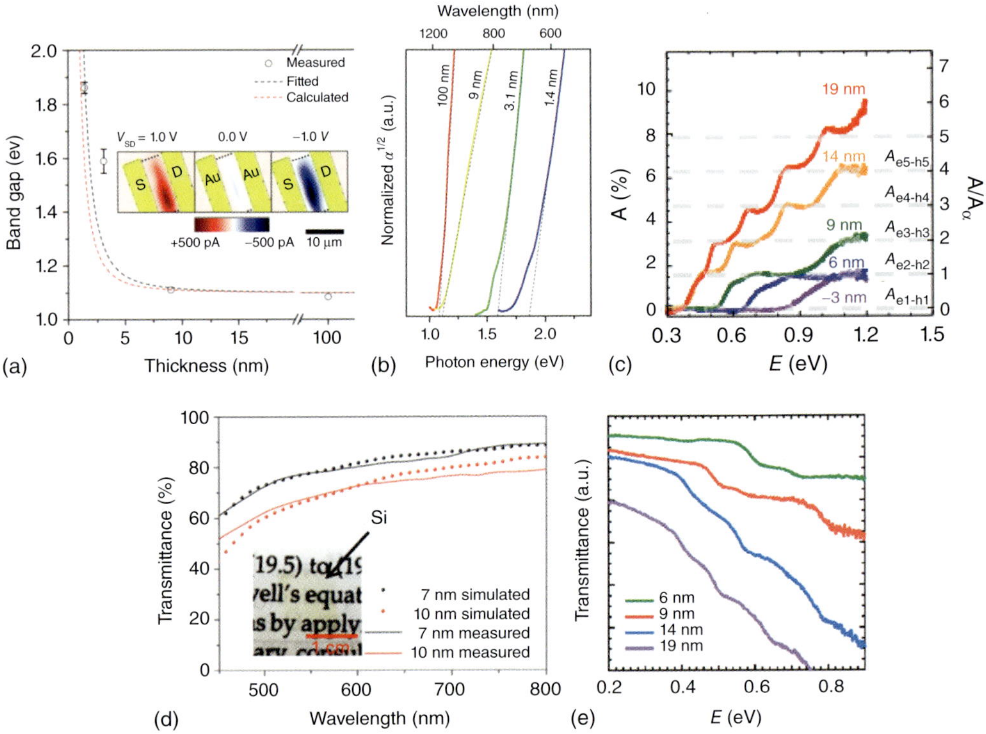

Figure 6.2 (a) The measured (circle), fitted curve (black dash), and theoretically calculated (red dash) bandgap of the Si as a function of thickness. The inset shows I_{ph} images of the ultrathin Si NMs with various drain voltages, which shows that the photocurrent is generated mainly within the Si itself [39]. (Reproduced with permission. Copyright © 2013, American Chemical Society.) (b) Normalized $\alpha^{1/2}$ as a function of phonon energy to show the absorption edges of the ultrathin Si NMs with various thicknesses [39]. (Reproduced with permission. Copyright © 2013, American Chemical Society.) (c) Absorbance spectra of 2D InAs quantum membranes (QMs) with thickness from 3 to 19 nm. Dashed line indicates 1.6% of absorbance [38]. (Reproduced with permission. Copyright © 2013, National Academy of Sciences.) (d) Simulated (dotted line) and measured transmittance (solid line) of a structure of native oxide/Si NM/epoxy/PET assembly. The thickness of the Si NM is 10 (red) and 7 nm (black) [39]. (Reproduced with permission. Copyright © 2013, American Chemical Society.) (e) Transmission spectra for various InAs QM thicknesses, illustrating clear steps, corresponding to the optical transitions between the 2D subbands [41]. (Reproduced with permission. Copyright © 2011, American Chemical Society.)

propagating distance of the light in the sample and I_0 is the intensity of the incident light. As the thickness of the NM approaches a few nanometers, the transmittance increases exponentially [39].

The transmittance values for 7- and 10-nm Si NMs were evaluated as 78% and 68%, respectively, at a wavelength of 550 nm, which matches well with the simulated transmittance as illustrated in Figure 6.2d. Figure 6.2e shows the transmittance characteristics of InAs NMs with various thicknesses as a function of the energy bandgap. The semitransparent nature of ultrathin NMs can be applied to fabricate transparent electronic devices.

6.2.2 Fabrication of Inorganic NMs for Flexible Electronics

6.2.2.1 Selective Etching

Organic materials are good candidates for flexible and large-area electronics because of their intrinsic flexibility, stretchability, and compatibility with solution processes [4, 9, 42]. But poor electrical properties of organic materials make their applications difficult in high-performance electronic system. On the other hand, many studies on NMs have found the potential of inorganic materials such as single-crystalline Si or compound semiconductors in the manufacture of flexible, high-performance electronic systems. To form the various NMs, a highly selective etching process to remove the sacrificial layer underneath the NM has been introduced [43]. Figure 6.3 shows the schematic structure of a wafer with a sacrificial layer and an NM stacked together, which is necessary to apply selective etching. An important feature of this structure is that it is compatible with high-temperature processes so that the quality of the NMs can be improved prior to the formation of flexible membranes. This process begins by stacking films of the required materials on proper sacrificial layers/substrates. Several representative materials for NMs and sacrificial layers, as well as proper etchants for selective etching, are summarized in Table 6.1.

Single-crystal Si on a SiO_2 sacrificial layer, the so-called SOI, is a good platform for single-crystal Si NMs with thicknesses from few micrometers down to 10 nm [44]. Brittle oxide materials such as ZnO and compound semiconductors such as GaAs and GaN can be grown on the corresponding sacrificial layers such

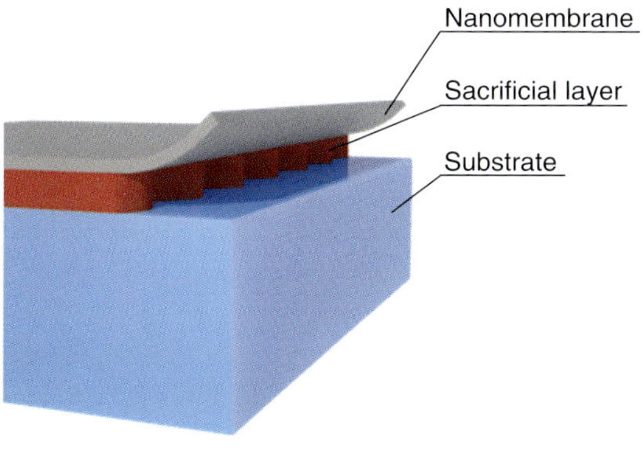

Figure 6.3 Schematic diagram of a structure for fabricating a NM released by removing a sacrificial layer underneath the NM.

Table 6.1 The representative combination of nanomembrane/sacrificial layer and their etchants to fabricate inorganic nanomembranes.

	Nanomembrane	Sacrificial layer	Substrate	Etchant
1	Single-crystal Si	SiO_2	Si wafer	HF
2	ZnO	GeOx	Si wafer	H_2O
3	GaAs	AlAs	GaAs wafer	HF
4	GaN	Si	Si wafer	TMAH
5	PZT	Amorphous Si	Si wafer	SF_6 RIE

as GeO_x, AlAs, and Si, respectively. After high-temperature processing (up to 350 °C) to form highly crystalline films, the sacrificial layers (GeO_x, AlAs, and Si) are removed by etchants such as H_2O, HF, and TMAH, generating high-quality ZnO, GaAs, and GaN NMs, respectively [45–47].

To implement this method in specific applications, the generated NMs need to be transferred onto arbitrary substrates throughout the wet or dry transfer process. Note that recent progress in the dry transfer method presents a "roll-to-roll" process which enables the transfer of NMs continuously over a large area [48–50]. In contrast to the previous methods, PZT ($PbZr_xTi_{1-x}O_3$, lead zirconate titanate), a ceramic perovskite material that exhibits the piezoelectric effect, NMs can be generated on an amorphous Si sacrificial layer by a dry etching process such as reactive ion etching (RIE) using sulfur hexafluoride (SF_6) plasma, instead of a wet etching process [51, 52].

6.2.2.2 Anisotropic Etching

Studies on anisotropic wet etching of single-crystal Si reveal that the etching rate can vary by as much as 400 times depending on the crystalline orientation [53]. Anisotropic etching process enables the generation of high-quality single-crystal Si NMs from a bulk single-crystal Si wafer, which has been widely used in semiconductor industry for half a century [53, 54].

Figure 6.4a shows the generation of Si NMs from bulk (100) and (111) wafers. Metal or Si_3N_4 is commonly used as a mask for the wet etching process using TMAH or KOH solution [43]. When it comes to anisotropic etching behavior, the slower etching rate (of nearly 400 times) along the <111> direction compared to the <110> direction directs the etching process into NMs to generate Si NMs. The resulting NMs from a bulk (100) wafer are developed into triangular and plate-like shapes from a bulk (111) wafer as a result of its crystal direction, as shown in Figure 6.4b.

NMs generated from (100) wafers have advantages in electrical properties, including higher mobility and low subthreshold voltage. On the other hand, NMs from (111) wafers are more favorable to large-area flexible applications because the NMs of triangular shape from (100) wafers maintain their narrow ribbon shapes.

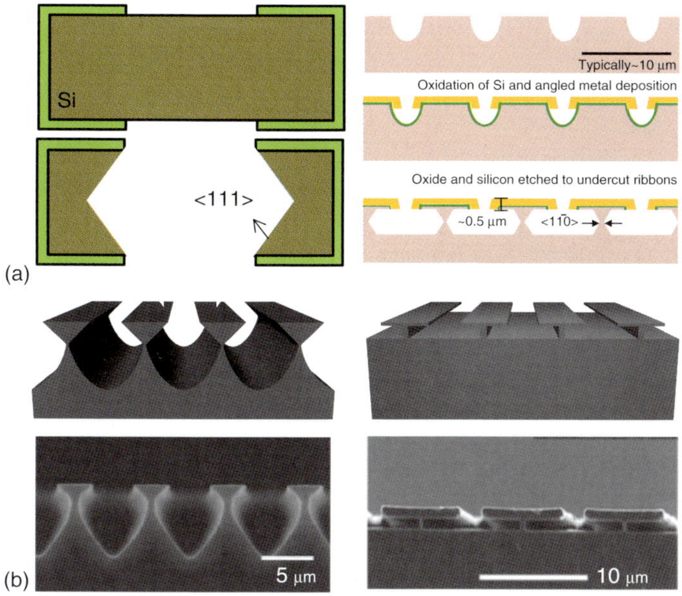

Figure 6.4 (a) Schematic diagram of the anisotropic etching phenomenon in Si (100) wafer (left) and (111) wafer (right) [43, 54]. (Reproduced with permission. Copyright © 2012, IOP Publishing, Ltd., 2010, AIP Publishing LLC.) (b) Schematic diagram of fabricated NMs and the corresponding SEM images in Si (100) wafer (left) and (111) wafer (right) [54, 55]. (Reproduced with permission. Copyright © 2010, AIP Publishing LLC, 2007, WILEY-VCH Verlag GmbH & Co. KGaA, Weinheim.)

6.2.2.3 **Mass Production of Inorganic NMs**

Although the use of a sacrificial layer is essential to yield high-quality inorganic NMs, this process involves a high cost to produce the structure. A repeatable transfer process can be one of the solutions to lower the production cost, thus suggesting the possibility to meet industrial requirements [55–57]. As can be seen in Figure 6.5a, the process consists of the repetitive growth of a sacrificial layer (AlAs) and a compound semiconductor layer (GaAs). The selective etching of AlAs using HF can release as many GaAs layers as necessary from the substrate. After the transfer process, the costly GaAs substrates can be recycled. Figure 6.5b shows the secondary ion mass spectroscopy (SIMS) data from the structure, and Figure 6.5c demonstrates partially fabricated stacked GaAs NMs during the selective etching process. An image of mass-produced GaAs NMs is shown in Figure 6.5d. Moreover, the fabrication of NMs through the anisotropic etching process has an advantage over the selective etching process because anisotropic etching is highly compatible with the established industry standards. Figure 6.5e illustrates the schematic of the mass-production process. The rippled wall is developed through inductively coupled plasma reactive ion etching (ICP-RIE), which induces a partially

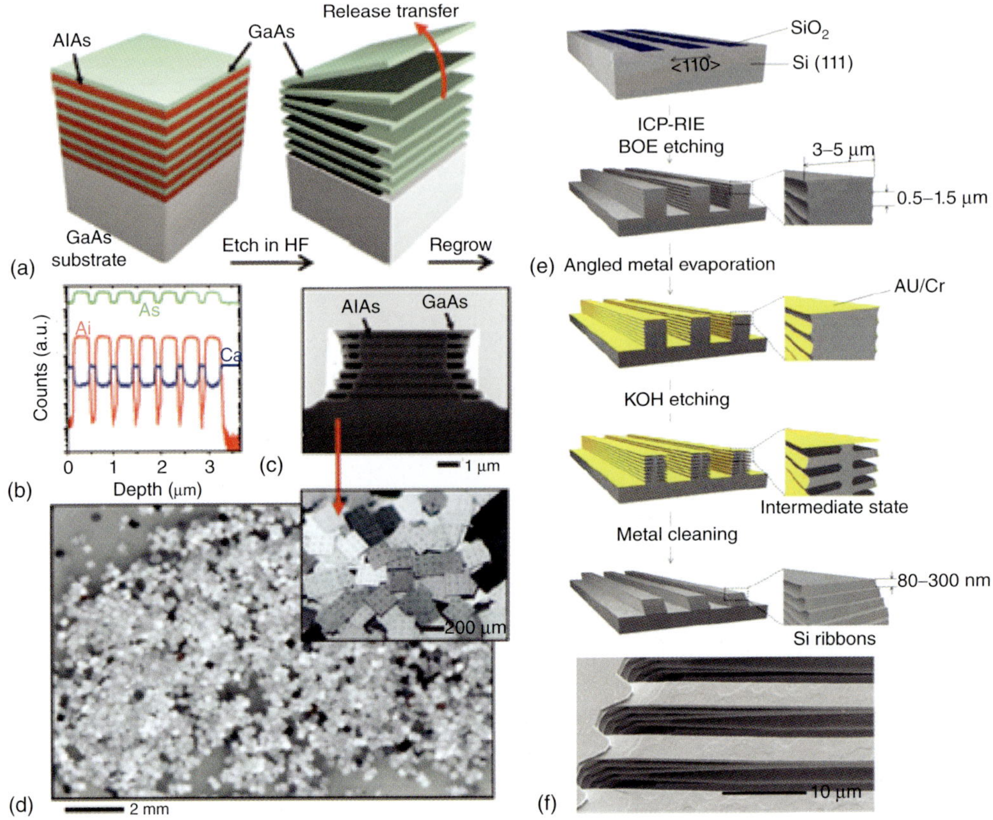

Figure 6.5 (a) Schematic diagram of the repeatable transfer method by repetitive stacking of AlAs and GaAs films. (b) Depth profile of the structure for the components including Al, As, and Ga using SIMS. (c) SEM image of the structure in the middle of etching process for the sacrificial layer. (d) Image of mass-produced GaAs NMs. (a–d) [58]. (Reproduced with permission. Copyright © 2010, Nature Publishing Group.) (e) Schematic diagram of the procedure for the mass production of Si NMs. (f) SEM image of the source wafer after extracting one Si NM. (e–f) [56]. (Reproduced with permission. Copyright © 2006, American Chemical Society.)

uncovered region during the angled metal deposition process. The deposited Cr/Au layer is used as a mask for the etchant (i.e., KOH), and the uncovered region is anisotropically etched to form multistacked single-crystal Si NMs. Figure 6.5f shows the bulk wafer after the transfer process of the top-placed Si NMs, indicating the capability of repetitive generation of Si NMs. Once all the Si NMs are transferred, the wafer can be reused after a polishing process.

6.2.2.4 Transfer Process

After fabricating inorganic NMs by the method discussed above, the membranes would be transferred to or suspended on arbitrary substrates or structures. Especially, when it comes to crystalline materials, a proper transfer process is required. Here, we introduce several transfer methods for inorganic NMs using various kinds of polydimethylsiloxane (PDMS) stamps. These stamps are used to transfer NMs from a mother substrate to a target substrate.

The transfer method consists of two steps: extracting NMs from a mother substrate, and releasing them onto a target substrate. The second step is far more challenging because the available transfer methods are restricted owing to the target substrates which are soon to be completed as the final form.

Since the report of the first trial using an adhesive layer, various methods have been developed to transfer NMs without any adhesive layer, as shown in Table 6.2. Transfer process 1 is the first demonstration of a dry transfer method using adhesive layers. However, the use of an adhesive layer limits its applicability in the transfer process because it does not facilitate electrical connection, the stacking process, and multiple transfer.

Therefore, the transfer processes 2–6 suggest various methods to release NMs without any adhesive layers. Process 2 utilizes patterned PDMS stamps of which the contact area to the NM is reduced so that the adhesive force between the stamp and the NM is reduced accordingly. Process 3 demonstrates a method to adjust the adhesive force by controlling the kinetics. This process manipulates the energy release rate, which directly corresponds to the adhesive force between a substrate and an NM by adjusting the separation speed. Process 4 also offers patterned PDMS stamps as in process 2, but this method overcomes the problems caused from the extracting process by using dynamically transformable PDMS. The adhesive force in process 2 is relatively weak, so that the transfer yield is low in the extracting process. In contrast, process 5 requires the application of a shear force to release the NM, taking advantage of decrease in the pull-off force under shear stress. This approach is a good method for transferring NMs onto hard substrates without using any adhesive layers. But it is not suitable for transfer onto soft substrates, because of the possible deformation of the substrate. Process 6 makes use of the thermal energy generated by a laser. The wavelength of the laser should

Table 6.2 The representative transfer processes for NMs using various release mechanisms.

Process	Adhesive layer	Release mechanism	Remarks	References
1	O	Adhesion layer	First demonstrated	[1]
2	X	Adhesion area control	Rectangular shape	[2]
3	X	Kinetic control	Separation speed	[3]
4	X	Adhesion area control	Pyramid shape	[4]
5	X	Shear force	Decrease in pull-off force	[5]
6	**X**	Heating by laser	—	[6]

be within a specific range which is transparent to PDMS but not the NMs. Then, the laser light would be absorbed at the interface between the membrane and the PDMS stamp. With the thermal energy generated by the laser, the NM will expand, thus releasing it from the PDMS stamp. All the demonstrated methods can be combined together to realize a high transfer yield in a specific environment of the target substrates used.

6.3 Applications in Flexible Electronics

The recent progress in the synthesis and fabrication of NMs has paved the way for NMs to be integrated into flexible electronics and has made it possible to overcome the technical challenges in various systems such as electronic, biomimetic, and mechanical [28, 33, 59]. The approaches to regulate the configuration of NMs anisotropically provide (i) a weak induced strain at the surface of the NMs upon bending; (ii) an extremely low stiffness, leading to easy bending and easy sticking to randomly curved surfaces; and (iii) easy deformation by the in-plane forces [27, 55]. In the following section, we will introduce representative applications, reflecting the various advantages as stated above.

6.3.1 Flexible Electronics

As introduced in the previous section, NMs have reduced strain on their surface, and are bendable enough for flexible electronic applications. In this section, we will introduce various applications of NMs, such as flexible solar cells, flexible memories, and flexible RF devices.

6.3.1.1 Flexible Solar Cell

Solar cells have been intensively studied as one of the environment-friendly energy harvesting systems with reliable efficiency, especially with regard to their cost and weight reduction [60, 61]. The good quality, thinness, and potential for mass production of NMs constitute some of the solutions to the current technical challenges [57, 58, 62].

Figure 6.6a shows a flexible solar cell made using (111) single-crystalline Si NMs. The fabrication of the cell begins with selective doping by injecting phosphorus and boron to make an optimized p–n junction for the solar cell. Anisotropic etching, as introduced in the previous section, can be carried out with KOH solution to form Si NM structures. The generated Si NMs are transferred onto a plastic substrate by the dry transfer method using a PDMS stamp to form arrays of micro cells. The NMs are now attached tightly on the substrate by using a photocurable polymer (NOA61) which is of 30 μm thickness, and they are interconnected with gold lines. Although thin NMs are useful to reduce the strain on their surface,

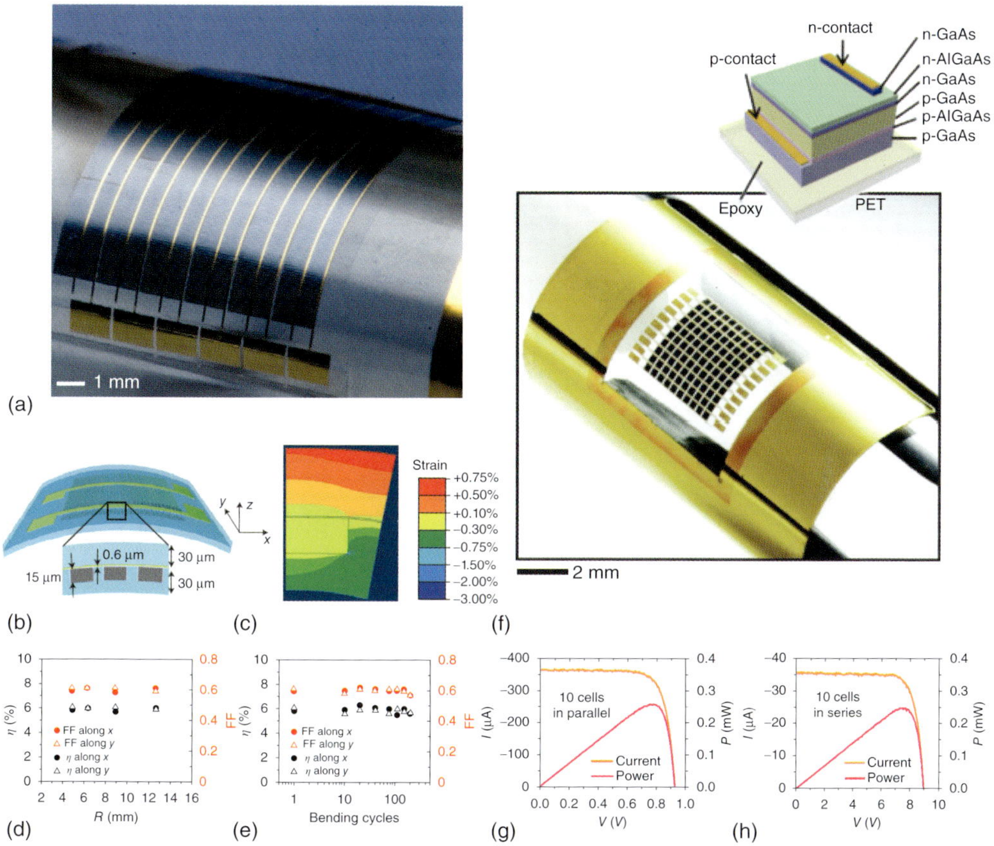

Figure 6.6 (a) Optical image of a flexible single-crystal Si solar cell. (b) Schematic illustration of a solar cell designed to place a neutral mechanical plane at the single-crystal Si NMs. (c) Measured current density as a function of driving voltage under AM: Air Mass 1.5 illumination in a flat state (blue) and bent state along the *x*-axis (red) and the *y*-axis (green). (d) Efficiency (η) and fill factor (FF) as a function of the bending radius (R) and (e) bending cycles up to 200 times at a bending radius of 4.9 mm [57]. ((a–e) Reproduced with permission. Copyright © 2008, Nature Publishing Group.) (f) Optical image of a flexible solar module on PET consisting of a 10 × 10 array of GaAs solar cells. The inset shows the structure of GaAs-based quantum well. (g) Light current and power as a function of voltage measured from GaAs solar modules in parallel interconnection and (h) in series interconnection of 10 cells in a flat state [58]. ((f–h) Reproduced with permission. Copyright © 2010, Nature Publishing Group.)

a thick substrate would increase the strain, on the other hand. It is a good solution to control and place the neutral mechanical plane in the middle of the NMs. Figure 6.6b shows a schematic diagram of a flexible solar cell. Si NMs are placed at the neutral mechanical plane covered with an epoxy layer. Strain analysis using the finite element method (FEM) suggests that negligible strain (<0.1%) is present

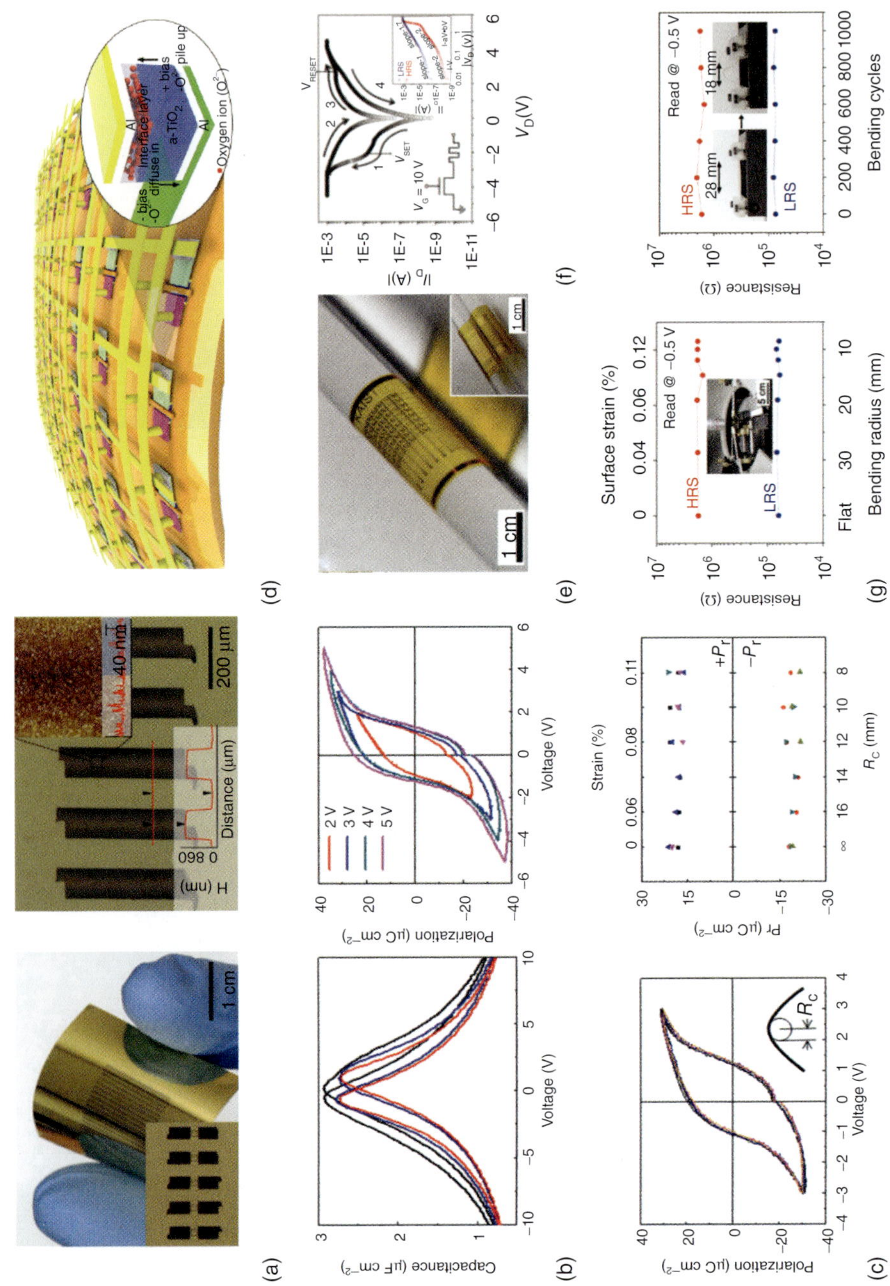
(a)
1 cm
200 μm
40 nm
H (nm)
Distance (μm)
(b)
Capacitance (μF cm−2)
Voltage (V)
Polarization (μC cm−2)
2 V
3 V
4 V
5 V
(c)
Strain (%)
Pr (μC cm−2)
+Pr
−Pr
RC (mm)
(d)
Al
Interface layer
a-TiO2
−bias
+bias
Oxygen ion (O2−)
pile up
diffuse in
(e)
1 cm
(f)
VG = 10 V
VSET
VRESET
VD(V)
LRS
HRS
(g)
Surface strain (%)
Read @ −0.5 V
HRS
LRS
Resistance (Ω)
Bending radius (mm)
Flat
28 mm
18 mm
Bending cycles

at the Si NM, which is the most brittle part in this device geometry, as shown in Figure 6.6c.

The module's performance is evaluated under AM 1.5 illumination, and it shows robust, consistent operation with bending radii of 12.6, 8.9, 6.3, and 4.9 mm. The module efficiency (η) is ~6.0% and the fill factor (FF) is ~0.6 even at the minimum bending radius, and these values are quite consistent even after 200 bending cycles (Figure 6.6d,e). The same strategy is applicable to compound semiconductors such as GaAs for high-performance electronic applications. The heterojunction structure of multiple GaAs NMs is transferred onto a poly(ethyleneterepthalate) (PET) substrate using epoxy as an adhesive layer by the selective etching process, as discussed previously (Figure 6.6f).

The fabricated solar cell is bendable enough to be wound around a curved surface of glass because of the thin GaAs active layer which is the most brittle component in the cell structure. As shown in Figure 6.6g,h, Yoon *et al.* demonstrated the voltage and current enhancement by connecting 10 solar cells in parallel and series, respectively, and the structure still maintained its flexibility.

6.3.1.2 **Flexible Memory**

As introduced in the previous section, it is worthwhile making brittle inorganic semiconductor materials into flexible forms by controlling the strain at the surface of NMs using thinning methods. This strategy has beneficial influences not only on semiconducting materials but also on inorganic dielectric materials which are essential components in transistor and memory applications [51, 63]. For flexible memory applications, organic ferroelectric materials such as poly(vinylidenefluoride-*co*-trifluoroethylene) (P(VDF-TrFE)), which is intrinsically flexible, have been investigated in the last few decades. However, the low degree of crystallinity and relatively poor performance hinder these

Figure 6.7 (a) Optical image of PZT capacitor arrays on a polyimide substrate (left). The inset shows a magnified image after fabrication. The optical microscope image of PZT ($PbZr_xTi_{1-x}O_3$) ribbons (right). The inset presents the surface roughness and thickness of PZT ribbons. (b) Representative C–V curve of the PZT capacitance measured on a bulk wafer (black), a PDMS (blue), and a plastic substrate (red) (left). The polarization hysteresis loop of a PZT capacitor on a plastic substrate with various driving voltages (right). (c) Hysteresis loop of a capacitor in a bent state (black: flat; orange: $R_c = 8$ mm; navy: 10 mm; dark yellow: 12 mm; magenta: 14 mm; blue: 16 mm) (left) and the change in terms of P_r values at different bending radii (right) [51]. ((a–c) Reproduced with permission. Copyright © 2010, Institute of Electrical and Electronics Engineers.) (d) Schematic illustration of a flexible 8 × 8 RRAM array on a plastic substrate. The inset shows the schematic structure and model for resistive switching of an a-TiO_2-based memristor. (e) Optical image of a flexible RRAM array wrapped around a quartz stick. The inset shows the possibility of conformal contact on a curvilinear surface. (f) Drain current versus voltage characteristics of the RRAM device on a flexible substrate. The left inset displays a circuit diagram and the right inset shows double-logarithmic plot in the negative sweep region. (g) Resistance ratio between the HRS and the LRS as a function of surface strain and bending radius (left) and as a function of bending times up to 1000 (right) [63]. ((d–g) Reproduced with permission. Copyright © 2011, American Chemical Society.)

organic materials from being used for practical applications [64]. In this section, flexible memory using inorganic NMs such as PZT and TiO_2 will be introduced to elucidate improved electrical properties from inorganic NMs rather than organic materials. Among the various materials used for memory, PZT is a good candidate for large-area and high-performance nonvolatile memory devices owing to its simple fabrication by the sol–gel method and to its high ferroelectric characteristics. A PZT layer of 360 nm thickness is spin-coated on the Pt/Ti bottom electrodes which are deposited on the amorphous Si sacrificial layer. As shown in Figure 6.7a, the patterned PZT ribbon arrays can be transferred onto plastic substrates by the dry transfer method after removing the sacrificial layer by NF_3 or XeF_4 plasma. In Figure 6.7b, the capacitor-type ferroelectric random access memory (FeRAM) shows negligible change in the capacitance even after the transfer process. In addition, the remnant polarization (P_r) increases from ~17 μC cm^{-2} at 2 V to 25 μC cm^{-2} at 5 V. These ferroelectric properties provide robust operation and consistent performance upon bending, as shown in Figure 6.7c. The minimum bending radius (R_c) of 8 mm displayed in the P_r vesus R_c plot corresponds to 0.11% strain. In the meantime, as the integration density of memory devices increase, cell-to-cell interference becomes a critical issue. However, this can be overcome by integrating the memory components in transistor structures, such as a single-transistor type memory device.

Kim *et al.* [63] successfully demonstrated NOR-type flexible resistive random access memory (RRAM) by fabricating high-performance single-crystal Si transistors with titanium oxide-based memristors. Figure 6.7d illustrates a schematic diagram of an RRAM with integrated titanium oxide-based memristors. The structure shown in the inset contains a 14-nm-thick a-TiO_2 layer deposited between the aluminum electrodes and a transistor with a channel length 10 μm, width 200 μm, and thickness 100 μm. The flexibility of an RRAM is demonstrated in Figure 6.7e, where the memory is wrapped around a glass bar. The mobility of the Si transistor is as high as 340 cm^2 V^{-1} s^{-1}, which cannot be achieved by organic-based devices. Moreover, RRAMs with a low threshold voltage require a low operation voltage (I_D ~ 300 μA at V_D = 1 V, and V_G = 3 V). In Figure 6.7f, the device experiences a transition from the high-resistance state (HRS) to the low-resistance state (LRS) as the drain voltage is swept from 0 V to the set voltage. The LRS is retained in a region below the positive reset voltage and is switched to the HRS in the region above reset voltage. Both HRS and LRS show significantly uniform resistance upon bending, with the bending radius of 28.6 mm down to 8.4 mm and even after 1000 cycles (Figure 6.7g).

6.3.1.3 Flexible High-Frequency Transistor

Transistors are basic components of logic devices to be operated at high frequencies. For flexible transistors, organic-based materials have been intensively studied with respect to their bending capability. However, the poor electrical property of organic-based transistors is a major obstacle for their operation at high frequencies because of their low carrier mobility. Fabricating thin Si and III–V compound semiconductor devices with plastic substrates enables the realization of flexible transistors operating in the high-frequency regime [65, 66]. There have been reports on flexible devices using Si for which the cut-off frequency was demonstrated as ~5 GHz without any scaling process, so that, with optimization, the stage delay of a 100-stage ring oscillator was reduced by as much as ~20 ps [67].

Figure 6.8a shows an optical image of flexible high-frequency transistor arrays in which single-crystal Si NMs are successfully integrated. A schematic diagram of a two-finger structure to enhance the operating frequency is shown at the bottom of the figure. The NMs as strain-released forms maintain their mobility in the active layer within 20% with a bending radius of 3 mm, which corresponds to over 0.4% strain for both compressive and tensile stress, as shown in Figure 6.8b.

In turn, stable maximum oscillation frequency (f_{max}) and cut-off frequency (f_T) are demonstrated in Figure 6.8c, which is attributed to the stable mobility in the transistor even when a strain is applied between −0.08% and 0.23%. In addition, passive components, such as inductors and capacitors, and active devices, such as transistors, are important for the whole system to be operational at high frequencies. Figure 6.8d shows the fabrication process of a flexible inductor and capacitor for RF applications, including metal deposition and dielectric layer deposition on a plastic substrate. The fabricated flexible inductor and capacitor arrays on a plastic substrate are displayed in Figure 6.8e. The 4.5-turn inductor shows stable inductance (L) around 6 nH between 45 MHz and ~5 GHz with a resonant frequency (f_{res}) at 9.1 GHz and a peak quality factor (Q) of 14.6 at 3.45 GHz (Figure 6.8f). The capacitance is as high as 0.45 pF at 5 GHz and f_{res} of 13.5 GHz in the flat condition (blue), as shown in Figure 6.8g. Additionally, the inductance and capacitance were measured under bent conditions with a minimum bending radius of 28.5 mm, showing the promising aspects of integrated circuits under a flexible configuration. This was also validated in active and passive devices that are operational at high frequencies.

6.3.1.4 **Foldable Transistor Using Ultrathin Si NMs**

The flexibility can be fully implemented by controlling the thickness of each component of an electronic device, such as the active layer, the dielectric, the electrodes, and the substrate. Jang *et al.* demonstrated ultrathin Si transistors by combining ultrathin Si NMs (7 nm) with a self-assembled monolayer (SAM, 4 nm) and graphene (0.3 nm). The total thickness of the ultrathin Si transistor was <12 nm on an SU-8 (a common negative photoresist based on epoxy) substrate (500 nm).

Figure 6.9a shows a TEM image (top) of the heterogeneous structure of three different materials and a schematic diagram of the transistor (bottom) [39]. Because of its ultrathin nature, the bending-induced strain is released dramatically so that the entire device can be easily folded at the edge of PET substrates (thickness: 25 μm) corresponding to a bending radius of 12.5 μm (Figure 6.9b). The transistor maintains its performance without significant change, as can be seen in Figure 6.9c. The device arrays are attached strongly to the edge of PET (bottom-left in Figure 6.9b) because they can be bent or even folded with a small

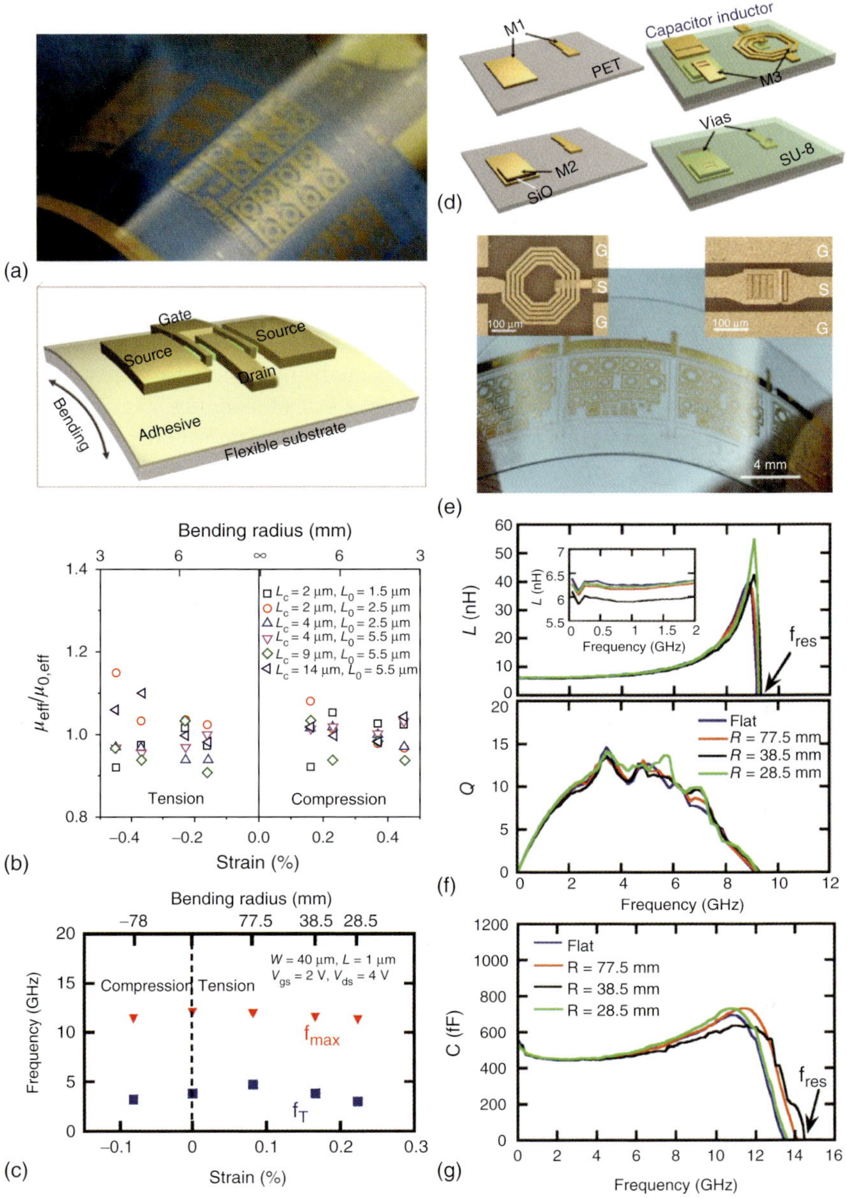
(a)
Gate
Source
Source
Drain
Bending
Adhesive
Flexible substrate
(b)
Bending radius (mm)
3 6 ∞ 6 3
L_c = 2 μm, L_0 = 1.5 μm
L_c = 2 μm, L_0 = 2.5 μm
L_c = 4 μm, L_0 = 2.5 μm
L_c = 4 μm, L_0 = 5.5 μm
L_c = 9 μm, L_0 = 5.5 μm
L_c = 14 μm, L_0 = 5.5 μm
$\mu_{eff}/\mu_{0,eff}$
Tension
Compression
Strain (%)
(c)
Bending radius (mm)
−78 77.5 38.5 28.5
W = 40 μm, L = 1 μm
V_{gs} = 2 V, V_{ds} = 4 V
Compression
Tension
Frequency (GHz)
f_{max}
f_T
Strain (%)
(d)
M1
PET
Capacitor inductor
M3
M2
SiO
Vias
SU-8
(e)
G
S
G
100 μm
4 mm
(f)
L (nH)
Frequency (GHz)
f_{res}
Q
Flat
R = 77.5 mm
R = 38.5 mm
R = 28.5 mm
(g)
C (fF)
Flat
R = 77.5 mm
R = 38.5 mm
R = 28.5 mm
f_{res}
Frequency (GHz)

Figure 6.8 (a) Optical image of RF transistors with two finger system (top) and schematic illustration of the device under bending test (bottom). (b) Normalized effective mobility as a function of bending radius and bending-induced strain for devices with various dimensions. (c) Measured f_T and f_{max} variations under different bending radii. (d) Schematic illustration of fabrication steps for flexible spiral inductors and metal–insulator–metal (MIM) capacitors. (e) Fabricated inductors and capacitors on a bent PET substrate. The left inset shows an optical microscope image of a 4.5-turn spiral inductor and the right inset shows an MIM capacitor. (f) Measured L values (top) and Q values (bottom) of the flexible inductor as a function of frequency with various bending radii from flat to 28.5 mm. The inset shows the magnified region between 0 and 2 GHz. (g) Measured capacitance from an MIM capacitor as a function of frequency in the flat and bent state [65]. (Reproduced with permission. Copyright © 2012, IOP Publishing, Ltd.)

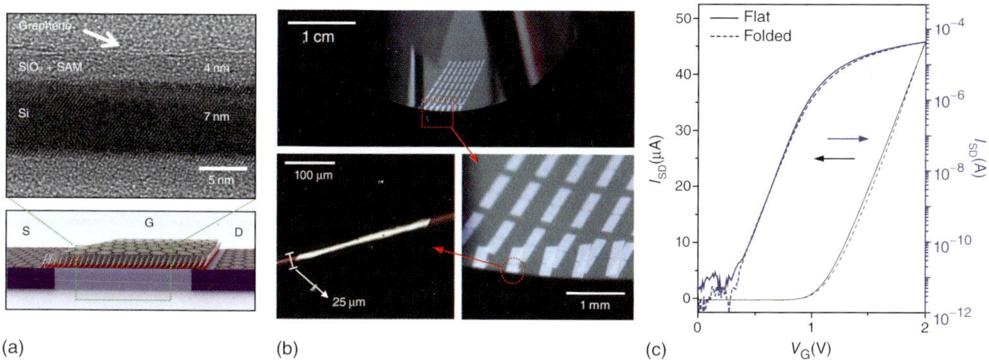

Figure 6.9 (a) Schematic illustration of an ultrathin transistor and a TEM image of the channel region. (b) Fabricated arrays of U-TFTs on an epoxy sheet of 300 nm thickness are folded around the edge of a PET substrate of 25 μm thickness (bending radius <12.5 μm). The magnified images (bottom) show folded U-TFTs at the edge without cracks. (c) Transfer characteristics of a U-TFT with a channel width of 200 μm and length of 20 μm at 0.1 V of drain voltage in linear (left axis) and log scale (right axis) before (solid line) and after folding (dashed line) [39]. (Reproduced with permission. Copyright © 2013, American Chemical Society.)

amount of force such as, for example, the surface tension of water. Low stiffness of the device structure makes conformal contact possible over rough surfaces, and this will be discussed in the next section.

6.3.2
Conformal Device

The ultrathin configuration of NMs offers not only high flexibility but also low bending stiffness. The stiffness, implying the resistance to the deformation of materials, is an important factor in determining the capability of materials when

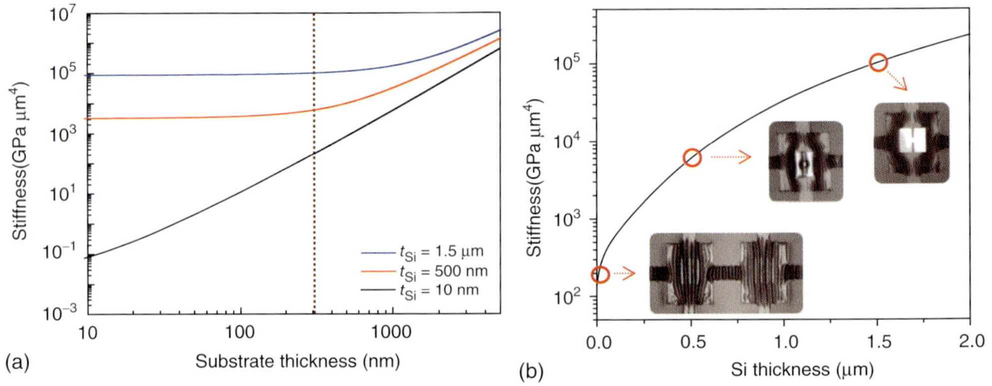

Figure 6.10 (a) Plot of the calculated bending stiffness versus substrate thickness with various thicknesses of Si in the log–log scale. (b) Calculated stiffness of U-TFT-based inverters with various thicknesses of Si NMs. A thin encapsulation layer (~150 nm) of epoxy on top of the inverters on PDMS places the devices at the neutral mechanical plane. The inset images correspond to each Si thickness marked with red circles [39]. (Reproduced with permission. Copyright © 2013, American Chemical Society.)

transformed [39]. Figure 6.10a presents the calculated stiffness of Si/SU-8 as a function of the substrate (SU-8) thickness with three different thicknesses of Si NMs in the log–log scale. This simulated stiffness is well matched with the experimental data for a wrinkled Si on a prestretched PDMS substrate, as shown in the inset of Figure 6.10b. The wrinkled structure can be obtained when the compressive strain from the prestrained substrate is large enough to transform the attached NMs. When the NM bends, the bending energy corresponding to the bending stiffness must be smaller than the adhesive force between the attached NM and the substrate so that sliding or separation of the NM from the substrate does not occur. A low stiffness is necessary for the NM to store compressive strain or to be attached onto rough surfaces by deforming its shape to fit the surface.

6.3.2.1 **Conformal Biomonitoring System**

It is a worthwhile strategy to integrate a device on the surface of the brain to monitor biosignals without causing any damage to it, in contrast to the preliminary research in which metal needles were usually inserted into the brain. However, the rough and soft surfaces of organs provide a difficult environment for electronic components to be placed on them. Furthermore, it would be even more challenging to utilize inorganic-based materials in a biomonitoring system. On the other hand, water, a major component of the organs, can exhibit a surface energy of 10 mJ m^{-2}, thus exerting an adhesive force [28]. This surface energy is enough to deform the device arrays of NMs along the rough surface of organs so that the arrays can be stuck closely to the organs to read out the electrical signals generated by the organs, especially the brain. As a result, devices in the NM form can make intimate contact with the surface of organs without damaging them. In fact, this is not the case for the conventional tip-type monitoring system which is rigid and brittle [28, 68].

Figure 6.11a illustrates electrode arrays on various substrates of different thicknesses on an artificial brain. The panels clearly demonstrate that the electrode arrays on a 25-μm-thick substrate can conformally cover the surface of the brain compared to thicker versions. Kim *et al.* [28] developed a novel process to fabricate electrode arrays on ultrathin substrates (2.5 μm) using dissolvable silk fibroin. Figure 6.11b shows the electrode arrays placed on a feline brain using dissolvable silk fibroin, measuring spatially resolved evoked potential response. However, the spatial resolution of the monitoring system becomes limited as the packing density of the switching circuits and amplifiers reaches its limit. To observe the evoked potential response of a brain with higher spatial resolution, Viventi *et al.* [69] demonstrated flexible logic circuits based on Si NMs. Figure 6.11c shows an optical image of electrode arrays for recording evoked potentials at the brain surface. Effectively, integrated circuits with a switching system are essential to increase not only the spatial resolution but also the temporal resolution, as shown in Figure 6.11d.

In Figure 6.11e, an *in vivo* experiment is demonstrated using these devices on the curvilinear surface of a heart to monitor the pulsation and visualize the cardiac motion. The 288-electrode array can analyze the activation patterns from the heart by mapping the cardiac motions. A series of cardiac signals during multiple beating cycles are shown in Figure 6.11f with a spatial resolution of ~500 μm and time resolution of 3.2 ms. Voltage signals from all channels are spatially constructed with color levels, and the noise level of recording is as low as 0.3 mV (Figure 6.11g).

6.3.3 Stretchable Electronics

As discussed in the previous section, materials in the NM form provide great opportunity for them to be integrated into flexible electronic devices. Furthermore, to realize human-friendly interface systems such as wearable computers and stretchable electronics, stretchability of the integrated circuits is critical. Strictly speaking, the materials in NM form are not suitable for stretchable electronics because of the high stress exerted within a small area when an in-plane force is applied. Therefore, extra effort should be made to improve stretchability so that even a wrinkled structure can store compressive strain. In this section, several representative studies will be addressed on stretchable electronics utilizing NMs.

6.3.3.1 Stretchable Logic Circuit

Transistors and logic circuits in flexible or even foldable forms have been demonstrated by using various NMs so that integrated circuits can be transformed with little external force as discussed in the previous section. The features of NMs are highly desirable to form wrinkled structures not only in one dimension but also

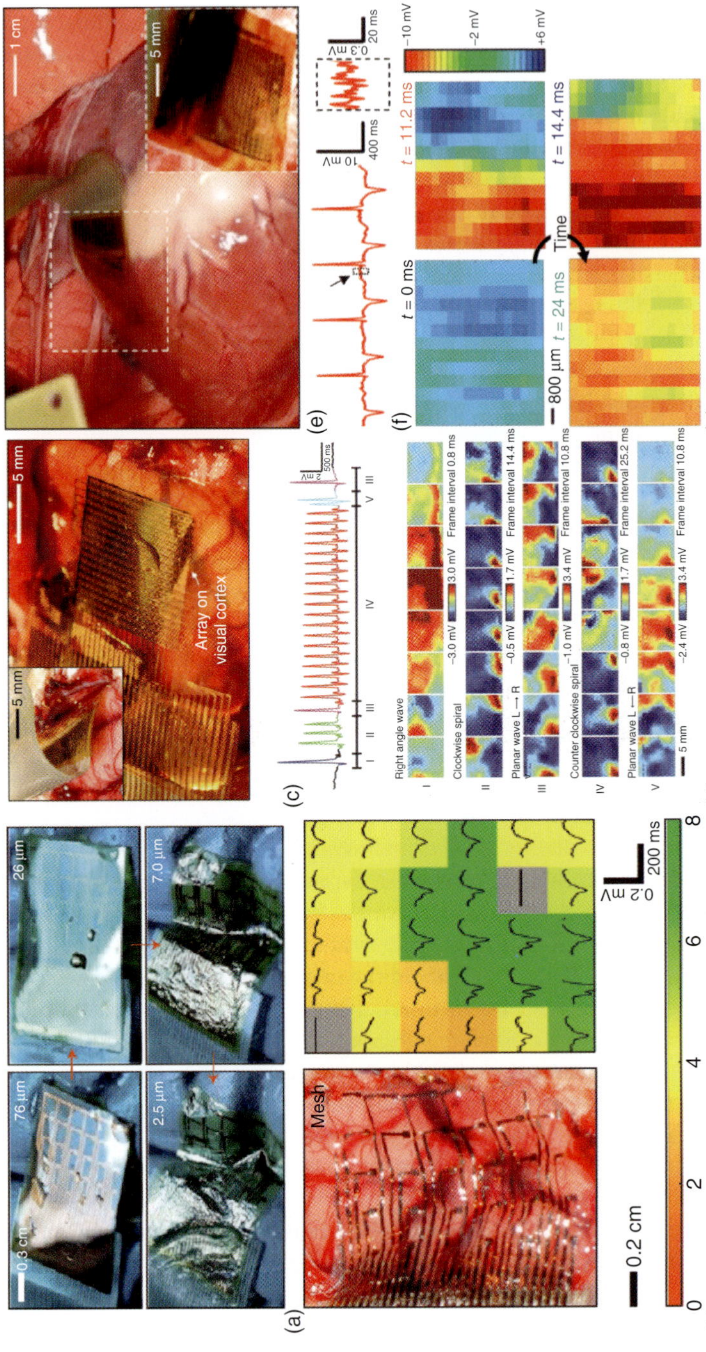
(a)
0.3 cm
76 µm
26 µm
2.5 µm
7.0 µm
(b)
Mesh
0.2 cm
0.2 mV
200 ms
0
2
4
6
8
(c)
5 mm
Array on visual cortex
I
II
III
IV
V
500 ms
(d)
Right angle wave
Clockwise spiral
Planar wave L → R
Counter clockwise spiral
Planar wave L ← R
Frame interval 0.8 ms
Frame interval 14.4 ms
Frame interval 10.8 ms
Frame interval 25.2 ms
Frame interval 10.8 ms
5 mm
(e)
1 cm
5 mm
10 mV
400 ms
0.3 mV
20 ms
(f)
t = 0 ms
t = 11.2 ms
800 µm
Time
(g)
t = 24 ms
t = 14.4 ms
−10 mV
−2 mV
+6 mV

Figure 6.11 (a) Series of pictures illustrating how the substrates of different thicknesses influence their conformal contact on the surface of a model brain [28]. (b) Optical image of conformal contact for electrode arrays on a brain (left), and clearly recorded signals generated from the brain (right). The two gray blocks are used as reference electrodes [28]. ((a, b) Reproduced with permission. Copyright © 2010, Nature Publishing Group.) (c) Flexible and high-density active electrode arrays placed on the visual cortex. The inset demonstrates the inserted electrode arrays in an inter-hemispheric fissure [70]. (d) μECoG signal from a representative channel of the electrode arrays during a short electrographic seizure (top panel). Movie frames (bottom) show varied spatial and temporal μECoG voltage patterns from all 360 electrodes during the labeled time intervals from the top panel. The frame intervals and color scale are provided for each set of eight movie frames, and the color scale is saturated at the 2nd and 98th percentile over eight frames to improve the visual display [70]. ((c, d) Reproduced with permission. Copyright © 2011, Nature Publishing Group.) (e) Photograph of flexible devices conforming to the cardiac tissue via surface tension. (Inset) Magnified image at another viewing angle. (f) Representative single voltage trace without external pacing. (Inset) Magnified view of system noise from the region indicated with an arrow. The signal-to-noise ratio (SNR) of the recorded signal is ~34 dB. (g) Time evolution of representative voltage distributions from all electrodes presenting the normal cardiac wavefront propagation. Color-coded scale is given on the right corner [6]. ((e–g) Reproduced with permission. Copyright © 2010, American Association for the Advancement of Science.)

in two dimensions, known as the *herringbone structure* [33, 71]. A 2D stretchable inverter array is shown in Figure 6.12a. Each inverter is fabricated with two transistors, and the inverter arrays in the NM form on an arbitrary substrate are transferred onto a prestrained elastomeric substrate. The total thickness including the substrate is <1.7 μm. Upon releasing the prestrained substrate, compressive strain is applied to form the herringbone pattern on the device so that the structure would store in-plane compressive strain. This provides stretchability to the wrinkled structure when a tensile strain is applied. Moreover, the stretchability strongly depends on the pre-strain. Figure 6.12b illustrates stretchable inverters with pre-strains of 2.7%, 3.9%, and 5.7% from left to right at the top, respectively. A 3D FEM and an SEM image of an inverter compressed with pre-strain of 3.9% are shown at the bottom. The formation of wrinkles mostly occurs in the interconnection area, which indicates regional differences in stiffness along the device due to the changes in stacking thickness and material type. The performance of the inverters is presented in Figure 6.12c. From the $V_{out} - V_{in}$ plot (upper panel), gain (red) is calculated from the output voltages (blue). The gain approaches 100, indicating the high performance of the stretchable inverter. The performance of p-MOS and n-MOS is also shown in the inset. The calculated and measured threshold voltage (V_M) with tensile strain along the x-axis (red) and y-axis (black) stays within the error range of 20%, as shown at the bottom of Figure 6.12c.

Kim *et al.* [33] demonstrated a highly integrated differential amplifier circuit by combining nine transistors. The optical images in Figure 6.12d show that the stretchable differential amplifier undergoes tensile strain of 0% (top) and 5% (bottom). As can be seen in Figure 6.12e, the stretchable differential amplifier

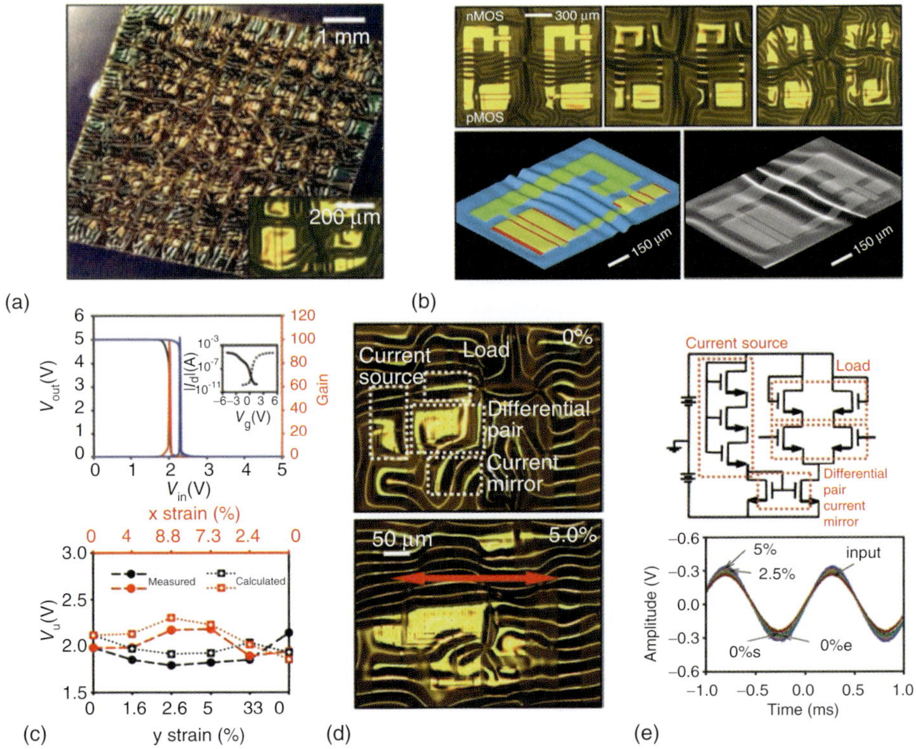

Figure 6.12 (a) Optical image of circuits in a wavy configuration on PDMS and magnified views of a single CMOS inverter (inset). (b) Wavy Si-CMOS inverters on PDMS, formed with various levels of pre-strain (top left, $\varepsilon_{pre} = 2.7\%$; top center, $\varepsilon_{pre} = 3.9\%$; top right, $\varepsilon_{pre} = 5.7\%$). Structural configuration determined by full 3D FEM of a system formed at $\varepsilon_{pre} = 3.9\%$ (bottom left) and perspective scanning electron micrograph of a sample fabricated with a similar condition (bottom right). (c) Simulated (blue) and measured (red and black) transfer characteristics from wavy inverters (top), and n- and p-channel metal–oxide–semiconductor field-effect transistors (MOSFETs) (solid and dotted lines, respectively, in the inset). Measured (solid circles) and simulated (open squares) threshold voltages from the inverters in terms of applied *x* strains (top, red) and *y* strains (bottom, black). (d) Optical images of a wavy differential amplifier in its as-fabricated state (top) and under an applied strain (5.0%) in a direction along the red arrow (bottom). (e) Circuit diagram of a differential amplifier (top) and output characteristics at various strain values (bottom) [33]. (Reproduced with permission. Copyright © 2008, American Association for the Advancement of Science.)

consists of a current source, a differential pair, a current mirror, and a load part. Stable output with various tensile strains up to 5% is demonstrated in Figure 6.12e.

6.3.3.2 Stretchable Light-Emitting Diode

Even though organic light-emitting diodes (OLEDs) retain an intrinsically flexible property which is advantageous for flexible displays, the light-emitting performance in terms of brightness, lifetime, and efficiency is still not comparable to those of inorganic-based LEDs [72]. The major challenge for the inorganic-based LEDs is to make them into flexible circuits. One of the strategies to realize this is to adopt inorganic materials in NM forms and transfer the NMs onto flexible substrates using a wrinkled structure to store the compressive strain. As discussed above, adopting materials in the NM form and transferring them onto flexible substrates paved the way for overcoming the challenges in flexible electronics [47].

Figure 6.13a and the insets display a 16×16 array of bendable quantum-well structure transferred on a polyurethane substrate. The inorganic light-emitting diode (ILED) consists of AlInGaP, AlGaInP, GaAs, and AlAs and gives out red light. Each cell in the array can be selectively operated by driving circuits through an interconnected metal mesh. The devices exhibit high current (>0.1 mA) at small driving voltages (<4 V) even after the transfer process, suggesting the possibility of stable performance with proper optimization (Figure 6.13b).

Wavelengths of the emitted light before and after the transfer process are comparable, as shown in Figure 6.13c. When the active regions and the interconnects are encapsulated with polymers, the neutral mechanical plane would be located at the center of the stacked structure. The encapsulated devices with pre-strain are transformed to be stretched on a rubber substrate, as can be seen in Figure 6.13d. These stretchable devices operate with a small driving voltage of 9.2 V and generate high current levels, as high as 100 μA. In addition, the devices exhibit stable performance under high stretching levels, even up to 22% strain (Figure 6.13e). When the strain applied is up to 22%, the ILED device and quantum-well structure experience at most 0.17% and 0.026% of strain, respectively, with respect to the interconnects in which most strain is concentrated. Reliability of the stretching property for the devices can be investigated by stretching cycle tests, as shown in Figure 6.13f. Stretchable ILED devices encapsulated by polymers maintain their operating voltage even after a few hundred stretching cycles.

6.3.3.3 Photodetector

When conventional optoelectronics becomes stretchable, there will be many advantages in unusual biomimetic systems derived from visual organs of human and even of arthropods [59, 73]. For instance, photodetector arrays mounted on a flat surface are limited in terms of the focusing area as well as the tunability of magnification. This limitation can be overcome

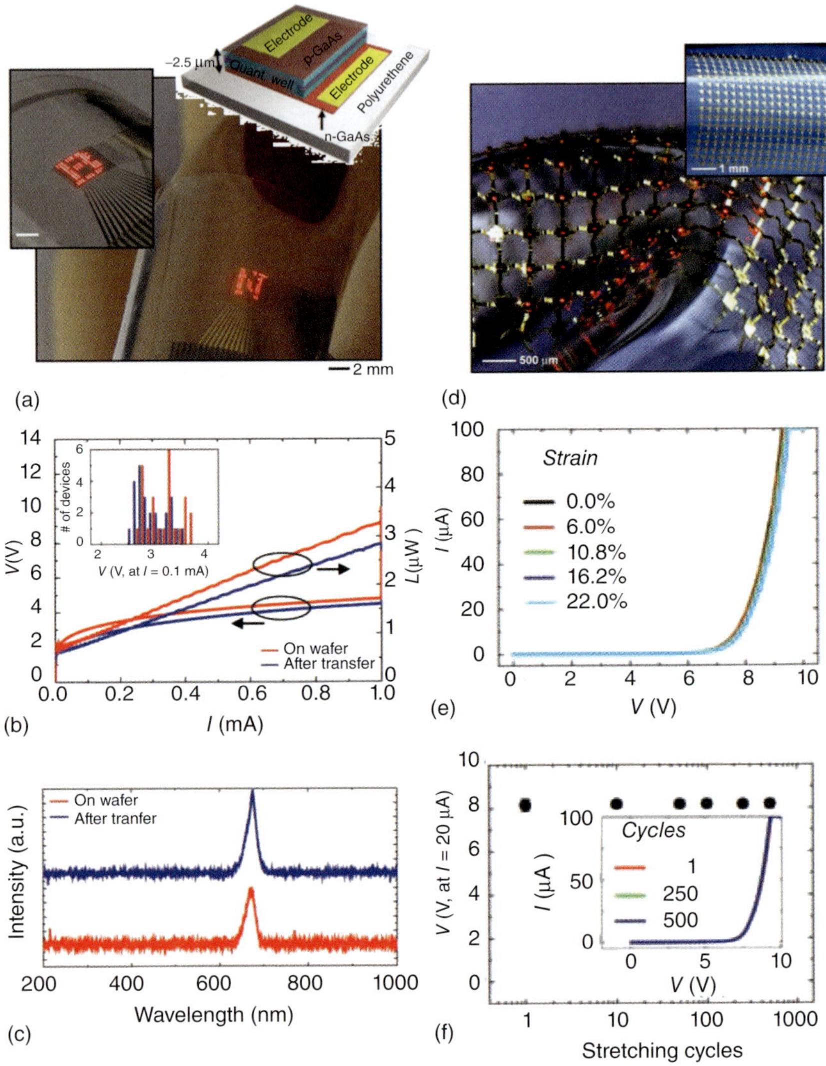
~2.5 μm
Electrode
p-GaAs
Quant. well
Electrode
Polyurethene
n-GaAs
2 mm
(a)
500 μm
1 mm
(d)
V(V)
L(μW)
# of devices
V (V, at I = 0.1 mA)
On wafer
After transfer
I (mA)
(b)
Strain
0.0%
6.0%
10.8%
16.2%
22.0%
I (μA)
V (V)
(e)
Intensity (a.u.)
On wafer
After tranfer
Wavelength (nm)
(c)
V (V, at I = 20 μA)
Cycles
1
250
500
I (μA)
V (V)
Stretching cycles
(f)

Figure 6.13 (a) Optical image of an ILED array on PET, wrapped around a finger of mannequin and a cylindrical glass tube (inset). (b) Current–voltage–emission characteristics of a GaAs LED before (red) and after transfer (blue). The inset shows a histogram of the number of devices versus the required voltage to produce currents of 0.1 mA. (c) Spectral characteristics of emission from a typical device on the wafer (red) and after transfer printing (blue). (d) Passive-matrix stretchable ILED display that uses a noncoplanar mesh configuration on a rubber substrate. The interconnect lines between adjacent devices are supported by arc-shaped bridge structures which can be deformed in response to an applied strain. Both the main panel and the inset image are collected from an automated camera system that combines pictures at different focal depths to provide a sharp, composite image. (e) Measured *I*–*V* curve from a representative ILED in the array with different applied strains. (f) Voltage needed to generate a current of 20 mA measured after stretching cycles up to 500 times at an applied strain of 22%. The inset shows the *I*–*V* behaviors after the cycling tests [47]. (Reproduced with permission. Copyright © 2009, American Association for the Advancement of Science.)

by fabricating photodetector arrays on a curvilinear surface, just like the human eye.

As can be seen in Figure 6.14a, stretchable photodetector arrays on a thin elastomeric PDMS membrane make a dynamically tunable detector. The radius of the hemisphere of the detector array can be controlled by liquid pressure, whereas it maintains its initial condition when strain is absent. This stable state is possible because the stored compressive strain would compensate the newly generated tensile strain in the array.

In Figure 6.14b, a large amount of strain (up to 10%) is concentrated in the middle of the interconnects of the photodetector arrays. Clear differences between planar and hemispherical configurations of the arrays are illustrated in Figure 6.14c. The image from the planar arrays reveals that only the corner areas are in focus while the central area is generally out of focus. In contrast, the images from the hemispherical arrays are uniformly in focus both at the corner and at the center. Therefore, it is possible to obtain focused images over the entire exposed area by using stretchable hemispherical arrays.

The strategy for systems mimicking the human eye using stretchable photodetector arrays can be also applied to mimic arthropods. Different from humans, arthropods have evolved to have a complex imaging system, including a wide field of view, low aberration, high acuity to motion, and infinite depth of field. Therefore, it is difficult to understand arthropods' visualizing mechanism because of its utterly disparate nature. Building photodetector arrays on a hemisphere would facilitate a biomimetic imaging system which imitates the visual organs of arthropods, as can be seen in Figure 6.14d. The photodetector arrays are arranged on the hemisphere with blocking diodes covered with an elastomeric microlens array with its maximum strain analyzed up to 50%. However, because of the strain concentration and the strain compensation by stored compressive strain, the applied strain is negligible. When an image of a soccer ball is placed on the left (−50°), on

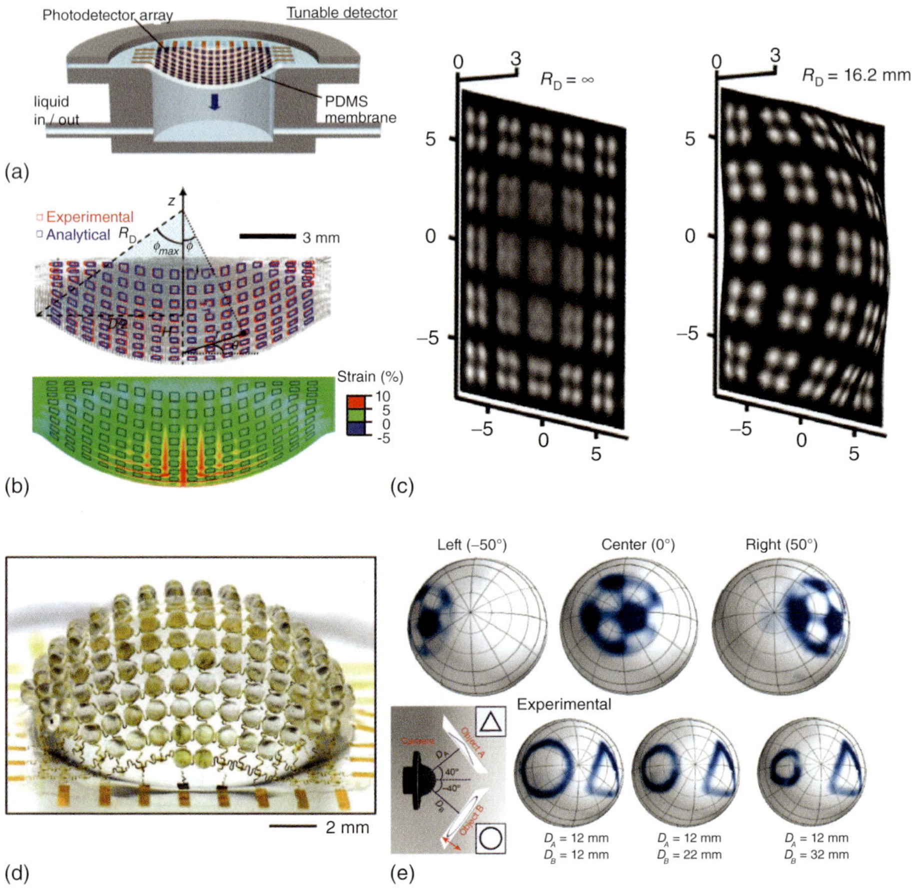
Photodetector array
Tunable detector
liquid
in / out
PDMS
membrane
(a)
Experimental
Analytical
R_D
ϕ_{max}
ϕ
3 mm
Strain (%)
10
5
0
-5
(b)
$R_D = \infty$
R_D = 16.2 mm
(c)
2 mm
(d)
Left (−50°)
Center (0°)
Right (50°)
Experimental
Camera
Object A
Object B
40°
−40°
D_A = 12 mm
D_B = 12 mm
D_A = 12 mm
D_B = 22 mm
D_A = 12 mm
D_B = 32 mm
(e)

Figure 6.14 (a) Schematic illustration of a camera including a tunable lens (top) and tunable detector (bottom) modules. The tunable lens is placed between a thin PDMS membrane (0.2 mm) and a glass window (1.5 mm), forming a plano-convex lens with 9 mm diameter. And the tunable detector consists of an array of interconnected Si photodiodes and blocking diodes (16 × 16 pixels) mounted on a thin PDMS membrane (0.4 mm) based on a mechanically optimized, open-mesh, serpentine design. This "lens-detector" sheet is mounted on a fluid-filled cavity. Regulating the pressure transforms the sheet into concave or convex hemispherical shapes with well-defined, tunable levels of curvature. (b) 3D rendering of the profile of the deformed surface measured by a laser scanner. The shape is close to a hemisphere with a radius (R_D) of 13.3 mm and a maximum deflection (H_D) of 2.7 mm. Calculated (blue) and measured (red) unit cell positions appearing as squares on this rendered surface (top). 3D rendering of circumferential strains in the Si devices (squares) and the PDMS membrane determined by finite element analysis (bottom). (c) Images of a test pattern of bright circular disks, acquired by the device in flat (left) and deformed hemispherical configurations (right), collected using a glass plano-convex lens (diameter: 9 mm; focal length: 22.8 mm). These images are rendered on the surfaces that match to those of the detector array. The distance between the lens and the source image is 75 mm. In the flat configuration (left), only the peripheral regions of the image are in focus. The image in the curved region (right) was acquired simply by actuating the detector into this shape without changing other aspects of the setup. This deformation brings the entire field of view to be focused as a result of matching the detector shape to the Petzval surface [73]. ((a–c) Reproduced with permission. Copyright © 2011, National Academy of Sciences.) (d) Image of lens arrays transformed on a hemispherical surface. (e) Imaging of a soccer ball captured at three different polar angles relative to the center of a camera: −50° (left, upper), 0° (center, upper), and 50° (right, upper). Schematic illustration of an experimental setup (bottom) to demonstrate key imaging characteristics. One object (object A, triangle) lies at an angular position of 40° and distance D_A; another object (object B, circle) at −40° and distance D_B. Pictures of these objects are collected at different values of D_B [59]. (Reproduced with permission. Copyright © 2013, Nature Publishing Group.)

the right (50°), and at the center (0°) of a camera, the arrays demonstrate the possibility of a wide field of view. Clear imaging of circular and triangular shapes which are 80° apart is illustrated at the bottom of Figure 6.14e. Sophisticated imaging system of the arthropod is feasible in the biomimetic system by integrating stretchable optoelectronic components.

6.3.4 Utilizing Deformation of NMs

As discussed in the previous section, materials in the NM form are suitable for flexible, conformal, and stretchable electronics because they can be easily deformed or bent with a perpendicularly applied force without inducing strain at the NM. On the other hand, an in-plane force can generate strain easily because the stress (i.e., the applied force per unit area) becomes larger with smaller thickness. In this section, strategies for integrating NMs into a strain-induced system will be discussed, including a nanogenerator and an RF transistor using strained Si.

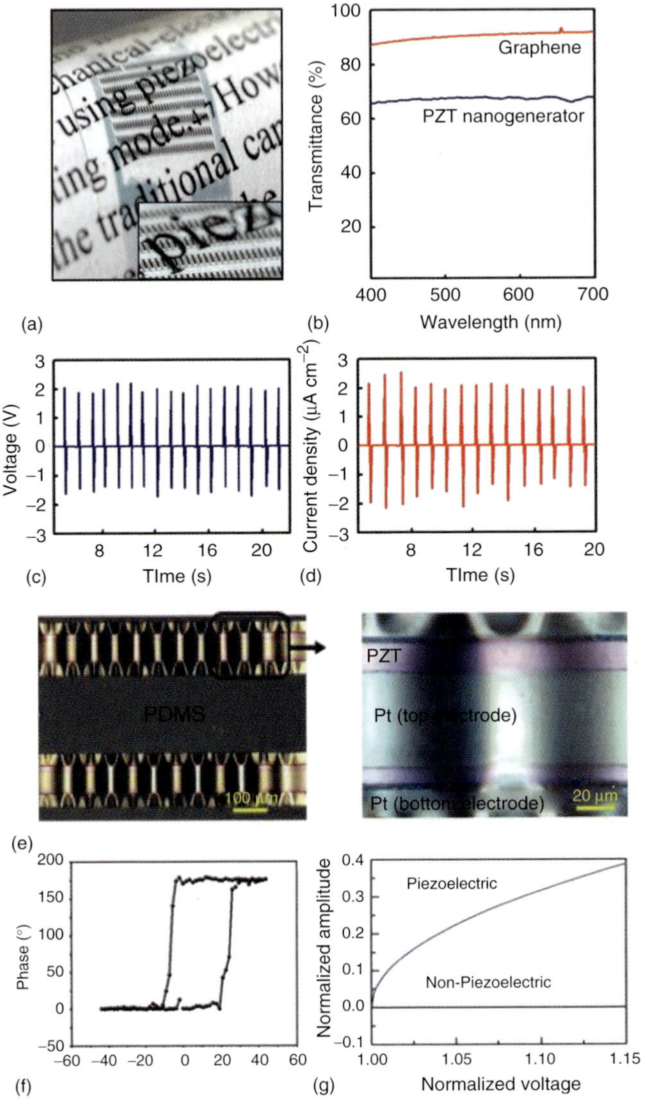

(a)
(b)
Transmittance (%)
Wavelength (nm)
Graphene
PZT nanogenerator
(c)
Voltage (V)
TIme (s)
(d)
Current density (μA cm−2)
TIme (s)
(e)
PDMS
100 μm
PZT
Pt (top electrode)
Pt (bottom electrode)
20 μm
(f)
Phase (°)
(g)
Normalized amplitude
Normalized voltage
Piezoelectric
Non-Piezoelectric

6.3.4.1 Nanogenerator and Actuator

Nanogenerators have attracted intense attention in energy harvesting because of key applications such as energy harvesters for artificial organs. The mechanism of the energy harvester involves the utilization of their piezoelectric characteristics which convert mechanical energy into electrical energy and vice versa. The voltage or current generators using this mechanism can fit well into the human heart. It is possible not only to fabricate various portable electronic devices consuming small power using the generated energy but also to stimulate heart using these nanogenerators as actuators when problems such as arrhythmia or heart attack occur within a heart.

At the early stage of the research, wurtzite-structured inorganic nanowires of ZnO, CdS, and GaN had been used because of their piezoelecric properties [74]. However, the performance of these nanowire devices degrades gradually as a result of the cracks and fractures generated in the active nanowires [75]. On the other hand, devices based on sol–gel PZT and polyvinylidene fluoride (PVDF) film are introduced by simply applying spin-coating and annealing [52, 76]. These devices show reliable and excellent nanogenerating properties even though they are not aligned vertically. Since PZT films have a larger piezoelectric constant than other piezoelectric materials, a thin PZT film can realize high-performance flexible nanogenerators on plastic substrates [31]. However, it is difficult to deposit PZT films by the sol–gel method on plastic substrates because a high-temperature process (>500 °C) is required for its crystallization. To solve this problem, the PZT films are initially deposited on a conventional SiO_2 wafer, and then crystallized through a high-temperature process. Then, the crystallized PZT films are etched like ribbon arrays by a top-down process. Once the bottom sacrificial layer is etched by lateral dry etching, ribbon arrays become suspended so that delamination and transfer onto a PET substrate can be easily effected with a polymer adhesive layer. As a result, a flexible PZT nanogenerator can be fabricated on a PET substrate, as can be seen in Figure 6.15a [52]. Especially, when graphene rather than a metal is introduced as the electrode, it provides semitransparency to the structure. As shown in Figure 6.15b, the transmittance of the nanogenerator exceeds 60%. This nanogenerator develops a high output voltage of ~2 V and a current density of ~2 μA cm^{-2} (Figure 6.15c,d). To generate

Figure 6.15 (a) Optical image of a fabricated nanogenerator with a combination of transferred PZT ribbon arrays on a PET substrate using NOA as an adhesive layer and graphene electrodes. The inset shows a magnified image. (b) The measured transmittance from the nanogenerator and the graphene electrodes. (c) The generated voltage and (d) current density from the PZT nanogenerator as a function of time when a load is applied to the generator by touch [52]. ((a–d) Reproduced with permission. Copyright © 2012, The Royal Society of Chemistry.) (e) Optical image of a PZT NM forming a wavy structure on prestretched PDMS (left) and its magnified image (right). (f) Piezoresponse force microscopy (PFM) hysteresis amplitude loop of a wavy PZT ribbon on PDMS. (g) Calculated buckling amplitude versus normalized voltage from wavy ribbons with and without piezoelectric effects [77]. ((e–g) Reproduced with permission. Copyright © 2011, American Chemical Society.)

this level of electrical signal, the force of 0.9 kgf is applied to the device upon bending. The matched strain value is found to be 4×10^{-4}%. In addition, the devices show reliable operation even after 7 days.

The PZT nanogenerator also works in reverse way: that is, converting electrical energy to mechanical deformation like an actuator. Nonplanar layouts of piezoelectric system have been developed on elastomeric substrates [77]. The PZT-based piezoelectric material can be delaminated from a wafer and transferred onto a PDMS substrate, as shown in Figure 6.15e. The amplitude of buckling can be modulated by applying electric fields, and about 1000 times larger values than that of the planar layout of piezoelectric systems can be achieved. Pre-strain is applied to the elastomeric substrate before the PZT thin ribbons are transferred. Uniform buckling is established on the PZT ribbons because of the compressive strain, and the top surface of the ribbons experiences tensile strain in the bent state as shown in Figure 6.15e. These transferred PZT ribbons on a PDMS substrate show piezoresponses in the presence of the driving voltages and they demonstrate stable phase conversion behavior (Figure 6.15f). The normalized buckling amplitude as a function of the normalized voltage is plotted in Figure 6.15g. The applied voltage induces buckling in a piezoelectric ribbon (blue), but there is no amplitude in a nonpiezoelectric film (black). These new nonplanar layouts of piezoelectric films on an elastomeric substrate suggest the possibilities of novel applications in piezoelectric systems.

6.3.4.2 RF Device Using Strained NMs

The strategy of using external stress is applicable to RF devices requiring high mobility, which can be generally improved as a tensile strain is applied. The heterogeneous structure of Si and Ge has been intensively studied for the suggested purpose. However, the method of generating strain and dislocation-based relaxation causes critical problems like developing rough surfaces, which are difficult to solve. Introducing the NM form can be one of the solutions when it is prepared with suspended NMs because strain can be easily induced by a thin Si–Ge strained layer.

Roberts *et al.* [34] proposed high-quality Si NMs with a very thin Si–Ge strained layer without dislocation, as discussed in the previous section. And Zhou *et al.* [78] presented fast flexible electronics using high-quality Si NMs. Figure 6.16a presents a schematic process for fabricating RF devices using a strained Si NM by transferring it onto a flexible substrate and forming a dielectric layer and the electrodes. The optical image of the RF device is shown on the left of the figure. The fabricated device can be bent because bending-induced strain is reduced when the material becomes thin enough, as discussed earlier. In Figure 6.16b, significant mobility enhancement is clearly displayed from a strained Si (>400 $cm^2 V^{-1} s^{-1}$) compared to an unstrained Si (270 $cm^2 V^{-1} s^{-1}$). And this device operates with an additional strain of up to 1%. As a result, both the maximum oscillation frequency (f_{max}) and cut-off frequency (f_T), as measures of the RF device performance, are improved from 11.5 and 3.5 GHz (unstrained silicon) to 15 and 5 GHz (strained silicon), respectively (Figure 6.16c).

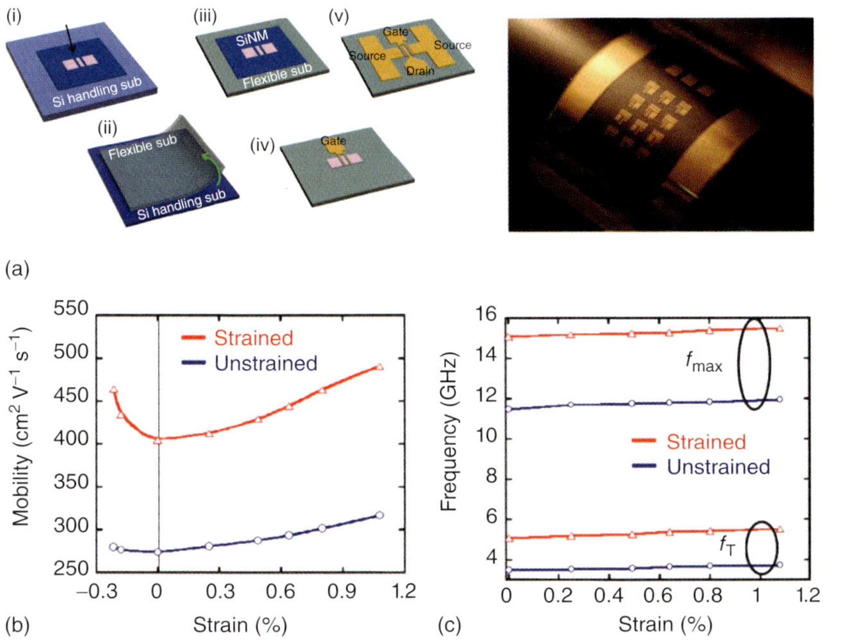

Figure 6.16 (a) Schematic illustration of fabrication process for devices using strained Si NMs (left) and the optical image of fabricated devices in bent state (right). (i) Release of a strained Si NM sitting on a handling substrate. (ii,iii) NM is flip-transferred to a PET substrate. (iv) Gate-stacking by lift-off techniques. (v) Source/drain and interconnect metallization. (b) Plot of the calculated mobility from transconductance measurement versus bending-induced strain for both unstrained and strained devices. (c) f_T and f_{max} of both unstrained (blue) and strained (red) devices as a function of bending-induced external strain [78]. (Reproduced with permission. Copyright © 2013, American Association for the Advancement of Science)

6.3.5
Transparent Transistor

Transparent electronic circuits have attracted interests in various fields, including military and industrial electronic systems. Oxide-based materials such as those of zinc and indium have been intensively studied because of their large intrinsic bandgap, optical transmittance, and good electrical properties [46, 79]. However, the relatively small carrier mobility and poor reproducibility of oxide-based materials are major obstacles for their applications. Jang *et al.* [39] demonstrated a novel way to overcome these challenges posed by oxide-based materials by thinning Si NMs below 10 nm. The completed structure of the ultrathin transistors is the integration of heterogeneous ultrathin materials: an ultrathin Si NM as an active layer, a SAM as a dielectric layer, and graphene as an electrode. As a result, ultrathin transistors exhibit not only promising mechanical flexibility and low stiffness but also outstanding optical transmittance

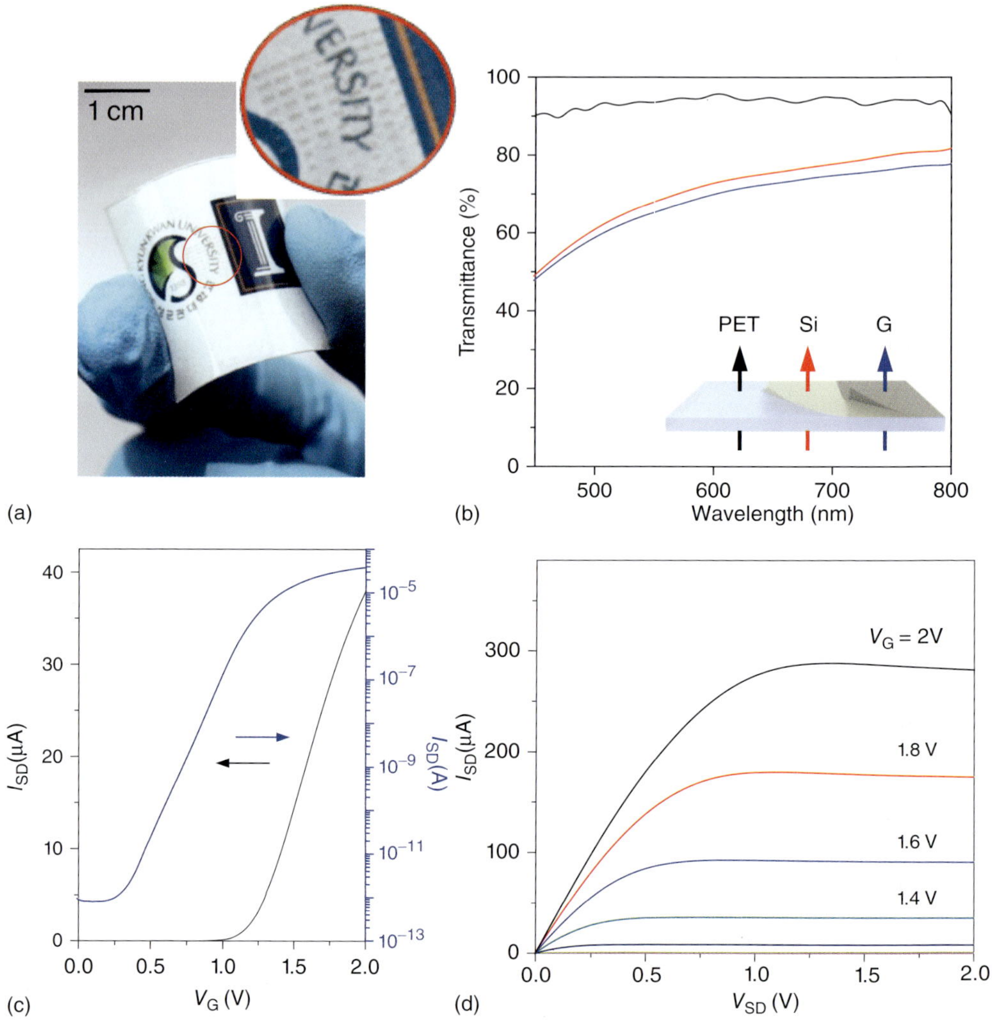

Figure 6.17 (a) Optical image of transparent and flexible transistor arrays using ultrathin Si NMs. The inset shows a magnified image to present the Si NM clearly. (b) Measured transmittance from PET substrate, PET/Si, and PET/Si/graphene. (c) Transfer characteristics of the transistor arrays with a channel width of 200 μm and length of 25 μm at 0.1 V of drain voltage in linear (left axis) and logarithmic (right axis) scales. (d) Full current–voltage characteristics at a gate bias from 0 V (bottom) to 2 V (top) in 0.2-V step [39]. (Reproduced with permission. Copyright © 2013, American Chemical Society.)

which directly comes from the nanometer-level thickness and indirect bandgap structure of Si.

Figure 6.17a displays an optical image of device arrays on a bent plastic substrate through which the level of optical transmittance can be estimated. Transmittance through a substrate, a Si NM, and a graphene electrode is given with the schematic diagram in Figure 6.17b. The transmittance of a 10-nm-thick Si NM reaches ~68% and that of the graphene-covered region ~65.3%, suggesting that the transmittance is decreased because of graphene by 2.7%. The device with a channel length of 25 μm and a width of 200 μm demonstrates a mobility of 110 $cm^2\ V^{-1}\ s^{-1}$, on/off ratio of over 10^7, and subthreshold voltage of 135 mV per decade, as shown in Figure 6.17c,d. These values are superior to those of oxide- or carbon-based systems and are comparable to those of a conventional Si-based electronic system without optimization. The current versus voltage characteristics with different gate biases show the expected current saturation behavior without significant adverse effects from the Schottky contacts at the source and drain.

6.4 Concluding Remarks

This chapter reviewed inorganic semiconductor nanomaterials and their integration methods by using transfer printing for large-area flexible and stretchable electronics. Inorganic semiconductor-based flexible electronics is receiving a lot of attention owing to the many benefits it offers over organic-based electronics, such as high performance, simple fabrication, and good compatibility with conventional electronics. These materials and techniques have successfully demonstrated a variety of high-performance electronics including TFTs, memory, RF devices, and integrated circuits on plastic or rubber substrates. Although printable inorganic semiconductor nanomaterials have various technological issues to overcome, they can provide interesting possibilities for flexible and stretchable electronics, which will significantly grow by the introduction of new products such as smart and interactive packaging and wearable electronics.

References

1. Chen, Y., Au, J., Kazlas, P., Ritenour, A., Gates, H., and McCreary, M. (2003) *Nature*, **423**, 136.
2. Kerpa, O., Weiss, K., and Worn, H. (2003) Development of a flexible tactile sensor system for a humanoid robot. Proceedings on 2003 IEEE/RSJ International Conference on Presented at Intelligent Robots and Systems (IROS 2003), October 27–31, 2003.
3. Lisong, Z., Wanga, A., Wu, S.-C., Jie, S., Park, S., and Jackson, T.N. (2006) *Appl. Phys. Lett.*, **88**, 083502.
4. Sekitani, T., Nakajima, H., Maeda, H., Fukushima, T., Aida, T., Hata, K., and Someya, T. (2009) *Nat. Mater.*, **8**, 494.
5. Chimot, N., Derycke, V., Goffman, M.F., Bourgoin, J.P., Happy, H., and Dambrine, G. (2007) *Appl. Phys. Lett.*, **91**, 153111.

6. Gu, G., Burrows, P.E., Venkatesh, S., Forrest, S.R., and Thompson, M.E. (1997) *Opt. Lett.*, **22**, 172.
7. Kim, B.J., Jang, H., Lee, S.-K., Hong, B.H., Ahn, J.-H., and Cho, J.H. (2010) *Nano Lett.*, **10**, 3464.
8. Sekitani, T., Zschieschang, U., Klauk, H., and Someya, T. (2010) *Nat. Mater.*, **9**, 1015.
9. Someya, T., Kato, Y., Sekitani, T., Iba, S., Noguchi, Y., Murase, Y., Kawaguchi, H., and Sakurai, T. (2005) *Proc. Natl. Acad. Sci. U.S.A.*, **102**, 12321.
10. Bryllert, T., Wernersson, L.E., Froberg, L.E., and Samuelson, L. (2006) *IEEE Electron Device Lett.*, **27**, 323.
11. Han, S., Jin, W., Zhang, D., Tang, T., Li, C., Liu, X., Liu, Z., Lei, B., and Zhou, C. (2004) *Chem. Phys. Lett.*, **389**, 176.
12. Lin, Y.-M., Jenkins, K.A., Valdes-Garcia, A., Small, J.P., Farmer, D.B., and Avouris, P. (2008) *Nano Lett.*, **9**, 422.
13. Liu, J., Rinzler, A.G., Dai, H., Hafner, J.H., Bradley, R.K., Boul, P.J., Lu, A., Iverson, T., Shelimov, K., Huffman, C.B., Rodriguez-Macias, F., Shon, Y.-S., Lee, T.R., Colbert, D.T., and Smalley, R.E. (1998) *Science*, **280**, 1253.
14. Plombon, J.J., Obrien, K.P., Gstrein, F., Dubin, V.M., and Jiao, Y. (2007) *Appl. Phys. Lett.*, **90**, 063106.
15. Chen, F., Ramayya, E.B., Euaruksakul, C., Himpsel, F.J., Celler, G.K., Ding, B., Knezevic, I., and Lagally, M.G. (2010) *ACS Nano*, **4**, 2466.
16. Fuechsle, M., Mahapatra, S., Zwanenburg, F.A., Friesen, M., Eriksson, M.A., and Simmons, M.Y. (2010) *Nat. Nanotechnol.*, **5**, 502.
17. Harneit, W. (2002) *Phys. Rev. A*, **65**, 032322.
18. Huang, M., Ritz, C.S., Novakovic, B., Yu, D., Zhang, Y., Flack, F., Savage, D.E., Evans, P.G., Knezevic, I., Liu, F., and Lagally, M.G. (2009) *ACS Nano*, **3**, 721.
19. Kaempgen, M., Duesberg, G.S., and Roth, S. (2005) *Appl. Surf. Sci.*, **252**, 425.
20. Kang, S.J., Song, Y., Yi, Y., Choi, W.M., Yoon, S.-M., and Choi, J.-Y. (2010) *Carbon*, **48**, 520.
21. Zhang, P., Tevaarwerk, E., Park, B.-N., Savage, D.E., Celler, G.K., Knezevic, I., Evans, P.G., Eriksson, M.A., and Lagally, M.G. (2006) *Nature*, **439**, 703.
22. Baca, A.J., Ahn, J.-H., Sun, Y., Meitl, M.A., Menard, E., Kim, H.-S., Choi, W.M., Kim, D.-H., Huang, Y., and Rogers, J.A. (2008) *Angew. Chem. Int. Ed.*, **47**, 5524.
23. Park, S.-I., Ahn, J.-H., Feng, X., Wang, S., Huang, Y., and Rogers, J.A. (2008) *Adv. Funct. Mater.*, **18**, 2673.
24. Kim, D.-H., Xiao, J., Song, J., Huang, Y., and Rogers, J.A. (2010) *Adv. Mater.*, **22**, 2108.
25. Lagally, M.G. (2007) *MRS Bull.*, **32**, 57.
26. Gleskova, H., Cheng, I.C., Wagner, S., Sturm, J.C., and Suo, Z. (2006) *Sol. Energy*, **80**, 687.
27. Cavallo, F., Grierson, D.S., Turner, K.T., and Lagally, M.G. (2011) *ACS Nano*, **5**, 5400.
28. Kim, D.-H., Viventi, J., Amsden, J.J., Xiao, J., Vigeland, L., Kim, Y.-S., Blanco, J.A., Panilaitis, B., Frechette, E.S., Contreras, D., Kaplan, D.L., Omenetto, F.G., Huang, Y., Hwang, K.-C., Zakin, M.R., Litt, B., and Rogers, J.A. (2010) *Nat. Mater.*, **9**, 511.
29. Khang, D.-Y., Jiang, H., Huang, Y., and Rogers, J.A. (2006) *Science*, **311**, 208.
30. Khang, D.-Y., Rogers, J.A., and Lee, H.H. (2009) *Adv. Funct. Mater.*, **19**, 1526.
31. Kwon, J., Sharma, B.K., and Ahn, J.-H. (2013) *Jpn. J. Appl. Phys.*, **52**, 06GA02.
32. Rogers, J.A., Lagally, M.G., and Nuzzo, R.G. (2011) *Nature*, **477**, 45.
33. Kim, D.-H., Ahn, J.-H., Choi, W.M., Kim, H.-S., Kim, T.-H., Song, J., Huang, Y.Y., Liu, Z., Lu, C., and Rogers, J.A. (2008) *Science*, **320**, 507.
34. Roberts, M.M., Klein, L.J., Savage, D.E., Slinker, K.A., Friesen, M., Celler, G., Eriksson, M.A., and Lagally, M.G. (2006) *Nat. Mater.*, **5**, 388.
35. Yuan, H.-C., Ma, Z., Roberts, M.M., Savage, D.E., and Lagally, M.G. (2006) *J. Appl. Phys.*, **100**, 013708.
36. Fischetti, M.V., Gamiz, F., and Hansch, W. (2002) *J. Appl. Phys.*, **92**, 7320.
37. Lockwood, D.J. (2008) *Proc. SPIE*, **7099**. doi: 10.1117/12.804577
38. Fang, H., Bechtel, H.A., Plis, E., Martin, M.C., Krishna, S., Yablonovitch, E., and Javey, A. (2013) *Proc. Natl. Acad. Sci. U.S.A.*, **110**, 11688.

39. Jang, H., Lee, W., Won, S.M., Ryu, S.Y., Lee, D., Koo, J.B., Ahn, S.-D., Yang, C.-W., Jo, M.-H., Cho, J.H., Rogers, J.A., and Ahn, J.-H. (2013) *Nano Lett.*, **13**, 5600.
40. Agrawal, B.K. and Agrawal, S. (2000) *Appl. Phys. Lett.*, **77**, 3039.
41. Takei, K., Fang, H., Kumar, S.B., Kapadia, R., Gao, Q., Madsen, M., Kim, H.S., Liu, C.-H., Chueh, Y.-L., Plis, E., Krishna, S., Bechtel, H.A., Guo, J., and Javey, A. (2011) *Nano Lett.*, **11**, 5008.
42. Krebs, F.C. (2009) *Org. Electron.*, **10**, 761.
43. Kim, H.-S., Kim, J.-M., Bang, Y.-S., Song, E.-S., Ji, C.-H., and Kim, Y.-K. (2012) *J. Micromech. Microeng.*, **22**, 095014.
44. Menard, E., Lee, K.J., Khang, D.Y., Nuzzo, R.G., and Rogers, J.A. (2004) *Appl. Phys. Lett.*, **84**, 5398.
45. Lee, K.J., Meitl, M.A., Jong-Hyun, A., Rogers, J.A., Nuzzo, R.G., Kumar, V., and Adesida, I. (2006) *J. Appl. Phys.*, **100**, 124507.
46. Park, K., Lee, D.-K., Kim, B.-S., Jeon, H., Lee, N.-E., Whang, D., Lee, H.-J., Kim, Y.J., and Ahn, J.-H. (2010) *Adv. Funct. Mater.*, **20**, 3577.
47. Park, S.-I., Xiong, Y., Kim, R.-H., Elvikis, P., Meitl, M., Kim, D.-H., Wu, J., Yoon, J., Yu, C.-J., Liu, Z., Huang, Y., Hwang, K.-C., Ferreira, P., Li, X., Choquette, K., and Rogers, J.A. (2009) *Science*, **325**, 977.
48. Bae, S., Kim, H., Lee, Y., Xu, X., Park, J.-S., Zheng, Y., Balakrishnan, J., Lei, T., Ri Kim, H., Song, Y.I., Kim, Y.-J., Kim, K.S., Ozyilmaz, B., Ahn, J.-H., Hong, B.H., and Iijima, S. (2010) *Nat. Nanotechnol.*, **5**, 574.
49. Kim, S.H., Yoon, J., Yun, S.O., Hwang, Y., Jang, H.S., and Ko, H.C. (2013) *Adv. Funct. Mater.*, **23**, 1375.
50. Sharma, B.K., Jang, B., Lee, J.E., Bae, S.-H., Kim, T.W., Lee, H.-J., Kim, J.-H., and Ahn, J.-H. (2013) *Adv. Funct. Mater.*, **23**, 2024.
51. Rho, J., Kim, S.J., Heo, W., Lee, N.-E., Lee, H.-S., and Ahn, J.-H. (2010) *IEEE Electron Device Lett.*, **31**, 1017.
52. Kwon, J., Seung, W., Sharma, B.K., Kim, S.-W., and Ahn, J.-H. (2012) *Energy Environ. Sci.*, **5**, 8970.
53. Mack, S., Meitl, M.A., Baca, A.J., Zhu, Z.T., and Rogers, J.A. (2006) *Appl. Phys. Lett.*, **88**, 213101.
54. Lee, S.-K., Jang, H., Hasan, M., Koo, J.B., and Ahn, J.-H. (2010) *Appl. Phys. Lett.*, **96**, 173501.
55. Baca, A.J., Meitl, M.A., Ko, H.C., Mack, S., Kim, H.S., Dong, J., Ferreira, P.M., and Rogers, J.A. (2007) *Adv. Funct. Mater.*, **17**, 3051.
56. Ko, H.C., Baca, A.J., and Rogers, J.A. (2006) *Nano Lett.*, **6**, 2318.
57. Yoon, J., Baca, A.J., Park, S.-I., Elvikis, P., Geddes, J.B., Li, L., Kim, R.H., Xiao, J., Wang, S., Kim, T.-H., Motala, M.J., Ahn, B.Y., Duoss, E.B., Lewis, J.A., Nuzzo, R.G., Ferreira, P.M., Huang, Y., Rockett, A., and Rogers, J.A. (2008) *Nat. Mater.*, **7**, 907.
58. Yoon, J., Jo, S., Chun, I.S., Jung, I., Kim, H.-S., Meitl, M., Menard, E., Li, X., Coleman, J.J., Paik, U., and Rogers, J.A. (2010) *Nature*, **465**, 329.
59. Song, Y.M., Xie, Y., Malyarchuk, V., Xiao, J., Jung, I., Choi, K.-J., Liu, Z., Park, H., Lu, C., Kim, R.-H., Li, R., Crozier, K.B., Huang, Y., and Rogers, J.A. (2013) *Nature*, **497**, 95.
60. Chen, T., Wang, S., Yang, Z., Feng, Q., Sun, X., Li, L., Wang, Z.-S., and Peng, H. (2011) *Angew. Chem. Int. Ed.*, **50**, 1815.
61. O'Regan, B. and Gratzel, M. (1991) *Nature*, **353**, 737.
62. Lee, J., Wu, J., Shi, M., Yoon, J., Park, S.-I., Li, M., Liu, Z., Huang, Y., and Rogers, J.A. (2011) *Adv. Mater.*, **23**, 986.
63. Kim, S., Jeong, H.Y., Kim, S.K., Choi, S.-Y., and Lee, K.J. (2011) *Nano Lett.*, **11**, 5438.
64. Naber, R.C.G., Asadi, K., Blom, P.W.M., de Leeuw, D.M., and de Boer, B. (2010) *Adv. Mater.*, **22**, 933.
65. Zhang, K., Seo, J.-H., Zhou, W., and Ma, Z. (2012) *J. Phys. D: Appl. Phys.*, **45**, 143001.
66. Sun, Y., Menard, E., Rogers, J.A., Kim, H.-S., Kim, S., Chen, G., Adesida, I., Dettmer, R., Cortez, R., and Tewksbury, A. (2006) *Appl. Phys. Lett.*, **88**, 183509.
67. Shahrjerdi, D. and Bedell, S.W. (2013) *Nano Lett.*, **13**, 315.
68. Kim, D.-H., Ghaffari, R., Lu, N., and Rogers, J.A. (2012) *Annu. Rev. Biomed. Eng.*, **14**, 113.

69. Viventi, J., Kim, D.-H., Moss, J.D., Kim, Y.-S., Blanco, J.A., Annetta, N., Hicks, A., Xiao, J., Huang, Y., Callans, D.J., Rogers, J.A., and Litt, B. (2010) *Sci. Transl. Med.*, **2**, 24ra22.
70. Viventi, J., Kim, D.-H., Vigeland, L., Frechette, E.S., Blanco, J.A., Kim, Y.-S., Avrin, A.E., Tiruvadi, V.R., Hwang, S.-W., Vanleer, A.C., Wulsin, D.F., Davis, K., Gelber, C.E., Palmer, L., Spiegel, J.V.D., Xiao, J.W.J., Huang, Y., Contreras, D., Rogers, J.A., and Litt, B. (2011) *Nat. Neurosci.*, **14**, 1599.
71. Kim, D.-H. and Rogers, J.A. (2008) *Adv. Mater.*, **20**, 4887.
72. Krames, M.R., Amano, H., Brown, J.J., and Heremans, P.L. (2002) *IEEE J. Sel. Top. Quantum Electron.*, **8**, 185.
73. Jung, I., Xiao, J., Malyarchuk, V., Lu, C., Li, M., Liu, Z., Yoon, J., Huang, Y., and Rogers, J.A. (2011) *Proc. Natl. Acad. Sci. U.S.A.*, **108**, 1788.
74. Bernardini, F., Fiorentini, V., and Vanderbilt, D. (1997) *Phys. Rev. B*, **56**, R10024.
75. Wang, Y., Joyce, H.J., Gao, Q., Liao, X., Tan, H.H., Zou, J., Ringer, S.P., Shan, Z., and Jagadish, C. (2011) *Nano Lett.*, **11**, 1546.
76. Bae, S.-H., Kahya, O., Sharma, B.K., Kwon, J., Cho, H.J., Özyilmaz, B., and Ahn, J.-H. (2013) *ACS Nano*, **7**, 3130.
77. Feng, X., Yang, B.D., Liu, Y., Wang, Y., Dagdeviren, C., Liu, Z., Carlson, A., Li, J., Huang, Y., and Rogers, J.A. (2011) *ACS Nano*, **5**, 3326.
78. Zhou, H., Seo, J.-H., Paskiewicz, D.M., Zhu, Y., Celler, G.K., Voyles, P.M., Zhou, W., Lagally, M.G., and Ma, Z. (2013) *Sci. Rep.*, **3**, 1291.
79. Nomura, K., Ohta, H., Takagi, A., Kamiya, T., Hirano, M., and Hosono, H. (2004) *Nature*, **432**, 488.

7
Dielectric Materials for Large-Area and Flexible Electronics

Sungjun Park, Sujin Sung, Won-June Lee, and Myung-Han Yoon

7.1
Introduction

Along with the semiconducting channel and conductive electrode materials, the gate dielectric material is one of the key components to realize high-performance thin-film transistor (TFT) devices. Particularly, the gate dielectrics play a critical role in terms of lowering the device's operating voltage and minimizing gate leakage current, and have been typically classified into inorganic and organic materials. Inorganic-based dielectrics, which mainly consist of silicon dioxide and various metal oxides (MOs) such as AlO_x, ZrO_x, HfO_x, and TiO_x, exhibit high areal capacitance, high dielectric strength, and low gate leakage current. These excellent figures of merit, however, are guaranteed only after the high-quality film formation, which typically requires expensive vacuum-deposition and/or high-temperature annealing processes. For these reasons, it is not trivial to fabricate inorganic dielectric films on top of organic semiconductors and plastic substrates that show relatively low melting or glass-transition temperature (typically $<200\,°C$). Moreover, most of the thick inorganic films are too brittle to be deposited on flexible plastic substrates.

Organic dielectric materials, which mainly consist of a variety of polymers, have shown several advantages in terms of versatile processability in solution and decent compatibility with organic/polymer semiconductors for the application of large-area and flexible electronics owing to their solubility in various solvents and their mechanical flexibility. Most of the solution-processed polymers, however, require high thickness for use as a TFT dielectric because a thin layer of polymeric dielectric usually contains many pinholes and, consequently, passes substantial amount of gate leakage.

In this chapter, we will focus on the recent progress in developing solution-processed organic, hybrid, and inorganic dielectric materials from the perspective of realizing large-area and flexible electronics composed of TFT arrays either by reinforcing the aforementioned weakness of organic/inorganic dielectrics or by searching for a new class of hybrid dielectric materials. The general class of gate

Large Area and Flexible Electronics, First Edition. Edited by Mario Caironi and Yong-Young Noh.

dielectric materials for the TFT application has been extensively reviewed, and detailed information can be found in several review articles [1–3].

7.2
General Polymer Dielectrics

There are several requirements for organic gate dielectric materials because many TFT characteristics are directly related to the properties of the gate dielectric. First, low gate leakage current can be achieved by minimizing pinholes on polymeric films. Second, the operation voltage can be lowered by increasing the areal capacitance of a dielectric layer with high dielectric constant or low film thickness. Furthermore, they also require good solvent resistance and relatively high glass-transition/melting temperature for additional layer deposition via the solution process and device performance enhancement via thermal annealing, respectively. The interfacial parameters such as the chemical functionality on and the surface roughness of the dielectric–semiconductor interface are also very important for efficient charge transport at the channel region.

Commonly used organic dielectrics mainly consist of polymers such as poly (vinyl phenol) (PVP), poly(vinyl alcohol) (PVA), polyimide (PI), poly(methyl methacrylate) (PMMA), polystyrene (PS), and poly(perfluorobutenylvinylether) (CYTOP) (Figure 7.1). Among them, a few representative polymeric insulators that are frequently reported in the latest research articles on organic thin-film transistors (OTFTs) are addressed in the following.

CYTOP is an amorphous perfluorinated polymer with high optical transparency and excellent chemical, thermal, and electrical properties. Moreover, its thermoplastic characteristics and good solubility in certain organic solvents have made many researchers opt this polymer as a popular TFT dielectric layer. Note

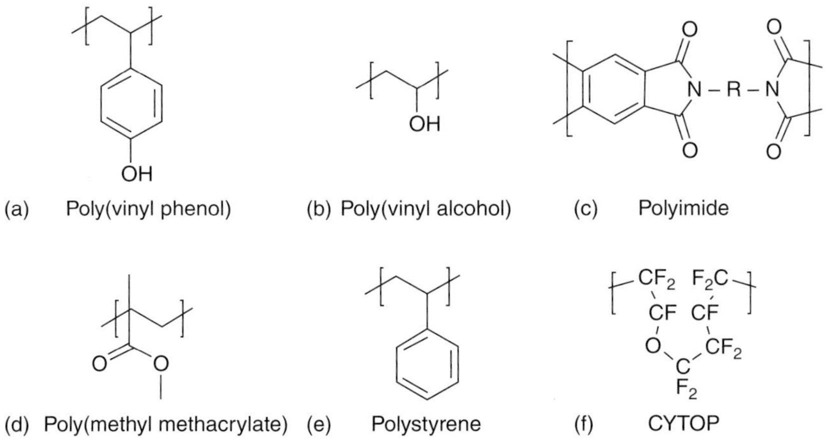

Figure 7.1 (a–f) Chemical structures of representative polymeric gate dielectric materials.

that most of the CYTOP-based TFTs are fabricated with the top-gate configuration because of CYTOP's poor wettability [4]. Additionally, it is well known that TFT devices constructed in the configuration of top-gate dielectric and electrode generally exhibit low contact resistance and improved environmental stability, with the upper dielectric layer acting as a passivation layer for the active channel.

Another kind of polymeric dielectric is PI. This polymer is regarded as a promising gate dielectric material because of its excellent mechanical and electrical properties and, in particular, thermal and chemical stability. Yi and coworkers have developed a variety of gate dielectric materials based on PI derivatives and successfully demonstrated their applications in TFTs. PI is typically prepared from polyamic acid (PAA) through a thermally activated imidization reaction. This reaction typically needs to be carried out above 300 °C, which prevents this material from being used on thermal-shock-sensitive materials. Ahn *et al.* [5] reduced such a high temperature for imidization down to 200 °C by adding 1,8-diazabicyclo[5.4.0]undec-7-ene (DBU) as a catalyst while the imidization efficiency was pushed to almost 100%. Recently, Jang *et al.* [6] developed a photo-patternable PI derivative with excellent electrical resistance. The authors introduced a photo-cross-linkable moiety into the polymer structure so that the cross-linking due to cyclobutane formation proceeded by activating cinnamate upon UV exposure and the unexposed areas were rinsed away during development.

7.3
Cross-Linked Polymer Dielectrics

The common polymers used for insulating purposes require a thick film (>300 nm) to achieve pinhole-free coatings with low gate leakage current. The resultant films provide low gate capacitance and, therefore, high operating voltages once TFT devices are constructed on them. If a given gate dielectric layer is to have high areal capacitance, a high carrier density can be induced at the interface between the semiconductor and the dielectric layer and the accumulation-mode TFT can be operated at a relatively low bias. In fact, most polymer-based dielectrics typically exhibit a low dielectric constant as a result of chemical bonds with low dipole moment and the amorphous structures in the film. Two approaches have been taken to obtain the high areal capacitance in polymer-based gate dielectric films while taking advantage of the aforementioned "polymeric" benefits, that is, excellent solution processability, mechanical flexibility, and low-temperature annealing. One is to reduce the film thickness, and the other is to use high-*k* polymers, both of which will be covered in the following.

The most popular way to reduce the thickness of polymer dielectric films while minimizing pinholes and retaining low leakage current is by cross-linking the polymer chains in the film during or after the polymer deposition. Several groups have proposed the construction of polymer dielectrics cross-linked by adding reactive cross-linkers. Especially, the chemically cross-linked dielectric film is even more robust during the solvent treatment, which is one of the desired properties for all solution-based TFT fabrication. Yoon *et al.* [7] utilized an organosilane as a cross-linker which is reactive to the OH functional group in both vinylphenol and silanol, and induced *in situ* cross-linking of PVP chains with chlorosilane molecules under ambient moisture conditions, as shown in Figure 7.2a. In the case of cross-linked poly(vinyl phenol) (cPVP), the insulating film thickness was reduced to 10–20 nm with areal capacitance of $\sim$300 nF cm^{-2} and a leakage current density of $\sim 10^{-7}$ A cm^{-2}. The same type of silane-based *in situ* cross-linking chemistry was successfully utilized in other kinds of polymer dielectric systems for minimizing pinholes, and thereby the leakage current, at the very thin layer. The Sirringhaus and Noh groups demonstrated excellent gate dielectrics based on PMMA [9] and CYTOP [10], respectively, of <100 nm thickness owing to the cross-linking reaction among the silane molecules. Note that both polymers lack OH functional groups but can be still densified with siloxane-based cross-linked network. Indeed, it is generally known that pinholes are formed when a CYTOP film is deposited below 200 nm thickness. In contrast, a pinhole-free CYTOP film (<50 nm) can be formed in the matrix of cross-linked siloxane network and exhibit good insulating properties for low-voltage thin-film organic transistor application [10].

Even though alkyl trichlorosilane is an efficient cross-linker with high reactivity even under ambient conditions, the acidic byproduct (i.e., HCl) can be an issue for transition to large-scale manufacturing. To avoid the generation of this harmful gas, Bao and coworkers demonstrated a similar type of *in situ* cross-linking reaction using chemical linkers containing dianhydride or diacyl chloride instead of bis-trichlorosilane (Figure 7.2b,c) [8]. In their work, PVP was chemically cross-linked with bifunctional anhydride molecules via esterification between the OH from the PVP chains and anhydride from the cross-linker molecules without harmful byproduct formation. With the same dielectric material, Roberts *et al.* [11] successfully fabricated pentacene-based OTFTs stably operating below 1 V in an aqueous solution. Besides the aforementioned "chemically cross-linked" dielectrics, various photo-cross-linked gate dielectric materials also have been reported [12, 13].

7.4 High-*k* Polymer Dielectrics

As mentioned above, high-*k* dielectric materials are used to obtain high areal capacitance without reducing the dielectric film thickness which generally introduces more pinholes and leakage current in polymeric films. Poly(vinylidene fluoride-*co*-trifluoroethylene), P(VDF-TrFE) has been proposed as a dielectric material with a high dielectric constant ($k > 10$), but this particular polymer exhibits ferroelectric properties with remnant polarization, and OTFTs using P(VDF-TrFE) as dielectric layer usually show large hysteresis in their I–V

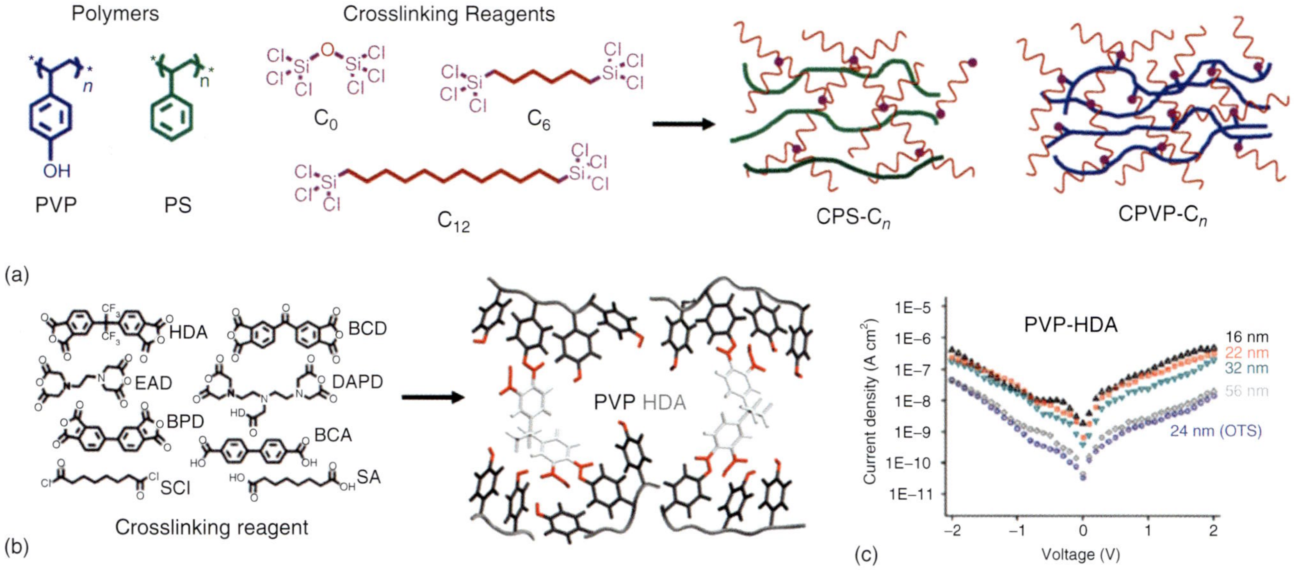

Figure 7.2 Cross-linked polymer systems used as gate dielectrics. Chemical structures of (a) cross-linked PVP/PS polymers and siloxane cross-linkers [7]. (Reproduced with permission. Copyright © 2005, American Chemical Society.) (b) Various cross-linking reagents and the cross-linked PVP with 4,4′-(hexafluoroisopropylidene)diphthalic anhydride (HDA). (c) *I*–*V* characteristics of PVP-HDA films [8]. ((b,c) Reproduced with permission. Copyright © 2009, American Chemical Society.)

characteristics. Thus, its application has been limited mostly to the memory devices. Several research groups have reported that blending P(VDF-TrFE) with a certain paraelectric polymer such as PVA or PMMA is effective in reducing the hysteresis effect from P(VDF-TrFE) through the depolarization effect of the blended paraelectric polymer chains while retaining the high dielectric constant. Baeg *et al.* [14] found that a 7 : 3 ratio of P(VDF-TrFE) and PMMA in the polymer blend is optimal for low-voltage OTFT performance with pentacene as an active channel (Figure 7.3). Based on a similar approach, the authors could effectively reduce the hysteresis of F8T2-based OTFT by suppressing the crystalline β-phase formation, which is the origin of strong ferroelectricity in P(VDF-TrFE) films [15]. They also reported that the ambipolar $I-V$ characteristics become prominent by modulating the charge transport with C–F dipoles in the channel region [16].

Instead of blending the existing high-k polymer with paraelectric amorphous polymers, Li *et al.* [17] attempted to develop for the first time a high-k relaxor ferroelectric terpolymer poly(vinylidene fluoride-trifluoroethylene-chlorofluoroethylene), P(VDF-TrFE-CFE), with a very high dielectric constant (Figure 7.4). This particular terpolymer has an extremely high dielectric constant of ~60 at room temperature, which is very rare in the class of polymer-based dielectric materials. The proposed mechanism of such a high dielectric property is that highly organized dipoles or ferroelectric nanodomains in the P(VDF-TrFE) unit can be aligned inside the CFE unit or the paraelectric matrix. Interestingly, distinct memory characteristics that may originate from the highly polarized P(VDF-TrFE) domains were not observed despite the extremely high dielectric constant and large areal capacitance value. pPBTTT-C16 OTFTs with this polymer as a dielectric layer ($C_i = 330\,\text{nF cm}^{-2}$ at 160 nm thickness) were operated at very a low voltage (below 3 V) with only marginal hysteresis.

7.5 Electrolyte Gate Dielectrics

Electrolyte-based dielectrics rely on the electric double-layer (EDL) capacitance of the order of $10\,\mu\text{F cm}^{-2}$, which enables very low voltage operation of organic transistors. When a potential is applied to the gate electrode, the ultrathin EDL of several nanometers is formed at the interface between the semiconductor and the dielectric layers while ionic redistribution takes place inside the electrolyte dielectric. Subsequently, the voltage applied to the gate electrode drops down mostly over the EDL, resulting in an extremely high capacitance value. In other words, a very large number of charge carriers accumulate on the semiconducting channel even with a very low gate bias. Since there is almost no voltage drop inside "conductive" electrolyte dielectrics, this double-layer capacitance is essentially independent of the dielectric layer thickness or the distance between the semiconductor and the gate electrode.

Ionic liquids, ion gels, polyelectrolytes, and polymer electrolytes are the commonly used materials for electrolyte dielectrics. Thiemann *et al.* investigated

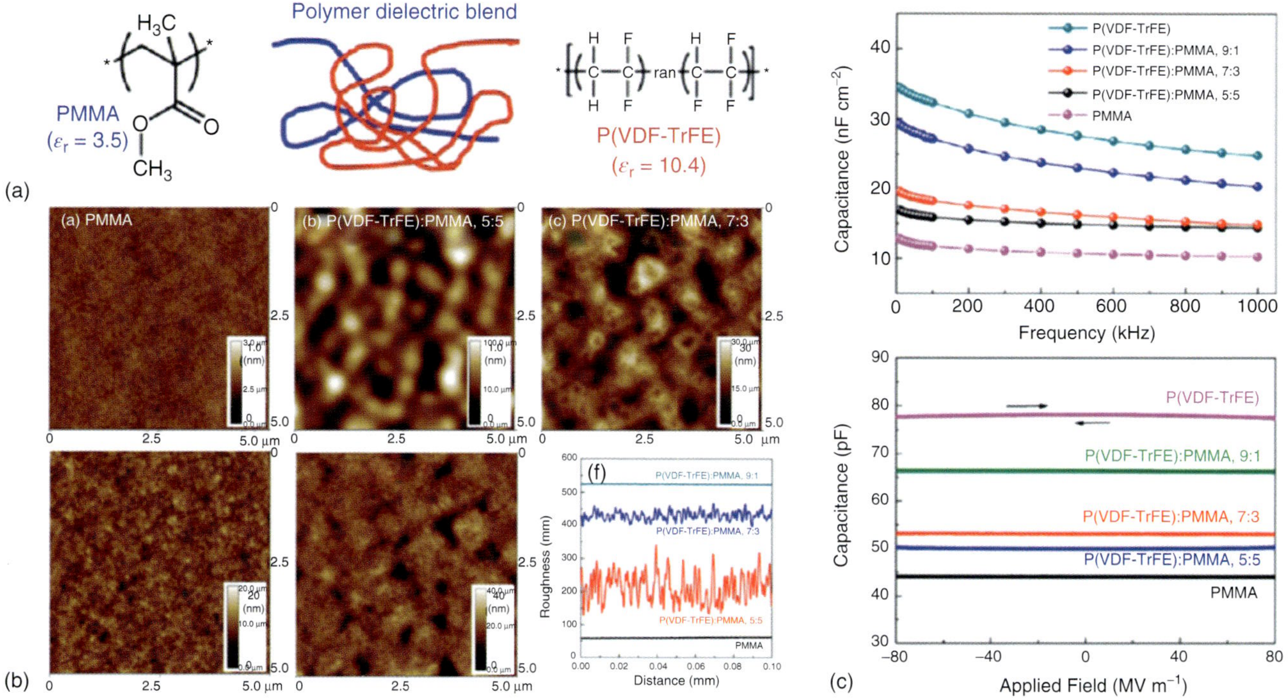

Figure 7.3 High-*k* polymer blends as gate dielectrics. (a) Chemical structures. (b) AFM morphology images and film roughness profiles. (c) Capacitance versus frequency (*C*–*f*) at 10 V bias and capacitance versus voltage (*C*–*V*) at 100 kHz frequency of PMMA, PMMA:P(VDF-TrFE) blend, and P(VDF-TrFE), respectively [14]. (Reproduced with permission. Copyright © 2012, American Chemical Society.)

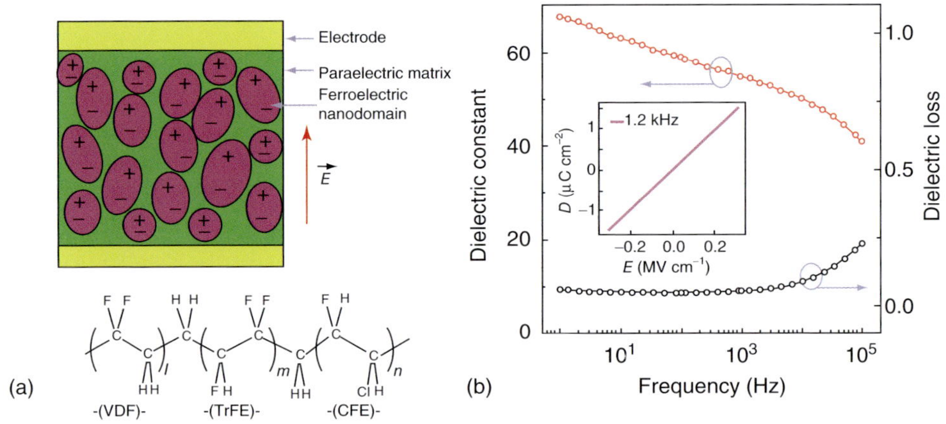

Figure 7.4 High-*k* polymer used as gate dielectric. (a) A schematic diagram of a capacitor based on P(VDF-TrFE-CFE) terpolymer, a relaxor ferroelectric polymer showing ferroelectric nanodomains embedded in paraelectric matrix, and its molecular structure. (b) *k* values (dielectric constant) and dielectric loss of its film as functions of frequency. The inset shows the electrical displacement versus electric field (*D*–*E*) loop of the film [17]. (Reproduced with permission. Copyright © 2012, WILEY-VCH Verlag GmbH & Co. KGaA, Weinheim.)

the effect of the chemical structure of a variety of imidazolium-based ionic liquids on their properties as electrolytes. The device characteristics of electrolyte-gated TFTs based on zinc oxide semiconductors showed a decrease of TFT mobility, which correlated well with the ionic liquid's capacitance [18]. Frisbie and coworkers pioneered in introducing various ionic gels as electrolyte gate dielectrics to improve the TFT switching speed which had been regarded as a weakness earlier. In fact, ionic liquids are promising materials for electrolyte dielectrics because of the former's electrochemical stability, large specific capacitance ($\sim 10\,\mu F\,cm^{-2}$), high ionic conductivity, and fast TFT switching speed (several kilohertz). Nonetheless, the ionic liquid itself is not mechanically robust because of its liquid form. To resolve this issue, many different types of ion gel, where an ionic liquid and a triblock copolymer are blended to give mechanical stability and high ion mobility, have been thoroughly examined (Figure 7.5c) [19]. The same group successfully demonstrated printed OTFTs based on ion gel dielectrics which reliably operated at very low voltages [20].

Despite their several advantages, ion gel films become unstable during temperature changes because of their volume change. Moreover, those films can be easily delaminated when the substrate is bent or a thermal shock is applied, because of their poor interfacial adhesion with organic semiconductors. Lee *et al.* [21] demonstrated patterned ion gel dielectrics via photo-cross-linking to achieve mechanical robustness. The ion gel, composed of an ionic liquid and the PS–PMMA–PS triblock copolymer, has been successfully used to operate graphene transistors at low bias [22]. Furthermore, a free-standing ion gel film with the highest areal capacitance ($\sim 10\,\mu F\,cm^{-2}$), which is comparable to that of

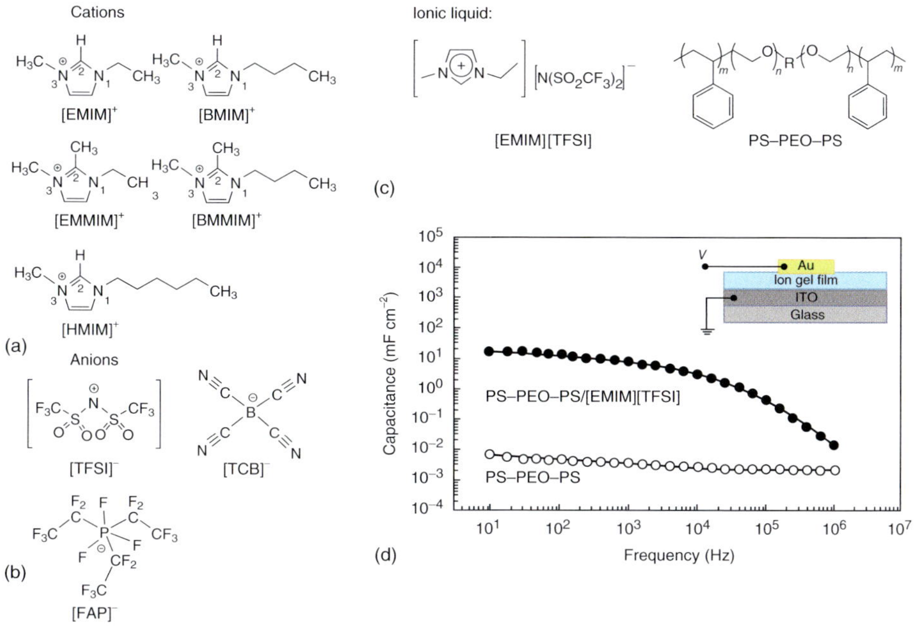

Figure 7.5 Poly-electrolytes used as gate dielectric. Molecular structures of cations (a) and anions (b) of ionic liquids [18], (c) ionic liquid and triblock copolymer: 1-ethyl-3-methylimidazolium bis(trifluoromethylsulfonyl)imide ([EMIM][TFSI]) and poly(styrene-block-ethylene oxide-block-styrene) (PS-PEO-PS), respectively, and (d) frequency dependence of the specific capacitance of triblock copolymer and ion-gel film. The inset shows the test device structure as either ITO/triblock copolymer or ion-gel film/Au [20]. ((a,b) Reproduced with permission. Copyright © 2012, American Chemical Society; (c,d) Reprinted with permission. Copyright © 2008, Macmillan Publishers Ltd.)

the conventional electrolyte ion gel, was developed recently with a simple and solvent-free route [23]. Because of its high mechanical (tensile) strength, this electrolyte-based dielectric film can be simply delaminated and transferred by using a pair of tweezers with minimal chance of solvent contamination.

The Berggren group used polyelectrolytes composed of a charged polymer chain and mobile counterions as gate dielectrics, as shown in Figure 7.6. By using p-type polyanions and n-type polycations as the gate insulator materials in the p- and n-channel transistors, respectively, the penetration of small counterions into the corresponding semiconducting layer was prevented through the polarity-matched polyelectrolyte-based gate dielectrics. Therefore, the electrochemical doping, which is one of the undesired phenomena in the TFT devices based on electrolyte gate dielectrics, in the semiconductor is blocked [24].

Several groups have also demonstrated solution-processed polymer electrolytes as gate dielectrics, for instance, lithium salts dissolved in a matrix of poly(ethylene oxide) (PEO) (called PEO/$LiClO_4$) [25]. However, this type of dielectric materials has a critical drawback, namely low switching speed (a few hertz), which is

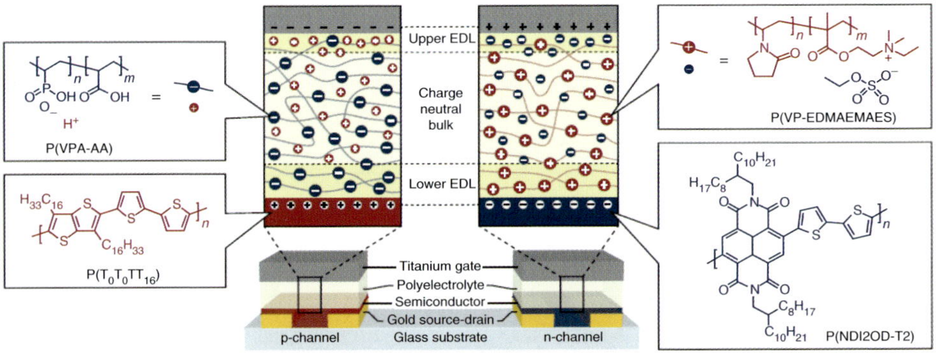

Figure 7.6 A schematics of polyelectrolyte-gated transistors showing p-type polyanion with positive counterions in the p-channel, and n-type polycation with negative counterions in the n-channel. The chemical structures of organic dielectric and semiconducting materials are displayed on the sides [24]. (Reproduced with permission. Copyright © 2011, WILEY-VCH Verlag GmbH & Co. KGaA, Weinheim.)

caused by the relatively slow polarization time and therefore very slow ion diffusion through the polymer chain, although these materials exhibit high polarization and large specific capacitance.

7.6
Self-Assembled Molecular Layer Dielectrics

In general, two major factors that determine the areal capacitance of a dielectric layer are its thickness, a physical parameter, and the dielectric constant, a chemical/material parameter. In this regard, both the self-assembled monolayers (SAMs) and self-assembled nanodielectrics (SANDs) are very effective in increasing the areal capacitance of gate dielectrics and thereby dramatically reduce the TFT operation voltages, because, as mentioned, their thickness of several nanometers enables very high areal capacitance while the highly organized SAMs guarantee minimal gate leakage current. Since the seminal research on the alkyl silane-based molecular insulator by Vuillaume and coworkers [26], SAM dielectrics have been intensively investigated. By using a SAM of tetradecylenyltrichlorosilane as the dielectric layer, the same research group demonstrated that the scaling limit for organic electronics could be pushed down to 2 and 30 nm for dielectric thickness and channel length, respectively, and the resulting sexithiophene TFT based on this dielectric materials showed a mobility of $3.6 \times 10^{-4}\,cm^2\,V^{-1}\,s^{-1}$ at V_D and $V_G = -2\,V$ [27]. Later, Halik *et al.* reported on a reliable method for fabricating low-voltage-operated pentacene TFTs based on 2.5-nm-thick (18-phenoxyoctadecyl)trichlorosilane SAM as an ultrathin dielectric layer [28]. In this work, the $\pi-\pi$ attractive phenoxy SAM dielectric enables densely packed molecular ordering of organic electronic materials and high transistor performance ($\mu \sim 1\,cm^2\,V^{-1}\,s^{-1}$). Their method led to substantial reduction of the power dissipation in a variety of OTFTs by using molecular-scale silane-based dielectric layers. Although OTFTs based on phenoxy-functionalized SAM showed good device performance, alkyl SAM dielectrics suffer from several drawbacks, including (i) low dielectric constant originating from the low-k alkyl silane materials, (ii) no chemical functionality on surface for additional layer deposition, and (iii) leakage-inducing defective sites on the SAM surface leading to the low device yield.

To increase the dielectric constant and minimize the defective sites, SANDs were proposed by Yoon *et al.* [29] using the cross-linked $\sigma-\pi$ organosilane multilayered dielectrics. In SANDs, each stacked chemisorbed layer is designed to play a distinct role: for instance, a highly organized alkyl layer (Alk) for blocking the main leakage current, a highly polarizable stilbazolium salt layer (Stb) for high dielectric constant matrix and as an additional leakage barrier, and an octachlorotrisiloxane layer (Cap) as a cement for filling the defects/pinholes and for regenerating the OH functional groups for the subsequent chemical modification on top. The same research group demonstrated that SAND is compatible even with inorganic semiconductors such as indium oxide and zinc oxide [30].

Generally, the silane-based surface reaction for the dielectric layer deposition is very sensitive to moisture and the reaction condition requires anhydrous organic solvents and an inert atmosphere. To overcome these issues, a vacuum deposition method of self-assembled nanodielectric (v-SAND) was reported [31]. An alternative route has been demonstrated by utilizing solution-processable SAMs with functional groups insensitive to moisture. Ha *et al.* [32] reported a hybrid type of multilayer dielectric composed of organophosphonic acid and inorganic zirconium oxide layers with a total dielectric thickness of 12 nm, capacitance of 450 nF cm^{-2}, gate leakage current of 3×10^{-7} A cm^{-2} at 2 MV cm^{-1}, and pentacene/ZnO TFT mobility of 0.38/3.5 cm^2 V^{-1} s^{-1}.

Even though silane-based SAMs, which are very reactive to the native silicon oxide surface, show excellent dielectric characteristics on the highly doped silicon for the gate electrode and substrate, their limited reactivity to other gate dielectric materials and processing sensitivity to the ambient moisture led to the search for moisture-insensitive and robust self-assembly chemistry other than silane-based ones. Klauk *et al.* [33] demonstrated that the alkylphosphonic acid SAM on the aluminum oxide created by plasma treatment of vacuum-deposited aluminum electrode showed excellent insulating properties. The n-octadecylphosphonic acid SAM on ultrathin alumina exhibits very high areal capacitance (~700 nF cm^{-2}), very low leakage current density (5×10^{-8} A cm^{-2}), and good compatibility with both p-type (0.6 cm^2 V^{-1} s^{-1}) and n-type organic semiconductors (0.02 cm^2 V^{-1} s^{-1}). With such excellent electrical performance, the same dielectric layer was successfully applied for a flexible memory circuit [34].

Jen and coworkers extended the idea of the simple alkylphosphonic acid SAM-Al_2O_3 hybrid to the aryl-terminated alkylphosphonic acid SAM on ultrathin sol–gel MOs such as HfO_x, AlO_x, and SiO_x (Figure 7.7d). In this scheme, while the key chemical groups are retained, including the long alkyl chain for blocking leakage current and the terminal phosphonic acid for chemisorption

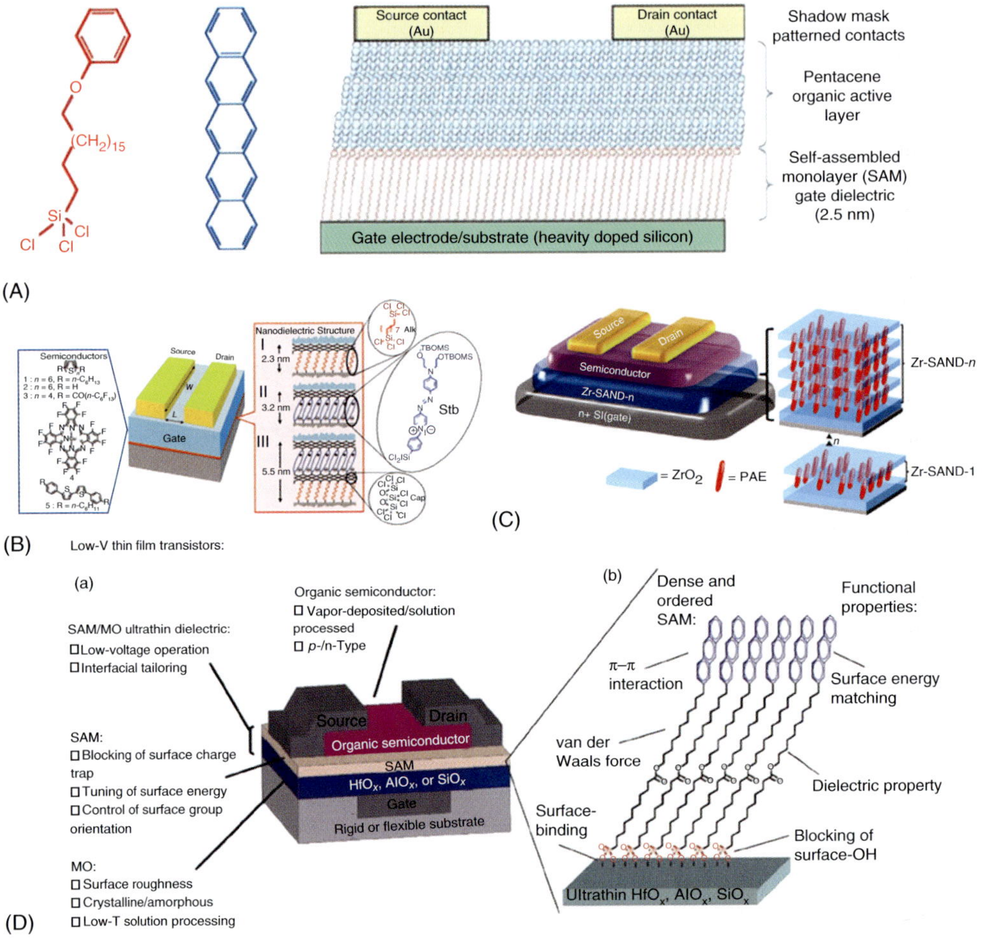

Figure 7.7 Self-assembled monolayer dielectrics. (A) SAM-based low-voltage organic TFT. Chemical structures of (18-phenoxyoctadecyl) trichlorosilane (PhO-OTS) used as a gate dielectric and pentacene as a semiconductor, respectively, and a schematic of the device structure [28]. (Reprinted with permission. Copyright © 2004, Macmillan Publishers Ltd.) (B) SAND-based low-voltage OTFT. Various organic semiconductors used (left) and components of SAND (right) [29]. (Reproduced with permission. Copyright © 2005, The National Academy of Sciences.) (C) A schematic of the device structure with zirconia-SAND [32]. (Reproduced with permission. Copyright © 2011, American Chemical Society.) (D) A comprehensive schematic of the device architecture composed of solution-processed metal oxide dielectrics and synthesized phosphonic acid SAM [35]. (Reproduced with permission. Copyright © 2012, Royal Society of Chemistry.)

to the substrate, the additional terminal aryl group controls the surface energy and modulates the interfacial trap, and the underlying sol–gel MO permits low-temperature solution processing without sacrificing the dielectric properties [35]. Furthermore, the enhanced reactivity of phosphonic acid to the hydroxyl groups on the underlying MO compared to silanol groups (Si–OH) on the native silicon dioxide induces strong metal–oxygen–metal bonds even after a simple spin-coating [36]. In a similar manner as above, a SAM containing both dielectric and semiconducting functionalities was deposited and its field-effect transistor (SAMFET, self-assembled monolayer field-effect transistor) activity was demonstrated by the same group despite a low transistor mobility of $\sim 1.0 \times 10^{-6}\,cm^2\,V^{-1}\,s^{-1}$. Later, Schmaltz *et al.* [37] demonstrated a further optimized SAMFET operation at as low as 3 V using a thin hybrid dielectric of 5.1 nm thickness, with a charge carrier mobility of $1.3 \times 10^{-2}\,cm^2\,V^{-1}\,s^{-1}$ due to the effective 2D lateral $\pi-\pi$ interaction even on the rough flexible substrate.

Although a variety of bilayers using SAMs and solution-processed metal oxide-based dielectric materials have been demonstrated by Jen and coworkers, their sol–gel-based metal oxide dielectrics typically require a high-temperature (>500 °C) annealing process [38]. Salleo and coworkers [39] reported on polymer semiconductor OTFTs based on bilayer dielectrics composed of alkylphosphonic acid and sol–gel zirconium oxide cured by UV irradiation at room temperature, exhibiting a capacitance of $320-522\,nF\,cm^{-2}$. In this case, the additional alkyl SAM is essential for effectively reducing the gate leakage current due to the very leaky ultrathin metal oxide layer which functions as tethering site for the SAM as well as the high-*k* inorganic dielectric.

7.7 Hybrid Dielectrics

Hybrid dielectric materials using both organic (polymeric) and inorganic dielectrics together have been intensely investigated in the past decades. The basic rationale is combining the advantages of both types of materials: solution processability and mechanical flexibility from organic dielectrics, and high dielectric constant and low leakage current from the inorganic counterpart. In this section, the hybrid dielectric materials are classified into three types: (i) organic–inorganic laminated bilayers or multilayers, (ii) organic polymeric/inorganic nanoparticle (NP)/hybrid composites, and (iii) organosiloxane- and organozirconia-based hybrid dielectrics.

7.7.1 Organic–Inorganic Laminated Bilayers/Multilayers

A very thin layer of a high-*k* MO exhibits very high capacitance because of its ultralow film thickness and high dielectric constant. However, it typically shows a very high leakage current and contains a large density of traps especially in the case of sol–gel metal oxide dielectrics. To resolve these problems, extra coatings of the organic layer have been proposed on top of thin metal oxide dielectrics. The Jen and Cho groups developed PS-coated high-*k* HfO_x [40] and TiO_2 dielectrics [41], respectively, to improve the quality of the dielectric interface with low gate leakage current and highly crystalline organic semiconductor structure. The additional PS coating has been known to suppress the leakage current, smooth the top metal oxide surface, and passivate the interfacial traps. The additional thermal treatment of PS/HfO_x at 120 °C improved the surface properties by flattening the phenyl group orientation, matching the interfacial surface energy, and inducing intimate connectivity between the PS interlayer and pentacene molecular layer [40]. These OTFTs on PS/HfO_x exhibit high crystallinity in the pentacene film and operate at low voltages (below 3 V) with large on/off ratios. In a similar way, a PS/TiO_x bilayer was also developed [41]. The nonpolar PS coating on an ultrathin TiO_2 layer was effective in covering the highly polar surface and smoothing out the roughness (Figure 7.8).

Reversed hybrid bilayers were also reported. Jang and coworkers [42, 43] modified the top surface of a thermally stable PI gate insulating layer with alumina and yttrium oxide interlayer. They argued that the improved crystallinity of the semiconducting layer and charge transfer at the semiconductor–dielectric interface were due to the metal oxide-based buffering interlayer deposited on top of thermally stable PI dielectrics.

7.7.2 Organic Polymeric/Inorganic Nanoparticle and Nanocomposites

As mentioned in Section 7.1, many research groups have devoted themselves to developing novel dielectric materials that exhibit excellent electrical performance and mechanical flexibility and incur low manufacturing cost. Polymer-based composites embedded with high-*k* dielectric NPs have been regarded as promising insulating materials with a high dielectric constant for large-area and flexible electronics applications. In general, the constituent NP composition, NP concentration in the matrix, and NP surface modification have been considered among the important control parameters to enhance the permittivity of well-known insulating polymers, including PS, poly-4-vinylphenol, PMMA, and poly(vinylidene fluoride/trifluoroethylene) copolymer. For this purpose, PVP has been chosen most frequently by many research groups since it contains OH functional groups that can be reactive to and cross-linkable with metal oxide NPs.

Jang *et al.* [44] investigated the dielectric properties of the insulating films based on PVP filled with barium strontium titanate (BST) NPs and demonstrated that the electrical performance of organic-based devices are modulated by the variation of the dielectric constant and surface roughness of the hybrid composite layers. More recently, Roy and coworkers reported on different types of polymer–NP composites by incorporating barium zirconate (BZ) [45] and *n*-octadecylphosphonic acid-functionalized aluminum titanate (AT) [46] NPs as high-*k* dielectric fillers into the PVP polymer matrix. They showed that

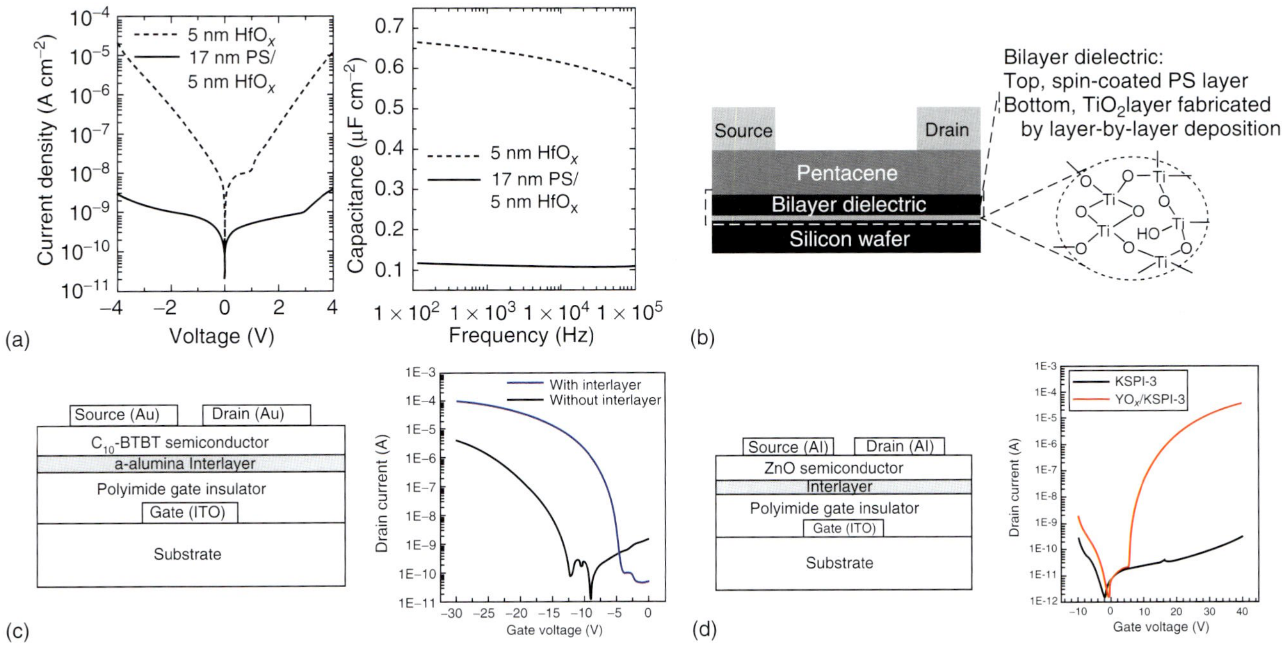

Figure 7.8 Bilayer hybrid dielectrics. Organic interlayer on inorganic dielectrics. (a) Characteristics of HfO_x and PS/HfO_x dielectrics [40]. (Reprinted with permission. Copyright © 2009, American Institute of Physics.) (b) Layer-by-layer deposition of TiO_2 bilayer on PS film [41]. (Reprinted with permission. Copyright © 2009, American Institute of Physics.) Inorganic interlayer on organic dielectrics: transfer characteristics of TFTs with (c) AlO_x [42]. (Reproduced with permission. Copyright © 2013, American Chemical Society.) and (d) YO_x on PI dielectrics [43]. (Reproduced with permission. Copyright © 2013, American Chemical Society.)

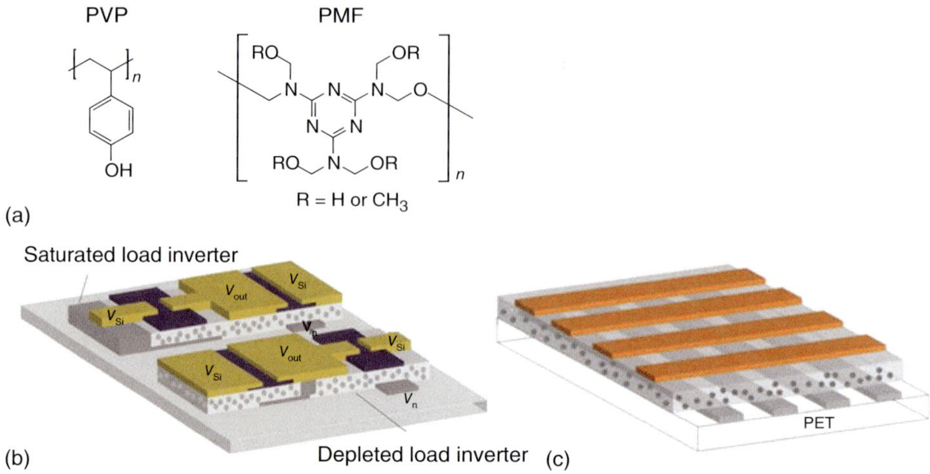

Figure 7.9 (a) The chemical structure of PVP and PMF (poly(melamine-*co*-formaldehyde)) in common hybrid composite dielectric materials [46] and devices. 3D schematic architecture of hybrid nanocomposite dielectric (b) inverter [45] and (c) capacitor arrays [46]. ((a,c) Reproduced with permission. Copyright © 2012, Royal Society of Chemistry; (b) Reproduced with permission. Copyright © 2012, Royal Society of Chemistry.)

the interfacial morphologies between pentacene-based semiconductor and dielectrics consisting of hybrid nanocomposites are responsible for the resulting device performance. To improve mechanical and electrical performance, the top surface of the insulating layer composed of the polymer and inorganic NPs was heavily cross-linked to allow solvent resistance for the next layer deposition process on top. More recently, Oh *et al.* [47] prepared a hybrid nanocomposite dielectric film by blending sol–gel-derived sodium beta alumina (SBA) and PVP to improve the gate dielectric strength and mechanical properties needed for solution-processable flexible TFTs. Their result showed a smooth and hydrophilic dielectric surface which made the semiconducting film uniform (Figure 7.9).

7.7.3
Hybrid Dielectrics Based on Organosiloxane and Organozirconia

Another distinct route to hybrid dielectric materials was recently reported that relies on a unique curing system exhibiting enormous potential to boost the synergetic effect from organic and inorganic components. The ideal polymer gate dielectric should be favorable for solution-based deposition, insoluble during the upper (semiconducting/conductive) layer coating, devoid of hydroxyl groups for minimal trap density, and so on. In this sense, unlike PI and polycarbonate, polymers that require low curing temperatures (below 200 °C) for making them insoluble are desirable to be used with conventional plastic substrates.

Nagase, T. *et al.* investigated the use of poly(methyl silsesquioxane) (PMSQ) for an organosiloxane-based dielectric, which can be thermally cured at relatively low temperatures (150 °C) via the condensation reaction between silanol (Si–OH) groups without adding curing agents [48]. The authors found the optimal solvent for PMSQ synthesis considering that the synthetic reaction condition strongly influences the Si–OH concentration, electrical resistivity, and TFT hysteresis in the resultant PMSQ dielectric film (Figure 7.10).

Hwang and coworkers reported on polyhedral oligomeric silsesquioxane (POSS) derivatives including the photo-curable chemical moieties such as methacryl (POSS-MA) [54], methylmethacryl (POSS-MMA) [50], and propylcinnamate (POSS-CYNNAM) [50] as gate insulators. POSS has a three-dimensional cubic silica structure (1–3 nm) in which it is possible to introduce additional functional groups at eight cubic edges and which shows good stability against external environmental factors. Thin films of POSS-MA and POSS-MMA were thermally cross-linked and subsequently solidified under UV irradiation upon adding the selected photo-radical initiators or photo-acid generators. In the case of POSS-CYNNAM, photo-cross-linking proceeds without any photo-additives. Compared to the PVP-based insulating films, POSS-based dielectrics showed excellent insulating properties and device performance with the capability of photo-patterning.

Moon and coworkers [55–57] intensively studied TFT gate dielectric fabrication based on organosiloxane-based hybrid materials using the sol–gel precursors including methyltriethoxysilane (MTES) and tetraethylorthosilicate (TEOS). They also demonstrated another path to make a hybrid gate dielectric film via UV curing-based patterning with the presence of a cross-linking promoter, hexa(methoxymethyl)melamine (HMMM) [58]. Additionally, superb hybrid dielectric characteristics were observed at a sufficiently low temperature (150 °C), which is well suited for device fabrication on flexible plastic substrates [51].

Bae and coworkers synthesized hybrid poly-condensate materials, called *hybrimers,* where the oligosiloxane nanoclusters are grafted by photo-polymerizable MA radicals using a simple nonhydrolytic sol–gel reaction between methacryloxypropyl trimethoxysilane (MPTMS) and diphenylsilanediol (DPSD). After exposure to UV light and subsequent thermal curing at 150 °C for removing residual solvent/impurities and densifying the as-spun film, these materials were applied as gate dielectric in TFT devices which showed low leakage current and no hysteresis in the $I-V$ characteristics [52, 59–61].

Instead of thermal curing only, Kim and coworkers searched for UV-curable hybrid dielectrics benefiting from both zirconium [53]/titanium [62] alkoxides for sol–gel cross-linking and cross-linkable materials such as acrylic monomers and dipentaerythritol penta-acrylate (DPHA) for UV cross-linking at 160 °C. The resultant ZrO_2 hybrid sols showed long-term stability, and their films (25–150 nm) with smooth and hydrophobic surfaces were capable of giving low leakage currents (10^6-10^7 A cm^{-2}) and high dielectric constants (~4).

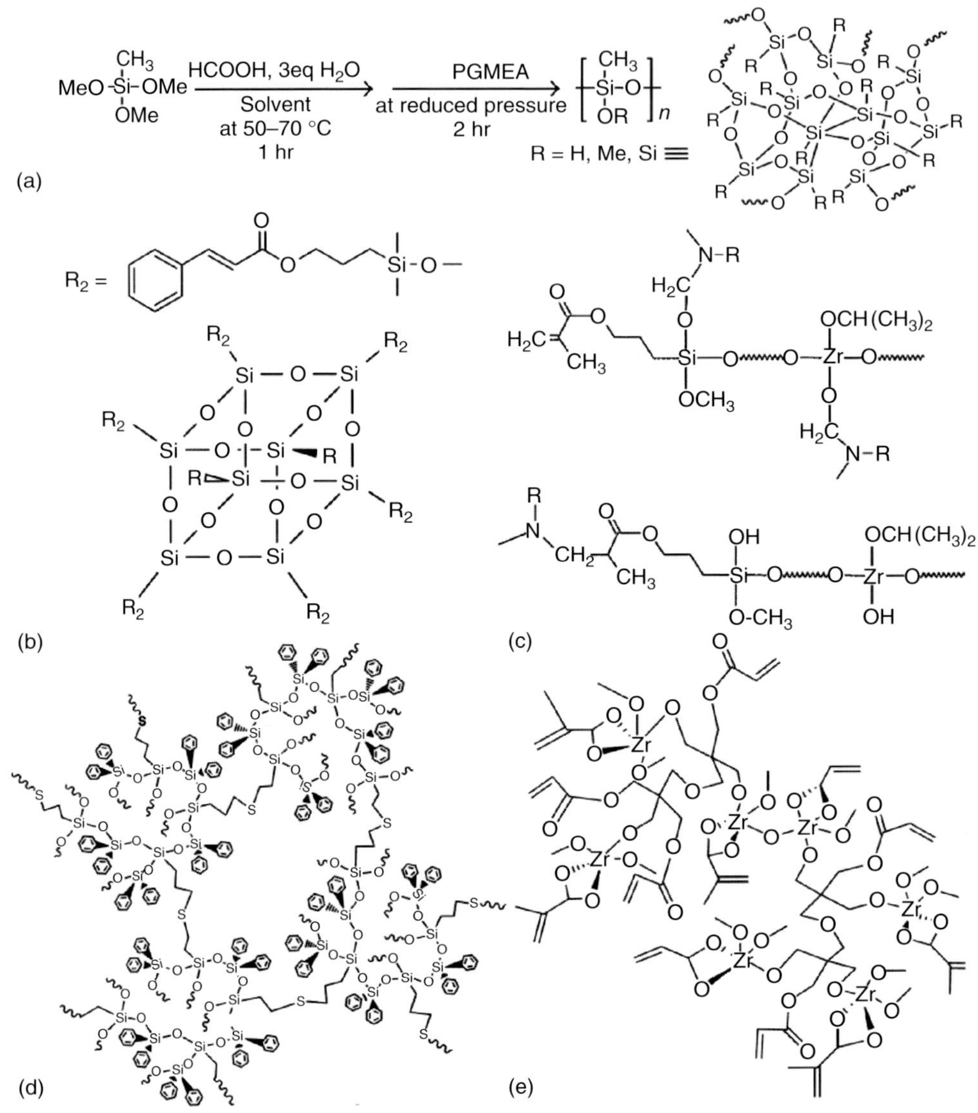

Figure 7.10 Chemical structure in organosiloxane- and organozirconia-based hybrid dielectrics. (a) Polymethylsilsesquioxane (PMSQ) with its synthetic recipe [49]. (Reproduced with permission. Copyright © 2012, Kimihiro Matsukawa *et al.*) (b) POSS-CYNAMM [50]. (Reprinted with permission. Copyright © 2013, Elsevier.) (c) Siloxane networks according to (up) o-alkylation and (down) c-alkylation of HMMM [51]. (Reproduced with permission. Copyright © 2011, Royal Society of Chemistry.) (d) Phenyl-sulfur hybrimer [52]. (Reproduced with permission. Copyright © 2013, ECS – The Electrochemical Society.) (e) Zirconium-based hybrimer [53]. (Reprinted with permission. Copyright © 2012, Elsevier.)

7.8 Sol–Gel High-*k* Inorganic Dielectrics

Silicon dioxide has been the most common dielectric material in industry and academia owing to its reliable dielectric performance and compatibility with silicon semiconductors and well-established fabrication platform. For the same reason, this classic dielectric material has been studied most frequently in conjunction with organic semiconductors and metal oxide semiconductors. For practically useful large-area and flexible circuits using organic or metal oxide semiconductors, the accompanying dielectric layer must have a larger capacitance since the TFT mobilities of these new types of semiconductors are relatively low compared to that of the conventional single-crystalline or polycrystalline silicon semiconductors. Because of its low dielectric constant ($k = 3.9$), however, silicon dioxide needs to be employed with a much thinner film to afford a larger capacitance, but a thinner silicon dioxide film means more leakage for the same bias. Besides, when solution-processed for large-area deposition at the relatively low-temperature process conditions compatible with flexible substrates, the silicon dioxide film that is spun on glass shows even poorer dielectric performance. To satisfy the above-mentioned requirements – in other words, large areal capacitance without sacrificing the gate leakage – many high-*k* inorganic materials (Figure 7.11) [63] have been introduced as the dielectric layer in unconventional TFTs based on organic/polymeric and sol–gel metal oxide semiconductors.

Although MO film formation from an organic solution was suggested first in 1969 [64], the availability of high-quality dielectric films for TFT application has been delayed because of practical issues such as film disruption/discontinuity,

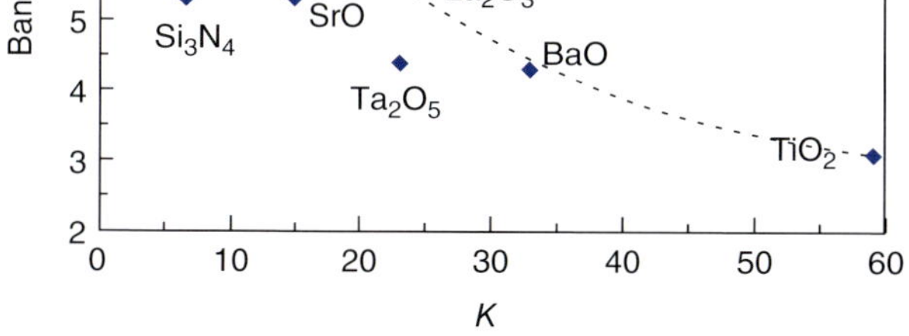

Figure 7.11 Relation between dielectric constant and bandgap of conventional high-*k* dielectric materials [63]. (Reproduced with permission. Copyright © 2006, Copyright © IOP publishing.)

grain boundaries, considerable porosity, rough surface morphology, and uneven crystallization, especially during the high-temperature annealing process. Later, researchers realized that appropriate selection of precursor materials, optimization of film deposition, and low-temperature post-annealing are very critical in large-area and possibly mass fabrication systems such as direct printing, spraying, and stamping. The various solution-processed MO dielectrics (Al_2O_3, CeO_2, HfO_2, Ta_2O_5, TiO_2, Y_2O_3, ZrO_2) for TFT application are summarized in Table 7.1. Note that all data in this table were collected from articles on all-metal oxide-based devices for comparison.

Keszler and coworkers pioneered in the demonstration of various solution-processed high-*k* dielectric films for TFT application. Solution-processed hafnium oxide sulfate [66], zirconium oxide sulfate [66], and aluminum oxide phosphate dielectric films [65, 70] were fabricated in combination with sputtered zinc indium oxide (ZIO), zinc tin oxide (ZTO), and zinc oxide (ZnO) semiconductors. The extracted breakdown field and dielectric constant were up to $4\,MV\,cm^{-2}$ and 12, respectively [66]. In the case of aluminum oxide phosphate, the authors also proved that lanthanum doping of the dielectric layer enhanced the dielectric constant from 4.8 up to 8.5, opening the possibility to control dielectric property with ionic composition in the precursor solution [65].

Recently, another approach to increase the dielectric constant in solution-processed MO dielectrics was demonstrated by several research groups [68, 74, 85]. Katz and coworkers [68] fabricated a sol–gel SBA, which provides high capacitance perpendicular to the planes with negligible leakage current due to the lack of charge carriers and limited interlayer mobility of sodium ions. The sol–gel SBA layer, which is compatible with organic and inorganic semiconducting materials, showed a maximum dielectric constant of as high as 170 and breakdown voltage over $3.5\,MV\,cm^{-2}$.

Although the various high-*k* dielectric materials have been successfully demonstrated in combination with many of organic semiconductors, the vacuum deposition or high-temperature processes (e.g., sputtering, chemical vapor deposition, atomic layer deposition, rapid thermal annealing) are not generally favorable for large-area and flexible electronics application. As an alternative route, the sol–gel process has been thoroughly investigated to fabricate metal oxide thin films for solution-processed TFTs. Despite the simplicity in solution preparation and film deposition, the post-annealing process at high temperature has prevented MO from being used in combination with organic semiconductors and plastic substrates, the properties of which are disturbed by the high temperature of typically over 200 °C. The various efforts to reduce the processing temperature to form high-quality MO film after annealing can be categorized into two approaches: (i) controlling the metal oxide precursor solutions by varying precursor/solvent compositions or adding additives [73, 92–94] and (ii) introducing an extra energy source during the post-annealing process, for example, treating with UV/ozone or irradiating with microwave/UV [86, 95–98]. Nonetheless, most of these efforts have so far been focused on metal oxide semiconductors.

Table 7.1 Recent developments in sol–gel-based metal oxide.

Dielectric	Method	Process temperature (°C)	*D* (nm)	C_i (nF cm^{-2})	Dielectric constant (*k*)	V_b (MV cm^{-1})	Semi-conductors	Maximum mobility	On/off ratio	Year	References
$AlPO_x$	SC	350	148–185	—	4.8	~10	ZnO	~3.5	~10^6	2007	[65]
$HafSO_x$ $ZircSO_x$	SC	325	150–300	—	9–12	~6	IZO	1	~10^6	2007	[66]
ZrO_x	SC	300	149	—	10	—	CdS	~48	~10^5	2009	[67]
AlO_x;Na	SC	600	75	~2000	170	~5.3	ZTO	28	~10^4	2009	[68]
AlO_x	SC	600	30	~56	~10	~2	ZnGaO	4.7	~10^3	2010	[69]
$AIPO_x$	SC	350	~180	—	5.2	~5	IGZO	4.5	~10^9	2011	[70]
AlO_x	SP	400	—	~50	~9.2	~1.8	ZnO	34	~10^5	2011	[71]
AlO_x	SC IJ	300	70	~80	~6.3	~4	ZTO	SP: 33 IJ: 24	~10^8	2011	[72]
AlO_x	SC	250 200	38 41	~188 ~173	—	—	InO	40 13	~10^7	2011	[73]
AlO_x;Na	SC	830	80	100	170	—	ZTO	18	~10^4	2011	[74]
ZrO_2	SP	450	~98	126	~14	~1.6	ZnO:Li	~85	~10^6	2011	[75]
HfO_x	SC	400	~120	—	~13	~5.5	IGZO	13.1	~10^7	2011	[76]
ZrO_x	SC	150 (UV)	5	~830	8.79	~18	ZnO	0.45	~10^5	2012	[77]
ZrO_x	SC	300	35	~240	24.5	—	ZTO	4	~10^5	2012	[78]
HfO_y/AlO_x bilayer	SC	400	~85	~150	—	—	ZTO	3.84	~10^5	2012	[79]
HfO_x	SC	300	65	~190	~14	—	ZTO	1.05	~10^5	2012	[80]
AlO_x	SC	350	242	143.7	—	~4	ZnO:Li	46.9	~10^5	2013	[81]
AlO_x;Li (Na, K)	SC	500	80	~200	9.2	~2	ZTO	19.6	~10^4	2013	[82]
ZrO_x	SC	400	100	—	—	—	SnO_2	~100	~10^5	2013	[83]
ZrO_x:B	SC	250	100	—	12.1	~3.94	In_2O_3	39.3	~10^7	2013	[84]
ZrO_x	SC	350	210	—	14.8	~2.8	IZO	7.21	~10^6	2013	[85]
ZrO_x/AlO_x bilayer	SC	180 (UV)	~35	~235	~9	—	ZnO	~11	~10^4	2013	[86]
ZrO_x	SC	500	90	~240	—	—	ZTO	~10	~10^9	2013	[87]
ZAZ bilayer	SC	350	250	45.8	14.8	~3.4	IZO	3.27	~10^7	2013	[88]
$TiO_2/AlPO_x$ bilayer	SC	350	150	—	8	~5.4	IGZO	3.2	~10^5	2013	[89]
Gd_2O_3	SC	500	100	146	9–14	~3.5	ZITO	~1.9	~10^3	2013	[90]
Gd_2O_3	SC	400	100	—	10–12	~3.5	ZTO	~2.53	~10^5	2013	[91]

SC, Spin-coating; SP, Spray Pyrolysis; IJ, Ink-jet; ZAZ, ($ZrO2/Al2O_3/ZrO_2$) multi-layer.

Very recently, several research articles have been published on the fabrication of metal oxide dielectrics at low temperatures (below 200 °C) which could be compatible with flexible substrates. Xu *et al.* [77] fabricated very thin zirconium oxide films (5 nm) below 150 °C under UV irradiation and successfully demonstrated all-solution-processed MO TFTs with the ZnO semiconductor. Anthopoulos and coworkers also employed UV irradiation in combination with an aqueous carbon-free MO precursor formulation to form MO semiconductors and dielectrics at low temperature between 80 and 180 °C [99]. In this work, the authors used a bilayer MO dielectric composed of aluminum oxide and zirconium oxide to ensure low gate leakage current and finally demonstrated all-solution-processed ZnO TFTs on a flexible polyethylene naphthalate (PEN) substrate with mobility as high as $10\ cm^2\ V^{-1}\ s^{-1}$. Baik and coworkers developed another low-temperature, high-*k* dielectric based on boron-doped peroxo-zirconium oxide [84]. To reduce the thermal annealing temperature to as low as 200 °C, the authors used an aqueous metal precursor solution to exclude the carbonyl group and boron dopant to form a strong B–O bond and thereby block the leakage current due to the increased density, and finally demonstrated all-solution-processed In_2O_3 TFTs on a flexible PI substrate with mobility $\sim 4\ cm^2\ V^{-1}\ s^{-1}$. Park and Yoon also reported on a high-performance, metal oxide-based semiconductor via deep ultraviolet (DUV) irradiation [98] and recently extended the feasibility of low-temperature DUV photoactivation to various MO dielectric materials. They investigated the photoactivation mechanism using various spectroscopic techniques proving the radical-involved reactions, and successfully fabricated all-solution-processed metal oxide-based electronic circuits such as seven-stage ring oscillators on flexible substrates.

7.9
Summary and Outlook

In this chapter, we reviewed various dielectric materials for low-voltage TFTs, particularly aiming at the solution-processed, large-area, and flexible electronics application. The polymeric dielectrics, which exhibit strength in terms of solution processability and mechanical flexibility, have been developed toward lowering the thickness via chemical cross-linking and increasing the dielectric constant via blending them with high-*k* polymers while maintaining low leakage current. Another route to high-performance polymeric dielectrics is the development of polymer-based electrolyte-gated dielectrics mostly focusing on the ion gel composed of an ionic liquid and block copolymers. Additionally, self-assembled molecular or multilayer dielectrics have been successfully demonstrated to dramatically reduce the operation voltages of OTFTs with their very high capacitance even on ultra-flexible substrates. Furthermore, SAM-based dielectric layers have been combined with various sol–gel MO films to realize the ultrathin hybrid dielectrics with high *k* and low leakage current. Most hybrid dielectrics have been developed in the direction of organic–inorganic laminated bi-/multilayers, polymer–high-*k*–NP composites, and organo-siloxane/organo-zirconia to combine the advantages of both organic and inorganic dielectrics. Finally, high-*k* MO dielectrics with low-temperature solution processability was reported very recently, showing tremendous potential to broaden the applications to large-area and ultra-flexible electronics.

References

1. Veres, J., Ogier, S., Lloyd, G., and de Leeuw, D. (2004) Gate insulators in organic field-effect transistors. *Chem. Mater.*, **16**, 4543–4555. doi: 10.1021/cm049598q
2. Facchetti, A., Yoon, M.H., and Marks, T.J. (2005) Gate dielectrics for organic field-effect transistors: new opportunities for organic electronics. *Adv. Mater.*, **17**, 1705–1725. doi: 10.1002/adma.200500517
3. Ortiz, R.P., Facchetti, A., and Marks, T.J. (2009) High-k organic, inorganic, and hybrid dielectrics for low-voltage organic field-effect transistors. *Chem. Rev.*, **110**, 205–239.
4. Sakanoue, T. and Sirringhaus, H. (2010) Band-like temperature dependence of mobility in a solution-processed organic semiconductor. *Nat. Mater.*, **9**, 736–740, *http://www.nature.com/nmat/journal/v9/n9/abs/nmat2825.html#supplementary-information* (accessed 18 July 2014).
5. Ahn, T., Choi, Y., Jung, H.M., and Yi, M. (2009) Fully aromatic polyimide gate insulators with low temperature processability for pentacene organic thin-film transistors. *Org. Electron.*, **10**, 12–17. doi: 10.1016/j.orgel.2008.09.002
6. Jang, K.-S. *et al.* (2012) Direct photopatternable, low-temperature processable polyimide gate insulator for pentacene thin-film transistors. *Org. Electron.*, **13**, 1665–1670. doi: 10.1016/j.orgel.2012.05.024
7. Yoon, M.-H., Yan, H., Facchetti, A., and Marks, T.J. (2005) Low-voltage organic field-effect transistors and inverters enabled by ultrathin cross-linked polymers as gate dielectrics. *J. Am. Chem. Soc.*, **127**, 10388–10395. doi: 10.1021/ja052488f
8. Roberts, M.E. *et al.* (2009) Cross-linked polymer gate dielectric films for low-voltage organic transistors. *Chem. Mater.*, **21**, 2292–2299. doi: 10.1021/cm900637p
9. Noh, Y.-Y., Zhao, N., Caironi, M., and Sirringhaus, H. (2007) Downscaling of self-aligned, all-printed polymer thin-film transistors. *Nat. Nanotechnol.*, **2**, 784–789.
10. Cheng, X. *et al.* (2010) Air stable cross-linked cytop ultrathin gate dielectric for high yield low-voltage top-gate organic field-effect transistors. *Chem. Mater.*, **22**, 1559–1566. doi: 10.1021/cm902929b
11. Roberts, M.E. *et al.* (2008) Water-stable organic transistors and their application in chemical and biological sensors. *Proc. Natl. Acad. Sci. U.S.A.*, **105**, 12134–12139. doi: 10.1073/pnas.0802105105
12. Lee, T.W., Shin, J.H., Kang, I.N., and Lee, S.Y. (2007) Photocurable organic gate insulator for the fabrication of high-field effect mobility organic transistors by low temperature and solution processing. *Adv. Mater.*, **19**, 2702–2706. doi: 10.1002/adma.200601752
13. Kim, S.H. *et al.* (2010) Photo-curable polymer blend dielectrics for advancing organic field-effect transistor applications. *Adv. Mater.*, **22**, 4809–4813. doi: 10.1002/adma.201000904
14. Baeg, K.-J. *et al.* (2012) Controlled charge transport by polymer blend dielectrics in top-gate organic field-effect transistors for low-voltage-operating complementary circuits. *ACS Appl. Mater. Interfaces*, **4**, 6176–6184. doi: 10.1021/am301793m
15. Soon-Won, J. *et al.* (2010) Low-voltage-operated top-gate polymer thin-film transistors with high capacitance poly(vinylidene fluoride-trifluoroethylene)/poly(methyl methacrylate) dielectrics. *J. Appl.*

Phys., **108**, 102810–102810-5. doi: 10.1063/1.3511697
16. Baeg, K.-J. *et al.* (2012) Remarkable enhancement of hole transport in top-gated N-type polymer field-effect transistors by a high-k dielectric for ambipolar electronic circuits. *Adv. Mater.*, **24**, 5433–5439. doi: 10.1002/adma.201201464
17. Li, J., Sun, Z., and Yan, F. (2012) Solution processable low-voltage organic thin film transistors with high-k relaxor ferroelectric polymer as gate insulator. *Adv. Mater.*, **24**, 88–93. doi: 10.1002/adma.201103542
18. Thiemann, S., Sachnov, S., Porscha, S., Wasserscheid, P., and Zaumseil, J. (2012) Ionic liquids for electrolyte-gating of ZnO field-effect transistors. *J. Phys. Chem. C*, **116**, 13536–13544. doi: 10.1021/jp3024233
19. Lee, J., Panzer, M.J., He, Y., Lodge, T.P., and Frisbie, C.D. (2007) Ion gel gated polymer thin-film transistors. *J. Am. Chem. Soc.*, **129**, 4532–4533. doi: 10.1021/ja070875e
20. Cho, J.H. *et al.* (2008) Printable ion-gel gate dielectrics for low-voltage polymer thin-film transistors on plastic. *Nat. Mater.*, **7**, 900–906, *http://www.nature.com/nmat/journal/v7/n11/suppinfo/nmat2291_S1.html* (accessed 18 July 2014).
21. Lee, S.W. *et al.* (2009) Periodic array of polyelectrolyte-gated organic transistors from electrospun poly(3-hexylthiophene) nanofibers. *Nano Lett.*, **10**, 347–351. doi: 10.1021/nl903722z
22. Kim, B.J., Lee, S.-K., Kang, M.S., Ahn, J.-H., and Cho, J.H. (2012) Coplanar-gate transparent graphene transistors and inverters on plastic. *ACS Nano*, **6**, 8646–8651. doi: 10.1021/nn3020486
23. Lee, K.H. *et al.* (2012) "Cut and stick" rubbery ion gels as high capacitance gate dielectrics. *Adv. Mater.*, **24**, 4457–4462. doi: 10.1002/adma.201200950
24. Herlogsson, L., Crispin, X., Tierney, S., and Berggren, M. (2011) Polyelectrolyte-gated organic complementary circuits operating at low power and voltage. *Adv. Mater.*, **23**, 4684–4689. doi: 10.1002/adma.201101757
25. Panzer, M.J. and Frisbie, C.D. (2007) Polymer electrolyte-gated organic field-effect transistors: low-voltage, high-current switches for organic electronics and testbeds for probing electrical transport at high charge carrier density. *J. Am. Chem. Soc.*, **129**, 6599–6607. doi: 10.1021/ja0708767
26. Boulas, C., Davidovits, J.V., Rondelez, F., and Vuillaume, D. (1996) Suppression of charge carrier tunneling through organic self-assembled monolayers. *Phys. Rev. Lett.*, **76**, 4797.
27. Collet, J., Tharaud, O., Chapoton, A., and Vuillaume, D. (2000) Low-voltage, 30 nm channel length, organic transistors with a self-assembled monolayer as gate insulating films. *Appl. Phys. Lett.*, **76**, 1941–1943.
28. Halik, M. *et al.* (2004) Low-voltage organic transistors with an amorphous molecular gate dielectric. *Nature*, **431**, 963–966.
29. Yoon, M.-H., Facchetti, A., and Marks, T.J. (2005) σ-π molecular dielectric multilayers for low-voltage organic thin-film transistors. *Proc. Natl. Acad. Sci. U.S.A.*, **102**, 4678–4682.
30. Wang, L. *et al.* (2006) High-performance transparent inorganic-organic hybrid thin-film n-type transistors. *Nat. Mater.*, **5**, 893–900. doi: 10.1038/nmat1755
31. DiBenedetto, S.A., Frattarelli, D., Ratner, M.A., Facchetti, A., and Marks, T.J. (2008) Vapor phase self-assembly of molecular gate dielectrics for thin film transistors. *J. Am. Chem. Soc.*, **130**, 7528–7529.
32. Ha, Y.-G. *et al.* (2011) Solution-deposited organic–inorganic hybrid multilayer gate dielectrics. Design, synthesis, microstructures, and electrical properties with thin-film transistors. *J. Am. Chem. Soc.*, **133**, 10239–10250.
33. Klauk, H., Zschieschang, U., Pflaum, J., and Halik, M. (2007) Ultralow-power organic complementary circuits. *Nature*, **445**, 745–748.
34. Sekitani, T. *et al.* (2009) Organic non-volatile memory transistors for flexible sensor arrays. *Science*, **326**, 1516–1519.
35. Ma, H., Acton, O., Hutchins, D.O., Cernetic, N., and Jen, A.K.-Y. (2012)

Multifunctional phosphonic acid self-assembled monolayers on metal oxides as dielectrics, interface modification layers and semiconductors for low-voltage high-performance organic field-effect transistors. *Phys. Chem. Chem. Phys.*, **14**, 14110–14126.
36. Acton, O. *et al.* (2011) Spin-cast and patterned organophosphonate self-assembled monolayer dielectrics on metal-oxide-activated Si. *Adv. Mater.*, **23**, 1899–1902.
37. Schmaltz, T. *et al.* (2013) Low-voltage self-assembled monolayer field-effect transistors on flexible substrates. *Adv. Mater.*, **25**, 4511–4514.
38. Hutchins, D.O. *et al.* (2012) Spin cast self-assembled monolayer field effect transistors. *Org. Electron.*, **13**, 464–468.
39. Park, Y.M., Daniel, J., Heeney, M., and Salleo, A. (2011) Room-temperature fabrication of ultrathin oxide gate dielectrics for low-voltage operation of organic field-effect transistors. *Adv. Mater.*, **23**, 971–974.
40. Wang, Y. *et al.* (2009) Low-voltage high-performance organic thin film transistors with a thermally annealed polystyrene/hafnium oxide dielectric. *Appl. Phys. Lett.*, **95**, 243302. doi: 10.1063/1.3268455
41. Lu, Y., Lee, W.H., Lee, H.S., Jang, Y., and Cho, K. (2009) Low-voltage organic transistors with titanium oxide/polystyrene bilayer dielectrics. *Appl. Phys. Lett.*, **94**, 113303. doi: 10.1063/1.3097010
42. Yoon, J.-Y. *et al.* (2013) Enhanced performance of solution-processed organic thin-film transistors with a Low-temperature-annealed alumina interlayer between the polyimide gate insulator and the semiconductor. *ACS Appl. Mater. Interfaces*, **5**, 5149–5155. doi: 10.1021/am400996q
43. Jang, K.-S. *et al.* (2013) Surface modification of a polyimide gate insulator with an yttrium oxide interlayer for aqueous-solution-processed ZnO thin-film transistors. *Langmuir*, **29**, 7143–7150. doi: 10.1021/la401356u
44. Jang, Y. *et al.* (2009) High field-effect mobility pentacene thin-film transistors with nanoparticle polymer composite/polymer bilayer insulators. *Appl. Phys. Lett.*, **94**, 183301. doi: 10.1063/1.3124660
45. Zhou, Y., Han, S.-T., Xu, Z.-X., and Roy, V.A.L. (2012) Polymer-nanoparticle hybrid dielectrics for flexible transistors and inverters. *J. Mater. Chem.*, **22**, 4060–4065. doi: 10.1039/c2jm15256h
46. Zhou, Y. *et al.* (2012) Functional high-k nanocomposite dielectrics for flexible transistors and inverters with excellent mechanical properties. *J. Mater. Chem.*, **22**, 14246–14253. doi: 10.1039/c2jm31813j
47. Oh, J.-Y. *et al.* (2013) Organic–inorganic hybrid gate dielectric for solution-processed ZnO thin film transistors. *J. Vac. Sci. Technol., B*, **31**, 050603. doi: 10.1116/1.4817499
48. Nagase, T. *et al.* (2010) Low-temperature processable organic-inorganic hybrid gate dielectrics for solution-based organic field-effect transistors. *Adv. Mater.*, **22**, 4706–4710. doi: 10.1002/adma.201001871
49. Matsukawa, K., Watanabe, M., Hamada, T., Nagase, T., and Naito, H. (2012) Polysilsesquioxanes for gate-insulating materials of organic thin-film transistors. *Int. J. Polym. Sci*, **2012**, Article ID 852063, 10 pp.
50. Kim, Y. *et al.* (2013) Photocurable propyl-cinnamate-functionalized polyhedral oligomeric silsesquioxane as a gate dielectric for organic thin film transistors. *Org. Electron.*, **14**, 2315–2323. doi: 10.1016/j.orgel.2013.05.030
51. Jung, Y. *et al.* (2011) Direct photopatternable organic-inorganic hybrid gate dielectric for solution-processed flexible ZnO thin film transistors. *J. Mater. Chem.*, **21**, 11879–11885. doi: 10.1039/c1jm10791g
52. Kim, J.-S., Kim, Y., Ko, J.-H., and Bae, B.-S. (2013) Thiol-ene reaction derived sol-gel hybrid dielectric layer for organic thin film transistors. *ECS Trans.*, **50**, 83–88. doi: 10.1149/05004.0083ecst
53. Wang, Y. and Kim, H. (2012) UV-curable organic–inorganic hybrid gate dielectrics for organic thin film transistors. *Org. Electron.*, **13**, 2997–3003. doi: 10.1016/j.orgel.2012.08.014

54. Kim, Y. *et al.* (2012) Organic thin-film transistors using photocurable acryl-fuctionalized polyhedral oligomeric silsesquioxanes as gate dielectrics. *Synth. Met.*, **162**, 1798–1803. doi: 10.1016/j.synthmet.2012.08.015
55. Jeong, S., Kim, D., Lee, S., Park, B.K., and Moon, J. (2007) Organic thin-film transistors using thin ormosil-based hybrid dielectric. *Thin Solid Films*, **515**, 7701–7705. doi: 10.1016/j.tsf.2006.11.143
56. Jeong, S., Kim, D., Park, B.K., Lee, S., and Moon, J. (2007) Influence of silanol groups on the electrical performance of organic thin-film transistors utilizing organosiloxane-based organic–inorganic hybrid dielectrics. *Nanotechnology*, **18**, 025204.
57. Jeong, S., Kim, D., Lee, S., Park, B.-K., and Moon, J. (2006) Organic-inorganic hybrid dielectrics with low leakage current for organic thin-film transistors. *Appl. Phys. Lett.*, **89**, 092101. doi: 10.1063/1.2338753
58. Jeong, S., Lee, S., Kim, D., Shin, H., and Moon, J. (2007) Photopatternable organosiloxane-based inorganic – organic SiO2 – ZrO2 hybrid dielectrics for organic thin film transistors. *J. Phys. Chem. C*, **111**, 16083–16087. doi: 10.1021/jp0761463
59. Eo, Y.-J., Kim, J.H., Ko, J.H., and Bae, B.-S. (2005) Optical characteristics of photo-curable methacryl-oligosiloxane nano hybrid thick films. *J. Mater. Res.*, **20**, 401–408. doi: 10.1557/JMR.2005.0045
60. Choi, C.G. and Bae, B.-S. (2007) Organic–inorganic hybrid materials as solution processible gate insulator for organic thin film transistors. *Org. Electron.*, **8**, 743–748. doi: 10.1016/j.orgel.2007.06.006
61. Choi, C.G. and Bae, B.-S. (2009) Organic–inorganic hybrid gate dielectrics for low-voltage pentacene organic thin film transistors. *Synth. Met.*, **159**, 1288–1291. doi: 10.1016/j.synthmet.2009.02.029
62. Wang, Y., Kim, Y., Lee, E., and Kim, H. (2012) Fabrication of organic thin-film transistors based on cross-linked hybrid dielectric materials. *Jpn. J. Appl. Phys.*, **51**, 09MJ02.
63. Robertson, J. (2006) High dielectric constant gate oxides for metal oxide Si transistors. *Rep. Prog. Phys.*, **69** (2), 327. doi: 10.1088/0034-4885/69/2/R02
64. Schroeder, H. (1969) Oxide layers deposited from organic solutions. *Phys. Thin Films*, **5**, 87–141.
65. Meyers, S.T. *et al.* (2007) Solution-processed aluminum oxide phosphate thin-film dielectrics. *Chem. Mater.*, **19**, 4023–4029.
66. Anderson, J.T. *et al.* (2007) Solution-processed HafSOx and ZircSOx inorganic thin-film dielectrics and nanolaminates. *Adv. Funct. Mater.*, **17**, 2117–2124.
67. Seon, J.-B., Lee, S., Kim, J.M., and Jeong, H.-D. (2009) Spin-coated CdS thin films for n-channel thin film transistors. *Chem. Mater.*, **21**, 604–611.
68. Pal, B.N., Dhar, B.M., See, K.C., and Katz, H.E. (2009) Solution-deposited sodium beta-alumina gate dielectrics for low-voltage and transparent field-effect transistors. *Nat. Mater.*, **8**, 898–903.
69. Xia, D. and Xu, J. (2010) High mobility and low operating voltage ZnGaO and ZnGaLiO transistors with spin-coated Al2O3 as gate dielectric. *J. Phys. D Appl. Phys.*, **43**, 442001.
70. Kim, K.M. *et al.* (2011) Competitive device performance of low-temperature and all-solution-processed metal-oxide thin-film transistors. *Appl. Phys. Lett.*, **99**, 242109–242109-3.
71. Adamopoulos, G., Thomas, S., Bradley, D.D., McLachlan, M.A., and Anthopoulos, T.D. (2011) Low-voltage ZnO thin-film transistors based on Y2O3 and Al2O3 high-k dielectrics deposited by spray pyrolysis in air. *Appl. Phys. Lett.*, **98**, 123503.
72. Avis, C. and Jang, J. (2011) High-performance solution processed oxide TFT with aluminum oxide gate dielectric fabricated by a sol–gel method. *J. Mater. Chem.*, **21**, 10649–10652.
73. Kim, M.-G., Kanatzidis, M.G., Facchetti, A., and Marks, T.J. (2011) Low-temperature fabrication of high-performance metal oxide thin-film electronics via combustion processing. *Nat. Mater.*, **10**, 382–388.

74. Zhang, B., Liu, Y., Agarwal, S., Yeh, M.-L., and Katz, H.E. (2011) Structure, sodium Ion role, and practical issues for β-alumina as a high-k solution-processed gate layer for transparent and low-voltage electronics. *ACS Appl. Mater. Interfaces*, **3**, 4254–4261.
75. Adamopoulos, G. *et al.* (2011) High-mobility low-voltage ZnO and Li-doped ZnO transistors based on ZrO2 high-k dielectric grown by spray pyrolysis in ambient air. *Adv. Mater.*, **23**, 1894–1898.
76. Jiang, K. *et al.* (2011) Low-energy path to dense HfO2 thin films with aqueous precursor. *Chem. Mater.*, **23**, 945–952.
77. Xu, X., Cui, Q., Jin, Y., and Guo, X. (2012) Low-voltage zinc oxide thin-film transistors with solution-processed channel and dielectric layers below 150 °C. *Appl. Phys. Lett.*, **101**, 222114–222114-3.
78. Lee, C.-G. and Dodabalapur, A. (2012) Solution-processed high-k dielectric, ZrO2, and integration in thin-film transistors. *J. Electron. Mater.*, **41**, 895–898.
79. Kim, Y.G., Avis, C., and Jang, J. (2012) Low voltage driven, stable solution-processed zinc-tin-oxide TFT with HfOy and AlOx stack gate dielectric. *ECS Solid-State Lett.*, **1**, Q23–Q25.
80. Avis, C., Kim, Y.G., and Jang, J. (2012) Solution processed hafnium oxide as a gate insulator for low-voltage oxide thin-film transistors. *J. Mater. Chem.*, **22**, 17415–17420.
81. Park, J.H. *et al.* (2013) Water adsorption effects of nitrate ion coordinated Al2O3 dielectric for high performance metal-oxide thin-film transistor. *J. Mater. Chem. C*, **1**, 7166–7174.
82. Liu, Y., Guan, P., Zhang, B., Falk, M.L., and Katz, H.E. (2013) Ion dependence of gate dielectric behavior of alkali metal Ion-incorporated aluminas in oxide field-effect transistors. *Chem. Mater.*, **25**, 3788–3796.
83. Jang, J. *et al.* (2013) Transparent High-Performance Thin Film Transistors from Solution-Processed SnO_2/ZrO_2 Gel-like Precursors. *Adv. Mater.*, **25**, 1042–1047.
84. Park, J.H. *et al.* (2013) Boron-doped peroxo-zirconium oxide dielectric for high-performance, low-temperature, solution-processed indium oxide thin-film transistor. *ACS Appl. Mater. Interfaces*, **5**, 8067–8075.
85. Park, J.H. *et al.* (2013) Low-temperature, high-performance solution-processed thin-film transistors with peroxo-zirconium oxide dielectric. *ACS Appl. Mater. Interfaces*, **5**, 410–417.
86. Lin, Y.H. *et al.* (2013) High-performance ZnO transistors processed via an aqueous carbon-free metal oxide precursor route at temperatures between 80-180 degrees C. *Adv. Mater.*, **25**, 4340–4346. doi: 10.1002/adma.201301622
87. Ha, T.-J. and Dodabalapur, A. (2013) Photo stability of solution-processed low-voltage high mobility zinc-tin-oxide/ZrO2 thin-film transistors for transparent display applications. *Appl. Phys. Lett.*, **102**, 123506–123506-3.
88. Park, J.H. *et al.* (2013) All-solution-processed, transparent thin-film transistors based on metal oxides and single-walled carbon nanotubes. *J. Mater. Chem. C*, **1**, 1840–1845.
89. Jiang, K., Meyers, S.T., Anderson, M.D., Johnson, D.C., and Keszler, D.A. (2013) Functional ultrathin films and nanolaminates from aqueous solutions. *Chem. Mater.*, **25**, 210–214.
90. Choi, S. *et al.* (2013) Low voltage operated, sol–gel derived oxide thin film transistor based on high-k Gd2O3 gate dielectric. *Mater. Chem. Phys.*, **138**, 1–4. doi: 10.1016/j.matchemphys.2012.11.015
91. Choi, S., Park, B.-Y., and Jung, H.-K. (2013) Growth and characterization of sol–gel prepared Gd2O3 films as gate insulators for Zn–Sn–O thin film transistors. *Thin Solid Films*, **534**, 291–295. doi: 10.1016/j.tsf.2013.03.023
92. Jeong, S., Ha, Y.G., Moon, J., Facchetti, A., and Marks, T.J. (2010) Role of gallium doping in dramatically lowering amorphous-oxide processing temperatures for solution-derived indium zinc oxide thin-film transistors. *Adv. Mater.*, **22**, 1346–1350.
93. Jeong, S., Lee, J.-Y., Lee, S.S., Choi, Y., and Ryu, B.-H. (2011) Impact of metal salt precursor on low-temperature annealed solution-derived Ga-doped In2O3 semiconductor for thin-film transistors. *J. Phys. Chem. C*, **115**, 11773–11780.

94. Hwang, Y.H. *et al.* (2013) An 'aqueous route' for the fabrication of low-temperature-processable oxide flexible transparent thin-film transistors on plastic substrates. *NPG Asia Mater.*, **5**, e45.
95. Han, S.-Y., Herman, G.S., and Chang, C.-H. (2011) Low-temperature, high-performance, solution-processed indium oxide thin-film transistors. *J. Am. Chem. Soc.*, **133**, 5166–5169.
96. Umeda, K. *et al.* (2013) Impact of UV/O3 treatment on solution-processed amorphous InGaZnO4 thin-film transistors. *J. Appl. Phys.*, **113**, 184509. doi: 10.1063/1.4804667
97. Jun, T. *et al.* (2011) High-performance low-temperature solution-processable ZnO thin film transistors by microwave-assisted annealing. *J. Mater. Chem.*, **21**, 1102–1108.
98. Kim, Y.-H. *et al.* (2012) Flexible metal-oxide devices made by room-temperature photochemical activation of sol-gel films. *Nature*, **489**, 128–132.
99. Lin, Y.H. *et al.* (2013) Correction: high-performance ZnO transistors processed via an aqueous carbon-free metal oxide precursor route at temperatures between 80–180 °C. *Adv. Mater.*, **25**, 4689.

8
Electrolyte-Gating Organic Thin Film Transistors

Moon Sung Kang, Jeong Ho Cho, and Se Hyun Kim

8.1
Introduction

Organic thin-film transistors (OTFTs) have received significant attention over the past decade for their potential to enable the fabrication of low-cost, printable, flexible electronic applications [1–6]. Progress in the field has relied on multilateral approaches, including (i) the synthesis of new electronic materials, (ii) the development and engineering of device fabrication processes, and (iii) the advancement in our understanding of the charge-transport physics in organic semiconductors [7–32]. High-performance electronic circuits based on OTFTs, such as ring oscillators and logic gates, are now achievable through simple and inexpensive processes. Practical applications of OTFTs and related devices such as portable displays, sensors, and radio frequency identification (RFID) tags driven by thin-film batteries will require the reduction of the devices' operating voltage. This is a critical remaining issue in OTFT research.

Efforts to reduce the OTFT operation voltage to acceptable levels have focused on the development of novel gate dielectric materials having a high specific capacitance (C'). Dielectrics with high capacitance, in principle, permit the accumulation of mobile charge carriers in semiconductors under a low applied gate voltage. The specific capacitance is proportional to κ/d (where κ is the dielectric constant of the insulator and d is the dielectric thickness), which suggests two alternative approaches to reducing the operation voltage. First, a high-κ gate dielectric material allows an OTFT to be driven at a few volts. Inorganic metal oxide films such as barium zirconate titanate (BZT) [33], barium strontium titanate (BST) [33], aluminum oxide (Al_2O_3) [34], tantalum pentoxide (Ta_2O_5) [35], titanium dioxide (TiO_2) [36], hafnium oxide (HfO_2) [37], zirconium oxide (ZrO_2) [38], cerium oxide (CeO_2) [39], and gadolinium oxide (Gd_2O_3) [40] exhibit high-κ values that are several times higher than those of conventional silicon dioxide or polymer gate dielectrics. These materials also display excellent dielectric strength which enables reliable device operation. Unfortunately, however, the high surface energy and roughness properties of these dielectric materials often damage the organic

Large Area and Flexible Electronics, First Edition. Edited by Mario Caironi and Yong-Young Noh.

semiconductor layer in contact with the dielectric material [41–43]. These dielectric materials are typically deposited using expensive vacuum processes that are incompatible with the cost-saving motivations underlying OTFT research.

As an alternative, ultrathin dielectrics can enable the preparation of OTFTs that operate under low driving voltages. Such approaches typically employ organic dielectric materials. Self-assembled monolayers [44–47]/multilayers [48–51], cross-linked polymers [52–54], polymer brush layers [55–57], and organic/inorganic hybrid layers [58–63] are common examples of ultrathin gate dielectrics that exhibit a high value of C' of several hundred nanofarad per square centimeter. High-performance, low-voltage OTFTs and other related functional devices (nonvolatile memory devices and sensors) can now be prepared using ultrathin organic dielectric materials on flexible substrates which benefit from the high compatibility of these materials and organic semiconductors [64–66]. Despite the significant progress made thus far, the attainment of uniform, defect-free, highly dense film structures from these materials remains a challenging issue.

Electrolytes have recently received considerable attention as alternative high-capacitance gate dielectric materials for use in OTFTs. Although the concept of utilizing electrolytes to modulate the current density in a semiconductor layer is not new, recent developments in printable solid electrolytes have attracted research interest in the use of this class of materials to drive OTFTs and operate electrical circuits on plastic substrates, textiles, or paper [67–71]. These materials exhibit a high specific capacitance which is several times higher than the values of the above-mentioned high-κ ultrathin gate dielectrics. Such a large specific capacitance permits a material to accumulate a very large charge density within the semiconductor material ($\sim 10^{14}$ carriers cm^{-2}). Accordingly, electrolyte gating not only improves the device operation voltage but also provides a versatile platform for investigating charge transport in semiconductors at large carrier densities. Electrolytes are, in general, solution-processable and are easily used in printing processes. Since variation of the electrolyte layer does not influence the charge storagability of the material drastically, many of the reproducibility issues that may arise from thickness variation during the printing processes can also be alleviated. The resulting devices display efficient charge injection from the electrode (a low source–drain contact resistance). Electrolytes with the above-mentioned characteristics are expected to show promise as gate dielectrics in printed OTFTs, provided that several technical issues are resolved: (i) the compatibility of these electrolytes with other contact materials must be improved, (ii) the materials display a slow polarization response time, (iii) the voltage stability windows of these materials tend to be narrow, (iv) the materials have high power consumption requirements because of the introduction of a high leakage current, and (v) device stability tends to be poor.

This chapter describes recent progress in the development of electrolyte-gated OTFTs toward the achievement of low-driving-voltage, flexible, and solution-processable electronics. This discussion begins with a basic explanation of the electrolyte-gated transistor operation mechanism, followed by an introduction of the various electrolyte materials employed in these devices. The electrical properties of electrolyte-gated transistors and their circuit applications are subsequently introduced.

8.2 Electrolyte-Gated OTFT Operation Mechanisms

Figure 8.1 shows several schematic diagrams of an electrolyte-gated OTFT during operation. Two distinct gating mechanisms can drive the generation of a large current in the semiconducting layer in contact with an electrolyte gate dielectric material: electric double-layer gating or electrochemical gating. The application of a bias to the gate electrode induces the accumulation of mobile ions at the electrolyte/electrode interface in order to compensate for the electric charges present at the gate electrode. The balanced layer of ions and countercharges is referred to as the *electric double layer* (*EDL*). The applied bias at the gate electrode also herds the counterions toward the electrolyte/semiconductor interface. Depending on the permeability of the counterions into the semiconductor layer, the counterions can either (i) form another EDL at the latter interface to accumulate counter charges in the semiconductor layer or (ii) penetrate into the semiconductor to dope/dedope the semiconductor electrochemically.

Electrolyte-gated transistors that operate via the former mechanism, in which two EDLs form at the gate/electrolyte and electrolyte/semiconductor interfaces, are often classified as *electrical double-layer transistors* (*EDLTs*), and the mechanism by which these devices function is called the *electric double-layer gating process* (the left panel in Figure 8.1). This process is likely to manifest in organic single crystals that are impermeable to ion penetration [72–75]. The polarization of an electrolyte involves ion migration under a gate bias, and carrier accumulation in the semiconductor layer is an electrostatic process similar to the process involved in typical field-effect transistor operations. The specific capacitance of an electrolyte layer containing two EDLs ($C'_{\text{electrolyte}}$) is determined by the sum of the respective EDL-specific capacitances (C'_{EDL1} and C'_{EDL2}) for EDLs connected in series ($1/C'_{\text{electrolyte}} = 1/C'_{\text{EDL1}} + 1/C'_{\text{EDL2}}$). The specific capacitance of each double layer can be estimated using the Helmholtz equation: $C'_{\text{EDL}} = \kappa\varepsilon_0/\lambda$, where ε_0 is the vacuum permittivity and λ is the Debye screening length or the thickness of the double layer (~1 nm). Because the specific capacitance of the EDL formed at the electrolyte/semiconductor interface is typically smaller than that at the gate/electrolyte interface, the former value dominates the total specific capacitance of the electrolyte and, in turn, yields a value well above 1 μF cm^{-2}. We emphasize that the large specific capacitance obtained using a common electrolyte is orders of magnitude larger than that of conventional dielectrics such as SiO_2 [76].

Alternatively, electrolyte-gated transistors that operate on the basis of electrochemical doping/dedoping processes are called *electrochemical transistors* (*ECTs*) (the right panel in Figure 8.1). These processes require a semiconductor permeable to ions, and most common polymeric semiconductor films can satisfy

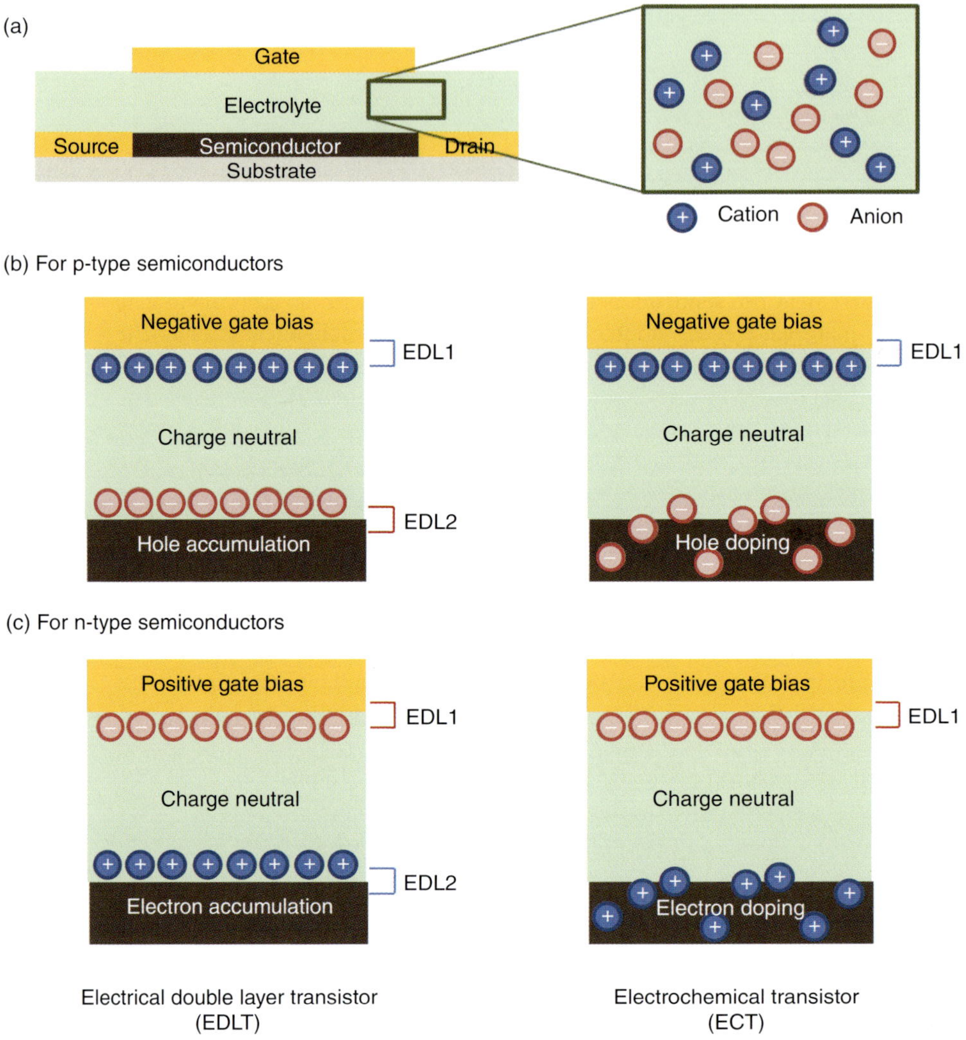

Figure 8.1 (a) Schematic diagram of an electrolyte-gated thin-film transistor in a top-gate, bottom-contact configuration. (b) Snapshots of hole accumulation and electrochemical doping processes for a p-type semiconductor in EDLT (left panel) and ECT (right panel). (c) Snapshots of electron accumulation and electrochemical doping processes for an n-type semiconductor in EDLT (left panel) and ECT (right panel).

this criterion. Ions adjacent to the permeable semiconductor layer that can readily diffuse into the semiconductor layer play a key role in ECT operation. Penetrating ions, which would have formed another EDL at the electrolyte/semiconductor interface in an EDLT, cause reversible oxidation/reduction of semiconductors under a gate bias [77]. Under a positive (negative) gate bias for example, penetrating cations (anions) in an n-(p-)type semiconductor layer induce the delivery of electrons (holes) into the semiconductor layer from the source electrode, thereby reducing (oxidizing) the semiconductor (doped semiconductor). Removal of the gate bias depletes the cations (anions) that have penetrated the semiconductor layer and neutralizes the semiconductor layer (to form a dedoped semiconductor). Such doping/dedoping processes are reversible as long as the semiconductor layers retain their neutral status upon removal of the gate bias. Structural distortions of the semiconductor layer are inevitable during multiple electrochemical doping/dedoping processes and eventually deteriorate the OTFT performance.

Regardless of the mechanisms described above, the operation of an electrolyte-gated OTFT relies on the migration of ions in an electrolyte layer. Because ion migration is slower than polarization in typical insulator-gate dielectrics, the operational speed or the switching speed of an electrolyte-gated OTFT is relatively low. A figure of merit used to describe the switching speed of an electrolyte gate dielectric is its *RC* time constant, which is the product of the electrolyte resistance (R) and the double-layer capacitance (C). The value of the *RC* time constant indicates the characteristic timescale in which an electrolyte can respond to a voltage change. A small *RC* time constant implies that ions in the electrolyte react promptly. The *RC* time constant can be expressed in terms of the intrinsic electrolyte parameters: $RC = C'l\sigma = \kappa\varepsilon_0 l/d\sigma$, where l is the entire electrolyte thickness, σ is the ionic conductivity, and d is the effective thickness of the double layer. Inserting the optimal values for the electrolytes, namely $\sigma = 10^{-2}$ S cm^{-1} and $C = 10$ μF cm^{-2} [78, 79], into the relation reveals that an electrolyte film of 1 μm thickness can have an *RC* constant as small as 0.1 μs. This, in turn, implies that an OTFT operating as fast as 10 MHz is, in principle, achievable if an electrolyte is employed as the gate dielectric layer. Of course, other factors, such as the device geometry, operation voltage, and semiconductor mobility, also contribute to the determination of the OTFT operation speed.

8.3 Electrolyte Materials

This section provides an overview of the different types of solid electrolyte that have been developed in an effort to achieve high-performance electrolyte-gated OTFTs. The critical electrolyte characteristics considered for the material design include (i) a high ionic conductivity for highly responsive devices, (ii) a large electrochemical window that can be sustained at high gate voltages, and (iii) retention of the solid-state integrity for practical device applications.

Solid polymer electrolytes, for example, poly(ethylene oxide) (PEO) doped with a lithium salt, such as $LiClO_4$, have been used extensively in many electrolyte gating studies [80]. The ionic motions in these systems couple to the polymer backbone relaxation motions because the oxygen lone pairs on the PEO chains coordinate to lithium ions in solution (Figure 8.2a). A flexible backbone is critical for enhancing the ion conductivity in such systems, which is expected to be influenced by the glass-transition temperature (T_g) of the polymer ($T_g = -60\,°C$

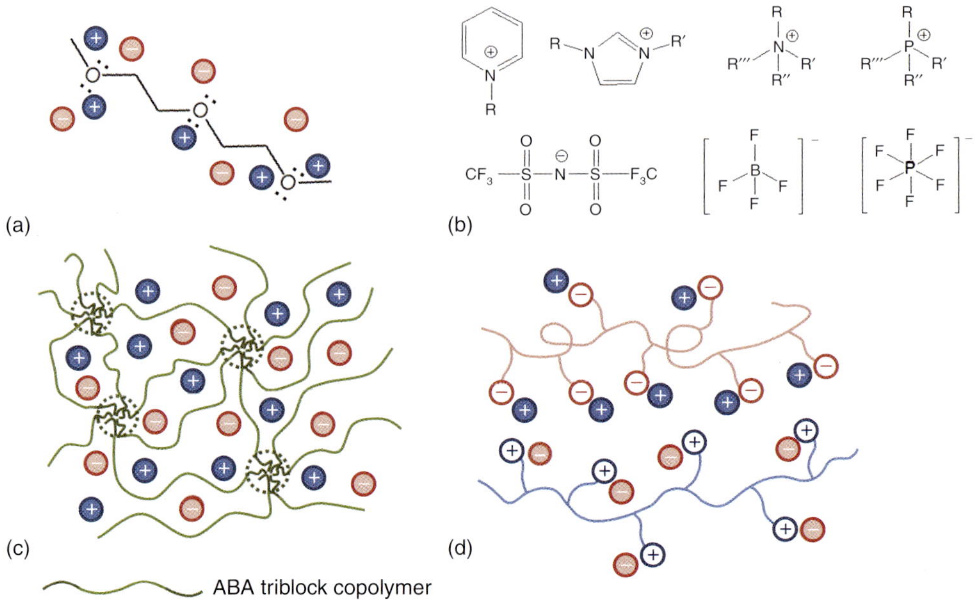

Figure 8.2 Electrolyte materials for organic transistors. (a) Schematic of PEO chain doped with salt. (b) Chemical structures of representative cations and anions for ionic liquids used as electrolyte-gate dielectric materials. (c) Schematic of ion gel network formed with ABA triblock copolymer and ionic liquids. (d) Schematic of polyelectrolytes used as electrolyte-gate dielectric materials.

for PEO [81]). Despite the amplified motions of the PEO backbone above T_g, the room-temperature ionic conductivities of the PEO/Li salt systems remain in the range 10^{-5} to 10^{-4}, which is low for fast-switching OTFT applications.

One approach to improving the ionic conductivity in an OTFT involves increasing the concentration of Li salts in the system because the ionic conductivity is, in principle, proportional to the density of ions present in the system. The addition of salts can increase the T_g of the system by enhancing the Lewis acid–base interactions between Li^+ ions and the PEO backbone to slow ion transport [82]. Experimental results show that the maximum ion conductivity is observed in PEO systems upon the addition of Li salts. Alternative methods have been devised to improve the ion conductivity in systems by replacing traditional PEO homopolymers with a cross-linked polymer or block/graft copolymer containing short-chain PEO, which can suppress polymer crystallization [83, 84]. Several inorganic/organic additives, including aluminum oxide (Al_2O_3), titanium oxide (TiO_2), silica (SiO_2), lithium aluminate ($LiAlO_2$), zeolites, and clay, have been added in an effort to tune the crystallinity of the electrolyte [85–88].

Recently, several types of ionic liquid have been employed as gate dielectrics in OTFTs [89–92]. Figure 8.2b shows representative cations and anions that have been used as ionic liquids. Ionic liquids are useful for gating OTFTs. First, the ion conductivity must exceed the conductivity of other electrolyte systems. For example, a typical ionic liquid composed of 1-ethyl-3-methylimidazolium bis(trifluoromethylsulfonyl)amide (<abbrev[EMI][TFSA]) exhibits an ion conductivity as high as $9.2\,\text{mS}\,\text{cm}^{-1}$ at room temperature [93]. These classes of materials display a large electrochemical window which is useful for accumulating high carrier densities in a semiconductor layer. The materials are negligibly volatile, nonflammable, and have exceptional thermal/chemical stability [94, 95]. The selection of combinatorial mixtures of ion components provides flexibility in materials design to fulfill the required physical properties.

The characteristics of ionic liquids that benefit the development of high-performance OTFTs are frequently accompanied by fluid properties that reduce the practicality of using an ionic liquid in an electronic device. To address the problems associated with ionic liquids, a clever method has been developed for providing ionic liquids with mechanical integrity without significantly compromising the excellent ionic conductivity of the material. A small amount of gelator [96, 97] and polymer [82] may be added to an ionic liquid to induce gelation and yield a solid electrolyte consisting mostly of ionic liquids, termed *ion gels*. Typical gelators include molecules that can form strong intermolecular interactions with an ionic liquid [96, 97] or a colloidal dispersion of nanoparticles [98]. Gelation in polymeric ion gels is driven through either chemical [99, 100] or physical [101, 102] cross-linking of the additive polymers. Watanabe *et al.* [99] reported the first example of a chemically cross-linked polymer ion gel in which a small amount of poly(methyl methacrylate) (PMMA) (as low as 6 wt%) was added to [EMI][TFSA]. Chemical cross-linking has also been achieved through the polymerization of vinyl monomers in the presence of a cross-linker [99, 103] or through a polyaddition reaction among (macro)monomers bearing multifunctional reactive groups [100, 104]. Alternatively, block copolymers with tunable material characteristics may be used to induce physical gelation through hydrogen bonding [102] or self-assembly wherein the self-assembly is driven by material compatibility differences [105]. Soft gels comprising ionic liquids may be prepared by adding as little as a 4 wt% ABA triblock copolymer bearing ionic liquid-incompatible A blocks and ionic liquid-compatible B blocks (Figure 8.2c) [106]. We emphasize that ion transport in an ion gel electrolyte is not significantly hampered by the formation of a gel matrix because only a small quantity of the gelation additive is present in the system. This is not the case for a PEO/lithium salt electrolyte (the lithium salt serves as a dopant, and most of the material forms the PEO backbone). Consequently, a relatively high ionic conductivity, as high as $10^{-3}\,\text{S}\,\text{cm}^{-1}$, may be obtained using a solid ion gel electrolyte at room temperature [107].

The third class of solid electrolytes is the polyelectrolytes, which are polymers bearing repeating units that contain ionizable groups. Figure 8.2d shows the chemical structures of several representative polyelectrolytes. Because ionic transport in this class of materials involves the migration of a large ionic backbone, the ion conductivity of the polyelectrolyte typically remains low, below $10^{-5}\,\text{S}\,\text{cm}^{-1}$, at room temperature [108]. Despite these drawbacks, polyelectrolytes are useful for EDLT applications because, unlike the other electrolytes discussed above, the large ionic backbone cannot penetrate the semiconductor layer, and the electrochemical doping process in the electrolyte-gated semiconductor is avoided [109].

Poly(ionic liquid)s or polymerized ionic liquids were recently developed for use as novel polyelectrolytes [110, 111]. As with conventional polyelectrolytes, the ionic conductivity of a poly(ionic liquid) is determined by the chemical nature of the polymer backbone, the identity of the counterion, T_g, and the humidity. It should be noted that poly(ionic liquid)s composed of polyanions generally have a higher ion conductivity than those composed of polycations. The ionic conductivity of a poly(ionic liquid) is low in most cases, around $\sim 10^{-4}$ S cm^{-1} at room temperature [110]. Accordingly, several methods for blending poly(ionic liquid)s with ionic liquids have been tested in an effort to increase the ion conductivity. Conductivities in the range 10^{-3} to 10^{-5} S cm^{-1} at room temperature have been achieved. These values are appropriate for use in electrolyte-gated OTFT applications.

8.4
OTFTs Gated with Electrolyte Dielectrics

The concept of coupling an electrolyte with electrical transport in a semiconductor was first developed in the 1970s to obtain an amplified electrical response from a small ionic signal [112]. Later, Wrighton *et al.* [113–116] conducted a series of elegant experiments that demonstrated the capacity for modulating the conductivity of a conjugate polymer by tuning the electrolyte composition. This concept has recently attracted interest as a result of the emergence of cheap, flexible OTFTs that require low operating voltages. This section highlights recent achievements in solid electrolyte-gated OTFTs and electronic circuits prepared using these transistors.

During the early stages of organic field-effect transistor (OFET) research, PEO systems dissolved with lithium salts ($LiClO_4$ or LiTFSA) were widely used as solid polymer electrolyte gate dielectrics [117–120]. The exceptionally high C' values of these materials, around 15 μF cm^{-2}, induced a very large charge carrier density that exceeded 10^{14} carriers cm^{-2} at the semiconductor/electrolyte interface. These values enabled the low voltage operation of OTFTs based on single organic crystals [119–121], carbon nanotubes (CNTs) [122], or thin films of molecules and polymers [117, 123]. PEO/$LiClO_4$ electrolytes provided an impressive hole mobility of 0.7 cm^2 V^{-1} s^{-1} from a benchmark p-type organic semiconductor, poly(3-hexylthiophene) (P3HT), at an operation voltage of <3 V. These systems resulted in a large room-temperature conductivity of 1000 S cm^{-1} and a very low conductance activation energy of 0.7–4 meV, without undergoing the structural rearrangements associated with chemical doping processes [124]. These devices inevitably suffered from a poor dynamic response to the applied electric field because of the low ionic conductivity of the electrolytes (10^{-4} to 10^{-5} S cm^{-1}). Consequently, switching frequencies exceeding 1 Hz were inaccessible, thereby restricting the practical applications of these devices [37]. Methods for boosting the switching speeds of electrolyte-gated devices have been a critical issue in this field.

To address this issue, Frisbie *et al.* and Lodge *et al.* [125, 126] demonstrated the preparation of a variety of OTFTs using various ion gel electrolytes. As mentioned above, ion gels are mixtures of ionic liquids and self-assembled block copolymers which form a physically cross-linked network in an ionic liquid. The high ionic conductivity of these ionic liquids (10^{-3} S cm^{-1}) compared to the conductivity in PEO-based electrolytes (10^{-4} to 10^{-5} S cm^{-1}) suggests that ion gel-gated OTFTs could yield a fast polarization response while preserving the high C' value critical for low-voltage device operation. As an example, ion gel-gated OTFTs based on various p-type polymer semiconductors were found to yield hole mobilities as high as 2 cm^2 V^{-1} s^{-1} at low voltages below 3 V, along with a high on/off current ratio exceeding 10^5 [89, 125]. The ion gel-gated P3HT TFTs displayed a good transient response at a 1 kHz switching frequency, nearly a 1000-fold improvement over the record value obtained from PEO-based electrolyte-gated OTFTs.

Flexible OTFTs and circuits based on printed ion gel electrolytes have been demonstrated using commercial aerosol jet printing techniques [125]. As shown in Figure 8.3, these devices were fabricated using a Au nanoparticle ink to print the source and drain electrodes on a flexible polyimide substrate, P3HT; an ion gel composed of [EMI][TFSA] and poly(styrene-*b*-methylmethacrylate-*b*-styrene) (PS-PMMA-PS) as the semiconductor and gate dielectric, respectively; and a conducting polymer ink composed of

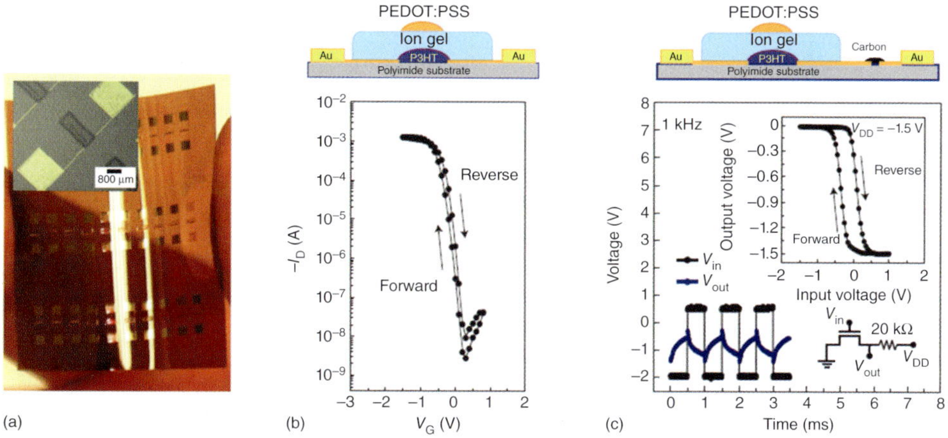

Figure 8.3 (a) Optical image of an aerosol-printed ion gel-OTFT array fabricated on a polyimide substrate. (b) Transfer characteristic of ion gel-gated P3HT OTFT at drain voltage (V_D) of −1 V. The top panels in (a) and (b) show the device schematic diagrams. (c) Output voltage response of a resistor loaded ion gel-gated OTFT inverter at the supply voltage (V_{DD}) of −1.5 V when gate voltage (V_G) is pulsed at 1 kHz. The input–output voltage characteristics are shown in the inset [125]. (Reproduced with permission. Copyright © 2008, NPG.)

poly(3,4-ethylenedioxythiophene):poly(styrene sulfonate) (PEDOT:PSS) to serve as the gate electrode. The resulting devices yielded a high hole mobility of $1.8\,cm^2\,V^{-1}\,s^{-1}$ and a large on/off ratio of 10^5. Optimization of the P3HT morphology and processing conditions (such as the post-annealing temperature) further improved the electrical performance, including the on/off current ratio, threshold voltage, hysteresis, dynamic response, bias stress resistance, and mechanical durability [127].

The Berggren group employed the polyanionic proton-conducting electrolyte poly(vinylphosphonic acid-*co*-acrylic acid) P(VPA-AA) as a gate dielectric in P3HT TFTs. This system took advantage of the facile movement of the smallest ionic species, the proton. The operation speed of the device could be further enhanced by fabricating submicrometer channel lengths [109] because the switching time is, in principle, proportional to the channel length. Interestingly, the short-channel effect [128, 129] could be suppressed by a high vertical electric field of $10^9\,V\,m^{-1}$ at 1 V, generated by P(VPA-AA). Here, the short-channel effect refers to the degraded device performance relative to the performance of an ideal long-channel device under conditions in which the lateral electric field between the source and drain electrodes exceeded the electric field in the vertical

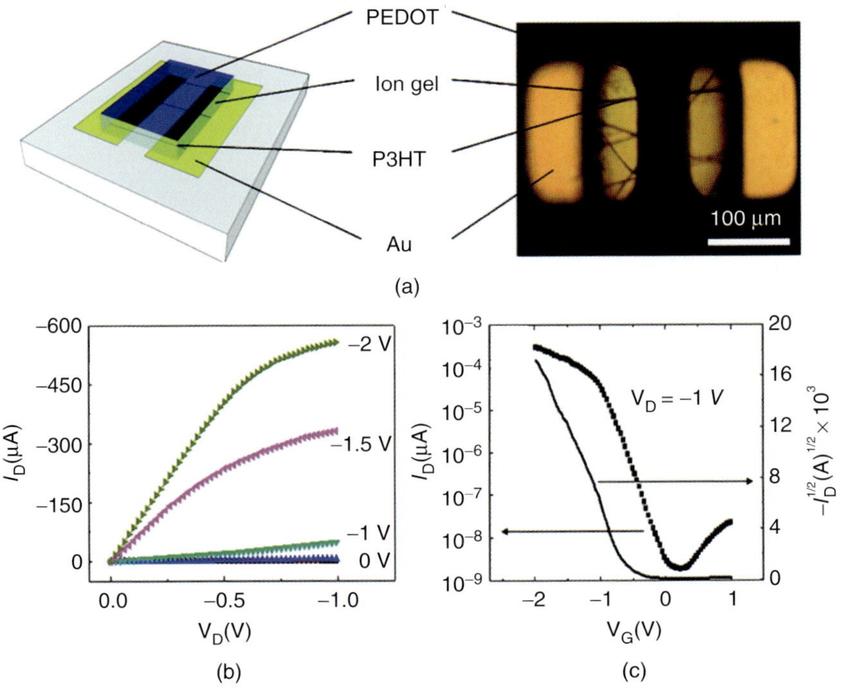

Figure 8.4 (a) Device structure scheme (left) and optical image (right) of ion gel-gated OTFTs. (b) Output characteristics. (c) Transfer characteristics [130]. (Reproduced with permission. Copyright © 2009, ACS.)

direction. The resulting P(VPA-AA)-gated OTFTs prepared with short channels could function at a switching frequency of 10 kHz.

The Cho and Jung groups have reported high-performance OTFTs based on photo-patternable ion gels and electrospun P3HT nanofibers on flexible polymer substrates treated with self-assembled monolayer (3-(trichlorosilyl)propyl methacrylate) [130]. To prepare the ion gels, [EMI][TFSA] ionic liquid was mixed with poly(ethylene glycol) diacrylate (PEG-DA) monomers and 2-hydroxy-2-methylpropiopenone (HOMPP) photoinitiator, which could be micro-patterned by UV irradiation through a shadow mask. Also, the UV irradiation induced chemical bonding between the PEG-DA and the self-assembled monolayer, yielding physical and mechanical tolerances against film delamination that may be caused during bending of the substrate. The resulting OTFT showed high device performance below 2 V: average field-effect mobility of $\sim 2\,cm^2\,V^{-1}\,s^{-1}$ and on/off current ratio of $\sim 10^5$ (Figure 8.4).

8.5 Circuits Based on Electrolyte-Gated OTFTs

Thanks to several notable developments in the field of OTFT technologies and device physics, basic logic circuits may now be fabricated by integrating multiple unit transistors. Future applications (e.g., flexible and portable electronic devices) will require these circuits to operate beyond a clock frequency of at least 100 Hz [131] and at an operating voltage less than a few volts so that simple power sources may be used, such as thin-film batteries or other energy harvesting systems. Along these lines, Xia *et al.* [132] reported the fabrication of a printed P3HT integrated circuit that operated at 2 V and included inverters, ring oscillators, NAND gates, and D flip-flops. In this work, all printed OTFTs and circuits were fabricated on flexible poly(ethylene naphthalate) (PEN) substrates. PEDOT:PSS (source, drain electrode), P3HT (semiconductor), [EMI][TFSA]/PS-PMMA-PS ion gel (gate dielectric), and PEDOT:PSS (gate electrode) were deposited using commercial aerosol jet-printing techniques. The ion gel-gated OTFTs were operated successfully at low voltages of ± 1 V with a very low subthreshold swing (SS) of 200 mV per decade and negligible hysteresis. The printed transistor-loaded inverter was optimized to increase the switching frequency to at least 10 kHz. Figure 8.5 shows an image of the device and a circuit diagram of the five-stage ring oscillator prepared using six transistor-loaded inverter-based ion gel-gated P3HT OTFTs. This ring oscillator operated at a frequency of 150 Hz, which corresponds to a propagation delay of 0.7 ms per stage. Printed D flip-flops consisting of eight NAND gates and three inverters were demonstrated. Flip-flops are critical building blocks present in virtually every integrated circuit application, from microprocessors to RFIDs and liquid crystal displays.

Berggren *et al.* [134] has demonstrated the fabrication of complementary integrated circuits based on OTFTs gated by polyelectrolytes. Complementary logic circuits in which n- and p-channel TFTs function in concert have several

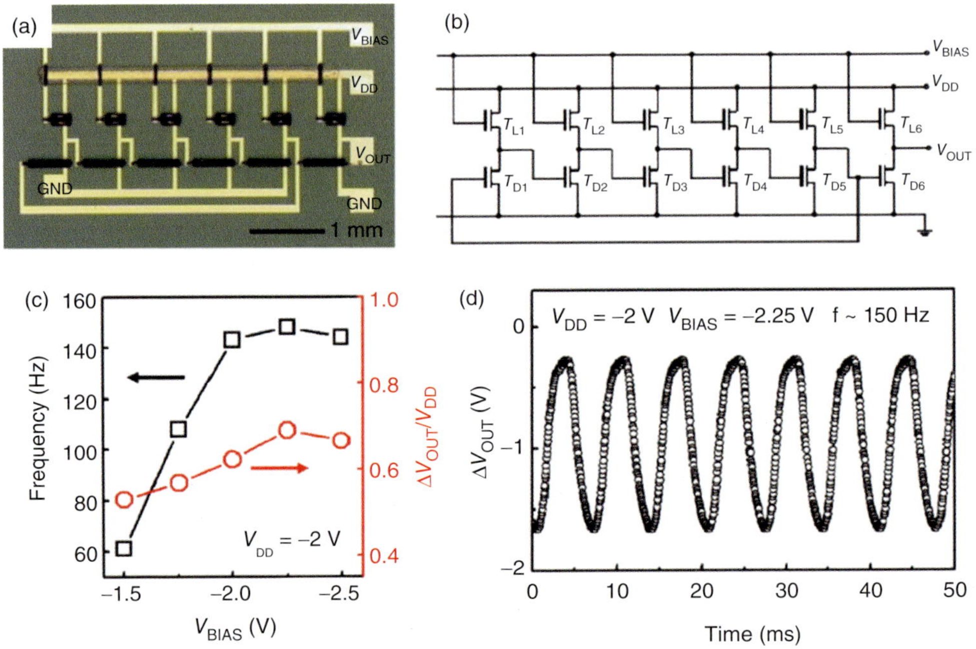

Figure 8.5 (a) Optical image of five-stage ring oscillator based on printed gel-OTFTs on a SiO_2/Si substrate. (b) Circuit diagram. (c) Frequency and output range of the ring oscillator as a function of load transistor gate voltage, V_{BIAS}. (d) Output characteristics of the ring oscillator with oscillation frequency as high as 150 Hz [133]. (Reproduced with permission. Copyright © 2010, Wiley-VCH.)

advantages over single-polarity circuit designs, including low power dissipation, fast operating speed, and stable operation. In this work, the polyanionic P(VPA-AA) or polycationic poly[(1-vinylpyrrolidone)-*co*-(2-ethyldimethylammonioethyl methacrylate ethyl sulfate)] (P(VP-EDMAEMAES)) electrolyte was used to form a gate dielectric in p- or n-channel transistors, respectively. Poly(2,5-bis(2-thienyl)-3,6-dihexadecylthieno[3,2-*b*]thiophene) (P(T0T0TT16)) and poly[*N*,*N*′-bis(2-octyldodecyl)-naphthalene-1,4,5,8-bis(dicarboximide)-2,6-diyl]-*alt*-5,5′-(2,2′-bithiophene) (P(NDI2OD-T2)) were used as p-type and n-type semiconductors, respectively (Figure 8.6). The resulting p- or n-channel OTFTs, respectively, provided carrier mobilities of 0.03 or 0.007 $cm^2\ V^{-1}\ s^{-1}$, a V_{th} of −0.5 or 0.1 V, and an SS of 0.13 or 0.07 V $decade^{-1}$. The resulting complementary integrated circuits ($W/L = 1000\ \mu m/2.5\ \mu m$) resulted in rail-to-rail output swings and signal gains of up to 17.5 for supply voltages between 0.2 and 1 V. The cut-off frequencies for the p- or n-channel transistors were estimated to be 16 or 6.7 kHz, respectively, under a supply voltage of 1 V. The static power dissipation was <2.5 nW per logic gate. Complementary ring oscillators displayed stable oscillations with a nearly full amplitude swing for supply voltages as low as 0.2 V and as high as 1.5 V. The corresponding propagation delays per stage at those voltages were 8 and 0.5 ms, respectively.

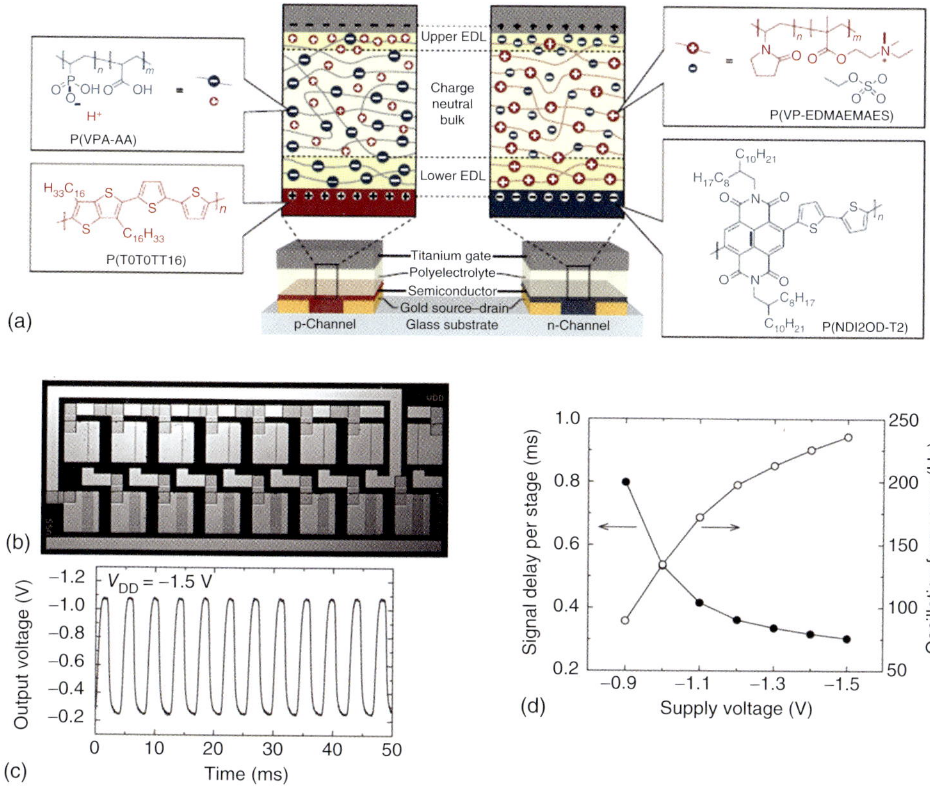

Figure 8.6 (a) Chemical structures of the semiconductor materials and device structure scheme for complementary polyelectrolyte-gated transistors. Schematic cross sections illustrate the charge distribution within the polyelectrolyte gate insulator layers in p- and n-channel transistors which are gated with negative and positive gate voltages, respectively. (b) Optical microscopy image of a seven-stage complementary ring oscillator circuit with output buffer. (c) Output voltage signal at the supply voltage (V_{DD}) of −1.5 V. (d) Oscillation frequency and calculated signal propagation delay as a function of V_{DD} [134]. (Reproduced with permission. Copyright © 2010, Wiley-VCH.)

High-speed organic printed circuits may be fabricated using carbon nanomaterials. For example, Ha *et al.* [133] investigated the performance of aerosol jet-printed TFT devices prepared using an ion gel electrolyte on a semiconducting CNT network. A variety of circuits beyond a unit transistor were demonstrated, including an inverter, a ring oscillator, and a NAND gate. The top-gate, bottom-contact CNT transistor was fabricated through the sequential printing of CNTs (semiconductor), an [EMI][TFSA]/PS-PMMA-PS ion gel (dielectric), and a PEDOT:PSS (electrode) layer onto a prepatterned Au electrode on SiO_2 and polyimide plastic substrates using aerosol jet-printing techniques. The channel length and width were 500 and 75 μm, respectively. The CNT transistors exhibited clear ambipolar behavior with an on/off ratio of 10^4 and effective mobilities of

$cm^2\,V^{-1}\,s^{-1}$ (hole) and $20\,cm^2\,V^{-1}\,s^{-1}$ (electron). The complementary inverter prepared by combining two identical ambipolar CNT transistors yielded an output swing of up to 1.2 V and a maximum inverter gain of 58 under a supply voltage of 1.5 V. The five-stage ion gel-gate CNT ring oscillators, in particular, displayed high performance at low operating voltages (<3 V). The oscillation frequencies reached values of up to 2.3 and 1.9 kHz under supply voltages of 1.5 and 2.5 V for the devices printed on SiO_2 and polyimide substrates, respectively, which yielded the corresponding minimum delay times per stage of 42 and 53 μs, respectively. The differences in the circuit performance of the two devices originated from the low mobility and the high parasitic capacitance due to the rough surface of the polyimide substrate [133]. More recently, the same group demonstrated the preparation of ion gel-gated CNT five-stage ring oscillators with an operating frequency exceeding 20 kHz and a stage delay of <5 μs under supply voltages below 3 V [135]. The transient behavior of the CNT circuits was systematically characterized as a function of the transistor channel length and the ionic conductivity of the ion gel electrolyte in order to establish the optimal conditions for boosting the circuit performance. A record ring oscillator performance was obtained from the printed organic circuits, yielding a delay time of 1.2 μs at 2 V. These performances were obtained by (i) reducing the channel length down to 1 μm, (ii) minimizing parasitic capacitance due to source–gate and drain–gate areal overlap, and (iii) using a highly conductive ion gel with a minimal thickness (Figure 8.7).

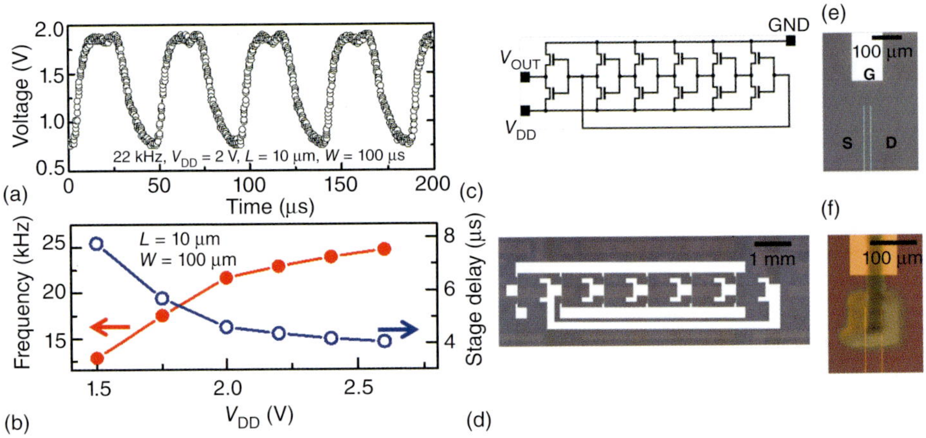

Figure 8.7 (a) Output signal of a CNT ring oscillator with a frequency of 22 kHz and stage delays of 4.5 μs at a supply voltage (V_{DD}) of 2 V. (b) Frequency and stage delay times of the ring oscillator as functions of V_{DD}. (c) Circuit diagram of the ring oscillator. (d) Photograph of the Au contacts in the ring oscillator before printing. (e) Photograph of the Au source, drain, and gate contacts for a single TFT before printing. (f) Picture of a single TFT after all layers were printed, including semiconducting CNT network, ion gel dielectric, and PEDOT:PSS gate electrode [135]. (Reproduced with permission. Copyright © 2013, Wiley-VCH.)

8.6 Conclusions

The need for attaining organic transistors driven by low voltages has led to the development of a variety of solid electrolytes that can be utilized as a high-capacitance dielectric layer. These solid electrolytes have been examined to expand our understanding of the physicochemical properties of these materials. Complementary circuits based on organic semiconductors, display devices, or sensors based on unit electrolyte-gated OTFT building blocks may be prepared through versatile printing processes. Although several issues have yet to be resolved, including increasing the device switching time, we expect that the unique advantages of this class of novel electrolytes, which display extremely high capacitance values and which may be readily processed using solution-based methods, will lead to the commercialization of electrolyte-based electronic devices.

References

1. Horowitz, G. (1998) Organic field-effect transistors. *Adv. Mater.*, **10** (5), 365–377.
2. Allard, S., Forster, M., Souharce, B., Thiem, H., and Scherf, U. (2008) Organic semiconductors for solution-processable field-effect transistors (OFETs). *Angew. Chem. Int. Ed.*, **47** (22), 4070–4098.
3. Loo, Y.-L. and McCulloch, I. (2008) Progress and challenges in commercialization of organic electronics. *MRS Bull.*, **33** (07), 653–662.
4. Braga, D. and Horowitz, G. (2009) High-performance organic field-effect transistors. *Adv. Mater.*, **21** (14-15), 1473–1486.
5. Holliday, S., Donaghey, J.E., and McCulloch, I. (2013) Advances in charge carrier mobilities of semiconducting polymers used in organic transistors. *Chem. Mater.* doi: 10.1021/cm402421p.
6. Dong, H., Fu, X., Liu, J., Wang, Z., and Hu, W. (2013) Key points for high-mobility organic field-effect transistors. *Adv. Mater.*, **25** (43), 6158–6183.
7. Rogers, J.A., Bao, Z.N., Makhija, A., and Braun, P. (1999) Printing process suitable for reel-to-reel production of high-performance organic transistors and circuits. *Adv. Mater.*, **11** (9), 741–745.
8. Bao, Z.N., Rogers, J.A., and Katz, H.E. (1999) Printable organic and polymeric semiconducting materials and devices. *J. Mater. Chem.*, **9** (9), 1895–1904.
9. Katz, H.E., Lovinger, A.J., Johnson, J., Kloc, C., Siegrist, T., Li, W., Lin, Y.Y., and Dodabalapur, A. (2000) A soluble and air-stable organic semiconductor with high electron mobility. *Nature*, **404** (6777), 478–481.
10. Sirringhaus, H., Kawase, T., Friend, R.H., Shimoda, T., Inbasekaran, M., Wu, W., and Woo, E.P. (2000) High-resolution inkjet printing of all-polymer transistor circuits. *Science*, **290** (5499), 2123–2126.
11. Meng, H., Bao, Z.N., Lovinger, A.J., Wang, B.C., and Mujsce, A.M. (2001) High field-effect mobility oligofluorene derivatives with high environmental stability. *J. Am. Chem. Soc.*, **123** (37), 9214–9215.
12. Calvert, P. (2001) Inkjet printing for materials and devices. *Chem. Mater.*, **13** (10), 3299–3305.
13. Rogers, J.A., Bao, Z., Baldwin, K., Dodabalapur, A., Crone, B., Raju, V.R., Kuck, V., Katz, H., Amundson, K., Ewing, J., and Drzaic, P. (2001)

Paper-like electronic displays: large-area rubber-stamped plastic sheets of electronics and microencapsulated electrophoretic inks. *Proc. Natl. Acad. Sci. U.S.A.*, **98** (9), 4835–4840.

14. Afzali, A., Dimitrakopoulos, C.D., and Breen, T.L. (2002) High-performance, solution-processed organic thin film transistors from a novel pentacene precursor. *J. Am. Chem. Soc.*, **124** (30), 8812–8813.
15. Loo, Y.L., Willett, R.L., Baldwin, K.W., and Rogers, J.A. (2002) Additive, nanoscale patterning of metal films with a stamp and a surface chemistry mediated transfer process: applications in plastic electronics. *Appl. Phys. Lett.*, **81** (3), 562–564.
16. Rogers, J.A. and Bao, Z. (2002) Printed plastic electronics and paperlike displays. *J. Polym. Sci. Part A: Polym. Chem.*, **40** (20), 3327–3334.
17. Stutzmann, N., Friend, R.H., and Sirringhaus, H. (2003) Self-aligned, vertical-channel, polymer field-effect transistors. *Science*, **299** (5614), 1881–1884.
18. Facchetti, A., Mushrush, M., Katz, H.E., and Marks, T.J. (2003) n-type building blocks for organic electronics: a homologous family of fluorocarbon-substituted thiophene oligomers with high carrier mobility. *Adv. Mater.*, **15** (1), 33–38.
19. Mushrush, M., Facchetti, A., Lefenfeld, M., Katz, H.E., and Marks, T.J. (2003) Easily processable phenylene-thiophene-based organic field-effect transistors and solution-fabricated nonvolatile transistor memory elements. *J. Am. Chem. Soc.*, **125** (31), 9414–9423.
20. Ong, B.S., Wu, Y.L., Liu, P., and Gardner, S. (2004) High-performance semiconducting polythiophenes for organic thin-film transistors. *J. Am. Chem. Soc.*, **126** (11), 3378–3379.
21. Sundar, V.C., Zaumseil, J., Podzorov, V., Menard, E., Willett, R.L., Someya, T., Gershenson, M.E., and Rogers, J.A. (2004) Elastomeric transistor stamps: reversible probing of charge transport in organic crystals. *Science*, **303** (5664), 1644–1646.
22. Sirringhaus, H. (2005) Device physics of solution-processed organic field-effect transistors. *Adv. Mater.*, **17** (20), 2411–2425.
23. Briseno, A.L., Mannsfeld, S.C.B., Ling, M.M., Liu, S., Tseng, R.J., Reese, C., Roberts, M.E., Yang, Y., Wudl, F., and Bao, Z. (2006) Patterning organic single-crystal transistor arrays. *Nature*, **444** (7121), 913–917.
24. McCulloch, I., Heeney, M., Bailey, C., Genevicius, K., Macdonald, I., Shkunov, M., Sparrowe, D., Tierney, S., Wagner, R., Zhang, W.M., Chabinyc, M.L., Kline, R.J., McGehee, M.D., and Toney, M.F. (2006) Liquid-crystalline semiconducting polymers with high charge-carrier mobility. *Nat. Mater.*, **5** (4), 328–333.
25. Gundlach, D.J., Royer, J.E., Park, S.K., Subramanian, S., Jurchescu, O.D., Hamadani, B.H., Moad, A.J., Kline, R.J., Teague, L.C., Kirillov, O., Richter, C.A., Kushmerick, J.G., Richter, L.J., Parkin, S.R., Jackson, T.N., and Anthony, J.E. (2008) Contact-induced crystallinity for high-performance soluble acene-based transistors and circuits. *Nat. Mater.*, **7** (3), 216–221.
26. Sekitani, T., Noguchi, Y., Zschieschang, U., Klauk, H., and Someya, T. (2008) Organic transistors manufactured using inkjet technology with subfemtoliter accuracy. *Proc. Natl. Acad. Sci. U.S.A.*, **105** (13), 4976–4980.
27. Yan, H., Chen, Z., Zheng, Y., Newman, C., Quinn, J.R., Dotz, F., Kastler, M., and Facchetti, A. (2009) A high-mobility electron-transporting polymer for printed transistors. *Nature*, **457** (7230), 679–U671.
28. Sirringhaus, H. (2009) Reliability of organic field-effect transistors. *Adv. Mater.*, **21** (38-39), 3859–3873.
29. Anthony, J.E., Facchetti, A., Heeney, M., Marder, S.R., and Zhan, X. (2010) n-type organic semiconductors in organic electronics. *Adv. Mater.*, **22** (34), 3876–3892.
30. Rogers, J.A., Someya, T., and Huang, Y. (2010) Materials and mechanics for stretchable electronics. *Science*, **327** (5973), 1603–1607.
31. Singh, M., Haverinen, H.M., Dhagat, P., and Jabbour, G.E. (2010) Inkjet

printing-process and its applications. *Adv. Mater.*, **22** (6), 673–685.

32. Sirringhaus, H., Bird, M., Richards, T., and Zhao, N. (2010) Charge transport physics of conjugated polymer field-effect transistors. *Adv. Mater.*, **22** (34), 3893–3898.
33. Dimitrakopoulos, C.D., Purushothaman, S., Kymissis, J., Callegari, A., and Shaw, J.M. (1999) Low-voltage organic transistors on plastic comprising high-dielectric constant gate insulators. *Science*, **283** (5403), 822–824.
34. Majewski, L.A., Grell, M., Ogier, S.D., and Veres, J. (2003) A novel gate insulator for flexible electronics. *Org. Electron.*, **4** (1), 27–32.
35. Tate, J., Rogers, J.A., Jones, C.D.W., Vyas, B., Murphy, D.W., Li, W., Bao, Z., Slusher, R.E., Dodabalapur, A., and Katz, H.E. (2000) Anodization and microcontact printing on electroless silver: solution-based fabrication procedures for low-voltage electronic systems with organic active components. *Langmuir*, **16** (14), 6054–6060.
36. Majewski, L.A., Schroeder, R., and Grell, M. (2005) One volt organic transistor. *Adv. Mater.*, **17** (2), 192–196.
37. Cho, S.W., Jeong, J.G., Park, S.H., Cho, M.H., Jeong, K., Whang, C.N., and Yi, Y. (2008) The characteristics and interfacial electronic structures of organic thin film transistor devices with ultrathin (HfO(2))(x)(SiO(2))(1-x) gate dielectrics. *Appl. Phys. Lett.*, **92** (21), 213302–213302-3.
38. Javey, A., Kim, H., Brink, M., Wang, Q., Ural, A., Guo, J., McIntyre, P., McEuen, P., Lundstrom, M., and Dai, H. (2002) High-[kappa] dielectrics for advanced carbon-nanotube transistors and logic gates. *Nat. Mater.*, **1** (4), 241–246.
39. Kim, C.S., Jo, S.J., Lee, S.W., Kim, W.J., Baik, H.K., and Lee, S.J. (2007) Surface-modified high-k oxide gate dielectrics for low-voltage high-performance pentacene thin-film transistors. *Adv. Funct. Mater.*, **17** (6), 958–962.
40. Kang, S.J., Chung, K.B., Park, D.S., Kim, H.J., Choi, Y.K., Jang, M.H., Noh, M., and Whang, C.N. (2004) Fabrication and characterization of the pentacene thin film transistor with a Gd2O3 gate insulator. *Synth. Met.*, **146** (3), 351–354.
41. Cheng, S.-S., Yang, C.-Y., Ou, C.-W., Chuang, Y.-C., Wu, M.-C., and Chu, C.-W. (2008) Pentacene thin-film transistor with PVP-capped high-k MgO dielectric grown by reactive evaporation. *Electrochem. Solid State Lett.*, **11** (5), H118–H120.
42. Shin, K., Yang, C.W., Yang, S.Y., Jeon, H.Y., and Park, C.E. (2006) Effects of polymer gate dielectrics roughness on pentacene field-effect transistors. *Appl. Phys. Lett.*, **88** (7).
43. Yang, C., Shin, K., Yang, S.Y., Jeon, H., Choi, D., Chung, D.S., and Park, C.E. (2006) Low-voltage organic transistors on a polymer substrate with an aluminum foil gate fabricated by a laminating and electropolishing process. *Appl. Phys. Lett.*, **89** (15).
44. Vuillaume, D., Boulas, C., Collet, J., Davidovits, J.V., and Rondelez, F. (1996) Organic insulating films of nanometer thicknesses. *Appl. Phys. Lett.*, **69** (11), 1646–1648.
45. Collet, J., Tharaud, O., Chapoton, A., and Vuillaume, D. (2000) Low-voltage, 30 nm channel length, organic transistors with a self-assembled monolayer as gate insulating films. *Appl. Phys. Lett.*, **76** (14), 1941–1943.
46. Halik, M., Klauk, H., Zschieschang, U., Schmid, G., Ponomarenko, S., Kirchmeyer, S., and Weber, W. (2003) Relationship between molecular structure and electrical performance of oligothiophene organic thin film transistors. *Adv. Mater.*, **15** (11), 917–922.
47. Woebkenberg, P.H., Ball, J., Kooistra, F.B., Hummelen, J.C., de Leeuw, D.M., Bradley, D.D.C., and Anthopoulos, T.D. (2008) Low-voltage organic transistors based on solution processed semiconductors and self-assembled monolayer gate dielectrics. *Appl. Phys. Lett.*, **93** (1), 013303.
48. Hur, S.-H., Yoon, M.-H., Gaur, A., Shim, M., Facchetti, A., Marks, T.J., and Rogers, J.A. (2005) Organic nanodielectrics for low voltage carbon nanotube thin film transistors and complementary logic gates. *J. Am. Chem. Soc.*, **127** (40), 13808–13809.

49. Ju, S., Lee, K., Janes, D.B., Yoon, M.-H., Facchetti, A., and Marks, T.J. (2005) Low operating voltage single ZnO nanowire field-effect transistors enabled by self-assembled organic gate nanodielectrics. *Nano Lett.*, **5** (11), 2281–2286.
50. Yoon, M.-H., Yan, H., Facchetti, A., and Marks, T.J. (2005) Low-voltage organic field-effect transistors and inverters enabled by ultrathin cross-linked polymers as gate dielectrics. *J. Am. Chem. Soc.*, **127** (29), 10388–10395.
51. Sekitani, T., Yokota, T., Zschieschang, U., Klauk, H., Bauer, S., Takeuchi, K., Takamiya, M., Sakurai, T., and Someya, T. (2009) Organic nonvolatile memory transistors for flexible sensor arrays. *Science*, **326** (5959), 1516–1519.
52. Yoon, M.-H., Kim, C., Facchetti, A., and Marks, T.J. (2006) Gate dielectric chemical structure-organic field-effect transistor performance correlations for electron, hole, and ambipolar organic semiconductors. *J. Am. Chem. Soc.*, **128** (39), 12851–12869.
53. Walser, M.P., Kalb, W.L., Mathis, T., and Batlogg, B. (2009) Low-voltage organic transistors and inverters with ultrathin fluoropolymer gate dielectric. *Appl. Phys. Lett.*, **95** (23).
54. Noh, Y.-Y. and Sirringhaus, H. (2009) Ultra-thin polymer gate dielectrics for top-gate polymer field-effect transistors. *Org. Electron.*, **10** (1), 174–180.
55. Li, L., Zhang, Y., Li, H., Tang, Q., Jiang, L., Chi, L., Fuchs, H., and Hu, W. (2009) Battery drivable organic single-crystalline transistors based on surface grafting ultrathin polymer dielectric. *Adv. Funct. Mater.*, **19** (18), 2987–2991.
56. Park, K., Park, S.H., Kim, E., Kim, J.-D., An, S.-Y., Lim, H.S., Lee, H.H., Kim, D.H., Ryu, D.Y., Lee, D.R., and Cho, J.H. (2010) Polymer brush as a facile dielectric surface treatment for high-performance, stable, soluble acene-based transistors. *Chem. Mater.*, **22** (18), 5377–5382.
57. Pinto, J.C., Whiting, G.L., Khodabakhsh, S., Torre, L., Rodriguez, A.B., Dalgliesh, R.M., Higgins, A.M., Andreasen, J.W., Nielsen, M.M., Geoghegan, M., Huck, W.T.S., and Sirringhaus, H. (2008) Organic thin film transistors with polymer brush gate dielectrics synthesized by atom transfer radical polymerization. *Adv. Funct. Mater.*, **18** (1), 36–43.
58. Hwang, D.K., Kim, C.S., Choi, J.M., Lee, K., Park, J.H., Kim, E., Baik, H.K., Kim, J.H., and Im, S. (2006) Polymer/YOx hybrid-sandwich gate dielectrics for semitransparent pentacene thin-film transistors operating under 5 V. *Adv. Mater.*, **18** (17), 2299.
59. Cao, Q., Xia, M.-G., Shim, M., and Rogers, J.A. (2006) Bilayer organic-inorganic gate dielectrics for high-performance, low-voltage, single-walled carbon nanotube thin-film transistors, complementary logic gates, and p-n diodes on plastic substrates. *Adv. Funct. Mater.*, **16** (18), 2355–2362.
60. Guo, N., DiBenedetto, S.A., Kwon, D.-K., Wang, L., Russell, M.T., Lanagan, M.T., Facchetti, A., and Marks, T.J. (2007) Supported metallocene catalysis for in situ synthesis of high energy density metal oxide nanocomposites. *J. Am. Chem. Soc.*, **129** (4), 766–767.
61. Acton, O., Ting, G., Ma, H., Ka, J.W., Yip, H.-L., Tucker, N.M., and Jen, A.K.Y. (2008) pi-sigma-phosphonic acid organic monolayer/sol-gel hafnium oxide hybrid dielectrics for low-voltage organic transistors. *Adv. Mater.*, **20** (19), 3697.
62. Sekitani, T., Zschieschang, U., Klauk, H., and Someya, T. (2010) Flexible organic transistors and circuits with extreme bending stability. *Nat. Mater.*, **9** (12), 1015–1022.
63. Jung, B.J., Hardigree, J.F.M., Dhar, B.M., Dawidczyk, T.J., Sun, J., See, K.C., and Katz, H.E. (2011) Naphthalenetetracarboxylic diimide layer-based transistors with nanometer oxide and side chain dielectrics operating below one volt. *ACS Nano*, **5** (4), 2723–2734.
64. Jung, S., Ji, T., and Varadan, V.K. (2010) Pentacene-based low-voltage strain sensors with PVP/Ta(2)O(5) hybrid gate dielectrics. *IEEE Trans. Electron Devices*, **57** (2), 391–396.
65. Klauk, H., Zschieschang, U., Pflaum, J., and Halik, M. (2007) Ultralow-power organic complementary circuits. *Nature*, **445** (7129), 745–748.

66. Roberts, M.E., Mannsfeld, S.C.B., Queralto, N., Reese, C., Locklin, J., Knoll, W., and Bao, Z. (2008) Water-stable organic transistors and their application in chemical and biological sensors. *Proc. Natl. Acad. Sci. U.S.A.*, **105** (34), 12134–12139.
67. Hamedi, M., Forchheimer, R., and Inganas, O. (2007) Towards woven logic from organic electronic fibres. *Nat. Mater.*, **6** (5), 357–362.
68. Hamedi, M., Herlogsson, L., Crispin, X., Marcilla, R., Berggren, M., and Inganas, O. (2009) Fiber-embedded electrolyte-gated field-effect transistors for e-textiles. *Adv. Mater.*, **21** (5), 573–577.
69. Said, E., Andersson, P., Engquist, I., Crispin, X., and Berggren, M. (2009) Electrochromic display cells driven by an electrolyte-gated organic field-effect transistor. *Org. Electron.*, **10** (6), 1195–1199.
70. Tao, X.Y., Koncar, V., and Dufour, C. (2011) Geometry pattern for the wire organic electrochemical textile transistor. *J. Electrochem. Soc.*, **158** (5), H572–H577.
71. Braga, D., Erickson, N.C., Renn, M.J., Holmes, R.J., and Frisbie, C.D. (2012) High-transconductance organic thin-film electrochemical transistors for driving low-voltage red-green-blue active matrix organic light-emitting devices. *Adv. Funct. Mater.*, **22** (8), 1623–1631.
72. Panzer, M.J., Newman, C.R., and Frisbie, C.D. (2005) Low-voltage operation of a pentacene field-effect transistor with a polymer electrolyte gate dielectric. *Appl. Phys. Lett.*, **86** (10), 103503–103503.
73. Takeya, J., Yamada, K., Hara, K., Shigeto, K., Tsukagoshi, K., Ikehata, S., and Aoyagi, Y. (2006) High-density electrostatic carrier doping in organic single-crystal transistors with polymer gel electrolyte. *Appl. Phys. Lett.*, **88** (11), 112102–112103.
74. Dhoot, A.S., Israel, C., Moya, X., Mathur, N.D., and Friend, R.H. (2009) Large electric field effect in electrolyte-gated manganites. *Phys. Rev. Lett.*, **102** (13), 136402.
75. Ye, J.T., Inoue, S., Kobayashi, K., Kasahara, Y., Yuan, H.T., Shimotani, H., and Iwasa, Y. (2009) Liquid-gated interface superconductivity on an atomically flat film. *Nat. Mater.*, **9** (2), 125–128.
76. Fedorov, M.V. and Kornyshev, A.A. (2008) Towards understanding the structure and capacitance of electrical double layer in ionic liquids. *Electrochim. Acta*, **53** (23), 6835–6840.
77. Lee, J., Kaake, L.G., Cho, J.H., Zhu, X.Y., Lodge, T.P., and Frisbie, C.D. (2009) Ion gel-gated polymer thin-film transistors: operating mechanism and characterization of gate dielectric capacitance, switching speed, and stability. *J. Phys. Chem. C*, **113** (20), 8972–8981.
78. MacCallum, J.R. and VIncent, C.A. (1987, 1989) *Polymer Electrolyte Review 1 and 2*, Elsevier.
79. Bard, A. J., Faulkner L. R. (2001) *Electrochemical Methods: Fundamentals and Applications*, (2nd edn) John Wiley & Sons, Inc., New York.
80. Fenton, D.E., Parker, J.M., and Wright, P.V. (1973) Wright reported the preparation of crystalline complexes of sodium and potassium salts with poly(ethylene oxide. *Polymer*, **14**, 589.
81. Brandrup, J., Immergut, E.H., and Grulke, E.A. (1999) *Polymer Handbook*, Wiley-Interscience, New York.
82. Ueki, T. and Watanabe, M. (2008) Macromolecules in ionic liquids: progress, challenges, and opportunities. *Macromolecules*, **41** (11), 3739–3749.
83. Gavelin, P., Jannasch, P., and Wesslen, B. (2001) Amphiphilic polymer gel electrolytes. I. Preparation of gels based on poly(ethylene oxide) graft copolymers containing different ionophobic groups. *J. Polym. Sci. Part A-Polym. Chem.*, **39** (13), 2223–2232.
84. Kaskhedikar, N., Burjanadze, M., Karatas, Y., and Wiemhofer, H.D. (2006) Polymer electrolytes based on cross-linked cyclotriphosphazenes. *Solid State Ion.*, **177** (35-36), 3129–3134.
85. Weston, J.E. and Steele, B.C.H. (1982) Effects of inert fillers on the mechanical and electrochemical properties of lithium salt poly (ethylene-oxide) polymer electrolytes. *Solid State Ion.*, **7** (1), 75–79.

86. Adebahr, J., Best, A.S., Byrne, N., Jacobsson, P., MacFarlane, D.R., and Forsyth, M. (2003) Ion transport in polymer electrolytes containing nanoparticulate TiO_2: the influence of polymer morphology. *Phys. Chem. Chem. Phys.*, **5** (4), 720–725.
87. Cheung, I.W., Chin, K.B., Greene, E.R., Smart, M.C., Abbrent, S., Greenbaum, S.G., Prakash, G.K.S., and Surampudi, S. (2003) Electrochemical and solid state NMR characterization of composite PEO-based polymer electrolytes. *Electrochim. Acta*, **48** (14-16), 2149–2156.
88. Sandi, G., Carrado, K.A., Joachin, H., Lu, W., and Prakash, J. (2003) Polymer nanocomposites for lithium battery applications. *J. Power Sources*, **119**, 492–496.
89. Cho, J.H., Lee, J., He, Y., Kim, B., Lodge, T.P., and Frisbie, C.D. (2008) High-capacitance ion gel gate dielectrics with faster polarization response times for organic thin film transistors. *Adv. Mater.*, **20** (4), 686–690.
90. Ono, S., Seki, S., Hirahara, R., Tominari, Y., and Takeya, J. (2008) High-mobility, low-power, and fast-switching organic field-effect transistors with ionic liquids. *Appl. Phys. Lett.*, **92** (10), 103313.
91. Chen, F., Qing, Q., Xia, J.L., Li, J.H., and Tao, N.J. (2009) Electrochemical gate-controlled charge transport in graphene in ionic liquid and aqueous solution. *J. Am. Chem. Soc.*, **131** (29), 9908–9909.
92. Ye, J.T., Inoue, S., Kobayashi, K., Kasahara, Y., Yuan, H.T., Shimotani, H., and Iwasa, Y. (2010) Liquid-gated interface superconductivity on an atomically flat film. *Nat. Mater.*, **9** (2), 125–128.
93. Tokuda, H., Hayamizu, K., Ishii, K., Susan, M., and Watanabe, M. (2005) Physicochemical properties and structures of room temperature ionic liquids. 2. Variation of alkyl chain length in imidazolium cation. *J. Phys. Chem. B*, **109** (13), 6103–6110.
94. Welton, T. (1999) Room-temperature ionic liquids. Solvents for synthesis and catalysis. *Chem. Rev.*, **99** (8), 2071–2083.
95. Macfarlane, D.R., Forsyth, M., Howlett, P.C., Pringle, J.M., Sun, J., Annat, G., Neil, W., and Izgorodina, E.I. (2007) Ionic liquids in electrochemical devices and processes: managing interfacial electrochemistry. *Acc. Chem. Res.*, **40** (11), 1165–1173.
96. Kimizuka, N. and Nakashima, T. (2001) Spontaneous self-assembly of glycolipid bilayer membranes in sugar-philic ionic liquids and formation of ionogels. *Langmuir*, **17** (22), 6759–6761.
97. Hanabusa, K., Fukui, H., Suzuki, M., and Shirai, H. (2005) Specialist gelator for ionic liquids. *Langmuir*, **21** (23), 10383–10390.
98. Ueno, K., Hata, K., Katakabe, T., Kondoh, M., and Watanabe, M. (2008) Nanocomposite ion gels based on silica nanoparticles and an ionic liquid: ionic transport, viscoelastic properties, and microstructure. *J. Phys. Chem. B*, **112** (30), 9013–9019.
99. Susan, M.A., Kaneko, T., Noda, A., and Watanabe, M. (2005) Ion gels prepared by in situ radical polymerization of vinyl monomers in an ionic liquid and their characterization as polymer electrolytes. *J. Am. Chem. Soc.*, **127** (13), 4976–4983.
100. Klingshirn, M.A., Spear, S.K., Subramanian, R., Holbrey, J.D., Huddleston, J.G., and Rogers, R.D. (2004) Gelation of ionic liquids using a cross-linked poly(ethylene glycol) gel matrix. *Chem. Mater.*, **16** (16), 3091–3097.
101. He, Y.Y. and Lodge, T.P. (2007) A thermoreversible ion gel by triblock copolymer self-assembly in an ionic liquid. *Chem. Commun.*, **26**, 2732–2734.
102. Noro, A., Matsushita, Y., and Lodge, T.P. (2008) Thermoreversible supramacromolecular ion gels via hydrogen bonding. *Macromolecules*, **41** (15), 5839–5844.
103. Jana, S., Parthiban, A., and Chai, C.L.L. (2010) Transparent, flexible and highly conductive ion gels from ionic liquid compatible cyclic carbonate network. *Chem. Commun.*, **46** (9), 1488–1490.
104. Matsumoto, K. and Endo, T. (2008) Confinement of ionic liquid by networked polymers based on multifunctional epoxy resins. *Macromolecules*, **41** (19), 6981–6986.
105. He, Y. and Lodge, T.P. (2008) Thermoreversible ion gels with tunable

melting temperatures from triblock and pentablock copolymers. *Macromolecules*, **41** (1), 167–174.
106. He, Y.Y., Boswell, P.G., Buhlmann, P., and Lodge, T.P. (2007) Ion gels by self-assembly of a triblock copolymer in an ionic liquid. *J. Phys. Chem. B*, **111** (18), 4645–4652.
107. Lee, K.H., Zhang, S.P., Lodge, T.P., and Frisbie, C.D. (2011) Electrical impedance of spin-coatable ion gel films. *J. Phys. Chem. B*, **115** (13), 3315–3321.
108. Meyer, W.H. (1998) Polymer electrolytes for lithium-ion batteries. *Adv. Mater.*, **10** (6), 439–448.
109. Herlogsson, L., Noh, Y.Y., Zhao, N., Crispin, X., Sirringhaus, H., and Berggren, M. (2008) Downscaling of organic field-effect transistors with a polyelectrolyte gate insulator. *Adv. Mater.*, **20** (24), 4708–4713.
110. Lu, J.M., Yan, F., and Texter, J. (2009) Advanced applications of ionic liquids in polymer science. *Prog. Polym. Sci.*, **34** (5), 431–448.
111. Mecerreyes, D. (2011) Polymeric ionic liquids: broadening the properties and applications of polyelectrolytes. *Prog. Polym. Sci.*, **36** (12), 1629–1648.
112. Bergveld, P. (1970) Development of an ion-sensitive solid-state device for neuro- physiological measurements. *IEEE Trans. Biomed. Eng.*, **17**, 70.
113. Moss, S.D., Janata, J., and Johnson, C.C. (1975) Potassium ion-sensitive field effect transistor. *Anal. Chem.*, **47**, 2238.
114. White, H.S., Kittlesen, G.P., and Wrighton, M.S. (1984) Chemical derivatization of an array of three gold microelectrodes with polypyrrole: fabrication of a molecule-based transistor. *J. Am. Chem. Soc.*, **106**, 5375.
115. Thackeray, J.W., White, H.S., and Wrighton, M.S. (1985) Poly(3-methyltiophene) coated electrodes: optical and electrical properties as a function of redox potential and amplification of electrical and chemical signals using Poly(3-methyltiophene) based microelectrochemical transistors. *J. Phys. Chem.*, **89**, 5133.
116. Ofer, D., Crooks, R.M., and Wrighton, M.S. (1990) Potential dependence of the conductivity of highly oxidized polythiophenes, polypyrroles, and polyaniline: observation of finite potential windows of high conductivity. *J. Am. Chem. Soc.*, **112**, 7869.
117. Dhoot, A.S., Yuen, J.D., Heeney, M., McCulloch, I., Moses, D., and Heeger, A.J. (2006) Beyond the metal-insulator transition in polymer electrolyte gated polymer field-effect transistors. *Proc. Natl. Acad. Sci. U.S.A.*, **103** (32), 11834–11837.
118. Panzer, M.J. and Frisbie, C.D. (2005) Polymer electrolyte gate dielectric reveals finite windows of high conductivity in organic thin film transistors at high charge carrier densities. *J. Am. Chem. Soc.*, **127** (19), 6960–6961.
119. Shimotani, H., Asanuma, H., Takeya, J., and Iwasa, Y. (2006) Electrolyte-gated charge accumulation in organic single crystals. *Appl. Phys. Lett.*, **89** (20), 203501.
120. Takeya, J., Yamada, K., Hara, K., Shigeto, K., Tsukagoshi, K., Ikehata, S., and Aoyagi, Y. (2006) High-density electrostatic carrier doping in organic single-crystal transistors with polymer gel electrolyte. *Appl. Phys. Lett.*, **88** (11), 112102.
121. Panzer, M.J. and Frisbie, C.D. (2006) High charge carrier densities and conductance maxima in single-crystal organic field-effect transistors with a polymer electrolyte gate dielectric. *Appl. Phys. Lett.*, **88** (20), 203504.
122. Ozel, T., Gaur, A., Rogers, J.A., and Shim, M. (2005) Polymer electrolyte gating of carbon nanotube network transistors. *Nano Lett.*, **5** (5), 905–911.
123. Panzer, M.J., Newman, C.R., and Frisbie, C.D. (2005) Low-voltage operation of a pentacene field-effect transistor with a polymer electrolyte gate dielectric. *Appl. Phys. Lett.*, **86** (10), 103503.
124. Panzer, M.J. and Frisbie, C.D. (2006) High carrier density and metallic conductivity in poly(3-hexylthiophene) achieved by electrostatic charge injection. *Adv. Funct. Mater.*, **16** (8), 1051–1056.
125. Cho, J.H., Lee, J., Xia, Y., Kim, B., He, Y., Renn, M.J., Lodge, T.P., and Frisbie,

C.D. (2008) Printable ion-gel gate dielectrics for low-voltage polymer thin-film transistors on plastic. *Nat. Mater.*, 7 (11), 900–906.

126. Lee, J., Panzer, M.J., He, Y., Lodge, T.P., and Frisbie, C.D. (2007) Ion gel gated polymer thin-film transistors. *J. Am. Chem. Soc.*, **129** (15), 4532–4533.

127. Kim, S.H., Hong, K., Lee, K.H., and Frisbie, C.D. (2013) Performance and stability of aerosol-jet-printed electrolyte-gated transistors based on poly(3-hexylthiophene). *ACS Appl. Mater. Interfaces*, **5** (14), 6580–6585.

128. Austin, M.D. and Chou, S.Y. (2002) Fabrication of 70 nm channel length polymer organic thin-film transistors using nanoimprint lithography. *Appl. Phys. Lett.*, **81** (23), 4431–4433.

129. Haddock, J.N., Zhang, X.H., Zheng, S.J., Zhang, Q., Marder, S.R., and Kippelen, B. (2006) A comprehensive study of short channel effects in organic field-effect transistors. *Org. Electron.*, **7** (1), 45–54.

130. Lee, S.W., Lee, H.J., Choi, J.H., Koh, W.G., Myoung, J.M., Hur, J.H., Park, J.J., Cho, J.H., and Jeong, U. (2009) Periodic array of polyelectrolyte-gated organic transistors from electrospun poly(3-hexylthiophene) nanofibers. *Nano Lett.*, **10** (1), 347–351.

131. Böhm, M., Ullmann, A., Zipperer, D., Knobloch, A., Glauert, W.H., and Fix, W. (2006) Printable electronics for polymer RFID applications. IEEE International Solid-State Circuits Conference 2006, Digest of Techinical Papers, p. 270.

132. Xia, Y., Zhang, W., Ha, M.J., Cho, J.H., Renn, M.J., Kim, C.H., and Frisbie, C.D. (2010) Printed sub-2 V Gel-electrolyte-gated polymer transistors and circuits. *Adv. Funct. Mater.*, **20** (4), 587–594.

133. Ha, M., Xia, Y., Green, A.A., Zhang, W., Renn, M.J., Kim, C.H., Hersam, M.C., and Frisbie, C.D. (2010) Printed, sub-3 V digital circuits on plastic from aqueous carbon nanotube inks. *ACS Nano*, **4** (8), 4388–4395.

134. Herlogsson, L., Crispin, X., Tierney, S., and Berggren, M. (2011) Polyelectrolyte-gated organic complementary circuits operating at low power and voltage. *Adv. Mater.*, **23** (40), 4684.

135. Ha, M., Seo, J.-W.T., Prabhumirashi, P.L., Zhang, W., Geier, M.L., Renn, M.J., Kim, C.H., Hersam, M.C., and Frisbie, C.D. (2013) Aerosol jet printed, low voltage, electrolyte gated carbon nanotube ring oscillators with sub-5 μs stage delays. *Nano Lett.*, **13** (3), 954–960.

9
Vapor Barrier Films for Flexible Electronics

Seok-Ju Kang, Chuan Liu, and Yong-Young Noh

9.1
Introduction

The protection of electronic and optoelectronic devices made from environmentally sensitive materials is increasingly important, since it mainly determines the device shelf-life. Conventionally, high barrier properties to protect the packed materials from the diffusion of various external permeants, especially oxygen and water vapor, are a crucial requirement in food packaging and many other industries [1]. To obtain high barrier properties of over 10^{-3} cm^{-2} g^{-1} day^{-1}, metal and ceramic materials have advantages over their polymer counterparts because of the former's high density (3 – 10 g cm^{-3}). However, they are not suitable for flexible optoelectronic devices such as organic light-emitting diodes (OLEDs), photovoltaic cells, and photodiodes, since all the metals and some ceramics are optically opaque and ceramic materials have much less flexibility. Alternative approaches include the use of a multilayered structure composed of a film with transparent flexible polymers and less flexible ceramic barrier layers such as SiO_x or AlO_x. Thin ceramic layers (~50 nm) offer high transparency and gas barrier properties without losing flexibility, while polymer layers (~500 nm thichness) provide excellent mechanical properties under bending conditions. Therefore, transparent oxides provide high barrier properties to oxygen and moisture in the ambient environment, and polymer layers provide flexibility. Vitex Systems first reported such a multilayer system called Barix™ for high-performance flexible barrier film in organic electronics and optoelectronic devices [2]. They demonstrated a remarkably low oxygen vapor transfer rate (OVTR) and water vapor transfer rate (WVTR) as low as 10^{-6} cm^{-2} g^{-1} day^{-1} by vacuum-sputtered Al_2O_3 layers and solution-deposited polymer layers. Now, many companies are using multilayered structures with different materials and processing techniques. Thin-film encapsulation has a history that spans decades. Since the early 1970s, thin aluminum films evaporated onto polymeric substrates have been applied commercially as gas barriers in food packaging. Single-layer metallic barrier coatings [3–5] were followed by single-layer transparent barrier coatings [6–9] and multilayer composite (oxide or nitride/polymer) barrier coatings.

Large Area and Flexible Electronics, First Edition. Edited by Mario Caironi and Yong-Young Noh.

Today, metallized plastic foil is widely used for packaging food, medical supplies, and sensitive electronic components. An aluminum film of 10–100 nm thickness reduces the permeability by atmospheric gases to 1/1000 that of bare polymeric foil. For microwaveability and product visibility, transparent oxide barrier coatings applied by sputtering or chemical vapor deposition entered development in the 1980s. To achieve flexibility, thin layers (<50 nm) of brittle oxides are routinely coated on polymeric substrates. Typical inorganic barrier materials such as SiO_2, SiN_x, and Al_2O_3 are highly impermeable to atmospheric gases. Ideally, a perfect thin layer could reduce gas or vapor permeation to nondetectable values. However, microscopic defects provide diffusion paths for gas and vapor, particularly through granular film growth, cracks above occluded dust particles, and pinholes [3, 9]. To slow down the ingress of atmospheric gas molecules, the length of their diffusion path is increased by stacking and alternating thin inorganic barrier layers and thick polymeric layers in a multilayer barrier. The polymeric layers stop the upward propagation of defects in the inorganic barrier layers and make the composite structure more flexible than a single, thick inorganic barrier layer. The entire multilayer barrier is typically several micrometers thick. In 1994, aluminum layers as barriers alternating with acrylate polymer layers were found to reduce the oxygen permeation rate through a polypropylene substrate by four orders of magnitude [10]. In 1996, oxygen and water permeation rates below the detection limit of $<0.015\ cm^{-2}\ g^{-1}\ day^{-1}$ were obtained using a three-layer barrier structure of 1-μm acrylate/25-nm AlO_x/0.24-μm acrylate on a PET (polyethylene terephthalate) substrate [11]. Impermeable thin-film barrier coatings have become a major challenge in making flexible OLED displays (Figure 9.1). Polymeric substrates need a barrier on the substrate side ("passivation") and on the top ("encapsulation") of the device to ensure long life. In general, the barrier in the path of light emission needs to meet the following additional criteria: (i) high optical transparency of 85–90% over the visible spectrum; (ii) control of optical microcavity effects; (iii) if on top of the OLED, low stress to avoid shear damage to the OLED; (iv) dense and conformal coating to avoid through-layer and edge leakage; and (v) low process temperature. Coating multilayer organic/inorganic barriers has been proven

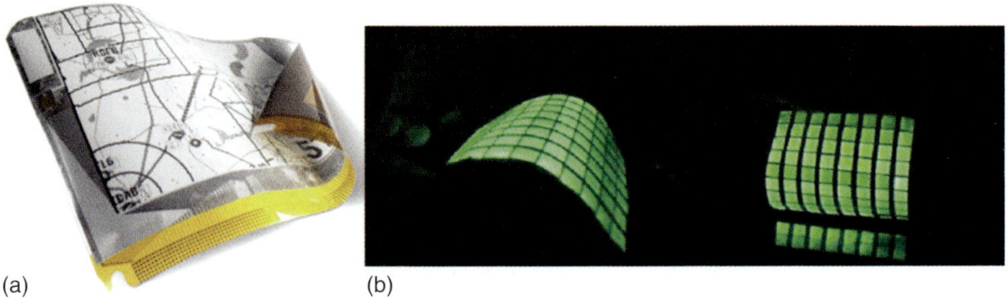

(a) (b)

Figure 9.1 Applications that need high-end flexible barrier films. (a) Active-matrix E-paper displays with polymer transistor for driving circuits. (b) Flexible OLED lighting device of 15 cm square on high-heat Lexan® polycarbonate substrate with graded barrier.

to dramatically extend the operating life of OLEDs [12–17]. For the purpose of cost reduction and barrier complexity, single dense inorganic barrier layers of aluminum oxide made by atomic-layer deposition (ALD) are being studied [18, 19].

9.2 Thin-Film Permeation Barrier Layers

Lack of impermeability to moisture and oxygen is a serious deficiency of all plastic substrates for flexible display applications, which are shown in Figure 9.1 [1, 20]. All display media including LCDs (liquid crystal displays), EPDs (electrophoretic displays), and OLEDs degrade when exposed to oxygen and moisture in ambient conditions, although at different rates, with OLEDs having the most sensitivity. For example, for the protection of an OLED display, the plastic substrate (barrier layer) must have a permeability of $<10^{-6}\,g\,m^{-2}\,day^{-1}$ for moisture and $10^{-5}\,ml\,m^{-2}\,day^{-1}$ for oxygen. In comparison, LCD displays have a barrier requirement of $10^{-2}\,g\,m^{-2}\,day^{-1}$ for both oxygen and moisture, which is significantly less stringent than that for OLEDs. The base plastic substrates typically have transmission rates of only about $10\,g\,m^{-2}\,day^{-1}$ for both oxygen and moisture, implying a great need for an extra barrier layer.

In principle, a thin layer of an inorganic film such as SiO_2, SiN_x, or Al_2O_3 deposited on the flexible plastic substrate can serve as a barrier layer with the required impermeability to oxygen and moisture. However, in practice, multilayer barrier film structures are required to counter the effects of pinholes or cracks in single-layer barrier layers. Several organizations are currently developing optically transparent multilayer barrier coatings for flexible OLED displays [21]. Vitex Systems [2] invented a barrier film called Barix, which employs alternating layers of polymer and ceramic inorganic layers for application in a vacuum, as shown in Figure 9.5. The inorganic films serve as barrier films for oxygen and moisture while the organic layers serve as a planarization/smoothing layer, and the multilayer structure provides redundancy against pinhole defects in the barrier films. The Barix layer is found to be an effective barrier layer to minimize the detrimental effects of pinholes and diffusion at grain boundaries. The film is about 3 μm thick and consists of about 500 Å thick Al_2O_3 ceramic films sandwiched between UV-cured acrylate films. Barix films were found to have water permeability in the range of $10^{-6}\,g\,m^{-2}\,day^{-1}$.

Most electronic devices require a high barrier against water vapor diffusion because of the sensitivity to moisture of the several integrated components. For example, water vapor barrier films are essential for OLEDs, electronic paper, and organic thin film transistors (TFTs) [22, 23]. This is because almost all high-performance semiconducting organic compounds show degraded performance when exposed to environmental moisture, which calls for effective encapsulation to prevent the ingression of water and oxygen to achieve adequate lifetimes. Also, new bendable substrates (plastic or metal-coated polymer films) for making light and flexible displays present weaker barriers to moisture compared to

Table 9.1 Water vapor and oxygen transmission rates of various polymer materials [4, 12, 27–29].

Polymer	WVTR ($g\,m^{-2}\,day^{-1}$)	OTR ($cm^3\,m^{-2}\,day^{-1}$)
Polyethylene	1.2–5.9	70–550
Polypropylene (PP)	1.5–5.9	93–300
Polystyrene (PS)	7.9–40	200–540
Poly(ethylene terephthalate) (PET)	3.9–17	1.8–7.7
Poly(ethersulfone) (PES)	14	0.04
PEN	7.3	3.0
Polyimide	0.4–21	0.04–17
15 nm Al/PET	0.18	0.2–2.9
SiO_x/PET	—	0.007–0.03
OLED requirement (estimate)	1×10^{-6}	1×10^{-5} to 1×10^{-3}

conventional glass. The permeation requirements vary for different materials and applications. Generally, devices that use chemically reactive electrodes, such as OLEDs, have the highest sensitivity to moisture, and degradation is often observed at the organic layer/electrode interface [24, 25]. Particularly, the cathodes are typically a low-work-function metal such as Li, Mg, In, Ca, or Ba, or alloys, which facilitates the injection of electrons into the light-emitting organic layer. These low-work-function metals are reactive and therefore require careful encapsulation. For example, Al is often used as an additional protective layer. Encapsulation of the whole device and isolation of the organic/cathode materials from the atmosphere have proven to be effective for prolonging the lifetime of OLEDs [26]. Unlike glass, polymer substrates do not provide sufficient protection to permeants as shown in Table 9.1, and an extra thin-film barrier is required on the substrate. Traditional encapsulation involves attaching a glass or metal lid to the substrate using a low-permeation epoxy. This approach is not compatible with flexible devices, so thin-film encapsulation is required. Moreover, at least one of these encapsulating layers must be transparent. Besides their use in electronics, water vapor barrier film with good optical transparency can also be applied to other fields, such as food packaging and encapsulation for medical, electronic, and optoelectronic devices.

The barrier properties of polymeric films are described using two measured quantities: the gas permeability (P) and the vapor transmission rate (VTR). For the diffusion of gases through solids, permeability ($P = g\,cm^{-1}\,s^{-1}\,atm^{-1}$) is the product of the diffusivity ($D = cm^3\,s^{-1}$) and the solubility ($S = g\,cm^{-3}\,atm^{-1}$). The WVTR is commonly reported in units of gram per square meter per day and must be qualified by specifying the temperature and relative humidity (%RH).

The widely quoted [30] requirement for the WVTR for an OLED lifetime of >10 000 h is $1\times10^{-6}\,g\,m^{-2}\,day^{-1}$. This value was originally estimated by

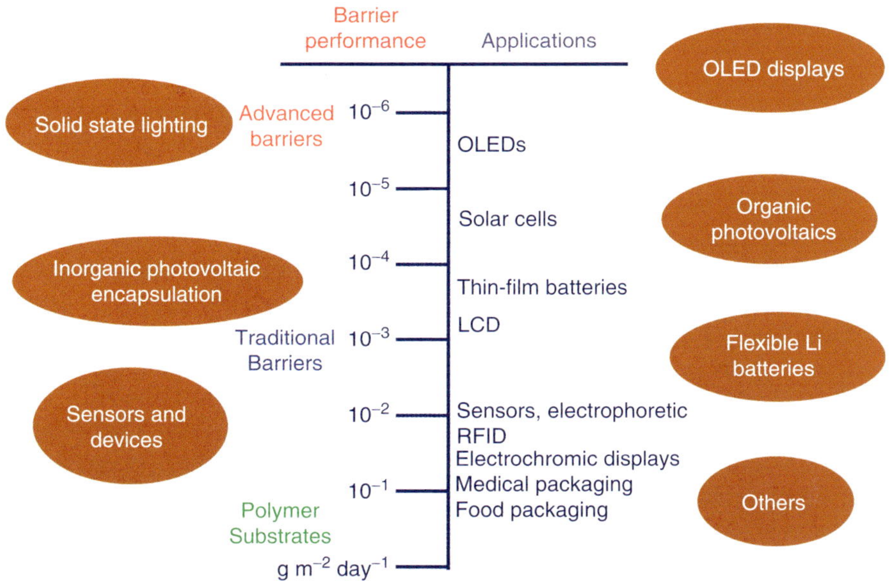

Figure 9.2 Water and oxygen vapor transfer rate requirements for various optoelectronic applications and the barrier performance provided by available materials (OLED, organic light-emitting diode; PV, photovoltaic; LCD, liquid crystal display; E-paper, electrophoretic paper display).

calculating the amount of oxygen and water needed to degrade the reactive cathode [31]. For organic TFT applications, the higher performing active layer materials also tend to have greater stability problems. The WVTR requirements for TFTs [32] are not as stringent as those for OLEDs, and are generally in the range of 1×10^{-3} to 1×10^{-1} g m^{-2} day^{-1}. For LCDs and EPDs [29, 33], the WVTR requirement is $\sim 1 \times 10^{-1}$ g m^{-2} day^{-1}. Figure 9.2 shows the order of magnitude of protection required for various organic electronic devices and those provided by various materials.

New developments also call for new probing techniques for measuring permeation barriers, because the permeation barrier requirements for OLEDs already exceed the minimum sensitivity of 5×10^{-4} g m^{-2} day^{-1} that can be detected by traditional permeation rate measurement techniques by over two orders of magnitude. Now, the most widely used approach is a Ca thin film of typically 50–100 nm deposited on the barrier-coated substrate [26, 34, 35]. The Ca film is encapsulated and becomes transparent as oxygen and moisture penetrate the barrier film. Monitoring the film optically allows the measurement of the Ca depletion rate, which can be converted to the permeation rate. Alternative approaches use radioactive tritiated water (T_2O or ^3H_2O) as a permeant to mimic moisture permeation [36, 37] or an ultrahigh vacuum residual gas analyzer to monitor the permeation of various species [38].

9.3 Permeation through Inorganic Thin Films

Thin-film permeation barriers have traditionally been formed from Al or oxides of Al or Si. Bulk oxides and Al are effectively impermeable to oxygen and water [12, 20, 39], as are perfect SiO_2 films [31, 40]. But traditional thin-film single barrier layers provide at best only two to three orders of magnitude improvement over the oxygen transmission rates (OTRs) of polymer substrates, whether deposited by plasma-enhanced chemical vapor deposition (PECVD), sputtering, or evaporation [4, 9, 40–44]. Generally, the oxygen permeation rate for single-layer barriers on polymer substrates changes by no more than an order of magnitude, irrespective of the coating method or material [9]. This limit is the result of permeation through defects or nanoscale pores rather than the bulk of the barrier film.

Even though the surface area fraction of defects is small, lateral diffusion causes the total permeation rate through many small pinholes to be much higher than if the pinholes are combined into fewer but larger defects [45, 46]. Lateral diffusion results in a large concentration gradient of permeates near the pinholes, so that increasing the barrier performance of the polymer adjacent to a pinhole defect will significantly improve the overall barrier performance [4, 47].

Figure 9.3 shows the correlation of film defect density with OTR for SiO_x, SiN_x, and Al-coated PET demonstrated by da Silva Sobrinho *et al.* [40]. The source of defect-driven permeation has been primarily attributed to pinhole defects [9, 40, 46]. Some recent studies have shown that, even in the absence of pinhole

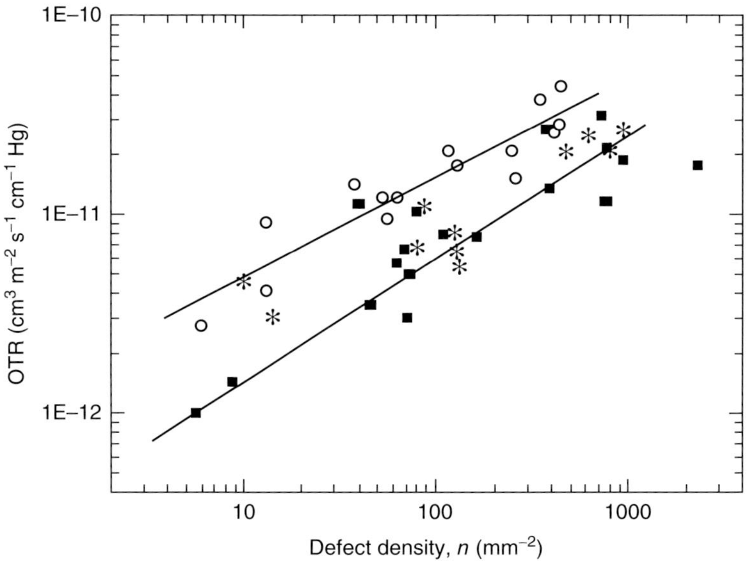

Figure 9.3 OTR as a function of defect density for Al (○), SiO_x (*), and SiN_x (■) films. Note that 1.5×10^{-12} $cm^3\ m^{-2}\ s^{-1}\ cm^{-1}$ Hg is the equivalent of 0.1 $cm^3\ m^{-2}\ day^{-1}$ [40]. (Reprinted with permission. Copyright © (2000), American Vacuum Society.)

defects, permeation rates are still reduced by only less than three orders of magnitude compared to the substrate alone [43]. The remaining permeation is shown to be the result of pores in the subnanometer to several nanometer range produced by surface roughness and/or low density of the films [43, 48–51].

Many barrier studies measure only OTR or WVTR, but it is important to bear in mind that barrier performance can be dependent on the permeant species. Permeation in polymers occurs by different mechanisms for nonpolar, noninteracting molecules such as oxygen compared to polar and condensable molecules such as water, which may interact with the polymer [43, 52]. Moreover, there is evidence that the permeation pathways through barrier films can be different for oxygen and moisture [51].

Two approaches to significantly improve the barrier performance have been explored: increasing the film density and producing multilayer structures. The first approach has been shown to reduce permeation through nanoscale pores [51]. Symmorphix has been demonstrated, using a proprietary high-plasma-density sputtering technique, to generate amorphous films that exhibit exceptionally high density. From these single-layer films, WVTR rates of 8×10^{-5} g m^{-2} day^{-1} have been measured using the Ca test under accelerated conditions of 60 °C/90% RH, indicating excellent performance at room temperature [53, 54]. Similarly, DuPont has reported excellent permeation barrier performance from ultrathin Al_2O_3 barrier films (~25 nm) grown by ALD methods on polyethylene naphthalate (PEN) substrates, featuring a WVTR of 1.7×10^{-5} g m^{-2} day^{-1} at 38°C/85% RH [19]. Also, researchers from the University of Colorado have reported Al_2O_3 films deposited by low-temperature ALD with a WVTR of 1×10^{-3} g m^{-2} day^{-1} [37].

However, it is technically difficult to produce a single-layer permeation barrier with sufficiently high film density and sufficiently low defect density. Because various organic electronic devices, including OLEDs, consist of films of a few hundred nanometers thickness, the smoothness of the substrate surface is very important. Even the smallest surface defects cause non-emitting areas called *dark spots*. This issue is very important even in glass substrate devices, but even more crucial in plastic substrate devices. Typically, plastic substrates have many small spikes and dimples on its surface, and these defects make holes in the moisture barrier film. Moisture penetrates into the devices through these holes, and then dark spots grow during storage. Alternatively, the multilayer structure comprising alternating polymer and inorganic layers has proven to be efficient and technically less demanding. In this system, a polymer layer could planarize and fill defects in the inorganic layers and subsequently provide a tortuous path for vapor molecules.

The polymer multilayer process was first demonstrated by GE, and polymer smoothing layers were subsequently developed elsewhere [11]. The smoothing layers were shown to be effective in reducing defects related to surface flaws for several types of films, including reactively sputtered and electron-beam-evaporated Al_2O_3 [11, 56]. The smoothing effect of the polyacrylate is a result of the deposition process in which the liquid acrylate monomer condenses on the substrate surface and is subsequently cured. This reduces the effect of surface flaws in the polymer substrate [56, 57]. A similar planarization has been achieved

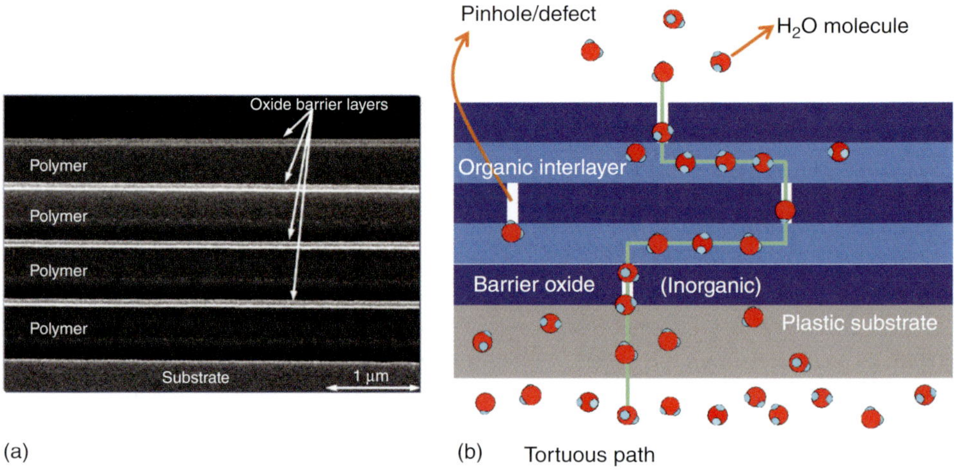

Figure 9.4 (a) Organic–inorganic multilayer structure from Vitex Systems. (Reproduced courtesy of Vitex Systems. Copyright © Vitex Systems) (b) Schematic multilayer barrier with extremely long effective diffusion path length due to large spacing between defects in inorganic layers.

with spin-coated resins [15, 58]. To avoid substrate defects, the plastic substrate could be coated with a UV-curable resin as the smoothing layer before depositing the inorganic moisture barrier film (e.g., SiON) ranging from several hundred nanometers to 1 μm. Definitely, the smoothing layer reduced the dark spots in the emitting areas, indicating that the emission after storage had improved by introducing a hybrid barrier structure consisting of an organic layer as a smoothing layer and an inorganic layer as a moisture barrier film.

Multilayer barrier-coated polymers from Vitex have demonstrated an estimated WVTR of $2 \times 10^{-6}\,\mathrm{g\,m^{-2}\,day^{-1}}$ in ambient conditions using the Ca test [26]. A cross-sectional micrograph of a Barix sample from Vitex is shown in Figure 9.4a,b. The company has also shown a combination of barrier-coated substrate and top encapsulation with a WVTR of $8 \times 10^{-6}\,\mathrm{g\,m^{-2}\,day^{-1}}$ in ambient conditions [59]. In other efforts, GE has reported a permeation barrier structure deposited by PECVD with graded junctions between multiple organic and inorganic regions with WVTR in the range of 10^{-5} to $10^{-6}\,\mathrm{g\,m^{-2}\,day^{-1}}$ calculated from accelerated measurements at 50 °C/95% RH [60].

The mechanism by which multilayer structures improve barrier performance has been a subject of wide discussion. The proposed explanations include the smoothing effect of the polymer layers, the "decoupling" of defects in neighboring films by the polymer layers, the increase of the diffusion path length, and the improved likelihood of having one "perfect" layer in the structure when multiple layers are used. Recently, Graff *et al.* [16] showed theoretically that multiple barrier layers should not improve the steady-state permeation rate by orders of magnitude. Rather, using reasonable figures for defect densities, their calculations show that the lag time could increase to several years. This implies that multiple layers delay a significant amount of permeation, which may be sufficient for many applications. This explanation seems to be supported by recent measurements [55].

9.4
Time-Resolved Measurements on Barrier Layers

Time-resolved permeation measurements combined with Fickian models have identified the mechanism of vapor permeation in multilayer structures and allow us to propose the design criteria for ultra-barrier films [16]. The calculated parameters shown in Figure 9.5 suggest that defects in the AlO_x layers of ultra-barrier stacks are so far apart that the vapor diffusion vector is mostly parallel to the plane of the polymer layer, resulting in a very long effective diffusion path between the AlO_x and substrate layers. For a polymer physical thickness t and defect spacing s, the effective diffusion path l is approximately shown in Figure 9.5. Indeed, the multilayer series-resistance model predicts minimal improvement in the steady-state flux (F_{SS}), even when considering the increased path length afforded by the multilayer stacking geometry. To explain the observed barrier performance, the effect of the multilayer structure on the lag time must be considered.

$$l = \left[t^2 + \left(\frac{s}{2}\right)^2\right]^{\frac{1}{2}} \sim \frac{s}{2} \tag{9.1}$$

1) Permeation through ultra-barriers is controlled by defects, and the defect size and spatial density in the vacuum-deposited inorganic layers are critical parameters that define the barrier performance of both single and multilayer vapor barriers.
2) The long apparent diffusion path induced by sandwiching thin polymer layers between low-defect-density inorganic layers causes a pronounced increase in lag time but a much less significant decrease in steady-state flux. The increased lag time is primarily responsible for the orders-of-magnitude improvement in the barrier performance as repeating organic/inorganic dyads are added to the multilayer barrier structure.

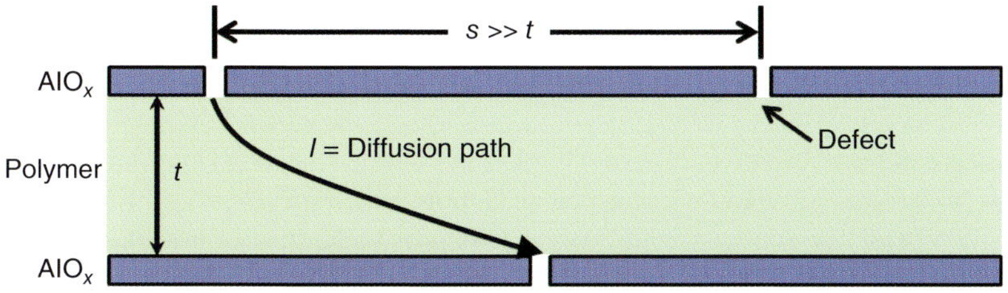

Figure 9.5 Schematic illustration of the effective diffusional thickness l of the polymer layer (approximately half the AlO_x defect spacing). This results in a path length which is much longer than the polymer's physical thickness (t) [16]. (Reprinted with permission. Copyright © 2004, AIP Publishing LLC.)

3) The demonstration of long-life plastic OLEDs using multilayer barriers is possible because of the extended lag time and not equilibrium diffusion effects. The attainment of steady-state diffusion likely does not occur in the typical test periods reported in the literature. Once the lag time is exceeded, the water vapor flux will reach a steady-state value that is substantially higher ($F_{ss} = 10^{-3}$ to 10^{-4} g m^{-2} day^{-1}) than the WVTR calculated during the transient period.
4) Reducing the diffusivity (D) and solubility (S) of the polymer interlayers will improve barrier performance. This may be accomplished by polymer selection (hydrophobic moieties or organic/inorganic copolymers), physical modifications (such as ion bombardment or crosslinking), or chemical modification (reactive etch or plasma surface treatment). The range of improvement by polymer selection/modification may be less than that by the inorganic layer, since the effective D of the inorganic layer is at least four orders of magnitude lower than that of the polymer interlayers.
5) The models also suggest that permeation can be reduced by thinning the polymer interlayers. However, this must be accomplished without sacrificing the smoothing properties (asperity/debris coverage) of the polymer.

9.5 Mechanical Limitations of Inorganic Films

Many demonstrations of electronic devices on flexible substrates involve the fabrication of traditional devices substituting only the substrate material. Often, smoothing layers are employed and, if necessary, the processing temperatures are reduced. However, there has been relatively little attention paid to the limits of flexibility of these devices. Mechanical failure in a device is determined by the weakest component. For devices that use inorganic thin films, these brittle layers are likely to be the source of failure. More recently, there have been efforts to develop devices with no brittle components whatsoever [61–64]. The fabrication of barrier layers should also take into account the important mechanical limitations.

9.6 Mechanics of Films on Flexible Substrates

The mechanics and brittle failure mechanisms of single and multilayered films on flexible substrates have been described in detail elsewhere [28, 65–72], and we only give a brief introduction here. When a sheet is bent, the outer surface experiences tensile stress and the inner surface undergoes compressive stress, while a plane inside the sheet, defined as the neutral plane (in the middle if the sheet is homogeneous), experiences no uniaxial stress at all. Films deposited on one or both sides shift the stress and strain distribution, but for many flexible electronic devices where the films are very thin relative to the substrate, one can make the

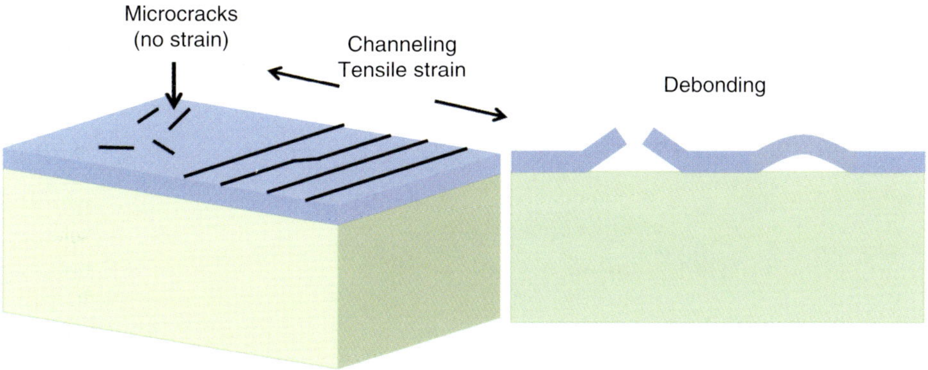

Figure 9.6 Primary failure illustration for brittle encapsulation layers on plastic substrates.

following simple approximation for the relationship between film strain and the radius of curvature:

$$\varepsilon_{\rm f} = \frac{d}{2r} \tag{9.2}$$

where d is the thickness of the substrate and r is the radius of curvature. To make a rough estimate of the radius of curvature at failure, we can see that for a 100-μm substrate and a typical failure strain of 0.5–1.0% for brittle films in tension, the minimum radius of curvature is 5–10 mm.

Failure modes of layered materials involve the growth of microcracks under stress [73], and the details of the fracture mode depend on the substrate modulus, film adhesion, and film cohesion, among other things. The most common failure modes for brittle films on flexible substrates are film cracking/channeling and debonding [71], as depicted in Figure 9.6. The former is more common for films in tension with good adhesion between layers, while the second mechanism is more common for films in compression or when adhesion is poor.

Early failure in brittle films is caused by preexisting defects. For oxide films on polymer substrates, these are often caused by defects in the substrate morphology. The failure curvature can therefore be reduced by minimizing such defects with a smoothing layer. In addition, defects can occur from scratches or edge defects. It is preferable to keep films from extending to the substrate edge, as stress can be concentrated near the edges.

Figure 9.7 shows several possibilities in thin-film bending structures. If the device layers are on a convex surface, the layers are placed under tensile strain when the device is bent and are therefore most prone to failure. If the device layers are on a concave surface, they are under compressive strain, which is preferred. Often, a convex or concave structure is determined not by mechanical considerations but by the device structure (e.g., top- or bottom-emitting OLEDs). In contrast, if the active device layers on the flexible substrate are encapsulated with the ductile barrier films,the condition for the device layers is different from the previous two cases, because the active layers are encapsulated bya second polymer barrier film that is comparable to the substrate. In this structure, the

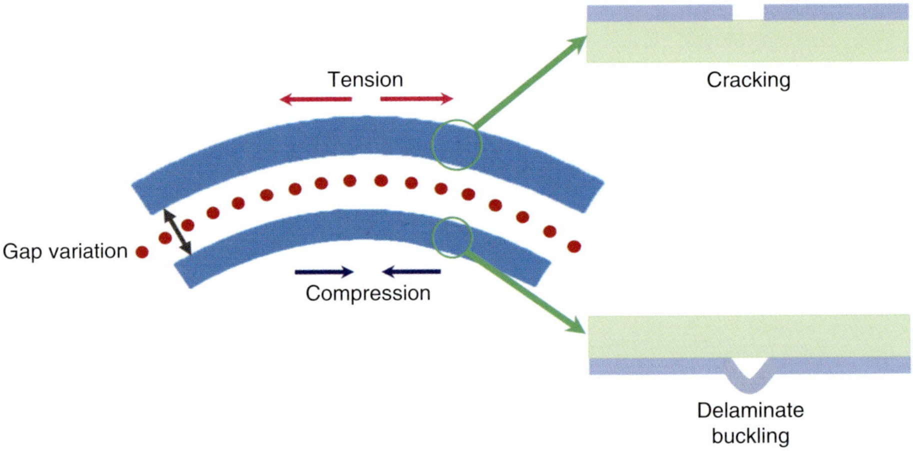

Figure 9.7 Geometries for flexible electronic devices. Devices (shaded) experience tensile strain, compressive strain, or shear stress depending on the position of bending geometry in the substrate.

device layers are in the neutral plane and can therefore be curled to a very tight radius [72]. However, for a higher modulus, the substrate shear stress can be significant and challenging for interlayer adhesion.

9.7
Summary

This chapter presented a thin-film barrier technology for flexible electronics, as well as the factors influencing the barrier's ability to stop permeable gases in air. For the future development of organic electronics, fabricating organic devices on flexible substrates, particularly on plastic, brings with it new challenges in order to make them commercially viable because plastic materials alone cannot provide a sufficiently hermetic barrier and especially for an OLED device which requires a most stringent level of barrier protection.

Specifically, one critical challenge is to design a means of hermetic packaging of the organic electronic device because many such devices exhibit a very short shelf-life if not protected from the ingress of environmental permeants such as oxygen and water vapor. However, inorganic barrier films show only limited performance as gas barriers because of the presence of macro and micro pinholes or defects. Moreover, organic coatings themselves have barrier properties inferior to those of inorganic coatings, but when inorganic layers are separated by organic coatings, there is then no defect that spans the whole thickness of the coating. As a result, to overcome the limitation of inorganic and organic barrier film, multiple advanced barrier technologies were developed through multiple layers of repeated inorganic and organic coatings for improving gas permeability and mechanical flexibility, which could be commercially beneficial to R2R fabrication processes. Therefore, the challenge is to demonstrate similar barrier performance using commercially viable solutions and R2R fabrication processes. From the increasing effort over the last decade, it seems likely that flexible thin-film barrier technology will advance and be developed progressively to enable flexible electronics applications.

References

1. Wong, W.S. and Salleo, A. (2009) *Flexible Electronics: Materials and Applications*, Springer-Verlag.
2. Moro, M., Chu, X., Hirayama, H., Krajewski, T., and Visser, R.J. (2006) A mass manufacturing process for Barix encapsulation of OLED displays. IMID/IDMC, 23 August.
3. Prins, W. and Hermans, J. (1959) Theory of permeation through metal coated polymer films. *J. Phys. Chem.*, **63**, 716–719.
4. Jamieson, E.H.H. and Windle, A.H. (1983) Structure and oxygen-barrier properties of metallized polymer film. *J. Mater. Sci.*, **18**, 64–80.
5. Weiss, J., Leppin, C., Mader, W., and Salzberger, U. (1989) Aluminum metallization of polyester and polypropylene films: properties and transmission electron microscopy microstructure investigations. *Thin Solid Films*, **174**, 155.
6. Klemberg-Sapiepha, J.E., Martinu, L., Kuttel, O.M., and Wertheimer, M. (1993) Transparent gas barrier coatings by dual-frequency PECVD. 36th Annual Technical Conference Proceedings of the Society of Vacuum Coaters, p. 445.
7. Philips, R.W., Markantes, T.M., and LeGallee, C. (1993) Evaporated dielectric colorless films on PET and OPP exhibiting high barriers towards moisture and oxygen. 36th Annual Technical Conference Proceedings of the Society of Vacuum Coaters, pp. 293–301.
8. Brody, A.L. (1994) Glass-coated flexible films for packaging: an overview. *Pack. Technol. Eng.*, **3**, 44.
9. Chatham, H. (1996) Review oxygen diffusion barrier properties of transparent oxide coatings on polymeric substrates. *Surf. Coat. Technol.*, **78**, 1–9.
10. Shaw, D.G. and Langlois, M.G. (1994) Use of vapor deposited acrylate coatings to improve the barrier properties of metallized film. 37th Annual Technical Conference Proceedings of the Society of Vaccum Coaters, p. 240.
11. Affinito, J.D., Gross, M.E., Coronado, C.A., Graff, G.L., Greenwell, E.N., and Martin, P.M. (1996) A new method for fabricating transparent barrier layers. *Thin Solid Films*, **290–291**, 63–67.
12. Lewis, J.S. and Weaver, M.S. (2004) Thin-film permeation-barrier technology for flexible organic light-emitting devices. *IEEE J. Sel. Top. Quantum Electron.*, **10**, 45–57.
13. Weaver, M.S., Michalski, L.A., Rajan, K., Rothman, M.A., Silvernail, J.A., Burrows, P.E., Graff, G.L., Gross, M.E., Martin, P.E., Hall, M., Mast, E., Bonham, C., Bennett, W., and Zumhoff, M. (2002) Organic light-emitting devices with extended operating lifetimes on plastic substrates. *Appl. Phys. Lett.*, **81**, 2929–2931.
14. Chwang, A.B., Rothman, M.A., Mao, S.Y., Hewitt, R.H., Weaver, M.S., Silvernail, J.A., Rajan, K., Hack, M., Brown, J.J., Chu, X., Moro, L., Krajewski, T., and Rutherford, N. (2003) Thin film encapsulated flexible OLED display. *Appl. Phys. Lett.*, **83**, 413–415.
15. Yoshida, A., Fujimura, S., Miyake, T., Yoshizawa, T., Ochi, H., Sugimoto, A., Kubota, H., Miyadera, T., Ishizuka, S., Tsuchida, M., and Nakada, H. (2003) 3-inch Full-color OLED display using a plastic substrate. *Proc. Soc. Inform. Display Symp. Dig. Tech. Papers*, **34**, 856–859.
16. Graff, G.L., Williford, R.E., and Burrows, P.E. (2004) Mechanisms of vapor permeation through multilayer barrier films: lag time versus equilibrium permeation. *J. Appl. Phys.*, **96**, 1840–1849.
17. Kim, T.W., Yan, M., Erlat, A.G., McConnelee, P.A., Pellow, M., Deluca, J., Feost, T.P., Duggal, A.R., and Schaepkens,

M. (2005) Transparent hybrid inorganic/organic barrier coatings for plastic organic light-emitting diode substrates. *J. Vac. Sci. Technol., A*, **23**, 971–977.
18. Ghosh, A.P., Gerenser, L.J., Jarman, C.M., and Fornalik, J.E. (2005) Thin-film encapsulation of organic light-emitting devices. *Appl. Phys. Lett.*, **86**, 223503.
19. Carcia, P.F., McLean, R.S., Reilly, M.H., Groner, M.D., and George, S.M. (2006) Ca test of Al_2O_3 gas diffusion barriers grown by atomic layer deposition on polymers. *Appl. Phys. Lett.*, **89**, 031915.
20. Lewis, J. (2006) Material challenge for flexible organic devices. *Mater. Today*, **9**, 38–45.
21. Graff, G., Burrows, P.E., Williford, R.E., and Praino, R.F. (2005) Barrier layer technology for flexible displays, in *Flexible Flat Panel Displays* (ed. G.P. Crawford), John Wiley & Sons, Inc., New York.
22. Lin, I.-L., Chen, C.-Y., Guo, Y.-B., Liu, C.-M., and Hong, F.C.-N. (2013) Transparent ultrahigh-strength moisture barrier film fabricated by lamination process. *Jpn. J. Appl. Phys.*, **52**, 01AC10.
23. Boehme, C. and Lucovsky, G. (2002) Origins of silicon solar cell passivation by SiNx:H anneal. *J. Non-Cryst. Solids*, **299–302**, 1157.
24. Papadimitrakopoulos, F. *et al* (1998) Chemical and morphology stability of aluminum Tris(8-Hydroxyquinoline)(Alq3): effects in light-emitting devices. *IEEE J. Sel. Top. Quantum Electron.*, **4**, 49.
25. Liew, Y.-F. *et al* (2000) Investigation of the sites of dark spots in organic light-emitting devices. *Appl. Phys. Lett.*, **77**, 2650.
26. Nisato, G. *et al* (2003) Thin-film encapsulation for OLEDs: evaluation of multi-layer barriers using the Ca test. *Proc. Soc. Inform. Display Symp. Dig. Tech. Papers*, **34**, 550.
27. Andrew, W. (1995) *Permeability and Other Film Properties of Plastics and Elastomers*, 1st edn, Plastics Design Library, Norwich, NY.
28. Leterrier, Y. (2003) Durability of nanosized oxygen-barrier coatings on polymers. *Prog. Mater. Sci.*, **48**, 1–55.
29. Lueder, E. (1998) Passive and active matrix liquid crystal displays with plastic substrates. *Electrochem. Soc. Proc.*, **98–22**, 336–354.
30. Burrows, P.E. *et al.* (2001) Gas permeation and lifetime test on polymer-based barrier coatings. *Proc. SPIE*, **4105**, 75.
31. Deal, B.E. and Grove, A.S. (1965) General relationship for the thermal oxidation of silicon. *J. Appl. Phys.*, **36**, 3770–3776.
32. Rutherford, N. (2005) Flexible substrates and packaging for organic displays and electronics. *Info. Display*, **21**, 20.
33. Kim, J.-Y. *et al.* (2001) Polymer-based multi-layer conductive electrode film for plastic LCD applications. *Appl. Phys. A*, **72**, 699.
34. Nisato, G. *et al.* (2001) Evaluating high performance diffusion barriers: the calcium test. Proceedings of the International Display Workshop/Asia Display, p. 1435.
35. Kumar, R.S. *et al.* (2002) Low moisture permeation measurement through polymer substrates for organic light emitting devices. *Thin Solid Films*, **417**, 120.
36. Dunkel, R. *et al.* (2005) Method of measuring ultralow water vapor permeation for OLED displays. *Proc. IEEE*, **93**, 1478.
37. Groner, M.D. and George, S.M. (2005) Gas diffusion barriers on polymers using Al_2O_3 atomic layer deposition. 48th Annual Technical Proceedings of the Society of Vacuum Coaters, p. 169.
38. Ranade, A. *et al.* (2005) High sensitivity gas permeability measurement system for thin plastic films. *Rev. Sci. Instrum.*, **76**, 013902.
39. Barrer, R.M. (1941) *Diffusion in and Through Solids*, Cambridge University Press, New York.
40. da Silva Sobrinho, A.S. *et al.* (2000) Defect-permeation correlation for ultrathin transparent barrier coatings on polymers. *J. Vac. Sci. Technol., A*, **18**, 149–157.
41. da Silva Sobrinho, A.S., Latreche, M., Czeremuszkin, G., Klemberg-Sapieha, J.E., and Wertheimer, M.R. (1998) Transparent barrier coatings on polyethylene terephthalate by single- and dual-frequency plasma-enhanced chemical

vapor deposition. *J. Vac. Sci. Technol., A*, **16**, 3190–3198.
42. da Silva Sobrinho, A.S. *et al.* (1999) Study of defect numbers and distributions in PECVD SiO_2 transparent barrier coatings on PET. *Mater. Res. Soc. Symp. Proc.*, **544**, 245.
43. Erlat, A.G. *et al.* (1999) SiOx gas barrier coatings on polymer substrates: morphology and gas transport considerations. *J. Phys. Chem. B*, **103**, 6047–6055.
44. Tropsha, Y.G. and Harvey, N.G. (1997) Activated rate theory treatment of oxygen and water transport through silicon oxide/poly(ethylene terephthalate) composite barrier structures. *J. Phys. Chem. B*, **101**, 2259–2266.
45. Rossi, G. (1993) Effect of local flaws in polymeric permeation reducing barriers. *J. Appl. Phys.*, **74**, 5471–5475.
46. Hanika, M., Langowski, H.-C., Moosheimer, U., and Peukert, W. (2003) Inorganic layers on polymeric films-Influence of defects and morphology on barrier properties. *Chem. Eng. Technol.*, **26**, 605–614.
47. Beu, T.A. and Mercea, P.-V. (1990) Gas transport through metallized polymer membranes. *Mater. Chem. Phys.*, **26**, 309–322.
48. Garcia-Ayuso, G. *et al.* (1996) Atomic force microscopy (AFM) morphological surface characterization of transparent gas barrier coatings on plastic films. *Surf. Coat. Technol.*, **80**, 203.
49. Henry, B.M. *et al.* (1999) A microstructural study of transparent metal oxide gas barrier films. *Thin Solid Films*, **355–356**, 500.
50. Henry, B.M. *et al.* (1998) Microstructural characterization of transparent SiO_2 permeation barrier coatings on PET. Proceedings of the 41st Annual Technical Conference Society of Vacuum Coaters, p. 434.
51. Erlat, A.G. *et al.* (2004) Mechanism of water vapour transport through PET/AlOxNy gas barrier films. *J. Phys. Chem. B*, **108**, 883.
52. Ouyang, H. *et al.* (2004) The study of mass transport of acetone in polycarbonate. *J. Appl. Phys.*, **96**, 7066.
53. Narasimhan, M. (2005) Progress on barrier on plastic substrates. USDC Flexible Displays Microelectronics Conference, Phoenix, AZ.
54. TRU Group *www.symmorphix.com* (accessed 17 July 2014).
55. *http://displayproducts.ga.com/thinfilm.html* (accessed 17 July 2014).
56. Affinito, J.D. *et al.* (1997) PML/oxide/PML barrier layer performance differences arising from use of UV or electron beam polymerization of the PML layers. *Thin Solid Films*, **308-309**, 19.
57. Shaw, D.G. and Langlois, M.G. (1993) Some performance characteristics of evaporated acrylate coatings. Proceedings of the 7th International Conference on Vacuum Web Coating, p. 268.
58. Yoshida, A. *et al.* (2001) Organic light emitting devices on polymer substrates. *J. Photopolym. Sci. Technol.*, **14**, 327.
59. Rutherford, N. (2005) Flexible displays – a low cost substrate/encapsulation packaging solution. USDC Flexible Displays and Microelectronics Conference, Phoenix, AZ.
60. Erlat, A.G. *et al.* (2005) Ultra-high barrier coatings for flexible organic electronics. Proceedings of the 48th Annual Technical Conference Society of Vacuum Coaters, p. 116.
61. Parashkov, R. *et al.* (2004) All-organic thin-film transistors made of poly(3-butylthiophene) semiconducting and various polymeric insulation layers. *J. Appl. Phys.*, **95**, 1594.
62. Lee, M.S. *et al.* (2005) Flexible all-polymer field effect transistors with optical transparency using electrically conducting polymers. *Thin Solid Films*, **477**, 169.
63. Klauk, H. *et al.* (2003) Pentacene organic transistors and ring oscillators on glass and on flexible polymeric substrates. *Appl. Phys. Lett.*, **82**, 4175.
64. Eder, F. *et al.* (2004) Organic electronics on paper. *Appl. Phys. Lett.*, **84**, 2673.
65. Leterrier, Y. *et al.* (1997) Adhesion of silicon oxide layers on poly(ethylene terephthalate). II: Effect of coating thickness on adhesive and cohesive strengths. *J. Polym. Sci., Part B: Polym. Phys.*, **35**, 1463.
66. Yanaka, M. *et al.* (1999) Thickness effect on cracking phenomena and mechanical

properties of submicron glass thin films deposited on a polymer substrate. *Mater. Res. Soc. Symp. Proc.*, **555**, 33.
67. Rochat, G. *et al.* (2003) Mechanical analysis of ultrathin oxide coatings on polymer substrates in situ in a scanning electron microscope. *Thin Solid Films*, **437**, 204.
68. Yanaka, M. *et al.* (1998) Cracking phenomena of brittle films in nanostructure composites analysed by a modified shear lag model with residual strain. *J. Mater. Sci.*, **33**, 2111.
69. Andersons, J. *et al.* (2003) Analysis of the initial fragmentation stage of oxide coatings on polymer substrates under uniaxial tension. *Thin Solid Films*, **434**, 203.
70. Leterrier, Y. *et al.* (2001) Biaxial fragmentation of thin silicon oxide coatings on poly(ethylene terephthalate). *J. Mater. Sci.*, **36**, 2213.
71. Hutchinson, J.W. and Suo, Z. (1992) Mixed mode crackings in layered materials. *Adv. Appl. Mech.*, **29**, 63.
72. Suo, Z. *et al.* (1999) Mechanics of rollable and foldable film-on-foil electronics. *Appl. Phys. Lett.*, **74**, 1177.
73. Servati, P. and Nathan, A. (2005) Orientation-dependent strain tolerance of amorphous silicon transistors and pixel circuits for elastic organic light-emitting diode display. *Appl. Phys. Lett.*, **86**, 033504.

10
Latest Advances in Substrates for Flexible Electronics

William A. MacDonald

10.1
Introduction

Since the early 2000s, there has been an explosion in interest in printed electronics both in terms of exploring the potential to develop new business opportunities and also in terms of developing the technology base. The opportunity to exploit flexible substrates in R2R (roll-to-roll) production has been the one that has excited the interest of the plastic films and associated processing and coating industries.

To replace glass, however, a plastic substrate should be able to offer some or all of the properties of glass, that is, clarity, dimensional stability, thermal stability, barrier, solvent resistance, and low coefficient of thermal expansion (CTE), coupled with a smooth surface. In addition, a further functionality such as a conductive layer might be required. No plastic film offers all these properties, so any plastic-based substrate replacing glass will almost certainly be a multilayer composite structure [1–3]. However, not all applications require such a demanding property set, and over the past decade plastic films have found application in areas broader than organic light-emitting diode (OLED) displays initially envisaged in the early days of the technology development. These include applications such as electrophoretic displays driven by thin-film transistor (TFT) arrays printed on plastic film, printed memory, and sensors. In addition, there has been a smearing of the boundary between flexible devices and printed electronics and/or flexible substrates in rigid devices, an example being the use of conductive films in touchscreens incorporated into smart phones and tablets.

This chapter will review the progress made in developing plastic substrates for flexible electronics over the past decade, with polyester films used as the main examples.

Large Area and Flexible Electronics, First Edition. Edited by Mario Caironi and Yong-Young Noh.

10.2 Factors Influencing Film Choice

10.2.1 Application Area

The requirements for the different applications areas envisaged for printable electronics are very different and will require substrates with different property sets.

This is summarized in Figure 10.1, and this classification is divided into "simple" organic circuitry, for example, radio frequency identification (RFID), organic-based active matrix backplanes, inorganic TFTs and OLED displays. As one moves up the list, the substrate requirements in terms of properties such as dimensional stability, surface smoothness, low CTE, conductive layers, and barrier become increasingly complex and more demanding, and this will be reflected in the cost of the substrate. Substrates with low temperature stability, including coated papers and oriented polypropylene film with good surface quality, are adequate for some applications such as RFID, but a film with high temperature stability and dimensional reproducibility coupled with excellent surface quality will be required for more complex devices such as active-matrix backplanes and OLED displays. Obviously, a substrate with a property set and price commensurate with the application should be chosen.

10.2.2 Physical Form/Manufacturing Process

The physical form of the display and whether it is (i) flat but exploiting light weight and ruggedness (ii) conformable, one time fit to nonflat surface, (iii)

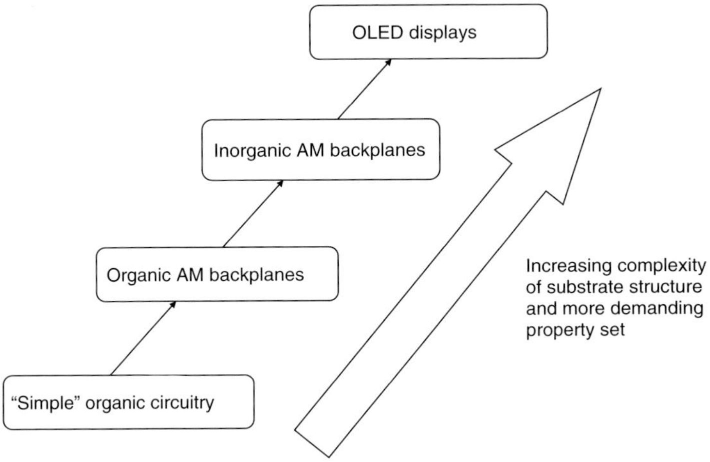

Figure 10.1 Illustration of the changing property requirements of different applications.

flexible and handlable, for example, electronic newspaper, and (iv) rollable will also influence film choice. Initially in the early 2000s, the vision was of flexible OLED displays and rollable displays in particular. Over the past decade, however, with the emergence of smart phones and tablets, the general public have become used to the portability and user interface of such devices, and the interest in a rollable or foldable display device has waned. Instead, the focus is now more on exploiting flexible electronics to give robustness to a rigid device, and the general trend on film substrates is to go thinner to reduce the bulkiness of a device.

Similarly, whether the film is manufactured batchwise or R2R can influence the film selection. Although R2R processing can be used for specific stages of device manufacture, for example, barrier or conductive coatings and for less complex printed electronic applications, for the most part more complex device manufacture such as active-matrix backplanes is carried out by a batch process on a rigid carrier. This fits with existing semiconductor manufacturing tooling equipment, and also the rigid carrier can be exploited to control the dimensional stability. However, batch processing introduces new challenges, such as bowing of the rigid carrier due to a mismatch in CTE and the shrinkage behavior between the film and the rigid carrier. The major factors influencing this are the rigidity of the carrier (more rigid gives less bow) and the thickness of the film (less bow with thinner film). A second issue is the release of the film from the carrier without damaging the printed circuitry, and this remains an active area of research.

10.3 Film Property Set

10.3.1 Polymer Type

Polyethylene terephthalate (PET), for example, DuPont Teijin Film (DTF) Melinex® polyester film, and polyethylene naphthalate (PEN), for example, DTF, Teonex® polyester film, are biaxially oriented semicrystalline films [4, 5]. The difference in chemical structure between PET and PEN is shown in Figure 10.2.

The substitution of the phenyl ring of PET by the naphthalene double ring of PEN has very little effect on the melting point (T_m), which increases by only a few degrees. There is, however, a significant effect on the glass-transition temperature (T_g), the temperature at which a polymer changes from a glassy state to a rubbery state, which increases from 78 °C for PET to 120 °C for PEN [2]. PET and PEN films are prepared by a process wherein the amorphous cast is drawn both in the machine direction (MD) and in the transverse direction (TD). The biaxially oriented film is then heat-set to crystallize the film [4, 5].

The success of polyester film in general applications is due to the properties derived from the basic polymer coupled with the manufacturing process of

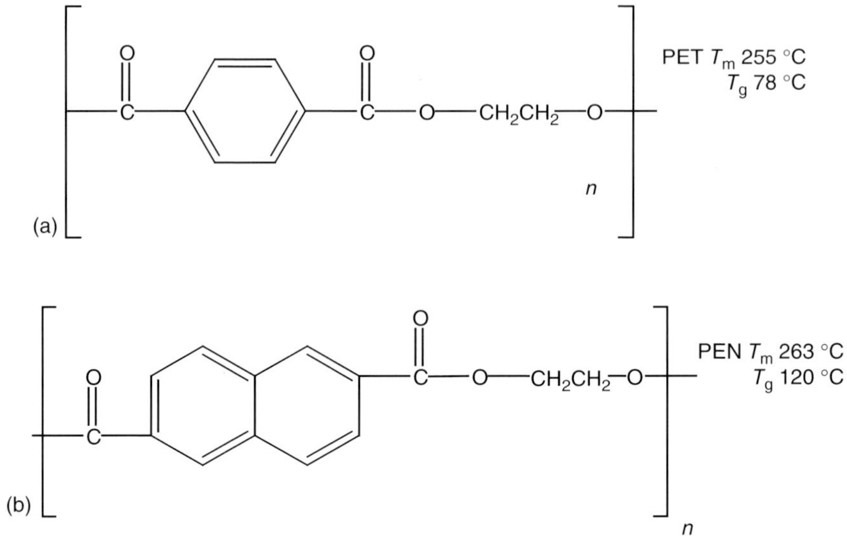

Figure 10.2 Chemical structures of (a) PET and (b) PEN films.

biaxial orientation and heat setting described above. These properties include high mechanical strength, good resistance to a wide range of chemicals and solvents, low water absorption, excellent dielectric properties, good dimensional stability, and good thermal resistance in terms of shrinkage and degradation of the polymer chains. Fillers can be incorporated into the polymer to change the surface topography and opacity of the film. The film surface can also be altered by the use of pretreatments to give a further range of properties, including enhanced adhesion to a wide range of inks, lacquers, and adhesives. These basic properties have resulted in PET films being used in a wide range of applications, from magnetic media and photographic applications, where optical properties and excellent cleanliness are of paramount importance, to electronics applications such as flexible circuitry and touch switches, where thermal stability is the key. More demanding polyester film markets, which exploit the higher performance and benefits of PEN, include magnetic media for high-density data storage and electronic circuitry for hydrolysis-resistant automotive wiring [6]. This property set provides the basis on which one can now build to meet the demands of the printed electronics market.

It is interesting to contrast these films with the other films that are currently being considered for flexible electronics applications. The main candidates are shown in Figure 10.3, which lists the substrates in terms of increasing glass transition (T_g).

Polymers can be further categorized into films that are semicrystalline (PET and PEN as mentioned above), amorphous and thermoplastic, and amorphous but solvent cast. Polymers with T_g's higher than 150 °C that are semicrystalline tend generally to have melting points that are too high to allow the polymers to be melt-processed without significant degradation. Victrex® poly(ether ether

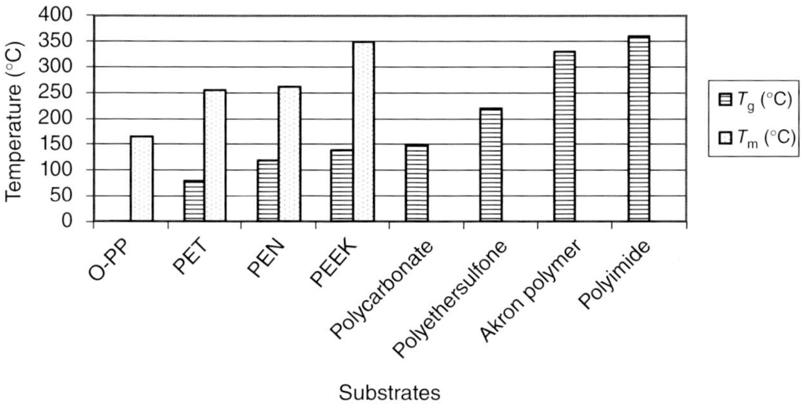

Figure 10.3 Glass transitions of film substrates of interest for flexible electronics applications.

ketone) (PEEK) film ACTIV®[1] is the highest performance semicrystalline material available in film form. The next category comprises polymers that are thermoplastic but noncrystalline and these range from polycarbonate (PC), for example, Teijins PURE-ACE®[2] and GE's Lexan®[3] with a T_g of ~150 °C, to polyethersulfone (PES), for example, Sumitomo Bakelite's Sumilite®[4] with a T_g of ~220 °C. Although thermoplastic, these polymers may also be solvent-cast to give high optical clarity. The third category includes high-T_g materials that cannot be melt-processed and include substrates based on Akron Polymer Systems (APS) resins,[5] and polyimide (PI), for example, DuPont's Kapton®.[6]

PET and PEN, by virtue of being semicrystalline, biaxially oriented, and heat stabilized (see Section 10.3.4), have a property set different from that of the amorphous polymers, and for simplicity these two basic categories will be used when comparing and contrasting the properties of the film types and the importance to the property set required for flexible electronic application.

10.3.2 Optical Clarity

The clarity of the film is important for bottom-emissive displays where one views through the film. A total light transmission (TLT) of >85% over 400–800 nm

1) APTIV™ Films, Victrex Technology Centre, Hillhouse International Thornton Cleveleys, Lancashire FY5 4QD, UK.
2) PURE-ACE™, Teijin Chemicals Ltd., 1-2-2 Uchisaiwai-cho, Chiyoda-ku, Tokyo, Japan.
3) Lexan™, GE Plastics, One Plastics Avenue, Pittsfield, MA 01201, USA.
4) Sumilite™, Sumitomo Bakelite Co. Ltd., Ten-Nouzu Parkside Blgd., 5-8,2-Chome, Higashi-Shinagawa, Shinagawa-Ku, Tokyo 140-0002, Japan.
5) Akron Polymer Systems, Inc., 62 N. Summit Street, Akron, OH 44308 USA.
6) Kapton™, DuPont High Performance Films, P.O. Box 89, Route 23 South and DuPont Road, Circleville, OH 43113, USA.

coupled with a haze of <0.7% is typical of what is required for this application. The polymers listed above meet this requirement apart from PI, which is yellow. For many display applications such as top-emissive displays and for electrophoretic display applications, clarity of the base substrate is not essential.

10.3.3 Birefringence

Biaxially oriented films such as PET and PEN are birefringent. For liquid crystal (LC) displays, which depend on light of known polarization, this means that birefringent films, which would change the polarization state, are unlikely to be used as base substrates. That said, polyester films are used extensively to enhance LC display performance, for example, brightness-enhancing and diffuser films. Films based on amorphous polymers are not birefringent and are more suitable for the base substrates for LC displays. Birefringence is not an issue with OLED, electrophoretic displays, and, indeed, some LC displays.

10.3.4 The Effect of Thermal Stress on Dimensional Reproducibility

PET and PEN films (Melinex® and Teonex®) are produced using a sequential biaxial stretching technology, which is widely used for semicrystalline thermoplastics [4, 5]. The process involves stretching the film in MD and TD and heat-setting at an elevated temperature. As a consequence, a complex semicrystalline microstructure develops in the material, which exhibits remarkable strength, stiffness, and thermal stability. Various studies have been made of the biaxial structure of polyester film manufactured in this way, and many descriptions have been written [7, 8]. The film comprises a mosaic of crystallites or aggregated crystallites, accounting for nearly 50 wt% of its material, which tend to align along the directions of stretch. Adjacent crystallites may not, however, share similar orientations. Crystallites show only a small irreversible response to temperature, which may take the form of growth or perfection. The noncrystalline region also possesses some preferred molecular orientation, which is a consequence of its connectivity to the crystalline phase. Importantly, the molecular chains residing in the noncrystalline region are, on average, slightly extended and therefore do not exist in their equilibrium Gaussian conformation.

Standard PET and PEN film will shrink between 1% and 3% at temperatures above the T_g. PET and PEN films can be further exposed to a thermal relaxation process, in which the film is transported under low tension through an additional heating zone at ~150 °C for PET and 180–200 °C for PEN. Some additional shrinkage is seen, which signifies a relaxation of the molecular orientation in the material [1–3, 8]. Fundamental measurements of fibers and films indicate that the relaxation occurs exclusively in the noncrystalline regions [9].

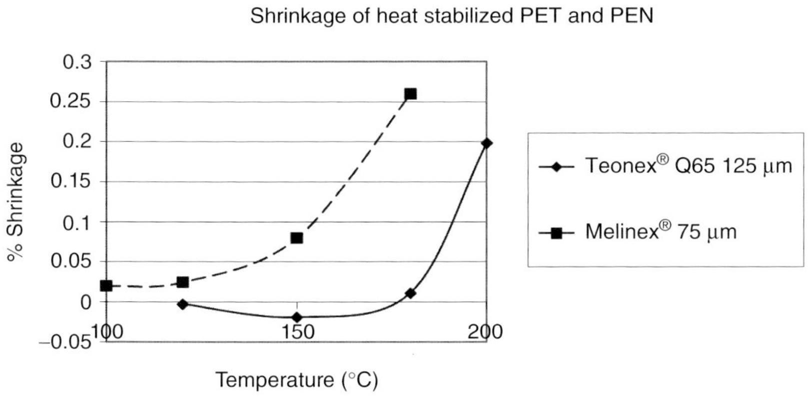

Figure 10.4 Shrinkage of heat-stabilized PET and PEN film versus temperature.

Figure 10.4 illustrates how the shrinkage of heat-stabilized PET (Melinex ST504/506) and PEN (Teonex Q65) change with temperature. For applications requiring low shrinkage above 140 °C, Teonex® Q65 is the preferred option, and there is continuing work within DTF to optimize the heat-stabilization process without having a detrimental effect on final film properties.

In a batchwise device production process, it is also possible to anneal the film at temperatures around 200 °C prior to processing it. It has been shown that it is possible to achieve a level of shrinkage down to 25 ppm at 150 °C with Teonex Q65 [10].

The second factor that impacts on the dimensional reproducibility is the natural expansion of the film as the temperature is cycled, as measured by the CTE. A low CTE (typically <20 ppm °C^{-1}) is desirable to match the thermal expansion of the base film to the layers that are subsequently deposited. A mismatch in thermal expansion means that the deposited layers can become strained and cracked under thermal cycling. In the temperature range from room temperature up to the T_g, the typical CTE of PEN is 18–20 ppm and of PET is 20–25 ppm (note that the T_g of PEN is 40 °C higher than PET). Above T_g, the natural expansion of PET and PEN films that have not been heat-stabilized is dominated by the shrinkage the films undergo as the internal strains in the film relax, as discussed above. However, the heat-stabilized films discussed above show only a very small increase in CTE in the temperature range from T_g to the temperature at which they were heat-stabilized [1–3]. This compares favorably with the quoted coefficient of expansion of amorphous polymers, which is typically 50 ppm °C^{-1} [2, 3] below the T_g but can increase by a factor of three times above the T_g.

In addition to dimensional stability, another important factor to be considered is the upper processing temperature (T_{max}) that a film can be subjected to. Although, as outlined above, the T_g does not define T_{max} with semicrystalline polymers, it largely does with the amorphous polymers.

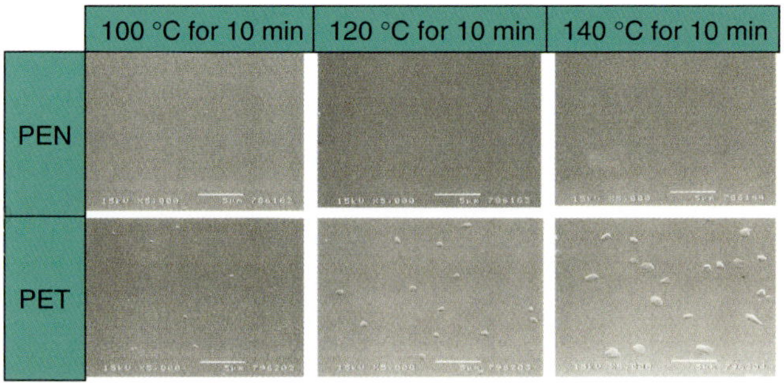

Figure 10.5 Growth of cyclic oligomers on PET and PEN film with time at 100, 120, and 140 °C.

10.3.5 Cyclic Oligomers

PET film contains 1.1 – 1.4 wt% cyclic oligomer, which can migrate to the surface if the film is held at elevated temperatures for tens of minutes [11, 12] as can occur, for example, in the manufacture of devices exploiting touch screens. The presence of these oligomers on the surface gives rise to "bloom" and increases the haze of the films, which is clearly undesirable. PEN with 0.3 wt% cyclic content has significantly lower cyclic oligomer content compared with PET and there is significantly less migration to the surface of the film (Figure 10.5).

One strategy to reduce this hazing effect is to exploit planarizing coatings (see Section 10.3.8) or hardcoats to act as a barrier to cyclic oligomer migration. Figure 10.6 shows a PET film with and without a planarizing coating annealed at 120 °C for 30 min. The presence of the cyclic oligomers can be seen on the surface of the noncoated film, but the coated film surface is clear.

Control of the polymerization process to yield a low oligomer polymer, coupled with careful control of the filming process, offers another strategy to a "low bloom" film. These low bloom films typically have a haze ≤1% after aging at 150 °C

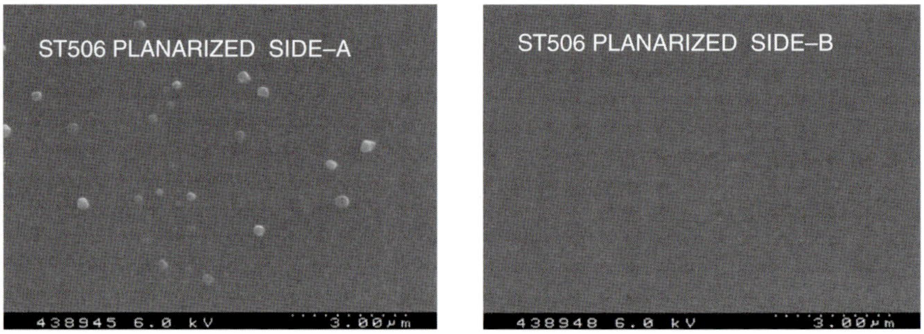

(a) Nonplanarized PET: 30 min/120 °C (a) Planarized PET: 30 min/120 °C

Figure 10.6 (a,b) Effect of planarizing coatings on acting as a barrier to blooming.

for 60 min. As a further refinement on the low bloom films, refractive index (RI) matching coatings have been developed which reduce the iridescence that can be observed with hard-coated films, which arises because of an RI mismatch between the pretreat and the hard coat.

10.3.6 Solvent and Moisture Resistance

A wide range of solvents and chemicals can potentially be used when laying down the various layers in displays depending on the processing steps involved. Amorphous polymers in general have poor solvent resistance compared to semicrystalline polymers. The solvent resistance of PET and PEN biaxially oriented films are shown in Table 10.1.

This deficiency is overcome by the application of a hardcoat to the amorphous resins, which significantly improves the solvent and chemical resistance to solvents such as *N*-methylpyrrolidinone (NMP), isopropyl alcohol (IPA), acetone, methanol, tetrahydrofuran (THF), ethyl acetate, 98% sulfuric acid, glacial acetic acid, 30% hydrogen peroxide, and saturated bases such as sodium hydroxide [13]. With PET and PEN films, a hardcoat is not required to give solvent resistance.

The residual shrinkage of stabilised PEN film after 30 min at 150 °C is of the order 500 ppm but can be reduced to below 100 ppm or better by careful process control – this will be discussed later. In addition to this residual shrinkage, the processing environment, in effect the prevailing humidity, must be taken into account because semicrystalline PEN will absorb up to 1500 ppm of moisture under ambient conditions yet readily lose this at higher processing temperatures.

Table 10.1 Table of solvent resistance.

	Unit	PEN	PET
Ketone	Acetone	Good	Good
	MEK	Good	Good
Alcohol	Methanol	Good	Good
	Ethanol	Good	Good
	Isopropanol	Good	Good
	Butanol	Good	Good
Ester	Ethyl acetate	Good	Good
Hydrocarbon	Formalin	Good	Good
	Tetrachloroethane	Good	Good
Acid	10% HCI	Good	Good
	10% HNO_3	Good	Good
	10% H_2SO_4	Good	Good
	Acetic acid	Good	Good
Alkali	10% NaOH	Good	Fair

It has been shown previously [14] that for every 100 ppm moisture absorbed, the film expands by ~45 ppm isotropically in the three film dimensions.

With the knowledge of the solubility level of moisture in PEN film and its rate of diffusion as a function of temperature, it is possible to model the impact of various environmental conditions on moisture content changes and hence volumetric changes in the different thicknesses of the film. PET and PEN will typically pick up about 1000 ppm moisture at an RH of 40% at 20 °C, and it can take up to 12 h to reach equilibrium depending on the thickness. Although PET and PEN will reach the same equilibrium level, PET picks up moisture at about three times the rate of PEN, as shown in Figure 10.7.

Films can typically take over 12 h to reach equilibrium. Figure 10.8 shows the impact of RH on this process and indicates that the final (equilibrium) moisture levels will change significantly with varying RH. At an RH of 20%, the film reaches an equilibrium level of about 500 ppm, but at 60% RH an equilibrium level of about 1500 ppm.

PET and PEN films will lose moisture upon subsequent heating, but now at a much faster rate. At 150 °C, the film takes 6 min to reach an equilibrium moisture level of 5 ppm, but at 90 °C the film takes 30 min to reach the level of 40 ppm (Figure 10.9).

Upon removal from the heated environment back into ambient conditions, the film initially picks up to 50% of its equilibrium moisture level while cooling in less than a minute or within a few seconds, but the remaining moisture reaches equilibrium at a much lower rate.

Taking into account the variation in moisture content that can result depending upon temperature, RH, and film thickness, which can be of the

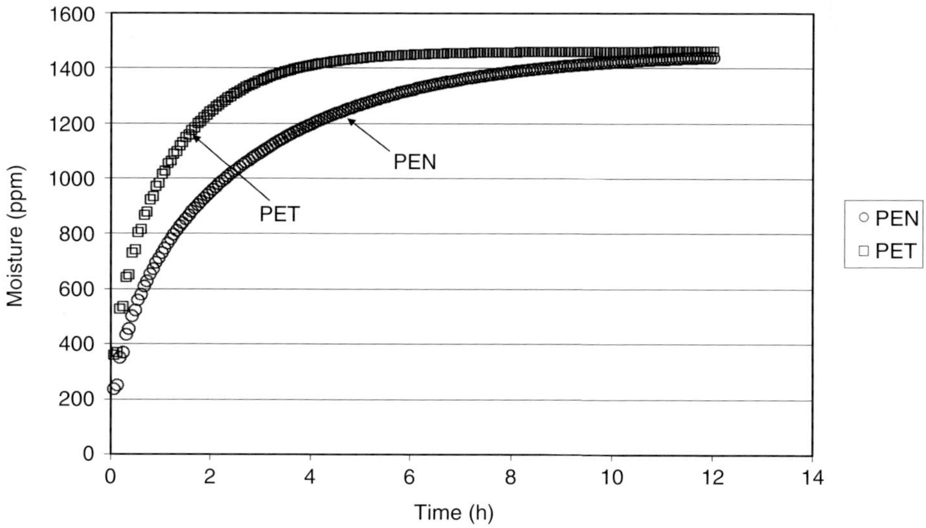

Figure 10.7 Moisture pickup versus time for PET and PEN.

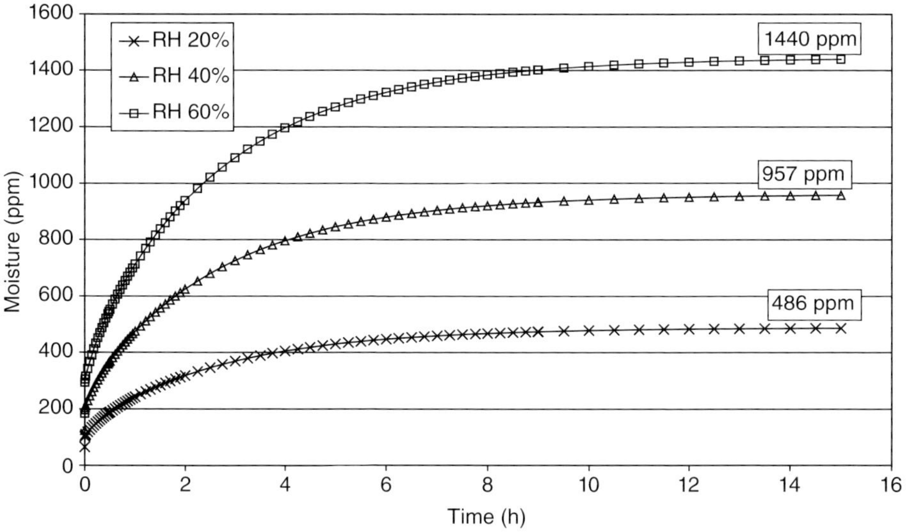

Figure 10.8 Effect of RH on moisture pickup of PEN film at 20 °C.

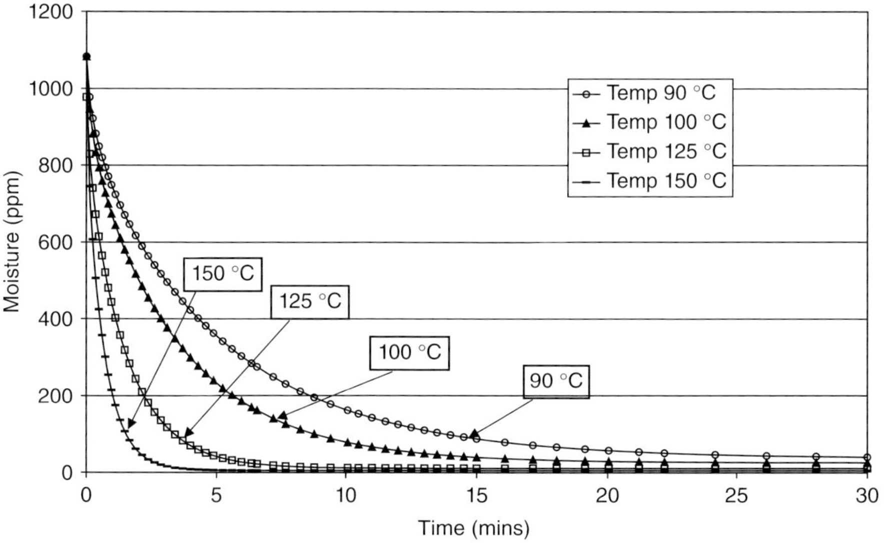

Figure 10.9 Moisture loss of PEN film versus time on heating to different temperatures at RH of 40%/20 °C.

order of several hundreds of parts per million, it can be seen that uncontrolled moisture pickup will have a significant influence on dimensional reproducibility. This will be particularly significant at lower processing temperatures where there is minimal film shrinkage but where the film takes a longer time to reach equilibrium moisture level than the time taken to carry out the processing step. Ideally, the film should be allowed to equilibrate before any registration points are established for laying down subsequent layers.

By carrying out modeling studies of the type described above, the impact of the moisture change can be minimized.

10.3.7
The Effect of Mechanical Stress on Dimensional Reproducibility

One of the main drivers in moving to plastic substrates is possibility for R2R processing and the consequent processing and economic advantages. Under these conditions, a winding tension will clearly be present, and polymer film substrates with low moduli will be susceptible to internal deformation, particularly at elevated process temperatures. Figure 10.10 shows a comparison between PET and PEN films.

The storage modulus E' is recorded using dynamic mechanical thermal analysis, and, as the temperature is increased, the stiffness of both materials is seen to fall. However, in the region 120–160 °C, PEN is significantly stiffer and stronger, with a modulus almost twice that of PET.

Any small tension or load applied to the film at that temperature will therefore impose a strain in the material, which will be "frozen-in" upon cooling and reappear as shrinkage upon reheating. The implication of this is that, if the film is constrained through a thermal processing cycle, for example, in an R2R process or by lamination to a rigid carrier, the strain introduced into the film may result in subsequent shrinkage on reheating.

Measures of stiffness such as Young's modulus are thickness-independent and do not indicate how the rigidity will change with thickness. Rigidity (D) can be defined by [14, 15].

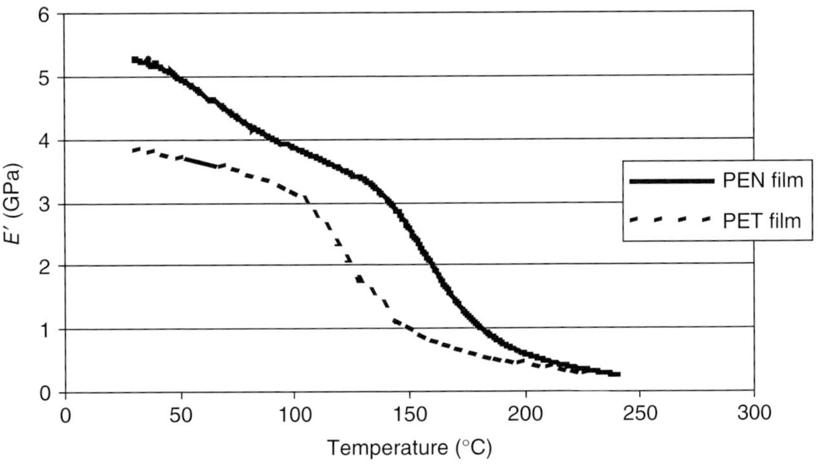

Figure 10.10 Change in storage modulus of PET and PEN with temperature.

Table 10.2 Comparative rigidity of amorphous and semicrystalline films.

Material	Thickness (μm)	Rigidity (N m ×10^{-4})	Rigidity relative to 125 μm amorphous film
Amorphous	125	5	1
Amorphous	200	20	4
PEN (Teonex® Q65)	125	15	3
PEN (Teonex® Q65)	200	61	12

$$D = \frac{Et^3}{12(1-\nu)}$$

where E is the elastic modulus (6.1 GPa), t is the thickness, and ν is the Poisson ratio (0.33).

Assuming a Young's modulus of 2 GPa for amorphous films and 6 GPa for the PEN film, it can be seen that (Table 10.2) a 200-μm PEN film is 4 times more rigid than a 125-μm PEN film and 12 times more rigid than a 125-μm amorphous film. This stiffness may prove to be an advantage in a batchwise display manufacturing process.

This extra rigidity can be exploited to reduce the stress and the resulting distortion, as evidenced by the difference in distortion measured through an a-Si TFT process on 125 μm (5 mil) PEN compared to 200 μm (8 mil) PEN where the distortion through the process is reduced from about 175 to 100 ppm (Figure 10.11).

10.3.8 Surface Quality

The surface quality of the film is an essential parameter for the more demanding display applications, and any surface defects or debris that could protrude through subsequent conductive or barrier layers should be removed. Surface quality can be divided into the following items:

10.3.8.1 **Inherent Surface Smoothness**

PET and PEN have an inherent surface smoothness of <1 nm, in terms of both roughness average (R_a) and root-mean-square roughness (R_q), as measured using white-light interferometric methods [16]. However, the use of R_a and R_q can be misleading, as they tend to be measured over a small sampling area and do not capture the occasional peak that occurs on the sample size of a display. The surface smoothness is largely determined by the internal cleanliness of the film substrate and typified by the presence of peaks of size tens to hundreds of nanometers.

DTF have worked with a leading surface metrology tool manufacturer to develop a large area metrology "LAM" tool – the first of its kind [17]. This

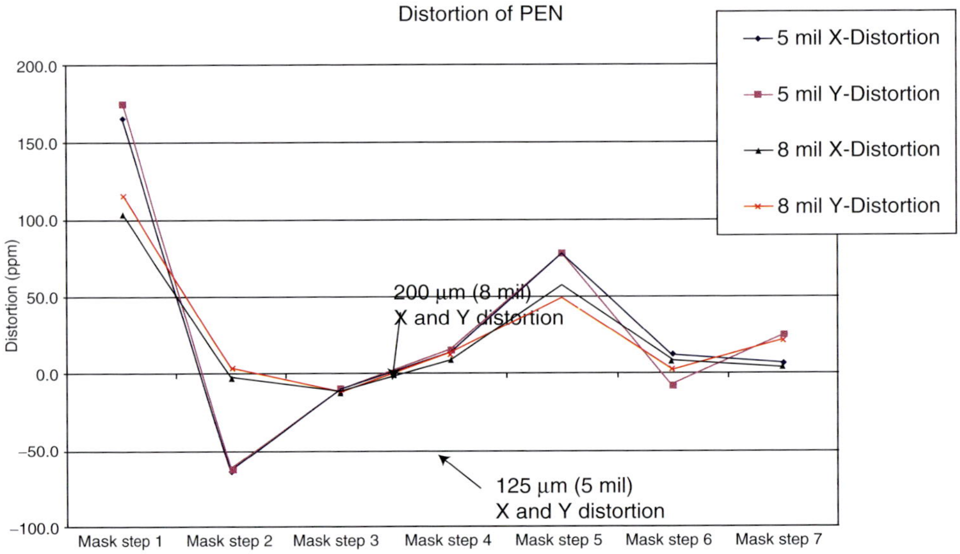

Figure 10.11 Distortion through a-Si TFT deposition. (slide courtesy of Flexible Electronic Display Centre, Arizona State University. Copyright © Flexible Electronic Display Centre, Arizona State University.)

device has the capability to map the defects on the surface of a film quickly and precisely, achieving optimum lateral resolution obtainable from white-light interferometry. The equipment was designed in addition to help distinguish between both intrinsic (polymeric based) and extrinsic (external, i.e., air-borne debris) defects.

Typical features of the tool are as follows:

1) Sampling area = 35 cm × 35 cm measured in around 2 h.
2) Map of intrinsic and extrinsic defects, precise locations registered for compositional analysis.
3) Full $X'Y'Z$ measurements of the surface defects, enabling significance testing.

Figure 10.12 shows a LAM surface analysis of a 50-μm uncoated PEN film, illustrating the current manufacturing quality. The graph is an exponential decay curve, expressing the intrinsic defects (area) at a given height threshold of interest. One can imagine the height threshold as an imaginary Z-plane bisecting the total $X'Y$ field of view. The propensity of defects or the opportunity to encounter one is quantified as a six-sigma term DPMOs (defects per million opportunities) or parts per million. Since such values are effectively a fraction of the total sampled area, it can be scaled up or extrapolated for much larger and meaningful surfaces areas. For example, to put this into perspective, 0.01 ppm defects of heights >0.5 μm relates to 1 mm^2 area within a total of 100 m^2 area. In other words, the intrinsic defects are not the issue dominating surface quality.

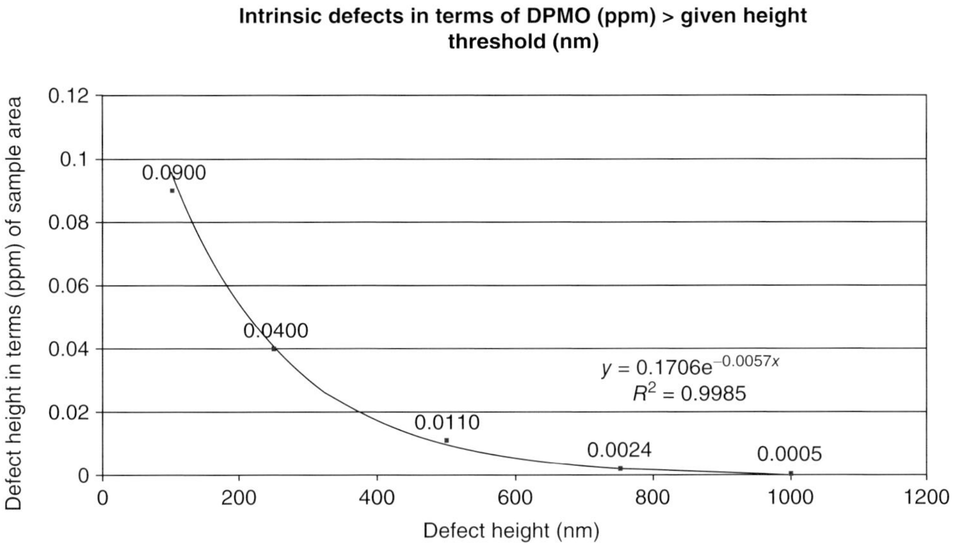

Figure 10.12 Frequency distribution of particles for PEN film (Teonex Q65FA).

10.3.8.2 Surface Cleanliness

In this chapter, surface cleanliness is taken to mean surface defects that are dominated by dust and surface scratches. These defects can range in size up to tens of micrometers both laterally and vertically, and, if not controlled, can have a significant impact on device fabrication. A recent analysis of a film entering a clean room carried out with CPI Printable Electronics Centre[7] and Teknek Ltd.[8] showed that the external debris on the surface was predominantly that associated with human, packaging, and process interaction – this is largely unavoidable in a film that is handled and slit in a non-clean-room environment. There are several strategies to address this, some of which are as follows:

1) Hygiene issues associated with handling transport and packaging of films, for example, use of plastic rather than cardboard cores, sharp blades to minimize debris on slitting, the double bagging of the film to ensure dust is not trailed into the clean room.
2) Control of static electricity as the film is unwound from a reel.
3) Contact and noncontact cleaning technologies.
4) The use of protecting films to protect the surface of the film from both external debris and surface scratching.
5) Planarizing coatings, coated in a clean room environment [2, 3] (see Section 10.4). These coatings have the dual functions of providing surface smoothness over a wide area and also imparting a certain surface hardness to prevent scratches on the surface during film processing.

7) CPI Printable Electronic Centre, NETPark Thomas Wright Way, Sedgefield, County Durham TS21 3FG, UK.
8) Teknek Ltd., River Drive, Inchinnan Business Park, Renfrewshire PA4 9RT, Scotland, UK.

This subject of minimizing the defects on the surface and at the same time being able to measure defects during actual film line processing is of considerable activity at the present. Perfection is unlikely, and, given that device manufactures now claim the same yields on films that they achieve on glass (see Section 10.5), it is probably not required. Understanding, however, which defects are critical and targeting to eliminate them are the current focus, and this is an achievable target.

10.4 Summary of Key Properties of Base Substrates

The main properties of heat-stabilized PET (Melinex® ST504/506) and PEN (Teonex® Q65HA) relevant to flexible electronics are summarized in Figure 10.13.

Unstabilized PET and PEN films exhibit the same property set except for the shrinkage and upper processing temperature, which for these films are largely dictated by the T_g.

It is difficult to directly compare the property sets of different film types, as information available in the public domain may not represent the films targeted for printed electronics applications. Table 10.3 shows the author's attempt to contrast the property set of the different substrates under consideration for flexible electronics, which should be used only as a rough guide.

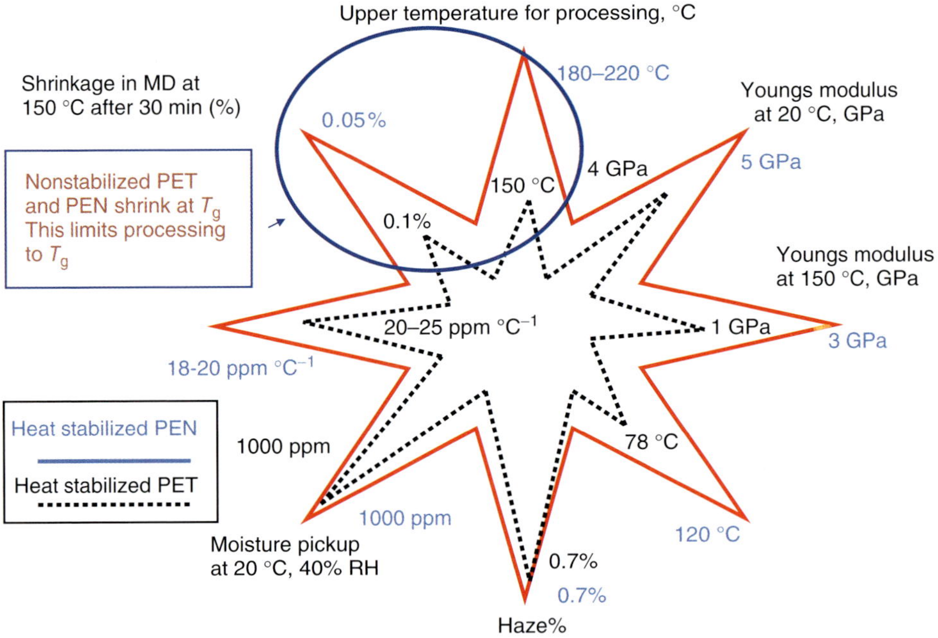

Figure 10.13 Summary of PET and PEN properties relevant to flexible electronics.

Table 10.3 Comparison of the key properties relevant to flexible electronics.

Property	Biaxially oriented heat stabilized PET	Biaxially oriented heat stabilized PEN	Biaxially oriented PEEK	PC	PES	Akron APS	Polyimide
CTE (−55 to 85 °C) (ppm °C^{-1})	√√	√√	√√	√	√	√	√√
%Transmission (400–700 nm)	√√	√√	√√	√√√	√√	√√	X
Water absorption (%)	√√	√√	√√	√	X	Unknown	X
Young's modulus (GPa)	√√	√√	√√	√	√	√	√
Tensile strength (MPa)	√√	√√	√√	√	√	√	√
Solvent resistance	√√	√√	√√	X	X	√√	√√
Upper operating temperature	√	√√	√√√	√	√√	√√√	√√√
Availability at commercial scale	√√√	√√√	√	√√√	√√√	√	√√√

√√√ is best, X is poor.

The general properties have been discussed in this chapter and also in previous papers [2, 3, 14]. Table 10.3 shows that both heat-stabilized PET (e.g., Melinex® ST504) and heat-stabilized PEN (Teonex® Q65HA) have an excellent balance of the key properties required for flexible electronics, and it is this property set coupled with commercial availability in grades tailored for flexible electronics that is leading both Melinex® ST grades and Teonex® Q65HA emerging as the leading films for the base substrate of the more demanding flexible and printed electronics applications. Victrex PEEK film is commercially available as a cast film, but to unlock the full potential of this high-performance polymer the material would have to be available in biaxially oriented form: although development quantities are available, the cost of scaling up remains a challenge. The Akron film development faces the same hurdle.

The ideal substrate for high-performance display applications would have the key attributes obtained with semicrystalline and biaxially oriented polymers, that is, low CTE, excellent solvent resistance, thermal stability of amorphous polymers such as PI, and high clarity. In addition to these, the film should be commercially available and also with the film surface quality that is only really achieved by manufacture at volume scale. Although research is being carried out on new materials, the development and production costs of developing a new film based on a new polymer, possibly requiring novel raw materials coupled with access to both polymer and film-making facilities, are prohibitively expensive for what is currently a small-volume market. The time taken to commercialize a new film is also lengthy. In the author's opinion, this is not an attractive proposition for a substrate supplier, and a more likely route forward is to push existing commercially available films to their limits and marry this with the advances that are being made in low-temperature processing of inorganic materials.

Prototype displays and backplanes have in the past been produced on flexible stainless steel substrates, but there is virtually no reported activity on this at present. Flexible glass obviously offers the advantages of glass with respect to barrier (although edge sealing remains a problem common with plastic-based barrier materials), but the availability of the flexible glass and the issue of major investment to commercialize manufacture for a relatively small market remain challenges to this industry.

10.5 Planarizing Coatings

The surface quality of the film can be controlled to an extent via recipe control, plant hygiene, web cleaning techniques, and film line processing conditions. The film however, is typically not manufactured in an environment that is clean enough to give a level of extrinsic contamination suitable for the more demanding applications in flexible electronics.

While a number of cleaning techniques can reduce extrinsic contamination, it is difficult to remove it completely. To achieve the surface quality required for most electronic devices, one option taken by DTF has been to apply a planarizing coating. This coating covers all surface defects and comprises a low-viscosity liquid that flows easily over the film imperfections to create a glass-smooth surface with fewer defects over a wide area. The planarizer coating ideally is carried out in a cleanroom to ensure that the surface has low intrinsic and extrinsic contamination. The planarizer coating also has anti-scratch properties, preventing scratching of the film during subsequent processing and handling.

The choice of chemistry and the coating thickness of the planarizing coating have an impact on the physical properties of the planarized film. Factors to consider include the maximum processing temperature during film processing and the chemistry of the layers that will be deposited on the planarizer surface. Suitable chemistry types that offer pencil hardness >2H range from predominantly inorganic materials, for example, sol–gel-derived coatings based on siloxane chemistry, to organic materials, for example, polyfunctional acrylate coatings.

Figures 10.14 and 10.15 illustrate the impact of the planarizing coating. These surface peaks, in terms of propensity and height, are clearly seen to reduce because of the "smoothing" effect of the planarizing coating. In Figure 10.14, it can be seen that there are no peaks above 250 nm in a 5 cm × 5 cm area with the planarized sample, whereas the "raw" PEN film surface in this particular study has peaks up to 600 nm.

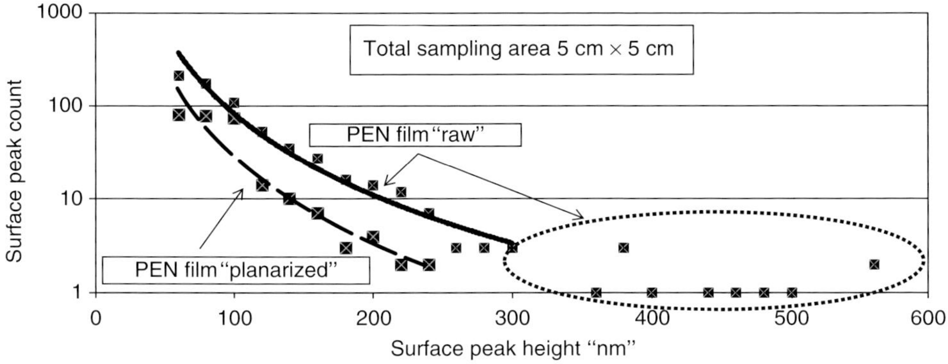

Figure 10.14 Effectiveness of planarizing coating on surface roughness of PEN film as measured by white-light interferometry.

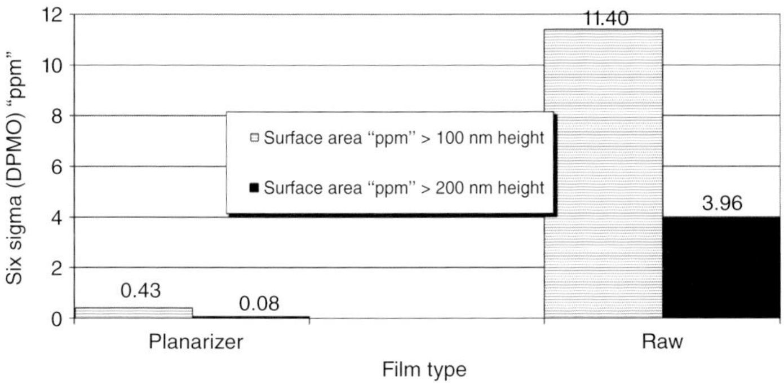

Figure 10.15 Intrinsic surface defect rate expressed as six-sigma DPMO ppm for both "raw" and planarized PEN film.

Figure 10.15 shows that there is a factor of 50 reduction in the occupied surface area of peaks >200 nm in height for planarized PEN films compared to standard PEN films.

Another key benefit from planarizing the surface is the reduction in the surface slopes associated with such peak heights, that is, rendering them less "sharp" and so less disruptive to subsequent coatings during device manufacture.

The thermal-mechanical analyses carried out on heat-stabilized PEN and an inorganic–organic hybrid planarized, coated, and heat-stabilized PEN using a thermomechanical analyzer (TMA-7, Perkin-Elmer) are shown in Figure 10.16. This figure shows that planarized PEN film's dimensional response to temperature cycling is very similar to that of uncoated PEN: that is, the planarized film remains a dimensionally reproducible substrate up to 180 °C. This is within the performance requirements of a flexible substrate for a display based on an inorganic TFT backplane.

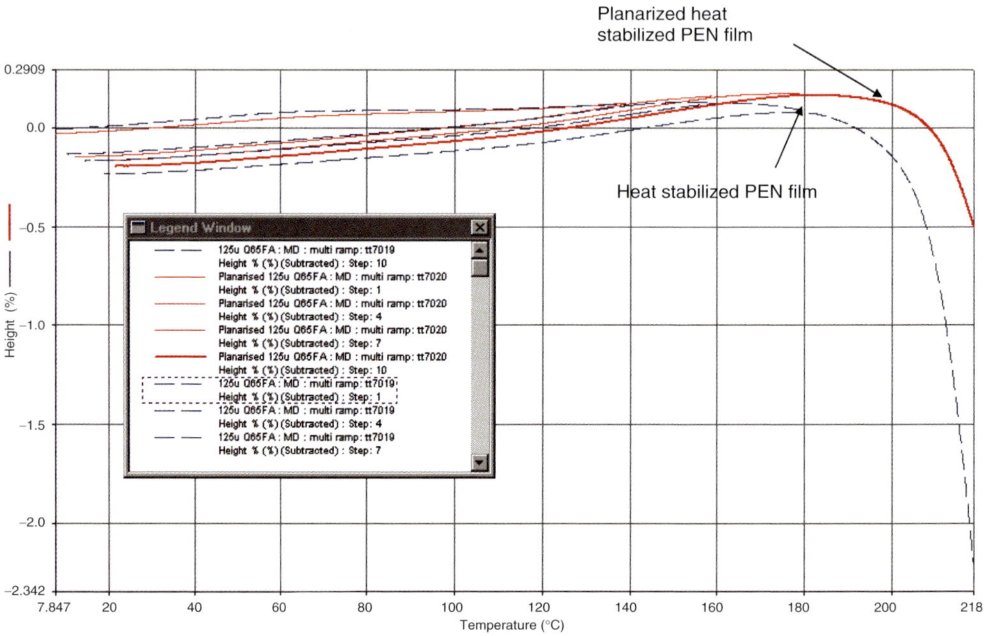

Figure 10.16 Thermal-mechanical analysis of heat-stabilized PEN and planarized heat stabilized PEN.

The planarized-coated PET and PEN films show excellent chemical resistance to common solvents used in photolithography processes and solution-based organic semiconductors.

10.6 Examples of Film in Use

There have been significant advances in the development of electronic paper-type applications based on electrophoretic displays driven by organic active-matrix backplanes. One of the pioneers in this area, Plastic Logic Ltd.,[9] claims to achieve the same yields on polyester film as on glass, and displays of a commercial quality are available at the time of writing this chapter, as evidenced from different application areas such as bus information signs (exploiting low power, readability in bright light) and second screen displays for smart phones that are always on (exploiting low power and robustness).

The key film features required for these applications are as follows:

1) Dimensional stability at the processing temperatures to obtain dimensional reproducibility through multilayer processing (discussed in Section 10.3.4).

9) Plastic Logic Ltd., 34 Cambridge Science Park, Cambridge CB4 0FX, UK.

2) Excellent surface smoothness to ensure that surface artifacts do not interfere with the organic TFTs (discussed in Sections 10.3.8 and 10.4).
3) Solvent resistance – the films must be able to withstand the solvents used in TFT manufacture (discussed in Section 10.3.6).

The relatively low processing temperatures of organic semiconductor materials are an advantage for plastic films such as PET and PEN.

Similarly, significant advances are being made on reducing inorganic deposition temperatures to allow fabrication of inorganic TFTs on plastic films at temperatures below 200 °C [18] In this case, the property set required of the base substrate is more demanding, in particular with respect to dimensional reproducibility at elevated temperatures, and the higher temperature performance of PEN is required. The Flexible Electronic Display Centre (FEDC)[10)] at ASU has also demonstrated the ability to manufacture mixed-oxide TFT arrays on PEN and has increased the processing area size in the manufacture of mixed-oxide TFT arrays from 150 mm round carriers to 370 mm × 470 mm scale without any yield losses due to bond failure or bow warpage. This technology has been exploited to manufacture prototype flexible X-ray detectors and flexible color OLED devices.

The exceptional performance above demonstrates that the film can be pushed beyond its data-sheet specification by careful control of the key factors discussed in this paper and by matching and optimizing device fabrication against the property set of a given film. The factors are as follows:

- Thermal stress (offline stabilization and reducing processing temperatures) (discussed in Section 10.3.4).
- Environment to minimize the effect of moisture (discussed in Section 10.3.6).
- Mechanical stress (exploiting rigidity through a rigid carrier) (discussed in Section 10.3.7).

The plastic substrates that are under consideration as flexible substrates typically have barriers of the order $1-10\,\mathrm{g\,m^{-2}\,day^{-1}}$ for water vapor transmission rate and about $1-10\,\mathrm{ml\,m^{-2}\,day^{-1}}$ for oxygen. OLED displays will require water vapor transmission rates of $<10^{-6}\,\mathrm{g\,m^{-2}\,day^{-1}}$ and oxygen transmission rates of $<10^{-5}\,\mathrm{ml\,m^{-2}\,day^{-1}}$, and from the start of interest in flexible electronics, achieving barrier films of this performance on flexible substrates has remained one of the key challenges. Detailed discussion of barrier technology is outside the scope of this chapter, but significant progress has been achieved over the past decade exploiting multilayer organic and inorganic coatings, and barrier materials with water vapor transmission of the order $10^{-3}\,\mathrm{g\,m^{-2}\,day^{-1}}$ are commercially available. Higher performance barrier materials are claimed. However, the limited availability of barrier films of the order $10^{-6}\,\mathrm{g\,m^{-2}\,day^{-1}}$ has slowed down the development of flexible OLED devices. Achieving a cost-effective high-barrier film still remains a key target for this industry and an area of active development.

10) Flexible Electronic and Display Center at Arizona State University, Arizona State University Research Park, 7700 South River Parkway, Tempe, AZ 85284-1808,USA.

10.7 Concluding Remarks

This chapter has addressed only the base substrate on which subsequent processing, including conductive and barrier coatings in addition to electronic device fabrication, will be carried out. The demanding property set required for flexible electronics is pushing the existing commercially available films to the limits of their performance, but, as has been discussed in this chapter, the approach of the substrate developer working closely with the device manufacturer, with each understanding the limitations of their respective technologies, is enabling significant progress in fabricating flexible electronics devices.

Acknowledgments

The author would like to acknowledge the significant contributions of his colleagues Robert Eveson, Duncan Mackerron, Kieran Looney, and Karl Rakos.

References

1. Crawford, G. (ed.) (2005) *Flexible Flat Panel Displays*, John Wiley & Sons, Ltd, Chichester.
2. MacDonald, W.A. (2004) Engineered films for display application. *J. Mater. Chem.*, **14**, 4.
3. MacDonald, W.A., Rollins, K., MacKerron, D., Eveson, R., Rakos, K., Adam, R., Looney, M.K., and Hashimoto, K. (2007) *J. SID*, **15/12**, 1075–1083.
4. MacDonald, W.A., Mackerron, D.H., and Brooks, D.W. (2002) in *PET Packaging Technology* (ed. D.W. Brookes), Sheffield Academic Press.
5. MacDonald, W.A. (2003) *Polyester Film, Encyclopedia Polymer Science and Technology*, 3rd edn, John Wiley & Sons, Inc, Hoboken, NJ.
6. MacDonald, W.A., Mace, J.M, and Polack, N.P. (2002) New developments in polyester films for display applications. 45th Annual Technical Conference Proceedings of the Society of Vacuum Coaters, 2002, p. 482.
7. Adlen, M.J., Kuusipalo, J., MacKerron, D.H., and Savijarvi, A.-M. (2004) *J. Mater. Sci.*, **39**, 6909–6919.
8. MacDonald, W.A., Rollins, K., Eveson, R., Rustin, R.A., and Handa, M. (2003) Plastic Displays-new developments in polyester films for plastic electronics. Society For Information Display, Digest of Technical Papers, 2003, pp. 264–267.
9. Gohil, R.M. (1994) *J. Appl. Polym. Sci.*, **52**, 925–944.
10. Sarma, K.R., Chanley, C., Dodd, S., Roush, J., Schmidt, J., Srdanov, G., Stevenson, M., Wessel, R., Innocenzo, J., Yu, G., O'Regan, M., Macdonald, W.A., Eveson, R., Long, K., Gleskova, H., Wagner, S., and Sturm, J.C. (2003) Active matrix OLED using 150 °C a-Si TFT backplane built on flexible plastic substrate. Proceedings from SPIE Aerosense, Techologies and Systems for Defense and Security, April 22–25, 2003.
11. Shiono, S. (1979) *J. Polym. Sci. Polym. Chem. Ed.*, **17**, 4123–4127.
12. Holland, B.J. and Hay, J.N. (2002) *Polymer*, **43**, 1797–1804.
13. Angiolini, S., Avidano, M., Bracco, R., Barlocco, C., Young, N.G., Trainor, M., and Zhao, X.-M. (2003) High performance plastic substrates for active matrix flexible FPD. Society For Information Display, Digest of Technical Papers, 2003, pp. 1325–1327.

14. MacDonald, W.A. (2006) in *Organic Electronics: Materials, Manufacturing and Applications*, Chapter 7 (ed H. Klauk), Wiley-VCH Verlag GmbH, Weinheim, pp. 163–179.
15. MacDonald, W.A., Rollins, K., MacKerron, D., Eveson, R., Rustin, R., Adam, R., Looney, M.K., and Hashimoto, K. (2005) Eurodisplay 2005. Proceedings of the 25th International Display Research Conference, 2005, pp. 36–41.
16. International Standard Organization (2013) ISO25178-604. *Geometrical Product Specifications (GPS)-Surface Texture: Areal. Part 604 Nominal Characterisitics of Non-contact (Coherence Scanning Interferometry) Instruments*, International Standard Organization.
17. Rakos, K. (2014) Metrology method. WO Patent WO2014/045038 A1.
18. Wong, W.S., Lujan, R., Daniel, J.H., and Limb, S. (2006) Digital lithography for large-area electronics on flexible substrates. *J. Non Cryst. Solids*, **352**, 1981–1985.

Part II
Devices and Applications

11
Inkjet Printing Process for Large Area Electronics

Sungjune Jung, Steve D. Hoath, Graham D. Martin, and Ian M. Hutchings

11.1
Introduction

Several features of inkjet printing make it particularly attractive as a manufacturing process in the context of large-area electronics.

First, it is a digital process. The location of each droplet of "ink" (i.e., the material being deposited) can be predetermined on a two-dimensional grid. If necessary, the location can be changed in real time, for example, to adjust for distortion or misalignment of the substrate, or to ensure that a certain height of the final deposit is achieved. Because it is a digital process, each product in a sequence can easily be made different from every other, in small or even in major ways; bespoke products are generated just as readily as multiple replicas of the same design. Because the pattern to be printed is held as digital data, there may be significant cost savings over processes that involve the use of a physical mask or template.

Second, it is a noncontact method; the only forces that are applied to the substrate result from the impact of very small liquid drops. Thus fragile and/or delicate substrates can be processed that would not survive more conventional printing methods. Material can be deposited on to nonplanar (rough or textured) substrates, because the process can be operated with a stand-off distance between the print head and substrate of at least 1 mm. In conventional contact-based printing, the printed material may also be transferred by accidental contact, potentially causing poor quality or contamination; such problems are avoided in a noncontact process.

Third, a wide range of materials can be deposited. By selection of an appropriate print head, liquids with viscosities from 1 to 50 mPa s or higher can be printed. Several different methods can be used to generate printed structures. Multiple combinations of materials can be used, and inkjet printing can also be combined with other process steps, so that in principle complex heterogeneous and composite structures can be produced, with different materials distributed in all three dimensions.

A further benefit is that inkjet printing is modular and scalable. Multiple print heads can be assembled to print in tandem, for example, by placing them side by

Large Area and Flexible Electronics, First Edition. Edited by Mario Caironi and Yong-Young Noh.

side to print a wider pattern, or one after the other to print several materials in sequence. These concepts are standard for graphical printing, where four or more colors are commonly used, but can also be readily extended to the manufacturing context.

In this chapter, we shall discuss the processes by which small drops of liquid are generated and delivered in inkjet printing, the rheological properties of the often complex fluids that are involved, the behavior of liquid drops during impact and spreading on a surface, and current applications of inkjet printing to large-area electronics.

11.2
Dynamics of Jet Formation

Inkjet printing involves the production of small drops of liquid and their deposition at precise locations on a substrate. Many methods for drop generation have been devised, but currently the most important methods used for inkjet printing are described as either "continuous inkjet" (CIJ) or "drop-on-demand" (DOD). In both the CIJ and DOD methods, the liquid flows through a small orifice (usually called a *nozzle*). The essential difference between the two lies in the nature of the flow through the nozzle: in CIJ, as the name implies, the flow is continuous, while in DOD it is impulsive. For the printing of active and passive electronics, DOD has been preferred by most users.

Except for those used for experimental or research purposes, nearly all DOD print heads contain multiple nozzles. Early commercial examples typically contained a few tens of nozzles, while modern heads contain hundreds or even thousands of nozzles, and several heads are often combined to cover a wider area. Each nozzle contributes to one or a few lines (depending on print configuration) of printed spots, making up the complete printed image.

As each nozzle generates a drop of printing fluid only when required, most of the nozzles are inactive for much of the time. This means that changes in the fluid at the open nozzle (for example, by evaporation of a solvent) can, after some time, affect the performance of that nozzle when it is next fired. In the worst case, the nozzle can fail to fire. Clearly this is to be avoided, and various strategies are employed to mitigate the effect, ranging from constraints on the ink formulation to purging into waste collectors during operation, and capping the nozzles when shut down.

DOD print heads require an actuation mechanism that will transfer enough energy to a suitable volume of liquid to enable it to form a drop (creating new surface area) and to travel to the substrate at a reasonable velocity. Various mechanisms have been proposed and patented, but the two most common techniques are thermal (or bubble) actuation and piezoelectric actuation.

Thermal actuation is used in many home and office inkjet printers. Within a cavity behind the nozzle is a small resistive heater in good thermal contact with the ink. Rapid heating causes superheating and vaporization of a thin layer of ink adjacent to the heating element. A vapor bubble quickly expands, producing fluid displacement and the kinetic energy necessary to force a drop from the nozzle. Once the drop is ejected and the heat source is switched off, the bubble rapidly collapses, drawing fresh ink in to the cavity, and the cycle is ready to start again once any residual pressure fluctuations and meniscus movements have died away. Clearly, the ink itself needs to be able to support and withstand the heating and vaporization process, and this limits the types of fluid that can be printed by this method. One problem to be avoided is the deposition of ink components on to the heater or elsewhere within the nozzle ("kogation"), and so attention is required in the ink formulation. These constraints on ink formulation have made thermal DOD technology less popular than the piezoelectric method discussed below for printing the wide range of materials required for electronics applications.

Piezoelectric actuation (PA) was one of the first DOD techniques to be described [1, 2]. PA-based inkjet print heads are used in home and office inkjet systems, and PA has dominated in those print heads that have been developed for industrial application, both for graphical printing and for digital fabrication. Lead zirconate titanate (PZT), a ceramic material that exhibits a strong piezoelectric effect, is used as the active component in most PA systems. An electric field applied to the ceramic causes it to change shape, and this phenomenon is used to eject the fluid.

Unlike thermal actuation, PA allows a negative gauge pressure (i.e., a pressure below atmospheric) to be achieved if desired, by arranging the movement of the ceramic actuator to increase the volume of the nozzle chamber. Some print head manufacturers use this to improve performance. For example, an initial negative pressure pulse when reflected and inverted from the end of the chamber can be combined with a later positive pressure pulse to increase the maximum driving pressure [3]. Drive waveforms can be modified to accommodate printing fluids with different characteristics.

Two physical properties dominate the behavior of the liquid jets and drops involved in inkjet printing: *surface tension* and *viscosity*. We shall remind ourselves briefly of their definitions.

The surface tension of a liquid reflects the fact that atoms or molecules at a free surface have a higher energy than those in the bulk. There is a cost in terms of energy in creating new surface area. For a free droplet of liquid, the shape with the lowest surface area, and therefore the lowest surface energy, is a sphere, and in the absence of other influences such as electrostatic or aerodynamic forces, that is the shape which a free drop will adopt. If the liquid is in contact with a solid surface, for example, after it has been printed on to a substrate, then we need to consider not only the energy of its free surface (which is usually in contact with air or solvent vapor) but also the energy of the interface between the liquid and the solid substrate. In that case (and where, as in inkjet printing, the effect of gravity is negligible), the equilibrium shape of the drop becomes a spherical cap.

The surface tension in a liquid causes a force to act in the plane of the free surface perpendicular to a free edge in that surface, which can be measured directly by various experimental arrangements. The force is proportional to the length of the edge, and the surface tension σ can therefore be defined as the force per unit length. Because by moving the edge in the direction of the force we will increase the area of the surface and also do work on the system against the surface tension, we can also treat the force per unit length as a *surface energy:* the work done in creating unit area of new surface. The two are equivalent, as are the SI units of newton per meter and joule per square meter. Most liquids of practical interest for inkjet deposition have surface tensions, γ, of the order of tens of millinewton per meter (or millijoule per square meter). For pure water at 20 °C, $\gamma = 72.5\ \text{mN m}^{-1}$, while for many organic liquids (which have smaller intermolecular energies than water) σ lies in the range from 20 to 40 mN m^{-1}. Liquid metals, on the other hand, have much higher values, typically several hundred mN m^{-1} for those with low melting points.

The tendency of a liquid to form the shape with the lowest total energy, which in the case of a free drop causes it to become a sphere, is crucial to the processes of inkjet printing. A continuous stream of liquid may initially be cylindrical, but that shape is unstable: under the action of surface tension, disturbances in the shape of the cylinder will grow and the jet will eventually break up into a series of spherical drops. This is the classical Rayleigh–Plateau instability: Rayleigh showed in 1879 that for a liquid cylinder of radius r, the wavelength λ of the disturbance that grows most rapidly is given by $\lambda \approx 9r$ [4].

The forces that resist the contraction of a liquid jet through the action of surface tension have two origins: the inertia of the liquid, and its viscosity. Inertial forces are those associated with a change in a body's momentum: in the case of a moving liquid, they are proportional to its density and the rate of change of velocity. Viscous forces arise from the interactions between molecules of the liquid and act between regions of liquid moving relative to each other. A simple example, which forms the basis for the definition of viscosity, is given by a shear flow as shown in Figure 11.1a. A region of liquid is defined by two parallel surfaces a distance d apart. The lower surface is stationary, and the upper surface moves relative to it at

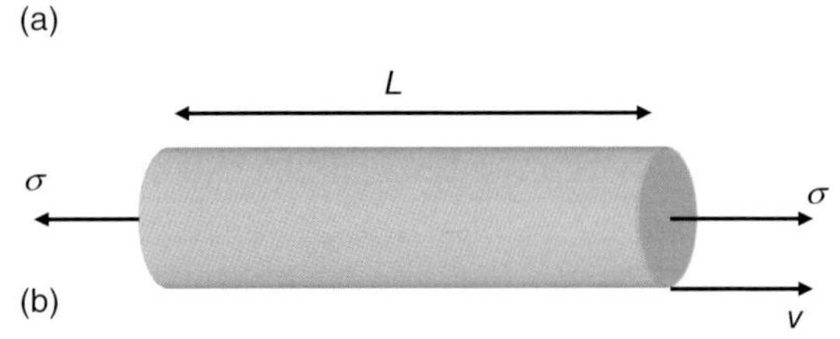

Figure 11.1 Geometry used to define (a) shear flow and (b) extensional flow.

a constant velocity v. We assume that this generates a linear gradient of velocity through the liquid, normal to the upper and lower surfaces. The shear strain rate ($\dot{\gamma}$) is given by

$$\dot{\gamma} = \frac{v}{d}$$

and for many liquids the shear stress τ acting on the upper and lower surfaces is proportional to shear strain (γ):

$$\tau = \eta\gamma$$

The constant η is the *dynamic viscosity* of the liquid, and if η is independent of $\dot{\gamma}$, then this behavior is termed *linear* or *Newtonian*. The simple term *viscosity*, without further qualification, usually means the dynamic viscosity as defined above for a shear flow. The SI unit for dynamic viscosity is pascal second (Pa s). The older (cgs) unit of viscosity, the poise (symbol P) or more commonly centipoise (cP), is still in widespread use and the conversion is straightforward: 1 mPa s = 1 cP. Water has a viscosity of almost exactly 1 mPa s at 20 °C, while fluids that are used in inkjet printing typically have viscosities in the range from ~2 to ~50 mPa s. The viscosity of most liquids falls rapidly with increasing temperature, and this is often exploited in inkjet printing: by varying the temperature of the print head, it is possible to optimize the viscosity of the ink for drop generation.

The shear flow shown in Figure 11.1a represents a particular pattern of liquid flow, but it does not represent well the flow that occurs in the formation or collapse of a jet. We may also need to consider the extensional or elongational flow, which is shown in idealized form in Figure 11.1b. If a cylindrical column of liquid of length L is stretched along its axis at a velocity v, the uniaxial strain rate ε is given by

$$\varepsilon = \frac{v}{L}$$

and the axial stress σ is linearly proportional to the strain rate:

$$\sigma = \eta_T\varepsilon$$

The constant of proportionality η_T is called the *extensional viscosity*, and for a Newtonian liquid η_T is three times the shear viscosity η. The ratio η_T/η is called the *Trouton ratio*, and can be significantly greater than 3 for non-Newtonian liquids that are *viscoelastic*. Viscoelastic liquids exhibit a time-dependent elastic response as well as viscosity, due, for the example, to the presence of polymer molecules in solution. Organic solvents are usually Newtonian, as is water, but practical printing fluids often exhibit some degree of viscoelasticity. The behavior of non-Newtonian liquids is discussed further in the following section.

Surface tension, inertia, and viscosity play key roles in the formation and behavior of liquid jets and drops. Two dimensionless numbers can be used to characterize the relative importance of these: the Reynolds number and the Weber number. The Reynolds number (*Re*) represents the ratio between the inertial and viscous forces in a moving fluid, and is defined by

$$Re = \frac{\rho d V}{\eta}$$

where ρ is the density of the fluid, V is its velocity, η is its viscosity, and d is a characteristic length: typically, the diameter of the jet, nozzle, or drop. The Weber number (We) depends on the ratio between inertia and the surface tension:

$$We = \frac{\rho d V^2}{\gamma}$$

where γ is the surface tension.

The influence of velocity in these two dimensionless groups can be removed by forming a further group, the Ohnesorge number (Oh), defined by

$$Oh = \frac{\sqrt{We}}{Re} = \frac{\eta}{\sqrt{\gamma \rho d}}$$

The value of the Ohnesorge number, which reflects only the physical properties of the liquid and the size scale of the jet or drop but is independent of the driving conditions (which control the velocity), turns out to be closely related to the behavior of a jet emerging from a nozzle, and thus to the conditions in DOD printing. If the Ohnesorge number is too high ($Oh > {\sim}1$), then viscous forces will prevent the separation of a drop, while if it is too low ($Oh < {\sim}0.1$), the jet will form a large number of satellite droplets. Satisfactory performance of a fluid in DOD inkjet printing thus requires an appropriate combination of physical properties, which will also depend on the droplet size and velocity (through the value of the Reynolds or Weber number), as shown in Figure 11.2 [5, 6]. Some authors use the symbol Z for the inverse of the Ohnesorge number ($Z = 1/Oh$). The range over which Newtonian liquids can be printed is often quoted as $10 > Z > 1$ (e.g., Reis *et al.* [7]), although another work has suggested that $14 < Z < 4$ may be more appropriate [8].

The ranges of Ohnesorge number noted above provide some bounds to the "printability" of the liquid, but other factors must also be considered: the jet must possess enough kinetic energy to be ejected from the nozzle (leading to the solid diagonal line in Figure 11.2 corresponding to $Re = 2/Oh$), and it is also desirable to avoid splashing of the drop on impact with the substrate (which leads to the broken diagonal line for which $OhRe^{5/4} = 50$) [5].

11.3
Ink Rheology: Non-Newtonian Liquids

Many functional inks for inkjet printing contain particles suspended or dispersed in a carrier liquid. In such cases, the viscosity of the liquid phase usually plays an important role in determining the flow properties of the ink. Dispersions have wide variations in performance depending on particle size, shape, concentration, and any interactions between the particles and with the continuous phase in which they are suspended. When there is a repulsive force between particles, they tend not to settle rapidly but instead form a network structure that will stabilize the suspension if undisturbed. Shearing forces during inkjet printing can destroy this delicate structure and break down the fluid viscosity. Polymers may

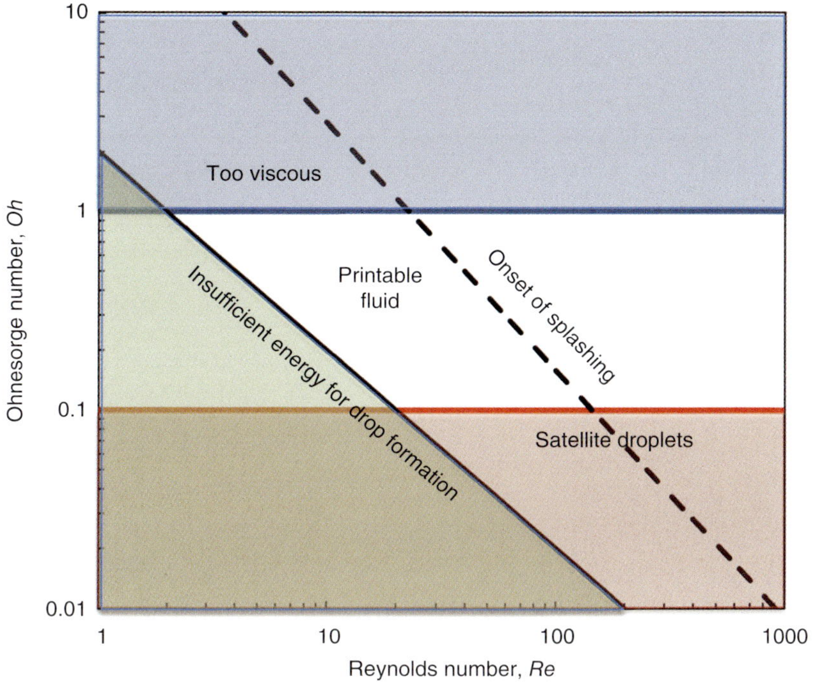

Figure 11.2 Schematic diagram showing the operating regime for stable operation of drop-on-demand inkjet printing in terms of Ohnesorge and Reynolds numbers [6]. (Reproduced with permission. Copyright © 2011, American Institute of Physics.)

be present, either to assist the dispersion of solid particles or as a key component of the fluid. Complex fluids may experience "hysteresis" or delays in regaining their initial properties, resulting in history-dependent behavior in materials testing and in jetting.

Newtonian behavior – where there is a simple linear relationship between the applied stress and the fluid flow as discussed in the previous section – is rarely found for functional inks, and *rheology* is concerned with determining how such complex fluids (Franck) actually flow, often expressing this in a mathematical form. Here we consider the rheology of non-Newtonian fluids, and then indicate where nonlinear behavior may affect functional material printing. Mathematical treatments of rheology may be found in specialist journals on rheology and (non-Newtonian) fluid mechanics, but they rarely address practical deposition issues.

Table 11.1 shows a classification of fluid types by their behavior under an increasing applied stress such as might be associated with entry into the inkjet nozzle or with drop impact on the substrate. The fluid may respond with or without a finite yield stress, and with "instantaneous" or finite time responses.

Table 11.1 Ink rheology.

Response\stress	Instantaneous		Finite time
	Linear	Non-linear	
Zero yield stress	Simple fluids Newtonian fluids	Pseudoplastic Non-Newtonian Shear thinning fluids (STF) Dilatancy (shear thickening)	— Non-Newtonian Thixotropy (STF hysteresis) Rheoplexy (reversible thickening)
Finite yield stress	Bingham fluids	—	—

There is a further subdivision of fluid types for either linear (instant) or nonlinear (instant or delayed) response.

The table shows that shear thinning fluids (STFs) are nonlinear but respond instantly without a yield stress, whereas thixotropic fluids are similar to STF with hysteresis; that is, they respond nonlinearly and without yield stress but take a finite time to change their bulk properties. A functional fluid of particular interest for electronics applications is aqueous poly(3,4-ethylenedioxythiophene):polystyrene sulfonate (PEDOT:PSS), which is an STF with rather weak elasticity and jets very well [9].

Rheology is important across the very wide range of timescales involved in inkjet printing. Times range from a few microseconds (μs) in high-frequency print heads up to milliseconds for droplet spreading on the substrate, and possibly seconds for solvent evaporation; less obviously, fluid might rest overnight in ink storage reservoirs at low or "zero" flow rates prior to daytime use in inkjet printing. Ink might contaminate nozzle exit surfaces for some minutes before remedial nozzle cleanups, or may show hysteresis effects after experiencing transient high-shear-rate conditions in the nozzle or after impact with the substrate.

Thus, the timescales relevant for ink rheology can easily span 12 decades (10^{-7} to 10^{5} s). No single rheological instrument covers this whole range of timescales, so many users still rely on measuring "zero" or low shear-rate viscosity (at times from 10^{-2} to 10^{2} s). Commercial instruments such as controlled-strain parallel plate rheometers (Rheometrics ARES (Advanced Rheometric Expansion System)) and research instruments such as filament stretching multipass rheometers (Caber MPR) can together span shear rates from 10 to 10^{5} s^{-1} (i.e., timescales from 10^{-5} to 10^{-1} s), as illustrated in Figure 11.3 for solutions of polystyrene in diethyl phthalate. Diffusive wave spectrometers useful for testing transparent particulate inks can access times from 10^{-6} to 1 s.

Steady flows of fluid through nozzles in inkjet print heads require a steady pressure difference across the nozzle length, which depends on the nozzle cross-sectional area and the fluid viscosity. Fluid flow is assumed to be stationary (the "no-slip condition") at the inner boundary wall and greatest along the central

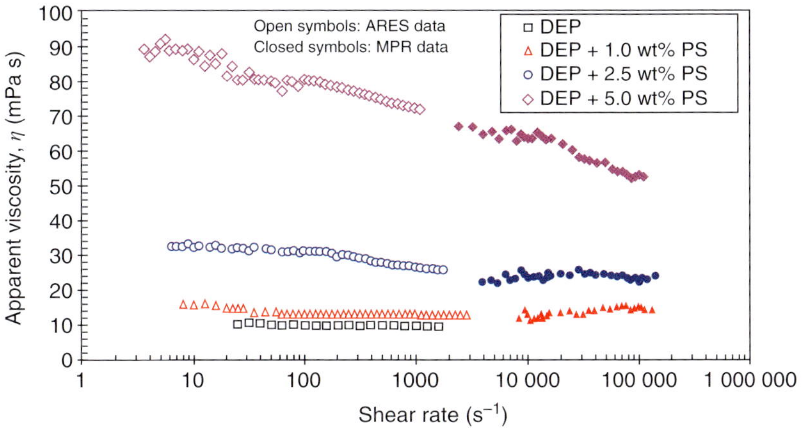

Figure 11.3 Apparent viscosity for various solutions of polystyrene (PS) in diethyl phthalate (DEP) as measured by ARES (open symbols) and MPR (closed symbols) at 25 °C for a range of shear rates [10]. (Reproduced with permission. Copyright © 2008, Elsevier.)

axis of the nozzle. As the inkjet fluids behave as if they were incompressible (apart from supporting acoustic waves), the value of the shear stress on the material is greatest at the nozzle walls.

Theoretical values for the maximum stress at the nozzle wall for a specified jetting speed suggest that there is an upper frequency limit of ~8 U/D, where the jet speed is U and nozzle diameter is D, for the generation of drops. This implies that ink rheology should always be assessed up to this frequency for the jetting performance to be predicted with any degree of confidence. Unfortunately, most practitioners still assess ink viscosity only at low shear rates, often many decades below the conditions pertinent to inkjet printing. One reason for this is that many organic solvents and simple aqueous solutions have a linear (Newtonian) rheological response that is independent of frequency or shear rate [3], so that a low-frequency viscometer is sufficient to characterize these inks. Another pragmatic reason is that the upper frequency limits are very high: at a jet speed $U = 2\,\text{m}\,\text{s}^{-1}$ and nozzle diameter $D = 50\,\mu\text{m}$, the upper frequency limit of $3 \times 10^5\,\text{s}^{-1}$ or $2 \times 10^6\,\text{rad}\,\text{s}^{-1}$ lies decades beyond the capability of many mechanical testing systems.

Materials can be usefully transferred to surfaces by inkjet printing, provided that they flow reliably through printing nozzles and deposit, dry, or bond suitably onto the substrates. Liquid inks are often formulated with this material in a suitable carrier fluid, for example, in a suspension, as a dispersion, or as a solution in one or more solvents. Other additives such as surfactants or defoamers that improve process reliability, jetting, drying, or deposition quality may be incorporated into specialist inks intended for functional applications such as large-area electronics. Even small quantities of such additives can have a significant influence on ink flow and inkjet reliability.

The viscosity and surface tension of liquids are significantly temperature-dependent, so that ink flow through nozzles can be facilitated by heating the ink reservoir and nozzle to suitable temperatures (usually ~40 – 50 °C) to reduce the dynamic viscosity. Evaporative loss (depending on vapor pressure and environment) from the exposed liquid meniscus and wetted nozzle face can significantly influence jetting reliability and limit the usable maximum print head temperature for complex inks.

Ink rheology is commonly assessed at 25 °C, so that the measured viscosity values may need some conversion factor to account for the print head operating temperature. Pragmatic practitioners merely note the range of assessed values that seem to jet well without making absolute viscosity conversions, which can complicate intercomparisons with results from other workers using different hardware and materials. An intermediate approach is to use the measured temperature dependence of the carrier fluid to correct the ink viscosity to values effective at the print head temperature.

STFs provide a means to reduce the viscosity of the fluid at the nozzle walls during actuation while rebuilding viscosity once the stress has been significantly reduced. Aqueous PEDOT:PSS has been used for many years in organic electronics applications. As seen in Figure 11.4, it happens to possess a near-Carreau-type response that shear-thins by an order of magnitude over the inkjet frequency regime ($10^5 – 10^6$), which aids in jetting it. Other additives possess far greater shear-thinning ranges: for example, aqueous Laponite gels that shear-thin by

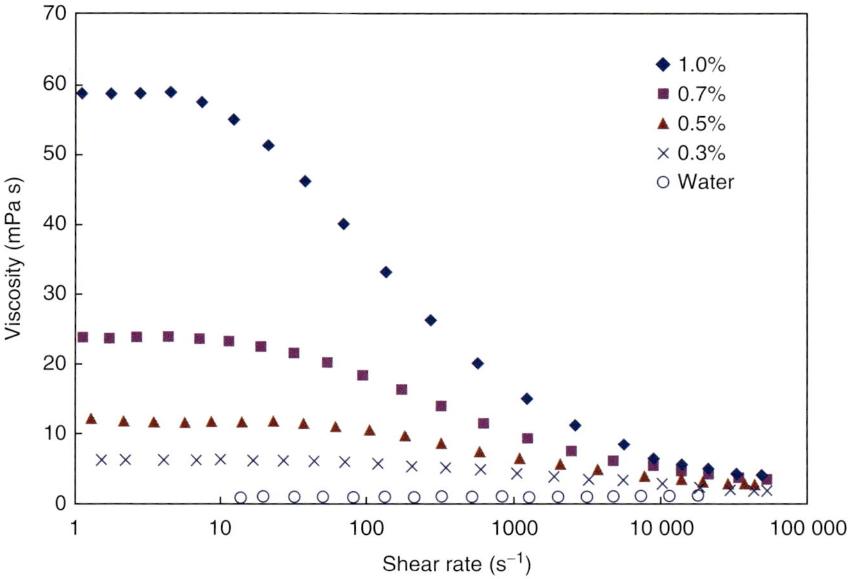

Figure 11.4 Shear viscosity versus shear rate for water and aqueous PEDOT:PSS at various weight percent loadings [9]. (Reproduced with permission. Copyright © 2012, Elsevier.)

several orders of magnitude can be readily jetted from inkjet nozzles. This is of interest for researching quality control of particulate depositions.

The structural rebuild time for shear-thinning materials may prove a key parameter, although shearing in inkjet occurs mainly at the outer surface of the jet. The observed persistence of long ligaments of PEDOT:PSS may arise from the high-viscosity core fluid rather than significant rebuild of the sheared surface fluid (although simulations imply thixotropy is below 10 μs). PEDOT:PSS drop impact would then start with a predominantly high viscosity bulk fluid which may also reduce spreading on deposition.

Shear-thickening fluids prove far harder to jet and are often dispensed by dripping. They can produce a deposition region that is insensitive to the drop impact velocity, and rheoplectic fluids might also improve drop ejection if they have ~1000 μs timescales. Figure 11.3 gives typical results from rheological studies of model polymer solutions using a parallel plate (ARES) rheometer and a filament-stretching (MPR) device [10].

11.4 Dynamics of Drop Impact and Spreading

The dynamics of a fluid drop impacting on a solid nonporous surface is a classical subject of interfacial hydrodynamics, which occurs in many industrial and environmental situations such as coating, rapid spray-cooling of hot surfaces, quenching of aluminum alloys and steels, motor jet, rain drop, pesticides, and inkjet printing [11, 12]. The impact of liquid drops on dry surfaces creates various flow patterns depending on the properties of the liquid and the surface. Liquids vary in density, viscosity, elasticity, and surface tension. The velocity and size of a droplet also have a crucial influence on the resulting behaviors. The solid surface may be rough or smooth, hydrophobic or hydrophilic, chemically homogeneous or heterogeneous, planar or nonplanar, and normal or oblique. Rioboo *et al.* [13] identified six possible consequences of a droplet falling on to a dry surface: deposition, prompt splash, corona splash, receding breakup, partial rebound, and complete rebound, as seen in Figure 11.5.

Inkjet printing for organic semiconductor electronics involves the spreading of a liquid drop on a smooth, dry, solid surface which can be described as a sequence of five successive phases: kinematic, spreading, relaxation, wetting, and equilibrium, as illustrated in Figure 11.6 [14]. Details of the drop's behavior depend on its impact speed, the liquid properties, and the surface wettability. As the drop collides with the surface, the initial shape of the region of the drop that is out of contact remains largely unchanged in the initial (kinematic) stage. This phase lasts approximately until the contact diameter reaches the original drop diameter. The spreading phase then follows, in which the contact line expands radially and nearly all the drop's initial kinetic energy is consumed. The fluid properties and the impact conditions all play a role in this phase. For example, a higher impact speed or larger drop size lead to faster spreading, whereas a higher surface tension

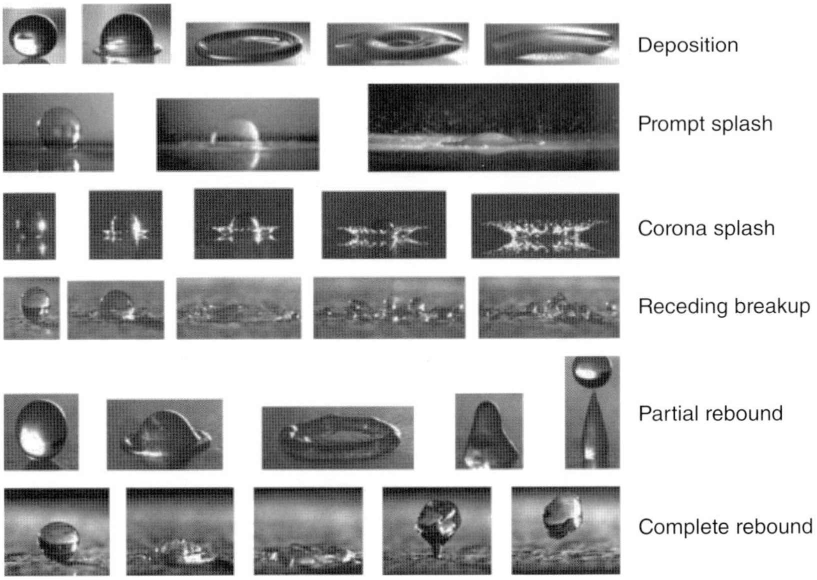

Figure 11.5 A variety of morphologies of liquid drop impact on to a dry surface [13]. (Reproduced with permission. Copyright © 2001, Begell House.)

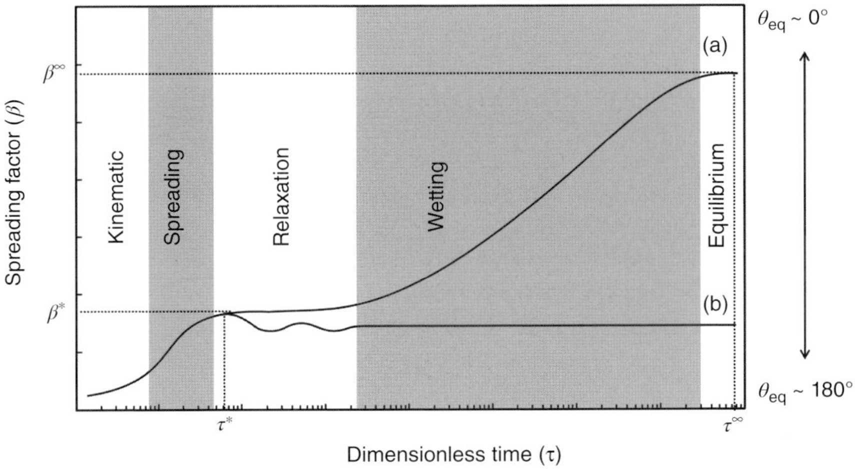

Figure 11.6 Schematic illustration of the evolution of the spreading factor (β) as a function of dimensionless time (τ) for drops impacting on three surfaces with different wettability. The maximum spreading factor (β^*) and the dimensionless time to reach $\beta^*(\tau^*)$ as well as the final spreading factor (β^∞) and dimensionless time to reach $\beta^\infty(\tau^\infty)$ for a highly wettable surface are indicated. The two curves correspond to (a) $\beta^* < \beta^\infty$ (highly wettable surface) and (b) $\beta^* > \beta^\infty$ (nonwettable surface) [15]. (Reproduced with permission. Copyright © 2012, Royal Society of Chemistry.)

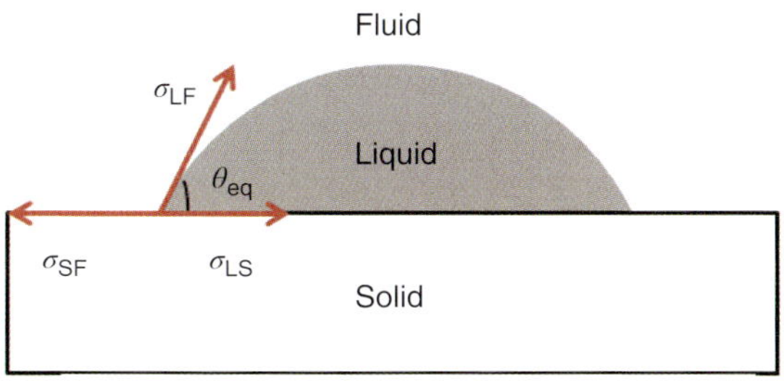

Figure 11.7 A liquid drop on a solid with an ideal contact angle (θ).

or viscosity results in slower expansion. After spreading to a maximum extent, the drop may experience relaxation or oscillation of shape, and possibly retraction of the contact line. A drop impacting on a hydrophobic surface may exhibit a receding contact line or even rebound if the initial kinetic energy is high enough; rebound is common from super-hydrophobic surfaces. On the other hand, a drop on a hydrophilic surface may oscillate or show a momentary pause in its expansion, and then undergo further spontaneous spreading due to capillary effects until its size reaches a final equilibrium state.

The extent to which a drop wets the surface is usually described by its equilibrium contact angle (θ_{eq}), which is defined as the angle between the liquid/vapor interface as it meets the solid surface (Figure 11.7). The liquid drop takes the shape that minimizes the free energy of the system, that is, minimizing the surface area of the drop in the absence of gravity. Gibbs demonstrated that minimizing the free energy requires the minimization of the sum (ψ) of three energies contributed by the three interfaces:

$$\psi = \sigma_{LF}A_{LF} + \sigma_{SL}A_{SL} + \sigma_{SF}A_{SF} \tag{11.1}$$

where σ is surface tension, A is area, and the subscripts LF, SL, and SF refer to liquid–fluid, solid–liquid, and solid–fluid interfaces, respectively. For a plane homogeneous surface, the minimization yields

$$\cos\theta = \frac{\sigma_{SF} - \sigma_{SL}}{\sigma_{LF}} \tag{11.2}$$

This is known as *Young's equation*. The wetting behavior of a liquid drop according to the contact angle is illustrated in Figure 11.8.

In an ideal situation of a liquid spreading on a uniform plane solid, there is only one equilibrium contact angle (θ_{eq}). But, in practice a number of stable angles can be measured [16]. Two relatively reproducible angles are the largest, called the *advancing contact angle* (θ_a), and the smallest called the *receding contact angle* (θ_r). The advancing angle can be measured by pushing the periphery of a drop over a surface, and the receding angle can be measured by pulling it back. The difference between the two angles ($\theta_a - \theta_r$) is the contact angle hysteresis.

A substantial number of experimental, numerical, and theoretical studies have been conducted to identify the important parameters influencing the wetting process and the final outcome of drop impact, for practical applications such as

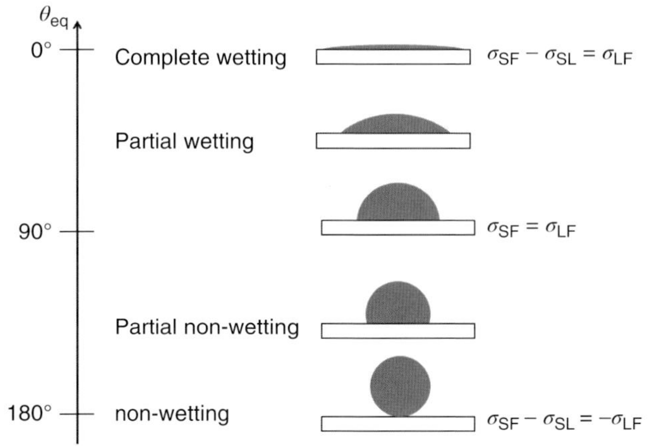

Figure 11.8 Wetting behavior of a liquid drop according to the contact angle (θ).

coating, painting, rapid spray cooling of hot surfaces, and the splat-quenching of metallic alloys [17]. Many of these applications require a more comprehensive understanding of the wetting process (e.g., to predict how quickly a deposited drop will wet a given area of the substrate). It is commonly reported that in the wetting stage the contact diameter D increases slowly with time t according to a power law ("Tanner's law"):

$$D = kt^n \tag{11.3}$$

where the coefficient k depends on the competition between surface tension and viscosity [18]. The value of the index n has been shown to be ~0.1 both on theoretical grounds and from experiments conducted with millimeter-sized drops [19].

With further development of high-speed imaging techniques, researchers have also been able to study the early stages of drop impact. These studies have attempted to quantify the influence of the impact parameters and liquid properties, and to develop predictive models [20–23]. While most of the earlier experimental studies of drop impact were undertaken with millimeter-sized drops, the understanding of the impact of much smaller drops (typically <100 μm in diameter) on to a dry solid surface has grown in importance with the development of inkjet printing technology. Some studies of the spreading of small drops, with diameters in the range ~20–100 μm relevant to inkjet printing, have been published, as summarized in Table 11.2. Van Dam and Clerc studied the spreading of water drops ejected from an inkjet print head. Interestingly, they observed and modeled the entrapment of a small air bubble inside the drop. Although they reported that they produced drops with diameters down to 36 μm, their paper only includes images of drops with a diameter >60 μm [24]. Dong and his colleagues [25] examined the impact of inkjet-printed drops (40–50 μm in diameter) on various smooth solid substrates with different static contact angles (6° – 107°) at a wide range of speeds (2–12 m s^{-1}) and studied the effects of impact speed and equilibrium contact angle θ_{eq} on the time evolution of spreading

Table 11.2 Comparison between the conditions used in experimental studies of inkjet-printed drop deposition.

Author	Fluid	D_0 (μm)	V_0 (m s^{-1})	$Z = 1/Oh$	θ_{eq}	Maximum record time (ms)	Minimum time resolution (μs)
van Dam	Water	36–85	0.74–13.8	52.6–77	15–70	0.25	0.25
Dong	Water	40.9–50.5	2.2–12.2	60.6–67.6	6–107	10	1
Son	Water	46	<2	58.8	10–110	3.3	8.25
Perelaer	PS in toluene	80	1	23.5–66.8	<13	N/A	N/A
Hsiao	PCR, UV	50	5	1–10	8–25	130	1
Jung	DEP	25–28	3–8	2.7	4–78	10 000	0.1

diameter at the early stages (up to ~250 μs). Son and his colleagues explored the behavior of drop-spreading in the regime of low Weber ($We = \rho D_0 V_0^2/\sigma$) and Reynolds number ($Re = \rho D_0 V_0/\eta$) ($0.05 < We < 2$ and $10 < Re < 100$). Slowly falling water drops were deposited on to glass substrates which were treated by UV-ozone plasma in order to vary their wettability. They presented images of the deposited drops from the impact phase to the equilibrium state, reporting that all the drops reached their equilibrium state within 3.3 ms regardless of the equilibrium contact angle [26]. Hsiao and his colleagues investigated the spreading and wetting dynamics of a printed drop, particularly in the context of printed circuit board masking. Drops of two commercial inks, a UV-curable ink and a phase-change resist (with melting temperature of 70 °C), were printed from an inkjet print head and imaged over a wide range of timescales up to 200 ms. In their work, both flash photography with a 20-ns spark light source and continuous framing with a 27 000-fps high-speed camera were used to explore early stage impact-driven spreading and the later stage of wetting [27].

More recently, Jung and Hutchings (2012) explored the full range of drop motion from the initial kinematic phase ($t < 5$ μs) to the long-time equilibrium state ($t > 5$ s). Short-duration flash photography was used to observe the various stages, with images captured over a range of timescales from every 100 ns during the rapidly developing early stages (equivalent to 10 million fps) to every second for the final stage. Their experimental conditions were chosen to be representative of DOD ink jet printing, and the work constitutes the first study of small drops (<30 μm diameter) at these values of initial Reynolds and Weber numbers, with high temporal and spatial resolution, which extended over eight orders of magnitude in timescale.

Although most works have focused on the drop impact process of Newtonian fluids so far, the impact and spreading of non-Newtonian fluids on solid surfaces has also played a key role in many fields of activity such as cooling, painting, food product, tablet coating, and pharmaceutical sprays. Understanding the spreading dynamics of a sub-100 μm low-viscoelastic drop has grown in significance because inkjet printing is used as a patterning tool for organic semiconductor devices. Unlike that of Newtonian fluids, the impact dynamics of a viscoelastic fluid may be effectively controlled by its rheological properties as well as viscosity or surface tension. But the spreading behavior of a non-Newtonian fluid drop is much less understood than that of a Newtonian drop. Particularly, the mechanism of an inkjet-printed non-Newtonian drop impact on a wettable surface remains unknown to date.

Bergeron *et al.* first studied the control of drop deposition with polymer additives. They discovered that drop rebound on a hydrophobic surface is markedly inhibited by adding very small amounts of a poly(ethylene)oxide (PEO) polymer to water [28]. They argued that the non-Newtonian elongational viscosity provided a large resistance to drop retraction after impact, thereby suppressing drop rebound. But, the underlying physical mechanism at work still remains controversial. Their explanation on the cause of the anti-rebound effect cannot explain why it occurs only for the retraction phase and does not affect the spreading phase at all, because the elongational viscosity is dependent on the total deformation of the drop and thus should act in the same way for both the spreading and the retraction phase.

Bartolo *et al.* investigated the dynamics of non-Newtonian drops on hydrophobic surfaces for polymeric solutions exhibiting strong normal stresses. They found that the normal stresses resulting from the polymeric constitution of the drop fundamentally altered the behavior of the contact line and could considerably slow drop retraction. No noticeable effects were observed during the expanding phase [29]. More recently, Smith and Bertola [30] reported that bulk properties of the polymeric fluid played no role in the anti-rebound phenomenon, but the stretching of polymer chains produced an effective resistance to the motion of the contact line. Starov *et al.* theoretically investigated the wetting of non-Newtonian fluids on solid substrates with the following assumptions: (i) complete wetting, (ii) a low dynamic contact angle approximation, (iii) a low non-Newtonian Reynolds number ($Re_n \ll 1$), and (iv) the dependence of the rheological properties of the fluid on shear deformation rate. They proposed an explicit expression for the drop contour for both regimes of gravitational spreading and capillary spreading, and showed that a shear-thinning drop spreads more slowly than an identical volume drop of corresponding Newtonian viscosity (i.e., the power-law index of the Tanner's law is <0.1) [31].

All the studies mentioned above used 1–2 mm drops impacting at $<2\,\mathrm{m\,s^{-1}}$, and a few studies attempted to look at a sub-100 μm drop whose Newtonian and non-Newtonian fluid properties and impact conditions lie within a practical inkjet printing regime. Son and Kim [32] experimentally studied the impact dynamics of non-Newtonian inkjet-printed drops on solid substrates for the first time. They used PEO solutions in a mixture of water and ethylene glycol (50 : 50% by weight) with $Oh \ll 1$ and focused on the polymer effect on the impact dynamics for a relaxation phase over 20–140 μs after impact using a high-speed camera at the frame rate of 100 000 fps. They found that the distinct difference between the Newtonian solvent and non-Newtonian solutions lay in their oscillation motion during the relaxation phase, with the viscoelastic solutions taking longer time in oscillation. Perelaer *et al.* [33] reported that the diameter of a dried liquid film of a polystyrene solution in toluene decreased with increasing molecular weight of the polymer as a result of the increase in low-shear viscosity.

Recently, the entire deposition process of a sub-30-μm polymer-containing drop on wettable surfaces was investigated for the first time over more than eight orders of magnitude in timescale under conditions fully representative of inkjet printing [15]. They discovered that the drop deposition dynamics of a polystyrene solution on a highly or partially wettable surface is independent of the high-shear and extensional rheology of the fluid, while the final drop size is significantly affected by surface wettability. Although in the spreading phase the extensional rate was strong enough to stretch the polymer chains (maximum Weissenberg number >5), elastic stresses were still negligible for two reasons: (i) the extensional flow was momentarily very strong in the beginning of the spreading phase, but the high rate persisted for a much shorter time than the Zimm relaxation time of the polymer, and (ii) an elasticity number <1 suggested that inertia force dominated the phase and elasticity stresses could not overwhelm it. For the relaxation and wetting phase, because the lamella experienced very slow spreading, non-Newtonian fluids showed the same behavior as Newtonian fluids. It can therefore be concluded that, even for viscoelastic inks, the deposition process is mainly controlled by their Newtonian properties and the surface condition, and is independent of the high-shear rate rheology of the fluid.

11.5 Applications of Inkjet Printing for Large-Area Electronics

11.5.1 Light-Emitting Diodes

Since the startling discovery of organic electroluminescence from polymers in the Cavendish Laboratory, P-OLEDs (polymer organic light emitting diodes) have received significant attention as the basis for the most promising next-generation flat-panel displays. Compared to conventional displays such as LEDs (light-emitting diode)s or liquid crystal displays (LCD)s, the self-luminous display does not require a backlight so that it can be thinner and lighter, consumes less energy, and can offer higher brightness and contrast. Inkjet printing technology has been demonstrated to be well suited to the deposition of light-emitting polymer (LEP) solutions.

OLED substrates are typically patterned with an acrylate or polyimide photoresist in the form of banks to define pixel wells into which the organic material is to be printed. Prior to printing, the substrate is preconditioned in order to maximize wetting within the well and give a nonwetting (low surface energy) surface to the photoresist bank in order to confine the drying inks to within the pixel. In an ideal situation, the ink completely fills the pixel well, such that the ink pins

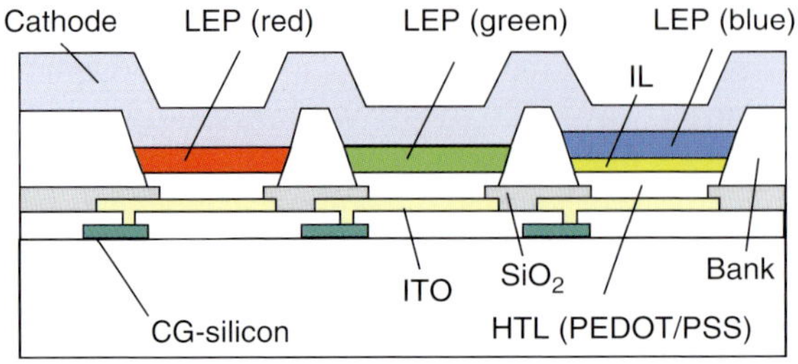

Figure 11.9 The cross sectional view of the device structure of a full-color AMOLED display by using inkjet technology [34]. (Reproduced with permission. Copyright © 2006, The Society for Information Display.)

at a point depending only on the volume of ink, rather than other factors, most notably the precise position of ink impact. Figure 11.9 illustrates a 3.6-in., 200-ppi, full-color active-matrix organic light emitting diode (AMOLED) display made by using inkjet technology [34]. Three organic layers are placed between the anode and the cathode; the hole transport layer, the interlayer, and the emission layer. All the layers are fabricated on a silicon active-matrix backplane by inkjet printing technology.

In 1998, Hebner and colleagues first used inkjet printing to directly deposit patterned luminescent-doped polymer (polyvinylcarbazole (PVK)) films, with coumarin 6 (C6) and coumarin 47 (C47) dyes, and fabricated P-OLEDs from inkjet-deposited doped polymers [35]. In the same year, Bharathan and Yang [36] also demonstrated a P-OLED device made with a conventional Epson printer by printing an aqueous solution of semiconducting PEDOT [36]. They also successfully fabricated dual-color light-emitting pixels by printing a 2 wt% aqueous solution of orange-emitting poly[5-methoxy-(2-propanoxy-sulfonide)-1,4-phenylene vinylene] (MPS-PPV) on a spin-coated film of blue-emitting poly[2,5-bis[2-(*N*,*N*,*N*-triethyl-ammonium)ethoxy]1,4-phenylene] (PPP-Net3+) [37].

A significant step to fabricate an active-matrix red–green–blue (RGB) multicolor panel was made by Cambridge University's Cavendish Laboratory in collaboration with Seiko Epson Corporation. Kobayashi *et al.* developed a system for depositing the LEP poly(*para*-phenylene vinylene) (PPV) for a green or red emitter with an Epson inkjet print head followed by spin-coating of poly(di-octyl fluorene) (F8) to form an electron-transferring layer or a blue emitter [38]. Using this system, they successfully patterned electroluminescent (EL) layers on a TFT (thin-film transistor) substrate and displayed an RGB multicolor image. In 2002, Dijksman and colleagues reported on the inkjet fabrication of a true full-color 80-ppi active and passive matrix display [39]. Much effort had been made in increasing the pixel resolution and improving the uniformity, longevity, and manufacturability. In 2007, Cambridge Display Technology (CDT), a company originally spun out of Cambridge University, announced an important step in the development of P-OLED display technology with the production of a 20.8-in. full-color television by using an inkjet printing process for the deposition of each color.

More recently, Singh and his colleagues demonstrated bright inkjet-printed OLEDs based on Ir-based phosphorescent macromolecules anchored on a polyhedral oligomeric silsesquioxane (POSS) molecular scaffolding used as a phosphorescent dye in a polymer inkjet containing a hole-transporting polymer, poly(9-vinylcarbazole), and an electron-transporting polymer, 2-4-biphenylyl-5-4-tertbutyl-phenyl-1,3,4-oxadiazole (PBD) [40]. A peak luminance of more than 6000 $cd\,m^{-2}$, a low turn-on voltage (6.8 V for 5 $cd\,m^{-2}$), and a relatively high quantum efficiency of 1.4% were achieved in their research. Through improvement in dye chemistry and print morphology, the authors were able to achieve a peak luminance of 10 000 $cd\,m^{-2}$. Bulovic's group demonstrated a simple and scalable printing method to achieve patterned pixels for flexible, full-color, large-area, AC-driven displays operating at video brightness [41]. They showed that a quantum dot–polymer composite could be inkjet-printed with stable ink solutions, and that it contributed to efficient and robust device architecture. They also reported that the inkjet printing technique was well suited for integration with metal oxide dielectric layers, which would enable improved optical and electrical performance.

11.5.2
Thin-Film Transistors

Organic thin-film transistors (OTFTs) have received much attention because their fabrication processes are far less complex than conventional silicon or other inorganic semiconductor technology. Field-effect transistors based on solution-processable organic semiconductors have experienced impressive improvements in both performance and reliability in recent years, particularly in the last decade. Printing-based manufacturing processes for integrated transistor circuits are being developed to realize low-cost, large-area electronic products on flexible substrates. OTFTs have already been demonstrated in applications such as electronic paper, sensors, and memory devices including radio frequency identification (RFID) tags [42].

One of the major obstacles to the development of organic transistors is the achievable minimum feature size by the inkjet printing process. The switching speed of a circuit relies on the mobility and channel length of the transistor. Commercial DOD inkjet printers can produce drops with a volume of some picoliters, which correspond to a drop diameter of typically 20–50 μm and a printing resolution of typically $\geq\pm5$ μm. The relatively poor resolution of the inkjet printing process limits the channel lengths achievable in printed OTFTs to 10–100 μm, resulting in a very low and insufficient switching speed of the transistor of 1–100 Hz [43].

Although inkjet printing had emerged as an attractive patterning technique for conjugated polymers in OLED displays in the late 1990s, it had not been applied to organic transistors until Sirringhaus [44] overcame this fundamental limitation in resolution in 2000. They demonstrated direct inkjet printing of complete transistor circuits, including via-hole interconnections based on solution-processed polymer conductors, insulators, and self-organizing semiconductors. In their work, a nonwettable polyimide pattern was defined by means of standard photolithography on a wettable glass substrate. After the pretreatment, they were able to deposit PEDOT:PSS for the source/drain and gate electrode by using a home-built piezoelectric inkjet printer and achieved high-resolution definition with a practical channel length of 5 μm.

Sekitani and his colleagues [45] demonstrated the feasibility of employing inkjet technology with sub-femtoliter drop volume and single-micrometer resolution for electronic device applications. They fabricated p-channel and n-channel organic TFTs with source/drain contacts prepared by sub-femtoliter inkjet printing of silver nanoparticles deposited directly on to the surface of the organic semiconductor layers, without the need for any photolithographic prepatterning or any surface treatment.

The Sirringhaus group further developed the additive printing technique for the fabrication of OTFTs. In 2005, they announced a novel bottom-up, self-aligned inkjet printing (SAP) process which was capable of defining sub-100-nm critical features with two simple additive printing steps using standard inkjet printing equipment without the need for any lithography or precise relative alignment [46]. The SAP technique comprises the following three steps (Figure 11.10): (i) patterning the first conductive polymer (PEDOT:PSS), (ii) modifying the surface of the first pattern to be of low surface energy without modifying the surface of the substrate, and (iii) printing a second conductive pattern such that it partially overlaps with the first conductive pattern. The droplets of the second pattern are repelled by and flow off the low-energy surface of the first pattern and dry with their contact line in close proximity to the edge of the first pattern, forming a small self-aligned gap. The typical channel gap attained by this method was shorter than 100 nm.

The low conductivity of conductive polymer materials such as PEDOT:PSS is another factor that limits the current flow in short-channel organic transistors. The Cambridge group extended the SAP printing method to the fabrication of functional conductive nanostructures with gold nanoparticle ink [47]. They printed the ink between two lithographically defined patterns to facilitate the study of the channel formation, which resulted in channel lengths from 4 μm to 60 nm achieved by controlling the surface tension and drying time of the ink.

This SAP process proved to be a capable tool for nanopatterning for OTFTs to overcome the limited resolution achievable by conventional inkjet printers. However, there were further needs for faster switching field-effect transistors (FETs) that required downscaling which the silicon industry has adopted for the last 40 years. In order to fully control the channel through the gate contact, it was necessary that the dielectric thickness was correspondingly reduced to below 1, while the greater channel length kept the gate leakage current low and limiting the overlap between the gate and the source and the drain contacts – by picoliter printing. In addition, a relatively high mobility is also required to obtain sufficient drain current.

Noh and coworkers [48] developed a printing process for sub-micrometer polymer TFTs which allowed basic scaling requirements to be met for downscaling and

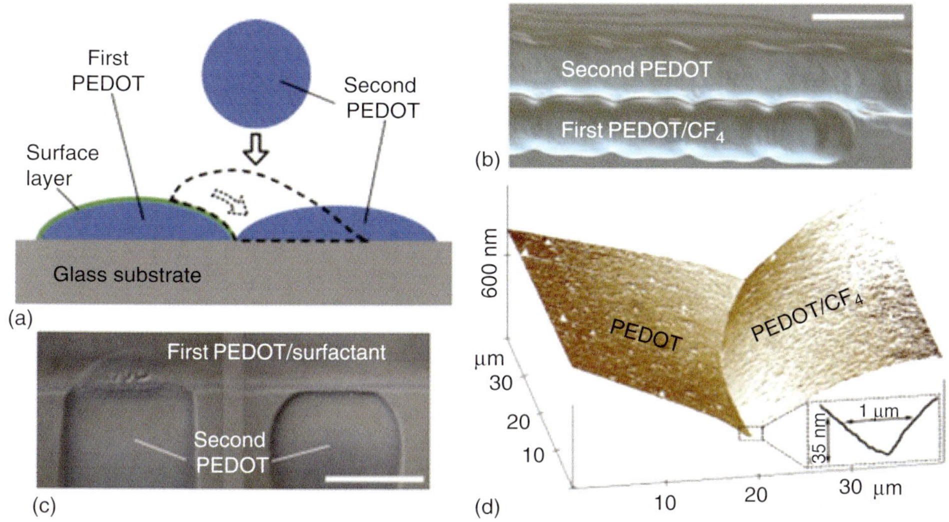

Figure 11.10 Self-aligned printing (SAP) process. (a) Schematic diagram of the self-aligned printing (SAP) process. The dashed line indicates initial position of printed PEDOT/PSS before dewetting from the hydrophobic surface layer. (b) Optical micrograph of the short channel between a CF_4-plasma-modified and a pure PEDOT/PSS line, formed by SAP on a glass substrate. (c) Optical image illustrating self-aligned PEDOT/PSS printed contacts: (left) incomplete dewetting and pinning without post-printing anneal of first pattern and (right) complete dewetting after post-printing anneal. (d) Atomic force microscopy (AFM) image of the short channel between two PEDOT/PSS lines as shown in (b); inset: smaller scale AFM scan of channel area. Scale bars in (b) and (c) represent 100 μm [46]. (Reproduced with permission. Copyright © 2005, John Wiley & Sons.)

at the same time remained compatible with the manufacturing requirements for large-area, flexible electronics. The authors used top-gate TFT architecture with source/drain electrodes fabricated by the SAP process. Sub-micrometer channel length was then defined by inkjet printing gold nanoparticles onto the surface of a previously printed first electrode pattern and causing the ink to flow off the surface of the first electrode pattern by modifying its surface with a hydrophobic self-assembled monolayer (SAM) such as 1*H*,1*H*,2*H*,2*H*-perfluoro-decanethiol (PFDT). After ink flow-off and drying, a small gap is formed between the two conducting electrodes, which is controlled in a range 60–400 nm by varying the process conditions.

The reduction of the channel length results in extraction of more current from a transistor. However, the maximum device operation frequency cannot be achieved unless the parasitic gate capacitance is minimized. A gate dielectric layer needs to be made sufficiently thin to meet mask scaling requirements. A thicker gate dielectric causes severe degradation of device performance owing to short-channel effects. To overcome this problem, Noh *et al.* [48] proposed the self-aligned gate printing (SAG) technique. The fabrication process of the

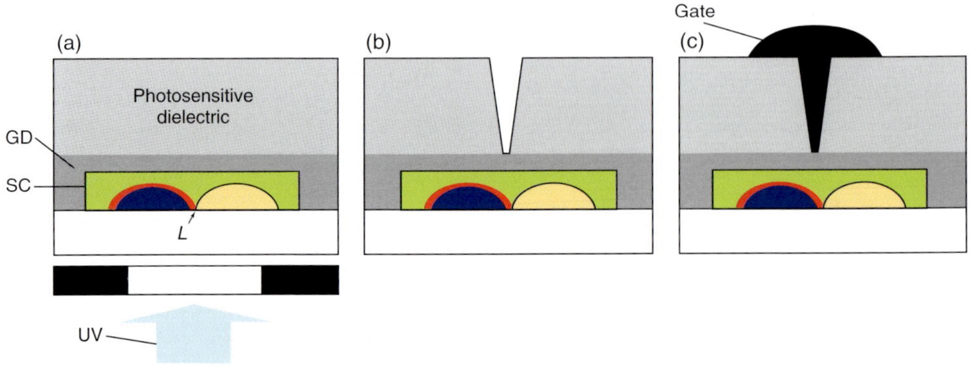

Figure 11.11 Schematic diagram of the self-aligned gate (SAG) process. (a) Deposition of a photosensitive second dielectric on top of the semiconducting (SC) and gate dielectric (GD) layer and UV irradiation through the back of substrate. (b) Development of the second dielectric to remove the exposed regions. (c) Inkjet printing of gate electrode [48]. (Reproduced with permission. Copyright © 2007, Nature Publishing Group.)

SAG structure follows four steps as shown in Figure 11.11: (i) depositing a thick (1–2 μm) UV-photosensitive dielectric layer on top of the gate dielectric gate, (ii) illuminating the substrate with light through the back, selectively exposing the channel region, (iii) developing the photoresist to obtain a trench structure self-aligned to the edges of the source and drain contacts, and (iv) inkjet printing a wide gate electrode with no need for precise alignment. The SAG configuration minimized parasitic overlap capacitance to values as low as 0.2–0.6 pF mm^{-1}, and allowed transition frequencies of $f_T = 1.6$ MHz to be reached.

Recently, a new T-shape configuration of electrodes was reported, based on single drop contacts, which achieved a high device yield of 94–100% on arrays with very low leakage current [48]. The authors also demonstrated that an inkjet-printable silver-complex-based ink, which can be sintered at low temperatures of 130 °C to achieve near-bulk silver conductivity, is fully compatible with integration with organic semiconductors dielectrics. This made it possible to fabricate organic semiconductors with a full solution process, without using any mask during the fabrication. Inkjet printing technique has also been developed to form single-crystal semiconductor films. Minemawari and his colleagues succeeded to inkjet-print single crystals of 2,7-dioctyl[1]benzothieno[3,2-*b*][1] benzothiophene (C8-BTBT), yielding TFTs with average carrier mobilities as high as 16.4 $cm^2\ V^{-1}\ s^{-1}$. This printing technique constitutes a major step toward the use of high-performance single-crystal semiconductor devices for large-area and flexible electronics applications.

Most organic semiconductor devices developed so far have been unipolar (typically p-type) and they have limitations in circuit integration because of low noise immunity and relatively high power consumption. An attempt has been made to develop complementary semiconductor devices composed of both p-type and n-type FETs. Baeg *et al.* first demonstrated inkjet-printed polymeric complementary metal–oxide–semiconductor (CMOS) inverters and ring oscillators, which are key building blocks for digital and analog integrated circuits. The inverters exhibited high voltage gains (>30) and the five-stage ring oscillators showed oscillator frequencies approaching 50 kHz [49]. More information on this subject can be found in Chapter 12.

11.5.3
Solar Cells

There is an urgent need to move toward clean and sustainable energy sources. World primary energy consumption has steadily grown during the last decade with the 10-year average of 2.6%, but renewables in power generation reached only 1.9% share of global primary energy consumption in 2012 [50]. Photovoltaic (PV) devices are one of the leading candidate technologies that could make a major contribution to solve the energy problem by harnessing a small fraction of the solar energy. Although current PV technology is predominantly based on inorganic silicon materials, high costs on materials and manufacturing processes limit its wide acceptance [51]. For this reason, organic photovoltaic devices (OPVs) have received much attention as one of the promising alternatives. A major advantage of OPVs is that they can be produced on a flexible substrate from solution by scalable printing technologies at low temperature and at low cost, without involving high-vacuum/high-temperature deposition processes (Figure 11.12).

Initial organic-based printed solar cells were successfully demonstrated in 2001 [52]. In that work, screen printing technology was used to fabricate bulk heterojunction plastic solar cells which showed 4.3% power conversion efficiency when

Figure 11.12 Inkjet-printed organic solar cells on a flexible substrate.

using an aluminum electrode and 488 nm illumination. Marin and colleagues [53] demonstrated inkjet printing of various electron donor/acceptor compositions to build solar cells with bulk heterojunction structures on a photoresist-patterned glass substrate. Polymer:fullerene blends were deposited by inkjet printing for active layers of organic solar cells, achieving a power conversion efficiency of 1.4% under simulated AM 1.5 solar illumination [54].

Brabec's group [55] reported a record power conversion efficiency of 3.5% for inkjet-printed poly(3-hexylthlophene):fullerene-based solar cells. They succeeded in their attempt to properly control the nanomorphology of the polymer blend in the inkjet printing process by adjusting the chemical properties of the poly(3-hexylthlophene) polymer donor. In their paper, they discussed the correlation between the material properties and performance of the devices fabricated with the inkjet printing patterning method. This morphology control of the polymer blends still remains the core crucial issue to achieve higher performance in the fabrication of solar cells [56]. More recently, Eom and his colleagues [57] fabricated high-efficiency polymer solar cells by inkjet printing both PEDOT:PSS and P3HT:PCBM (poly(3-hexylthiophene) and 1-(3-methoxycarbonyl)-propyl-1-phenyl-(6,6)C61) layers. They found that the inclusion of additives with high boiling points significantly influenced the film morphology, the optical and electrical properties, and device performances. ITO-free organic solar cells could be fabricated with inkjet-printed current-collecting grids and high-conductivity PEDOT:PSS as a composite anode [58]. In their work, the role of busbars and effects of grid geometry and conductivity on the performance of 2 cm × 2 cm organic bulk-heterojunction solar cells were quantitatively evaluated. All inkjet-printed and all-air-processed organic solar cells with an average power conversion efficiency of 2% were demonstrated for the first time [59]. In this work, highly efficient printed solar cells (~5%) with a simple three-layer structure, whose cathodes are the only nonprinted layer made by evaporation, were reported. Apart from the inkjet printing process, other roll-to-roll (R2R) printing processes are also being used to make complete polymer solar cell modules with no vacuum steps [60].

The same soft materials for a donor and acceptor layer and an anode/cathode in organic solar cells can be used to fabricate organic photodetectors. Azzellino *et al.* [61] showed a fully inkjet-printed organic photodetector with high quantum yield. As photoactive layer, they chose P3HT:PC_{61}BM blend which has been most widely used for organic solar cell fabrication. Arias and her colleagues [62] reported a polyfluorene photosensor for high-illuminance detection, which was also fully inkjet-printed.

11.6 Summary

Inkjet printing technology encompasses the generation, control, and deposition of sub-100-μm liquid drops. As we have discussed in this chapter, new opportunities for inkjet printing are starting to be exploited commercially in the manufacture of high-value, high-precision electronic devices and systems. By maximizing the major advantages of the inkjet process (e.g., noncontact, material-conserving, flexible, additive) to achieve the high speed, large coverage, and material diversity of an additive fabrication system, we can transform the conventional inkjet method as a graphic technology into one of major manufacturing potential. This transformation will not happen without a fundamental understanding of the physical phenomena involved in the rheology of complex fluids and the generation, impact, spreading, and drying of printed drops, which require further systematic studies. Thorough research on these phenomena will provide key inputs to fabricate various smart, large-area electronic devices and to establish innovative inkjet R2R manufacturing systems. Further information on the use of inkjet technology for digital fabrication can be found elsewhere [63].

References

1. Zoltan, S. (1972) Pulsed droplet ejecting system. US Patent 3,683,212.
2. Kyser, E.L. and Sears, S.B. (1976) Method and apparatus for recording with writing fluids and drop projection therefor. US Patent 3,946,398.
3. Wijshoff, H. (2010) The dynamics of the piezo inkjet printhead operation. *Phys. Rep.*, **491**, 77–177.
4. Rayleigh, L. (1879) On the capillary phenomena of jets. *Proc. R. Soc. London*, **29** (71-97).
5. Derby, B. (2010) Inkjet printing of functional and structural materials: fluid property requirements, feature stability, and resolution. *Annu. Rev. Mater. Res.*, **40**, 395–414.
6. McKinley, G.H. and Renardy, M. (2011) Wolfgang von Ohnesorge. *Phys. Fluids*, **23**, 127101.
7. Reis, N., Ainsley, C., and Derby, B. (2005) Ink-jet delivery of particle suspensions by piezoelectric droplet ejectors. *J. Appl. Phys.*, **97**, 094903.
8. Jang, D., Kim, D., and Moon, J. (2009) Influence of fluid physical properties on ink-jet printability. *Langmuir*, **25**, 2629–2635.
9. Hoath, S.D., Jung, S., Hsiao, W.-K., and Hutchings, I.M. (2012) How PEDOT:PSS solutions produce satellite-free inkjets. *Org. Electron.*, **13**, 3259–3262.
10. Tuladhar, T.R. and Mackley, M.R. (2008) Filament stretching rheometry and break-up behaviour of low viscosity polymer solutions and inkjet fluids. *J. Non-Newtonian. Fluid Mech.*, **148**, 97–108.
11. Rein, M. (1993) Phenomena of liquid drop impact on solid and liquid surfaces. *Fluid Dyn. Res.*, **12**, 61–93.
12. Yarin, A.L. (2006) Drop impact dynamics: splashing, spreading, receding, bouncing.... *Annu. Rev. Fluid Mech.*, **38**, 159–192.
13. Rioboo, R., Marengo, M., and Tropea, C. (2002) Time evolution of liquid drop impact onto solid, dry surfaces. *Exp. Fluids*, **33**, 112–124.
14. Jung, S. and Hutchings, I.M. (2012) The impact and spreading of a small liquid drop on a non-porous substrate over an extended time scale. *Soft Matter*, **8**, 2686.
15. Jung, S., Hoath, S.D., and Hutchings, I.M. (2012) The role of viscoelasticity in drop impact and spreading for inkjet printing of polymer solution on a wettable surface. *Microfluid. Nanofluid.*, **14**, 163–169.
16. Marmur, A. (2006) Soft contact: measurement and interpretation of contact angles. *Soft Matter*, **2**, 12.
17. De Gennes, P. (1985) Wetting: statics and dynamics. *Rev. Mod. Phys.*, **57**, 827–863.
18. Tanner, L.H. (1979) The spreading of silicone oil drops on horizontal surfaces. *J. Phys. D Appl. Phys.*, **12**, 1473–1484.

19. De Coninck, J., de Ruijter, M.J., and Voué, M. (2001) Dynamics of wetting. *Curr. Opin. Colloid Interface Sci.*, **6**, 49–53.
20. Kim, H.-Y. and Chun, J.-H. (2001) The recoiling of liquid droplets upon collision with solid surfaces. *Phys. Fluids*, **13**, 643.
21. Ukiwe, C. and Kwok, D.Y. (2005) On the maximum spreading diameter of impacting droplets on well-prepared solid surfaces. *Langmuir*, **21**, 666–673.
22. Attané, P., Girard, F., and Morin, V. (2007) An energy balance approach of the dynamics of drop impact on a solid surface. *Phys. Fluids*, **19**, 012101.
23. Vadillo, D., Desie, G., and Soucemarianadin, A. (2004) Spreading behavior of single and multiple drops. Proceeding of the 21st International Congress on Theoretical and Applied Mechanics, Warsaw, Poland, pp. 21–22.
24. Van Dam, D.B. and Le Clerc, C. (2004) Experimental study of the impact of an ink-jet printed droplet on a solid substrate. *Phys. Fluids*, **16**, 3403.
25. Dong, H., Carr, W.W., Bucknall, D.G., and Morris, J.F. (2007) Temporally-resolved inkjet drop impaction on surfaces. *AIChE J.*, **53**, 2606–2617.
26. Son, Y., Kim, C., Yang, D.H., and Ahn, D.J. (2008) Spreading of an inkjet droplet on a solid surface with a controlled contact angle at low Weber and Reynolds numbers. *Langmuir*, **24**, 2900–2907.
27. Hsiao, W.-K., Hoath, S.D., Martin, G.D., and Hutchings, I.M. (2009) Ink jet printing for direct mask deposition in printed circuit board fabrication. *J. Imaging Sci. Technol.*, **53**, 050304.
28. Bergeron, V., Bonn, D., Martin, J., and Vovelle, L. (2000) Controlling droplet deposition with polymer additives. *Nature*, **405**, 772–775.
29. Bartolo, D., Boudaoud, A., Narcy, G., and Bonn, D. (2007) Dynamics of non-newtonian droplets. *Phys. Rev. Lett.*, **99**, 174502.
30. Smith, M.I. and Bertola, V. (2010) Effect of polymer additives on the wetting of impacting droplets. *Phys. Rev. Lett.*, **104**, 154502.
31. Starov, V.M., Tyatyushkin, A.N., Velarde, M.G., and Zhdanov, S.A. (2003) Spreading of non-Newtonian liquids over solid substrates. *J. Colloid Interface Sci.*, **257**, 284–290.
32. Son, Y. and Kim, C. (2009) Spreading of inkjet droplet of non-Newtonian fluid on solid surface with controlled contact angle at low Weber and Reynolds numbers. *J. Non-Newtonian. Fluid Mech.*, **162**, 78–87.
33. Perelaer, J. *et al.* (2009) The spreading of inkjet-printed droplets with varying polymer molar mass on a dry solid substrate. *Macromol. Chem. Phys.*, **210**, 495–502.
34. Gohda, T. *et al.* (2006) 58.3: A 3.6-In. 202-ppi full-color AMPLED display fabricated by ink-jet method. SID Symposium on Digest of Technical Pappers, vol. 37, p. 1767.
35. Hebner, T.R., Wu, C.C., Marcy, D., Lu, M.H., and Sturm, J.C. (1998) Ink-jet printing of doped polymers for organic light emitting devices. *Appl. Phys. Lett.*, **72**, 519.
36. Bharathan, J. and Yang, Y. (1998) Polymer electroluminescent devices processed by inkjet printing: I. Polymer light-emitting logo. *Appl. Phys. Lett.*, **72**, 2660.
37. Yang, Y., Chang, S., Bharathan, J., and Liu, J.I.E. (2000) Organic/polymeric electroluminescent devices processed by hybrid ink-jet printing. *J. Mater. Electron.*, **11**, 89–96.
38. Kobayashi, H. *et al.* (2000) A novel RGB multicolor light-emitting polymer display. *Synth. Met.*, **111-112**, 125–128.
39. Dijksman, J.F. *et al.* (2007) Precision ink jet printing of polymer light emitting displays. *J. Mater. Chem.*, **17**, 511.
40. Singh, M., Haverinen, H.M., Dhagat, P., and Jabbour, G.E. (2010) Inkjet printing-process and its applications. *Adv. Mater.*, **22**, 673–685.
41. Wood, V. *et al.* (2009) Inkjet-printed quantum dot-polymer composites for full-color AC-driven displays. *Adv. Mater.*, **21**, 2151–2155.
42. Reese, C., Roberts, M., Ling, M., and Bao, Z. (2004) Organic thin film

transistors. *Mater. Today*, **7**, 20–27, *http://www.sciencedirect.com/science/article/pii/S1369702104003980* (accessed 22 July 2014).

43. Zielke, D. *et al.* (2005) Polymer-based organic field-effect transistor using offset printed source/drain structures. *Appl. Phys. Lett.*, **87**, 123508.
44. Sirringhaus, H. (2000) High-resolution inkjet printing of all-polymer transistor circuits. *Science*, **290**, 2123–2126.
45. Sekitani, T., Noguchi, Y., Zschieschang, U., Klauk, H., and Someya, T. (2008) Organic transistors manufactured using inkjet technology with subfemtoliter accuracy. *Proc. Natl. Acad. Sci. U.S.A.*, **105**, 4976–4980.
46. Sele, C.W., von Werne, T., Friend, R.H., and Sirringhaus, H. (2005) Lithography-free, self-aligned inkjet printing with sub-hundred-nanometer resolution. *Adv. Mater.*, **17**, 997–1001.
47. Zhao, N. *et al.* (2007) Self-aligned inkjet printing of highly conducting gold electrodes with submicron resolution. *J. Appl. Phys.*, **101**, 064513.
48. Noh, Y.-Y., Zhao, N., Caironi, M., and Sirringhaus, H. (2007) Downscaling of self-aligned, all-printed polymer thin-film transistors. *Nat. Nanotechnol.*, **2**, 784–789.
49. Baeg, K.-J. *et al.* (2011) High speeds complementary integrated circuits fabricated with all-printed polymeric semiconductors. *J. Polym. Sci., Part B: Polym. Phys.*, **49**, 62–67.
50. BP (2013) BP Statistical Review of World Energy June 2013, *bp.com/statisticalreview* (accessed 11 December 2013).
51. Li, G., Zhu, R., and Yang, Y. (2012) Polymer solar cells. *Nat. Photonics*, **6**, 153–161.
52. Shaheen, S.E., Radspinner, R., Peyghambarian, N., and Jabbour, G.E. (2001) Fabrication of bulk heterojunction plastic solar cells by screen printing. *Appl. Phys. Lett.*, **79**, 2996.
53. Marin, V. *et al.* (2005) Ink-jet printing of electron donor/acceptor blends: towards bulk heterojunction solar cells. *Macromol. Rapid Commun.*, **26**, 319–324.
54. Aernouts, T., Aleksandrov, T., Girotto, C., Genoe, J., and Poortmans, J. (2008) Polymer based organic solar cells using ink-jet printed active layers. *Appl. Phys. Lett.*, **92**, 033306.
55. Hoth, C.N., Choulis, S.A., Schilinsky, P., and Brabec, C.J. (2007) High photovoltaic performance of inkjet printed polymer:fullerene blends. *Adv. Mater.*, **19**, 3973–3978.
56. Chen, H. *et al.* (2009) Polymer solar cells with enhanced open-circuit voltage and efficiency. *Nat. Photonics*, **3**, 649–653.
57. Eom, S.H. *et al.* (2010) High efficiency polymer solar cells via sequential inkjet-printing of PEDOT:PSS and P3HT:PCBM inks with additives. *Org. Electron.*, **11**, 1516–1522.
58. Galagan, Y. *et al.* (2011) ITO-free flexible organic solar cells with printed current collecting grids. *Sol. Energy Mater. Sol. Cells*, **95**, 1339–1343.
59. Jung, S., Sou, A., Banger, K., Ko, D.-H., Chow, P. C. Y., McNeill, C. R. and Sirringhaus, H. (2014) All-inkjet-printed, all-air-processed organic solar cells. *Adv. Energy Mater.*, **4**: 1400432. doi: 10.1002/aenm.201400432.
60. Krebs, F.C. (2009) Fabrication and processing of polymer solar cells: a review of printing and coating techniques. *Sol. Energy Mater. Sol. Cells*, **93**, 394–412.
61. Azzellino, G. *et al.* (2013) Fully inkjet-printed organic photodetectors with high quantum yield. *Adv. Mater.*, **25**, 6829–6833.
62. Lavery, L.L., Whiting, G.L., and Arias, A.C. (2011) All ink-jet printed polyfluorene photosensor for high illuminance detection. *Org. Electron.*, **12**, 682–685.
63. Hutchings, I.M. and Martin, G.D. (2013) *Inkjet Technology for Digital Fabrication*, John Wiley & Sons, Ltd, Chichester.

12
Inkjet-Printed Electronic Circuits Based on Organic Semiconductors

Kang-Jun Baeg and Yong-Young Noh

12.1
Printed Organic Electronics

12.1.1
Printed Electronic Devices

Printed electronics is an emerging industry that takes advantage of conventional graphic art printing (GAP) technologies to manufacture electronic and opto-electronic circuits with various form factors that are flexible, rollable, and even stretchable [1]. The use of GAP methods, such as screen, flexography, gravure, reverse-offset, bar coating, spray, slot-die, and inkjet, enables the fabrication of various electronic components and products at highly efficient use of raw materials and very high volume compared to traditional photolithography-based electronics (Figure 12.1) [2]. Through this new technology, a whole new family of products can be realized in the near future, including flexible and stretchable displays [3–5], plastic solar cells [6], medical and consumer disposables [7, 8], electronic labels, radio frequency identification (RFID) tags [9, 10], and wearable smart devices [11].

The attraction of printed electronic technology mainly comes from the possibility of preparing stacks of microstructured electrical layers in much simpler and more cost-effective ways for large-area low-cost applications compared to conventional electronics. The selection of printing methods is determined by several requirements concerning the rheological properties of printable inks or pastes and the printed layers, such as printed feature dimensions, electrical functionalities, and economic considerations (productivity and cost) (Table 12.1). Among the traditional GAP methods, gravure, offset, flexography, screen, and inkjet printing are predominantly used in the fabrication of printed electronic products. Mass-printing methods (gravure and flexography) are commonly employed as roll-to-roll methods, while inkjet and screen printing are mostly used as sheet-fed methods [2].

Large Area and Flexible Electronics, First Edition. Edited by Mario Caironi and Yong-Young Noh.

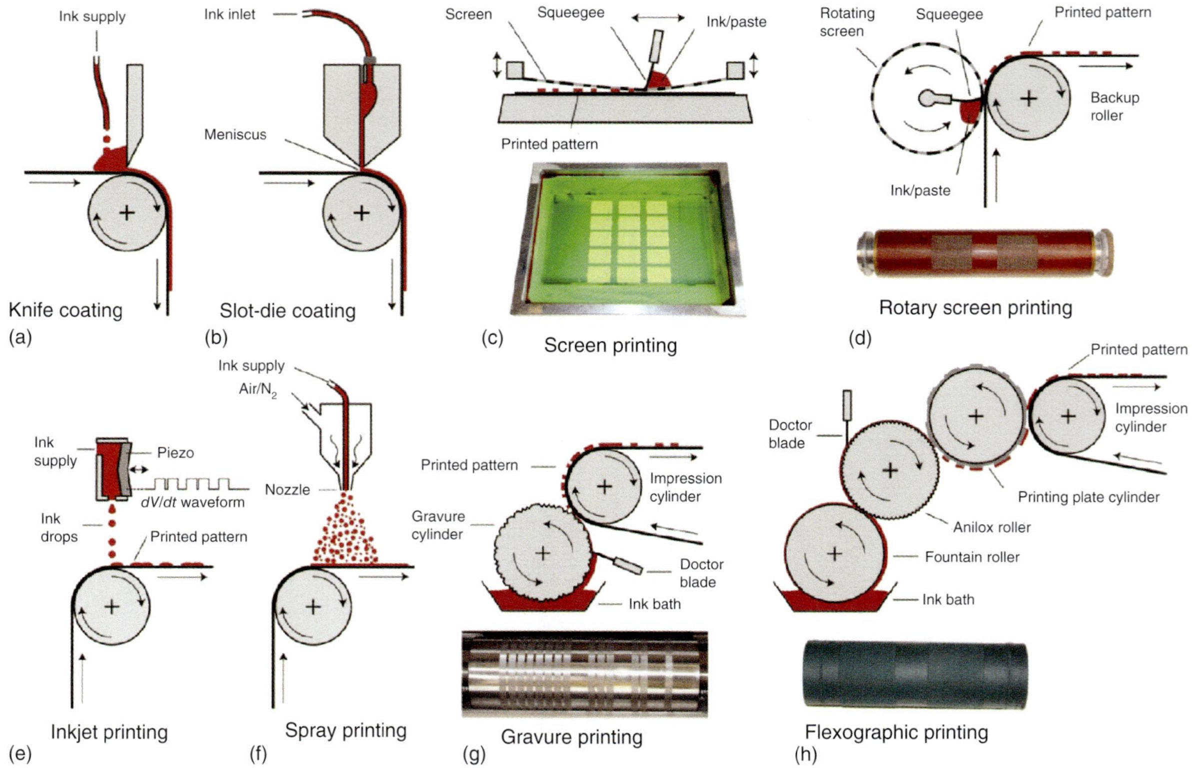

Figure 12.1 Schematic illustration of printing techniques: (a) knife coating, (b) slot-die coating, (c) screen printing, (d) rotary screen printing, (e) inkjet printing, (f) spray printing, (g) gravure printing, and (h) flexographic printing [2]. (Reproduced with permission. Copyright © 2012, John Wiley & Sons, Inc.)

Table 12.1 Characteristics of the various printing techniques.

Printing techniques	Viscosity (Pa s)	Thickness (μm)	Feature size (μm)	Throughput ($m^2 s^{-1}$)	Registration (μm)	Features
Flexography	0.05–0.5	0.04–2.5	80	3–30	<200	Inexpensive plate pattern, high throughput, thick layer/low-viscosity ink
Gravure	0.01–0.2	<0.1–8	75	3–60	>20	Fast printing, high resolution, relatively high plate cost, low dot gain
Offset	5–100	0.5–2	10–50	3–30	>10	High quality, high throughput, need for ink additives
Screen	0.5–50	0.015–100	20–100	2–3	>25	Robust, simple, thick layer, large feature size, high ink viscosity, slower speed
Inkjet	0.001–0.04	0.01–20	20–50	0.01–0.5	5–20	Noncontact, small ink quantities, digital printing, low-viscosity ink, slow speed

12.1.2 Inkjet Printing Technology

Inkjet printing is an attractive process for fabrication in many applications, such as organic and flexible electronics, printed circuit boards, nanotechnology, cell patterning for tissue engineering, and biosensors, because of its ability to precisely deposit picoliter volumes of solutions or suspensions in well-defined patterns [12–14]. The direct-write printability of the inkjet method enables cost savings by removal of the need for masks, allows efficient use of materials minimizing waste, and leads to minimum contamination

A typical piezoelectric-type drop-on-demand (DoD) inkjet print head is schematically represented in Figure 12.2. When pressure waves form as a result of a sudden volume change by the application of a voltage to the piezoelectric actuator, wave propagation begins throughout the capillary [15]. When a positive pressure wave approaches the nozzle, the fluid is pushed outward. A droplet is ejected when the amount of kinetic energy transferred outward is larger than the surface energy needed to form a droplet [15]. Notably, the velocity of the droplet depends on the transferred kinetic energy, and its initial velocity needs to be higher than several meters per second in order to overcome the decelerating action of ambient air; otherwise, droplets could deviate from their intended trajectory. Depending on the amplitude of the driving voltage, the volume and

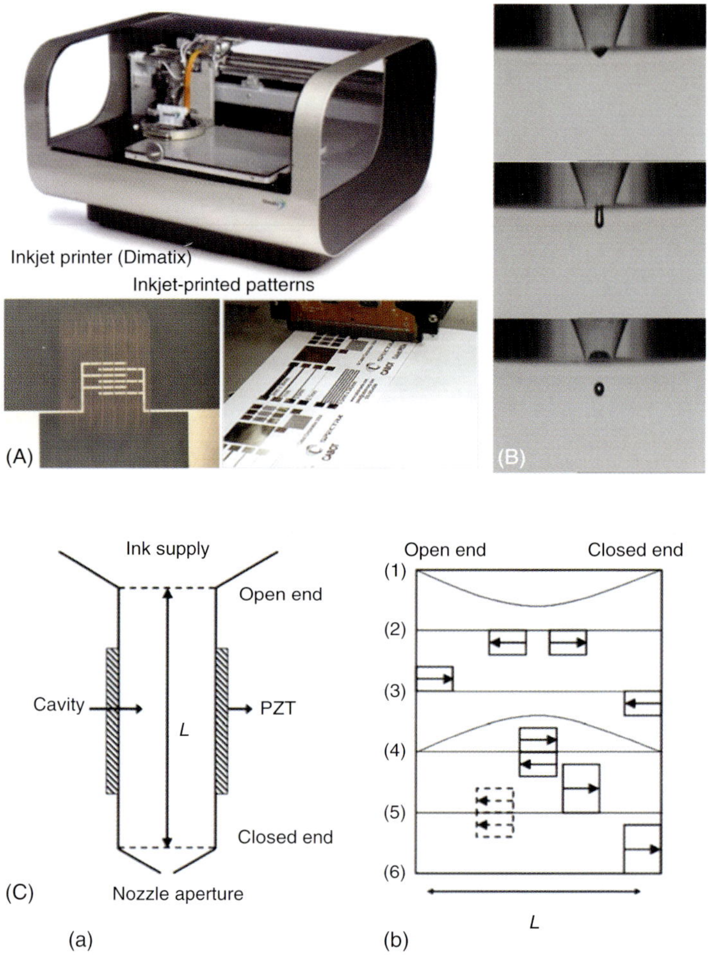

Figure 12.2 (A) Commercial inkjet printer for printed electronics devices and its printed features on flexible substrates. (B) Charge-coupled device (CCD) camera images of inkjet droplet formation at different propagation delays. (C) (a) Schematic diagram of a piezoelectric inkjet print head and (b) schematic representation of wave propagation and reflection in a piezoelectric tubular actuator [15]. (Reproduced with permission. Copyright © 2008, Royal Society of Chemistry.)

velocity of a droplet increase linearly, but they show a more complicated behavior with changing frequency and pulse width. The maximum droplet velocity is a function of the pulse width, but it does not depend on the driving voltage amplitude. Similar behavior is also observed for the droplet size [15]. At high frequencies, piezoelectric DoD print heads typically result in a chaotic droplet ejection behavior, since earlier pressure waves do not decay completely and interact with the next pressure waves that are generated [15]. The decay time of the residual waves is dependent on the ink composition.

For predicting the mechanism of inkjet droplet formation, Fromm [16] used the Z number grouping of fluid properties. This is equivalent to the inverse of the Ohnesorge number (Oh) as follows:

$$Z = \frac{1}{Oh} \tag{12.1}$$

$$Oh = \frac{\nu}{\sqrt{\rho \sigma L}} = \frac{\sqrt{We}}{Re} \sim \frac{\text{viscous forces}}{\sqrt{\text{inertia} \cdot \text{surface tension}}} \tag{12.2}$$

where ν is the liquid viscosity, ρ is the liquid density, σ is the surface tension, L is the characteristic length scale (i.e., the diameter of the nozzle aperture), Re is the Reynolds number, and We is the Weber number. Droplet formation in DoD inkjet systems was predicted to be possible for only $Z > 2$, and the droplet volume increases as the value of Z increases under a given pressure pulse condition [16].

Fromm's prediction was refined by enlarging the Z window where droplet production occurs (in the range of $1 < Z < 10$). The lower limit is determined by the ink's viscosity-dissipating the pressure pulse, while the upper limit represents the formation of satellite droplets rather than a single droplet [15]. Derby *et al.* also observed that the droplet volume increases when Z is in the range 1–14, which is consistent with the predictions of Fromm. Schubert *et al.* [17] found that a number of common solvents with low viscosities ranging from 0.4 to 2 mPa s and surface tensions of 23–73 mN m^{-1} could be successfully printed (the Z numbers for these printable solvents varied from 21 to 91). They indicated that the main factor affecting printability was solvent vapor pressure, resulting in unstable droplets, and that no droplets will be produced for solvents with vapor pressures higher than ~100 mmHg [15]. For practical applications of an inkjet printing system, the Ohnesorge numbers can be calculated from the fundamental parameters of the solvents typically used.

12.2 CMOS Technology

Complementary metal–oxide–semiconductor (CMOS) technology, which consists of complementary and symmetrical pairs of p-type and n-type transistors, is mostly used in Si-based microelectronics and integrated circuits (ICs) for logic functions [18]. The most basic and fundamental element in ICs is the inverter, which carries out the NOT logic function. As can be seen in Figure 12.3, unipolar inverters are typically constructed when the p-type or n-type semiconductors do not have equivalent transistor characteristics, such as when n-type or p-type behavior is dominant for amorphous oxide semiconductors or organic semiconductors, respectively [19, 20]. A complementary design is possible when both n-type and p-type transistors are available and show reasonably balanced electron and hole mobilities (μ) and threshold voltages (V_{Th}). "Complementary-like" logic can also be implemented with both the close patterning of two different unipolar n-type and p-type semiconductors and with

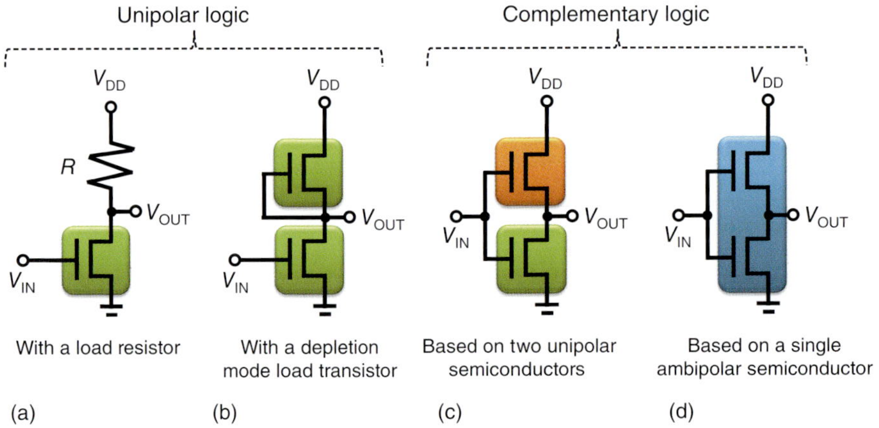

Figure 12.3 Various circuit configurations for the organic inverters: (a) and (b) are unipolar architectures, while (c) and (d) represent complementary "CMOS-like" configurations. p-Type only or n-type only semiconductor-based unipolar inverter with (a) a load resistor or (b) a depletion mode load transistor. (c) Complementary inverter consisting of two different p-type and n-type unipolar semiconductors. (d) Complementary inverter based on a single ambipolar semiconductor [21]. (Reproduced with permission. Copyright © 2013, John Wiley & Sons, Inc.)

a single ambipolar semiconductor [21]. Notably, the ambipolar semiconductors can enable the transport of holes and electrons simultaneously, but this frequently results in the deterioration of CMOS circuit characteristics in terms of small noise margin (NM) and large static power consumption. Therefore, unipolarization of the ambipolar semiconductor active channel via p-/n-doping and interface engineering of semiconductor contact and semiconductor dielectrics are used to achieve fully complementary logic circuits.

12.2.1 CMOS Inverters

Figure 12.4 shows a generic circuit diagram of a CMOS inverter and its basic operation mechanism assuming that the p-type and n-type transistors in the logic gate behave as perfect unipolar devices. For a positive supply bias ($V_{DD} > 0$ V), a positive input voltage ($V_{IN} > 0$ V; logic "1"; V_{IN} has to be large enough to turn off the p-type transistor) at the input would make the pull-up (p-type) transistor turn OFF and the pull-down (n-type) transistor turn ON. Therefore, the initial output voltage (V_{OUT}) of high V_{DD} changes to ground (GND; logic "0") (Figure 12.4e). On the other hand, at $V_{IN} = 0$ V (logic "0"), it will make the p-type transistor turn ON and the n-type transistor turn OFF, pulling the V_{OUT} close to high V_{DD}, that is, logic "1" (Figure 12.4d). These V_{IN} versus V_{OUT} voltage transfer (inverting) characteristic (VTC) curves are determined by the different operating regimes of p- and n-type transistors along the VTC curves, as highlighted by the switching points A–E in Figure 12.4b.

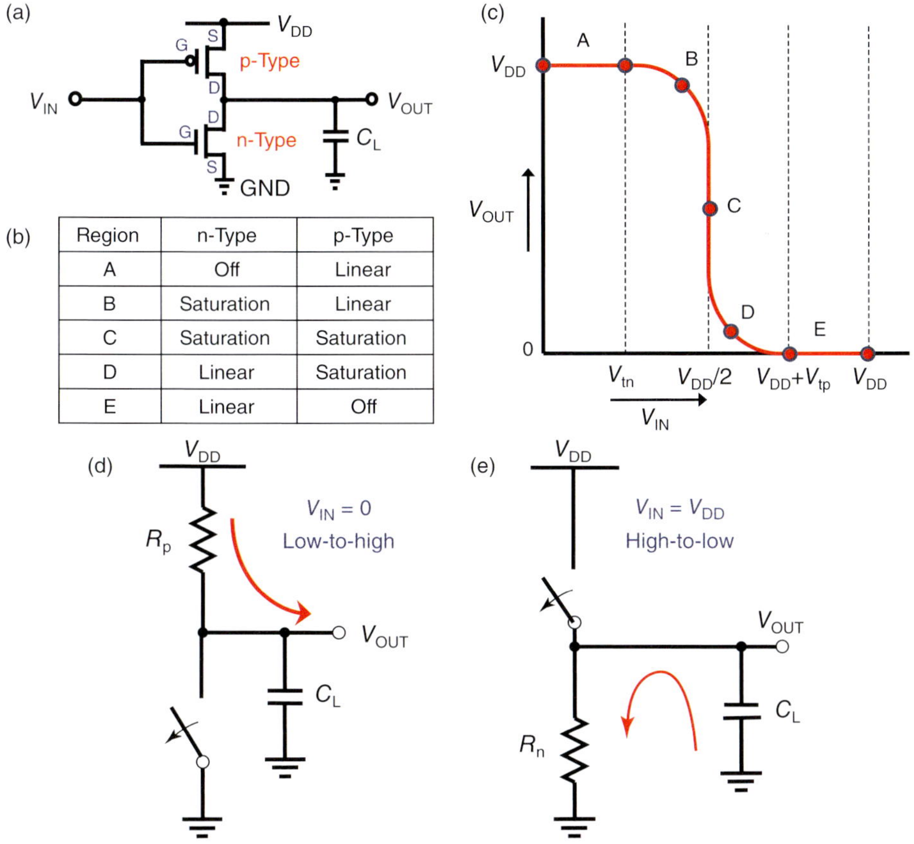

Region	n-Type	p-Type
A	Off	Linear
B	Saturation	Linear
C	Saturation	Saturation
D	Linear	Saturation
E	Linear	Off

Figure 12.4 CMOS inverter characteristics. (a) Circuit configuration, (b) characteristic operation regions, and (c) corresponding voltage transfer characteristics for a CMOS inverter. Schematic descriptions during the operation of CMOS inverter at (d) $V_{IN} = 0\,V$ and (e) $V_{IN} = V_{DD}$ [21]. (Reproduced with permission. Copyright © 2013, John Wiley & Sons, Inc.)

The switching threshold (V_m) of the generic inverter (Figure 12.5a) corresponds to the intersection point on the VTC curve when V_{IN} is equal to V_{OUT}. At this point, both the p-type and n-type transistors operate in the saturation region, where V_{IN} (or V_{OUT}) is called V_m or the inverter switching point voltage (V_{inv}). Therefore, the saturation current in each transistor must be equal at V_{inv}, which can be expressed by Eq. (12.3) [22]:

$$V_{inv} = \frac{\sqrt{\dfrac{\beta_n}{\beta_p}} \cdot V_{Th,n} + (V_{DD} - V_{Th,p})}{1 + \sqrt{\dfrac{\beta_n}{\beta_p}}} \tag{12.3}$$

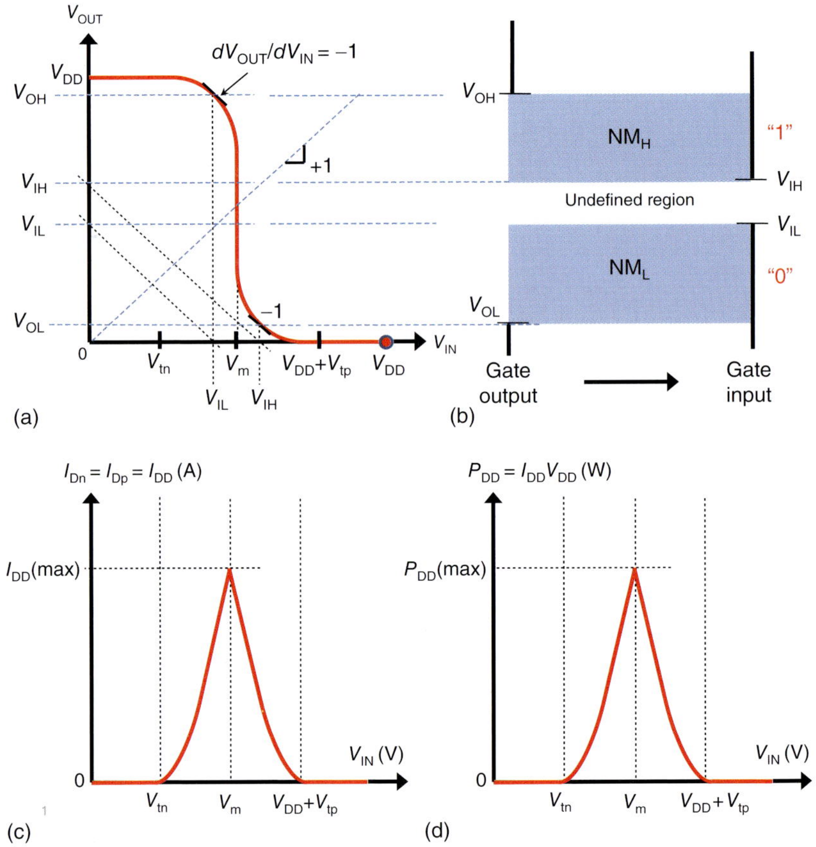

Figure 12.5 CMOS inverter characteristics. (a) VTC and (b) corresponding noise margins. (c) I_{DD} and (d) its power consumption during the operation of a CMOS inverter [21]. (Reproduced with permission. Copyright © 2013, John Wiley & Sons, Inc.)

where V_{Th} is the threshold voltage, $\beta = (W/L)\mu C_i$, is a design factor for adjusting the p-type and n-type transistor saturation currents, in which W indicates the field-effect transistor (FET) channel width, L is the channel length, μ is the field-effect mobility, and C_i is the dielectric capacitance. The subscripts p and n denote the semiconductor type.

The ideal CMOS inverter with equivalent p-type and n-type transistors shows V_{inv} at $1/2V_{DD}$. To yield an ideal inverter, if μ and V_{Th} are different, the design factor (β) has to be adjusted properly to obtain equivalent saturation currents. In contrast to Si-based technology, organic field-effect transistors (OFETs) still do not have a fixed and optimized material library. Notably, OFETs have shown a very broad range of electron and hole mobilities, from 10^{-4} to a few tens of $cm^2\ V^{-1}\ s^{-1}$ depending on the chemical structure of the π-conjugated organic semiconductor molecules and the thin-film deposition methods [21].

The polarity (p-type, n-type, or ambipolar) and charge transport mechanism are also significantly variable. Therefore, proper selection of the p-type and n-type semiconductor materials with balanced μ and V_{Th} is a primary consideration before adjusting β in an organic CMOS inverter. The other ways to achieve the equivalent p-type and n-type OFETs are by controlling the charge injection at the semiconductor-contact electrode and charge transport properties near the semiconductor–dielectric interface [23–25].

The functionality of the CMOS inverter is determined by a set of criteria. The first criterion is that the inverter gain (g_{inv}), which is obtained by calculating $-dV_{OUT}/dV_{IN}$ in the inverting region, must be >1. To work properly in a complex circuit, this gain should be as high as possible. In practical circuitry, values between 4 and 10 have been indicated as the minimum [26]. More importantly, a CMOS circuit has the two characteristic features of high noise immunity and very low power consumption in the static ON and OFF states. NM is the maximum noise signal that can be superimposed on a digital signal without causing a malfunction of the digital circuit element. NMs can be defined in several different ways [27]: (i) the maximum equal criterion, which is probably the most adopted one in the literature on organic electronics [28]. The NM is simply defined as the side of the largest square that can be embedded within the loops formed by the VTC curve and its inverse of an inverter pair. (ii) the "−1" slope criterion, Figure 12.5a,b, which most of the digital circuits textbooks use. It is likely due to the fact that it is easy to find an analytical expression for each quantity involved. The maximum/minimum input and output voltages (V_{IH}/V_{IL} and V_{OH}/V_{OL}) are defined by the critical points where $|dV_{OUT}/dV_{IN}| = 1$. The NMs at the high (NM_H) and low levels (NM_L) are described as

$$NM_H = V_{OH} - V_{IH} \tag{12.4}$$

$$NM_L = V_{IL} - V_{OL} \tag{12.5}$$

The other important feature of a CMOS inverter is negligible static power consumption. Significant amounts of current flow through both p-type (I_{Dp}) and n-type transistors (I_{Dn}), but only near the switching threshold (V_m) in a CMOS inverter (Figure 12.5c,d). In the static logic "1" (at $V_{IN} = V_{DD}$) and logic "0" (at $V_{IN} = 0\,V$) states, p-type or n-type transistors are fully turned OFF while the counter-transistor is turned ON for discharging and charging the load capacitor in a inverter circuit, thus minimizing the power consumption.

12.2.2 Ring Oscillators

A ring oscillator is a logic element composed of an odd number of inverters connected in a chain, where the output of the last inverter is fed back into the first one (Figure 12.6). Each gate inverts the logic value at its input, and the last inverter will negate the input of the first one. Given an odd number of gates, an oscillation of the circuit sets in, limited by the time needed by the signal to propagate through

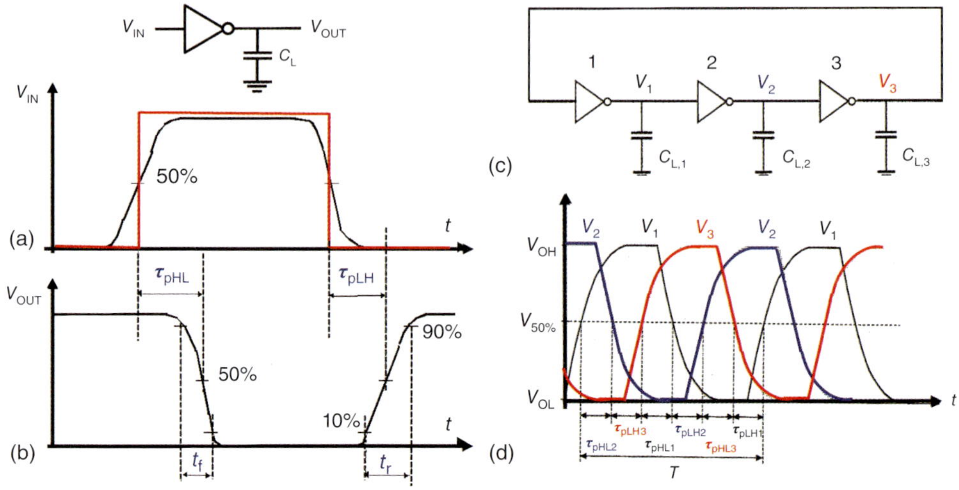

Figure 12.6 Circuit configuration of a complementary inverter and inverter dynamic characteristics of (b) V_{OUT} versus (a) V_{IN}, (c) circuit configuration of a three-stage ring oscillator, and (d) dynamic characteristics of the ring oscillators [21]. (Reproduced with permission. Copyright © 2013, John Wiley & Sons, Inc.)

the inverting chain [21]. This simple circuit typically is used to test the operation of inverters in a logic circuit. Notably, two main conditions must be satisfied to measure the exact circuit speed:

- The inverter should only delay the signal without distorting it.
- The load at the output should be equal to the input impedance of the inverter.

These conditions are met well by a ring oscillator. Therefore, this basic ring oscillator circuit element is frequently used to measure the operating frequency and stage delay time of digital circuits.

In a physical device, the gate input capacitance must be charged by the preceding gate output current in order to alter its voltage level. Because this delay adds up at each gate, adding more inverters to the ring oscillator chain would increase the total circuit delay time, reducing the oscillation frequency (f_{osc}). The f_{osc} of a ring oscillator circuit has a relationship with the propagation delay (τ_p) of the single inverter stage (Figure 12.6d), which is given by

$$f_{osc} = \frac{1}{T} = \frac{1}{2n\tau_p} \tag{12.6}$$

where n is the number of inverter stages and T is the total delay time of $T = \tau_{pHL1} + \tau_{pHL2} + \tau_{pHL3} + \tau_{pLH1} + \tau_{pLH2} + \tau_{pLH3}$. Thus, the obtained τ_p depends on the shape of the input signal and on the load at the output of each inverter stage in a digital circuit. The τ_p of the complementary inverter (Figure 12.6a,b) for the high-to-low (τ_{pHL}) and low-to-high (τ_{pLH}) transitions with an ideal rectangular input signal can be derived using first-order linear RC network analysis, as follows [29]:

$$\tau_{\text{pHL}} \approx \ln(0.5)R_{\text{n}}C_{\text{L}} \approx 0.69R_{\text{n}}C_{\text{L}} \tag{12.7}$$

$$\tau_{\text{pLH}} \approx \ln(0.5)R_{\text{p}}C_{\text{L}} \approx 0.69R_{\text{p}}C_{\text{L}} \tag{12.8}$$

where R_{n} and R_{p} are the very simplified equivalent ON-resistances of the n-type and p-type transistors, respectively. By assuming an identical equivalent load capacitance (C_{L}) both for the high-to-low and low-to-high transitions, the overall propagation delay (τ_{p}) of the inverter is defined as the average of the two values:

$$\tau_{\text{p}} = \frac{\tau_{\text{pHL}} + \tau_{\text{pLH}}}{2} \approx 0.69C_{\text{L}}\left(\frac{R_{\text{n}} + R_{\text{p}}}{2}\right) \tag{12.9}$$

From these simplified equations, it is possible to deduce strategies to minimize the propagation delay of a gate: (i) by limiting the load capacitance C_{L}, which mainly depends on the input capacitance of a similar gate and on the interconnections capacitance, or (ii) by reducing the transistor's ON-resistance, which depends on the charge carrier mobility (μ), the geometrical factors W/L, the specific capacitance (C_{g}), and V_{DD}.

Along with improving the charge-carrier mobility of printable semiconductors, the reduction of the dimensions (the channel length, L) is still a clear issue in printed and organic electronics because of the resolution limitations of common printing techniques. A longer W induces a higher gate capacitance, so this cannot be considered as a parameter for boosting the circuit speed, at least as long as the gate capacitance prevails on extrinsic contributions, such as the capacitance of interconnections. When the inverter is switched between the ON and OFF states $f_{0\to1}$ times per second, the dynamic power consumption is given by $P_{\text{dyn}} = C_{\text{L}}V_{\text{DD}}^2 f_{0\to1}$. Thus the V_{DD} should also be limited to reduce power consumption, leaving the increase of C_{g} as a possibility to improve the currents at the same V_{DD} [21].

12.3
High-Speed Organic CMOS Circuits

The state-of-the-art printed organic electronic devices and circuits are still limited by their operational speeds, which are far below a few megahertz [25, 30–35], but there has been remarkable progress in the last decade [36–39]. The transition (or cut-off) frequency (f_{T}) is a common figure of merit used to indicate the maximum operating frequency for a transistor. f_{T} in the saturation regime can be simply described as

$$f_{\text{T}} \approx \frac{\mu_{\text{app}}V_{\text{DD}}}{2\pi L(L + 2L_{\text{ov}})} \tag{12.10}$$

where μ_{app} is the apparent device mobility, $V_{DD} = (V_g - V_{Th})$, and L_{ov} is the overlap of equal sized gate/source and gate/drain electrodes. This gives the main parameters affecting the operation speed of a device. The operation speed of an organic transistor and its integrated circuitry can be improved by increasing the charge-carrier mobility of the active semiconductors, by realizing source/drain electrodes with a short channel length and negligible contact resistance, by using a small parasitic overlap capacitance, and by using a high dielectric capacitance for low-voltage operation (Figure 12.7).

12.3.1 High-Mobility Printable Semiconductors

In the last decades, the charge-carrier mobilities of organic semiconductors have been greatly improved from $\sim 10^{-5}$ to a few tens of $cm^2\ V^{-1}\ s^{-1}$ through a better understanding of the charge-carrier transport mechanism in π-conjugated organic molecules as well as by well-designed synthetic strategies and optimization of the device fabrication process. There have been a number of studies on the development of complementary electronic circuits based on high-mobility p-type and n-type organic semiconductors as well as on ambipolar organic semiconductors. Baeg *et al.* [25, 37, 39] reported high-speed (~80 kHz) complementary (or complementary-like) inkjet-printed electronic circuits using p-type and n-type (or ambipolar) polymeric semiconductors.

The current mobility of state-of-the-art organic semiconductors is considered valuable enough to successfully drive active-matrix organic light-emitting diode (OLED) displays and electronic papers, and even for operating printed RFIDs for item-level tagging (13.56 MHz frequency). Although organic semiconductors have many combinational advantages in comparison with their inorganic counterparts, such as mechanical flexibility, low-temperature easy processing with GAP methods for high-throughput mass production, and stability under ambient conditions, the apparent mobility has to be improved further for applications to high-end printed electronics, such as printed CPUs and microprocessors.

As an alternative, other classes of printable active materials have been developed and tested with the goal of enhancing the overall performance of printed electronics device, including nanocarbon materials (single-walled carbon nanotubes [40, 41] and nanoribbon graphene [42]), metal dichalcogenide nanosheets [43], amorphous oxide semiconductors [44–46], and printable silicon [47]. All of these materials have much higher intrinsic charge-carrier mobilities than organic semiconductors, ranging from a few tens of $cm^2\ V^{-1}\ s^{-1}$ (amorphous oxide semiconductors) to $\sim 10^3\ cm^2\ V^{-1}\ s^{-1}$ (for graphene transistors grown by chemical vapor deposition, despite a theoretical maximum value of $\sim 10^6\ cm^2\ V^{-1}\ s^{-1}$) [21]. Much interest has focused on developing new generations of printable active materials and their applications. However, many remaining hurdles have to be overcome, such as the sorting of chirality and semiconducting/metallic carbon nanotubes, the large amount of OFF current attributed to the zero-bandgap properties of graphene, and the rare p-channel properties and operational instabilities of the

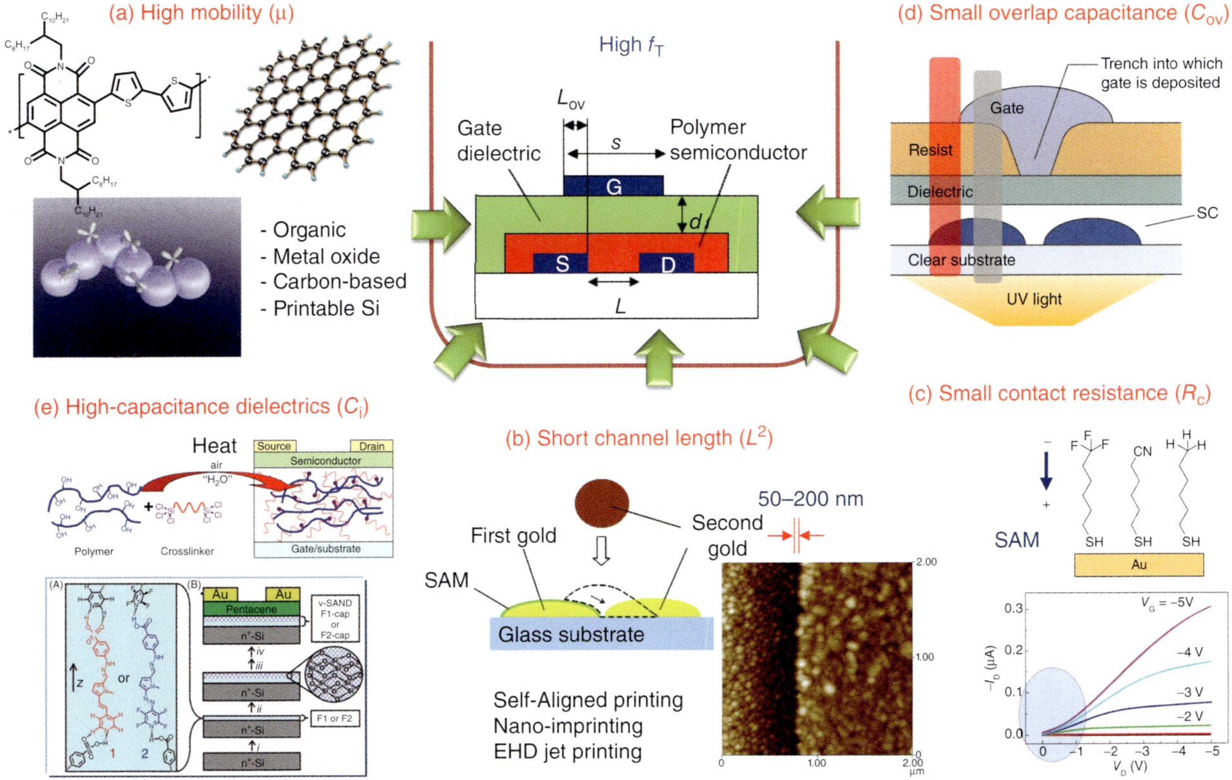

Figure 12.7 Strategies for developing high-performance organic electronic circuits via (a) active-channel materials with high charge-carrier mobility, (b) S/D electrodes with short channel length, (c) negligible contact resistance for charge-carrier injection, (d) small parasitic overlap capacitance, and (e) high-capacitance gate dielectrics for low-voltage operation [21]. (Reproduced with permission. Copyright © 2013, John Wiley & Sons, Inc.)

oxide semiconductors, before these technologies are adapted to commercialized products [21].

12.3.2
Downscaling of Channel Length

It is quite challenging to obtain down-scaled transistor channels with micrometer size or even submicrometer lengths using the well-developed standard GAP methods. This is because the printed feature sizes are typically very large, frequently more than a few tens of micrometers. There have been several approaches to realize down-scaled organic transistors with submicrometer channel lengths using printing techniques. Sirringhaus *et al.* [48, 49] demonstrated submicrometer channels of 50–200 nm in size produced using self-aligned printing (SAP) methods, obtained by controlling the receding contact-line motion of the liquid conductive ink (either metal nanoparticles or poly(3,4-ethylenedioxythiophene):polystyrene sulfonate (PEDOT:PSS) conducting polymer). These droplets were repelled by and flowed off the surface of a previously deposited electrode, with surface modification to change the wettability (Figure 12.8).

As an alternative to the SAP method of conventional inkjet printing combined with a unique surface modification technique to produce short-*L* electrodes, Murata *et al.* and Rogers *et al.* modified conventional inkjet printing techniques to demonstrate superfine inkjet printing [50] and electrohydrodynamic (EHD) jetting [51], respectively. The superfine inkjet system allowed minimum feature

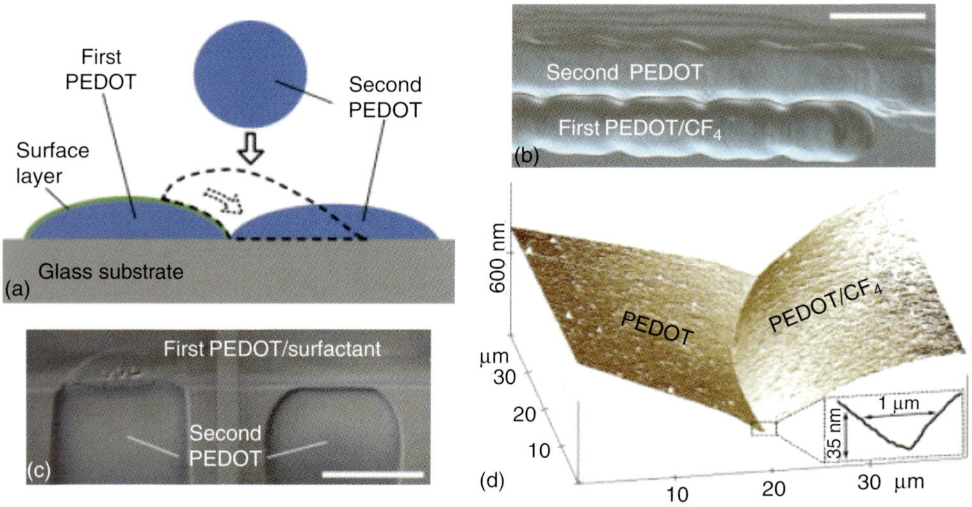

Figure 12.8 (a–d) Self-aligned inkjet printing of PEDOT:PSS ink for fabrication of source and drain electrodes for organic thin-film transistors [48]. (Reproduced with permission. Copyright © 2013, John Wiley & Sons, Inc.)

sizes of <1 μm. This made possible direct printing of ultrafine silver nanopaste wiring of a few micrometers in width on a glass plate, as well as site-selective growth of carbon nanotubes using a catalyst, and ultimately fabricating organic transistors. EHD jet printing uses electric fields rather than the thermal or acoustic energy used in conventional inkjet heads to overcome capillary forces in a small-diameter nozzle and create the fluid flows necessary for delivering a tiny volume of ink on to a substrate. This enables inkjet-printing-based electronic devices with submicrometer resolution by controlling the fluid flow using fine micro-capillary nozzles.

12.3.3 Reducing Contact Resistance

In OFETs, the contact resistance (R_c) tends to become comparable to or higher than the channel resistance (R_{ch}) as the device is down-scaled to the submicrometer regime. Therefore, the apparent device mobility (μ_{app}) in the saturation region drops significantly below the intrinsic mobility (μ_o) as given by Eq. (12.11) [52, 53]:

$$\mu_{app} \approx \mu_o \left[1 - \left(\frac{\mu_o C_i W R_c \left(V_g - V_{Th} \right)}{L + \mu_o C_g W R_c (V_g - V_{Th})} \right)^2 \right] \tag{12.11}$$

Theoretically, f_T values of 10 and 100 MHz are expected to be achievable at voltages of 3 and 30 V, respectively, when the OFET device has a μ_{app} of 1 cm^2 V^{-1} s^{-1} and lateral dimensions of 1 μm [30]. However, the measured f_T of a practical OFET device with these mobility and dimensions has been significantly low, typically below 1 MHz, because of the R_c effect. Currently, most of the reported short-channel OFETs are contact-limited. There have been a number of approaches for reducing R_c, including energy-level matching by varying the metal electrodes [54–56], controlling the interfacial dipoles using self-assembled monolayers (SAMs) [57–59], incorporating charge injection interlayers, and using polymer and other nanocarbon electrodes [60–62].

12.3.4 Reducing Parasitic Overlap Capacitance

As transistors are down-scaled and become highly integrated, parasitic capacitance, mainly due to the overlap L_{ov} between the gate and source/drain electrodes, becomes a critical issue that strongly limits f_T [21]. Most GAP tools for the fabrication of printed electronics devices typically have large feature sizes of more than a few tens of micrometers. The areal margins of printing machines apparently cause a large overlap area between the gate and source/drain electrodes. Therefore, novel approaches have to be introduced in the fabrication of printed transistors in order to minimize this parasitic overlap capacitance and to achieve the highest possible operation

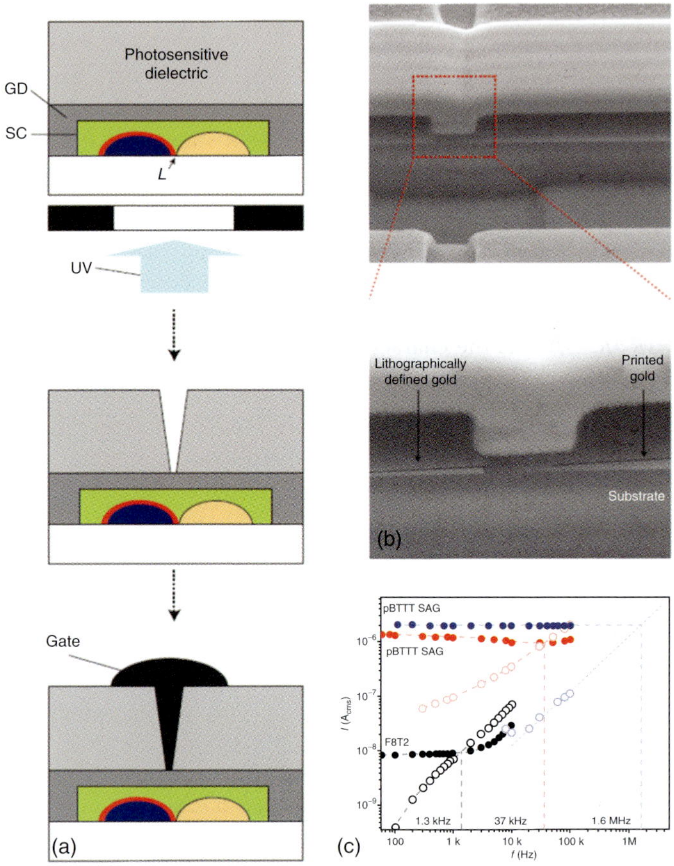

Figure 12.9 Self-aligned gate (SAG) architecture. (a) Schematic of the process to form a SAG structure. Top panel: Deposition of a photosensitive second dielectric on top of the semiconducting (SC) and gate dielectric (GD) layer and UV irradiation through the back of the substrate. Middle panel: Development of the second dielectric to remove the exposed regions. Bottom panel: Inkjet printing of gate electrode. (b) Cross-sectional view of the SAG structure measured by FIB-SEM. (c) Operation frequency evaluation of the SAG FETs [59]. (Reproduced with permission. Copyright © 2007, Nature Publishing Group.)

frequency [30, 59, 63]. Noh *et al.* [59] significantly reduced the parasitic overlap capacitance to as low as 0.2–0.6 pF mm^{-1} by using a self-aligned gate (SAG) technique. The SAG structure was fabricated using a photoresist developed selectively on the channel area via back-illuminated UV light, as can be seen in Figure 12.9.

For staggered organic transistors, it is important that L_{ov} be optimized to reduce both R_c and the parasitic overlap capacitance (C_{ov}) [21]. In contrast to coplanar OFET devices where the charge injection is limited to the narrow edge region near the active channel, staggered structures have much larger charge injection areas because the source/drain electrode is located opposite to the semiconductor layer [30]. Therefore, staggered OFET devices generally have smaller R_c. The W-normalized R_c is related to L_{ov} as [30]

$$R_c W = 2R_{sheet} L_T \coth\left(\frac{L_{ov} + L_{ext}}{L_T}\right) \tag{12.12}$$

where L_{ext} the extended contact length, which is a further contribution to the charge injection in a nonperpendicular direction between the source contact and the gate due to the electrostatic potential distribution in the organic semiconductor layer, $L_T = (R_y/R_{sheet})^{1/2}$ is the characteristic injection length or effective overlap length. Here, R_y is the apparent y-direction resistance normalized by the contact area (in ohms-square centimeter) and R_{sheet} is the sheet resistance of the semiconductor layer. Unfortunately, the requirements of small R_c and small parasitic capacitance are in competition, and a compromise between these two opposite constraints should be found [21].

For typical organic semiconductor FETs with $\mu_{app} = \sim 1\,cm^2\,V^{-1}\,s^{-1}$ and $R_c = \sim 0.6\,k\Omega\,cm$, L_T is measured to be around $10\,\mu m$. R_c increases as L_{ov} is decreased below $L_T \sim 10\,\mu m$, which decreases the effective mobility of the OFETs. By considering R_c and L_{ext}, Klauk *et al.* [30] showed that f_T reaches a plateau for $L_{ov} \approx L_T$, and then remains approximately constant when L_{ov} is reduced further. It was discovered that a minimization of L_{ov} may not be required for achieving a high f_T in staggered OFETs. Notably, L_T is mostly considered in a staggered device structure and determined by $(R_y/R_{sheet})^{1/2}$. This provides useful insight into the development of high-speed printed electronic circuits based on organic semiconductors. Therefore, the highest f_T can be obtained by considering such important parameters as charge-carrier mobilities, channel dimension, contact resistance, parasitic resistance, and the device configuration.

12.4
Conclusions

The most desirable patterning method for manufacturing OFETs should involve the direct patterning of active materials, in which the patterning and deposition are carried out in a single step, and should afford the ability to process large areas in a high-throughput process. To this end, inkjet printing is one of the most promising methods, and is often used in combination with photolithographic methods to decrease the feature size and improve the accuracy and resolution. The possibility of realizing large-area organic ICs using GAP technology represents a huge opportunity that would pave the way for ubiquitous adoption of cheap and flexible electronic applications, from microprocessors of printed near field communication (NFC) and RFID tags to plastic CPUs in wearable smart devices.

References

1. Parashkov, R., Becker, E., Riedl, T., Johannes, H.H., and Kowalsky, W. (2005) *Proc. IEEE*, **93**, 1321.
2. Søndergaard, R.R., Hosel, M., and Krebs, F.C. (2012) *J. Polym. Sci., Part B: Polym. Phys.*, **51**, 16.
3. Gelinck, G.H., Huitema, H.E.A., van Veenendaal, E., Cantatore, E., Schrijnemakers, L., van der Putten, J.B.P.H., Geuns, T.C.T., Beenhakkers, M., Giesbers, J.B., Huisman, B.-H., Meijer, E.J., Benito, E.M., Touwslager, F.J., Marsman, A.W., van Rens, B.J.E., and de Leeuw, D.M. (2004) *Nat. Mater.*, **3**, 106.
4. Jain, K., Klosner, M., Zemel, M., and Raghunandan, S. (2005) *Proc. IEEE*, **93**, 1500.
5. Sekitani, T., Nakajima, H., Maeda, H., Fukushima, T., Aida, T., Hata, K., and Someya, T. (2009) *Nat. Mater.*, **8**, 494.
6. Søndergaard, R., Hösel, M., Angmo, D., Larsen-Olsen, T.T., and Krebs, F.C. (2012) *Mater. Today*, **15**, 36.
7. Furtak, N.T., Skrzetuska, E., and Krucińska, I. (2013) *Fibres Text. East. Eur.*, **21**, 84.
8. Wei, Y., Torah, R., Yang, K., Beeby, S., and Tudor, J. (2013) *Sens. Actuators, A*, **203**, 241.
9. Koski, K., Koski, E., Virtanen, J., Bjorninen, T., Sydanheimo, L., Ukkonen, L., and Elsherbeni, A.Z. (2012) *Int. J. Adv. Manuf. Technol.*, **62**, 167.
10. Myny, K., Steudel, S., Vicca, P., Beenhakkers, M.J., van Aerle, N.A.J.M., Gelinck, G.H., Genoe, J., Dehaene, W., and Heremans, P. (2009) *Solid State Electron.*, **53**, 1220.
11. Pang, C., Lee, C., and Suh, K.Y. (2013) *J. Appl. Polym. Sci.*, **130**, 1429.
12. De Gans, B.J. and Schubert, U.S. (2003) *Macromol. Rapid Commun.*, **24**, 659.
13. De Gans, B.J., Duineveld, P.C., and Schubert, U.S. (2004) *Adv. Mater.*, **16**, 203.
14. Delaney, J.T., Smith, P.J., and Schubert, U.S. (2009) *Soft Matter*, **5**, 4866.
15. Tekin, E., Smith, P.J., and Schubert, U.S. (2008) *Soft Matter*, **4**, 703.
16. Fromm, J.E. (1984) *IBM J. Res. Dev.*, **28**, 322.
17. de Gans, B.-J., Kazancioglu, E., Meyer, W., and Schubert, U.S. (2004) *Macromol. Rapid Commun.*, **25**, 292.
18. Rabaey, J.M., Chandrakasan, A., and Nikolić, B. (2003) *Digital Integrated Circuits*, Prentice Hall, Upper Saddle River, NJ.
19. Facchetti, A. (2011) *Chem. Mater.*, **23**, 733.
20. Fortunato, E., Barquinha, P., and Martins, R. (2012) *Adv. Mater.*, **24**, 2945.
21. Baeg, K.-J., Caironi, M., and Noh, Y.-Y. (2013) *Adv. Mater.*, **25**, 4210.
22. Baker, R.J., Li, H.W., and Boyce, D.E. (1998) *CMOS Circuit Design, Layout, and Simulation*, IEEE Press, New York.
23. Baeg, K.J., Khim, D., Jung, S.W., Kang, M., You, I.K., Kim, D.Y., Facchetti, A., and Noh, Y.Y. (2012) *Adv. Mater.*, **24**, 5433.
24. Baeg, K.-J., Khim, D., Kim, J., Han, H., Jung, S.-W., Kim, T.-W., Kang, M., Facchetti, A., Hong, S.-K., Kim, D.-Y., and Noh, Y.-Y. (2012) *ACS Appl. Mater. Interfaces*, **4**, 6176.
25. Baeg, K.-J., Kim, J., Khim, D., Caironi, M., Kim, D.-Y., You, I.-K., Quinn, J.R., Facchetti, A., and Noh, Y.-Y. (2011) *ACS Appl. Mater. Interfaces*, **3**, 3205.
26. Schrom, G. and Selberherr, S. (1996) Ultra-low-power CMOS technologies. CAS'96 Proceedings, October 1996, p. 237.
27. Hauser, J.R. (1993) *IEEE Trans. Educ.*, **36**, 363.
28. Bode, D., Rolin, C., Schols, S., Debucquoy, M., Steudel, S., Gelinck, G.H., Genoe, J., and Heremans, P. (2010) *IEEE Trans. Electron Devices*, **57**, 201.
29. Kang, S.-M. and Leblebici, Y. (2003) *CMOS Digital Integrated Circuits: Analysis and Design*, McGraw-Hill, New York.
30. Ante, F., Kalblein, D., Zaki, T., Zschieschang, U., Takimiya, K., Ikeda, M., Sekitani, T., Someya, T., Burghartz, J.N., Kern, K., and Klauk, H. (2012) *Small*, **8**, 73.
31. Cho, J.H., Lee, J., Xia, Y., Kim, B., He, Y.Y., Renn, M.J., Lodge, T.P., and Frisbie, C.D. (2008) *Nat. Mater.*, **7**, 900.

32. Crone, B., Dodabalapur, A., Lin, Y.Y., Filas, R.W., Bao, Z., LaDuca, A., Sarpeshkar, R., Katz, H.E., and Li, W. (2000) *Nature*, **403**, 521.
33. Fix, W., Ullmann, A., Ficker, J., and Clemens, W. (2002) *Appl. Phys. Lett.*, **81**, 1735.
34. Klauk, H., Zschieschang, U., Pflaum, J., and Halik, M. (2007) *Nature*, **445**, 745.
35. Knobloch, A., Manuelli, A., Bernds, A., and Clemens, W. (2004) *J. Appl. Phys.*, **96**, 2286.
36. Ante, F., Kalblein, D., Zschieschang, U., Canzler, T.W., Werner, A., Takimiya, K., Ikeda, M., Sekitani, T., Someya, T., and Klauk, H. (2011) *Small*, **7**, 1186.
37. Baeg, K.-J., Jung, S.-W., Khim, D., Kim, J., Kim, D.-Y., Koo, J.B., Quinn, J.R., Facchetti, A., You, I.-K., and Noh, Y.-Y. (2013) *Org. Electron.*, **14**, 1407.
38. Kitamura, M., Kuzumoto, Y., Aomori, S., and Arakawa, Y. (2011) *Appl. Phys. Express*, **4**, 051601.
39. Baeg, K.-J., Khim, D., Kim, D.-Y., Jung, S.-W., Koo, J.B., You, I.-K., Yan, H., Facchetti, A., and Noh, Y.-Y. (2011) *J. Polym. Sci., Part B: Polym. Phys.*, **49**, 62.
40. Fukuda, K., Sekitani, T., Zschieschang, U., Klauk, H., Kuribara, K., Yokota, T., Sugino, T., Asaka, K., Ikeda, M., Kuwabara, H., Yamamoto, T., Takimiya, K., Fukushima, T., Aida, T., Takamiya, M., Sakurai, T., and Someya, T. (2011) *Adv. Funct. Mater.*, **21**, 4019.
41. Chen, P.C., Fu, Y., Aminirad, R., Wang, C., Zhang, J.L., Wang, K., Galatsis, K., and Zhou, C.W. (2011) *Nano Lett.*, **11**, 5301.
42. Burghard, M., Klauk, H., and Kern, K. (2009) *Adv. Mater.*, **21**, 2586.
43. Huang, X., Zeng, Z., and Zhang, H. (2013) *Chem. Soc. Rev.*, **42**, 1934.
44. Kim, M.-G., Kanatzidis, M.G., Facchetti, A., and Marks, T.J. (2011) *Nat. Mater.*, **10**, 382.
45. Banger, K.K., Yamashita, Y., Mori, K., Peterson, R.L., Leedham, T., Rickard, J., and Sirringhaus, H. (2011) *Nat. Mater.*, **10**, 45.
46. Kim, Y.H., Heo, J.S., Kim, T.H., Park, S., Yoon, M.H., Kim, J., Oh, M.S., Yi, G.R., Noh, Y.-Y., and Park, S.K. (2012) *Nature*, **489**, 128.
47. Harting, M., Zhang, J., Gamota, D.R., and Britton, D.T. (2009) *Appl. Phys. Lett.*, **94**, 193509.
48. Sele, C.W., von Werne, T., Friend, R.H., and Sirringhaus, H. (2005) *Adv. Mater.*, **17**, 997.
49. Caironi, M., Gili, E., Sakanoue, T., Cheng, X., and Sirringhaus, H. (2010) *ACS Nano*, **4**, 1451.
50. Murata, K. (2003) Super-fine ink-jet printing for nanotechnology. Proceedings of the International Conference on MEMS, Nano and Smart Systems, Banff, Canada, p. 346.
51. Park, J.U., Hardy, M., Kang, S.J., Barton, K., Adair, K., Mukhopadhyay, D.K., Lee, C.Y., Strano, M.S., Alleyne, A.G., Georgiadis, J.G., Ferreira, P.M., and Rogers, J.A. (2007) *Nat. Mater.*, **6**, 782.
52. Benor, A. and Knipp, D. (2008) *Org. Electron.*, **9**, 209.
53. Gundlach, D.J., Zhou, L., Nichols, J.A., Jackson, T.N., Necliudov, P.V., and Shur, M.S. (2006) *J. Appl. Phys.*, **100**, 024509.
54. Tiwari, S.P., Zhang, X.H., Potscavage, W.J., and Kippelen, B. (2009) *J. Appl. Phys.*, **106**, 054504.
55. Hamadani, B.H., Ding, H., Gao, Y., and Natelson, D. (2005) *Phys. Rev. B*, **72**, 235302.
56. Baeg, K.-J., Khim, D., Kim, D.-Y., Koo, J.B., You, I.-K., Choi, W.-S., and Noh, Y.-Y. (2010) *Thin Solid Films*, **518**, 4024.
57. Campbell, I.H., Rubin, S., Zawodzinski, T.A., Kress, J.D., Martin, R.L., Smith, D.L., Barashkov, N.N., and Ferraris, J.P. (1996) *Phys. Rev. B*, **54**, 14321.
58. Boudinet, D., Benwadih, M., Qi, Y.B., Altazin, S., Verilhac, J.M., Kroger, M., Serbutoviez, C., Gwoziecki, R., Coppard, R., Le Blevennec, G., Kahn, A., and Horowitz, G. (2010) *Org. Electron.*, **11**, 227.
59. Noh, Y.-Y., Zhao, N., Caironi, M., and Sirringhaus, H. (2007) *Nat. Nanotechnol.*, **2**, 784.
60. Pang, S.P., Hernandez, Y., Feng, X.L., and Mullen, K. (2011) *Adv. Mater.*, **23**, 2779.
61. Yun, D.J., Hong, K., Kim, S.H., Yun, W.M., Jang, J.Y., Kwon, W.S., Park, C.E.,

and Rhee, S.W. (2011) *ACS Appl. Mater. Interfaces*, **3**, 43.

62. Zhang, Y.Y., Shi, Y.M., Chen, F.M., Mhaisalkar, S.G., Li, L.J., Ong, B.S., and Wu, Y.L. (2007) *Appl. Phys. Lett.*, **91**, 223512.

63. Palfinger, U., Auner, C., Gold, H., Haase, A., Kraxner, J., Haber, T., Sezen, M., Grogger, W., Domann, G., Jakopic, G., Krenn, J.R., and Stadlober, B. (2010) *Adv. Mater.*, **22**, 5115.

13
Large-Area, Printed Organic Circuits for Ambient Electronics

Tsuyoshi Sekitani, Tomoyuki Yokota, and Takao Someya

13.1
Introduction

Realization of a sustainable society will require the development of industrial manufacturing processes that have minimal impact on the environment. The emerging field of printable electronics [1–7] has much to offer in this regard, as it has the potential to drastically reduce the ecological footprint of electronics manufacturing (that is, the energy consumed in manufacturing electronic products). In addition, developments in printing allow greater economy in the use of inks, thus minimizing material wastage in comparison to conventional manufacturing techniques.

However, a significant obstacle to the development of large-area printed electronics such as large-area displays and sensors has been the fundamental need to compromise between manufacturing efficiency, transistor performance, and power consumption, as past improvements in manufacturing efficiency achieved through the use of printing techniques have inevitably led to poor device performance and large power consumption, while attempts to improve performance or reduce power have led to higher process temperatures and increased manufacturing costs. In addition to these tradeoffs, large variations in electrical performance may result from crystal defects, grain boundaries, ion-based impurities, atmospheric degradation of organic semiconductors (OSCs), or geometric fluctuations that result when organic LEDs (light-emitting diodes) or organic transistors are printed onto plastics.

In this chapter, we describe how these fundamental issues can be avoided through the use of state-of-the-art screen printing technologies, printable organic materials, and circuit design, allowing the printing of large-area, ultraflexible, high-performance organic transistors and integrated circuits onto flexible, thin plastic substrates.

In order to demonstrate the feasibility of large-area printed electronics, we have developed three applications: a pressure sensor sheet; an intelligent medical catheter in which an ultraflexible pressure sensor sheet is wrapped around a very fine medical catheter to improve the spatial distribution of pressure; and an

Large Area and Flexible Electronics, First Edition. Edited by Mario Caironi and Yong-Young Noh.

organic LED pixel circuit that utilizes printed floating-gate organic transistors to drive organic LEDs. The last of these, a 230 mm × 230 mm printed active-matrix circuit composed of 64 × 64 screen-printed organic two-transistor-one-capacitor (2T1C) cells, can compensate LED brightness variations and degradation for more than 6 months.

We also discuss future prospects of printed large-area electronics and present a picture of the nature of the requirements – in terms of material science, electronic engineering, and mechanical engineering – for next-generation manufacturing processes.

The contents of this chapter are as follows: (13.1) Introduction, (13.2) Manufacturing Process and Electrical Characteristics: (13.2.1) Materials and Methods for Manufacturing Large-Area, Printed Active Matrix, (13.2.2) Electrical Characteristics, (13.2.3) Mechanical Characteristics, (13.2.4) Inverter Circuits and the Ring Oscillator, (13.2.5) Organic Floating-Gate Transistors; (13.3) Demonstrations; and (13.4) Future Prospects.

13.2 Manufacturing Process and Electrical Characteristics

13.2.1 Materials and Methods

Screen printing is one of the most effective methods for patterning materials onto large-area substrates with high throughput. Commercially available screen printing processes are supported by well-established large-size (3500 mm × 3500 mm) screen shadow masks with 30 μm repeat accuracy (Figure 13.1). Screen shadow masks with 20 μm resolution and very high tensile strength (3000 N mm^{-1}) can pattern high-viscosity Ag paste (Toyo-chem, Rex-α 156) and lead to line widths of 20 μm with 20 μm line spacing over a large area (Figure 13.1b–d). As Ag paste has a viscosity of 360 Pa s and uses very few organic solvents and binders, we can print low-resistivity patterns onto a plastic substrate at low drying temperatures; that is, 27 μΩ cm after drying at room temperature (RT) for 2 days, and 18 μΩ cm after drying at 150 °C for 30 min. These drying temperatures are low enough to enable the use of almost any plastic substrate, while the resistivities are low enough for use in printed electrodes.

13.2.2 Organic Transistors Manufactured Using Printing Technologies

Using source/drain electrodes formed through screen printing, we fabricated two types of organic transistor. Figure 13.2 shows the cross-sectional views of these: (i) an organic transistor with top-contact geometry in which an OSC layer of dinaphtho[2,3-*b*:2′,3′-*f*]thieno [3,2-*b*]thiophene (DNTT) [9] is

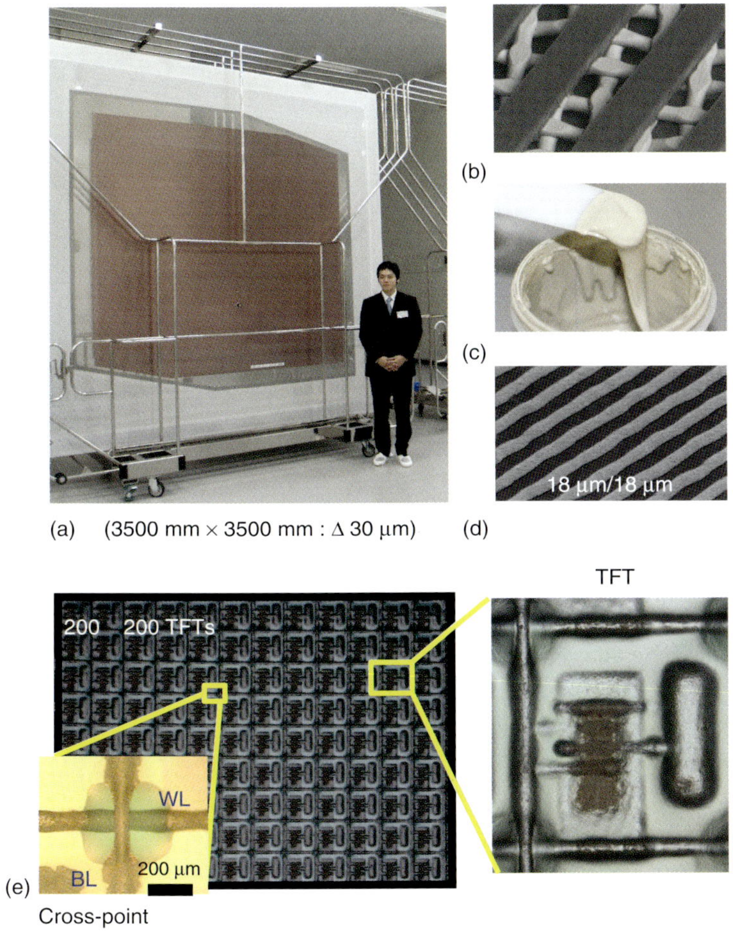

Figure 13.1 (a) Images of screen shadow mask. (b) Magnified image of twill-woven screen mesh and masking emulsion layers with line width 20 μm. (c) Image of high-viscosity Ag paste. (d) Image of printed Ag electrodes with 18 μm line width and 18 μm spacing. (e) Micrograph of organic transistor active matrix with a 1-mm pixel pitch. Magnified pictures of a printed transistor and a cross-point of a bit-line and a word-line are also shown here; channel length of the transistor is 50 μm, while printed linewidths are 100–200 μm. Insulating layers of 3 μm thickness are formed between bit-lines and word-lines to reduce parasitic capacitance and chance of leakage currents [8]. (Reproduced with permission. Copyright © 2012, Society for Information Displays.)

formed using a vacuum evaporation process (in this paper this is referred to as a *DNTT transistor*) and (ii) an organic transistor with bottom-contact geometry in which an OSC layer consisting of solution-processable polycrystalline material (Lisicon OSC) [10] is formed using dip coating (an all-printed transistor).

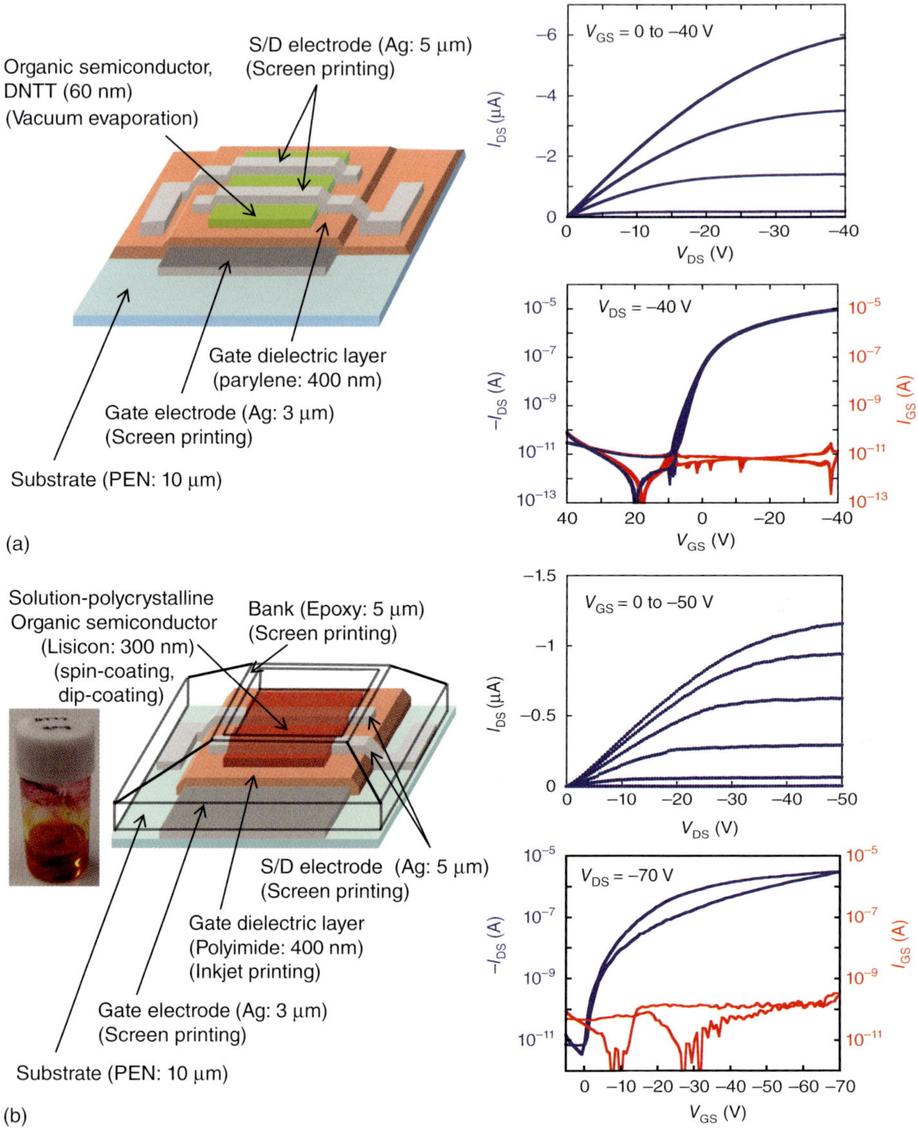

Figure 13.2 Structure and electrical characteristics of printed organic transistors. (a) Schematic illustration of organic transistors with top-contact geometry formed with organic semiconductor layer dinaphtho[2,3-*b*:2′,3′-*f*]thieno [3,2-*b*]thiophene (DNTT) [9] using vacuum evaporation. Typical output, transfer (blue), and leakage (red) transistor characteristics. (b) Schematic illustration of organic transistors with bottom-contact geometry with organic semiconductor layer of solution-processable polycrystalline (Lisicon OSC) material [10], formed using dip-coating process. Typical output, transfer (blue), and leakage (red) transistor characteristics [8]. (Reproduced with permission. Copyright © 2012, Society for Information Displays.)

13.2.2.1 Manufacturing Process for DNTT Transistors

Silver paste is screen printed and dried at 120 °C in air on a flexible substrate of 10-μm-thick polyethylene naphthalate (PEN) film to form 3-μm-thick gate electrodes. Using chemical vapor deposition (CVD), a parylene (diX-SR, Daisan-kasei) layer is formed, producing a 400-nm-thick gate dielectric. A masking resist (#503B-SH, Asahi Chemical) is then patterned by screen printing to form a shadow mask for the vacuum evaporation of OSCs. The patterned resist is baked at 150 °C for 90 s in ambient air for stiffening and the removal of solvents, and the base film with the printed shadow mask is rolled onto a 10-cm-diameter cylindrical bar, which is then rotated and subjected to vacuum evaporation to create a 60-nm-thick DNTT to serve as an organic semiconducting channel – a method that is well applicable to large-area material formation [11]. Following evaporation, the resist shadow mask is peeled off to allow patterning of the organic channel layer, another Ag paste is formed by screen printing to create printed source/drain electrodes, and, finally, the whole device is coated with parylene, which serves as an encapsulation layer.

13.2.2.2 Electrical Performance of DNTT Transistors

We then measured the characteristics of the printed transistors fabricated by the processes described above. As shown in Figure 13.2a, the source–drain current (I_{DS}) of the transistors was measured as a function of the source–drain voltage (V_{DS}) for a gate voltage (V_{GS}) varying from 0 to −40 V in steps of −10 V. The corresponding transfer curves for V_{GS} varying from +40 to −40 V at a constant V_{DS} of −40 V are also shown for this device. The hysteresis was very small (with the ON:OFF ratio exceeding 10^6), demonstrating that no significant leakage current flowed through the parylene layer. The average mobility in the saturation regime was 0.2 $cm^2\,V^{-1}\,s^{-1}$, indicating excellent transistor characteristics. Typical DNTT transistors with screen-printed electrodes remain "Normally-ON," that is, they show large I_{DS} at 0 V, which is mainly due to unintentional doping from atmospheric components such as oxygen and humidity.

13.2.2.3 Manufacturing Process for All-Printed Transistors

Using screen printing, a silver paste pattern is formed and dried at 120 °C in air to form 3-μm-thick gate electrodes on thin flexible substrates. A polyimide precursor is then formed with inkjet printing and cured at 120 °C to form a 400-nm-thick gate dielectric layer, and the Ag paste is again screen printed to create printed source/drain electrodes. Except for the channel area consisting of a 300-nm-thick polycrystalline semiconductor layer formed by dip coating in a solution, the entire surface is formed via screen printing. Finally, the whole device is coated with an encapsulation layer of parylene.

13.2.2.4 Electrical Performance of All-Printed Transistors

We then measured the electrical performance of the all-printed transistor. As shown in Figure 13.2b, I_{DS} of the transistor was measured as a function of V_{DS}, with V_{GS} varying from 0 to −50 V in steps of −10 V. The corresponding transfer curves are shown for V_{GS} varying from +5 to −70 V at a constant V_{DS} of −70 V. Again, hysteresis was very small, with the ON:OFF ratio exceeding 10^5. The average mobility in the saturation regime was 0.1 $cm^2 V^{-1} s^{-1}$, which is almost half that measured in the DNTT transistors.

Owing to the gas barrier characteristics of the parylene layers, the electrical performance of these devices does not show degradation even after 6 months of storage in air. This longevity suggests that the organic semiconducting layer will not be damaged by the residual organic solvents in our Ag paste, as the latter contains very small amounts of organic solvents and other materials – which in any event evaporate easily at low temperatures.

13.2.3 Mechanical Characteristics

Using a high-precision mechanical stage, the mechanical and electrical characteristics of the transistors were evaluated simultaneously. Figure 13.3 shows the pictures and the transfer characteristics of several printed-electrode DNTT transistors that were manufactured as previously described using three different substrates. By bending the transistor through a radius of 0.5 mm radius, the saturation currents decreased and the threshold voltages shifted in the negative direction, while the mobility decreased by 50%. Although these performance changes were

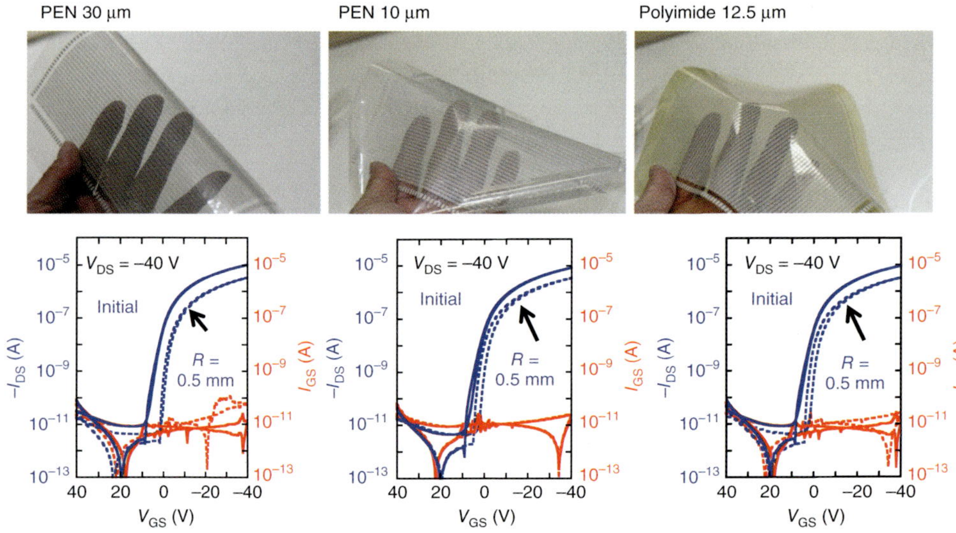

Figure 13.3 Mechanical and electrical characteristics of printed transistors. Pictures of the active matrices of printed organic transistors on three different plastic substrates. Transfer (blue) and leakage (red) characteristics are also shown, with measurements performed before (represented as "initial"; solid-line) and during bending to 0.5 mm bending radius (represented as "R = 0.5 mm"; dashed line) [8]. (Reproduced with permission. Copyright © 2012, Society for Information Displays.)

reversible through unbending, a further decrease in the bending radius resulted in irreversible degradations in performance owing primarily to the delamination of the source and drain electrodes in the DNTT layers. The observed changes from bending stress can be understood as an effect of altering the intergrain distance among the polycrystalline DNTT semiconductors [12, 13].

13.2.4 Inverter Circuits and Ring Oscillator Using Printed Transistors

Although we have developed two types of transistors, namely DNTT transistors with screen-printed electrodes and all-printed transistors, several circuits in the following discussion are based on DNTT transistors with screen-printed electrodes because device-to-device nonuniformity of all-printed transistors is much larger than that of DNTT transistors with screen-printed electrodes, resulting in instability of the circuits.

Using p-type DNTT transistors on plastic substrates, two types of inverters were constructed: unipolar (Figure 13.4a) and pseudo-complementary (Figure 13.4b). The unipolar inverter, which was constructed using two p-type transistors, operates at −40 V with a gain of only 2 (such a small gain is intrinsic to unipolar inverters [14]). The pseudo-complementary inverter consists of four p-type transistors, which were originally designed by our group [15]. Although this inverter consists of p-type unipolar transistors with no n-type transistors, it shows steep inversion characteristics and its gain exceeds 70.

The high-gain inverter above was used to construct a printed three-stage ring oscillator (Figure 13.4c) with an oscillation frequency of 8.4 Hz, corresponding to a propagation delay per stage of 19.8 ms. The frequency response (cut-off frequency: f_T) of this device is in excellent agreement with the conventional theoretical formula

$$f_T = \frac{\mu(V_{GS} - V_{th})}{2\pi L(L + \Delta L)}$$

where L is the channel length, ΔL is the overlap length (i.e., the line width of the source and drain electrodes), μ is mobility, V_{GS} is the gate voltage, and V_{th} is the threshold voltage. The observed slow frequency response of the oscillator is due mainly to the large L (100 μm) of the printed transistors; decreasing this value to 20 μm, will improve the frequency response to 1 ms per stage.

13.2.5 Printed Organic Floating-Gate Transistors

A floating-gate transistor is a field-effect transistor (FET) with two gate electrodes. In addition to the control gate seen in a regular transistor, a floating gate is embedded into the gate dielectric. If the dielectric is sufficiently thin, electronic charge can be brought onto the floating gate through quantum tunneling or thermal emission when a sufficiently large program voltage is applied between the control gate

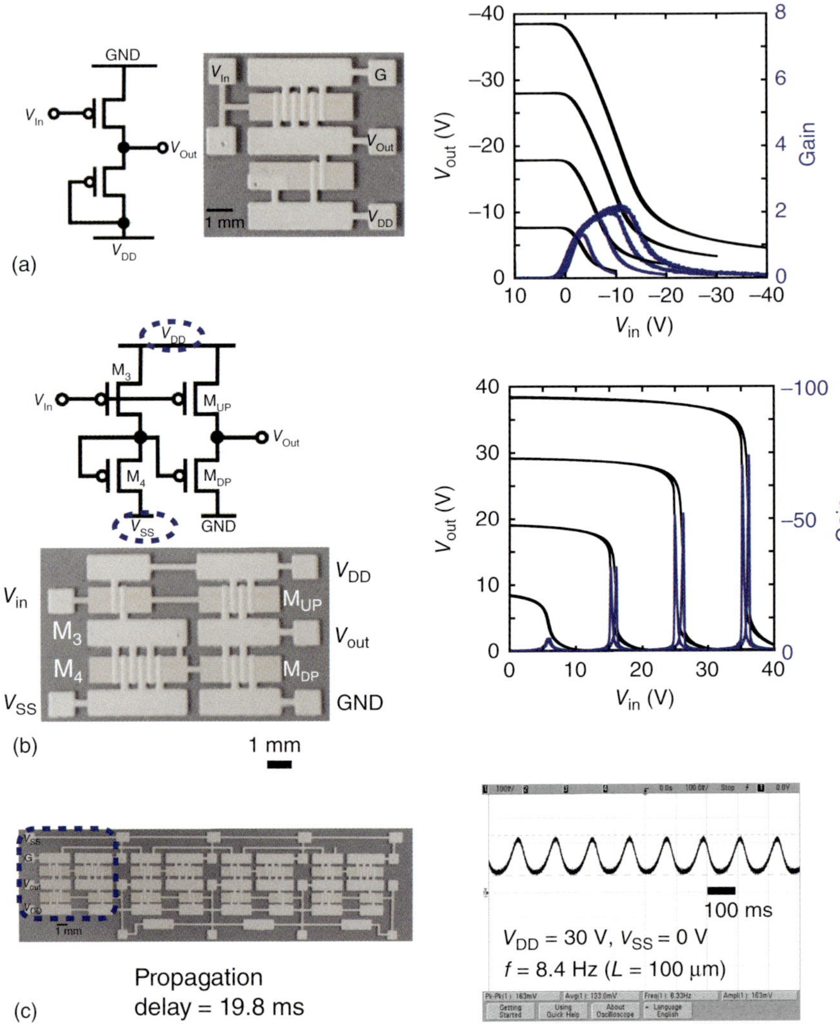

Figure 13.4 Inverter circuits and three-stage ring oscillator. Schematic circuit diagram, picture, and inverter characteristics (input voltage V_{in} – output voltage V_{out} and gain) of (a) unipolar inverter and (b) pseudo-complementary inverter. V_{DD} represents the power supply voltage, V_{SS} is the tuning voltage, and GND is ground. The four transistors (M_3, M_4, M_{UP}, and M_{DP}) forming the pseudo-complementary inverter are precisely designed with W/L ratios 1 : 5 : 3 : 3. (c) Picture and oscillation characteristic of a three-stage ring oscillator comprised of pseudo-complementary inverters. Channel length of the printed transistor is 100 μm, oscillation frequency is 8.4 Hz, and the corresponding propagation delay per stage is 19.8 ms [8]. (Reproduced with permission. Copyright © 2012, Society for Information Displays.)

and the source contact. Because the charge on the floating gate partially screens the electric field between the control gate and the semiconductor, charging the floating gate changes the transistor's threshold voltage; this shift can be detected by measuring the drain current at a certain gate–source voltage. Because the floating gate is completely isolated by the dielectric, the stored charge remains without the need for any applied voltage; this forms the basis of nonvolatile memory. To erase the memory, a voltage of opposite polarity can be applied, discharging the floating gate through the dielectric. The threshold voltage can be controlled by the program voltage, which is typically two or three times higher than normal operation (read-out) voltages. More details regarding organic floating-gate transistors can be seen in our previous report [16].

Taking full advantage of the characteristics of organic floating-gate transistors, we fabricated non-uniformity-compensator and organic active-matrix LED pixel circuits on plastic substrates (Figure 13.5).

13.2.5.1 Manufacturing Process

The organic floating-gate transistors were also fabricated using screen printing; Figure 13.5a shows a cross-sectional view of our organic floating-gate transistors with top-contact geometry. The manufacturing process is almost identical to that of the previously described DNTT transistors, with printed floating gates embedded into the gate dielectric layer. Figure 13.5b shows a photograph of an organic LED pixel circuit utilizing printed floating-gate thin-film transistors (FG-TFTs).

13.2.5.2 Electrical Performance

The threshold voltage (V_{th}) of a floating-gate transistor can be dynamically and systematically controlled by using a program voltage that is three times higher than the normal operational voltage to induce channel carrier injection into the floating gate; a schematic illustration of the operational mechanism can be seen in Figure 13.6a. As shown in Figure 13.6b–d, the electrical characteristics of a standalone printed floating-gate transistor were evaluated. Initially, V_{th} was between +9 and +10 V (see Figure 13.6b); for programming, a voltage of −80 V of 100-ms duration was then applied between the control gate and the source contact. This created a gate current of more than 1 μA, which charged the floating gate and shifted V_{th} to about −20 V. To erase this state, a voltage of +80 V was applied to discharge the floating gate and recover the initial threshold voltage. Cycling the gate–source voltage between +80 and −80 V produced a hysteresis. The exact shift in V_{th} (i.e., the threshold voltage window) depended on the voltage and the duration of the program pulse (see Figure 13.6c,d); as shown in Figure 13.6c, V_{th} was shifted by program pulses >−60 V (duration of 100 ms). Figure 13.6d shows the effect of the program pulse duration, indicating that a longer duration will generate a larger shift in V_{th}. Employing this systematic shift in V_{th}, we constructed organic LED pixel circuits with device-to-device nonuniformity compensators as shown in Figure 13.5. Figure 13.6e,f shows the compensation of device-to-device performance variations to limit the variation size by controlling V_{th}. Initially, the

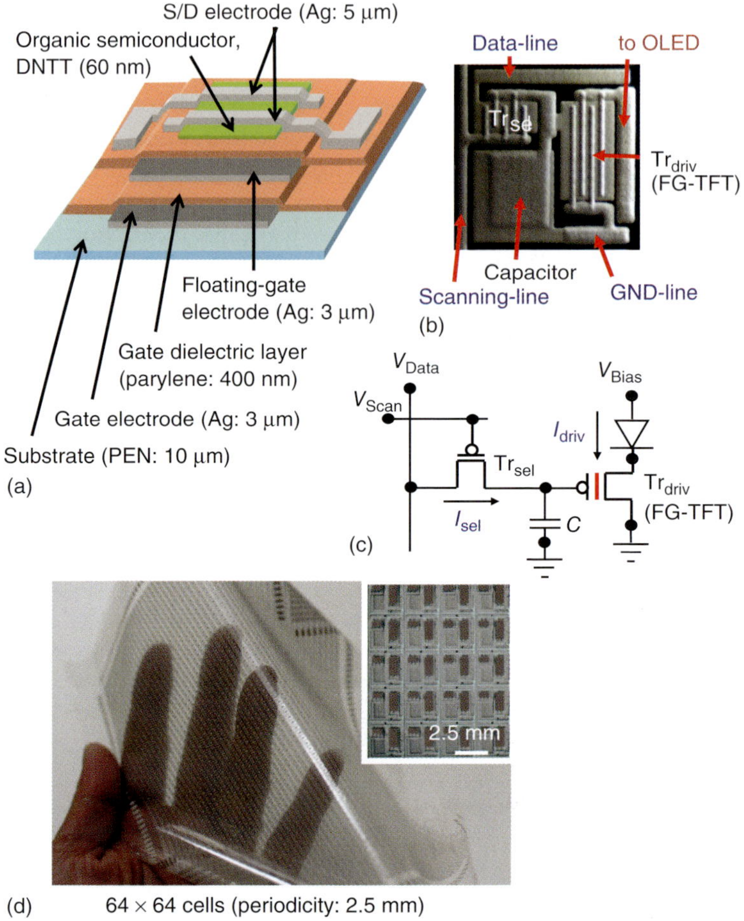

Figure 13.5 Floating-gate transistor and organic active-matrix LED pixel circuit. (a) Schematic cross-sectional illustration of printed organic floating-gate transistors. (b) Magnified image of one LED pixel circuit cell comprised of two transistors and one capacitor (2T1C) in which the drive transistor (Tr_{driv}) is formed using a floating-gate transistor (FG-TFT). (c) Circuit diagram of one LED pixel circuit cell (2T1C). (d) Picture of 64 × 64 active-matrix LED pixel circuit with pixel pitch 2.5 mm. Inset shows a magnified picture of the active-matrix array [8]. (Reproduced with permission. Copyright © 2012, Society for Information Displays.)

three-pixel circuit shown had large device-to-device nonuniformities of 56.4% and 29.4% in terms of drive current (I_{driv}, which determines the luminance of an organic LED) and V_{th}, respectively (Figure 13.6e). After compensation (i.e., providing an appropriate program pulse to each transistor), device-to-device nonuniformity decreased significantly to 9.8% and 4.8% in terms of I_{driv} and V_{th}, respectively (Figure 13.6f).

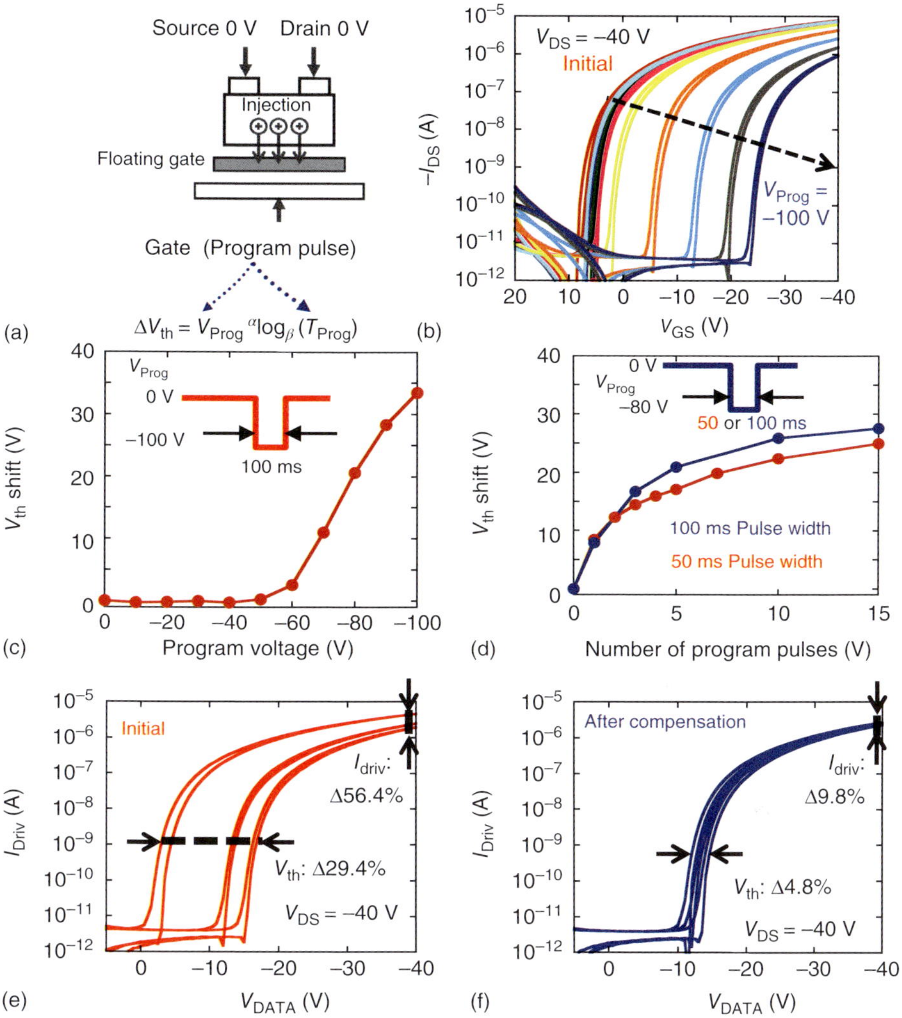

Figure 13.6 Electrical characteristics of printed floating-gate transistor and compensation of device-to-device nonuniformity using floating-gate-based 2T1C pixel circuit. (a) Schematic illustration of floating-gate charging mechanism. (b) Transfer characteristics of floating-gate transistors after various program voltage pulses (100 ms duration) changing from −10 to −100 V were applied to the bottom (control) gate electrode. When reading out the data after programming, voltage for the control gate (V_{GS}) was varied from +20 to −20 V by the application of a source–drain voltage (V_{DS}) of −40 V. (c,d) Threshold voltage (V_{th}) shift of organic floating-gate transistors with changing amplitude and program pulse voltage duration (V_{Prog}); (c) duration 100 ms, amplitude 0 to −100 V and (d) duration 50 or 100 ms, amplitude −80 V. (e,f) Compensation of performance variation. Pixel driving current (I_{Driv}) of three different floating-gate transistors (FG-TFTs) were modified using dynamic control of V_{th} shift to reduce performance variation [8]. (Reproduced with permission. Copyright © 2012, Society for Information Displays.)

13.3 Demonstration

13.3.1 Organic Active-Matrix LED Pixel Circuits

By taking full advantage of state-of-the-art high-definition screen printing, low-resistivity Ag paste with low drying temperature, and printed organic transistors, we were able to fabricate an ultraflexible active matrix of organic LED pixel circuits on 10-μm polyimide films for applications in medical sensing and treatment. This printed floating-gate organic transistor circuit has a 230 mm × 230 mm printed active-matrix circuit comprised of 64 × 64 screen-printed organic 2T1C cells with a periodicity of 2.5 mm (Figure 13.5) and can compensate LED brightness variations and degradation for more than 6 months. Because the substrate of the circuit is made of a 10-μm film, its critical bending radius is <0.5 mm. By applying feedback control to the threshold voltages of the floating-gate organic transistors and the organic LED driving currents, we are able to realize a nonuniformity of <5% at luminances >200 cd m^{-2}. Although this circuit design and system for feedback control were discussed in our previous SID (Society for Information Display) papers [8, 17], this paper has demonstrated for the first time the development of a large-area, flexible, active-matrix circuit using high-definition screen printing. Furthermore, by taking advantage of new device configurations for organic floating-gate transistors, we have clearly demonstrated the feasibility of dynamic threshold voltage control as a feedback protocol for reducing electrical performance variations in organic LEDs and printed organic transistors.

13.3.2 Large-Area Flexible Pressure Sensor Sheet

After producing an active transistor matrix of printed organic transistors, we coated its surface with 5-μm-thick parylene passivation layers. A CO_2 laser process machine was used to make via holes through the parylene on the drain electrodes, and Ag paste was then deposited using a microdispenser system to form Ag electrodes and interconnections to each electrode through the via holes. Finally, a pressure-sensitive rubber and a PEN film with a Cu electrode were laminated using an active matrix. A cross-sectional illustration, picture, and circuit diagram of this structure are shown in Figure 13.7a–c.

Using this sensor sheet, we developed a flexible pressure sensor that uses an organic transistor active matrix to read the spatial distribution of pressure and assessed its characteristics. By applying pressure to the sensors, the resistance of the pressure-sensitive rubber decreases, resulting in the activation of organic FETs from the application of a drain–drain voltage (V_{DD}) (the principles of this effect are explained further in Ref. [11]). As can be seen in Figure 13.7c, we were able to obtain a clear picture of the spatial distribution of the pressure applied to the

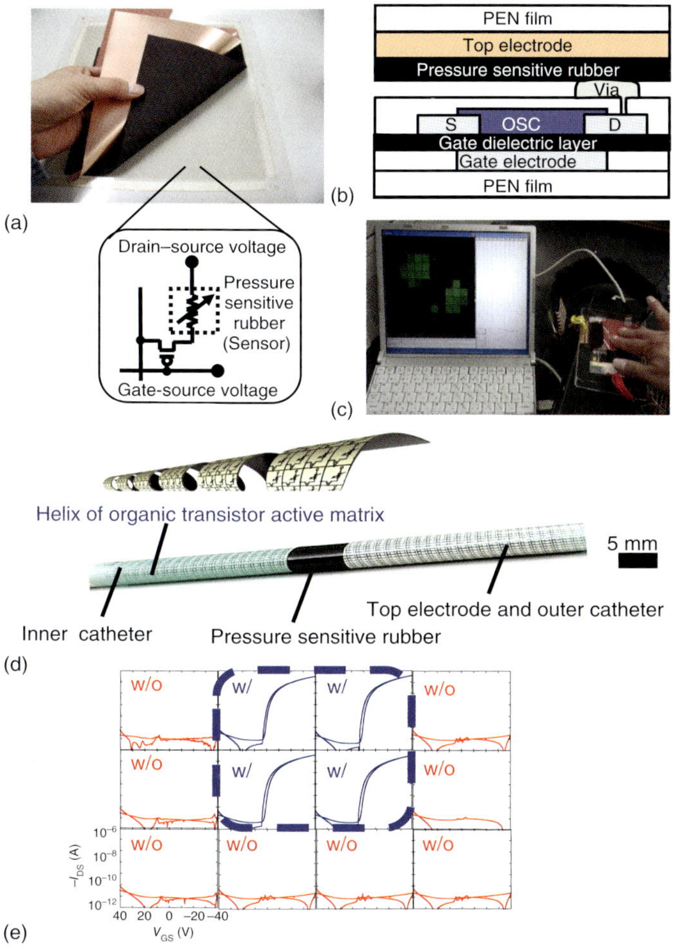

Figure 13.7 Demonstration of a pressure sensor sheet. (a) Development picture of a 300 mm × 300 mm pressure sensor sheet comprising (bottom) printed organic transistor active matrix, (middle) pressure-sensitive rubber, and (top) plastic film with Cu foil as common electrode. Schematic circuit diagram of one sensor cell is also shown. (b) Corresponding cross-sectional illustration of the pressure sensor. (c) Demonstration of real-time readout of spatial distributions of pressure applied by two fingers (left) and spatial distributions of pressures read out by the printed organic transistor active matrix (right). (Note that applying pressure to the pressure-sensitive rubber sheet creates a conducting path between the drain of the transistor and the counterelectrode. Thus, the potential of −40 V is present on the drain of the transistor only when pressure is applied, allowing the array to measure the spatial distribution of pressure). (d) Photograph of a tightly wound active-matrix sensor array for a catheter that measures the spatial distribution of pressure. The array was fabricated by laminating three sheets: (i) a foldable array of printed transistors, (ii) a pressure-sensitive rubber sheet, and (iii) a 10-μm-thick PEN sheet with printed Ag counterelectrodes. (e) Transfer characteristics of a 4 × 3 sensor cells measured at 10 kPa pressure. Transistors are activated only where mechanical pressure is applied [8]. (Reproduced with permission. Copyright © 2012, Society for Information Displays.)

sensor by a rubber object, thus demonstrating the feasibility of using a printed organic transistor active matrix as a read-out circuit for sensor applications.

13.3.3 Intelligent Sensor Catheter for Medical Diagnosis

To demonstrate a possible application for printed organic transistors that can operate in a bent state, we assembled a thin catheter to measure the spatial distribution of mechanical pressure. The sensor was fabricated by laminating three sheets (Figure 13.7d,e): a foldable 4×36 array of DNTT transistors, a pressure-sensitive rubber sheet, and a 10-μm-thick PEN film substrate with a printed Ag counterelectrode. The drain contacts of all 144 transistors used were connected to the rubber sheet, while the counterelectrode was in contact with the opposite surface of the rubber sheet. When mechanical pressure is exerted on the catheter, the electrical resistance between the top and bottom surfaces of the rubber decreases. At positions where pressure is applied, a resulting counterelectrode potential of −40 V is supplied to the transistors, allowing the spatial distribution of pressure to be obtained by interrogating the printed transistors in the active-matrix array.

13.4 Future Prospects

Future applications of large-area, flexible structures such as those developed in this work will enable the development of electronic circuits that can be mounted onto surfaces that have thus far been unsuitable for electrical functionalities. For example, inserting flexible organic circuits [18] into banknotes with large-area-compatible printing technologies would represent an ultimate anticounterfeit technology and tracking system. One of the most attractive features of flexible electronics is their compatibility with living bodies; in addition to wearable electronics, such as sensors and actuators that function from outside the body, disposable medical electronics that function inside the body are expected to play important roles in medical diagnostics, securing life and safety, healthcare, and smart welfare. Organic integrated circuits with exceptional flexibility and bending stability can be used to design new intelligent surfaces that can be used as human-friendly electronic interfaces. In the future, it is expected that new ways will be found to use such intelligent surfaces to interact with people, objects, and the environment.

Acknowledgments

We thank Mr. Y. Sano (SP-Solutions, Japan), Dr. T.-C. (Jim) Huang, Dr. K. Ishida, Prof. Takayasu Sakurai, Prof. Makoto Takamiya (University of Tokyo), Dr. Hiroyoshi Nakajima, Dr. Hiroki Maeda (Dai Nippon Printing Co., Ltd.), Nippon

Denki Kagaku Co., Ltd. and members of the Someya/Sekitani Lab for their technical support and valuable discussions. This study was supported by NEDO. We also thank Asada-Mesh Co., Ltd. for providing us with very fine shadow masks and Daisankasei Co., Ltd. for high-purity parylene (diX-SR).

References

1. Service, R.F. (2004) *Science*, **304**, 675.
2. Katz, H.E. (2004) *Chem. Mater.*, **16**, 4748.
3. Rogers, J.A. (2001) *Science*, **291**, 1502.
4. Hamedi, M. *et al.* (2007) *Nat. Mater.*, **6**, 357.
5. Garnier, F. *et al.* (1994) *Science*, **265**, 1684.
6. Rogers, J.A. *et al.* (2001) *Proc. Natl. Acad. Sci. U.S.A.*, **98**, 4835.
7. Sekitani, T. *et al.* (2007) *Nat. Mater.*, **6**, 413.
8. Sekitani, T., *et al.* (2012) Society for Information Displays (SID), Display Week, p. 32.3.
9. Yamamoto, T. and Takimiya, K. (2007) *J. Am. Chem. Soc.*, **129**, 2224.
10. Lloyd, G. *et al.* (2010) International Display Workshop (IDW)'10.
11. Noguchi, Y. *et al.* (2007) *Appl. Phys. Lett.*, **91**, 133502.
12. Sekitani, T. *et al.* (2005) *Appl. Phys. Lett.*, **86**, 073511.
13. Sekitani, T. *et al.* (2010) *Nat. Mater.*, **9**, 1015.
14. Klauk, H., Zschieschang, U., Pflaum, J., and Halik, M. (2007) *Nature*, **445**, 745.
15. Fukuda, K. *et al.* (2011) *IEEE Electron Device Lett.*, **32**, 1448.
16. Sekitani, T. *et al.* (2009) *Science*, **326**, 1516.
17. Huang, T.C., *et al.* (2011) Society for Information Displays (SID), Display Week, p. 13.2.
18. Zschieschang, U. *et al.* (2011) *Adv. Mater.*, **23**, 654.

14
Polymer and Organic Nonvolatile Memory Devices

Seung-Hoon Lee, Yong Xu, and Yong-Young Noh

14.1
Introduction

The main drive to develop organic nonvolatile memory is currently for applications in cost-effective, environmentally friendly, disposable/recyclable, lightweight, unbreakable, and easy-to-carry devices: printed and flexible (or stretchable) devices can satisfy these requirements. Flexible radio frequency identification (RFID) tags, near-field communication (NFC) devices, wearable electronics, and portable flexible displays are some examples of such wide electronic applications that can be realized in the near future. The items where today's integration of silicon chip does have economical merit, such as toys, cards, labels, badges, value paper, and medical disposables, can also be equipped with organic electronics including organic memory. Recently, numerous manufacturers such as Samsung, Sony, HP, and Nokia have been showing a great deal of interest in developing flexible display for e-readers, mobile phones, TVs, and many other consumer electronic items. For example, Sony illustrated a rollable OLED (organic light-emitting diode) display in 2010 driven by TFTs (thin-film transistors). Based on the flexible AMOLED (active-matrix organic light-emitting diode) display technology, Samsung Electronics demonstrated two prototype mobile devices at CES 2013 [1].

General memories can be divided into two main categories: volatile and nonvolatile. When the system power is off, volatile memory loses all data, whereas in the case of nonvolatile memory the stored information can be still retained. As shown in Figure 14.1, nonvolatile memory can be further divided into write-once read-many times (WORM) memory, hybrid nonvolatile and rewritable (flash) memory, and dynamic random access memory (DRAM). These are the types of organic memories that have been widely reported so far [2–4].

The simplest nonvolatile memory for such applications is the so-called hardwired memory, where binary data (bits) are defined as permanently present conductive lines during the memory production, for example, by printing conductive lines (additive) or ablating unwanted lines (subtractive). The second type is WORM memory , where the bits can be written on the site of use only

Large Area and Flexible Electronics, First Edition. Edited by Mario Caironi and Yong-Young Noh.

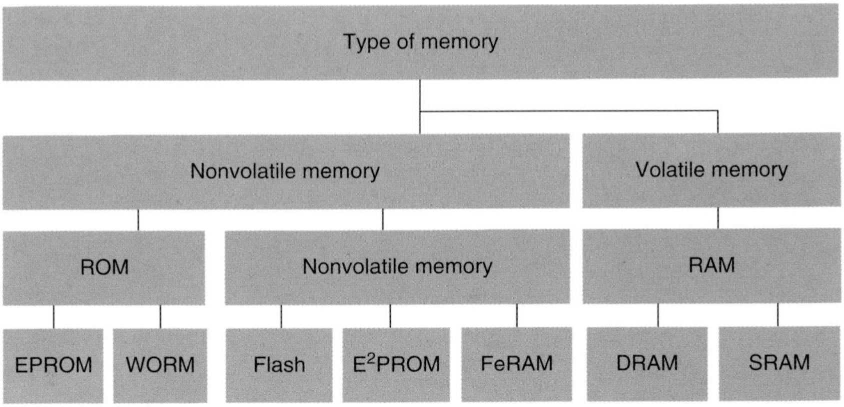

Figure 14.1 Category of memories, EPROM, electrically programmable read-only memory; WORM, write-once read-many; Flash, hybrid nonvolatile and rewritable; EPROM, electrically programmable read-only memory; E^2PROM, electrically erasable programmable read-only memory; FERAM, polymer ferroelectric memory, DRAM: dynamic random-access memory; and SRAM, static random-access memory.

once. WORM memory can be used to store archival standards, databases, and other massive data, where information is often kept accessible for a long term. Nowadays, the typical WORM memories are the conventional compact disc recordable (CDR), DVD $\pm$ R, or PROMs (programmable read-only memory). Reproducible and reliable technology by π-conjugated material has been presented by using amphorsulfonic-acid-doped polyaniline [5] and poly(3,4-ethylenedioxythiophene):poly(styrenesulfonate) (PEDOT:PSS) [6], showing the feasibility of fabricating organic memory [2, 7].

DRAM is a random access memory that saves each bit of data in a separate capacitor. As real-world capacitors inevitably leak charge, the stored information will finally vanish if the capacitor is not refreshed periodically. Because of this need for refreshing, it is dynamic memory and, because DRAM loses data after the power is switched off, it belongs to the volatile memory class. DRAM is nowadays the most important memory in most microcomputers. Interestingly, some organic memories also show volatile memory characteristics with the ability to write, read, erase, and refresh their electric states.

The most promising memory type should be reprogrammable, so that a mechanism of repetitively switching between different states is needed, and the most convenient mechanism is reprogrammability by a voltage or current, that is, the electrically erasable and programmable nonvolatile memory. In this chapter, we limit the discussion to this type of memory.

We, therefore, provide an overview of the device architectures that have been suggested in the past years to realize such electrically reprogrammable nonvolatile memory arrays, specifically using organic semiconductors and mostly compatible with organic TFTs. These principles have been pioneered in silicon technology, and are being or have been used there.

NAND flash memory is the leading nonvolatile memory produced by silicon technology, known for long retention and high density with low cost. The operation of a flash transistor-type memory is based on charge storage on a floating gate or in a gate dielectric layer of a transistor structure. A similar structure can be expected for organic electronics, since charge carriers can be equally well injected from the conjugated semiconductor into charge-storing gate dielectric (also called *electret*). Charge-trapping characteristics can be further optimized by incorporating specific trapping sites in the dielectric, for example, metal nanoparticles (NPs) or dispersed charge-accepting (and typically π-conjugated) molecules. Such concepts of transistor-based memory comprise a structure that is very similar to a normal transistor, and therefore this device architecture is beneficial for embedding memory in logic circuits and enables satisfying the requirements for a successful organic, nonvolatile memory technology.

In silicon technology, the scalability of flash memory devices toward nanosized transistors is approaching their physical limits. Contenders for future technology generations include diode-structured resistor-type bistable memory, which is sandwiched between two conductive electrodes (Figure 14.2). From Table 14.1, it can be seen that the prime concern for organic memories is not scalability to the nanometer size. However, the simplicity of crossbar arrays is highly attractive for achieving low cost. Thus, research to attain switchable resistor crossbar arrays for organic and printed electronics is an attractive field. Ultimate scaling of these devices could even conceivably lead to the use of a single molecular layer that features conductive switching.

Ferroelectric memories have been extensively researched as an alternative nonvolatile memory to be integrated with Si technology [7]. They are based on ferroelectric materials, which have two stable states of polarization, leading naturally to a nonvolatile binary (Boolean) system. Ferroelectric perovskite oxides such as lead zirconate titanate (PZT) and bismuth strontium tantalate (BST) were applied as dielectric layers of ferroelectric capacitors in one-transistor-one-capacitor (1T1C) memory cells, a commercial product today. Next-generation

Figure 14.2 Sketch of (a) an island and (b) a cross-bar resistive memory array with the memory material sandwiched between the bottom contacts (horizontal lines) and the top contacts (vertical lines).

Table 14.1 Requirements for thin-film nonvolatile memory arrays.

Parameter	Value
Retention	Several (10) years at maximum operation temperature
Write time	Micro to milliseconds
Read time	Micro to milliseconds
Memory size	Strongly application-dependent: Initial RFID item-level: 8 – 128 bit Expectations for smart packaging, cards, toys, badges, e-tickets, wireless sensor nodes: typically 1 – 64 kbit
Write voltage	<24 V
Read/write/erase endurance	Application-dependent: 10^3 to $>10^6$ write/erase cycles
Temperature range	Storage: −40 to +85 °C Operational: −20 to +50 °C
Density	Not critical up to 64 kbit: approximately square millimeter or even more per array
Integration with logic	Desirable for most applications
Cost	Critical per array, but not necessarily per bit
Extra features	Integrated on flexible substrates Transparent for some applications

Source: Reprinted with permission from Ref. [2]. Copyright © 2011, American Chemical Society.

ferroelectric memory structure uses a ferroelectric material in the gate stack of a so-called ferroelectric field-effect transistor (FeFET).

14.2 Resistive Switching Memories

Unlike the conventional Si memories that show device characteristics associated with specific structures (e.g., FETs), resistive switching memories show typically in the form of a crossbar or an island cell (Figure 14.2). Memory devices including switchable resistive materials are generically classified as resistor-type memory or resistive random access memory (RRAM). Different from the capacitor-type and transistor-type organic memories, which still need to be integrated into basic CMOS (complementary metal–oxide– semiconductor) logic circuits [2], the RRAMs are not based on silicon CMOS technology. The RRAMs are main contenders for scaled memory in silicon technologies.

14.2.1 Fundamentals of Resistive Switching Principles

Figure 14.3 shows the classification of the physical and chemical principles that can be used to obtain resistive switching properties [7]. Among these, electronic

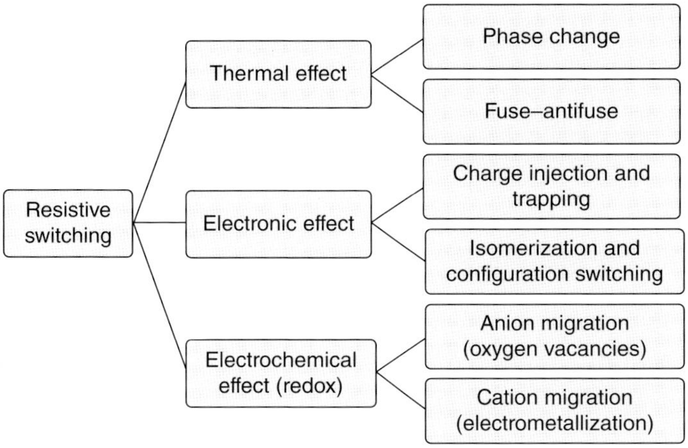

Figure 14.3 Classification of resistive switching effects for nonvolatile memory applications [2]. (Reprinted with permission. Copyright © 2011, American Chemical Society.)

effects and electrochemical effects have been shown with thin-film organic memories.

The programming/erasing operation of RRAM is to change their initially high-resistance OFF state (HRS)/low-resistance ON state (LRS) to LRS/HRS, and vice versa, by the application of a voltage or current across the electrodes. During the programming (or SET) process, the current abruptly increases from HRS to LRS as the voltage increases from 0 V to the critical point which is referred to as the *threshold voltage* (V_{Th}). On the other hand, the current level also abruptly decreases from LRS to HRS as the reset voltage (V_R) resets under the erasing (or RESET) process. Sweeping the gate voltage between the binary states of LRS and HRS repeatedly performs the SET and RESET processes.

Both unipolar and bipolar memory systems exhibit electrically reversible switching. Unipolar memory (Figure 14.4a) uses the same polarity of voltage to write in and erase data. The transition at V_1 from the low current state (i.e., the OFF state) to the high current state (i.e., the ON state) serves as the "SET" or "Writing" process (i.e., the first sweep). The RESET operation (transition from HIGH to LOW current state) is performed at voltage V_2 at which the current across the memory cell is higher than that during the SET process. Unipolar memory has several technical advantages; for instance, it enables rewritable switching properties in the architecture of one diode and one resistor (1D1R) in order to avoid crosstalk [8]. Moreover, the external circuits that are required in practical applications can be also simplified by using a power supply of only a single polarity. Nevertheless, signals of different polarities that are often required for bipolar (= opposite polarity switching) memory might be optionally used for unipolar memory; see Figure 14.4b. Bipolar memory applies a signal of one polarity for the SET operation and the opposite polarity for RESET operation. The current during the SET operation should be limited not to exceed the current compliance in order to prevent overprogramming.

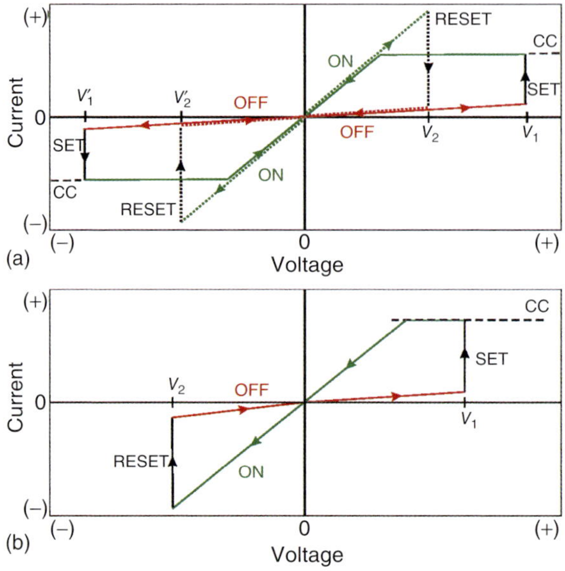

Figure 14.4 Basic *I*–*V* schemes of resistive switching memory cells (linear current and voltage scales). (a) Unipolar switching with the SET operation shown with full line and the RESET operation in dashed line (switching is shown for both polarities). (b) Example of bipolar switching in the case of a memory cell where the SET operation is performed at positive voltages and the RESET operation at negative voltages. CC denotes limitation of the current compliance by a current compliance [2]. (Reprinted with permission. Copyright © 2011, American Chemical Society.)

14.2.2 Mechanisms of Resistive Switching

14.2.2.1 Filamentary Conduction

Filamentary conduction is generally observed when the flow of ON-state current is confined to a highly localized small fraction of device area [9]. Two types of filamentary conduction are widely suggested for RRAM devices. One is the carbon-rich filaments formed by local degradation of organic films [10, 11]. The other one is the metallic filament that is related to the migration of the electrode through the organic film [12, 13]. For the metallic filament, there are three important parameters: electrode thickness, film thickness, and the nature of the "forming" atmosphere [14]. For example, Ag ions penetrating into organic layer under forward bias could result in filament formation in Ag/WPF-BT-FEO (poly[(9,9-bis(6′-(*N*,*N*,*N*-trimethylammonium) hexyl)-2,7-fluorene)-*co*-(9,9-bis(2-(2-methoxyethoxy)ethyl)-fluorene)-*co*-(2,1,3-benzothiadiazole)] dibromide)/heavily doped p-type poly-Si device, which was experimentally proved by transmission electron microscopy (TEM) analysis. In inorganic resistive memory, Lu's group [15] recently reported the lateral conductive Ag filaments through *in situ* TEM. In their study, the narrowest region of the filament was found to be near the dielectric/inert-electrode interface in the memory devices, irrespective of the growth direction.

14.2.2.2 **Space Charge and Traps**

If the contact between electrodes and organic material is ohmic and the organic material is trap free, the accumulation of charge carriers near the electrodes can build up a space charge region. Mutual repulsion between individual charges prohibits the total charges from being injected into the organic active layer, and the resultant current is called *space-charge limited current* (*SCLC*). All ohmic contacts do not always form the SCLC. The reason for the occurrence of SCLC is the difference in mobility from the electrode to the organic active layer. Figure 14.5 shows the typical bipolar $I-V$ characteristics and its log–log plot for SCLC analysis. In the low-voltage region (a), the slope of the logarithmic plot of the $I-V$ curve (see Figure 14.5b) is approximately 1 in all organic memory devices. This fits typical ohmic behavior induced by thermally generated free carriers. In the next (b) region, the trap-limited SCLC is found to fit well the $I-V$ curve for all memory devices. In the SCLC conduction model, the current density J can be described as [16–21]

$$J \propto d \left(\frac{V}{d^2} \right)^m$$

where V, d, and m are the applied voltage, the sample thickness, and the fitting parameter, respectively. If the insulator is ideal, which means that there are no free carriers in equilibrium and no bulk traps, m is equal to 2. On the other hand, in extrinsic materials containing traps, values of m ranging from 2 to 11 have been reported [16–21]. When a memory material has no or poor mobility, numerous factors can contribute to SCLC phenomenon: injection of electrons or holes from electrode, presence of ionized dopants in interfacial depletion regions, and accumulation of mobile ions at electrode/organic material interface.

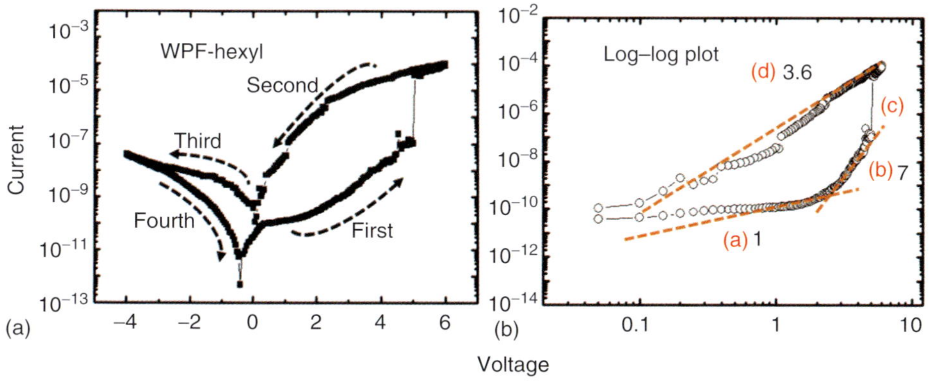

Figure 14.5 (a) Typical $I-V$ characteristics of bipolar memory devices. (b) A log–log plot of the current as a function of the applied voltage for (a).

14.2.2.3 Charge Transfer

A charge-transfer complex, which is defined as an electron donating moiety–electron acceptor complex, operates by electronic transition to the excited states accompanied by partial electronic charge transfer from the donor to the acceptor compound.

14.2.2.4 Ionic Conduction

Ionic conduction normally occurs in materials that contain ionic groups, such as polyelectrolytes, ionomers, or ionic liquids. Adsorption of water may play a key role because water can act as an ion source, a local structure modifier, or a dielectric impurity. A long transition time and high activation energy are the typical characteristics of ionic conduction compared to electronic conduction.

14.2.3 The Role of π-Conjugated Material in Switching Process

In some cases, a π-conjugated material plays a main role in the switching process compared to filament conduction or the SCLC mechanism. In the case of copper 7,7,8,8-tetracyanoquinodimethane (CuTCNQ) (where TCNQ is 7,7,8,8-tetracyano-*p*-quinodimethane) the metal-organic material was suggested to play the critical role in memory devices. In such a system, CuTCNQ acts as a solid ionic conductor containing mobile Cu^+ cations that form conductive (metallic) filaments by an electrochemical reaction (Figure 14.6). The Cu^+ ion of reduction occurs at the negative electrode, and the Cu filaments grow in a (porous) oxide layer at that electrode, for example, native aluminum oxide [22]. This suggested role of CuTCNQ also explains the reported switching of this material when probed with a scanning tunneling microscope tip: metallic filaments growing from the tip can connect the nanometer gap between the material and the tip, similar to what has been reported for Ag_2S "quantum point contact switch" [22]. The role of CuTCNQ is similar to that of Ag_2S because the organic material fulfills the requirements of good (cat)ionic conductors [22]: it has (i) strongly polarizing monovalent cations (as, for example, Ag^+ and Cu^+), (ii) highly polarizable anions (because the charge in the relatively large π-conjugated system $TCNQ^{\bullet-}$ is highly delocalized), and (iii) 4-, 3-, and perhaps 2-coordination of the cations by the anions (in CuTCNQ, the Cu^+ cations are tetra-coordinated by the nitrile groups of the $TCNQ^-$ anions). According to this set of requirements, it would be expected that the ionic conductivity of Cu^+ salts with π-conjugated organic anions without nitrile groups should be poorer, and therefore switching properties of compounds such as copper tetrachloro-*p*-benzoquinone should be worse than those of CuTCNQ.

π-Conjugated polymers with strongly coordinating heteroatoms such as N and S (as, e.g., poly(3-hexylthiophene), P3HT) enable the realization of reproducible electrical switching involving metallic filament formation in memory systems with Al bottom contacts and Cu top contacts [23]. When a bias is applied to the Cu electrode, mobile copper ions formed by electrochemical oxidation of the Cu top contact can migrate within the active polymer [23].

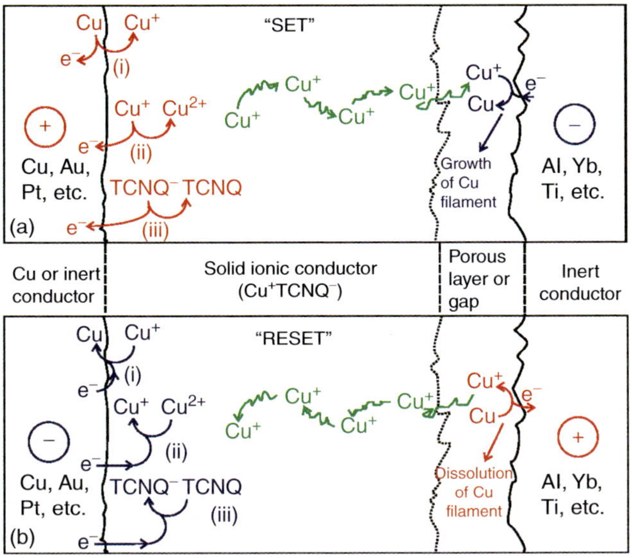

Figure 14.6 Switching mechanism of organic memories containing CuTCNQ. (a) The SET operation starts with the formation of a conductive Cu filament by electrochemical reduction of copper ions at the cathode and simultaneous electrochemical oxidation in another species (i), (ii), or (iii) at the anode. (b) The RESET operation under opposite polarization leading to the disappearance of the conductive channel [2]. (Reprinted with permission. Copyright © 2011, American Chemical Society.)

Like that of CuTCNQ, the switching of Cu/conjugated polymer with N or S atoms/Al memory devices takes place by reversible Cu filament formation within the thin AlO_x layer because it is impossible to prevent the formation of a thin native AlO_x layer in the case of Al bottom contacts.

14.2.4
Recent Flexible RRAM Based on Organic–Inorganic Bistable Materials

Two-dimensional nanomaterials have demonstrated a vast array of unique physical, optical, and electrical properties [24]. Their novel structure and unique properties are promising for future electronics and surely for memory applications. Recently, Liu *et al.* demonstrated a flexible RRAM device based on solution-processed MoS_2-PVP (polyvinylpyrrolidone) nanocomposites containing a structure of PET/rGO (polyethylene terephthalate/reduced graphene oxide)/MoS_2-PVP/Al by using a facile method for exfoliation and dispersion of MoS_2 with the aid of PVP [24b]. These memory devices showed an on:off ratio of around 10^2. In that paper, the charge trapping and detrapping behavior of the composite was proposed to explain the SCLC model (Figure 14.7).

Son *et al.* [18] also demonstrated a flexible organic bistable device by incorporating a graphene sheet into the insulating PMMA (poly(methyl methacrylate))

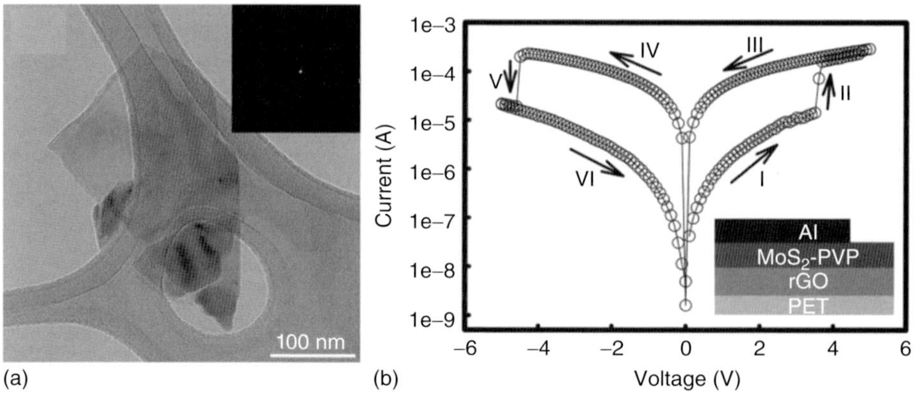

Figure 14.7 (a) TEM image of an exfoliated MoS_2-PVP nanosheet. Inset shows the selected area electron diffraction (SAED) pattern of the corresponding MoS_2 sheet. (b) The I–V characteristics of rGO/MoS_2-PVP/Al flexible memory device. Inset: Schematic diagram of the memory device structure [24]. (Reproduced with permission. Copyright © 2012, Wiley.)

layer to act as the active layer. The graphene sheet is sandwiched between PMMA polymer layers with the structure PET/ITO/PMMA/graphene/PMMA/Al (ITO, indium tin oxide). Those devices exhibited a maximum on:off ratio of 10^7, an endurance number of 10^5 cycles, and a retention time of 10^5 s. To verify the conduction mechanism of such memory devices, $I-V$ characterization and the log–log plot were examined. The main contribution to the bistable behavior was found to be by the conducting filament formed by the diffusion from Al electrode. The authors speculated that only those electrons trapped near the lowest unoccupied molecular orbital (LUMO) level of PMMA do not have the capability to generate a conducting filament unless the graphene layer with sufficient trap sites is inserted.

14.3
Charge Storage in Transistor Gate Dielectric

Research on developing organic nonvolatile memories has been almost exclusively based on organic field-effect transistors (OFETs) because these devices can be read without destroying their memory state ("nondestructive read-out"). These transistor-type memory devices have a manufacturing advantage because of their architectural compatibility with integrated circuits. Furthermore, the sneak current problem in cross-bar type devices can also be solved by using the integration of transistor devices.

Transistors comprise three electrodes: source, drain, and gate. Organic transistors use π-conjugated organic molecules as active material (semiconductor) for connecting the source and the drain electrodes and a thin gate dielectric layer for insulating the semiconductor film from the gate electrode. Charge carriers can be accumulated at or depleted from the semiconductor/dielectric interface by applying an appropriate gate bias (V_g). In the accumulation mode, the charge carriers create a conductive channel between the source and the drain electrodes, and upon applying a drain bias, a current (I_d) can flow though the semiconductor film. Therefore, the gate controls the OFET as a switch, and the threshold voltage (V_{Th}) is defined as the voltage at which a conductive channel begins to form. In the region far beyond V_{Th} (strong accumulation), the charge density per unit area becomes proportional to the difference between V_g and V_{Th}. Transistors can be easily integrated into logic gates and thus into more complex logic circuits [25]. The characteristics of π-conjugated organic molecules and the polarity of the applied biases determine whether the primary charge carriers are electrons (giving rise to so-called "n-type transistors") or holes ("p-type transistors").

14.3.1 Operation of Charge-Storage OFET Memory Devices

The accumulated charges in OFETs are basically "volatile," that is, only present as long as a gate voltage exceeding V_{Th} is applied. If either reversible charge trapping or detrapping of the gate dielectric layer occurs, such transistors can be used for memory application. Several types of gate dielectrics enable reversible trapping of charges upon application of a gate field. For example, polymer electrets are dielectrics with embedded metallic or semiconducting NPs or organic conjugated molecules, and ferroelectric gate insulators with permanent and/or switchable electrical dipoles can achieve the memory effect in OFETs [4]. In this section, we introduce the basic properties, operating mechanisms, and recent progress in memory devices based on OFETs with gate insulators that have charge-storage capacity.

Charge trapping in the gate dielectric causes a shift of the threshold voltage (ΔV_{Th}) [26]. The magnitude of ΔV_{Th} is proportional to the stored charge density per unit area Δn:

$$\Delta V_{Th} = \frac{e\Delta n}{C_i} \quad (14.1)$$

where e is the elementary charge, and C_i is the capacitance of the gate dielectric per unit area. This shift, that is, the V_{Th} difference between the fresh transistor and the one with charged gate dielectric, can be confirmed in the transfer curve of the transistor. This V_{Th} shift is called as the *memory window*. As shown in Figure 14.8, there are two different stable states of the drain current for each V_g, namely high I_d and low I_d representing "1" and "0" states of the memory, respectively, which is one of the core functions of modern computers. Here, it is important to distinguish the memory behaviors from the bias-stress effect which also leads to V_{Th} shifts due to the presence of defects and impurities in the semiconductor film and/or the gate dielectric [5]. The bias-stress effect would enhance the memory window but it cannot be controlled and is not permanent either. Nonvolatile memory means

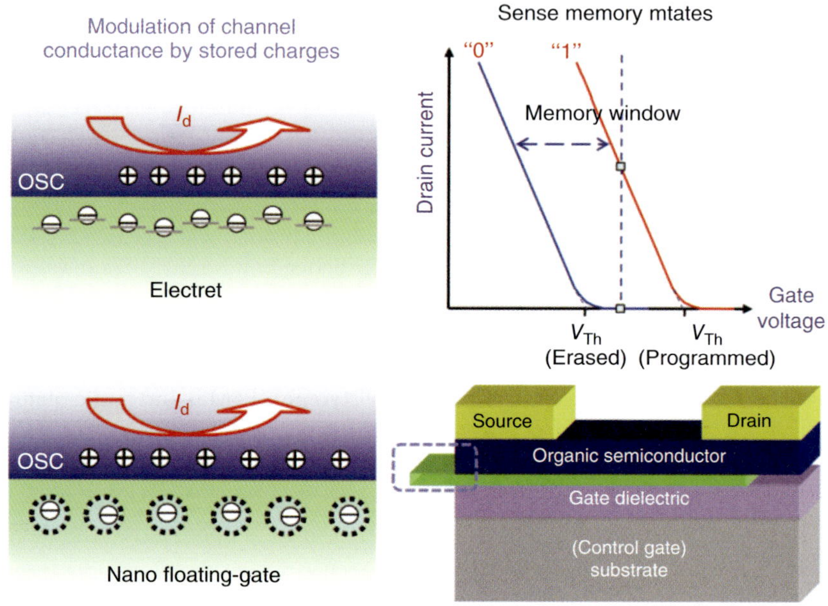

Figure 14.8 Schematic configuration and operating mechanism of an organic memory with a p-type transistor [2]. (Reprinted with permission. Copyright © 2011, American Chemical Society.)

that the V_{Th} shift is (quasi) permanent as long as the state of the device is not intentionally overwritten. If the programming state can be reversibly recovered to its original state by applying an appropriate bias, the memory transistor is rewritable. Such transistors are programmed by shifting their V_{Th} such that two different stable states are established at zero V_g, and, ideally, the two states correspond to positive and negative V_{Th}. This ensures that it allows the read-out to be maintained at state zero V_g without the requirement of applying a gate field, which is potentially destructive. Additionally, it enables low power consumption and reliable data storage. The charge-storage mechanisms in the gate dielectric layer OFET-based memory devices can be classified into three types: (i) charge-trap memory, (ii) floating-gate memory, and (iii) ferroelectric memory. Ferroelectric memory transistors will be described in the next section of this article, so only floating-gate and charge-trap-based OFET memory are discussed in this section.

14.3.2
Charge Storage in Polymer Electrets

An electret is defined as a dielectric material that has a quasi-permanent electric charge or dipolar polarization, including ferro-, piezo-, and pyroelectric polymers [27, 28]. Polymer electrets have been very useful in a wide variety of applications such as in transducers, electrophotography, optical display systems, gas filters, and motors because of their ability to store charges for a relatively long time

[27, 28]. Katz *et al.* [29] first demonstrated an organic transistor-type memory device with a polarizable gate insulator. Both p-type 1,4-bis(5-phenyl-2-thienyl)benzene (PTPTP) and n-type *N,N′*-bis(1*H*,1*H*-perfluorooctyl) naphthalene-1,4,5,8-tetracarboxylic diimide (F_{15}-NTCDI) OFETs were demonstrated. Switching V_{Th} from the accumulation mode to either zero or the depletion mode was demonstrated by the application of a depletion voltage (i.e., positive V_g for p-type semiconductors and negative V_g for n-type semiconductors) on the gate electrode. The PTPTP and F_{15}-NTCDI semiconductors were chosen because of their "Normally ON" property, that is, their initial V_{Th} is positive for the p-channel PTPTP and negative for the n-channel F_{15}-NTCDI OFETs. As chargeable gate dielectrics, two inorganic materials (SiO_2 and glass resin) and two hydrophobic polymers (poly(4-methylstyrene) (P4MS) and cyclic olefin copolymer) were used. The hydrophobic polymer dielectrics exhibited better memory effect as a result of their superior charge-storage (electret) properties. However, such memory devices need a relatively long switching time (∼10 min).

Baeg *et al.* [30] reported an advanced memory transistor with pentacene and dual-gate dielectrics composed of poly(α-methylstyrene) (PαMS) and SiO_2. This memory device had a very large memory window with a relatively long retention time of >100 h (Figure 14.9a,b). The long retention characteristic indicates that mobile charge carriers from pentacene are deeply trapped in the PαMS layer or at the interface between PαMS and SiO_2. The device also showed a very short switching time (up to microseconds). In experiments using various polymer electrets, the same group also reported the importance of using polymer electrets to realize stable memory behavior (Figure 14.9c) [31]. OFETs with nonpolar and hydrophobic polymer electrets, such as polystyrene (PS), PαMS, P4MS, and poly(2-vinyl naphthalene) (PVN), showed only the systematically controllable and reversible shifts in the transfer characteristics. Furthermore, the magnitude of the memory windows is inversely proportional to both the hydrophilicity and polarity of the dielectrics (Figure 14.9d). Recently, Debucquoy *et al.* [32] observed that holes trapped in the gate dielectric of p-type OFETs are more easily erased by overwriting trapped holes with injecting electrons from the semiconductor channel (at positive V_g) than by detrapping of holes. This means that reversible switching of the memory needs the (nominally p-type) semiconductor containing some ambipolarity; that is, some amount of mobility is required for both types of charge carriers, electrons and holes, in the semiconductor. Very recently, this group reported a flexible printed organic NAND (NOT AND) flash memory based on high-performance OFETs using bilayer polymeric dielectrics and a top-gated structure (Figure 14.9) [33]. Because of the smaller energy bandgap and highly localized deep trap density, these memory devices showed better charge-transfer and trapping properties of PVN than PS. Bilayer dielectrics of low *k* and high *k* can inject and trap electrons in an electret because the applied electric field is stronger in the low-*k* layer than in the high-*k* layer. The authors also designed a strategy for the expansion of monolithically inkjet-printed flexible organic NAND flash memory on polyethylene naphthalate (PEN) substrates

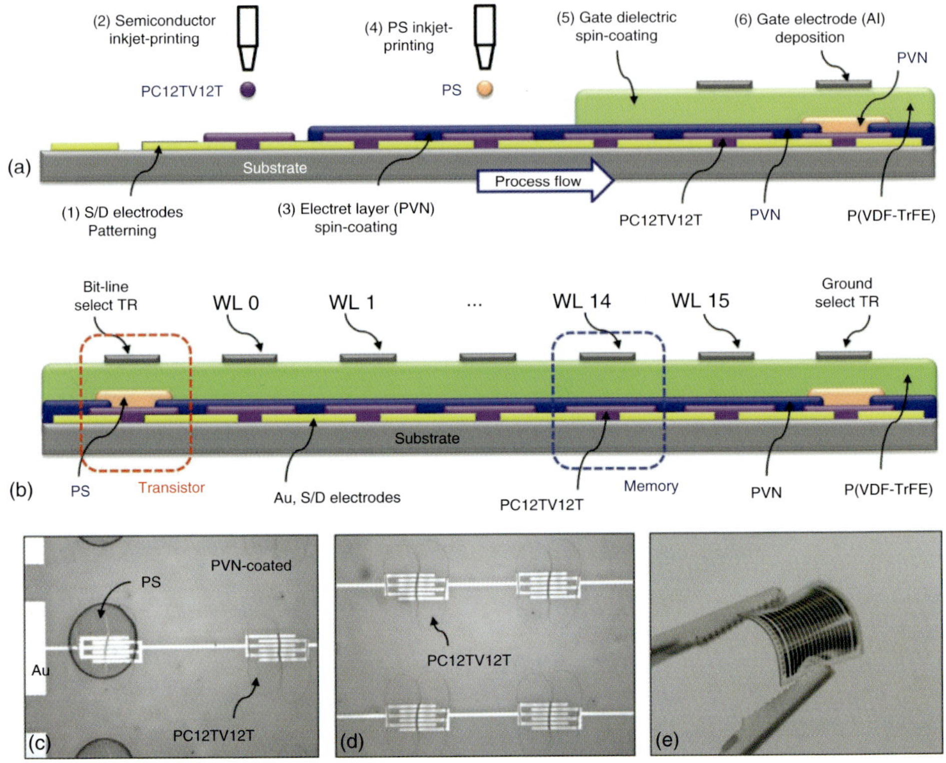

Figure 14.9 Facile strategy for developing flexible printed organic NAND flash memory. (a) Schematic diagram of fabrication process flow and (b) schematic illustration of a cross-sectional view of the 256-bit NAND-type organic flash memory. Charge coupled device (CCD) camera images (c) of PS inkjet-printed onto PVN and PC12TV12T layers and (d) inkjet-printed PC12TV12T active feature and spin-coated PVN layer above the PC12TV12T. (e) Digital camera image of the fabricated flexible printed organic flash memory on a PEN substrate [33]. (Reproduced with permission. Copyright © 2012, Wiley.)

through delicately selecting polymer dielectrics, where poly[(vinylidenefluoride-co-trifluoroethylene] (PVN/P(VDF-TrFE)) and PS/P(VDF-TrFE) devices were utilized as nonvolatile memory cells for ground-line and bit-line selecting transistors, respectively. Electrical simulations showed that the flexible printed organic NAND flash is simple to be programmed, read, and erased in the memory array repeatedly without disturbing the nonselected memory cells.

14.3.3 Nanoparticle-Embedded Gate Dielectrics

Another type of charge-trapping OFET memory is organic floating-gate memory [34]. In this device, charges are stored in a metal or semiconducting layer called the *floating gate*, which is located within the insulating gate dielectric and is completely surrounded by an insulator (Figure 14.8). Charges are injected into the floating gate from the transistor channel by applying a high V_g. As the floating gate traps charges in the insulator, the transistor's V_{Th} is shifted similar to the electret memory. To discharge the floating gate, a reverse bias is applied on the gate electrode. The magnitude of the V_{Th} shift depends on the density of the trapped charges in the floating gate [35]. A variety of insulating polymers can be used for floating-gate memory as the controlling and tunneling gate dielectrics.

In silicon technology, flash memory often uses floating gate, but this geometry has not been demonstrated for organic memory so far [4]. As pioneered in silicon, an alternative is to insert conducting NPs in an insulating gate dielectric matrix [36, 37]. Each particle is an independent charge-storage site isolated from the others. This is more robust against loss of charge as compared to a floating gate.

The major advantages of Au NPs are their great chemical stability and the fact that they can be formed and processed from solution [37]. Zhen *et al.* [38] demonstrated an all-organic NP floating-gate memory device with CuPc as the active channel. They fabricated floating gates using e-beam-deposited Au NPs inside a polyimide (PI) gate dielectric layer. Holes are injected from CuPc into the Au NPs by the application of a negative gate bias by Fowler–Nordheim tunneling. To erase data, the stored charges could be removed from the Au NPs back to the CuPc under a positive gate bias. Mabrook *et al.* [39] chose the solution-processed Au NPs formed by a self-assembled dipping method on PMMA to obtain the floating gate instead of using an e-beam. However, the memory devices showed poor retention time because the Au NPs were positioned directly in contact with the (pentacene) transistor channel, leading to spontaneous leakage of the stored charges. To separate the transistor channel (pentacene) from the Au NP floating gate, Kim *et al.* [40] incorporated Au NPs under a tunneling dielectric layer (PMMA or PVP/HfO_2). The Au NPs were deposited by using the layer-by-layer (LbL) method on positively charged poly(allylamine hydrochloride) (PAH), as part of a PAH/PSS multilayer dielectric. A tunneling layer of HfO_2 separated the Au NPs from the pentacene channel on top of this stack. This device exhibited certain shifts in V_{Th} and a large memory window of 3–5 V after application of a V_g of 50–70 V with a relatively long retention time (several months). A similar approach was used to solution-processable conjugated polymers. Liu *et al.* [41] reported a floating gate memory transistor using electrostatic LbL self-assembly of Au NPs. However, the retention time was as short as ~200 s because the poor insulating properties of the PVP barrier layer might not have worked well [31]. Baeg *et al.* [34] reported a solution-processed poly[[9,9-dioctylfluorenyl-2,7-diyl]-*co*-(bithiophene)] (F8T2) transistor with a top-gate/bottom-contact configuration (Figure 14.10) with Au NPs incorporated between the bilayered dielectrics PS and cPVP. A flexible polymer memory device was also fabricated on a PET plastic substrate, operating as well as those on glass.

GO is the insulating form of graphene, and surface carbonyl, carboxylic, and hydroxylic groups are introduced into graphene to create more charge-trapping

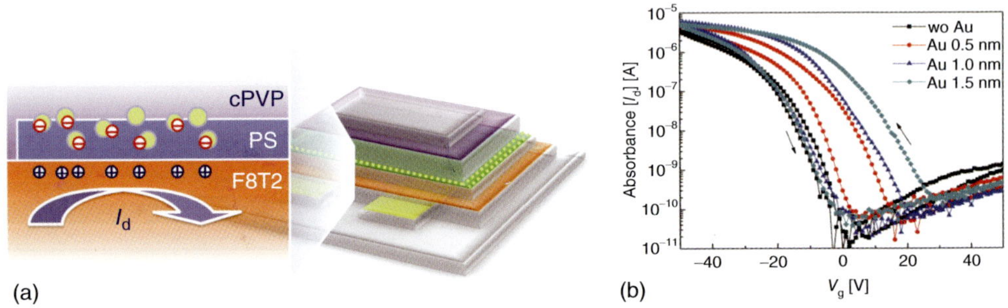

Figure 14.10 (a) Schematic device configuration of a top-gate/bottom-contact F8T2 floating-gate memory transistor and the charge-trapping mechanism in Au NPs. (b) Transfer characteristics of F8T2 OFET after thermal evaporation of Au of different thicknesses ranging from 0.5 to 1.5 nm, where V_g was varied from 90 to −90 V [34]. (Reproduced with permission. Copyright © 2010, Wiley.)

sites and destroy the conjugation after oxygen decoration [42]. GO has been successfully applied in the flexible charge-trapping transistor memory device [43]. Transparency is another attractive feature for the next-generation light-weight, flexible, and portable electronic systems. Yu *et al.* [43] demonstrated ultratransparent flexible memory devices based on graphene electrodes and an active channel of single-walled carbon nanotube (SWCNT) networks (Figure 14.11). Additional ozone treatment induced the oxidation of the graphene electrode to bond oxygen atoms to the surface of graphene as C–O–C, C=O, and C–OH to act as charge-trap sites. The memory device showed no transmittance reduction after oxidation, in good contrast with devices with Au and Al NP trap layers that produced 11.4% and 25% decrease of transmittance, respectively. The nonvolatile transistor-type memory device showed a mobility (μ) of ~44 $cm^2\ V^{-1}\ s^{-1}$ and an operation speed of 100 ns. The memory window was stable during bending tests, confirming their excellent mechanical flexibility and programmable properties. However, the short retention time of just 1000 s indicated that the devices might contain charge-trapping sites. The structure employed in this work with oxidized graphene electrode as charge-trapping element is unlike the traditional one in which the charge carriers are trapped by the insulating layer between the tunneling and blocking dielectric layers. The relatively short retention time was presumably due to the direct connection between the traps and the electrode, which leads to easy leakage of stored charges.

14.4 Polymer Ferroelectric Devices

Ferroelectrics are polar substances of either solid (crystalline or polymeric) or liquid crystal material, in which the spontaneously generated electric polarization can be reversed by inverting the external electric field. The alignment

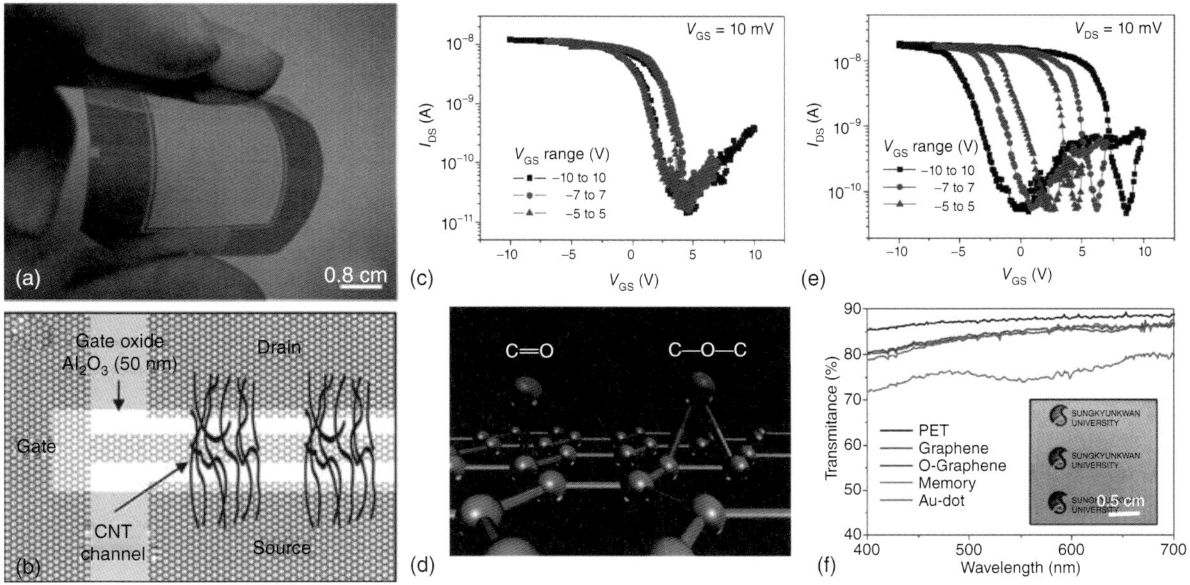

Figure 14.11 (a) Photograph of the transparent graphene–CNT integrated circuit array (16 × 18). (b) Schematic of the graphene electrode (source, drain, and gate) and SWCNT network channel on the PET film. Hysteresis in the (c) graphene gate electrode and (e) oxygen-decorated graphene gate electrode. (d) 3D model of the C=O and C–O–C bonds formed on the graphene surface. (f) Transmittances of the PET, graphene, oxygen-decorated graphene, memory device, and the Au NP-decorated device. The inset image shows the uniform transmittance of the fabricated memory device array on a PET film [43]. (Reproduced with permission. Copyright © 2011, Wiley.)

of intrinsic dipole moments inside the ferroelectric material induces spontaneous polarization. Mathematically, the equation describing the ferroelectric polarization is

$$\boldsymbol{D} = \varepsilon_r \boldsymbol{E} + \boldsymbol{P}$$

where $\boldsymbol{P}$, $\boldsymbol{D}$, and $\boldsymbol{E}$ are the polarization, dielectric displacement, and electric field, respectively, and ε_r is the relative permittivity of dielectric material. The spontaneous polarization $\boldsymbol{P}$ originates from the alignment of intrinsic dipole moments inside the material. Depending on the direction of the applied external electric field, the direction of these dipoles can be changed up or down.

The critical electric field for reversing the polarization of a ferroelectric material is referred to as the *coercive field*. The electric displacement as a function of field strength consequently draws a hysteresis curve ($\boldsymbol{D}-\boldsymbol{E}$ loop) between opposite polarities. These characteristics give an electric bistability. In particular, ferroelectric random access memory (FeRAM) and FeFETs are very promising for practical applications.

A standard method for measuring the $\boldsymbol{D}-\boldsymbol{E}$ loop is to use the Sawyer–Tower circuit (Figure 14.12) [44]. A sinusoidal voltage signal is applied to one of the electrodes of the ferroelectric capacitor, and the amount of charge displacement in the other electrode is measured using the voltage that is created on a reference capacitor connected in series. An idealized $\boldsymbol{D}-\boldsymbol{E}$ hysteresis curve is shown in

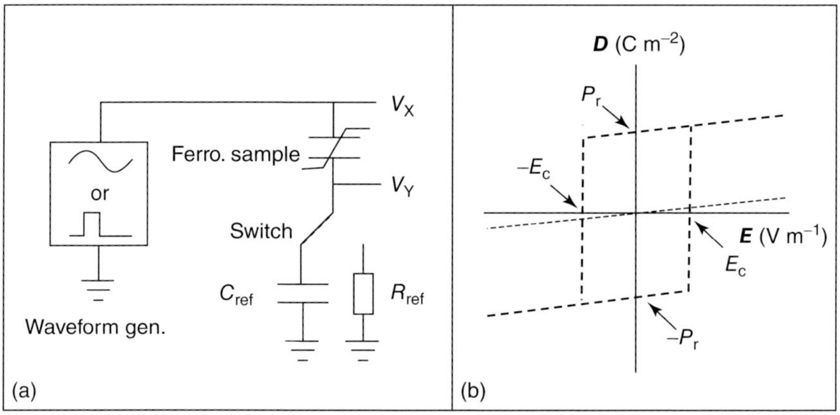

Figure 14.12 (a) Hysteretic loop of ferroelectric thin film between two electrodes. The displacement ($\boldsymbol{D}$) of the dielectric is given by the sum of free charges ($\varepsilon_r \boldsymbol{E}$) and polarization ($\boldsymbol{P}$), with $\boldsymbol{E}$ being the externally applied electric field. The slope of the line is the dielectric constant of the material and characterizes the charging of capacitance. In this idealized case, the remanent polarization P_r can adopt two opposing values (positive and negative), the absolute magnitudes of which are equal and constant. They are interchanged instantly at a particular externally applied field, called the coercive field E_c. For $-E_c < \boldsymbol{E} < E_c$ the capacitor may contain different charges (polarizations) at each $\boldsymbol{E}$. (b) Measurement of $\boldsymbol{D}-\boldsymbol{E}$ hysteresis and transient switching phenomena using the Sawyer–Tower setup [2]. (Reprinted with permission. Copyright © 2011, American Chemical Society.)

Figure 14.12 for a ferroelectric capacitor with two parallel electrodes. At low voltages, only the linear component is measured: the field is not large enough to affect $\boldsymbol{P}$. At higher fields, a hysteretic response develops, which saturates at even higher fields. P_r is the remanent polarization, defined as the polarization after removing the external electric field ($\boldsymbol{E}=0$), and E_c is the coercive field, which is the field needed to reduce the polarization to zero ($\boldsymbol{P}=0$). P_r and E_c are the key parameters describing the ferroelectric characteristics of a material. A ferroelectric with high coercive field yields a higher memory window ($2E_c \times d$ with d being the thickness of the ferroelectric), but this is at the expense of higher voltages required for switching. A ferroelectric with high remnant polarization P_r leads to higher switching currents in capacitors and currents in transistors [45] but cannot be fully polarized using low-voltage operation. Hence, subloops are used, and they are less stable.

14.4.1
Materials

In 1969, Kawai [46] first reported the piezoelectric effect in the polymer polyvinylidene fluoride (PVDF). Just 2 years later, the ferroelectric properties of the same polymer were reported. Throughout the 1970s and 1980s, these polymers were applied to data storage and retrieval. Since the discovery of PVDF, many other polymers have been sought for ferroelectric properties. Initially, different blend systems and copolymers of PVDF were studied, such as a PVDF with PMMA [47]. Other polymers were also discovered that possessed ferroelectricity, such as polytrifluoroethylene [48], copolymers of VDF and TrFE, and odd-numbered nylon [45, 49, 50].

Small-molecule organic ferroelectrics are very rare [51], probably because polarization reversal of a small molecule requires reorientation of the whole molecule within tight crystalline structures. A notable exception is liquid-crystalline materials. They switch at low electric fields and are therefore used in displays. For memory purposes, their low P_r value is unfavorable.

The ferroelectric parameters of the most studied organic ferroelectric materials are listed in Table 14.2. Also included are the parameters of the two inorganic ferroelectric materials that are currently studied most: $SrBi_2Ta_2O_9$ (SBT) and $PbZrTi_{1-x}O_3$ (PZT) both of which are used in memories. They both have the perovskite structure and a high dielectric permittivity (high k) of ~250 in thin films [52]. Hybrid organic–inorganic composite materials are also explored and have characteristics dependent on the loading of the inorganic NP.

From Table 14.2, it is clear that PVDF and the copolymers of VDF and TrFE are very promising candidates for ferroelectric memory because of their relatively large remanent polarization and short switching time. However, the switching time of ferroelectric nylons, for example, is longer by orders of magnitude at the same applied field [47]. Hence, the properties of PVDF and P(VDF-TrFE) will be further discussed.

Table 14.2 Parameters of ferroelectric materials.

Material class	Material	E_c (MV m^{-1})	P_r (mC m^{-2})	Switching time (s)
Inorganic perovskites	$SrBi_2Ta_2O_9$ $PbZrTi_{1-x}O_3$	<5	100–500	$<10^{-8}$
Polymer	Vinylidene difluoride based polymer	50–60	50–90	10^{-5}
	Odd-nylons	70	50–90	100
	Poly(thio)urea	40	300	100
Molecule	Liquid crystals	0.2	5	?
Composite	Polymer loaded with inorganic particles	Intermediate	Intermediate	Intermediate

Source: Reprinted with permission from Ref. [2]. Copyright © 2011, American Chemical Society.

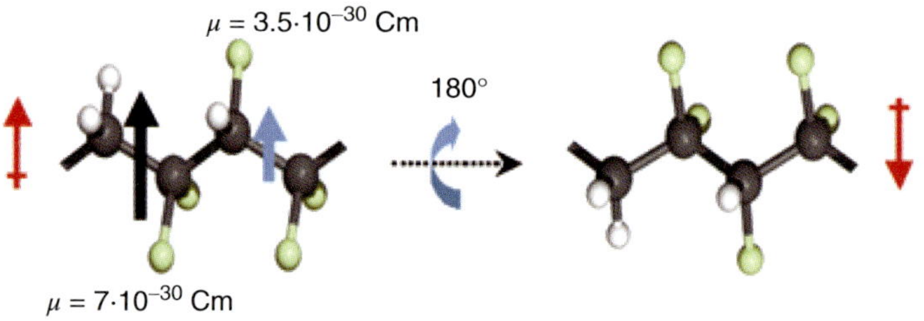

Figure 14.13 Ferroelectric switching in P(VDF-TrFE): dipole reorientation due to 180° chain rotation in crystalline regions [2]. (Reprinted with permission. Copyright © 2011, American Chemical Society.)

The ferroelectricity of PVDF and P(VDF-TrFE) stems from the dipole moments in the molecule that can be aligned under an applied field by a rotation of the polymer chain, as illustrated in Figure 14.13. Strongly electronegative fluorine atoms in the polymer induce the dipole moments. The dipole switching, which depends on temperature and the strength of applied field, involves rotation of the dipoles around the backbone [49]. This already implies that the preferred chain configuration is all-trans, and this phase is referred to as the *β-phase.* This phase gives the highest ferroelectric response because of an optimal alignment of all the F–H dipole moments in the crystal unit cell.

When PVDF is processed into a thin film, either from a melt or from a solution, it becomes the so-called α-phase: the polymer chains have alternating trans and gauche conformation [47], which cancels out the overall polarization originating from the "electronegative fluoride-electropositive hydrogen" (F–H) dipoles perpendicular to the backbone of the polymer. To impart ferroelectric properties to PVDF, it is usually mechanically stretched and/or electrically poled to orient the molecular chains in the all-trans conformation (β-phase). Other methods to realize this phase are the addition of hygroscopic salts, epitaxy on KBr, control of the cooling and heating rate, the solvent evaporation rate, and blending with poly(methylmethacrylate) [53–58].

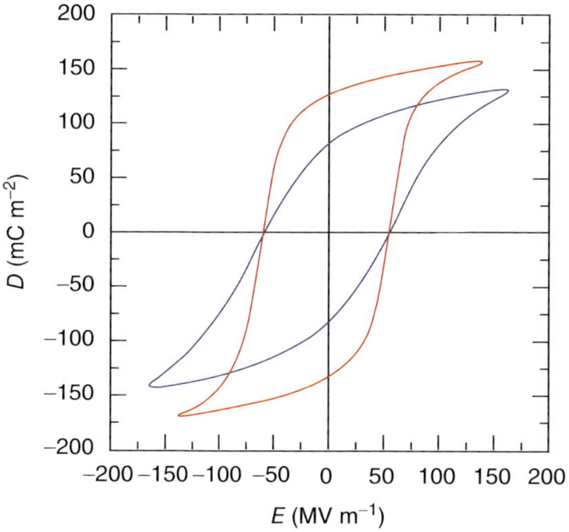

Figure 14.14 *D–E* hysteretic loop of P(VDF-TrFE) films with different VDF/TrFE ratios. In blue, the molar ratio VDF/TrFE is 50 : 50. In red, it is 80 : 20 [2]. (Reprinted with permission. Copyright © 2011, American Chemical Society.)

The copolymerization of PVDF and PTrFE is an efficient way to form the β-phase. The larger fluoride atom in the PTrFE monomer which replaces the smaller hydrogen atom in the PVDF monomer induces a strong steric hindrance, which is spontaneously more favorable the all-trans than the gauche conformation. Generally, the β-phase can be obtained for molar ratios 50–80% of VDF by introducing TrFE monomers directly. No additional processing is needed [59].

The crystalline polymorphs of P(VDF-TrFE) highly depend on the temperature. Ferroelectric compounds show ferroelectricity only below a certain phase transition temperature, called the *Curie temperature*, T_c, and are paraelectric above this temperature. T_c, which is lower than the melting point T_m, ranges from 70 to 140 °C. To increase the crystallinity and thereby the ferroelectric response, solution-cast P(VDF-TrFE) is thermally annealed in the paraelectric phase (above T_c) for some time. Thermal motion induces the molecular chains to rearrange their positions, forming a highly crystalline state [60].

Figure 14.14 manifests D-E hysteresis loops of random copolymers of VDF and TrFE containing different TrFE unit. In fact the one with less TrFE shows a higher Pr. Note that Ec is not influenced by a differing TrFE content.

14.4.2
Principles of Memory Operation

In ferroelectric-based memory, information is stored via the polarization state. Polarization in one or the other direction is referred to as "0" or "1." In ferroelectric random access memory (FRAM), the ferroelectric is used as a capacitor in the architecture shown in Figure 14.15. The FeFET is derived from the basic

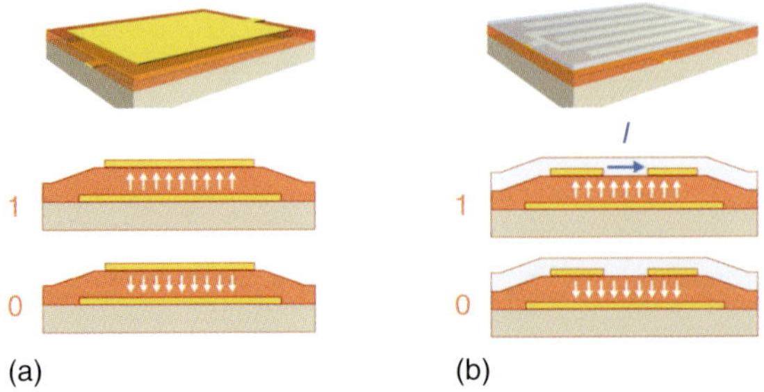

Figure 14.15 (a) FRAM and (b) FeFET basic cells [2]. (Reprinted with permission. Copyright © 2011, American Chemical Society.)

1T1C (one transistor–one capacitor) FRAM cell by inserting the ferroelectric in the gate stack of the select transistor below the gate electrode, thus resulting in a smaller basic cell. Because of the smaller cell, a higher integration density is possible. Furthermore, as will be shown in the following, this new structure enables a nondestructive read-out.

14.4.2.1 Capacitor

In its simplest form, a ferroelectric capacitor is composed of metal–ferroelectric layer–metal structure. By applying a sufficiently large voltage on metal electrode, the ferroelectric layer is polarized in one of two stable states. The polarization state can be read (or, better, inferred) by detecting the amount of charge build-up during a certain time period (t) by measuring the voltage over a reference capacitor with respect to ground using an integrating circuit. Depending on whether the polarization is aligned or not with the direction of the applied field, a low or a high charge displacement current response is measured. If the directions of the internal polarization and the applied field are opposite, the read-out changes the polarization state and the stored information is altered. Thus, it is necessary to rewrite the memory cell to its original state after the read-out. In other words, the read-out protocol for capacitor-based memories is destructive. It requires a relatively complex read-out circuitry, which becomes increasingly difficult when the capacitor size is downscaled. Furthermore, because the read-out is destructive, the memories should endure a high number of write–rewrite cycles (endurance).

14.4.2.2 Field-Effect Transistor

The simplest structure of an FeFET comprises a metal–ferroelectric–semiconductor layer stack (Figure 14.15), in which the ferroelectric layer serves as the gate dielectric. Sometimes, another dielectric layer can be inserted between the semiconductor and the ferroelectric. The ferroelectric layer, because of its remanent polarization, can choose either of two stable polarization states. These states are maintained when no bias is applied, that is, at power shutdown. Switching from one polarization state to another is accomplished by applying a sufficiently large gate voltage. Depending on the direction of the polarization, positive or negative countercharges accumulate in the semiconductor film close to the semiconductor–ferroelectric interface, effectively causing a positive or negative shift of the onset voltage. Therefore, a memory window, defined by the shifted onset voltage, exists wherein the drain current may have either of two levels depending on the actual polarization state of the ferroelectric gate dielectric. The corresponding drain current levels can be used to define the Boolean "0" and "1" states of nonvolatile memory with nondestructive read-out [45]. This is illustrated schematically in Figure 14.15. Nondestructive read-out offers the advantage that the memory is not subjected to destructive read/rewrite cycles which are commonly employed in ferroelectric capacitor memories, so that the lifetime of those devices is limited only by the number of writing times.

Multilevel memory cells can be realized based on either flash or resistive switching operation [61, 62]. Hwang *et al.* [63] demonstrated a reliable and mechanically flexible multilevel FeFET using a P(VDF-TrFE) ferroelectric

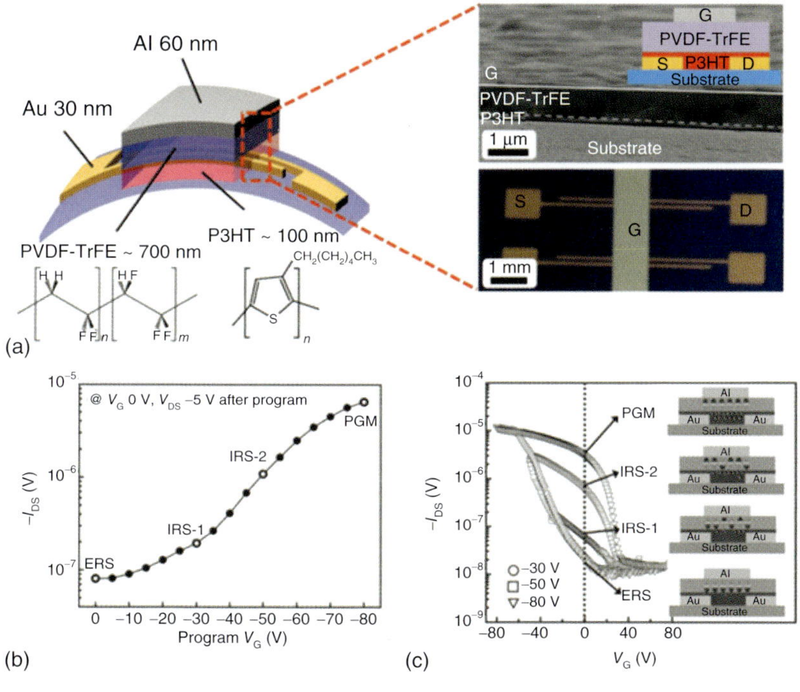

Figure 14.16 (a) Schematic of the top-gate/bottom-contact FeFET memory device with P3HT channel and ferroelectric PVDF-TrFE insulator. SEM image (top) and photograph (bottom) show the cross-sectional and plan views of the FeFET, respectively. (b) Nonvolatile I_{DS} values obtained at the read voltage ($V_G = 0$ V) after various program V_G values ranging from −5 to −80 V in the DC sweeping mode. (c) The multilevel $I_{DS}-V_G$ transfer curves of an FeFET after various program V_G values. The schematic in the inset shows nonvolatile I_{DS} levels controlled by the ferroelectric polarization at different program gate voltages [63]. (Reproduced with permission. Copyright © 2012, Wiley.)

dielectric layer. The top-gate/bottom-contact FeFET memory device incorporating Al gate electrode/P(VDF-TrFE)/P3HT/Au source/drain electrodes was constructed on both Si and flexible PI substrate (Figure 14.16). The multilevel storage memory device relies on the applied gate voltage which controls the degree of remnant polarization. An FeFET with four-level data storage was achieved by selecting writing voltages at +80, −30, −50, and −80 V. The subsequent gate voltage was swept through the OFF state, ±30, ±50, and ±80 V.

Recently, many research groups investigated organic electronic circuits on paper and banknotes to explore the possibility to authenticate money and stop counterfeiting. Among them, Alshareef *et al.* [64] demonstrated an all-polymer ferroelectric capacitor memory device and FeFET memories on banknotes. Figure 14.17 shows the fabrication process of ferroelectric memory on a banknote. First, the planarizing layer is deposited, which is then spin-coated with polydimethylsiloxane (PDMS) followed by the active layer deposition. Bottom gate electrodes are dimethyl sulfoxide (DMSO)-doped conducting PEDOT:PSS, and the ferroelectric layer is P(VDF-TrFE) of ~140 nm. The semiconducting layer is thermally evaporated pentacene, and Au serves as source and drain electrodes. The all-polymer devices showed performance with a maximum P_r of ~8.2 μC cm^{-2}, low E_c of 50 MV m^{-1}, high μ of ~0.12 cm^2 V^{-1} s^{-1}, low leakage of ~10^{-6} A cm^{-2}, large memory window of ~8 V, high I_{on}/I_{off} ratio of ~5 × 10^3, low turn-on voltage of ~0.5 V, low operating (after switching) voltages of ~4 V, and excellent retention characteristics up to 10^4 s.

(a) (b)

Figure 14.17 (a) Fabrication of polymer ferroelectric memory devices on banknotes: (i) PDMS-coated banknote, (ii) spin-coating PEDOT: PSS bottom-gate electrodes, (iii) spin-coating P(VDF-TrFE) ferroelectric layer ~140 nm, and (iv) thermally evaporated pentacene thin film (~60 nm). (b) Photograph of a 1-Saudi Riyal note covered with arrays of polymer ferroelectric memory devices [64]. (Reproduced with permission. Copyright © 2012, Wiley.)

Table 14.3 Comparison of the types of polymer memories classified by their primary circuit elements.

Type	MIM type	Transistor type (OFET)	Resistor type (MSM and FEdiode)
Physical description	Capacitor stores charges of opposite polarity on plate electrodes. Each bit is stored in a separate capacitor.	Charges storage and/or polarization at the dielectric–semiconductor interface modulate channel current of the transistor. Level of channel current indicates the bit level of the OFET memory.	Data storage is based on the high and low conductivity (bistability of resistance) of resistor in response to the applied electric field.
Device structure	a) Metal–ferroelectric–metal (MFM) b) Metal–insulator–ferroelectric–metal (MIFM).	a) Floating gate b) Charge trapping dielectric c) Ferroelectric as gate dielectric.	a) MIM = metal–insulator–metal b) MSM = metal–semiconductor–metal c) MISM = metal–insulator–semiconductor–metal d) M\S:FE\M = metal–semiconductor: ferroelectric blend–metal.
Materials	Ferroelectric polymers: • PVDF or P(VDF-TrFE) • odd nylons • cyano polymers • poly(thio)ureas • Ferroelectric liquid crystal (FLC) polymers.	a) FET semiconductor: • π-conjugated polymers/molecules • metal oxides b) Gate dielectric: Inorganic insulators, polymer dielectrics, polymer–polymer blends, composites with nanoparticles, insulators with blended donor or acceptor molecules, ferroelectric polymers.	• Interfacial oxides • Insulating polymers (also with nanoparticles) • Semiconducting polymers and small molecules (also with NPs) • Solid-state ionic conductors • Composite materials, ferroelectric:semiconductor blends.

(*continued overleaf*)

Table 14.3 (*Continued*)

Type	MIM type	Transistor type (OFET)	Resistor type (MSM and FEdiode)
Mechanism	Ferroelectric polymer maintains permanent electric polarization that can be repeatedly switched between two stable states by an external electric field.	Charge storage or polarization in OFET gate dielectric gives rise to a shift of V_{Th} or hysteresis, creating bistable transistor transfer curves.	Electrical bistability of resistance by (i) creation and annihilation of filaments (metallic or oxygen vacancy); (ii) redox doping/dedoping of a semiconductor; and (iii) switching of carrier injection in a semiconductor.
Memory cell	a) 1 (pass) transistor + 1 capacitor (1T1C) b) 1 (pass) transistor + 2 capacitors (1T2C) c) 2 transistors + 2 capacitors (2T2C).	a) 1 memory transistor (1T) "NAND flash, NOR flash, DRAM" b) 1 pass transistor + 1 memory transistor (2T).	a) 1 memory diode (1D) b) 1 diode + 1 memory resistor (1D1R) c) 1 pass transistor + 1 memory resistor (1T1R).
Performance factors	Polymer composition, crystallinity, film thickness, switching dynamics, film defects, metal electrodes, field pulses, fatigue characteristics.	Charge mobility, capacitance per area, maximum electric displacement, impurity, morphology, crystal packing, energy barrier, deposition conditions.	Filamentary conduction, filament rupture, electrode materials, control of interfaces and interfacial oxides, redox doping effects, carrier injection at electrodes.
Status	Close to commercialization	Research	Research
Technical limitations	a) Destructive read-out b) Material degradation (fatigue) c) Capacitor scaling.	a) Thickness and roughness control of dielectric layer b) Parasitic capacitance c) Depolarization and retention time.	a) Mechanisms unascertained b) Retention time and temperature stability of conductive filaments c) Complexity of programming of bipolar memories d) Sneak current paths in resistor arrays.

Source: Reprinted with permission from Ref. [2]. Copyright © 2011, American Chemical Society.

14.5 Conclusions

We summarize in Table 14.3 the main characteristics of the devices and architectures discussed in this chapter. Each of memory class has the potential, but there still remain several issues to be solved. Practical applications are expected to be achieved with ferroelectric capacitor memory banks first. Transistors are expected to simplify mainly the read-out, as they are able to remove the constraint of destructive read-out. Although many parameters of Table 14.1 have been shown separately for transistor-type memory device, the integration of all parameters in one FET technology is not proven. Also, the retention time of these devices needs to be improved.

Resistive-type memories with the diode structure are the most attractive for high-density integration. In this kind of devices, the sneak current is still a big problem. Corresponding technologies necessitate further work to acquire maturity, as indicated in Table 14.2.

To achieve advanced memory devices, integration with TFT circuitry for read-out and write is the next subject of future research. It is very promising in memory banks integrated into various applications, which are not possible by present-day silicon technology.

References

1. (a) AOL Inc. *http://www.engadget.com/2010/05/26/sonys-rollable-oleddisplay-can-wrap-around-a-pencil-our-heart/* (accessed 17 July 2014); (b) CNET *http://asia.cnet.com/samsung-shows-off-flexible-display-phones-at-ces-keynote-62220087.htm* (accessed 17 July 2014).
2. Heremans, P., Gelinck, G.H., Muller, R., Baeg, K.J., Kim, D.Y., and Noh, Y.Y. (2011) *Chem. Mater.*, **23**, 341.
3. Ouyang, J., Chu, C.W., Szmanda, C.R., Ma, L.P., and Yang, Y. (2004) *Nat. Mater.*, **3**, 918.
4. Ling, Q.D., Liaw, D.J., Zhu, C.X., Chan, D.S.H., Kang, E.T., and Neoh, K.G. (2008) *Prog. Polym. Sci.*, **33** (10), 917–978.
5. Marsman, A.W., Hart, C.M., Gelinck, G.H., Geuns, T.C.T., and de Leeuw, D.M. (2004) *J. Mater. Res.*, **19** (7), 2057.
6. de Brito, B.C., Smits, E.C.P., van Hal, P.A., Geuns, T.C.T., de Boer, B., Lasance, C.J.M., Gomes, H.L., and de Leeuw, D.M. (2008) *Adv. Mater.*, **20** (19), 3750.
7. Waser, R., Dittmann, R., Staikov, G., and Szot, K. (2009) *Adv. Mater.*, **21** (25), 2632.
8. Cho, B., Kim, T.-W., Song, S., Ji, Y., Jo, M., Hwang, H., Jung, G.-Y., and Lee, T. (2010) *Adv. Mater.*, **22**, 1228.
9. Dearnaley, G., Morgan, D.V., and Stoneham, A.M. (1970) *J. Non-Cryst. Solids*, **4**, 593.
10. Pender, L.F. and Fleming, R.J. (1975) *J. Appl. Phys.*, **46**, 3426.
11. Segui, Y., Ai, B., and Carchano, H. (1976) *J. Appl. Phys.*, **47**, 140.
12. Hwang, W. and Kao, K.C. (1974) *J. Chem. Phys.*, **60**, 3845.
13. Graves-Abe, T. and Sturm, J.C. (2005) Dynamics of write and erase mechanisms in a novel organic memory with extremely low ON resistance. Device Research Conference (DRC), Santa Clara, CA, IEEE, pp. 49–50.
14. Ling, Q.D., Liaw, D.J., Zhu, C.X., Chan, D.S.H., Kang, E.T., and Neoh, K.G. (2007) *Polymer*, **48**, 5182.

15. Yang, Y., Gao, P., Gaba, S., Chang, T., Pan, X., and Lu, W. (2012) *Nat. Commun.*, **3**, 732.
16. Lee, S-H, Oh, S-H, Ji, Y, Kim, J, Kang, R, Khim, D, Lee, S, Yeo, J-S, Lu, N, Kim, MJ, Ko, HC, Kim, T-W, Noh, Y-Y, Kim, D-Y, *Org. Electron.* 2014, doi. 10.1016/j.orgel.2014.02.027
17. Kuang, Y.B., Huang, R., Tang, Y., Ding, W., Yu, Z., Ma, Y., Zhang, L.J., Wu, D., Wen, Y.Q., and Song, Y.L. (2009) *IEEE Electron Device Lett.*, **30**, 931.
18. Son, D.I., Kim, T.W., Shim, J.H., Jung, J.H., Lee, D.U., Lee, J.M., Park, W.I., and Choi, W.K. (2010) *Nano Lett.*, **10**, 2441.
19. Kondo, T., Lee, S.M., Malicki, M., Domercq, B., Marder, S.R., and Kippelen, B. (2008) *Adv. Funct. Mater.*, **18**, 1112.
20. Arif, M., Yun, M., Gangopadhyay, S., Ghosh, K., Fadiga, L., Galbrecht, F., Scherf, U., and Guha, S. (2007) *Phys. Rev. B*, **75**, 195202.
21. Tu, C.H., Lai, Y.S., and Kwong, D.L. (2006) *IEEE Electron Device Lett.*, **27**, 354.
22. (a) Billen, J., Steudel, S., Müller, R., Genoe, J., and Heremans, P. (2007) *Appl. Phys. Lett.*, **91** (26), 263507; (b) Deleruyelle, D., Muller, C., Amouroux, J., and Müller, R. (2010) *Appl. Phys. Lett.*, **96** (26), 263504; (c) Terabe, K., Hasegawa, T., Liang, C., and Aono, M. (2007) *Sci. Technol. Adv. Mater.*, **8** (6), 536; (d) Armstrong, R.D., Bulmer, R.S., and Dickinson, T. (1973) *J. Solid State Chem.*, **8** (3), 219.
23. Joo, W.-J., Choi, T.-L., Lee, J., Lee, S.K., Jung, M.-S., Kim, N., and Kim, J.M. (2006) *J. Phys. Chem. B*, **110** (47), 23812.
24. (a) Butler, S.Z., Hollen, S.M., Cao, L., Cui, Y., Gupta, J.A., Gutiérrez, H.R., Heinz, T.F., Hong, S.S., Huang, J., and Ismach, A.F. (2013) *ACS Nano*, **7**, 2898–2926; (b) Liu, J., Zeng, Z., Cao, X., Lu, G., Wang, L.-H., Fan, Q.-L., Huang, W., and Zhang, H. (2012) *Small*, **8**, 3517.
25. Katz, H.E. (2004) *Chem. Mater.*, **16** (23), 4748.
26. Sze, S.M. and Ng, K.K. (2007) *Physics of Semiconductor Devices*, 3rd edn, John Wiley & Sons, Inc., Hoboken, NJ.
27. Broadhurst, M.G., Davis, G.T., Gross, B., Mascarenhas, S., Sessler, G.M., van Turnhout, J., and West, J.E. (1980) *Topics in Applied Physics: Electrets*, vol. **33**, Springer-Verlag, Berlin, Heidelberg, New York.
28. Sessler, G.M. (2001) *J. Electrostat.*, **51**, 137.
29. Katz, H.E., Hong, X.M., Dodabalapur, A., and Sarpeshkar, R. (2002) *J. Appl. Phys.*, **91** (3), 1572.
30. Baeg, K.-J., Noh, Y.-Y., Ghim, J., Kang, S.-J., Lee, H., and Kim, D.-Y. (2006) *Adv. Mater.*, **18** (23), 3179.
31. Baeg, K.-J., Noh, Y.-Y., Ghim, J., Lim, B., and Kim, D.-Y. (2008) *Adv. Funct. Mater.*, **18** (22), 3678.
32. Debucquoy, M., Rockelé, M., Genoe, J., Gelinck, G.H., and Heremans, P. (2009) *Org. Electron.*, **10** (7), 1252.
33. Baeg, K.-J., Khim, D., Kim, J., Yang, B.-D., Kang, M., Jung, S.-W., You, I.-K., Kim, D.-Y., and Noh, Y.-Y. (2012) *Adv. Funct. Mater.*, **22**, 2915.
34. Baeg, K.-J., Noh, Y.-Y., Sirringhaus, H., and Kim, D.-Y. (2010) *Adv. Funct. Mater.*, **20** (2), 224.
35. Kim, S.-J., Park, Y.-S., Lyu, S.-H., and Lee, J.-S. (2010) *Appl. Phys. Lett.*, **96** (3), 033302.
36. Tiwari, S., Rana, F., Hanafi, H., Hartstein, A., Crabbe, E.F., and Chan, K. (1996) *Appl. Phys. Lett.*, **68** (10), 1377.
37. Tsoukalas, D., Dimitrakis, P., Kolliopoulou, S., and Normand, P. (2005) *Mater. Sci. Eng., B*, **124-125**, 93.
38. Zhen, L., Guan, W., Shang, L., Liu, M., and Liu, G. (2008) *J. Phys. D: Appl. Phys.*, **41** (13), 135111.
39. Mabrook, M.F., Yun, Y., Pearson, C., Zeze, D.A., and Petty, M.C. (2009) *Appl. Phys. Lett.*, **94** (17), 173302.
40. Kim, Y.-M., Park, Y.-S., O'Reilly, A., and Lee, J.-S. (2010) *Electrochem. Solid-State Lett.*, **13** (4), H134.
41. Liu, Z.C., Xue, F.L., Su, Y., Lvov, Y.M., and Varahramyan, K. (2006) *IEEE Trans. Nanotechnol.*, **5** (4), 379.
42. Chen, Y., Zhang, B., Liu, G., Zhuang, X., and Kang, E.-T. (2012) *Chem. Soc. Rev.*, **41**, 4688.
43. Yu, W.J., Chae, S.H., Lee, S.Y., Doung, D.L., and Lee, Y.H. (2011) *Adv. Mater.*, **23**, 1889.

44. Sawyer, C.B. and Tower, C.W. (1930) *Phys. Rev.*, **35** (3), 269.
45. Scott, J.F. (2000) *Ferroelectric Memories in Advanced Microelectronics*, Springer-Verlag.
46. Kawai, H. (1969) *Jpn. J. Appl. Phys.*, **8** (7), 975.
47. Nalwa, H.S. (ed.) (1995) *Ferroelectric Polymers: Chemistry, Physics and Applications*, Marcel Dekker, Inc., New York.
48. Bune, A.V., Fridkin, V.M., Ducharme, S., Blinov, L.M., Palto, S.P., Sorokin, A.V., Yudin, S.G., and Zlatkin, A. (1998) *Nature*, **391** (6670), 874.
49. Dawber, M., Rabe, K.M., and Scott, J.F. (2005) *Rev. Mod. Phys.*, **77** (4), 1083.
50. Vizdrik, G., Ducharme, S., Fridkin, V.M., and Yudin, S.G. (2003) *Phys. Rev. B*, **68** (9), 094113.
51. Horiuchi, S. and Tokura, Y. (2008) *Nat. Mater.*, **7** (5), 357.
52. Tuttle, B.B.A., Voigt, J.A., Goodnow, D.C., Lamppa, D.L., Headley, T.J., Eatough, M.O., Zender, G., Nasby, R.D., and Rodgers, S.M. (1993) *J. Am. Ceram. Soc.*, **76** (6), 1537.
53. Salimi, A. and Yousefi, A.A. (2003) *Polym. Test.*, **22** (6), 699.
54. Scheinbeim, J., Nakafuku, C., Newman, B.A., and Pae, K.D. (1979) *J. Appl. Phys.*, **50** (6), 4399.
55. Lovinger, A.J. (1981) *Polymer*, **22** (3), 412.
56. Kang, S.J., Park, Y.J., Sung, J., Jo, P.S., Park, C., Kim, K.J., and Cho, B.O. (2008) *Appl. Phys. Lett.*, **92** (1), 012921.
57. Yang, D.C. and Thomas, E.L. (1987) *J. Mater. Sci. Lett.*, **6** (5), 593.
58. Leonard, C., Halary, J.L., and Monnerie, L. (1988) *Macromolecules*, **21** (10), 2988.
59. Furukawa, T. (1989) *Phase Transitions*, **18** (3-4), 143.
60. Fukuma, T., Kobayashi, K., Horiuchi, T., Yamada, H., and Matsushige, K. (2000) *Jpn. J. Appl. Phys.*, **39** (6B), 3830.
61. Lee, J.-S., Kim, Y.-M., Kwon, J.-H., Sim, J.S., Shin, H., Sohn, B.-H., and Jia, Q. (2011) *Adv. Mater.*, **23**, 2064.
62. Liu, M., Abid, Z., Wang, W., He, X., Liu, Q., and Guan, W. (2009) *Appl. Phys. Lett.*, **94**, 233106.
63. Hwang, S.K., Bae, I., Kim, R.H., and Park, C. (2012) *Adv. Mater.*, **24**, 5910.
64. Khan, M.A., Bhansali, U.S., and Alshareef, H.N. (2012) *Adv. Mater.*, **24**, 2165.
65. Prince, B. (2002) *Emerging Memories: Technologies and Trends*, Kluwer Academic Publishers, Boston, MA.

15
Flexible Displays

Chung-kun Song and Gi-Seong Ryu

15.1
Introduction

Emerging technologies represented by organic electronics and printing technology have stimulated ideas for new products – which cannot be realized by Si technology – using large-area and flexible substrates. Large, flexible electronic circuits are expected to give more freedom in the future, replacing hand-held devices with wearable electronics, solid square-shaped displays with flexible free-shaped displays that are installable anywhere, and radio frequency identification (RFIDs) attached on products with those embedded in and inline-processed with products.

Displays will be the first to utilize such innovations, because the display industry is facing saturation on the market and urgently needs alternative devices to break the stagnation and create new markets [1]. Many expect that flexible displays can meet the various requirements of future displays in that they can be installed on objects with any shape, such as rounded domed ceilings, cylindrical columns, large windows, and even walls [2–4]. In addition, the demand to access information anywhere and anytime in the future will require many displays in many places, and for this the display panels should become inexpensive. In conclusion, flexible displays can meet the key features of future displays, namely, flexibility and low price.

Therefore, in this chapter, the potential and limitations of the current technologies available for flexible displays will be reviewed, including flexible substrates, the various display modes feasible for flexible displays, and thin-film transistors (TFTs). In addition, the active-matrix organic light-emitting diode (AMOLED) panel, which is considered as "an ultimate display," will be discussed from design to fabrication by using the organic thin film transistor (OTFT) as a device driving pixel.

Large Area and Flexible Electronics, First Edition. Edited by Mario Caironi and Yong-Young Noh.

15.2
Flexible Substrates

Conventional glass substrates have a number of drawbacks for use in flexible displays. They are not robust (can break when dropped), rigid (no potential for roll-to-roll production), and relatively thick (heavy). However, the typical features of glass, including thermal stability, optical transparency, oxygen and moisture permeability, scalability, and size, must be met by substrates of flexible displays.

Plastics, thin glasses, and metal foils are considered for the substrates of flexible displays. Metal foils have the advantages of being impervious to oxygen and moisture and having high thermal and chemical resistance. However, the large surface roughness requires an extra process to smooth it. Furthermore, film coatings must be added to the metal foils to electrically isolate devices on the substrate from the conductive metal substrates. The most serious problem with metal foils is that once bent they cannot be restored to their original shape.

Thin glasses have high transparency and small surface roughness, are impervious to oxygen and moisture, and have high thermal and chemical resistance: the same as the glasses used for current displays. However, thin glasses are unbendable and easily broken, and thus are hard to handle during fabrication. In particular, thin glasses cannot be used in roll-to-roll processes, which is a highly desirable feature for flexible substrates.

Plastic substrates have attractive features such as physical durability, high flexibility which makes them good for roll-to-roll processing, light weight, and potentially low cost when produced in large volumes. Those features have motivated active research on flexible displays using plastic substrates. It is the roll-to-roll processing potential that makes plastics such a promising candidate for use as substrates of flexible displays. Plastic substrates provide opportunities to overcome the drawbacks of glasses in terms of robustness, continuous manufacturing, and form factors. Even so, there are a number of challenges that must be met in order to duplicate the most favorable features of glasses in plastic substrates. For example, for AMOLED, barrier against oxygen and moisture is essential for a long life time; thus the plastic substrate should have a smooth and clean surface, low outgassing, robustness, and low defects. For maximum light extraction, they need low haze, low optical defects, and high transparency. The backplane alignment is the most important factor for high resolution. Therefore, plastic substrates should have dimensional stability at high temperatures, low moisture absorption, low shrinkage, and low coefficient of thermal expansion (CTE). For use with AMOLED, plastic substrates should have the following properties; although the exact values are still controversial [5]:

Optical transmission: $>90\%$
Roughness: <2 nm rms (root mean square)
Stability under heat
Hardness: >6H
Moisture permeation: $<10^{-5}\ cm^3\ m^{-2}\ day^{-1}$ at 1 atm
Oxygen permeation: $<1\ \mu g\ m^{-2}\ day^{-1}$.

Table 15.1 Typical parameter values for plastic substrates.

Property	PET	PEN	PC	PES	PI (color)	PI (colorless)	FPR
Thickness (μm)	125	125	150	200	—	100	100
T_g (°C)	78	120	150	223	450	~340	>350
CTE (ppm °C^{-1})	15	13	70	54	30	~10	14
Transmission (%)	91	87	90	88	~30	>87	90
Density (g cm^{-3})	1.4	1.36	1.2	1.37	—	1.4	—
Water uptake (weight)	0.3	0.3	0.3	1.4	—	1.8	—
Young's modulus (GPa)	5.3	6.1	3	2.2	—	2.5	15
Tensile strength (MPa)	225	275	—	83	—	~200	—

Among the currently available plastic substrates, there is none that can satisfy all these requirements. Each candidate plastic has its own advantages and disadvantages. In Table 15.1, we summarize the typical values of the parameters of interest for candidate plastics that might be used as substrates.

15.2.1 Thermal Stability

It is a challenge for flexible displays to work with substrates that can tolerate only low thermal stress during processing. The level of thermal stress that flexible substrates must withstand is determined by the product of the length of time spent at high temperature and the peak temperature. For example, although plastics may be subjected to very high temperatures during processing by a pulsed laser, they will not be damaged if the time of irradiation is extremely short. For long processes, even at relatively low temperatures, it is sometimes difficult to determine a safe temperature. Typically, an upper limit is set by the melting temperature T_m. The glass transition temperature T_g is a common boundary beyond which the dimensional stability of plastic films begins to degrade. However, some films remain stable above T_g if they are thermally pretreated. The shrinkage of polyethersulfone (PES) at 180 and 200 °C is presented in Ref. [6]. The shrinkage characteristics can be explained by an activated process of gas desorption from the plastic substrate [7]. To prevent degassing of the plastic substrates, the films of SiO_2, SiN_x, AlO_x, AlN_x, and Ta_2O_5 have been used as gas barriers [8].

The CTE is also an important parameter to be considered, even if the plastics are stabilized by thermal pretreatment. Stress arises from differences in CTEs between plastic and the films deposited on them. The stressed films are susceptible to peeling, cracking, and bending, which become serious at higher deposition temperatures [8]. Therefore, peeling, cracking, and bending due to the large difference of CTEs are major problems to be overcome in flexible displays, and thus the deposition temperatures should be kept as low as possible. In order to prevent the plastic substrates from bending, both sides of such substrates are coated with the same thickness of gas-barrier materials. The typical CTE of plastics is 3–18-fold larger than that of glass (4 ppm °C^{-1}). Therefore, plastics are subject to thermal expansion whenever the temperature rises, which results in a significant difference between the expansions of the plastic substrate and the films deposited upon it. This can lead to poor interface integrity, resulting in cracking or misalignment during pattern registration. By the way, polycarbonate (PC) has the largest CTE (70 ppm °C^{-1}) among the plastics, as seen in Table 15.1, but it is almost completely restored to the original position after cooling. Therefore, its final variation due to temperature is the smallest among the plastic substrate candidates. Therefore, if a thermal process is performed at an early step in the processing and only once, PC can produce the minimum misalignment. Polyimide (PI) and fiber-reinforced plastic (FRP) exhibit relatively high T_g, which makes them important candidates for use as substrates for flexible displays.

15.2.2
Optical Transparency

Optical transparency must be better than 90% in the range of visible wavelengths (400–700 nm), and it varies from one material to another. Coatings are often used to provide resistance against absorption of chemicals and oxygen, which have a different transmission, depending on the wavelength, from that of the plastic substrates on which they are coated. Such coatings have transmission qualities that differ from those of the plastic substrates and can change the overall transmission [9]. The effective transparency of plastic substrates with film coatings should satisfy the criteria of 90% at least at the green wavelength of 550 nm. For LCD (liquid crystal display) applications, birefringence is another important criterion. Poly(ethylene naphthalate) (PEN) and poly(ethylene terephthalate) (PET) have a crystal structure that makes them unsuitable for LCDs because of their large birefringence. PES is better suited for Twisted Nematic (TN)-mode LCDs.

15.2.3
Permeation of Oxygen and Moisture

The high permeation of oxygen and moisture through plastic substrates is a serious problem and reduces the life time of displays. For example, 20 cm^3 m^{-2} day^{-1} of oxygen and 8 g m^{-2} day^{-1} of moisture are transmitted through uncoated PET. That is several orders of magnitude higher than can be tolerated by displays. For OLED (organic light-emitting diode) applications, oxygen and moisture penetration should be $<10^{-5}$ cm^3 m^{-2} day^{-1} and 1 μg m^{-2} day^{-1} at 1 atm, respectively [5]. These values are more stringent than for LCDs and EPDs (electrophoretic display)s, which are in the range of 10^{-3}–10^{-2} cm^3 m^{-2} day^{-1} [5]. Therefore, plastic substrates must be enclosed within coating materials that prevent the entry of oxygen and moisture in order to achieve acceptable lifetimes.

15.2.4 Chemical Resistance

Crystalline plastics (i.e., PET and PEN) exhibit strong resistance against most of the chemicals used in semiconductor processing. However, amorphous plastics such as PES and PC have poor chemical resistance. PES, in particular, is vulnerable to ketone solvents, and thus hard film coatings are required to provide protection.

15.2.5 Surface Roughness

Almost all plastics suffer from surface roughness and particulate contamination. Although the rms roughness of the typical industrial PEN is only 2.55 nm, there are many spikes more than 50 nm tall [5]. Such spikes are tall enough to penetrate several layers of an OLED structure. Spikes can be removed by polishing or by coating the surface with a smoothing layer. Planarization also increases the effectiveness of antipermeation coatings but the effective transparency can be degraded.

In conclusion, no single plastic substrate can satisfy all the requirements by itself. Therefore, hard film coatings must be added to the plastic film in order to overcome drawbacks such as high oxygen and moisture penetration, low chemical resistance, and roughness of surface. However, the properties of the plastic film must provide the basic qualities (i.e., high T_g, low CTE, and high transmission) which cannot be compensated by film coatings. In terms of providing these basic qualities, the colorless PI is the most promising candidate for plastic substrates for AMOLED flexible displays with stringent requirements.

15.3 Display Mode

Currently, three display modes are available for flexible displays: LCD, EPD, and OLED. The operating mechanism of each mode is briefly described, and their feasibility for application to flexible displays is discussed.

LCD consists of two glass substrates with a polarizing film and an liquid crystal (LC) between the glasses and light source (backlight). One glass substrate is placed perpendicular to the other in a polarizing direction, and thus light is visible on the opposite side to the backlight in normal conditions (V is off) because the molecules of the LC are gradually twisted as they approach the other side and are out of phase with 90° on the opposite side, as shown in Figure 15.1a. In this result, the light was screened to the polarizing direction of the opposite. When an electric field (V is on) is applied between the glasses, all molecules are in phase and block light transmission, as shown in Figure 15.1b, resulting in being invisible.

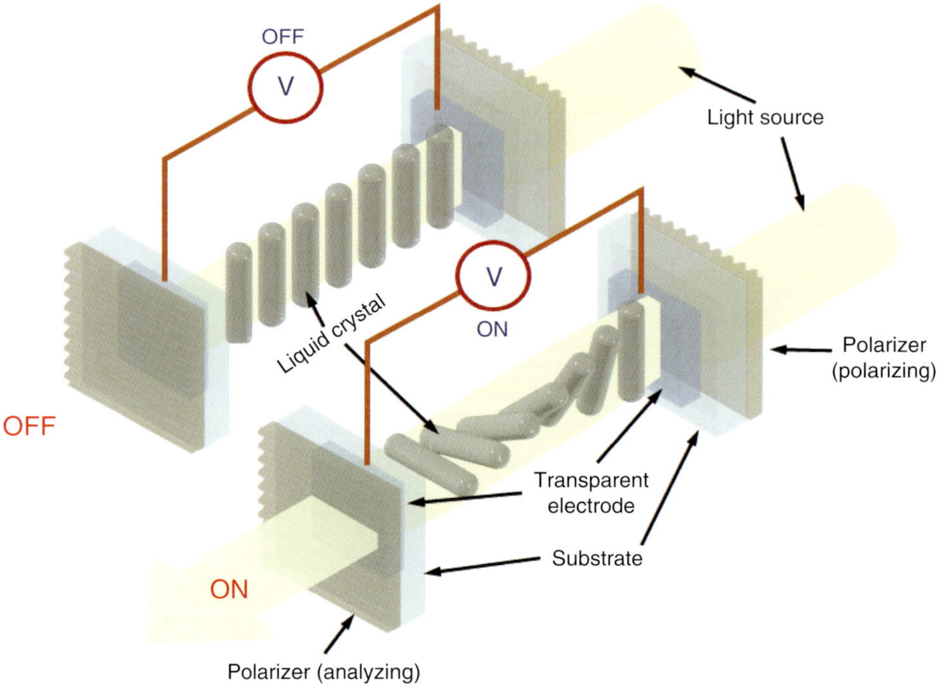

Figure 15.1 Conceptual picture of LCD operation. (a) The twisted LC molecules when the voltage if off, making it visible and (b) the aligned LC molecules when the voltage is on, making it invisible.

LCD technology has become sufficiently mature to make UHD (ultrahigh definition) displays. However, flexible LCD displays cannot be realized unless a flexible light source is developed and the cell gap between the two substrates is kept uniform over the entire area of a large flexible substrate and maintained even when bending.

EPD is a reflection-type display that controls the reflection of light with an applied voltage. EPD consists of two glass substrates and charged particles in the middle. One of the substrates should have a transparent electrode in order to reflect and transmit light. Two types of EPDs, namely capsule type and microcup type, are available at present. The operating principles of the two EPDs are identical; black and white charged particles are attracted to the electrode when a voltage with the polarity opposite to the particles is applied, and thus, the image is realized by the color associated with the attracted particles.

Capsule-type EPD has charged particles dispersed in a transparent solution within a capsule of a few hundred micrometers in diameter, as shown in Figure 15.2a. A distinctive feature of capsule-type EPD is the ease of fabrication of the display panel by simply applying a paste made of the capsules on one substrate and then laminating the other substrate on it. Of particular interest, there is no limitation for the application to flexible substrates at all. However, it has the

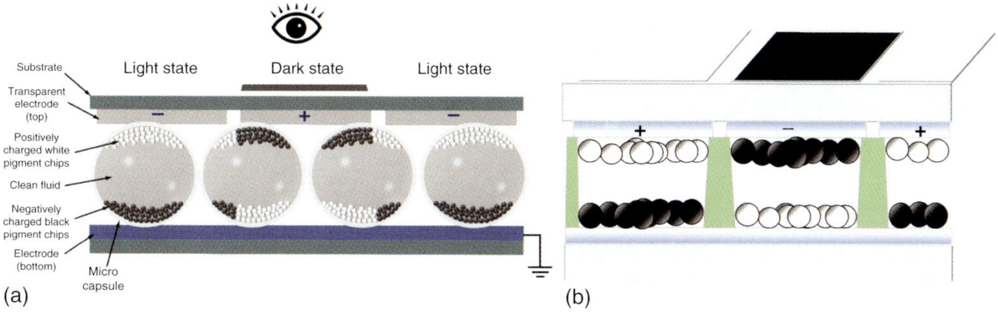

Figure 15.2 Schematic pictures of the two types of EPDs. (a) Capsule-type, and (b) micro-cup type.

disadvantage of having a low switching speed of a few hundred milliseconds because the particles move in the solution. In addition, there is no threshold for switching, so that the capsule-type EPD should be addressed by an active mode, which has a switching circuit at each pixel consisting of transistors with a threshold.

The micro-cup EPD is implemented by placing the charged particles inside micro-cups fabricated between substrates, as shown in Figure 15.2b. Therefore, for flexible displays, it is challenging to make micro-cups with uniform thickness over a large area and keep the upper substrate stick to the cups even when bent and to reduce the high operating voltage of a few tens of volts down to a few volts. The high switching speed of a few hundred microseconds is an advantage due to the particles moving in the air inside the cups. Another advantage is that there is a threshold for switching, and thus it is possible to make a panel operating with passive matrix mode without any switching circuit.

The common limitation of the two EPD types is the difficulty of colorization. Two methods of coloration are investigated: using color particles and color filters. However, both are not ready for commercialization yet. Color particles exhibiting clear R/G/B colors are not available. Color filters decrease the intensity of the reflected light, resulting in poor visibility. Therefore, their applications are limited to areas where color is not essential, such as e-books, price tags, and public advertisements.

OLED is a self-emitting device functioning through the recombination of electrons and holes. An OLED consists of several organic semiconductor (OSC) layers, basically the hole transport layer (HTL), the emitting layer (EML), and the electron transport layer (ETL), as shown in Figure 15.3. Several layers can be added to match hole transport with electron transport, to generate light only in the EML, and to control carrier injection from the electrodes. Because the device generates R/G/B light by itself, natural colors can be easily realized without color filters or light sources that are used for LCDs. Therefore, the structure of the OLED panel is very simple, and thus is light-weight. In particular, there are no limitations for its application to flexible displays because OLEDs are fabricated at room temperature and even with solution processes. Therefore, some say that "it is an ultimate

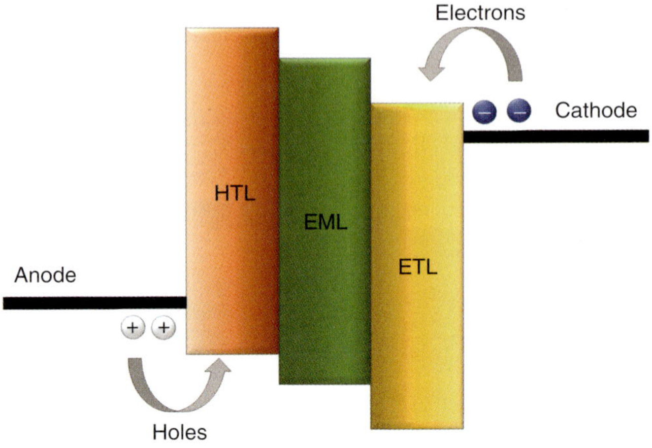

Figure 15.3 Structure of OLED.

display." However, it requires barrier films for long lifetimes, because OSCs are vulnerable to oxygen and moisture permeation.

15.4 Thin-Film Transistor

For large-area electronics, TFT is an adequate device because of the low-cost processing required by the production of large-area devices and because of its performance suitable for the required switching speed of active-matrix displays, which is in the range of tens of milliseconds. In this section, we give a brief history of TFT technology development and then examine the potential and limitations of the currently available TFTs with respect to flexible applications.

After the invention of TFT in 1925 [10–12] and the first production of TFT using polycrystalline CdS in 1962 [13–15], TFTs were largely ignored until the introduction of the "killer application" of TFTs proposed in 1971 [16]; that is, the so-called active-matrix mode of LCDs where a CdS TFT array was used for addressing pixels. In 1979, a new material, amorphous Si, was introduced as the active material in TFTs [17]. Their performance, with an on-current of several microamperes and an off-current of 1 nA, was recognized to fit the requirement to drive LCDs [18]. In particular, the amorphous nature provided a clue for resolving the drawbacks of polycrystalline CdS, which is influenced by the grain size, grain boundary interface states, stoichiometry, and sensitivity to ambient gases such as oxygen and moisture, because a-Si has no grain boundaries and is self-passivating. In 1980, poly-Si TFT was introduced [19] with a mobility of $50\,cm^2\,V^{-1}\,s^{-1}$, employing SiO_2 as the gate insulator grown thermally at 1050 °C. The performance of poly-Si TFT was improved with electron mobility of $400\,cm^2\,V^{-1}\,s^{-1}$ by using laser recrystallization of poly-Si

Table 15.2 Comparison between the available TFT technologies [24–28].

TFT properties	a-Si TFT	LTPS TFT	Organic TFT	Oxide TFT
Process temperature (°C)	<350	<500	RT	RT–350
Mobility ($cm^2 V^{-1} s^{-1}$)	0.5	50–100	1–10	1–100
Flexibility	Poor	Poor	Good	Medium
Process	Vacuum	Vacuum	Soluble/in-line	Vacuum/soluble
Substrate	Glass/metal foil	Glass/metal foil	Plastic	Plastic
Reliability	Low	High	Low in air	High
Manufacturing cost	Low	High	Low	Low

[20, 21]. Low temperature poly-Si (LTPS) TFTs emerged with mobilities of more than 20 $cm^2 V^{-1} s^{-1}$, similar to that of high-temperature poly-Si TFTs [22]. This was achieved by using solid-phase crystallization at 600 °C of a-Si deposited at 550 °C.

The year 1990 saw the introduction of a new class of OTFT, with OSC as the active layer material [23]. Evaporated hexathiophene was used as the active material, producing mobility comparable to that of a-Si. Organic TFTs are currently being pursued as candidates for integration onto flexible plastic substrates for future generations of displays.

In Table 15.2, the performance of the currently available TFTs is summarized and compared [24–28].

15.4.1
a-Si TFT

Today, a-Si TFT is a workhorse in the LCD industry and huge investment has been made because of its low-cost fabrication, adequate large-area processing, and the fact that its mobility of 0.5 $cm^2 V^{-1} s^{-1}$ is sufficient to drive LCD, which does not require high switching speed. However, the process temperature is still too high to be applied to plastic substrates. The high-temperature processes are associated with the deposition of a-Si thin film and gate dielectric, and also with controlling the hydrogen content in the a-Si thin film.

Hydrogenated amorphous silicon has a short-range order, which means that the coordination number, bonding angle, and bond length are close to those for single-crystalline Si within two or three atomic distances, but there is no periodicity in the long range. Because of this, there are a considerable number of localized states in the gap. The hydrogen in the a-Si:H reduces the number of dangling bonds by passivation. Therefore, the dangling bond density of $\sim 10^{20}$ cm^{-3} in vacuum-evaporated a-Si or sputtered a-Si is reduced to $10^{15}-10^{16}$ cm^{-3} in Plasma Enhanced Chemical Vapor Depostion (PECVD) a-Si:H, where the hydrogen content is 10–30%. Typically, undoped a-Si:H for TFT applications is prepared at a substrate temperature of 220–350 °C. The substrate temperature greatly affects the hydrogen content of the final film, which decreases with increasing substrate temperature because of the enhanced out-diffusion of hydrogen from the film. For application to flexible substrates, the substrate temperature should be kept as low as possible because of the dimensional changes of the plastic substrates during heating. A typical substrate temperature for a-Si:H on glass is ~300 °C. The temperature should be kept below 150 °C for plastic substrates. However, the a-Si:H TFTs processed at low temperatures produced a large gate leakage current through the SiN gate dielectric. One reason for this is that the content of hydrogen in SiN increases at low substrate temperatures [29]. Dilution of the silane/ammonia plasma with halogen or hydrogen in the deposition chamber could improve the quality of SiN even at low substrate temperatures [30].

15.4.2
LTPS TFT

The motivation of the development of poly-Si TFTs is the high mobility of 100 $cm^2 V^{-1} s^{-1}$, making it possible to integrate driving circuits consisting of poly-Si TFTs with LCDs on the same substrate. Typically, the poly-Si thin film is obtained from the amorphous-phase Si through a subsequent laser crystallization step. Although the pulsed laser step takes the plastics to a very high temperature during processing, the plastic will not be damaged if the time is extremely short. However, the dehydrogenation process cannot be performed without raising the substrate temperature to at least 400 °C. The high hydrogen content in the as-deposited a-Si:H films is explosively out-diffused as a result of the extremely high temperature gradients during laser annealing, resulting in the development of excessive porosity and surface roughness [31, 32]. To eliminate this problem, the hydrogen content of the film should be reduced to levels compatible with laser annealing, which is the so-called dehydrogenation process. This process can be accomplished by a temperature of 400–500 °C [33]. This is a relatively low temperature compared to the typical temperature of 600 °C for deposition of poly-Si but still too high to be applied to plastic substrates.

15.4.3
Oxide TFT

Oxide TFTs using a complex Indium Gallium Zinc Oxide (IGZO) single crystal as the active semiconductor layer were reported in 2003 with an impressive effective mobility of 80 $cm^2 V^{-1} s^{-1}$ [34]. A very high temperature of 1400 °C was necessary to attain this level of performance [34]. In 2004, a new class of oxide TFTs was introduced using pulse-laser deposition (PLD)-deposited amorphous IGZO as the semiconductor [35]. The mobility was 8.3 $cm^2 V^{-1} s^{-1}$, which was 10 times smaller than the previous ones, but they exhibited the potential of oxide TFTs by demonstrating a transparent TFT on a plastic substrate using near-room-temperature processing, as shown in Figure 15.4 [35]. The relatively high mobility despite the amorphous state originates from the electronic orbital structure of the material. As shown in Figure 15.5,

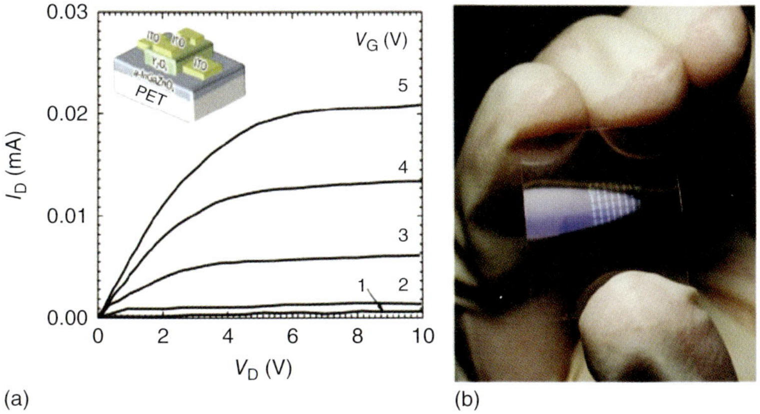

Figure 15.4 (a) Electrical characteristics and (b) the pictures of IGZO oxide TFTs fabricated on a plastic substrate [35]. (Reproduced with permission. Copyright © 2004, Nature Publishing Group.)

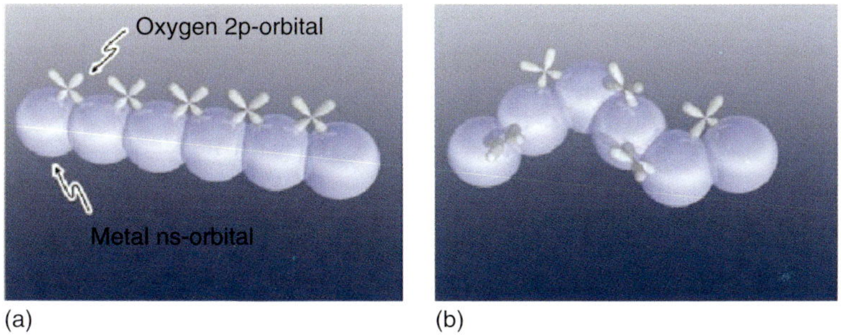

Figure 15.5 (a,b) Electronic orbital structure of crystalline and amorphous of IGZO [35]. (Reproduced with permission. Copyright © 2004, Nature Publishing Group.)

direct overlap between the 5s orbitals of neighboring indium atoms forms a conducting path for free electrons. The spherical symmetry of the 5s orbitals makes the conduction band isotropic and IGZO material insensitive to a distorted amorphous structure, maintaining high mobility in the amorphous state [35].

The results described above were obtained from an optimized PECVD SiO_2 process at 400 °C, but the temperature is too high for application to plastic substrates. Sputtering is an alternative method for low-temperature dielectric fabrication, but generally low-temperature sputtering induces a low sputtering rate of dielectrics, which require high-energy deposition processes and, thus, can degrade the film quality and their interfaces. Another method with low temperature is performed by using solution processes. The first research on the solution-processing of inorganic semiconductor materials was reported in 2001, achieving high levels of device performance [36]. The mobility was $0.2\,cm^2\,V^{-1}\,s^{-1}$, but it was processed at 900 °C. In 2007, zinc tin oxide (ZTO) and

IZO (indium zinc oxide) TFTs produced by spin coating exhibited a mobility of 16 $cm^2\,V^{-1}\,s^{-1}$ after annealing at 600 °C [37, 38]. The performance was enhanced with the mobility of 85 $cm^2\,V^{-1}\,s^{-1}$ by doping Li into ZnO TFTs using ambient spray pyrolysis and blends of precursor solutions and by annealing at 400 °C [39–41]. It is still challenging to obtain high performance in oxide TFTs without applying high temperature.

15.4.4 OTFT

The mobility of an organic crystal is in the range of 1–10 $cm^2\,V^{-1}\,s^{-1}$, depending upon the used processes and gate dielectric. The mobility of an intrinsic OSC is determined by two factors. The first is the width of the energy band of an organic crystal. In an organic crystal, the π orbitals of the molecules are overlapped to form a current path, resulting in a semiconductor-like energy band. However, the bandwidth is a few kilotesla to 100 meV because organic molecules are loosely bound by weak van der Waals force. The narrow bandwidth induces a large relaxation time, and therefore a small mobility, according to the relationship of $(1/\mu) \sim \tau_{rel} = (h/\Delta E) \sim 10^{-14}$ s, which is about 100 times larger than that of a single Si crystal. The second factor is the long band tails above the valence band and below the conduction band, which are associated with molecular disorder originating from the variation of bond lengths between molecules due to thermal agitation. The band tails hinder carrier transport by trapping and releasing carriers in the bands. Therefore, the strategy to increase mobility is to achieve appropriate ordering of the molecules to make a current path through the π orbitals in small molecules and in polymers, as shown in Figure 15.6.

Thin films with small molecules can be formed by thermal evaporation. The growth mechanism of an organic thin film with thermal evaporation will be described in brief, as well as way to produce appropriate ordering of molecules. A molecule evaporated onto a substrate diffuses around until it arrives at a

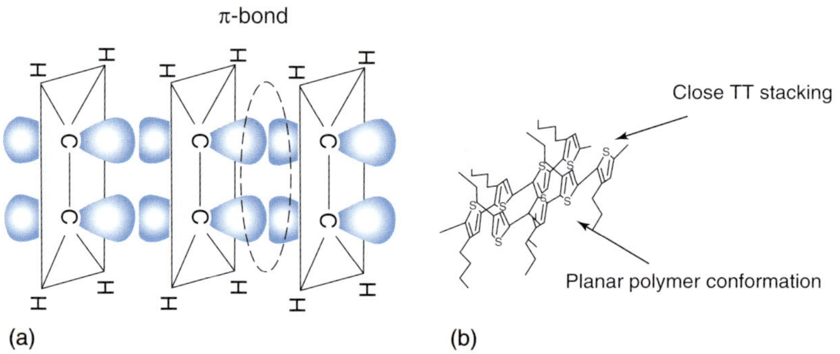

Figure 15.6 Molecular ordering with (a) small molecules and (b) polymer molecules. (Reproduced with permission. Copyright © 2002, American Chemical Society.)

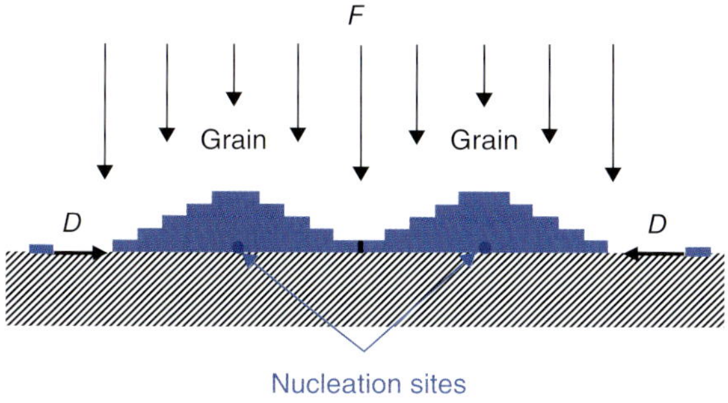

Figure 15.7 Conceptual picture to model the growth mechanism of organic thin film with small molecules.

thermodynamically low energy state to form a nucleation site. The molecules that arrive subsequently from the heating source aggregate at the site, and then a crystal grain grows, as illustrated in Figure 15.7. The size of a grain is determined by the density N of the nucleate sites. It is proportional to the deposition rate F, which is the number of molecules arriving at a substrate per second, and inversely proportional to the diffusion coefficient D at the substrate; that is, $N \sim F/D$. An organic thin film is generally a poly-crystal consisting of crystal grains. Thus, carrier transportation is largely affected by the trap states on grain boundaries in inter-grains in addition to the small bandwidth and tail states in intra-grains. Therefore, it is important to produce large grains or a small grain density for high mobility. In order to obtain large grains, N should be small so that F must be small and D large. F is controlled by the temperature of the heating source, and thus the temperature should be low enough for F to be about a few tens of angstroms per second. D has the relationship of $D = a^2 k_s$, where a is the distance from site to site and k_s is the hopping rate from site to site. And $k_s \propto \exp(-V_s/kT)$, where V_s is the potential barrier energy from site to site, k is the Boltzmann constant, and T is the temperature. In order to obtain a large D, V_s should be small and T large. Therefore, the substrate's surface must be kept hydrophobic with a low surface energy and heated to a temperature of less than 100 °C. In conclusion, in order to obtain a large grain, the deposition rate should be maintained as low as possible during evaporation, and the substrate should have a low surface energy with a hydrophobic surface and also heated to an appropriate low temperature.

For instance, the pentacene thin-film crystal, which is a representative small molecule exhibiting high mobility, consists of triclinic unit cells that are interlocked with a herringbone structure to increase bond strength, as shown in Figure 15.8a, and thus exhibits a high mobility of ~1 cm^2 V^{-1} s^{-1}. In Figure 15.8b, the AFM pictures of pentacene thin films are depicted for different deposition rates and substrate temperatures. The smaller the deposition rate and the larger the substrate temperature, the larger the grain size, as described above. In Figure 15.8c, the AFM pictures of a pentacene film are shown with and without

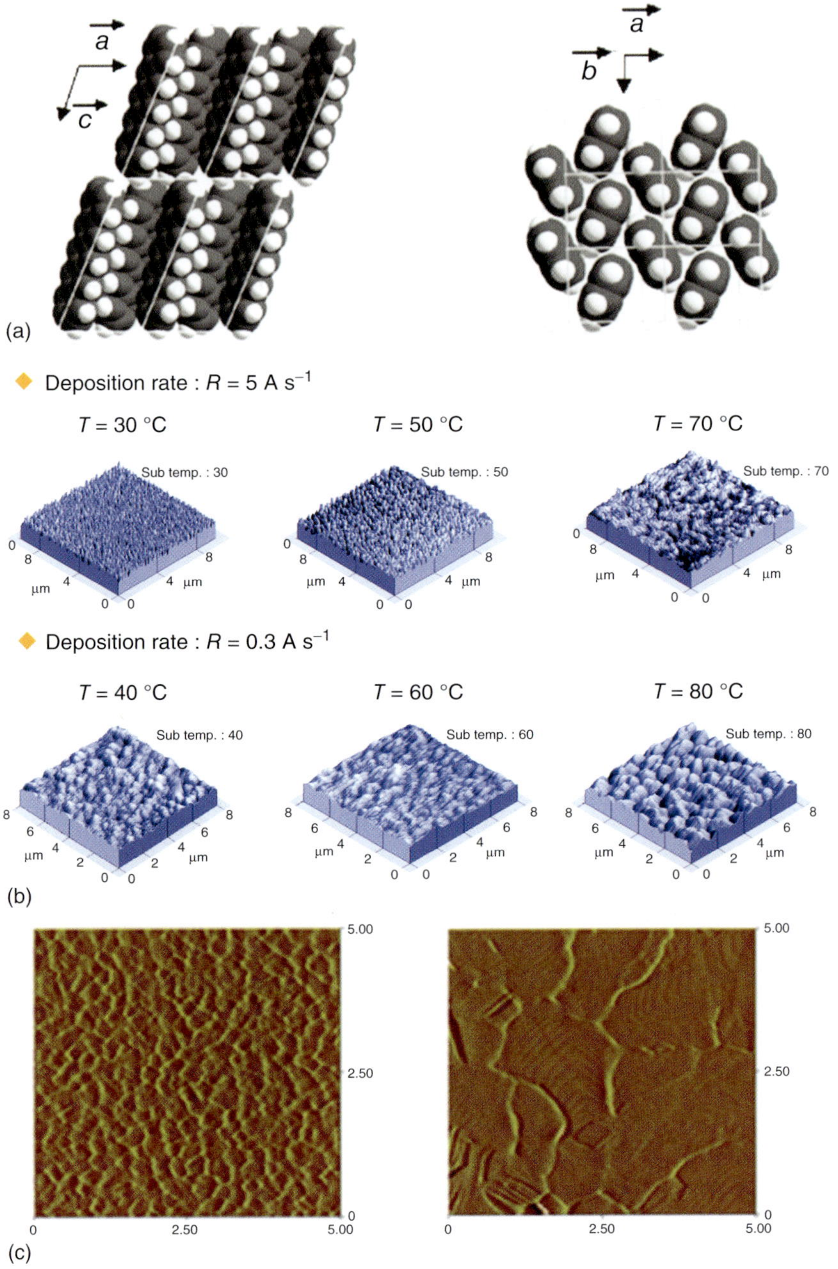

Figure 15.8 (a) Unit cell of pentacene crystal. (b) AFM pictures of pentacene thin film according to growth rate and substrate temperatures. (c) AFM pictures of pentacene thin film without and with OTS treatment on substrate; OTS treatment produces the larger grains. (Reproduced with permission. Copyright © 2012, American Chemical Society.)

surface treatment with OTS (octadecyltrichlorosilane). The grains became larger with OTS treatment because OTS provided the surface of the gate dielectric a low surface energy.

A polymer thin film is usually formed by a solution process such as dip coating because the molecule of the polymer is too heavy to be thermally evaporated. To obtain appropriate ordering of the molecules with polymers, self-organization is an effective method, which manipulates the molecules. For instance, the Poly(2,5-bis(3-hexadecylthiophen-2-yl)thieno[3,2-b]thiophene) (PBTTT) molecule has long side chains on the backbone, as shown in Figure 15.9a, and thus the molecules are self-organized within a layer and between layers by placing the side chains interdigitally [42]. The PBTTT film is prepared by dip coating, and the alignment is controlled by the dip-coating speed. At a low dip-coating speed of 2 μm s^{-1}, the polymer chains are uniaxially oriented, and long-range-aligned continuous polymer fibrous layers are obtained, as shown in Figure 15.9b, yielding pronounced structural anisotropy and a high average mobility of ~0.7 $cm^2 V^{-1} s^{-1}$ with the

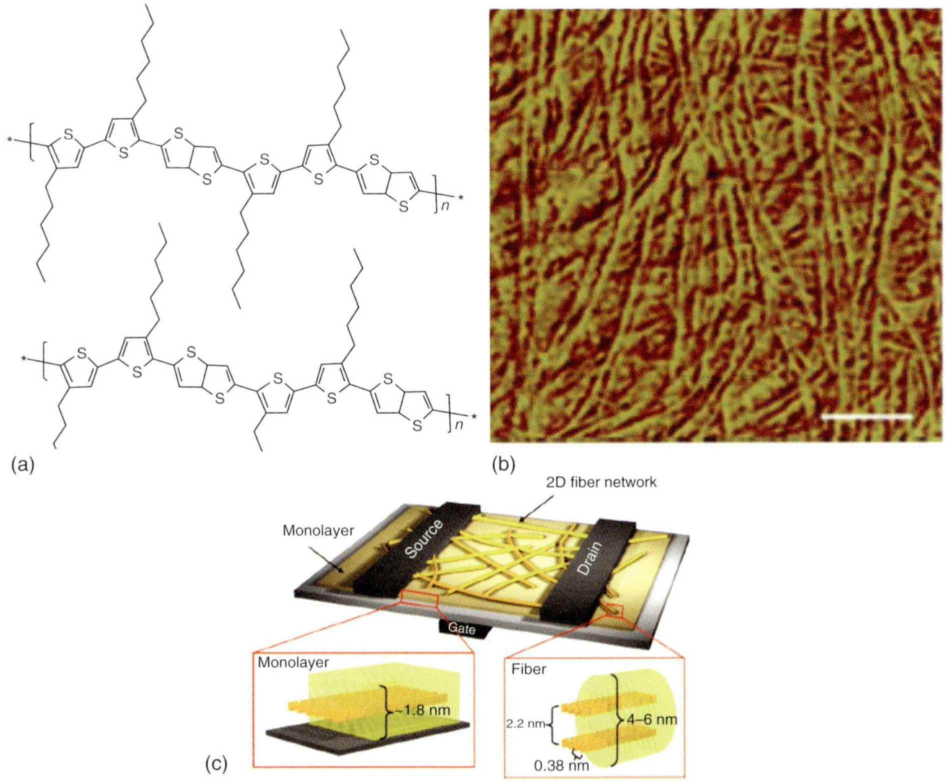

Figure 15.9 (a) Molecular ordering of PBTTT. (b) AFM images of PBTTT film prepared by a slow dip-coating speed of 2 μm s^{-1}; the polymer chains are uniaxially oriented. (c) Arrangement of PBTTT layers [43]. (Reproduced with permission. Copyright © 2012, American Chemical Society.)

highest mobility of 1.3 $cm^2 V^{-1} s^{-1}$ in the alignment direction. The mobility perpendicular to the alignment direction is only 6.2×10^{-2} $cm^2 V^{-1} s^{-1}$. This is a significantly larger anisotropy ratio of ~20. According to Grazing Incidence Wide Angle X-ray Scattering (GIWAXS) analysis, the model for the high mobility can be described as follows: The monolayer at the dielectric surface consists of polymer chains arranged edge-on. On top of this first layer, fibers are formed in which the edge-on polymer backbones are oriented along the fiber axis. This corresponds also to the fiber growth direction and faster charge-carrier migration. In these 4–6-nm-thick fibers, the out-of-plane chain-to-chain spacing of the polymer is 2.2 nm, which is in an identical range to that of the monolayer thickness. This means that, in the fiber, two or three polymer chains are arranged on top of each other, as shown in Figure 15.9c [43].

15.5 AMOLED Panel with Printing Technology

The OTFT/OLED combination is the most promising one as the driving transistor and display mode, respectively, for flexible displays in terms of process compatibility with flexible substrates made from a low-temperature process and solution processing capability. In this section, printing technologies to fabricate AMOLED panels using OTFTs as the driving devices will be discussed. A printing technology must be selected by considering the property of the layers of a panel. Because the gate electrodes of OTFTs and the scan bus lines and pads of the panels occupy most of the area of the substrate, a simple and cheap printing process such as screen printing is preferable. For the active layer of OTFT and OLED, expensive OSC materials are used, and thus ink jet printing is recommended in order to conserve the precious materials by jetting only a tiny amount of OSC solution on a specific area only. We investigated the feasibility of several printing technologies for the OTFT backplane of AMOLED panels.

15.5.1 Design and Fabrication of OTFT Backplane

A pixel consists of two OTFTs, which are the switching (SW) and driving (DR) OTFTs, one storage capacitor C_{st}, and one OLED, as shown in Figure 15.10a. The voltage signal V_{scan} of the scan line selects a pixel by applying a signal to the gate of the SW OTFT, and then the data on the data line V_{data} is transferred to the storage capacitor through the SW OTFT. The signal stored on C_{st} opens or closes the gate of the DR OTFT depending upon the polarity of the signal, the current from voltage source V_{dd} is supplied to the OLED through the DR OTFT, and then a light is switched on or off. It is important to reduce the off-state current I_{off} of the SW OTFT down to the level of picoamperes in order to minimize the leakage current from C_{st} through SW OTFT and thus to maintain the signal on C_{st} during

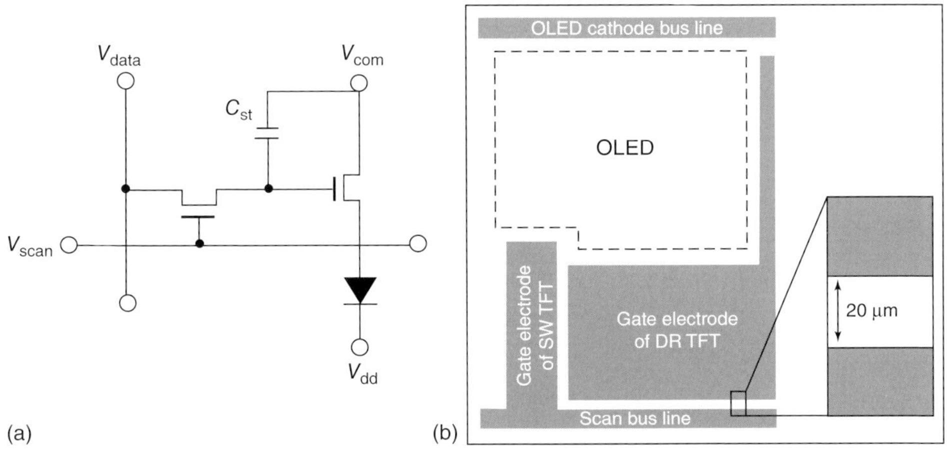

Figure 15.10 (a) Circuit diagram and (b) the layout of a pixel of AMOLED. (Reproduced with permission. Copyright © 2013, Elsevier.)

frame time. The DR OTFT should have the ability to supply a sufficient on-state current I_{on} to the OLED.

According to the luminescence versus current (J–I) characteristics of the OLED fabricated in this study, the OLED produces 100 cd m^{-2} at 0.7 mA cm^{-2}, as shown in Figure 15.17a. In order to handle the current, the channel lengths of the SW and DR OTFTs were determined to be 20 μm, and the widths to be 1040 and 1980 μm, respectively. A pixel was realized within a pitch of 1000 μm × 1000 μm by considering the limitation of the printing technology used in this study. The interdigitated structure for the source and drain (S/D) electrodes could implement a wide channel in a small area, and thus supply sufficient current to the OLED. The area of the OLED was 730 μm × 360 μm, resulting in an aperture ratio of 26.3%, as illustrated in Figure 15.10b. The diagonal length of the panel was 7.4 in., in which 17 500 pixels (140 × 125) were contained.

The schematic diagram of the fabrication processes is depicted in Figure 15.11. First, an IZO layer, which was predeposited on a glass substrate for the transparent anode electrode of the OLED, was patterned by photolithography and an etching process. Then, a silver (Ag) layer for the gate electrodes of the OTFTs and the scan bus lines and the contact pads of the panel were coated by screen printing with Ag ink over the whole area of the substrate and then patterned by photolithography. The process related to screen printing will be described in detail in the following subsection. Next, the gate insulator layer using a photo-patternable organic polymer was spin-coated and then patterned by photolithography. In this step, because the gate dielectric polymer itself was photo-patternable, the several steps for patterning the gate dielectric, such as photoresist coating and photolithography and etching of gate dielectric, which would be carried out if the conventional nonpatternable material were used, could be eliminated. Thus, we could simplify the process steps and avoid the performance degradation of the OTFTs associated with

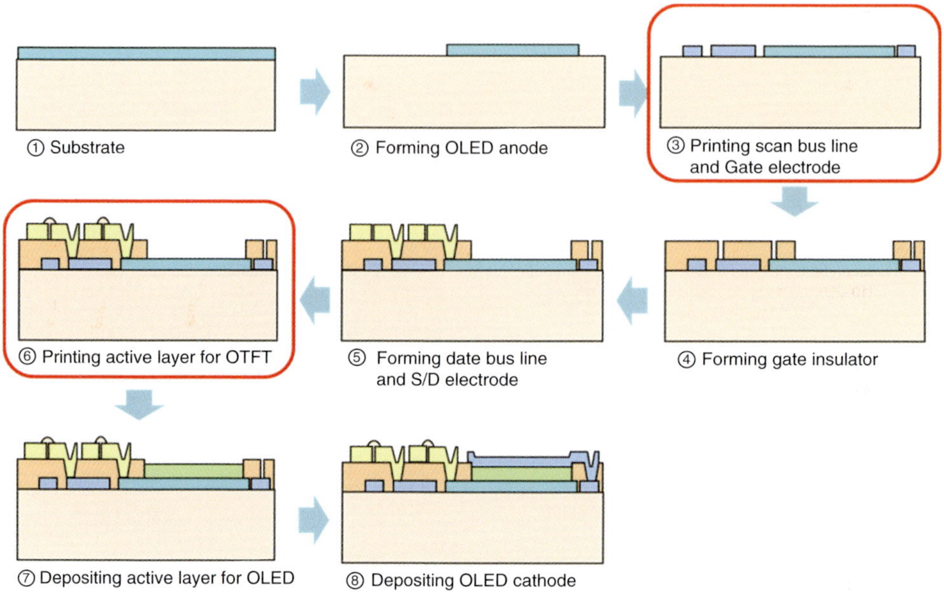

Figure 15.11 Schematic diagram of the fabrication process of AMOLED with OTFTs.

the effects of the photolithography process with a photoresist on the channels [44]. Subsequently, Au was thermally evaporated for the S/D electrodes and data bus lines, which was followed by inkjet printing of 6,13-Bis(triisopropylsilylethynyl) pentacene (TIPS-pentacene) for an OSC layer. Finally, the layers for the OLED and Al for the cathode electrode were continuously deposited by thermal evaporation through a shadow mask. A total of five masks, including one shadow mask, were used.

In this process, two printing technologies were employed: screen printing for the gate electrodes, and ink jet printing for the OSC. We will describe the printing processes in detail in the following section.

15.5.2
Screen Printing of the Gate Electrodes and Scan Bus Lines

In an AMOLED panel, the gate electrode of the SW OTFT is directly connected to a long scan line, as shown in Figure 15.10a, and the requirements are very stringent in terms of geometric dimensions and resistivity. The small resistance of the gate electrode and scan line is very important to reduce the signal delay along the long scan line. The resistance is expressed by $R = \rho(L/Wd)$, where ρ is the resistivity, L the length, W the width, and d the thickness of the scan line. The length is determined by the panel size, and thus a large panel can be a cause of high resistance. The width is also limited by resolution. Additionally, the thickness should be small for good step coverage of the gate dielectric on the gate electrode. Therefore, these are the key factors for reducing resistivity as much as possible for a

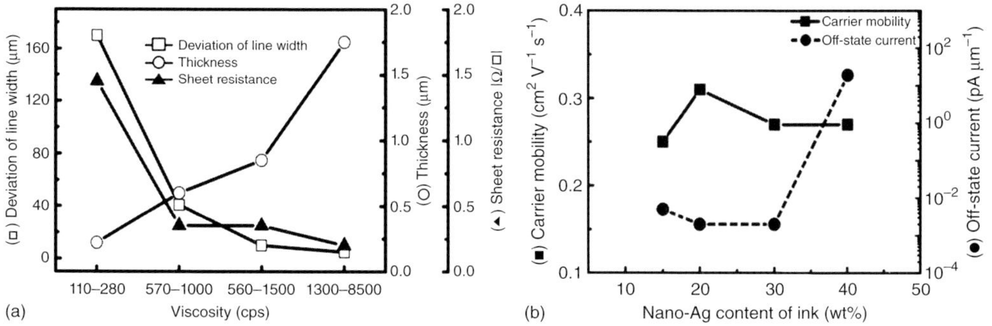

Figure 15.12 Electrical characteristics of nano-Ag film fabricated by screen printing according to (a) the viscosity and (b) the Ag content of ink. (Reproduced with permission. Copyright © 2013, Elsevier.)

high-performance panel. In this study, the AMOLED panel operating at 60 Hz frame frequency required a resistivity of less than $8 \times 10^{-4}\ \Omega\,cm$ for an allowable delay time of 1.3 μs.

One of the objectives of this study was to achieve a long, narrow, thin, and low-resistive gate electrode and scan bus lines with printing, which occupied most of the area of a pixel as shown in Figure 15.10b. We selected a screen printing process using nano-siver ink because of its simplicity and low cost. The above geometric and electrical properties of the Ag film are strongly dependent on the viscosity of the ink and hence by the Ag content of the ink, as shown in Figure 15.12. As the viscosity increases, the sheet resistance decreases and the deviation of the width also decreases, while the thickness increases. Moreover, the intersecting point of two oppositely dependent curves could not provide the appropriate viscosity to satisfy the requirement simultaneously. Additionally, the minimum feature size in a pixel was the gap space between the gate electrode of the DR OTFT and the scan line, as shown in Figure 15.10b, which was 20 μm. Such a narrow spacing cannot be produced by any of the currently available printing technologies alone. Therefore, in order to implement the above geometric and electrical requirements simultaneously, we adopted photolithography together with screen printing. First, an Ag thin film layer was deposited over the entire area of the substrate by screen printing. In this step, it was very important to make the thickness <0.1 μm, which was achieved by reducing the viscosity of the Ag ink to 80–120 cPs. Subsequently, the Ag film was cured at 200 °C for 30 min. Finally, photolithography and the etching process of the Ag film were carried out to obtain the final patterns.

The content of Ag particles in the Ag ink is an important factor in this screen printing. It determines the viscosity of the ink, and thus affects the thickness and resistivity of the gate electrode as well as the performance of the OTFTs. As shown in Figure 15.12b, OTFTs using the screen-printed gate electrodes exhibit Ag content dependence of mobility and the off-state current. The mobility reached the maximum value of $0.3\ cm^2\ V^{-1}\ s^{-1}$ at 20 wt% of Ag content. Above

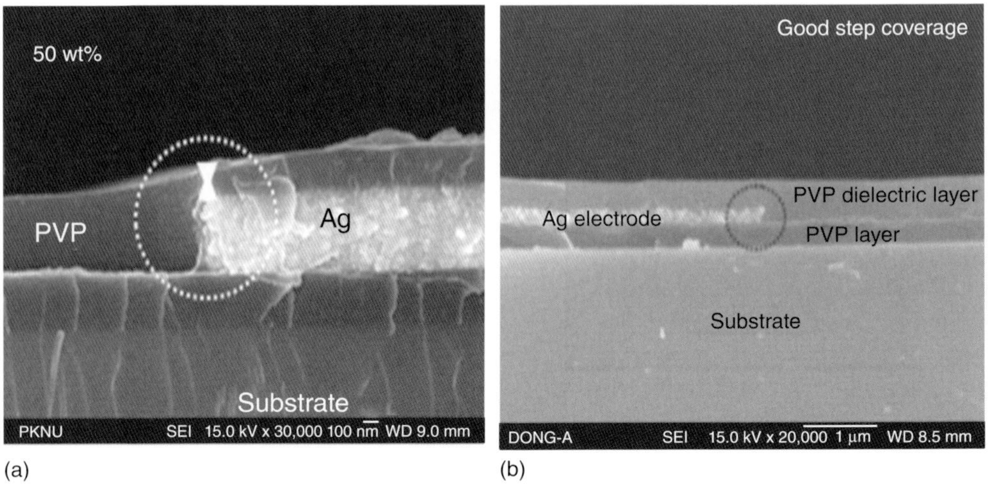

Figure 15.13 (a) Poor step coverage of PVP gate dielectric on the screen-printed gate electrode with high content of Ag. (b) Good step coverage with a proper content of Ag. (Reproduced with permission. Copyright © 2013, Elsevier, Copyright © 2010, Elsevier.)

that, the mobility slightly decreased. However, the off-state current increased rapidly above 30 wt%. The reason can be found in the large leakage current through the gate dielectric at the edge of the gate electrode caused by the poor step coverage of the gate dielectric on the thick gate electrode that was formed by a high Ag content of 30 wt%, as shown in Figure 15.13a. Therefore, 20 wt% Ag content was selected for screen printing of the gate electrodes to produce good step coverage, as shown in Figure 15.13b. The corresponding viscosity of the Ag ink was about 100 cP s.

With a viscosity of 100 cP s, the thickness and resistivity of the gate electrode were 60 nm and $3.0 \times 10^{-5}\ \Omega$ cm, respectively. Two requirements of the gate electrode – thickness of <1000 nm and resistivity of $<8.0 \times 10^{-4}\ \Omega$ cm – which exhibited opposite dependence on the viscosity, as shown in Figure 15.12a, were satisfied simultaneously with a viscosity of 100 cP s. The last requirement of the gate electrode, that is, a narrow line width, was achieved by photolithography. In Figure 15.14, the final printed gate electrodes and scan bus lines are shown. The figure exhibits the clear patterns of the gate electrodes, scan bus lines, and contact pads. In particular, the space between the DR OTFT gate electrode and the scan bus line, which is the minimum feature size of a pixel, is clearly seen. The line edge is very clear, compared to that with thermal evaporation. It originated from the nano-sized Ag particles.

Although this study was not for the sole utilization of screen printing, the significance of this study is in replacing the conventional thermal evaporation process with the cheap screen printing for deposition of the metal layer over the entire area. In future, photolithography can be replaced by other printing processes, such as reverse offset. However, the latter process requires alignment equipment on

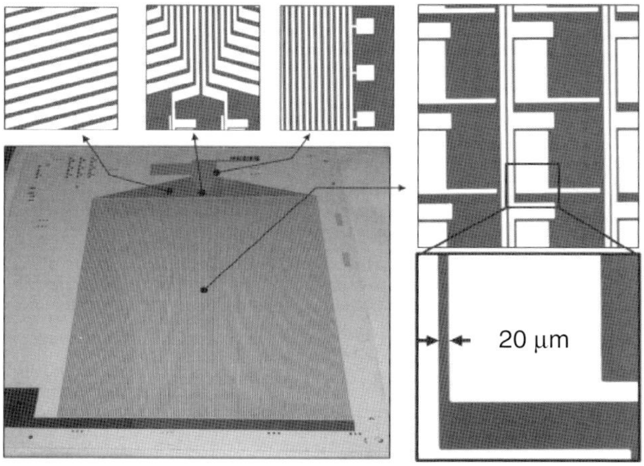

Figure 15.14 Gate electrodes and scan and data bus lines fabricated by screen printing combined with photolithography. (Reproduced with permission. Copyright © 2013, Elsevier.)

the printer because gate electrodes should be aligned with the IZO anode of the OLED. It will be difficult to align patterns during printing, especially for large-area printing.

15.5.3 Inkjet Printing of TIPS-Pentacene for OTFTs

Inkjet printing is an appropriate process for OSCs because the area of the OSC in an AMOLED is only a small channel region of the OTFT, and an inkjet can efficiently conserve the expensive OSC by dropping the OSC ink on a specific small area. However, the droplet of OSC usually produces the so-called "coffee ring," where the most of molecules accumulate on the contact line of the droplet, as shown in Figure 15.15b. In that case, the molecules are rare in the middle channel area, so that, in general, it is difficult to make high-performance OTFTs with inkjet printing. Thus, it is important to obtain a uniformly thick OSC layer without a coffee ring.

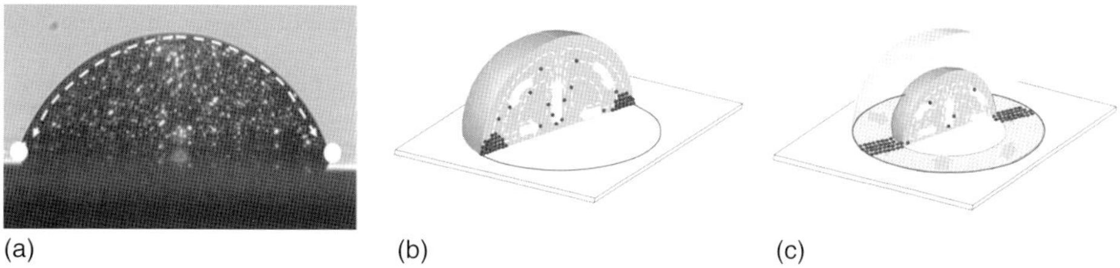

Figure 15.15 (a) Picture of particle movement in a droplet, and the conceptual figure to illustrate (b) coffee ring formation due to the contact line fixation and (c) the formation of uniform layer by making the contact line move. (Reproduced with permission. Copyright © 2013, Elsevier.)

We determined the mechanism of coffee ring formation and found a way to remove it. To identify the mechanism, we took serial pictures of the motion of the particles in a droplet over time. As shown in Figure 15.15a, the particles continuously moved from the central bottom to the top and then to the contact line along the surface because of the convective flow of the solvent. Furthermore, the contact line was fixed during drying, so that the particles continuously accumulated on the contact line, as shown in Figure 15.15b, because the contact line played the role of a thermally stable nucleate site, generating the coffee ring after complete drying. Based on this mechanism, a simple process was designed to eliminate the coffee ring and obtain uniform performance of the OTFTs over a large area. This process is described in the next section.

In order to eliminate the coffee ring and obtain a uniformly thick layer, the substrate was heated during jetting and drying. With heating, the contact line moved toward the central region during drying instead of being fixed, as shown in Figure 15.15c, because the solvent at the contact line evaporated faster than at the center. Because the contact line was a thermally stable nucleate site, the moving contact line made the crystal grains grow from the edge to the center, generating a uniform layer. However, the speed of the contact line movement should be properly matched to that of crystal formation; otherwise, the thickness would vary. Additionally, the contact line movement also depends upon the surface condition of the channel and the solvent type of the droplet. Therefore, a suitable temperature must be determined for each specific case.

For the TIPS-pentacene layer, which was mixed in anisole and jetted on a PVP gate dielectric, an appropriate temperature was 46 °C. This temperature was also applied for the photo-patternable gate dielectric used in this study. The details can be found in Refs [45, 46]. The attractiveness of this heating process is its simplicity and efficiency, which enables it to produce uniform performance over a large area just by keeping the whole area of the substrate at a specific temperature. Forty OTFTs randomly selected from the panel exhibited concentrated transfer curves within a small range, as shown in the inset of Figure 15.16. The mobility extracted from the 40 OTFTs was $0.31 \pm 0.05\,cm^2\,V^{-1}\,s^{-1}$. The deviation of 17% seems to be very small considering a panel area of $12\,in.^2$. The threshold voltage ranged from 0 to +1.5 V, and the off-state current was in the range of $3.6-9.6 \times 10^{-4}\,pA\,\mu m^{-1}$.

The number of droplets on a single channel also affected the performance of the OTFTs. The performance variation was examined with respect to the number of droplets. The number of droplets changed the final size of the TIPS-pentacene layer, which was restricted by the pixel pitch. Thus, the appropriate number of droplets was determined to satisfy the required performance as well as the limited resolution. Depending on the number of droplets on a single channel, the mobility also changed, as shown in Figure 15.16. At four drops, the mobility reached a maximum value of about $0.5\,cm^2\,V^{-1}\,s^{-1}$. However, the diameter of the droplet increased to about 400 μm, which was above the limit of the OTFT pitch of 380 μm. Therefore, three drops were chosen by compromising the

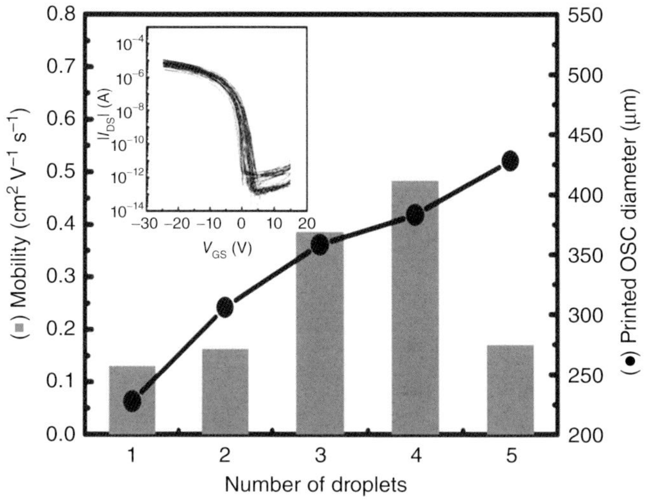

Figure 15.16 Variation of mobility and diameter of a droplet according to the number of droplets. (Reproduced with permission. Copyright © 2013, Elsevier.)

Table 15.3 Summary of the performance parameters of OTFTs fabricated by ink jet printing for TIPS-pentacene.

μ (cm^2 V^{-1} s^{-1})	I_{on}/I_{off}	V_{Th} (V)	SS (V s^{-1})	I_{off} (pA μm^{-1})
0.31 ± 0.05	$3.96 \pm 1.45 \times 10^7$	0.64 ± 0.85	0.47 ± 0.17	$6.61 \pm 3.08 \times 10^{-4}$

mobility with the pitch limit. For this case, the mobility was about 0.3 cm^2 V^{-1} s^{-1} and the diameter 350 μm. The performance parameters are summarized in Table 15.3.

15.6 Fabrication of the OLED and AMOLED Panel

In this study, a green OLED was employed to investigate the feasibility of using OTFTs for driving an AMOLED panel. The structure of the OLED is shown in the inset of Figure 15.17a. All the layers were deposited by thermal evaporation through a shadow mask. It is very important for the shadow mask to be strongly attached to the substrate in order to prevent defects from being induced by the shadow effect. The OLEDs emitted green light with a wavelength of 514 nm. The luminescence intensity varied linearly with the current density, as shown in Figure 15.17a, so that the gray level could be implemented. At a current density of 0.7 mA cm^{-2}, the luminescence intensity was 100 cd m^{-2}, which was brightly visible by the naked eye.

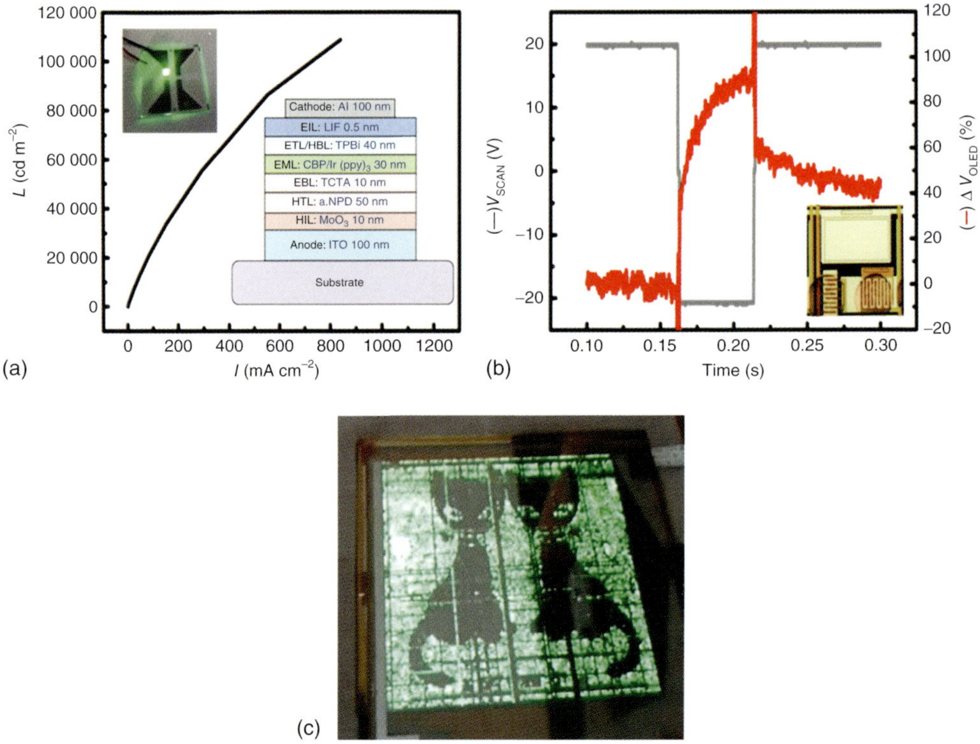

Figure 15.17 (a) Luminescence versus current characteristics of OLED. (b) Switching characteristics of a pixel of AMOLED. (c) AMOLED panel successfully displaying patterns. (Reproduced with permission. Copyright © 2013, Elsevier.)

The inset of Figure 15.17b shows the fabricated pixel array. The droplet on the SW OTFT seems larger than the channel and overlaps the data bus line, resulting in an electrical short-circuit. However, there was a gate dielectric layer between them, so the electrical short-circuit did not occur. From the pixel array, the waveforms of the scan bus line and the voltage across the OLED were measured, which are depicted in Figure 15.17b. The signal at each node of the pixel circuit could not be measured in a pixel because it was difficult to pick a node with a probe in the small area. Thus, we examined the voltage V_{OLED} across OLED by connecting a resistor across the OLED outside of the pixel. Thus, we could examine the operation of the SW and DR OTFT, and also the charging across the storage capacitor C_{st} in a pixel with the variation of V_{OLED}. As shown in Figure 15.17b, when a voltage of −20 V was applied to the scan bus line for 50 ms, V_{OLED} increased, reflecting the SW OTFT turning on, C_{st} being charged, and DR OTFT supplying current to the OLED. As the scan voltage returned to +20 V, V_{OLED} decreased by 30% from the peak, representing the kick-back voltage, and gradually decreased down to 50% of the peak during the off state of the SW OTFT. Based on the waveforms, we confirmed that the operation of the pixel circuit was successful.

Patterns were successfully displayed in the AMOLED panel, as shown in Figure 15.17c. However, several kinds of defects were found. The defects related to the fabrication process were generated by dropping TIPS-pentacene droplets in the wrong position and misalignment of the OLED pattern due to the drooping of the metal shadow mask and also to contaminations. The defects caused by degradation of the devices were associated with burning of the Al cathode of the OLED and generation of dark spots in the OLED due to operational stress.

15.7 Future Prospects

Technologies for flexible displays are currently being developed. Display-mode OLED will dominate the market because of its simple panel structure and fabrication processes resulting from its self-emitting nature. TFT technologies should be improved to reduce the process temperature and to be compatible with plastic substrates. Otherwise, the thermal properties of plastic substrates must be enhanced to withstand the high temperature of TFT backplane processes. In terms of low-temperature processes, OTFTs are the most suitable for flexible applications, but the performance and reliability need to make progress in mobility and in developing new materials. Because printing technologies are expected to enable the production of large-area and flexible electronics at low cost, TFTs made from solution processes, such as OTFTs, will be the key devices.

References

1. Tseng, F.M., Cheng, A.C., and Peng, Y.N. (2009) Assessing market penetration combining scenario analysis, Delphi, and the technological substitution model: the case of the OLED TV market. *Technol. Forecast. Soc. Change*, **76**, 897–909.
2. Allen, K.J. (2005) Reel to real: prospects for flexible displays. *Proc. IEEE*, **93**, 1394–1399.
3. Choi, M.C., Kim, Y.K., and Ha, C.S. (2008) Polymers for flexible displays: from material selection to device applications. *Polym. Sci.*, **33**, 581–630.
4. Crawford, G.P. (2005) *Flexible Flat Panel Display Technology*, John Wiley & Sons, Ltd, Chichester.
5. DISPLAYSEARCH and FlexTech (2008) 2008 Flexible Electronics and Displays Report.
6. Kagan, C.R. and Andry, P. (2003) *Thin-Film Transistors*, Chapter 2, Marcell Dekker, Inc.
7. Sorbrinho, A.S.d.S. and Czeremuszkin, G. (1999) A study of defects in ultra-thin transparent coatings on polymer. *Surf. Coat. Technol.*, **116-119**, 1204–1210.
8. Polach, S., Horst, D., Maier, G., Kallfass, T., and Lueder, E. (1999) Matrix of light sensors addressed by a-Si:H TFTs on a flexible plastic substrate. *Proc. SPIE*, **3649**, 31–39.
9. Yan, M., Kim, T.W., Erlat, A.G., Pellow, M., Foust, D.F., Liu, J., Schaepkens, M., Heller, C.M., McConnelee, P.A., Feist, T.P. and Duggal, A. (2005) A Transparent, High Barrier, and High Heat Substrate for Organic Electronics. *Proc. IEEE*, **93** (8), 1469.
10. Luo, F.C., Hester, W.A., and Brody, T.P. eds (1978) Alphanumeric and video performance of a 6"x6" 30 lines-per-inch thin-film-transistor -liquid crystal display panel. Digest 1978 SID International Symposium, San Jose, CA, p. 94.

11. Brody, T.P., Luo, F.C., Szepesi, Z.P., and Davies, D.H. (1975) A 6×6-in. 20-lpi electroluminescent display panel. *IEEE Trans. Electron Devices*, **ED-22**, 739.
12. Lee, M.J., Wright, S.W., and Judge, C.P. (1980) Electrical and slruciuri properties of cadmium selenide thin-film transistors. *Solid State Electron.*, **23**, 671.
13. Carlson, D.E. and Wronski, C.R. (1976) Amorphous silicon solar cells. *Appl. Phys. Lett.*, **28**, 671.
14. Luo, F.C., Chen, I., and Genovese, F. (eds) (1980) *Conference on Record Biennial Display Research Conference*, IEEE, New York, p. 111.
15. Kallfass, T. and Lueder, E. (1979) High-voltage thin-film transistors manufactured with photolighography and with Ta_2O_5 as the gate oxide. *Thin Solid Films*, **61**, 259.
16. Lechner, B.J., Marlowe, F.J., Nester, E.O., and Tults, J. (eds) (1971) Liquid crystal matrix displays. *Proc. IEEE*, **59**, 1566.
17. LeComber, P.G., Spear, W.E., and Ghaith, A. (1979) Amorphous-silicon field-effect device and possible application. *Electron. Lett.*, **15**, 179.
18. Brody, T.P., Asars, J.A., and Dixon, G.D. (1973) A 6. 6 inch 20 lines-per-inch liquid crystal display panel. *IEEE Trans. Electron Devices*, **ED-20**, 995.
19. Depp, S.W., Juliana, A., and Huth, B.G. (eds) (1980) *Proceeding of the 1980 International Electron Device Meeting*, IEEE, New York, p. 703.
20. Juliana, A., Depp, S.W., Huth, B. and Sedgwick, T. (eds) (1982) Thin-film polysilicon devices for flat panel display circuitry. Digest 1982 Society for Information Display International Symposium, San Jose, CA, p. 38.
21. Nishimura, T., Akasaka, Y., and Nakata, H. (1982) Characteristics of TFT fabricated in laser-recrystallized polysilicon for active LC display. SID International Symposium Digest of Technical Papers, p. 36.
22. Little, T.W., Koike, H., Takahara, K., Nakazawa, T., and Ohshima, H. (eds) (1991) *A 9.5-Inch 1.3-Mega-Pixel Low-Temperature Poly-Si TFT-LCD Fabricated by SPC of Very Thin Films and an ECR-CVD Gate Insulator, Conference Record 1991 International Display Research Conference*, IEEE, New York, p. 219.
23. Garnier, F., Peng, X., Horowitz, G., and Fichou, D. (1990) An all-organic "soft" thin film transistor with very high carrier mobility. *Adv. Mater.*, **2**, 592.
24. Park, J.S., Kim, T.S., Son, K.S., Lee, K.H., Jung, J.S., Maeng, W.J., Kim, H.S., Kim, E.S., Park, K.B., Seon, J.B., Kwon, J.Y., Ryu, M.K., and Lee, S. (2010) High-performance and stable transparent Hf–In–Zn–O thin-film transistors with a double-Etch-stopper layer. *IEEE Electron Device Lett.*, **31** (11), 1248.
25. Park, J.S., Kim, T.S., Son, K.S., Lee, K.H., Maeng, W.J., Kim, H.S., Kim, E.S., Park, K.B., Seon, J.B., Choi, W., Ryu, M.K., and Lee, S.Y. (2010) The influence of SiOx and SiNx passivation on the negative bias stability of Hf-In-Zn-O thin film transistors under illumination. *Appl. Phys. Lett.*, **96** (26), 262109.
26. Ahn, B.D., Shin, H.S., Kim, H.J., Park, J.S., and Jeong, J.K. (2008) Comparison of the effects of Ar and H2 plasmas on the performance of homojunctioned amorphous indium gallium zinc oxide thin film transistors. *Appl. Phys. Lett.*, **93** (20), 203506.
27. Sato, A., Abe, K., Hayashi, R., Kumomi, H., Nomura, K., Kamiya, T., Hirano, M., and Hosono, H. (2009) Amorphous In–Ga–Zn–O coplanar homojunction thin-film transistor. *Appl. Phys. Lett.*, **94**, 133502.
28. Lim, H., Yin, H., Park, J.S., Song, I., Kim, C., Park, J., Kim, S., Kim, S.W., Lee, C.B., Kim, Y.C., Park, Y.S., and Kang, D. (2008) Double gate GaInZnO thin film transistors. *Appl. Phys. Lett.*, **93** (6), 063505.
29. Gleskova, H. and Wagner, S. (2001) Low-temperature silicon nitride for thin-film electronics on polyimide foil substrates. *Appl. Surf. Sci.*, **175-176**, 12–16.
30. He, S., Nishiki, H., Hartzell, H.J., and Nakata, Y. (2000) Low-temperature PECVD a-Si:*H* TFT for plastic substrate. SID 2000 Digest, pp. 278–281.
31. Winer, K., Anderson, G.B., Ready, S.E., Bachrach, R.Z., Johnson, R.I., Ponce, F.A., and Boyce, J.B. (1990) Excimer-laser-induced crystallization of

hydrogenated amorphous silicon. *Appl. Phys. Lett.*, **57**, 2222.
32. Brotherton, S.D., McCulloch, D.J., Clegg, J.B., and Gowers, J.P. (1993) Excimer-laser-annealed poly-Si thin-film transistors. *IEEE Trans. Electron Devices*, **40**, 407.
33. Mei, P., Boyce, J.B., Hack, M., Lujan, R.A., Johnson, R.I., Anderson, G.B., Fork, D.K., and Ready, S.E. (1994) Laser dehydrogenation/crystallization of plasma-enhanced chemical-vapor-deposited amorphous silicon for hybrid thin-film transistors. *Appl. Phys. Lett.*, **64**, 1132.
34. Nomura, K., Ohta, H., Ueda, K., Kamiya, T., Hirano, M., and Hosono, H. (2003) Thin-film transistor fabricated in single-crystalline transparent oxide semiconductor. *Science*, **300**, 1269.
35. Nomura, K., Ohta, H., Takagi, A., Kamiya, T., Hirano, M., and Hosono, H. (2004) Room-temperature fabrication of transparent flexible thin film transistors using amorphous oxide semiconductors. *Nature*, **432**, 488.
36. Ohya, Y., Niwa, T., Ban, T., and Takahashi, Y. (2001) Thin film transistor of ZnO fabricated by chemical solution deposition. *Jpn. J. Appl. Phys. Part 1-Regul. Pap. Short Notes Rev. Pap.*, **40**, 297.
37. Chang, Y.J., Lee, D.H., Herman, G.S., and Chang, C.H. (2007) High-performance, spin-coated zinc tin oxide thin-film transistors. *Electrochem. Solid-State Lett.*, **10**, H135.
38. 38. Lee, D.H., Chang, Y.J., Herman, G.S., and Chang, C.H. (2007) A general route to printable high-mobility transparent amorphous oxide semiconductor. *Adv. Mater.*, **19**, 843.
39. Bashir, A., Wobkenberg, P.H., Smith, J., Ball, J.M., Adamopoulos, G., Bradley, D.D.C., and Anthopoulos, T.D. (2009) High-performance zinc oxide transistors and circuits fabricated by spray pyrolysis in ambient atmosphere. *Adv. Mater.*, **21**, 2226.
40. Adamopoulos, G., Bashir, A., Thomas, S., Gillin, W.P., Georgakopoulos, S., Shkunov, M., Baklar, M.A., Stingelin, N., Maher, R.C., Cohen, L.F., Bradley, D.D.C., and Anthopoulos, T.D. (2010) Spray-deposited Li-doped ZnO transistors with electron mobility exceeding 50 cm^2/Vs. *Adv. Mater.*, **22**, 4764.
41. Adamopoulos, G., Thomas, S., Wöbkenberg, P.H., Bradley, D.D.C., McLachlan, M.A., and Anthopoulos, T.D. (2011) High-mobility low-voltage ZnO and Li-doped ZnO transistors based on ZrO High-k dielectric grown by spray pyrolysis in ambient air. *Adv. Mater.*, **23**, 1894.
42. McCulloch, I., Heeney, M., Bailey, C., Genevicius, K., MacDonald, I., Shkunov, M., Sparrowe, D., Tierney, S., Wagner, R., Zhang, W., Chabinyc, M.L., Kline, R.J., McGehee, M.D., and Toney, M.F. (2006) Liquid-crystalline semiconducting polymers with high charge-carrier mobility. *Nat. Mater.*, **5**, 328.
43. Wang, S., Kiersnowski, A., Pisula, W., and Mullen, K. (2012) Microstructure evolution and device performance in solution-processed polymeric field-effect transistors: the key role of the first monolayer. *J. Am. Chem. Soc.*, **134**, 4015–4018.
44. Ryu, G.S., Jeong, S.H., Kwon, D.H., Kim, J.S., and Song, C.K. (eds) (2012) *AMOLED Panel Driven by TIPS-Pentacene OTFTs*, IMID 2012 Digest, p. 210
45. Kim, M.J., Heo, H.W., Suh, Y.K., and Song, C.K. (2011) Morphology control of TIPS-pentacene grains with inert gas injection and effects on the performance of OTFTs. *Org. Electron.*, **12**, 1170–1176.
46. Lee, M.W., Ryu, G.S., Lee, Y.U., Pearson, C., Petty, M.C., and Song, C.K. (2012) Control of droplet morphology for inkjet-printed TIPS-pentacene transistors. *Microelectron. Eng.*, **95**, 1–4.

16
Flexible Organic Solar Cells for Scalable, Low-Cost Photovoltaic Energy Conversion

Seunghyup Yoo, Jongjin Lee, Donggeon Han, and Hoyeon Kim

16.1
Overview of Organic Photovoltaic (OPV) Cells

16.1.1
Motivation for OPV Cells

The history of modern solar cells starts in the 1950s, when Chapin, Fuller, and Pearson at the Bell Laboratories developed the first solar cell with a practically meaningful efficiency using a crystalline Si p–n junction [1]. Since then, solar cells based on crystalline Si have shown a steady improvement and secured manufacturing technologies for the PV (photovoltaic) module, or an array of individual cells connected in series or in parallel to meet a certain specification for voltage and current output desired by a target application. Because of technological maturity and excellent stability, they soon dominated the global PV market just like modern integrated electronic devices. However, the proliferation of the Si PV technology has often been limited by its manufacturing cost rather than by the technology itself. In fact, the continued demand for driving down the cost has given birth to thin-film PV cells such as those based on a-Si or on compound semiconductors such as CdTe or $CuIn_{1-x}Ga_xSe_2$ (CIGS).

Note that the price of electricity generated by a given PV technology is determined by various factors, including (but are not limited to) the efficiency, lifetime, cost of PV modules, annual insolation, interest rate for capital investment, balance of system (BOS) cost (e.g., cost of land and power-conditioning system, batteries), and operation and management (O&M) overhead cost [2, 3]. Among them, the per-area cost and efficiency of a PV cell or module allow one to estimate the so-called "$\$/W_p$" or the cost to produce 1 W under the full sun condition. A value smaller than 1–2 $\$/W_p$ is often regarded, for a PV module based on crystalline Si technology, as a target to realize "grid parity", at which the price of electricity (given in dollar per kilowatthour) generated by a PV technology becomes comparable to that of electricity generated by conventional methods (=about 10–30 cents kWh^{-1} in most countries) [4] (see Figure 16.1) [5]. Note that the $\$/W_p$ value required to meet a target electricity generation cost depends on the

Large Area and Flexible Electronics, First Edition. Edited by Mario Caironi and Yong-Young Noh.

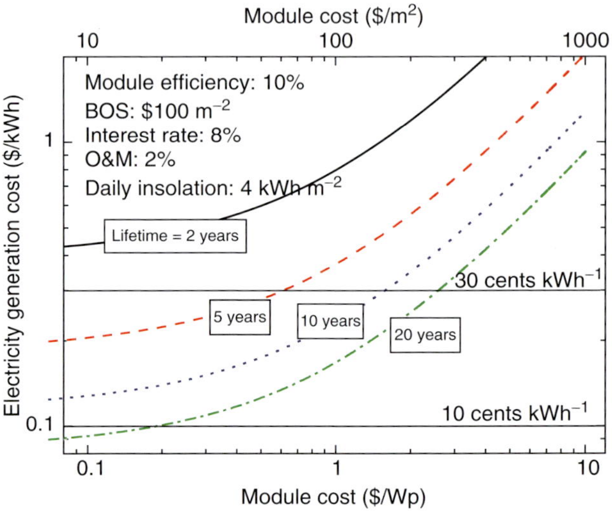

Figure 16.1 Electricity generation cost estimated as module cost ($/m^2 or $/W$_p$) for PVs with power conversion efficiency of 10% and lifetime of 2, 5, 10, and 20 years. Assumptions used for estimation are provided in the inset. See Refs [2, 3] for further details.

lifetime of a given PV module; in case of a PV module with a relatively short lifetime, the required $/W$_p$ could be far <1–2 $/W$_p$. Under reasonable assumptions specified in the inset of Figure 16.1, for example, a PV module with power conversion efficiency and lifetime of 10% and 5 years, respectively, requires a module cost to be smaller than ~0.6$/W$_p$ (equivalent to $60 m^{-2}) to realize the electricity generation cost below 30 cents kWh^{-1}, while a PV module with the same efficiency but with lifetime of 20 years can realize it with a module cost as large as 2.6$/W$_p$. Organic PV and several other emerging PV technologies are thus trying to meet the goal of grid parity by significantly reducing their module cost and increasing their lifetime while maintaining reasonably high efficiency in the range of 10–20%.

What makes OPV (organic photovoltaic) technologies promising as low-cost PV technologies lies in their low-temperature processability. It allows OPVs to be fabricated on various substrates including plastic substrates, which, in turn, enables high-throughput fabrication based on roll-to-roll (R2R) technologies, opening up ways to realize a significant reduction in the overall manufacturing cost by volume production [6]. In addition, a study done by Krebs and his coworkers [7] indicates that the low-temperature processability of OPVs coupled with R2R printing can result in a PV technology with energy payback time (EPBT) on the order of a few days so that the overall energy consumed to produce PV modules can be recovered very soon by the energy produced by those PV modules. This is in a sharp contrast with the case of conventional crystalline PV technology, whose the EPBT is typically on the order of a few years.

Another obvious benefit associated with their compatibility with plastic substrates and inherent mechanical flexibility of polymer and small-molecule thin films is the fact that solar cells can be made highly flexible and light so that they can be easily integrated into other products, enabling various opportunities based on creative ideas such as PVs in the form of portable rolls, wall papers, blinds, tents, and so on, thereby extending the applicability to areas that have not been readily available to conventional PV technologies.

16.1.2
Fundamentals of OPV Technologies

16.1.2.1 General Operation of PV Cells

In most cases, a PV cell can be conveniently modeled by an equivalent circuit which consists of a diode connected in parallel with a DC current source having a current that is equals to the photocurrent (I_{ph}) generated by the given PV cell (Figure 16.2a). The current versus voltage characteristics of a PV cell are then those of a diode but shifted downward by I_{ph} (Figure 16.2b). The fourth quadrant corresponds to the PV operation region, as it has negative power, indicating a situation where electrical power is provided by a PV cell to an external load. The power conversion efficiency (η) of a PV cell with active area A is then defined by the ratio of the maximum electrical power ($= P_{max}^{(elec)}$) that can be produced by the cell to the optical power ($= P_{in}^{(opt)}$) provided by a light incident on the cell:

$$\eta = \frac{P_{max}^{(elec)}/A}{P_{in}^{(opt)}/A} = \frac{|JV|_{max}}{E_{in}^{(opt)}} = \text{FF}\frac{J_{sc}V_{oc}}{E_{in}^{(opt)}} \quad (16.1)$$

where J_{sc} is short-circuit current density, V_{oc} is open-circuit voltage, and FF refers to fill factor, which is defined as $(JV)_{max}/J_{sc}V_{oc}$. For solar cells used in terrestrial

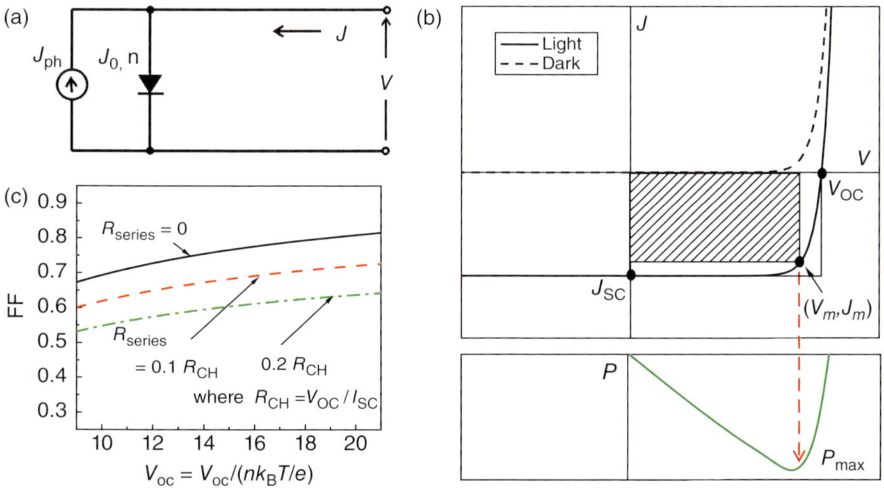

Figure 16.2 (a) Equivalent circuit of an ideal solar cell. (b) J–V and P–V characteristics. (c) FF of a solar cell described by the equivalent circuit in (a) with the diode based on a p–n junction. In (c), cases with a finite series resistance are also presented.

applications, the solar spectrum, specified as AM1.5G, is used as the standard illumination condition with the irradiance ($= E_{in}^{(opt)}$) of 1 sun, given by 100 mW cm^{-2}.

In the case provided by the equivalent circuit shown in Figure 16.2a with the diode based on an ideal p–n junction, the following relation holds at temperature T (in kelvin) for J_{sc} and V_{oc}:

$$J_{sc} = \frac{-I_{ph}}{A} = -J_{ph} \tag{16.2a}$$

$$V_{oc} = \frac{nk_B T}{e} \ln\left(1 + \frac{J_{ph}}{J_0}\right) \tag{16.2b}$$

where J_0 and n refer to diode reverse saturation current density and diode ideality factor, respectively, and k_B and e are the Boltzmann constant and electronic charge, respectively. For FF, there is no analytic expression, but the following empirical formula is known to be very useful, providing an accurate approximation to numerically obtained values (Figure 16.2c):

$$\mathrm{FF}_{\mathrm{empirical}} = \left.\frac{v_{oc} - \ln\left(v_{oc} + 0.72\right)}{v_{oc} + 1}\right|_{v_{oc} \equiv V_{oc}/(nk_B T/e)}, \quad (v_{oc} > 10) \tag{16.3}$$

In a practical solar cell, actual J–V characteristics can differ significantly from the ideal curves for various reasons. For example, J_{SC} and FF may get severely compromised by the presence of a large series resistance, which may come from the relatively high sheet resistance of transparent electrodes (TEs) (see Figure 16.2c). In the case of OPVs, working mechanisms different from that of conventional p–n junction type solar cells can also lead to nonideal J–V characteristics, although their characteristics can be fitted *phenomenologically* to the equivalent circuit in Figure 16.2a relatively well often with series and shunt resistance included. For further details, readers may refer to Refs [8–12] and references therein.

16.1.2.2 **Working Principle of OPV Cells**

The operation of OPVs relies on a photo-induced carrier generation process occurring mainly at donor–acceptor (D–A) heterojunctions [11, 13]. Absorption of photons in organic semiconductors (OSCs) creates bound electron–hole pairs, often referred to as *excitons*. Excitons in organic materials differ from those in inorganic crystalline semiconductors in that they have a binding energy on the order of a few hundred millielectronvolts and, thus, are not easily separable into independent electrons and holes in the room temperature range unless a sufficient external energy is provided; properly chosen donor–acceptor heterojunctions with a sufficient energy offset between their lowest unoccupied molecular orbital (LUMO) levels (ΔE_A in Figure 16.3a) or between their highest occupied molecular orbital (HOMO) levels (ΔE_H in Figure 16.3a) can provide an energetic means to dissociate them into free electrons and holes by a charge-transfer process. Those separated electrons and holes are then transported and collected selectively by a cathode and an anode, respectively [14].

It is noteworthy that excitons are electrically neutral, and thus have to reach the D/A junction by diffusion. Unfortunately, most OSCs reportedly have a very short diffusion length (L) in the range of a few nanometers to a few tens of nanometers. The total amount of photogenerated carriers in OPVs with a neat donor layer and an acceptor layer (Figure 16.3b) is therefore limited by this short exciton diffusion length in most cases. That is, one cannot make donor and/or acceptor layers arbitrarily thick to maximize the chance of photon absorption [15]. This is because excitons created at a distance larger than L from the D/A junction are most likely to recombine before even reaching the junction, simply wasting absorbed photons without contributing to photocarrier generation. To overcome this dilemma, the so-called bulk heterojunction (BHJ) geometry is widely adopted in OPVs, in which donor and acceptor materials are mixed together to form interpenetrating networks, as shown in Figure 16.3b (right, bottom). In a BHJ geometry, excitons do not have to travel far to reach the D/A junction because they can find a nearby D/A junction within a few nanometers or so in OPVs with a proper nanoscale intermixed D/A BHJ geometry. A layer with the proper BHJ geometry can be formed (i) by a carefully controlled co-evaporation if one uses thermal evaporation of small molecules, or (ii) simply by blending donor and acceptor materials into one solution and forming a layer from that solution with a coating process such as spin coating or spraying. Note that the BHJ geometry is not a perfect cure; the thickness of active layers in OPVs based on BHJs cannot be arbitrarily increased as well, because it is still often limited by the carrier collection length determined mainly by the carrier mobility of OSCs and the intermixing morphology. In that respect, efforts are being made to develop OSCs with high carrier mobility and to find a way to control the morphology of D/A networks, for example, by a thermal or solvent annealing process.

While the photocarrier generation process mentioned above is related mainly to J_{sc}, or the maximum current per device area that can be generated by a solar cell, another important parameter of a solar cell is its photovoltage output. The maxi-

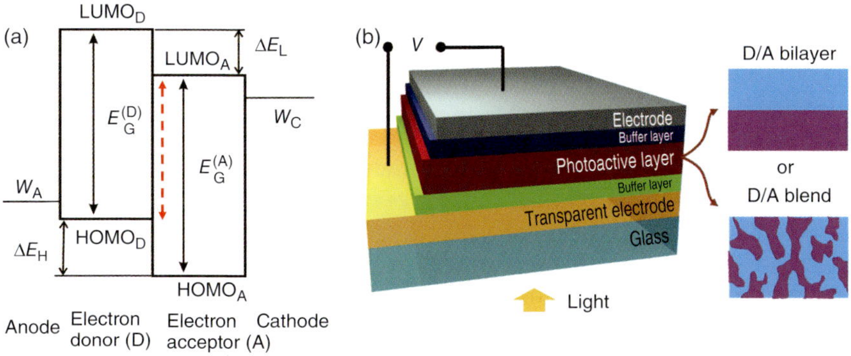

Figure 16.3 (a) Schematic energy diagram for a donor (D) acceptor (A) heterojunction. (b) Typical device geometry in OSCs.

mum voltage output that can be provided by a solar cell is $V_{oc,}$ and it should be as large as possible to achieve high power conversion efficiency. Many reports indicate a positive correlation between V_{oc} and the effective bandgap of a given D/A pair, which is defined by the energy difference between the LUMO of the acceptor material and the HOMO of the donor material used in an OPV cell. However, actual V_{oc} can be reduced for various reasons, including the presence of carrier leakage paths, recombination, energetic disorder or traps, imperfect energy-level matching between the HOMO and LUMO levels of the BHJ, and the electrode work function. Further details on V_{oc} can be found in Refs [11, 16, 17].

16.1.2.3 **Major Components and Various Configuration of OPV Cells**

Overview As shown in Figure 16.3b, OPV cells typically consist of a substrate, an anode, a cathode, and a photoactive layer (a D/A blend or a neat D/A bilayer), with one of the electrodes chosen to be optically transparent to allow incident light to enter the photoactive layer. Optionally, buffer layers can be inserted between each electrode and the photoactive layer. Depending on the location of the cathode with respect to the substrate, OPV cells are classified as "normal or conventional" (the cathode is the "top" electrode or the electrode that is away from a substrate) or "inverted" (the cathode is the "bottom" electrode or the electrode on the substrate). As the name suggests, the conventional geometry that is predominantly used has its anode on the substrate side; this is rather a natural choice, as ITO (indium tin oxide) – the *de facto* TEs in OPVs – has a relatively high work function (about 4.7 eV) suitable for anodes. Nevertheless, inverted OPVs are gaining attention, as they are regarded as advantageous in terms of device stability because cathodes, which are believed to be more vulnerable to air and humidity due to their low work function, are buried underneath other layers, making them better protected from ambient atmosphere. In the sections below, basic facts on components of OPVs are briefly summarized.

Photoactive Layers As mentioned in the previous section, photoactive layers of OPVs consist of organic electron donor (D) and acceptor (A) materials. These D/A materials are based on conjugated compounds, or organic compounds in which single and double bonds alternate. In such a compound, π electrons from sp^2 hybrid orbitals are highly delocalized, enabling rich electrical properties in organic materials, which would otherwise be a good insulator in most cases [18]. As the degree of π conjugation increases, the HOMO–LUMO energy gap (E_{HL}) tends to decrease. When E_{HL} is in the range of 1–4 eV, they can absorb photons in the UV-NIR (near-infrared) spectral region via $\pi-\pi^*$ transitions, making them especially attractive for PV or light-emitting applications.

Donor materials have a relatively small ionization potential, and acceptor materials have a relatively large electron affinity. In such a case, the energy diagram formed by D and A can be similar to that of the p–n junction in inorganic semiconductors, providing an ideal situation for solar cells, as was shown in the previous section (Figure 16.3a). Notable donor and acceptor materials are summarized in

Table 16.1. Acceptor materials used in a seminal work by Tang *et al.* were based on perylene derivatives, but fullerenes (C_{60}, C_{70}) or their solution-processable derivatives such as phenyl-C_{61}-butyric acid methyl ester (PCBM-60) or phenyl-C_{71}-butyric acid methyl ester (PCBM-70) have been predominantly adopted because Heeger and his coworkers [19] found photo-induced charge transfer from conjugated polymers to C_{60} to be highly efficient. Although the electrical property of C_{60} is better than that of C_{70}, C_{70} and its derivatives are also popular because of their superior optical absorption in the visible spectral range [20]. Recently, the indene-C_{60} bisadduct (ICBA) was adopted as an acceptor replacing PCBM-60 in some BHJ OPV cells [21]. Its electron affinity, which is smaller than that of PCBM-60, was shown to yield higher V_{oc} for OPV cells containing ICBA.

While acceptor materials were more or less fixed as fullerene derivatives, donor materials were based on various moieties. In the case of thermally evaporable small molecules, phthalocyanines and oligoacenes (tetracene, pentacene) have been popular. In case of solution-processable polymers, derivatives of poly(*p*-phenylene vinylene) (PPV), polythiophene, and polyfluorene have been popular. Recent trends in the development of donor polymeric materials adopt copolymers in which a donor unit (D_0) and an acceptor unit (A_0) alternate; the interaction between D_0 and A_0 can lower the HOMO–LUMO energy gap (E_G) so that the resultant polymer can absorb low-energy photons in the NIR region, which is often difficult to realize with a homopolymer-type conjugated polymer. A simple explanation for the E_G lowering can be provided, as shown in Figure 16.4a, using an argument similar to the energy splitting that occurs for a linear combination of atomic orbitals (LCAOs) [22]. Representative examples include alternating copolymers in which D_0 is based on a dithiophene group, a carbazole group, or a benzodithiophene group. Poly[2,6-(4,4-bis-(2-ethylhexyl)-4*H*-cyclopenta [2,1-*b*;3,4-*b*′] dithiophene)-*alt*-4,7 (2,1,3-benzothiadiazole)] (PCPDTBT), shown in Figure 16.4b, is a low-E_G donor polymer and was used in a historic work on polymeric tandem solar cells [23]. Poly[[4,8-bis[(2-ethylhexyl)oxy] benzo[1,2-*b*:4,5-*b*′]dithiophene-2,6-diyl][3-fluoro-2-[(2-ethylhexyl) carbonyl] thieno[3,4-*b*]thiophenediyl]] (PTB7) is a donor polymer used for OPV cells with efficiency in the range of 7–9%.

Electrodes Electrodes are an essential element in any PV cells, collecting photo-generated carriers. Because of the difficulty in carrier doping of OSCs and of their thin-film geometry, electrodes in OPVs typically cover the whole active region. Materials with high and low work functions are used for anodes and cathodes, respectively, to provide a built-in field with a proper polarity. One of the electrodes should be transparent so that light can enter the photoactive region. A TE commonly adopted in OPV cells is ITO. ITO has a high conductivity (about $10^4\ S\ cm^{-1}$) and optical transmittance and can be easily patterned with a conventional photolithography or laser patterning technique [24]. Furthermore, it has a relatively uniform surface with a low surface roughness unlike fluorine-doped tin oxide (FTO) commonly used in dye-sensitized solar cells [25, 26]. Having a low surface roughness on the order of a few nanometers or less is important because

Table 16.1 Example of donors (D) and acceptors (A) in organic solar cells.

Common name	Full name	Molecular structure	HOMO (eV)	LUMO (eV)
Donor				
Pentacene	Pentacene	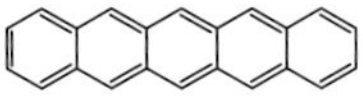	−4.9	−3.0
CuPc	Copper phthalocyanine	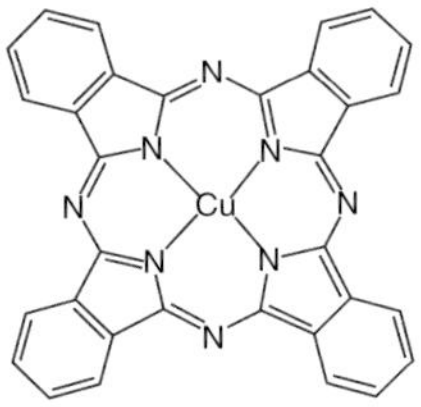	−5.2	−3.5
P3HT	Poly(3-hexylthiophene)	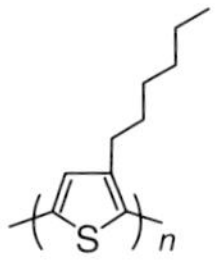	−5.2	−3.3
PCDTBT	Poly[*N*-9″-hepta-decanyl-2,7-carbazole-*alt*-5,5-(4′,7′-di-2-thienyl-2′,1′,3′-benzothiadiazole)]	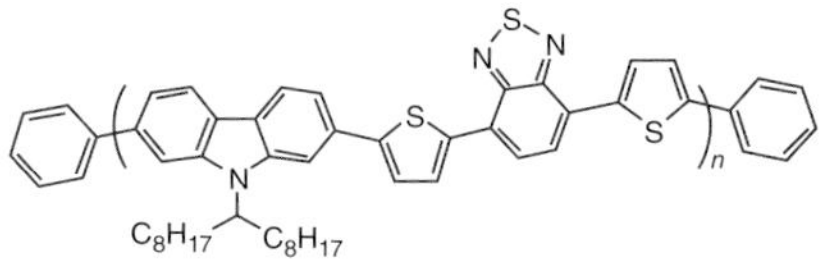	−5.45	−3.6
PCPDTBT	Poly[2,6-(4,4-bis-(2-ethylhexyl)-4*H*-cyclopenta[2,1-*b*;3,4-*b*′]-dithiophene)-*alt*-4,7-(2,1,3-benzothiadiazole)]	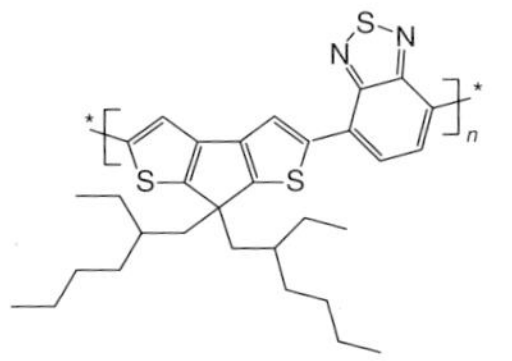	−5.3	−3.55
PTB7	Poly[[4,8-bis[(2-ethylhexyl)oxy] benzo[1,2-*b*:4,5-*b*′] dithiophene-2,6-diyl] [3-fluoro-2-[(2-ethylhexyl)carbonyl] thieno[3,4-*b*]thiophenediyl]]	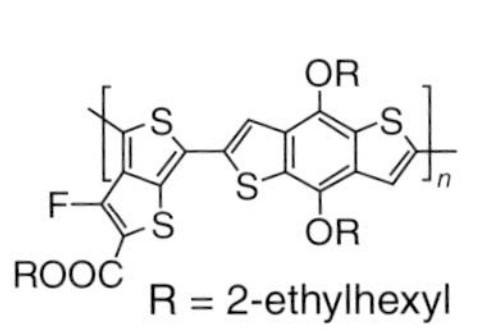	−5.15	−3.31

Table 16.1 (*Continued*)

Common name	Full name	Molecular structure	HOMO (eV)	LUMO (eV)
Acceptor				
PTCBI	3,4,9,10-Perylenetetra-carboxylic bisbenzimidazole	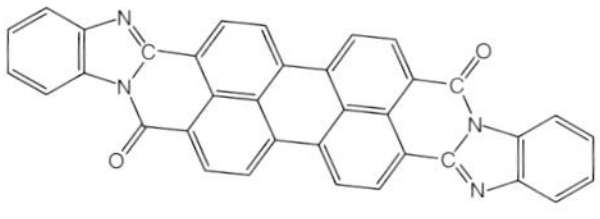	−6.2	−3.6
C_{60}			−6.2	−3.9
C_{70}		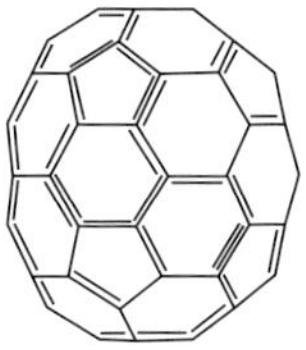	−6.1	−4.2
PCBM60	[6,6]-Phenyl-C_{61}-butyric acid methyl ester	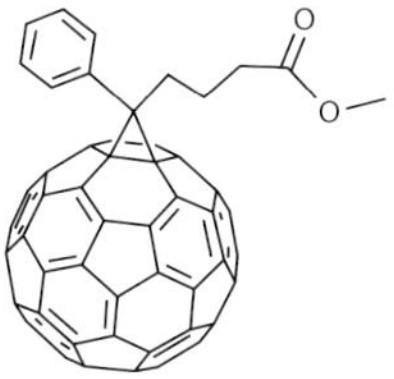	−6.1	−3.7
PCBM70	[6,6]-Phenyl-C_{71}-butyric acid methyl ester	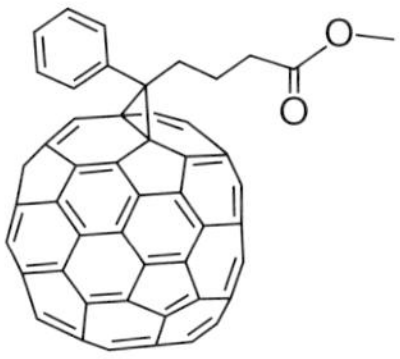	−6.1	−4.3
$IC_{60}BA$	Indene-C_{60} bisadduct	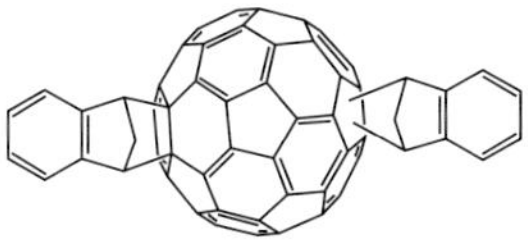	−5.94	−3.74
$IC_{70}BA$	Indene-C_{70} bisadduct		−5.61	−3.72

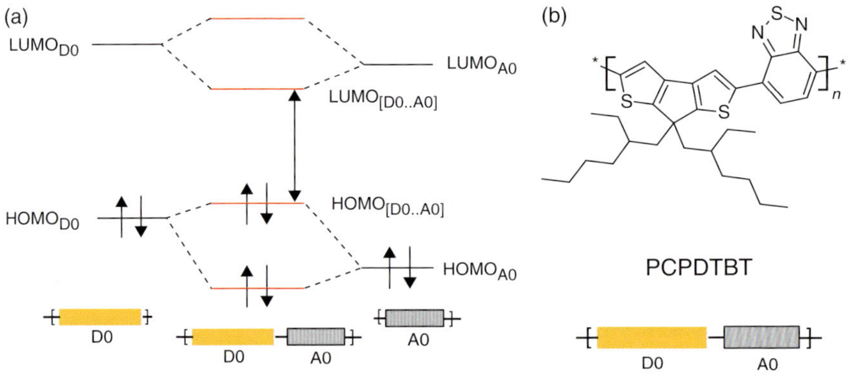

Figure 16.4 (a) Schematic energy diagram for a copolymer based on donor (D_0) and acceptor (A_0), explaining the origin of the lowered energy gap. (Images are redrawn from Ref. [22].) (b) Example of PCPDTBT.

the photoactive layers in OPV cells have a thickness in the range 50–200 nm. Most transparent conducting oxides (TCOs) including ITO are typically formed by means of a vacuum deposition technique other than thermal evaporation. For example, ITO or indium zinc oxide (IZO) layers are typically prepared by a sputtering technique. As organic layers can be easily damaged during the sputtering process, however, it is very difficult to deposit them onto organic layers [27]. Therefore, TCOs precoated on a substrate are typically used as the bottom electrode, and thermally evaporated metal layers are used as the top electrode. Recently, there have been several attempts to find an alternative to ITO because ITO has limited mechanical flexibility and a potential cost issue due to the unstable supply of indium [28–34]. A thin-metal-based semiTE is a good example; in such a case, a TE can be placed as the top electrode [35]. This is useful when OPVs have to be made on an opaque substrate or when they are desired to have a see-through property [35] so that they can be used for energy-generating windows in building-integrated photovoltaic (BIPV) applications.

Buffer Layers Buffer layers can be placed optionally in between photoactive layers and electrodes. They were shown to be very effective in improving the efficiency by playing one or more of the followings roles [15]: (i) an exciton blocking layer (EBL) reducing interfacial exciton quenching that can happen near electrodes by energy transfer to conducting layers; (ii) a layer physically blocking direct diffusion of metal into organic layers; (iii) a layer protecting against thermal or sputter damage to organic layers from a specific deposition process of top electrodes [36]; (iv) a layer controlling the effective work function of an electrode/buffer system so that high or low work function can be realized regardless of the kind of metal [37, 38]; (v) a layer controlling the selective carrier collection [39]; or (vi) an optical spacer layer [39] whose thickness can control the degree of constructive interference between the light incident on the back electrode and the light reflected from it such that photon absorption is maximized within the photoactive layers. Anodic buffer layers that were proven useful include NiO_x [40], MoO_x [41], WO_3 [42], and PEDOT:PSS (polyethylene dioxythiophene:poly(styrenesulfonate)), and cathodic buffer layers include TiO_x [39], ZnO [43], Cs_2CO_3 [44], and polyethylenimine ethoxylated (PEIE) [37]. Wide-gap OSCs such as bathocuproine (BCP) [15] and BPhen [45] were also shown to be useful as EBLs.

16.2 Efforts toward Realization of Flexible OSCs

16.2.1 Overview

The flexibility of organic thin films has been recognized as a feature that can differentiate organic electronic devices from conventional PV technologies. Flexible solar cells can offer the advantages of unbreakability, light weight, expandability (by rolling or folding), and/or easy integration onto other objects. Furthermore, the compatibility with R2R process will be critical in driving down the manufacturing cost.

Flexible OSCs have therefore been explored from the early days of OSC history [46], and promising results have been reported by several groups. This section provides a brief review on such efforts and addresses key issues critical for the development of flexible OSCs, including transparent conductors and the encapsulation structure.

16.2.2 Transparent Electrodes (TEs) for Flexible OSCs

The development of flexible TEs has been a key agenda for the realization of highly flexible OSCs. This is because TCOs – the most common TEs in OSCs – have very limited flexibility because of the brittleness of oxide-based materials [47]. In addition, the properties of TCOs prepared on plastic substrates are often inferior to those on glass substrates because of the high-temperature annealing typically required to obtain high-quality TCO films. In that respect, there have been intensive efforts to find or develop flexible TE alternatives to TCOs. As it is difficult to realize flexible TEs that meet all the requirements of mechanical, electrical, and optical properties, hybrid approaches are often employed which take advantages of two or more technologies. The following are good examples of this: (i) conducting polymers (or very thin TCOs) and metal grids or (ii) oxide–metal–oxide (OMO) or dielectric–metal–dielectric (DMD) type multilayer transparent electrodes (MTEs).

16.2.2.1 Metal Grids Combined with Other Transparent Electrodes

Metal grids or mesh-type electrodes can have a very low sheet resistance by taking advantage of the highly conductive nature of metals such as Ag, Cu, and Al. Considering that the loss of power in large-area cells comes largely from the excessive series resistance of the transparent conductor, metal grid electrodes can be an effective means to maintain the power conversion efficiency in large-area solar cells. Furthermore, the grid electrodes have much better flexibility than TCOs and can be prepared by various cost-effective methods such as rotary screen printing, gravure offset printing, and imprinting.

Once the width of a grid line is fixed at w_g, the transmittance of a metal grid electrode in a square-lattice configuration is determined by the ratio of the area not shadowed by the metal to the total electrode area, which is given for a spatial period of W_p by the following equation:

$$T_{grid} = \frac{(W_p - w_g)^2}{W_p^2} \approx 1 - 2\left(\frac{w_g}{W_p}\right) \quad \text{if } W_p \gg w_g \tag{16.4a}$$

Therefore, W_p should be roughly 20 (10) times w_g to obtain T_{grid} of 90% (80%). Note that each printing technique has a limited resolution, and thus the minimum feature size achievable in a given printing process is often limited. For example, the minimum feature size of gravure printing is typically about 50 μm and that of screen printing or offset printing is about 10–20 μm at best. With 20 μm (50 μm) taken as w_g, W_p leading to T_{grid} of 90% should be as large as 400 μm (1 mm). Such a length scale could be acceptable for most inorganic crystalline solar cells, but it could be too large a distance for carriers to travel in OSCs that have a limited transport property. In that respect, metal grids in OSCs are typically used together with an auxiliary transparent electrode covering the whole active area.

To this end, Galagan *et al.* demonstrated a hybrid TE consisting of Ag grids in combination with conducting polymers based on a complex of poly(3,4-ethylenedioxythiophene) (PEDOT) and polystyrene sulfonate (PSS) , commonly referred to as *PEDOT:PSS*. Conducting polymers are conjugated polymeric materials along which charge carriers can be conducted efficiently [48, 49]. Doping, which creates quasi-particles such as charged solitons and bipolarons, is regarded as an essential process that enables metallic conduction in conjugated organic compounds [48–50]. Among the various conducting polymers developed to date, PEDOT:PSS is the most popular because it can be easily processed from aqueous solutions and possesses relatively good environmental stability and optical transparency in the visible spectral range [51]. PEDOT:PSS formulations are commercially available in a wide range of conductivities so that they can be used optionally as a buffer layer or as a TE. Its standard grade has a nominal conductivity of $1\,\mathrm{S\,cm^{-1}}$, which is several orders of magnitude smaller than that of typical TCOs ($\sim 10^4\,\mathrm{S\,cm^{-1}}$). Therefore, there have been several methods proposed to enhance the conductivity of PEDOT:PSS to a significant degree. For example, a small addition of a solvent such as dimethyl sulfoxide (DMSO) or ethylene glycol (EG) [51], the addition of a fluoro-surfactant called *Zonyl* [52] or *hexafluoroacetone* (*HFA*), and the immersion of solidified PEDOT:PSS films in methanol [33] were shown to be very effective. With such methods, conductivity >1000 S cm^{-1} can now be achieved. Nevertheless, the conductivity is still 10 times smaller than that of commercially available ITO or FTO films; thus, it is challenging to use PEDOT:PSS as a stand-alone TE. Demonstration of decent performance can be made even with a TE having a relatively high sheet resistance in lab-scale small-area cells, but its negative effect soon appears as the size of the PV cells is scaled up. In that respect, PEDOT:PSS and metal grids may be regarded as an ideal match for each other. In particular, the planarizing capability of PEDOT:PSS can be useful because metal grids tend to provide organic active layers with local bumps that can induce unwanted leakage paths. In some cases, metal grids are embedded within substrates or other functional layers (e.g., encapsulating layers) to further ensure the smoothness of the hybrid TEs [53]. The high work function of PEDOT:PSS (~5 eV) is also useful in defining the hybrid electrode as an anode regardless of the kind of metal used for grids.

Recently, Kang and his coworkers [54] demonstrated hybrid electrodes composed of (i) Ag grids embedded in a polyimide (PI) layer and (ii) a very thin layer of amorphous ITO as a highly flexible TE (Figure 16.5). Ag grids (w_g = 39 μm; 1.3 μm thick) were initially printed on a glass substrate pretreated with a self-assembled monolayer (SAM), making its surface hydrophobic. A thick PI layer (~400 μm thick before curing) was then coated by doctor-blading over the Ag grid structure. After the curing process, the PI layer with an after-cure thickness of ~40 μm was detached from the glass, which served as a highly flexible substrate with the embedded Ag grids. A 20-nm-thick amorphous ITO layer was then sputtered onto the side on which the Ag grids were exposed. The rms surface roughness of the resultant electrode was as small as 0.79 nm. The transmittance and sheet resistance of the proposed electrode were approximately 93% and 13 Ω/sq., respectively. Large-area OSCs fabricated on the PI substrate with the proposed TE exhibited power conversion efficiency comparable to that of control cells built on glass with a TE based on crystalline indium tin oxide (c-ITO). The efficiency did not degrade even after 1000 bending cycles at a bending radius as small as 1 mm.

16.2.2.2 Other Flexible Transparent Electrodes

The high degree of flexibility of metal-grid-based TEs stems from the ductile nature of metal layers. Hence, a thin metal film also has the potential to serve as a flexible TE provided that there is an effective method to enhance its transmittance, which is typically low because of significant reflection and finite absorption. TEs based on a DMD or an OMO structure try to obtain high transparency by suppressing reflectance from their metal layer by means of destructive interference [29]. Essentially the same structure has been utilized for a transparent heat rejection film which transmits the visible part and reflects the infrared part of the sunlight. Many variants were proposed, including anodes of ZnS/Ag/WO_3 (see Figure 16.6a,b) [29], WO_3/Ag/WO_3 [55], MoO_x/Ag/MoO_x [30], ITO/Ag/ITO [56], and TeO_2/Ag/PEDOT:PSS [57], and a cathode of ZnS/Ag/TiO_x [58]. While Ag is the most popular material because of its low absorption in the visible

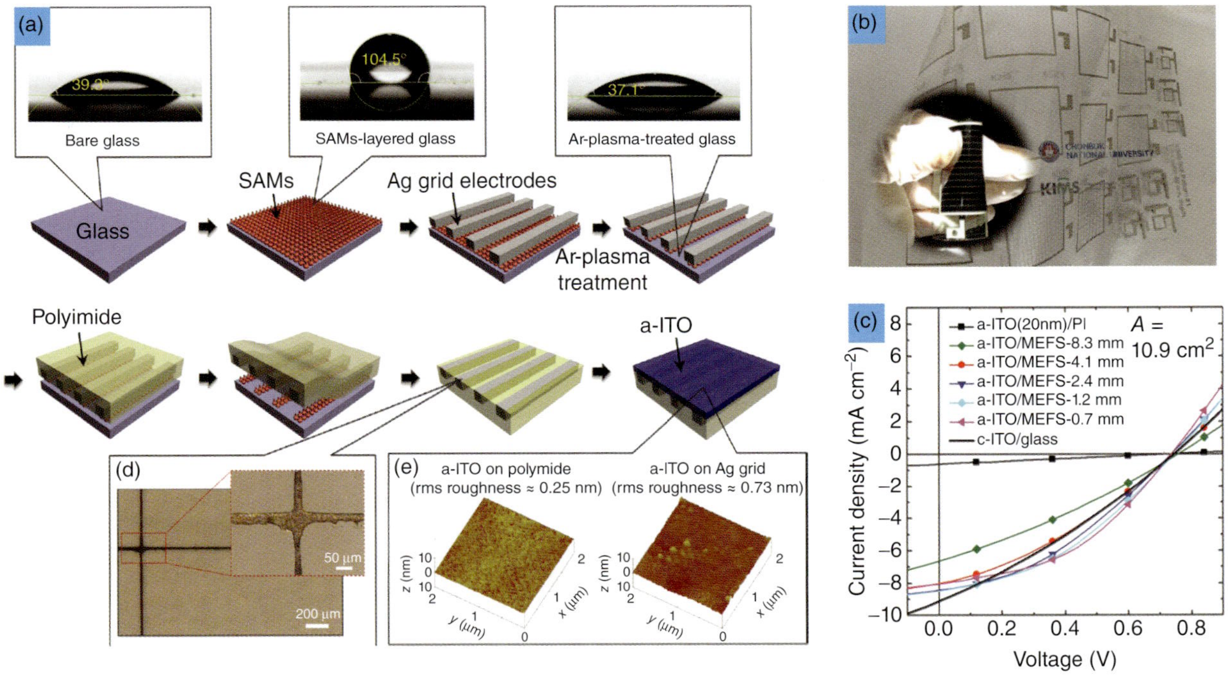

Figure 16.5 (a) Process flow for formation of hybrid electrodes based on embedded metal mesh (grid) electrodes and a-ITO. (b) Test pattern printed on plastic substrates. (c) *J–V* characteristics of large-area solar cells ($A = 10.9\ cm^2$). The inset in (b) shows the large-area cell. (d) The microscope image of the embedded Ag grids. (e) AFM images showing the surface morphologies of a-ITO grown on bare polyimide and on Ag grids embedded in polyimide. (Reprinted with permission from Ref. [54]. Copyright © 2014, John Wiley and Sons.)

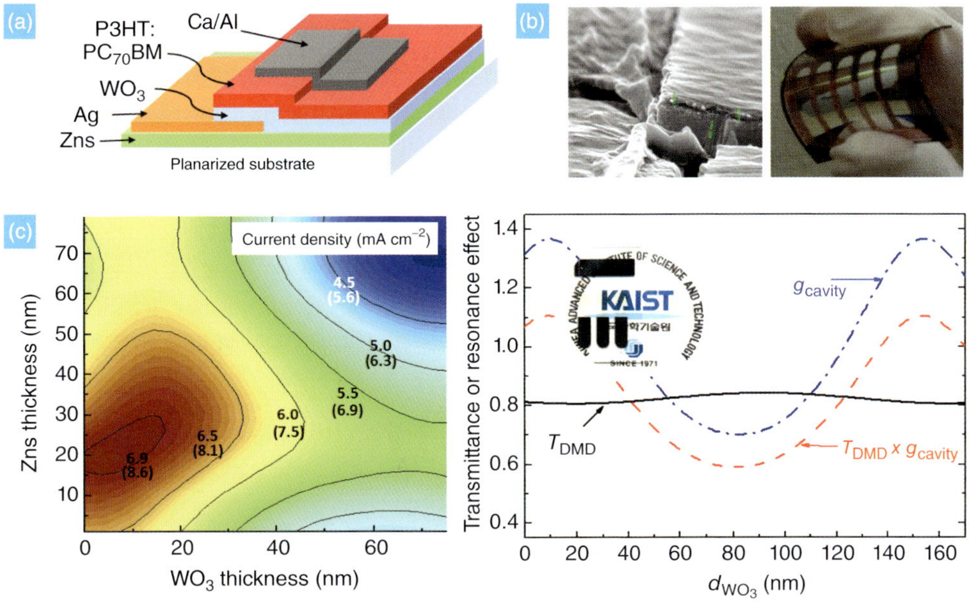

Figure 16.6 (a) Device structure of OSCs with ZnS-Ag-WO_3 (ZAW) electrodes. (b) The SEM image (left) showing the deep cracks formed by ITO after repeated bending. Photograph of flexible OSCs with ZAW electrodes. (c) (left) Optimization of OSC performance and (right) importance of the cavity resonance effect in ZAW-based OSCs. (Image in (c) is redrawn from Ref. [29].) The inset shows the photograph of a transparent organic device made by the dielectric–metal–dielectric electrodes.

spectrum, Cu [59] and Au [60] have also been tried. In particular, Cu-based DMD TEs may be regarded as a low-cost and flexible alternative to ITO electrodes. Performance comparable to that of ITO-based devices has been demonstrated in all the cases mentioned above. Considering the presence of a finite reflectance from DMD or OMO electrodes, Han *et al.* [29] and Sergeant *et al.* [30] showed that the optimization of OSCs containing those electrodes should be done by carefully monitoring the competition between electrode transmittance and the cavity resonance effect, as can be seen in Figure 16.6c [29, 30]. As the oxide or dielectric layers used in these DMD or OMO type electrodes are typically thinner than 50–60 nm, the overall flexibility of OSCs produced thereof has been demonstrated to be superior to that of ITO-based control cells, which tend to form ITO-induced cracks after repeated bending (Figure 16.6b). Upon finding the optimal deposition condition for metal films with thickness in the range of 10–20 nm, sheet resistance on the order of ~5 Ω/sq. is readily achievable in these electrodes [61], making them suitable for large-area OSCs as well. However, the reliance on vacuum-based deposition techniques may be regarded disadvantageous from a cost standpoint. Nevertheless, it may still be a viable option for vacuum-based small-molecular OSCs provided that vacuum R2R processes can be utilized [62].

Other promising TEs for flexible OSCs include those based on Ag nanowires (Ag-NWs) and carbon nanomaterials such as carbon nanotubes (CNTs) and graphene. Bae, Lee, and their coworkers [63] have recently demonstrated flexible OSCs based on Ag-NWs embedded into a transparent hybrimer reinforced with glass fibers. The rms roughness being most problematic for Ag-NW-based TEs, an embedding scheme similar to that described in the previous section was utilized to minimize the occurrence of unwanted electrical shorts, illustrating an effective way to fully utilize the benefits that Ag-NW technology can offer, namely, high transmittance, low sheet resistance, and excellent flexibility [63]. The process for CNT-based TEs has matured during the last decade, but its application appears to be still limited to areas that do not require a very low sheet resistance (e.g., touch sensor applications). Compared to CNTs, graphene-based TEs are still in their infancy. A T value of 90% and R_{sheet} of 30 Ω/sq. have recently been demonstrated for doped graphene [64]. At this moment, it is likely that the use of graphene-based TEs will also be limited to low-end applications, similar to the case of CNTs, or to applications requiring more extreme deformation, like stretching.

16.2.3
Encapsulation Issues

Flexible OSCs mostly rely on plastic substrates which are notorious for their high water vapor transmittance rate (WVTR) and oxygen transmittance rate (OTR). Because organic materials, buffer layers, and electrode materials degrade upon exposure to H_2O and O_2, it is very important to keep OSCs away from ambient air by providing a tight encapsulation. The presence of light can complicate the situation by enabling photo-induced degradation. While it could be acceptable in the high-end display market to have an encapsulation structure based on a sophisticated multilayer barrier [65, 66] with multiple dyads of alternating polymeric layers and inorganic oxide or nitride layers, the use of such expensive processes should be avoided or minimized, if possible, in OSCs targeting ultralow-cost applications. A study indicated that a thin Al_2O_3 layer conformally coated over OSCs by atomic layer deposition (ALD) can be highly effective in protecting OSCs from ambient air [67]. With a single ALD-based Al_2O_3 layer, pentacene/C_{60}-based OSCs were shown to maintain their initial performance even after being stored for 8000 h in ambient air [66]. Nevertheless, a single oxide layer is almost always subject to defect-driven propagation of O_2 and H_2O. Therefore, a multilayer barrier approach may not be completely avoidable in flexible OSCs. Otherwise, it may be necessary to find an internal structure that can significantly reduce the encapsulation requirements so that one or two dyads of barrier layers can be sufficient. An inverted OSC structure is a good example because cathodes, generally known to be more reactive, are buried underneath the other layers [68]. It will also be important to develop a cost-effective process for the formation of multilayer barrier structures.

16.3 Flexible OSCs for High-Throughput Production: A Printing-Based Approach to Low-Cost Solar Energy Conversion

16.3.1 Printing Technology Overview

High-throughput production such as R2R fabrication is essential in significantly reducing the cost of solar cells [69]. Furthermore, it will be highly cost effective if "monolithically" integrated PV modules (Figure 16.7a) can be produced by such R2R methods, particularly in an additive (mask-free) way. Organic solar cells are highly suitable in these respects because their solution-processability and flexibility make them compatible with R2R and/or printing-based technology [71–73].

Nevertheless, applying R2R printing technologies for the fabrication of OSC modules is a nontrivial task and requires careful engineering in many aspects. For example, a selected printing method must be compatible with the ink formulation of a given OSC material, which may have a wide range of viscosity (e.g., high viscosity for metal ink formulation and very low viscosity for ink formulation of OSCs [74].) Forming a very thin layer (typically ~100 nm thick) uniformly over a large area is also a challenging task. Maintaining high resolution and precise relative alignment among participating layers is another key issue in extracting the full potential a given PV technology can offer, because the finite inactive area ("$W_{inact} \times L$" in Figure 16.7a) can severely limit the area utilization ratio. As can be seen in Figure 16.7b, the problem may get lessened if the width of each strip ($= W_{unit} = W_{inact} + W_{act}$) increases [74]; however, W_{unit} cannot be made arbitrarily large because it will increase the series resistance coming from a transparent conducting layer. In this section, printing technologies for facile fabrication of OSC modules are reviewed, with a discussion on the issues that can be raised in each technology. As each material has a different ink formulation with a wide

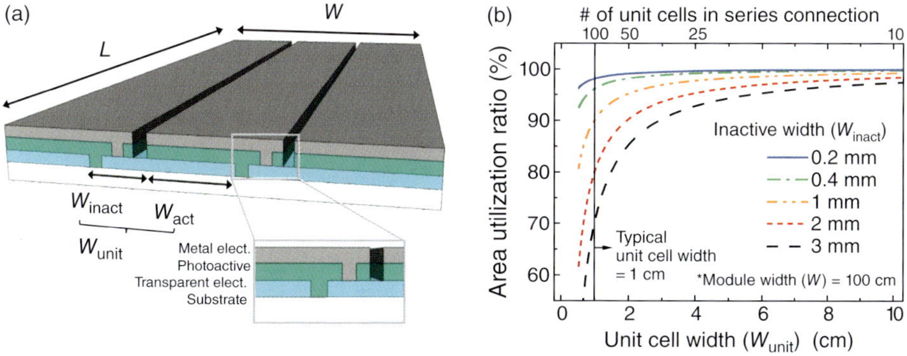

Figure 16.7 (a) Device structure of a monolithically integrated solar module. (b) Area utilization ratio of a monolithic module versus unit cell width for various inactive widths. (Image (b) is reprinted with permission from Ref. [74]. Copyright © 2011. Elsevier.)

range of viscosity, technologies will also be described in terms of suitability for a certain type of material.

16.3.2 Review of Printing Technologies Used for OSCs

Various printing methods have been used to fabricate OSCs. Most of the works are focused on the printing of active layers, while some of the works also try to the metal mesh grid indispensable for large-area thin film devices. Among the various printing technologies, we focus on those that are particularly useful for OSC fabrication.

16.3.2.1 Screen Printing

Screen printing is a widely used industrial technique in large-area printing such as metal mesh printing in silicon solar cells [70] and dye printing for T-shirts. High-viscosity inks are squeezed through the hole of a prepatterned mesh with a thickness a few tens of micrometers. This method was used in the early stages of printed OSC history because PPV-based polymers – the first-generation polymers in OSCs – had relatively high viscosities [75]. However, because the recent high-efficiency materials, including the widely studied polythiophene, have low viscosity in solution, they naturally flow through the mask; thus, this method is now considered mainly as a way to implement a metal mesh grid as used in silicon solar cell technology. Typically, screen printing is a batch process, but a rotary version is also available, as shown in Figure 16.8 [77]. Nevertheless, screen printing has a relatively low material utilization ratio and it is cumbersome to maintain its mask clean. Hence, alternative high-speed R2R printing methods based on gravure or flexographic printing are often preferred [76].

16.3.2.2 Droplet Coating and Printing

Droplet printing covers a wide area including various spray methods and on-demand-type inkjet printing (Figure 16.9a). As no spray method can form a pattern by itself, additional masks should be used to shape a pattern; on the other hand, the inkjet is made to print a pattern, as used in desktop printers. The spray method uses fine droplets of inks which are atomized in various ways such as pneumatic gas flow of air or nitrogen [79], sonication of inks [80], and electrostatic breakdown of Taylor cones of inks [81]. Because droplet printing relies on a continuous supply of tiny droplets and the OSCs use volatile organic solvents, pileups of individual droplet results in islands of coffee stains [78] and the merger of each droplet is often hard to obtain during the deposition process. There are a variety of methods to flatten these island coffee stains: solvent vapor annealing (Figure 16.9b) [82], additional spraying of solvents [78], and addition of high-boiling-point additives into ink formulation [78]. By these methods, photoactive layers with flat, uniform morphologies can be obtained, and the power conversion efficiency (PCE) of the OSCs made thereof are almost the same as those of the OSCs with active layers formed by the traditional spin-coating method [83].

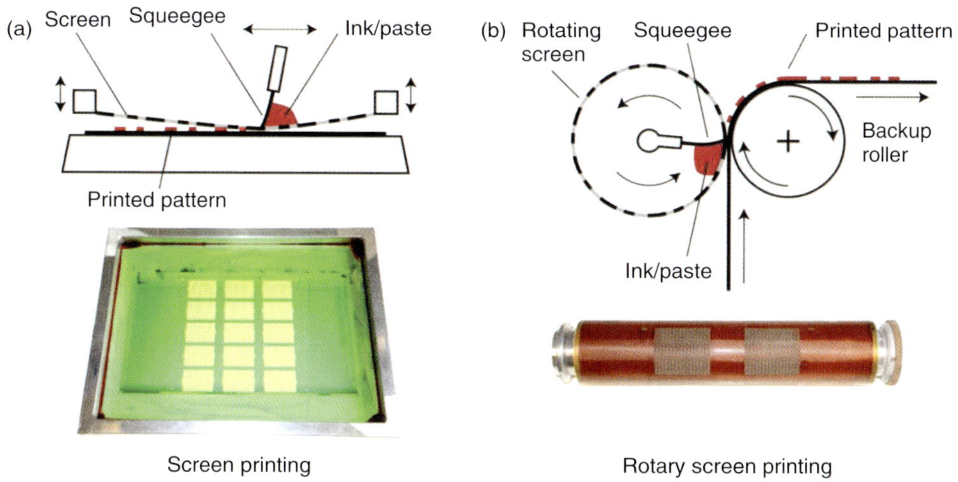

Figure 16.8 Diagrams for (a) screen printing (a) and (b) R2R-compatible rotary screen printing. Photographs of the screen in each technology are also shown. (Reprinted with permission from Ref. [76]. Copyright © 2012, Wiley Periodicals, Inc., A Wiley Company.)

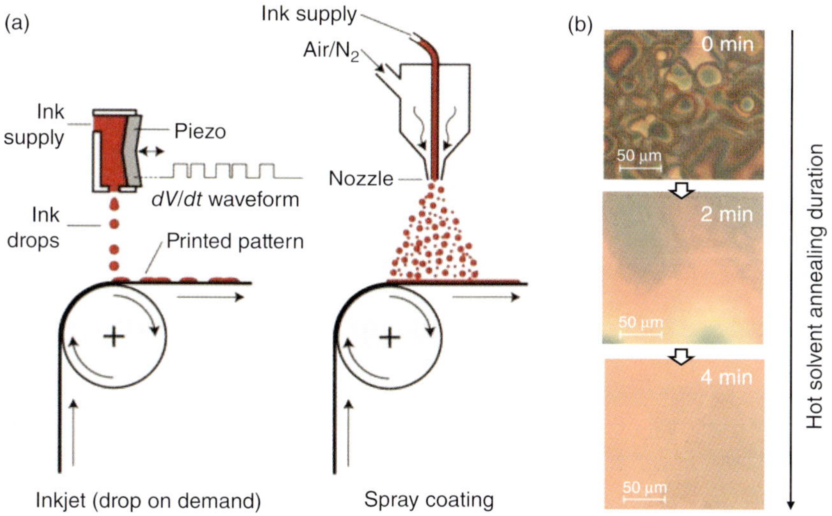

Figure 16.9 (a) Schematics of inkjet printing and spray coating. Spray header can be replaced by other types such as ultrasonic and electrostatic units. (Images are reprinted from Ref. [77] with permission from Journal of polymer science part B: polymer physics, Wiley Periodicals, Inc., A Wiley Company.) (b) Optical microscope images of spray-deposited films: as-deposited, 2 and 4 min after hot-solvent vapor annealing at 45 °C, respectively. (Reprinted with permission from Ref. [82]. Copyright © 2011, Elsevier.)

Although spray methods and inkjet printing use the same principle of droplet deposition, and both of them can be applied to R2R production, they have different processing features or target applications. Spraying methods are simple to implement and can easily deposit materials over a large area. Often, traditional hand-held spray and compressed air (or, better, nitrogen) are adequate to make functional films. Even the most sophisticated system involving an ultrasonic spray does not require a complex system except for a sonication header. Because spraying methods cannot define a pattern without a mask, OSC modules may be formed by a combinatorial method that involves (i) zero-dimensional coating by spraying and (ii) post-patterning such as laser ablation.

Inkjet printing is the most versatile printing method to change the pattern, if necessary, because it uses software-coded digital patterns rather than "hard-coded" physical patterns adopted in screen, slot-die, and gravure printings. Active area patterns in OSC modules can thus be defined easily not only into stripes but also into any arbitrary shape, if desired. Nevertheless, it is difficult to form thick layers with inkjet printing, which may be disadvantageous for metal printing, which generally prefers to have a thick layer for low sheet resistance.

16.3.2.3 Blade/Knife Edge Coating and Slot-Die Printing

The blade/knife edge (hereafter blade) method is a widely used industrial coating method that can make a thick polymer film through a gap between the knife edge and the substrate controlling the base material supplied in front of the edge [84]. The wet thickness is determined by the gap (d_{gap}) between the blade and the substrate and is typically $0.5-1.0d_{gap}$ [85]. Thus, this method is highly sensitive to the clearance of the knife and flatness of the substrate but not to the coating speed [71]. This insensitivity to coating speed makes it useful in regulating the amount of ink supply to the first roller in other R2R printers. However, it is quite challenging to attain large-area regulation of a submicrometer-scale gap distance which is critical for R2R production of OSCs. Although it cannot be used for a direct patterning process, being a zero-dimensional coating, it is adequate for a printability test onto a substrate for specific ink formulation. Early stages of OSC printing used this method for small-scale laboratory cells before implementing in large-area R2R production. Recent application of this method using a novel polymer of poly{(benzo-2,1,3-thiadiazol-4,7-diyl)-alt-(3′,4″di (2-octyldodecyl)-2,2′;5′,2″;5″,2‴-quaterthiophen-5,5‴-diyl)} (POD2T-DTBT) with $PC_{71}BM$ rendered the highest PCE of 6.74% without any post processing or additives [85]. Another issue with this method in its application to large-area R2R production of OSCs is that inks are fed in front of the knife edge and it is exposed to the open space. Because organic solvents of OSC active materials are highly volatile, concentrations of the inks can continuously vary in this case. To solve this problem and realize a certain level of printability, slot-die printing is actively used in many research groups.

Slot-die coating/printing is an extension of blade/knife edge coating and has been demonstrated in R2R configurations for both active and buffer layers. The ink feeding mechanism is similar to that in blade/knife edge coating, as it is supplied in front of the printing blade. In slot-die printing, however, another blade is placed in front of the regulating blade. Thus, the inks are fed through a small gap between these two blades (Figure 16.10a). This feature provides a significant advantage in OSC fabrication. Because inks are fed through the gap between two blades, volatile solvents cannot evaporate efficiently during the printing, and ink feeding can be easily regulated by an external pumping through this gap [85]. Although this method uses face-sandwiched knife edge blocks, the mechanism that determines the thickness is different from that of blade/knife edge coating [71]. Because the supplied amount of ink is fixed by an external pumping system, the rheological thickness determined by the coating head and substrate configuration must be matched [87]. If this condition is not met, defects and instabilities can occur. Another feature of this method is that it can handle a wide range of viscosities with a single apparatus from a few millipascal seconds (mP s) to several thousands of pascal seconds [85]. When it is used in printing high-viscosity inks, it operates in the "extrusion" coating mode (Figure 16.10b).

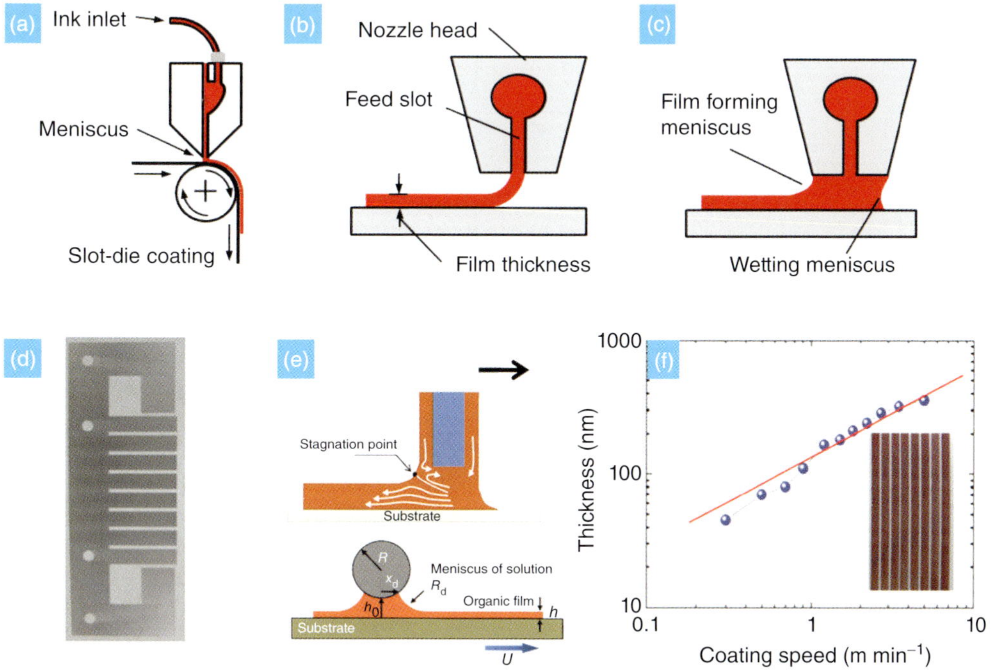

Figure 16.10 (a) Schematic of slot-die coating/printing. (b,c) Two different modes of operation which are determined by the viscosity of the inks. (Images in (a–c) are redrawn from Refs [81, 84].) (b) Extrusion coating mode and (c) slot coating mode. (d) Slot-die mask with protruded shims to pattern 1D stripe patterns. (e) Meniscus formed between the protruded mask shim and the substrate. (f) Thickness versus coating speed for the slot-coating mode. Inset: the photograph of an OSC module fabricated by a slot-die printing. (Images in (d–f) are reprinted with permission from Ref. [89]. Copyright © 2009, Elsevier.)

But when the viscosity is low enough for the viscous and capillary forces to be dominant, it operates in the "slot" coating mode (Figure 16.10c), which is the case for OSC material inks [88].

Another advantage of the slot-die method is that one can easily control the width of 1D stripe patterns by opening or closing the gap between two head blocks. In many OSC studies based on the slot-die method, mask shims (Figure 16.10d) protrude slightly from the bottom of the slot-die head. These protruded shims make another mode of printing in slot-die operation. Traditional slot coating uses the meniscus between both lips of its head block and a substrate, and the stagnation streamline of the ink is confined to this meniscus [85]. However, when one uses protruded mask shims, the fed inks are moved from the gap to the protruded mask shims by "adhesive" surface tension; thus, printed stripes follow the shape of protruded mask shims and not that of open slits (Figure 16.10e and the inset of Figure 16.10f) [89]. In this case, the traditional meniscus explanation between the lips of head block and a substrate cannot hold; instead, a new explanation of the meniscus between the mask shims and the substrate is required. The thickness dependence in this mode becomes essentially same as that of a horizontal dip-coating method, as shown in Figure 16.10e (bottom) [86]. In this case, the higher the web speed, the thicker the coated layer (Figure 16.10f).

16.3.2.4 **Gravure Printing**

Gravure printing and its variants are the most important high-throughput printing methods in the industrial world. It uses printing cylinders that have engraved cell patterns on the surface. Inks are fed into these engraved cells with the help of a doctor blade and then they are transferred to the substrate assisted by a soft impression cylinder [90] (Figure 16.11). Engraving parameters such as cell density, depth, width, and angles of the screen and stylus must be adjusted depending on ink properties. As gravure printing has many processing parameters, it is hard to optimize, but the throughput gain can be substantial once it is achieved. Kopola *et al.* [91] demonstrated the feasibility of adopting gravure printing as a layer deposition method for hole-transporting PEDOT:PSS and P3HT:PCBM photoactive materials. In their work, they could achieve a flat surface with a concentration as high as 130 mg ml^{-1} of polymer ink (the normally used concentration in spin coating is about 1 mg ml^{-1}) and a printing speed of 17 m min^{-1}. Recently, Yang *et al.* [90] tried to increase the coating speed up to 35 m min^{-1} with half the concentration used in Kopola's work. But at this time, gravure printing research is still in the early stages, and there is room for further optimization.

16.3.2.5 **Other Coating/Printing Methods**

Many other coating or printing methods have also been demonstrated for OSC fabrication. Some of them are traditional methods such as dip coating [92], bar coating, and brush coating [93], but their results were presented in small-sized laboratory cells. They need to be demonstrated, at least in large-area devices, to show their feasibility in large-area low-cost production. On the contrary, Guo *et al.* [94] demonstrated an unconventional coating method in the name of "ESSENCIAL" in

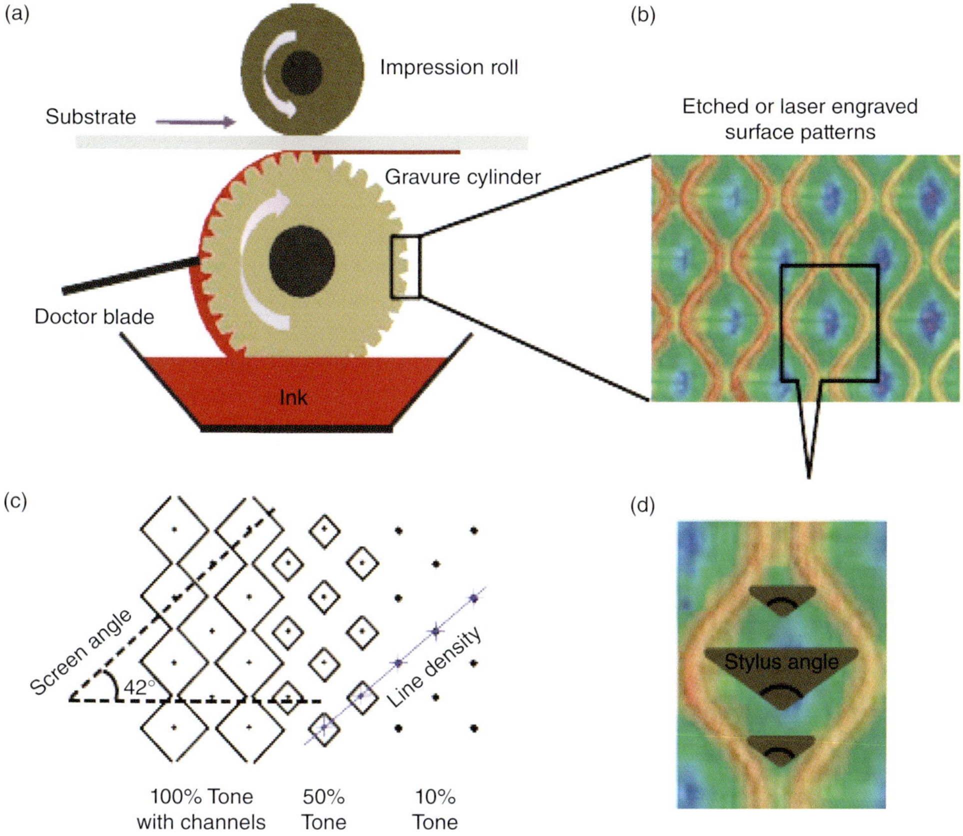

Figure 16.11 (a) Schematics of gravure printing. (b) Representative engraved surface patterns. (c,d) Pattern parameters tone, screen angle, line density, and stylus angle. (Reprinted with permission from Ref. [90]. Copyright © 2012, Elsevier.)

an R2R system which enhanced the polymer crystalline ordering by introducing a solvent annealing step during the deposition. In this method, another encapsulation web confines the deposited inks from the atmospheric environment; thus, the confined ink does not feel the surface energy difference during the solvent evaporation through the additional web.

While the present chapter mainly deals with solution processing, vacuum-based large-area coating and jet-printing method are also available: organic vapor deposition technique [95] is an example of a vacuum-based areal deposition method, and organic vapor jet printing [96, 97] may be regarded as a "vapor" version of inkjet printing. These methods are advantageous in that they do not require any solvent, which can be harmful for environment and health, but the use of heat for evaporation may tend to increase the overall amount of energy for fabrication.

16.3.3 Issues in Module Fabrication

By various large-area coating/printing methods, printing of large-area OSCs has been successfully demonstrated. However, several issues still need to be solved, and thus there is much room for further improvement. For example, the current slot-die coating used in OSC production adopts a meniscus coating principle [86, 90], where the thickness (d) increases with the coating speed (v) according to $d = v^{2/3}$. Most organic-based inks used for high-performance OSCs are required to be ~100 nm thick, whereas this value is hard to achieve in a high-speed slot-die coating except for solutions with very a low solid content. These highly diluted solutions, however, are prone to the coffee stain effect during the printing and drying processes [81], thus deteriorating the overall performance of the OSC modules through insecure series connection [78].

Not only a module fabrication process but also a well-engineered module architecture can play an important role. With the challenges specific to OSCs such as a very small film thickness (<100 nm), fluidic motion control during thin-film deposition, and multiple layer stacking [98], new large-area module architectures having high-fidelity series connection and robustness against variations in the process conditions are essential in enhancing the commercial viability of printed OSCs. Recently, such a process-friendly architecture called *SeleCT* (*seamless continuous terminal architecture*) has been demonstrated by arranging multiple cells with alternating polarities [99]. Although SeleCT was not fully solution-processed, it suggested a possibility of a unique architecture in printed large-area OSCs with high efficiency and relatively good processability (Figure 16.12).

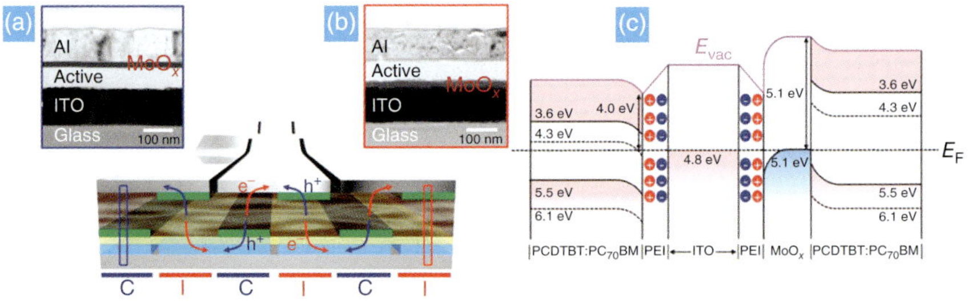

Figure 16.12 Schematics of seamless, continuous terminal architecture (SeleCT) module. (a,b) Corresponding electrode configuration utilizing buried tunneling junction. (c) Energy configuration in SeleCT modules. (Reproduced with permission from Ref. [99]. Copyright ©, The Royal Society of Chemistry.)

16.4 Summary and Outlook

Solar cells made of organic materials were reviewed as a next-generation PV device. Its hallmark is its low-temperature process which can lead to highly affordable solar cells with the additional benefit of flexibility. Brief descriptions were provided on the fundamental aspects of organic solar cells including their motivation, major components, typical structure, and operation mechanism. Then, issues in making highly flexible OSCs were introduced. In particular, TEs and encapsulation were noted as critical elements key to the successful realization of flexible OSCs. Hybrid TEs containing metal grids or thin metal films were introduced as a potential solution that can overcome the challenges posed by the mechanically brittle nature of ITO, which are the common TEs used in OSCs. Other TEs, such as conducting polymers, carbon nanomaterials, and Ag-NWs, were also introduced, and their pending issues were briefly summarized. The necessity of an affordable multilayer barrier structure for the reliable yet cost-effective encapsulation of flexible OSCs was also noted. As an example to utilize the flexibility of organic compounds and substrates for driving down the cost of OSCs, various printing technologies were introduced and discussed from the perspectives of high-throughput fabrication. The advantages and deficiencies of each technology were described, and major issues were discussed in terms of R2R OSC fabrication.

As existing Si PV technologies continue to reduce their production cost, and other thin-film PV technologies keep improving their performance, OSCs will need to clearly define their strategic position and continue to develop technologies for features that differentiate them from other competitors. To this end, attention must be paid not only to the efficiency but also to the practical viability of OSCs. The development of a truly low-cost process and device structure that can be easily scaled up and yet compatible with flexible OSCs should be achieved.

References

1. Sze, S.M. (1981) *Physics of Semiconductor Devices*, 2nd edn, John Wiley & Sons, Inc., New York.
2. Paish, O., MacNellis, B., and Derrick, A. (1994) in *Solar Electricity* (ed. T. Markvart), John Wiley and Sons, Ltd, Chichester.
3. Notton, G., Muselli, M., and Poggi, P. (1998) Costing of a stand-alone photovoltaic system. *Energy*, **23**, 289.
4. Sioshansi, P. (2013) Deutsche Bank Says Solar is Approaching Grid Parity, Breaking Energy (Sept. 2013), *http://breakingenergy.com/2013/09/05/* (accessed 21 July 2014).
5. Conibeer, G. (2007) Third-generation photovoltaics. *Mater. Today*, **10**, 42.
6. Morel, D.L., Ghoshj, A.K., Feng, T., Stogryn, E.L., Purwin, P.E., Shaw, R.F., and Fishman, C. (1978) High-efficiency organic solar cells. *Appl. Phys. Lett.*, **32**, 495.
7. Espinosa, N., Hösel, M., Angmo, D., and Krebs, F.C. (2012) Solar cells with

one-day energy payback for the factories of the future. *Energy Environ. Sci.*, **5**, 5117–5132.

8. Yoo, S., Domercq, B., and Kippelen, B. (2005) Intensity-dependent equivalent circuit parameters of organic solar cells based on pentacene and C60. *J. Appl. Phys.*, **97**, 103706.
9. Nelson, J., Kirkpatrick, J., and Ravirajan, P. (2004) Factors limiting the efficiency of molecular photovoltaic devices. *Phys. Rev. B*, **69**, 035337.
10. Kumar, A., Sista, S., and Yang, Y. (2009) Dipole induced anomalous S-shape I-V curves in polymer solar cells. *J. Appl. Phys.*, **105**, 094512.
11. Blakesley, J.C. and Neher, D. (2011) Relation between energetic disorder and open-circuit voltage in bulk heterojunction organic solar cells. *Phys. Rev. B.*, **84**, 075210.
12. Riedel, I., Parisi, J., Dyakonov, V., Lutsen, L., Vanderzande, D., and Hummelen, J.C. (2004) Effect of temperature and illumination on the electrical characteristics of polymer-fullerene bulk-heterojunction solar cells. *Adv. Funct. Mater.*, **14**, 38–44.
13. Kippelen, B. and Brédas, J.-L. (2009) Organic photovoltaics. *Energy Environ. Sci.*, **2**, 251.
14. Heremans, P., Cheyns, D., and Rand, B.P. (2009) Strategies for increasing the efficiency of heterojunction organic solar cells: material selection and device architecture. *Acc. Chem. Res.*, **42**, 1740–1747.
15. Peumans, P., Yakimov, A., and Forrest, S.R. (2003) Small molecular weight organic thin-film photodetectors and solar cells. *J. Appl. Phys.*, **93**, 3693.
16. Rand, B.P. and Burk, D.P. (2007) Offset energies at organic semiconductor heterojunctions and their influence on the open-circuit voltage of thin-film solar cells. *Phys. Rev. B*, **75**, 115327.
17. Potscavage, W.J., Yoo, S., and Kippelen, B. (2008) Origin of the open-circuit voltage in multilayer heterojunction organic solar cells. *Appl. Phys. Lett.*, **93**, 193308.
18. Sun, S.-S. (2008) Basic electronic structures and charge generation in organic optoelectronic materials, in *Introduction to Organic Electronic and Optoelectronic Materials and Devices* (eds S.-S. Sun and L. Dalton), CRC Press, Boca Raton, FL.
19. Yu, G., Gao, J., Hummelen, J.C., Wudl, F., and Heeger, A.J. (1995) Polymer photovoltaic cells: enhanced efficiencies via a network of internal donor-acceptor heterojunctions. *Science*, **270**, 1789.
20. Yoo, S., Potscavage, W.J., Domercq, B., Kim, J., Holt, J., and Kippelen, B. (2006) Integrated organic photovoltaic modules with a scalable voltage output. *Appl. Phys. Lett.*, **89**, 233516.
21. Zhao, G., He, Y., and Li, Y. (2010) 6.5% efficiency of polymer solar cells based on poly(3-hexylthiophene) and indene-C60 bisadduct by device optimization. *Adv. Mater.*, **22**, 4355.
22. Brabec, C.J., Heeney, M., McCulloch, I., and Nelson, J. (2011) Influence of blend microstructure on bulk heterojunction organic photovoltaic performance. *Chem. Soc. Rev.*, **40**, 1185.
23. Kim, J.Y., Lee, K., Coates, N.E., Moses, D., Nguyen, T.-Q., Dante, M., and Heeger, A.J. (2007) Efficient tandem polymer solar cells fabricated by all-solution processing. *Science*, **317**, 222.
24. Gordon, R.G. (2000) Criteria for choosing transparent conductors. *MRS Bull.*, **25**, 52–57.
25. O'Reagan, B. and Grätzel, M. (1991) A low-cost, high-efficiency dye-sensitized solar cells based on colloidal TiO2 films. *Nature*, **353**, 737.
26. Yang, F. and Forrest, S.R. (2006) Organic solar cells using transparent SnO2:F anodes. *Adv. Mater.*, **18**, 2018.
27. Hung, L.S. and Tang, C.W. (1999) Interface engineering in preparation of organic surface-emitting diodes. *Appl. Phys. Lett.*, **74**, 3209.
28. Cho, H., Yun, C., Park, J.-W., and Yoo, S. (2009) Highly flexible organic light-emitting diodes based on $ZnS/Ag/WO_3$ multilayer transparent electrodes. *Org. Electron.*, **10**, 1163.
29. Han, S., Lim, S., Kim, H., Cho, H., and Yoo, S. (2010) Versatile multilayer transparent electrodes for ITO-free and flexible organic solar cells. *IEEE J. Sel. Top. Quantum Electron.*, **16**, 1656.
30. Sergeant, N.P., Hadipour, A., Niesen, B., Cheyns, D., Heremans, P., Peumans, P., and Rand, B.P. (2012) Design of

transparent anodes for resonant cavity enhanced light harvesting in organic solar cells. *Adv. Mater.*, **24**, 728.

31. Ouyang, J., Chu, C.-W., Chen, F.-C., Xu, Q., and Yang, Y. (2005) High-conductivity poly(3,4-ethylenedioxythiophene) – poly(styrenesulfonate) film and its application in polymer optoelectronic devices. *Adv. Funct. Mater.*, **15**, 203.
32. Galagan, Y., Coenen, E.W.C., Sabik, S., Gorter, H.H., Barink, M., Veenstra, S.C., Kroon, J.M., Andriessen, R., and Blom, P.W.M. (2012) Evaluation of ink-jet printed current collecting grids and busbars for ITO-free organic solar cells. *Sol. Energy Mater. Sol. Cells*, **104**, 32.
33. Kim, Y.H., Sachse, C., Machala, M.L., May, C., Müller-Meskamp, L., and Leo, K. (2011) Highly conductive PEDOT:PSS electrode with optimized solvent and thermal post-treatment for ito-free organic solar cells. *Adv. Funct. Mater.*, **21**, 1076.
34. Lee, J.Y., Connor, S.T., Cui, Y., and Peumans, P. (2008) Solution-processed metal nanowire mesh transparent electrodes. *Nano Lett.*, **8**, 689.
35. Han, D., Kim, H., Lee, S., Seo, M., and Yoo, S. (2010) Realization of efficient semitransparent organic photovoltaic cells with metallic top electrodes: utilizing the tunable absorption asymmetry. *Opt. Express*, **18**, A513.
36. Meyer, J., Winkler, T., Hamwi, S., Schmale, S., Johannes, H.-H., Weimann, T., Hinze, P., Kowalsky, W., and Riedl, T. (2008) Transparent inverted organic light-emitting diodes with a tungsten oxide buffer layer. *Adv. Mater.*, **20**, 3839.
37. Zhou, Y., Fuentes-Hernandez, C., Shim, J., Meyer, J., Giordano, A.J., Li, H., Winget, P., Papadopoulos, T., Cheun, H., Kim, J., Fenoll, M., Dindar, A., Haske, W., Najafabadi, E., Khan, T.M., Sojoudi, H., Barlow, S., Graham, S., Brédas, J.-L., Marder, S.R., Kahn, A., and Kippelen, B. (2012) A universal method to produce low-work function electrodes for organic electronics. *Science*, **336**, 327.
38. Greiner, M.T., Helander, M.G., Tang, W.-M., Wang, Z.-B., Qiu, J., and Lu, Z.-H. (2012) Universal energy-level alignment of molecules on metal oxides. *Nat. Mater.*, **11**, 76.
39. Kim, J.Y., Kim, S.H., Lee, H.-H., Lee, K., Ma, W., Gong, X., and Heeger, A.J. (2006) New architecture for high-efficiency polymer photovoltaic cells using solution-based titanium oxide as an optical spacer. *Adv. Mater.*, **18**, 572.
40. Berry, J.J., Widjonarko, N.E., Bailey, B.A., Sigdel, A.K., Ginley, D.S., and Olsen, D.C. (2010) Surface treatment of NiO hole transport layers for organic solar cells. *IEEE J. Sel. Top. Quantum Electron.*, **16**, 1649.
41. Sun, Y., Takacs, C.J., Cowan, S.R., Seo, J.H., Gong, X., Roy, A., and Heeger, A.J. (2011) Efficient, air-stable bulk heterojunction polymer solar cells using MoOx as the anode interfacial layer. *Adv. Mater.*, **23**, 2226.
42. Han, S., Shing, W.S., Seo, M., Gupta, D., Moon, S.-J., and Yoo, S. (2009) Improving performance of organic solar cells using amorphous tungsten oxides as an interfacial buffer layer on transparent anodes. *Org. Electron.*, **10**, 791.
43. Sun, Y., Seo, J.H., Takacs, C.J., Seifter, J., and Heeger, A.J. (2011) Inverted polymer solar cells integrated with a low-temperature-annealed sol-gel derived ZnO film as an electron transport layer. *Adv. Mater.*, **23**, 1679.
44. Li, G., Chu, C.-W., Shrotriya, V., Huang, J., and Yang, Y. (2006) Efficient inverted polymer solar cells. *Appl. Phys. Lett.*, **88**, 253503.
45. Chan, M.Y., Lee, C.S., Lai, S.L., Fung, M.K., Wong, F.L., Sun, H.Y., Lau, K.M., and Lee, S.T. (2006) Efficient organic photovoltaic devices using a combination of exciton blocking layer and anodic buffer layer. *J. Appl. Phys.*, **100**, 094506.
46. Brabec, C.J., Padinger, F., Hummelen, J.C., Janssen, R.A.J., and Sariciftci, N.S. (1999) Realization of large area flexible fullerene – conjugated polymer photocells: a route to plastic solar cells. *Synth. Met.*, **102**, 861.
47. Wang, G.F., Tao, X.M., and Wang, R.X. (2008) Flexible organic light-emitting diodes with a polymeric nanocomposite anode. *Nanotechnology*, **19**, 145201.
48. Heeger, A.J. (2001) Semiconducting and metallic polymers: the fourth generation

of polymeric materials (Nobel Lecture). *Angew. Chem. Int. Ed.*, **40**, 2591–2611.
49. Shirakawa, H. (2001) The discovery of polyacetylene film: the dawning of an era of conducting polymers (Nobel Lecture). *Angew. Chem. Int. Ed.*, **40**, 2574–2580.
50. MacDiarmid, A.G. (2001) "Synthetic metals": a novel role for organic polymers (Nobel Lecture). *Angew. Chem. Int. Ed.*, **40**, 2581–2590.
51. Elschner, A., Kirchneyer, S., Lövenich, W., Merker, U., and Reuter, K. (2011) *PEDOT: Principles and Applications of an Intrinsically Conductive Polymer*, CRC Press, Boca Raton, FL.
52. Vosgueritchian, V., Lipomi, D.J., and Bao, Z. (2012) Highly conductive and transparent PEDOT:PSS films with a fluorosurfactant for stretchable and flexible transparent electrodes. *Adv. Funct. Mater.*, **22**, 421–428.
53. Galagan, Y., Rubingh, J.-E.J.M., Andriessen, R., Fan, C.-C., Blom, P.W.M., Veenstra, S.C., and Kroon, J.M. (2011) ITO-free flexible organic solar cells with printed current collecting grids. *Sol. Energy Mater. Sol. Cells*, **95**, 1339.
54. Jung, S., Lee, S., Song, M., Kim, D.-G., You, D.S., Kim, J.-K., Kim, C.S., Kim, T.-M., Kim, K.-H., Kim, J.-J., and Kang, J.-W. (2014) Extremely flexible transparent conducting electrodes for organic devices. *Adv. Energy Mater.*, **4**, 1300474.
55. Hong, K., Kim, K., Kim, S., Lee, I., Cho, H., Yoo, S., Choi, H.W., Lee, N.-Y., Tak, Y.-T., and Lee, J.-L. (2011) Optical properties of WO_3/Ag/WO_3 multilayer as transparent cathode in top-emitting organic light emitting diodes. *J. Phys. Chem. C*, **115**, 3453.
56. Yun, J., Wang, W., Bae, T.S., Park, Y.H., Kang, Y.-C., Kim, D.-H., Lee, S., Lee, G.-H., Song, M., and Kang, J.-W. (2013) Preparation of flexible organic solar cells with highly conductive and transparent metal-oxide multilayer electrodes based on silver oxide. *ACS Appl. Mater. Interfaces*, **5**, 9933.
57. Salinas, J.-F., Yip, H.-L., Chueh, C.-C., Li, C.-Z., Maldonado, J.-L., and Jen, A.K.-Y. (2012) Optical design of transparent thin metal electrodes to enhance in-coupling and trapping of light in flexible polymer solar cells. *Adv. Mater.*, **24**, 6362.
58. Han, D., Lee, S., Kim, H., Jeong, S., and Yoo, S. (2013) Cathodic multilayer transparent electrodes for ITO-free inverted organic solar cells. *Org. Electron.*, **14**, 1477.
59. Lim, S., Han, D., Kim, H., Lee, S., and Yoo, S. (2012) Cu-based multilayer transparent electrodes: a low-cost alternative to ITO electrodes in organic solar cells. *Sol. Energy Mater. Sol. Cells*, **101**, 170.
60. Wang, Z.B., Helander, M.G., Qiu, J., Puzzo, D.P., Greiner, M.T., Hudson, Z.M., Wang, S., Liu, Z.W., and Lu, Z.H. (2011) Unlocking the full potential of organic light-emitting diodes on flexible plastic. *Nat. Photonics*, **5**, 753.
61. Cho, H., Choi, J.-M., and Yoo, S. (2011) Highly transparent organic light-emitting diodes with a metallic top electrode: the dual role of a Cs_2CO_3 layer. *Opt. Express*, **19**, 1113.
62. Fraunhofer-Gesellschaft *http://www.comedd.fraunhofer.de/en/news/press/2013/2013-09-02.html* (accessed 21 July 2014).
63. Im, H.-G., Jin, J., Ko, J.-H., Lee, J., Lee, J.-Y., and Bae, B.-S. (2014) Flexible transparent conducting composite films using a monolithically embedded AgNW electrode with robust performance stablilty. *Nanoscale*, **6**, 711.
64. Bae, S., Kim, H., Lee, Y., Xu, X., Park, J.-S., Zheng, Y., Balakrishnan, J., Lei, T., Kim, H.R., Song, Y.I., Kim, Y.-J., Kim, K.S., Özyilmaz, B., Ahn, J.-H., Hong, B.H., and Iijima, S. (2010) Roll-to-roll production of 30-inch graphene films for transparent electrodes. *Nat. Nanotechnol.*, **5**, 574.
65. Lewis, J.S. and Weaver, M.S. (2004) Thin-film permeation-barrier technology for flexible organic light-emitting devices. *IEEE J. Sel. Top. Quantum Electron.*, **10**, 45.
66. Kim, N., Potscavage, W.J., Sundaramoothi, A., Henderson, C., Kippelen, B., and Graham, S. (2012) A correlation study between barrier film performance and shelf lifetime of encapsulated organic solar cells. *Sol. Energy Mater. Sol. Cells*, **101**, 140.

67. Potscavage, W.J., Yoo, S., Domercq, B., and Kippelen, B. (2007) Encapsulation of pentacene/C_{60} organic solar cells with Al2O3 deposited by atomic layer deposition. *Appl. Phys. Lett.*, **90**, 253511.
68. Jiang, C.Y., Sun, X.W., Zhao, D.W., Kyaw, A.K.K., and Li, Y.N. (2010) Low work function metal modified ITO as cathode for inverted polymer solar cells. *Sol. Energy Mater. Sol. Cells*, **94**, 1618.
69. Espinosa, N., Garcia-Valverde, R., Urbina, A., Lenzmann, F., Manceau, M., Angmo, D., and Krebs, F.C. (2012) Life cycle assessment of ITO-free flexible polymer solar cells prepared by roll-to-roll coating and printing. *Sol. Energy Mater. Sol. Cells*, **97**, 3–13.
70. Gatz, S., Hannebauer, H., Hesse, R., Werner, F., Schmidt, A., Dullweber, T., Schmidt, J., Bothe, K., and Brendel, R. (2011) 19.4%-efficient large-area fully screen-printed silicon solar cells. *Phys. Status Solidi (RRL)*, **5**, 147–149.
71. Wengeler, L., Schmitt, M., Peters, K., Scharfer, P., and Schabel, W. (2013) Comparison of large scale coating techniques for organic and hybrid films in polymer based solar cells. *Chem. Eng. Process. Process Intensif.*, **68**, 38–44.
72. Rowell, M.W. and McGehee, M.D. (2011) Transparent electrode requirements for thin film solar cell modules. *Energy Environ. Sci.*, **4**, 131–134.
73. Shah, A.V., Schade, H., Vanecek, M., Meier, J., Vallat-Sauvain, E., Wyrsch, N., Kroll, U., Droz, C., and Bailat, J. (2004) Thin-film silicon solar cell technology. *Prog. Photovoltaics Res. Appl.*, **12**, 113–142.
74. Kim, H., Lee, S., Han, D., and Yoo, S. (2014) High-density organic photovoltaic modules: mask-free fabrication using nozzle jet printing and oblique deposition. *Sol. Energy Mater. Sol. Cells*, **120**, 561–565.
75. Shaheen, S.E., Radspinner, R., Peyghambarian, N., and Jabbour, G.E. (2001) Fabrication of bulk heterojunction plastic solar cells by screen printing. *Appl. Phys. Lett.*, **79**, 2996–2998.
76. Yu, J.-S., Kim, I., Kim, J.-S., Jo, J., Larsen-Olsen, T.T., Søndergaard, R.R., Hösel, M., Angmo, D., Jørgensen, M., and Krebs, F.C. (2012) Silver front electrode grids for ITO-free all printed polymer solar cells with embedded and raised topographies, prepared by thermal imprint, flexographic and inkjet roll-to-roll processes. *Nanoscale*, **4**, 6032–6040.
77. Søndergaard, R.R., Hösel, M., and Krebs, F.C. (2013) Roll-to-roll fabrication of large area functional organic materials. *J. Polym. Sci., Part B: Polym. Phys.*, **51**, 16–34.
78. Lee, J.-H., Sagawa, T., and Yoshikawa, S. (2011) Morphological and topographical characterizations in spray coated organic solar cells using an additional solvent spray deposition. *Org. Electron.*, **12**, 2165–2173.
79. Vak, D., Kim, S.-S., Jo, J., Oh, S.-H., Na, S.-I., Kim, J., and Kim, D.-Y. (2007) Fabrication of organic bulk heterojunction solar cells by a spray deposition method for low-cost power generation. *Appl. Phys. Lett.*, **91**, 081102.
80. Girotto, C., Moia, D., Rand, B.P., and Heremans, P. (2011) High-performance organic solar cells with spray-coated hole-transport and active layers. *Adv. Funct. Mater.*, **21**, 64–72.
81. Kim, Y., Kim, G., Lee, J., and Lee, K. (2012) Morphology controlled bulk-heterojunction layers of fully electro-spray coated organic solar cells. *Sol. Energy Mater. Sol. Cells*, **105**, 272–279.
82. Huang, Y.-C., Chia, H.-C., Chuang, C.-M., Tsao, C.-S., Chen, C.-Y., and Su, W.-F. (2013) Facile hot solvent vapor annealing for high performance polymer solar cell using spray process. *Sol. Energy Mater. Sol. Cells*, **114**, 24–30.
83. Aernouts, T., Aleksandrov, T., Girotto, C., Genoe, J., and Poortmans, J. (2008) Polymer based organic solar cells using ink-jet printed active layers. *Appl. Phys. Lett.*, **92**, 033306.
84. Kistler, S.F. and Schweizer, P.M. (1997) *Liquid Coating Method*, Springer.
85. Lim, S.-L., Chen, E.-C., Chen, C.-Y., Ong, K.-H., Chen, Z.-K., and Meng, H.-F. (2012) High performance organic photovoltaic cells with blade-coated active layer. *Sol. Energy Mater. Sol. Cells*, **107**, 292–297.

86. Park, B. and Han, M.-y. (2009) Photovoltaic characteristics of polymer solar cells fabricated by pre-metered coating process. *Opt. Express*, **17**, 13830–13840.

87. Jakubka, F., Heyder, M., Machui, F., Kaschta, J., Eggerath, D., Lövenich, W., Krebs, F.C., and Brabec, C.J. (2013) Determining the coating speed limitations for organic photovoltaic inks. *Sol. Energy Mater. Sol. Cells*, **109**, 120–125.

88. Blankenburg, L., Schultheis, K., Schache, H., Sensfuss, S., and Schrödner, M. (2009) Reel-to-reel wet coating as an efficient up-scaling technique for the production of bulk-heterojunction polymer solar cells. *Sol. Energy Mater. Sol. Cells*, **93**, 476–483.

89. Hong, S., Lee, J., Kang, H., and Lee, K. (2013) Slot-die coating parameters of the low-viscosity bulk-heterojuction materials used for polymer solar cells. *Sol. Energy Mater. Sol. Cells*, **112**, 27–35.

90. Yang, J., Vak, D., Clark, N., Subbiah, J., Wong, W.W.H., Jones, D.J., Watkins, S.E., and Wilson, G. (2013) Organic photovoltaic modules fabricated by an industrial gravure printing proofer. *Sol. Energy Mater. Sol. Cells*, **109**, 47–55.

91. Kopola, P., Aernouts, T., Guillerez, T., Jin, H., Tuomikoski, M., Maaninen, A., and Hast, J. (2010) High efficient plastic solar cells fabricated with a high-throughput gravure printing method. *Sol. Energy Mater.*, **94**, 1673–1680.

92. Hu, Z., Zhang, J., Xiong, S., and Zhao, Y. (2012) Annealing-free, air-processed and high-efficiency polymer solar cells fabricated by a dip coating process. *Org. Electron.*, **13**, 142–146.

93. Kim, S.-S., Na, S.-I., Jo, J., Tae, G., and Kim, D.-Y. (2007) Efficient polymer solar cells fabricated by simple brush painting. *Adv. Mater.*, **19**, 4410–4415.

94. Park, H.J., Kang, M.-G., Ahn, S.H., and Guo, L.J. (2010) A facile route to polymer solar cells with optimum morphology readily applicable to a roll-to-roll process without sacrificing high device performances. *Adv. Mater.*, **22**, E247–E253.

95. Baldo, M. *et al.* (1998) Organic vapor phase deposition. *Adv. Mater.*, **10**, 1505.

96. Shtein, M., Peumans, P., Benziger, J.B., and Forrest, S.R. (2004) Direct, mask- and solvent-free printing of molecular organic semiconductors. *Adv. Mater.*, **16**, 1615.

97. Yun, C., Choi, J., Kang, H.W., Kim, M., Moon, H., Sung, H.J., and Yoo, S. (2012) Digital-mode organic vapor-jet printing (D-OVJP): advanced jet-on-demand control of organic thin-film deposition. *Adv. Mater.*, **24** (21), 2857–2861.

98. Zakhidov, A., Lee, J.-K., Fong, H.H., DeFranco, J.A., Chatzichristidi, M., Taylor, P.G., Ober, C.K., and Malliaras, G.G. (2008) Hydrofluoroethers as orthogonal solvents for the chemical processing of organic electronic materials. *Adv. Mater.*, **20**, 3481–3484.

99. Lee, J., Back, H., Kong, J., Kang, H., Song, S., Suh, H., Kang, S.-O., and Lee, K. (2013) Seamless polymer solar cell module architecture built upon self-aligned alternating interfacial layers. *Energy Environ. Sci.*, **6**, 1152–1157.

17
Flexible Inorganic Photovoltaics

Zhuoying Chen

17.1
Introduction

Photovoltaics represent a clean and sustainable way to generate electric power through converting renewable solar radiation to electricity. Even though the discovery of the photovoltaic (PV) effect dates back to the nineteenth century, it was not until 1950s that its development took off after the discovery of the $p-n$ junction in silicon through exposing naturally n-type silicon wafers to boron trichloride gas for the development of a p-type skin. Being the basic building blocks in silicon electronics, these $p-n$ junctions form the first modern solar cells exhibiting good rectifying behavior and PV effect with an initial efficiency of 6% [1]. These early solar cells were primarily used in space applications, where cost was generally not a concern. Cost became a concern starting from the 1970s, when a growing interest to secure renewable energy sources alternative to fossil fuels emerged. This resulted in a rapid expansion in the research and development of photovoltaics. Since then, there have been continuous efforts in searching for strategies to both increasing device efficiencies and reducing material and fabrication costs. To date, the power conversion efficiency of crystalline silicon (c-Si) single-junction solar cells has reached 25% in laboratories [2], which is approaching the theoretical Shockley–Queisser efficiency limit [3] which is calculated to be 30% for a $p-n$ junction solar cell of 1.1 eV bandgap under the assumption that the only recombination mechanism is radiative. In terms of cost per watt, over the last 40 years the price of c-Si solar modules has fallen steadily mainly owing to the increase of device efficiency, the establishment of large-scale manufacturing bases, and the resultant supply-chain benefits [4]. While photovoltaics are already competitive technologies in geographic regions where conventional electricity supply is expensive, their current contribution to the global energy generation is still rather limited. Besides the PV module cost, additional costs need to be considered when one would like to convert a household's electricity supply from conventional to PV technology. These additional costs are mainly related to installation and maintenance, among which the cost of the inverter for grid-connected systems and the cost of batteries for off-grid systems play important roles. As a result, without

Large Area and Flexible Electronics, First Edition. Edited by Mario Caironi and Yong-Young Noh.

government subsidies, there is still a lack of incentive for the majority of the public to install solar panels.

To lower the PV module and installation costs, various alternative PV materials, such as amorphous and microcrystalline silicon, other nanostructured or thin-film materials, photoelectrochemical junctions, organic semiconductors, and colloidal inorganic nanocrystals, have been developed. Compared to the heavy and rigid crystalline silicon and III–V compounds, these alternative materials are attractive not only because of their cost-reduction potential but also because of their light-weight, robust, and flexible characteristics which favor the diversity of the PV market. Being compatible with flexible substrates, these solar cells can be fabricated on a large scale and in a cost-effective manner such as through roll-to-roll printing or vapor deposition. Without the need of heavy maintenance, they can be integrated, rather than installed, into various surfaces such as buildings, fabrics, tents, and sails without any constraint on the flatness of the surface. The savings in building or fabric materials, together with their light-weight characteristics, lower their installation cost. Flexible solar cells can thus be integrated to act as multifunctional materials, such as roofing layers, both providing electricity and serving as a waterproof system [5]. Such multifunctional applications open up opportunities to a large variety of products demanded by the consumer market.

This chapter will mainly cover the progress and examples of *inorganic* photovoltaics which are compatible with flexible substrates. For other material systems applicable on flexible substrates, interested readers may refer to recent excellent reviews on organic [6–10] and hybrid [11–14] solar cells. This chapter divides the discussions on different inorganic PV materials into three sections depending on the methods required to deposit them onto flexible substrates. We will first start with crystalline solar cells which have to be transferred onto flexible substrates, then thin film solar cells grown directly onto flexible substrates by vapor deposition, and, finally, thin-film solar cells deposited from solution.

17.2
Thin Crystalline Solar Cells Transferred onto Flexible Substrates

Solar cells based on a few hundred micrometers of c-Si are normally not flexible because of the brittle nature of silicon wafers. But they can possibly undergo some degree of flexure when the thickness of the silicon layer is reduced to a few tens of micrometers [15, 16]. One of the most representative examples in this field is the Sliver® cells , invented and developed by researchers from the Australia National University [16–18]. Typically, an individual Sliver solar cell is a long (50–100 mm), narrow (1–2 mm), and thin (40–60 μm) crystalline silicon stripe, while a Sliver module contains thousands of these stripes. The fabrication and processing of such Sliver cells have been well developed: Starting from a 1–2-mm-thick silicon wafer, thin silicon stripes are obtained by forming deep narrow grooves all the way through the wafer by techniques such as laser scribing, dicing with a saw, or an anisotropic alkaline etching of (110) wafers. Cell processing (diffusion, oxidation, deposition) is completed while the silicon stripes are still supported by the silicon substrate at the edge of the wafer. They are subsequently removed from the wafer and laid on their sides on selected substrates. A silicon wafer of 1.5 mm thickness and 150 mm diameter can yield about 1000 Sliver cells with a combined surface area of about 1500 cm^2, signifying the large potential of this technology in the reduction of silicon usage. In addition, because of the reduced dimension of each stripe, Sliver cells can be deposited on flexible substrates. Blakers and Armour [16] have carried out experimental characterizations on the cell's behaviors under static and repeated flexure . The breaking radii are found to range from a few to a few tens of millimeters depending on the sample thickness. For a given radius of curvature, these thin c-Si solar cells either broke immediately or were able to sustain multiple flexing with only a small compromise in their power conversion efficiencies (maintaining about 95% of the efficiencies of unflexed samples).

More flexible c-Si cells can be achieved when the dimension of the silicon stripes is further reduced. One of these examples is represented by the work of Yoon *et al.* [15], who fabricated flexible ultrathin silicon microcells composed of c-Si stripes with lengths, widths, and thicknesses of 1.55 mm, 50 mm, and 15 μm, respectively. Starting from bulk silicon wafers, individual silicon microcells were defined (and selectively doped through a doping mask defined) by multiple photolithography and etching steps. These microcells were then transferred and printed on flexible substrates using a PDMS stamp by an automated machine. These silicon microcells typically achieved power conversion efficiency in the range of 6–8% and open-circuit voltage of 450–510 mV. At the bending radii of 12.6, 8.9, 6.3, and 4.9 mm and after 200 bending cycles, their module efficiency and fill factor remained impressively unchanged (Figure 17.1). Because the thickness of silicon in these examples is very small, efficient absorption of solar energy can, however, be a problem, leading to lower power conversion efficiencies. While light-trapping techniques such as backside reflectors can be used to partially alleviate this problem, there is also an increase in edge surface recombination in these microcells compared to their conventional bulk and inflexible counterparts due to the unpassivated surfaces.

Similar to crystalline silicon which becomes flexible when it is thin, flexible InAs/GaAs quantum dot (QD) solar cells composed of a 4.1-μm-thick compound semiconductor PV layer was demonstrated by Tanabe *et al.* [19] . Because of their high absorption coefficients, compound semiconductors of a few micrometers thickness are typically used in solar cells. Therefore, unlike in the case of c-Si, there is no particular need for cutting down the thickness of the active layer for them to become flexible in this case. The QD layer is first grown onto lattice-matched crystalline substrates (GaAs) and then transferred onto flexible plastic substrates. Even though detailed characterizations on the possible bending radii are not available, these solar cells show interesting PV characteristics equal to those of as-grown QDs on an inflexible GaAs substrate.

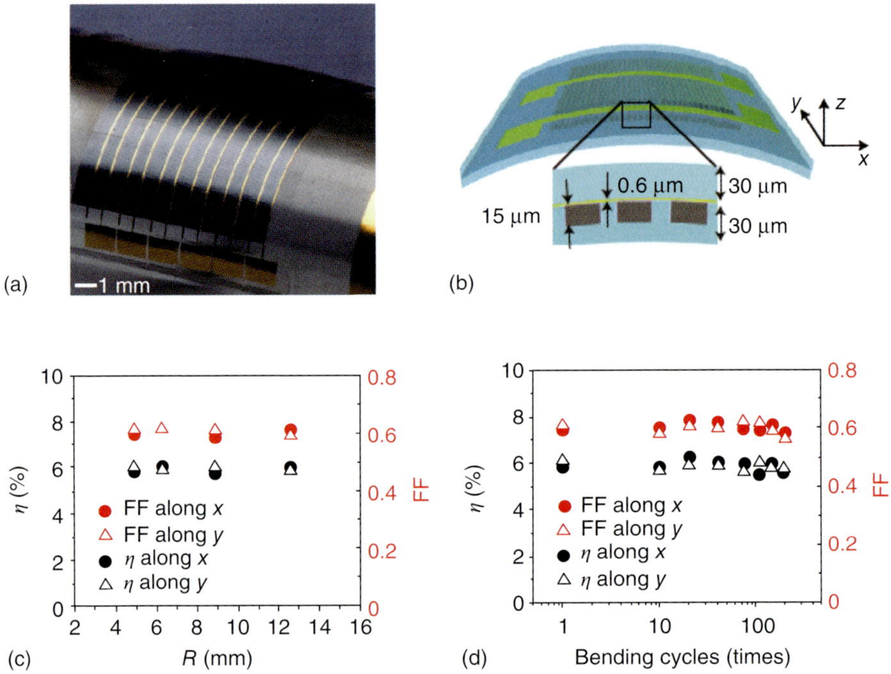

Figure 17.1 (a) Optical image, (b) schematic illustration, and (c,d) photovoltaic performance of mechanically flexible modules that incorporate arrays of interconnected c-Si microcells. (a) Optical image of a module bent along a direction parallel to the width of the microcells to a bending radius of 4.9 mm. (b) Schematic illustration of an optimized design in which the neutral mechanical plane is positioned near the center of the microcells (gray) through judicious choices of thickness for the polymer (blue) substrate and overcoat. The inset of (b) shows the cross sections of three microcells interconnected by an electrode. (c) Plot of power conversion efficiency and fill factor under AM 1.5 illumination for $R = 12.6$, 8.9, 6.3, and 4.9 mm. (d) Plot of power conversion efficiency and fill factor as a function of bending cycles up to 200 times at $R = 4.9$ mm [15]. (Reproduced with permission. copyright © 2008, Macmillan Publishers.)

In all the above-mentioned flexible solar cell examples, neither c-Si nor InAs/GaAs epitaxial QDs could be grown directly on flexible plastic films or metal foils. Instead, rather complicated etching and transfer processes are required to define them and put them onto flexible substrates, which may not be desirable for large-area and fast-throughput fabrications.

17.3 Thin-Film Solar Cells Grown Directly onto Flexible Substrates by Vapor Deposition

Unlike the crystalline examples of the previous section where rather complicated processes are needed to transfer them onto flexible substrates, there are thin-film

PV materials, with thicknesses ranging from a few hundred nanometers to a few micrometers, which are compatible with the mechanical requirements to become flexible. More importantly, they can be deposited directly onto large-area, flexible substrates by physical or chemical vapor deposition processes, which opens mass-production possibilities. Examples of these thin-film PV materials compatible with flexible substrates include amorphous silicon (a-Si), polycrystalline cadmium telluride (CdTe), and polycrystalline copper indium gallium diselenide ($CuIn_xGa_{(1-x)}Se_2$, CIGS).

In the case of a-Si , while the research to apply it to solar cells dates back to the late 1970s, its application to flexible solar cells started only about 20 years ago [20, 21]. Besides being cheaper to produce compared to c-Si, a-Si is a better photon absorber because of the loss of long-range order and the resultant relaxed selection rules for photon absorption. However, its amorphous nature means that there are plenty of unsatisfied "dangling" bonds, which inevitably leads to the existence of defect states inside the bandgap. These defect states trap charges and act as recombination centers. a-Si is therefore usually used in the form of thin films because charge transport will not be efficient in thick films. Because of the short exciton diffusion length in doped a-Si, a-Si solar cells adapt typically a $p-i-n$ or n–i–p junction design where photons are mainly absorbed in the central intrinsic region. To boost the photon absorption efficiency, tandem strategies have been applied, combining multiple undoped and doped a-Si solar cells of different bandgaps [22–24].

Notable examples of flexible a-Si solar cells at the individual and module levels have been demonstrated by companies such as Fuji and Uni-Solar on plastic sheets [20, 25] and metallic foils [21, 26, 27] exhibiting multijunction power conversion efficiency ranging from about 8% to 15%. In the flexible triple-junction example fabricated by Uni-Solar [21], there are three a-Si cells of $p-i-n$ configurations, the intrinsic (i) layer of which is either undoped for absorbing "blue" photons or doped with different amount of Ge to absorb "green" and "red" photons (Figure 17.2). The manufacture of such triple-junction solar cells is performed in a roll-to-roll manner. A roll of stainless steel moves continuously through different machines

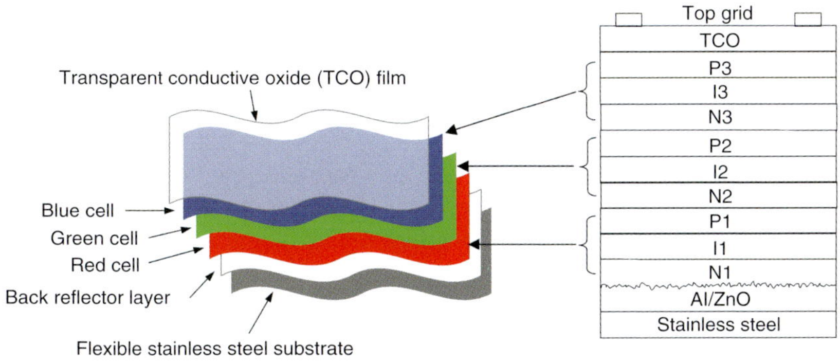

Figure 17.2 Schematic of a triple-junction structure containing amorphous silicon.

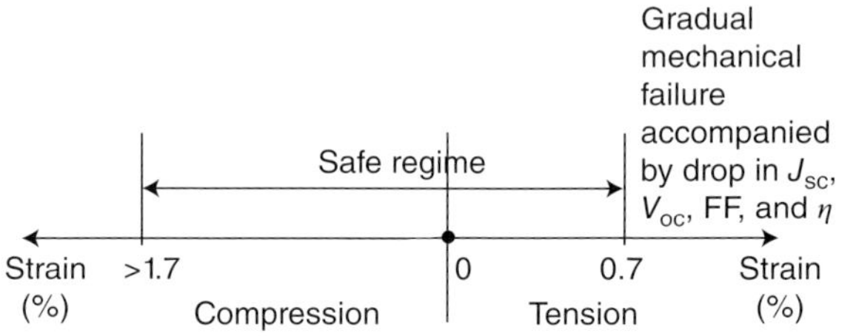

Figure 17.3 Summary of the response of a triple-junction amorphous silicon–germanium solar cell on stainless steel to mechanical strain [29]. (Reproduced with permission. Copyright © 2006, Elsevier.)

for substrate cleaning, back-reflector deposition, a-Si or a-SiGe alloy deposition, and, finally, indium tin oxide (ITO) deposition. The back-reflector layer typically is composed of a layer of metal (Al or Ag) and metal oxide (ZnO). It serves as a "buffer" layer to provide ohmic contact between the stainless steel and the a-Si alloy and facilitates multiple light reflections due to its textured surface. Upon optimizing this back-reflector, such flexible solar cells exhibit promising total current densities of 22.74 mA cm^{-2} and a quantum efficiency of about 0.21 at 850 nm. Applying a series of optimization steps on the tunnel junctions and the transparent conductive oxide layer, Uni-Solar further achieved more optimized triple-junction a-Si solar cells exhibiting a total photocurrent of 26.88 mA cm^{-2} [23]. Detailed mechanical studies have been performed on such flexible triple junction a-Si cells. Interestingly, they are much more robust under compression than under tension. Little deterioration in PV performance was found when the cells were under compression up to 1.7% strain, while the cell characteristics began to degrade at a tensile strain of 0.75% [28] (Figure 17.3). While the technology of flexible thin-film a-Si solar cells was originally designed toward lowering their cost compared to c-Si solar cells, the complicated processes involved in the fabrication of multijunctions (with each p–i–n junction containing an "i" layer of different Ge doping levels) may reduce their cost advantage.

Compared to thin-film a-Si solar cells built from p–i–n structures, thin-film polycrystalline CdTe and CIGS solar cells are mostly based on the formation of a heterojunction with another material that transports carriers of the opposite type. As compound semiconductors have relatively high absorption coefficients, single-junction solar cells based on CdTe and CIGS of a few micrometers thickness can achieve equivalent PV performance compared to multijunction a-Si thin-film solar cells. With a direct bandgap of 1.45 eV, polycrystalline CdTe thin films deposited by physical and chemical vapor deposition have been widely applied in solar cells during the last 40 years [30, 31]. Conventional CdTe cells typically utilize a "superstrate" structure (Figure 17.4a). In such a structure, different layers, including the front contact, CdS, CdTe, and the back contact, are deposited sequentially on a transparent, rigid bottom substrate (such as glass) through which light enters the solar cell. State-of-the-art power conversion efficiency of 19.6% on 1 cm^2 rigid substrates has been achieved by GE Global Research recently on single-junction thin-film CdTe cells [2]. There have also been intensive

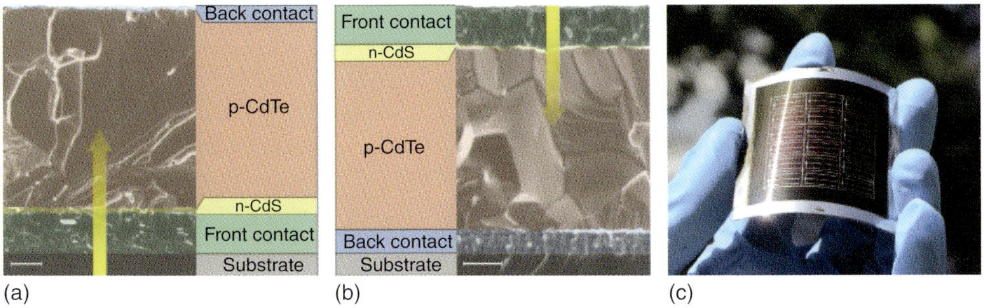

Figure 17.4 CdTe solar cell structures. (a,b) Scanning electron micrograph and schematic of cross section of a CdTe solar cell in (a) the conventional "superstrate" configuration and (b) the inverted "substrate" configuration which allows the use of opaque substrates such as metal foils. In the substrate configuration, Mo/MoO_x and i-ZnO/ZnO:Al can be used as electrical back and front contact, respectively. The scale bars correspond to 1 μm. The yellow arrows show the direction of illumination. (c) Photograph of a sample with several CdTe solar cells on a flexible metal foil fabricated by the Empa-Swiss Federal Laboratories for Materials Science and Technology [33]. (Reproduced with permission. Copyright © 2013, Macmillan Publishers.)

research efforts to achieve flexible CdTe solar cells of high efficiency. One general issue related to the use of flexible substrates is the required modification of the experimental procedures: for example, lower substrate temperature allowed, or a different (inverted) device structure required and the resultant compromised performance. For example, by applying a special polyimide film developed by DuPont as the bottom transparent substrate, researchers from the Empa-Swiss Federal Laboratories for Materials Science and Technology demonstrated in 2011 flexible CdTe solar cells showing a peak power conversion efficiency of 13.8% [32]. While the application of a conventional device structure (the "superstrate" structure) on polyimide films so far has allowed the highest efficiency in flexible CdTe solar cells, there has also been much research effort to use the inverted structure (the "substrate" structure, Figure 17.4b), which allows flexible opaque substrates such as common metal foils. In this "substrate" structure, light enters the solar cell from the top layer, which is the front contact. Initial studies on CdTe solar cells using this "substrate" structure did not demonstrate promising efficiencies probably because of the different device processing conditions required when one inverted the conventional structure and the resulting compromised electronic properties of the CdTe layer. Through a controlled doping of the CdTe layer by the evaporation of a sub-monolayer of copper and subsequent annealing, researchers from the Empa-Swiss Federal Laboratories obtained in 2013 a breakthrough in efficiency of flexible CdTe solar cells, reaching 10.9% and 11.5% with the use of flexible stainless steel and molybdenum foils as the bottom substrate, respectively (Figure 17.4c).

Compared to thin-film CdTe solar cells, chalcopyrite CIGS cells so far have exhibited slightly higher power conversion efficiencies [2, 34]. In fact, thin film CIGS cells are currently the best performing solar cells among the thin-film PV technologies [2]. The initial search for Ga-free copper indium selenide (CIS)

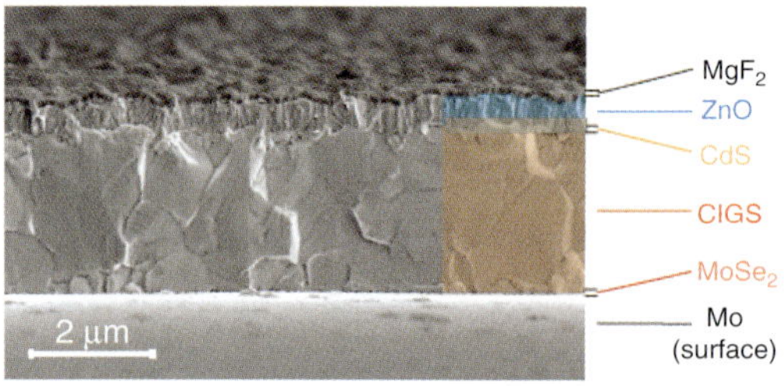

Figure 17.5 (a) Scanning electron micrograph of a crosssection obtained from a complete CIGS solar cell on flexible polyimide substrate with an efficiency >18%. The top layer of MgF_2 was used to minimize reflection losses. (b) Photograph of flexible CIGS solar cells on polyimide substrates [38]. (Reproduced with permission. Copyright © 2011, Macmillan Publishers.)

solar cells dates back to the 1970s [35]. It was subsequently identified that doping this material with gallium could allow the engineering of its bandgap, from 1.0 eV for pure CIS to 1.7 eV for pure indium-free copper gallium selenide (CGS). State-of-the-art CIGS solar cells typically have a ratio of gallium to the total amount of gallium and indium of about 0.3–0.4, which leads to a bandgap of around 1.2 eV. Most state-of-the-art CIGS solar cells are based on the formation of a heterojunction between CIGS and CdS. Similar to the device structures used in CdTe solar cells, CIGS solar cells have been built using both the conventional configuration on transparent substrates [36–38] (Figure 17.5a) and the inverted configuration on opaque substrates with a transparent top contact [39–41]. The CIGS layers are typically deposited through a multistage co-evaporation process,[1)] where a combination of In, Ga, and Se is deposited during the first and the third stage and Cu and Se are deposited during the second stage. On rigid substrates of 0.5 cm^2 area, 20.8% of power conversion efficiency has been demonstrated [34]. Metal and ceramic foils have been used to build flexible CIGS solar cells under the inverted configuration, showing 17.9% in power conversion efficiency [39, 40]. By comparison, flexible polyimide substrates can potentially induce less metallic impurities in the active layer; however, they can withstand only a lower substrate temperature, which imposes constraints in the absorber deposition process. After identifying that the composition gradient in the absorber layer was the major culprit for the poor PV performance and by adjusting it properly, Chirila *et al.* [38] demonstrated a power conversion efficiency as high as 18.7% in flexible CIGS solar cells using polyimide substrates (Figure 17.5b).

In summary, thin films based on materials such as a-Si, CdTe, and CIGS have been deposited by chemical or physical vapor deposition directly onto flexible substrates for functional solar cells. Very promising PV performance has been achieved with this technology, showing peak power conversion efficiency of 18.7% in flexible CIGS solar cells. This value is very close to that achieved in rigid CIGS

1) There are CIGS solar cells fabricated by the deposition of CIGS nanoparticle ink and subsequent annealing. These solution-processed examples will be discussed in the next section.

solar cells (20.8%), suggesting little performance limitation due to the application of flexible substrates.

17.4 Solution-Processed Thin-Film Solar Cells Deposited Directly onto Flexible Substrates

To attain more cost advantages in fabrication, flexible solar cells with their active layer deposited directly from solution have aroused much interest [42]. Advances in this field mainly rely on the progress of colloidal chemistry. Unlike those vacuum-deposited thin films described in the previous sections, nanoparticle solutions (nanoparticle "ink") of various compositions and nanoparticle precursor solution can be spin-coated, doctor-bladed, sprayed, or inkjet-printed on various substrates at room (or mild) temperature with and without the need of post-deposition annealing. Because of various reasons such as compromised charge transport in thick films related to the existence of grain boundaries or nanoparticle surfaces, these solution-processed thin films for solar cells typically have a thickness of no more than a few micrometers, allowing mechanical compatibilities with flexible applications.

Depending on the final goal of the thin film desired, there are mainly two subcategories in this technology: The first includes those solar cells with their active semiconducting layer in the form of poly or nanocrystalline thin films similar to the examples described in the previous section but deposited from solution. In this case, the nanoparticle or precursor solution is used because of the ease of fabrication, while it is not necessarily to keep the form of nanoparticles in the final device but to form dense polycrystalline thin films with the aid of post-deposition annealing. Notable examples in this subcategory include the "nano ink" developed by Nanosolar Inc. [43, 44]. Combined with a post-deposition annealing and selenization step, with nanoparticle inks, this company has achieved 17.1% power conversion efficiency in flexible CIGS solar cells compatible with roll-to-roll printing fabrication [45]. Similar solution processes were also used for kesterite ($Cu_2ZnSn(S,Se)_4$, (CZTS)) solar cells, which have cost-reduction advantages over CIGS cells because of the use of benign and earth-abundant elements. For example, by depositing an ink containing a mixture of copper and tin chalcogenide and Zn powder in hydrazine on substrates followed by a post-deposition heat treatment at 540 °C, Todorov *et al.* [46] achieved CZTS thin-film solar cells showing 9.6% power conversion efficiency. Without the use of the toxic hydrazine solvent, Cao *et al.* proposed an alternative ink containing a mixture of CuSnS, ZnS, and SnS nanoparticles. Thin films deposited from this nanoparticle ink could be converted to dense polycrystalline CZTS through thermal annealing, allowing 8.5% power conversion efficiency in solar cells [47]. Though the above-mentioned CZTS solar cells were demonstrated only on rigid substrates, there is no reason why these methods based on precursor and nanoparticle inks cannot be applied onto flexible substrates. Indeed, despite the to-be-improved PV performance, inks based on hydrothermally derived CZTS nanoparticles have already been successfully applied to achieve functional CZTS solar cells on flexible aluminum foils [48].

The second subcategory includes solar cells based on solution-processed colloidal inorganic semiconducting nanocrystals . Because of their compound semiconductor nature, colloidal nanocrystals of II–VI and VI–VI compositions typically exhibit high absorption coefficients (~10^5 cm^{-1}). Colloidal nanocrystals, when their dimensions fall in the quantum-confined regime, are attractive for various large-area and low-cost optoelectronic applications [49–57] mainly due to their readily tunable absorption and emission across the visible and the near-IR spectrum, as well as their solution processability. For photovoltaics, the size-tunable absorption characteristics of nanocrystals offer important opportunities in the design of single-junction solar cells capable of absorbing up to a specific wavelength. Such absorption tunability is very precious, particularly in the near-IR spectrum where conventional solar materials offer only limited choices. These absorption-specific nanocrystal solar cells can be applied in tandem cells in a second stage to maximally harvest the solar spectrum. In addition, in quantum-confined nanocrystals (QDs), there is the multiple exciton generation (MEG) phenomenon, a process generating two or more electron–hole pairs upon the absorption of high-energy photons of at least twice the bandgap energy. If the MEG process can be properly applied to nanocrystal solar cells, it can potentially boost the solar cell efficiency beyond the Shockley–Queisser limit. Solar cells in this subcategory typically involve no or only mild post-deposition annealing for the active layer in order not to destroy the quantum-confined optical properties specific to the nanocrystal dimension.

There have been many successful examples in applying colloidal nanocrystals in functional solar cells. For example, while being slightly out of the scope of this chapter focused on *inorganic* flexible solar cells, inorganic–organic hybrid solar cells using CdSe nanorods [58], tetrapods [59], and hyperbranched [60] nanocrystals have exhibited power conversion efficiencies of 2.6%, 2.8%, and 2.2%, respectively. Hybrid solar cells based on a planar heterojunction between *p*-type PbS QDs and *n*-type C_{60} have shown a promising efficiency of 5.2% [61]. For inorganic solar cells, in 2008, 1.6% power conversion efficiency was demonstrated on both rigid ITO/glass and flexible ITO-coated polyester substrates using a planar heterojunction structure formed between two different layers of colloidal nanocrystals of Cu_2S and CdS (Figure 17.6a) [62]. In these flexible cells, only slight efficiency degradation (~8%) was found after bending the device through 105°. While the exact dimensions of these samples were not reported, a curvature of 105° gives a bending radius of about 5 mm if the entire sample width under bending is 1 cm. Much research effort has also been devoted to the IV–VI PbS and PbSe systems because of their tunable near-IR absorption. For example, solar cells based on a Schottky barrier formed at the interface of a *p*-type PbS colloidal QD film and a metal cathode have shown power conversion efficiency up to 3.93% [63]. Bilayer inorganic solar cells based on PbS QDs and a TiO_2 layer have achieved an efficiency of 5.1% in 2010 [51]. Since then, significant progress in the device

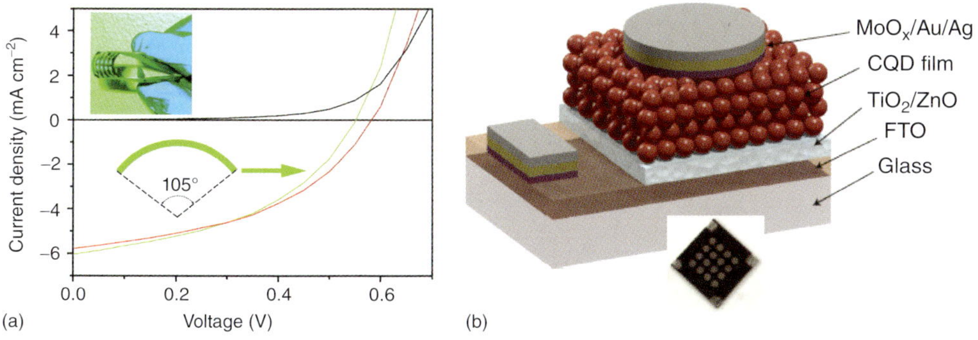

Figure 17.6 (a) Cu_2S–CdS nanocrystal photovoltaic devices on a plastic substrate. Current density versus voltage characteristics of the photovoltaic device under zero illumination (black curve) and standard illumination when flat (red curve) and released to flat after bent to a curvature of 105° (green curve). Inset: photograph of the Cu_2S–CdS nanocrystal plastic solar cell [62]. (Reproduced with permission. Copyright © 2008, American Chemical Society.) (b) Schematic of the depleted heterojunction colloidal QD solar cell containing a heterojunction formed between TiO_2 and PbS QDs [67]. (Reproduced with permission. Copyright © 2012, Macmillan Publishers.)

performance of PbSe and PbS colloidal QD solar cells has been achieved by optimizing both device structures and nanocrystal surface ligands [64–68]. A power conversion efficiency up to 8.5% has been recently demonstrated on cells with the depleted TiO_2/PbS QD heterojunction structure (Figure 17.6b) after combining a donor-supply electrode strategy and the use of hybrid ligands for QD surface passivation [69]. Tandem collidal quantum dot (CQD) solar cells were also realized based on two batches of PbS QDs, with visible-absorbing QDs for the front cell and near-IR-absorbing QDs for the back cell [64]. While most of these academic demonstrations were done on rigid substrates, the technology of colloidal solar cells is generally considered to be compatible with flexible substrates and the roll-to-roll fabrication process [70–72]. Toward flexible applications, graphene has been suggested as a promising alternative electrode to replace ITO [73].

In summary of this section, polycrystalline or nanocrystal thin films of various compositions can be deposited directly from nanoparticle or precursor solutions onto flexible substrates for functional solar cells. Despite being so far less performing compared to those solar cells deposited through vacuum processes described in the previous section, solution-processed solar cells offer additional cost advantages in fabrication. Promising PV performance has been achieved in solution-processed flexible CIGS and CZTS thin-film solar cells, with 17.1% power conversion efficiency in the CIGS system, which is very close to the efficiency obtained in co-evaporated CIGS (18.7%). On the other hand, being compatible with flexible applications, colloidal nanocrystal QDs have also been widely investigated for solution-processed solar cells. Representative colloidal examples were demonstrated using VI–VI QDs, achieving 8.5% power conversion efficiency. Despite the modest PV performance, which can be potentially improved through further tuning the device structure and nanocrystal surface chemistry, colloidal nanocrystals offer additional opportunities because of their absorption-tunable characteristics and the existence of a new physical mechanism (MEG process) which can further boost their PV performance.

17.5 Summary

Cost reduction and the development of nonconventional solar cell products meeting the need to market diversification are important motivations for flexible solar cells. To be compatible with flexible substrates, which enable the cost-effective roll-to-roll fabrication, one of the most important criteria is the thickness of the PV active layer. When the total thickness is only a few micrometers, crystalline and multicrystalline materials as well as nanocrystal films can all allow some degree of flexibility. So far, the PV performances of many materials on flexible substrates have been somewhat inferior to those on conventional rigid substrates. This is mainly due to various issues, such as inefficient absorption, the low deposition temperatures allowed, and the special device structures required, all associated with the use of flexible substrates. However, it has been shown that such issues can be partially or completely resolved after taking proper engineering measures on the material deposition and device configurations. Depending on the fabrication methods, flexible inorganic solar cells based on different material systems have been described in this chapter, together with the particular engineering measures necessary to boost their performance on flexible substrates. These different systems include crystalline examples transferred onto substrates, thin-film examples deposited directly onto substrates through vapor deposition, and solution-processed thin film examples. Notably, among these systems, there are cases (such as the CIGS flexible solar cells) where the PV performance achieved on flexible substrates is already very close to that achieved on conventional rigid substrates, suggesting promising perspectives for this field of flexible inorganic solar cells.

References

1. Chapin, D.M., Fuller, C.S., and Pearson, G.L. (1954) *J. Appl. Phys.*, **25**, 676.
2. Green, M.A., Emery, K., Hishikawa, Y., Warta, W., and Dunlop, E.D. (2013) *Prog. Photovoltaics Res. Appl.*, **21**, 827.
3. Schockley, W. and Queisser, H.J. (1961) *J. Appl. Phys.*, **32**, 510.
4. Goodrich, A.C., Powell, D.M., James, T.L., Woodhouse, M., and Buonassisi, T. (2013) *Energy Environ. Sci.*, **6**, 2811.
5. Gregg, A., Blieden, R., Chang, A., and Ng, H. (2005) Conference Record of the Thirty-first IEEE Photovoltaic Specialists Conference 2005, p. 1615.
6. Clarke, T.M. and Durrant, J.R. (2010) *Chem. Rev.*, **110**, 6736.
7. Zhou, H., Yang, L., and You, W. (2012) *Macromolecules*, **45**, 607.
8. Li, G., Zhu, R., and Yang, Y. (2012) *Nat. Photonics*, **6**, 153.
9. Ameri, T., Khoram, P., Min, J., and Brabec, C.J. (2013) *Adv. Mater.*, **25**, 4245.
10. Yeh, N. and Yeh, P. (2013) *Renew. Sustain. Energy Rev.*, **21**, 421.

11. Park, N.-G. (2013) *J. Phys. Chem. Lett.*, **4**, 2423.
12. Delcamp, J.H., Yella, A., Holcombe, T.W., Nazeeruddin, M.K., and Grätzel, M. (2013) *Angew. Chem. Int. Ed.*, **52**, 376.
13. Kamat, P.V. (2008) *J. Phys. Chem. C*, **112**, 18737.
14. Kamat, P.V., Tvrdy, K., Baker, D.R., and Radich, J.G. (2010) *Chem. Rev.*, **110**, 6664.
15. Yoon, J., Baca, A.J., Park, S.-I., Elvikis, P., Geddes, J.B.I., Li, L., Kim, R.H., Xiao, J., Wang, S., Kim, T.-H., Motala, M.J., Ahn, B.Y., Duoss, E.B., Lewis, J.A., Nuzzo, R.G., Ferreira, P.M., Huang, Y., Rockett, A., and Rogers, J.A. (2008) *Nat. Mater.*, **7**, 907.
16. Blakers, A.W. and Armour, T. (2009) *Sol. Energy Mater. Sol. Cells*, **93**, 1440.
17. Stocks, M.J., Weber, K.J., Blakers, A.W., Babaei, I., Everett, V., Neuendod, A., Kerr, M.J., and Verlinden, P.J. (2003) Proceeding of t3rd Wold Conference on Photovoltaic Energy Conversion, Osaka, Japan, May 11–18, 2003, 1P-D3-04, p. 184.
18. Weber, K.J., Blakers, A.W., Deenapanray, P.N.K., Everett, V., and Franklin, E. (2005) Sliver® solar cells. Proceeding of the Thirty-first IEEE Photovoltaic Specialists Conference, Lake Buena Vista, Florida, January 3–7, 2005, pp. 991–994.
19. Tanabe, K., Watanabe, K., and Arakawa, Y. (2012) *Appl. Phys. Lett.*, **100**, 192102.
20. Ichikawa, Y., Tabuchi, K., Kato, S., Takano, A., Sasaki, T., Tanda, M., Saito, S., Sato, H., Fujikake, S., Yoshida, T., and Sakai, H. (1994) A new structure A-Si solar cell with plastic film substrate. Presented at IEEE 1st World Conference on Photovoltaic Energy Conversion, 1994, Conference Record of the Twenty Fourth IEEE Photovoltaic Specialists Conference – 1994, December 5–9, 1994.
21. Guha, S., Yang, J., Banerjee, A., Hoffman, K., Sugiyama, S., Call, J., Jones, S.J., Deng, X., Doehler, J., and Izu, M., and Ovshinsky, H.C. (1997) Triple-junction amorphous silicon alloy PV manufacturing plant of 5 MW annual capacity. Presented at Photovoltaic Specialists Conference, 1997, Conference Record of the Twenty-Sixth IEEE, September 29–3 October, 1997.
22. Guha, S., Yang, J., Nath, P., and Hack, M. (1986) *Appl. Phys. Lett.*, **49**, 218.
23. Yang, J., Banerjee, A., and Guha, S. (1997) *Appl. Phys. Lett.*, **70**, 2975.
24. Gregg, A., Blieden, R., Chang, A., and Ng, H. (2005) Performance Analysis of Large Scale, Amorphous Silicon Photovoltaic Power Systems. Proceeding of the Thirty-first IEEE Photovoltaic Specialists Conference, Lake Buena Vista, Florida, January 3–7, 2005 pp. 1615–1618.
25. Beernink, K., Pietka, G., Noch, J., Wolf, D., Banerjee, A., Yang, J., Guha, S., and Jones, S.J. (2002) High Specific Power Amorphous Silicon Alloy Photovoltaic Modules. Proceeding of the Twenty-Ninth IEEE Photovoltaic Specialists Conference, New Orleans, Louisiana, May 19–24, 2002, pp. 998–1001.
26. van Cleef, M., Lippens, P., and Call, J. (2001) Superior Energy Yields of UNI-SOLAR® Triple Junction Thin Film Silicon Solar Cells compared to Crystalline Silicon Solar Cells under Real Outdoor Conditions in Western Europe. Proceeding of the 17th European Photovoltaic Solar Energy Conference and Exhibition, Munich, Germany, 22–26 October, 2001.
27. Yang, J., Banerjee, A., and Guha, S. (2003) *Sol. Energy Mater. Sol. Cells*, **78**, 597.
28. Jones, R., Johnson, T., Jordan, W., Wagner, S., Yang, J., and Guha, S. (2002) Effects of Mechanical Strain on the Performance of Amorphous Silicon Triple-Junction Solar Cells. Proceeding of the 29th IEEE Photovoltaic Specialists Conference, New Orleans, Louisiana, May 19–24, 2002, pp. 1214–1217.
29. Gleskova, H., Cheng, I.-C., Wagner, S., Sturm, J.C., and Suo, Z. (2006) *Sol. Energy*, **80**, 687.
30. Adirovich, E.I., Yuabov, Y.M., and Yagudaev, G.R. (1969) *Sov. Phys.: Semicond.*, **3**, 61.
31. Aliyu, M.M., Islam, M.A., Hamzah, N.R., Karim, M.R., Matin, M.A., Sopian, K., and Amin, N. (2012) *Int. J. Photoenergy*, **2012**, 1–10, Article ID 351381.

32. Photonics.com *http://www.photonics.com/Article.aspx?AID=47412* (accessed 21 July 2014).
33. Kranz, L., Gretener, C., Perrenoud, J., Schmitt, R., Pianezzi, F., La Mattina, F., Blosch, P., Cheah, E., Chirila, A., Fella, C.M., Hagendorfer, H., Jager, T., Nishiwaki, S., Uhl, A.R., Buecheler, S., and Tiwari, A.N. (2013) *Nat. Commun.*, **4**, 2306.
34. ZSW *http://www.zsw-bw.de/en/support/press-releases/press-detail/zsw-stellt-weltrekord-solarzelle-her.html* (accessed 21 July 2014).
35. Shay, J.L., Wagner, S., and Kasper, H.M. (1975) *Appl. Phys. Lett.*, **27**, 89.
36. Gabor, A.M., Tuttle, J.R., Albin, D.S., Contreras, M.A., Noufi, R., and Hermann, A.M. (1994) *Appl. Phys. Lett.*, **65**, 198.
37. Repins, I., Contreras, M.A., Egaas, B., DeHart, C., Scharf, J., Perkins, C.L., To, B., and Noufi, R. (2008) *Prog. Photovoltaics Res. Appl.*, **16**, 235.
38. Chirila, A., Buecheler, S., Pianezzi, F., Bloesch, P., Gretener, C., Uhl, A.R., Fella, C., Kranz, L., Perrenoud, J., Seyrling, S., Verma, R., Nishiwaki, S., Romanyuk, Y.E., Bilger, G., and Tiwari, A.N. (2011) *Nat. Mater.*, **10**, 857.
39. Ishizuka, S., Yamada, A., Matsubara, K., Fons, P., Sakurai, K., and Niki, S. (2008) *Appl. Phys. Lett.*, **93**, 124105.
40. Yagioka, T. and Nakada, T. (2009) *Appl. Phys. Express*, **2**, 072201.
41. Shi, C.Y., Sun, Y., He, Q., Li, F.Y., and Zhao, J.C. (2009) *Sol. Energy Mater. Sol. Cells*, **93**, 654.
42. Ghyselen, B., Hartmann, J.M., Ernst, T., Aulnette, C., Osternaud, B., Bogumilowicz, Y., Abbadie, A., Besson, P., Rayssac, O., Tiberj, A., Daval, N., Cayrefourq, I., Fournel, F., Moriceau, H., Di Nardo, C., Andrieu, F., Paillard, V., Cabie, M., Vincent, L., Snoeck, E., Cristiano, F., Rocher, A., Ponchet, A., Claverie, A., Boucaud, P., Semeria, M.N., Bensahel, D., Kernevez, B., and Mazure, C. (2004) *Solid-State Electron.*, **48**, 1285.
43. Yu, D., Fidanza, J., Roscheisen, M.R., and Sager, B.M. (2010) Solution–based fabrication of photovoltaic cell. US Patent 7663057 B2, Google Patents.
44. Sager, B.M., Yu, D. and Robinson, M.R. (2007) Coated nanoparticles and quantum dots for solution-based fabrication of photovoltaic cells. US Patent US7, 306, 823 B2.
45. Semiconductor Today*http://www.semiconductor-today.com/news_items/2011/OCT/NANOSOLAR_061011.html* (accessed 21 July 2014).
46. Todorov, T.K., Reuter, K.B., and Mitzi, D.B. (2010) *Adv. Mater.*, **22**, E156.
47. Cao, Y., Denny, M.S.J., Caspar, J.V., Farneth, W.E., Guo, Q., Ionkin, A.S., Johnson, L.K., Lu, M., Malajovich, I., Radu, D., Rosenfeld, H.D., Choudhury, K.R., and Wu, W. (2012) *J. Am. Chem. Soc.*, **134**, 15644.
48. Tian, Q., Xu, X., Han, L., Tang, M., Zou, R., Chen, Z., Yu, M., Yang, J., and Hu, J. (2012) *CrystEngComm*, **14**, 3823.
49. Greenham, N.C., Peng, X.G., and Alivisatos, A.P. (1996) *Phys. Rev. B: Condens. Matter*, **54**, 17628.
50. Ginger, D.S. and Greenham, N.C. (2004) in *Semiconductor and Metal Nanocrystals: Synthesis and Electronic and Optical Properties* (ed. V.I. Klimov), Marcel Dekker, New York, 232–277.
51. Pattantyus-Abraham, A.G., Kramer, I.J., Barkhouse, A.R., Wang, X., Konstantatos, G., Debnath, R., Levina, L., Raabe, I., Nazeeruddin, M.K., Grätzel, M., and Sargent, E.H. (2010) *ACS Nano*, **4**, 3374.
52. Huynh, W.U., Peng, X.G., and Alivisatos, A.P. (1999) *Adv. Mater.*, **11**, 923.
53. Caruge, J.M., Halpert, J.E., Wood, V., Bulovic, V., and Bawendi, M.G. (2008) *Nat. Photonics*, **2**, 247.
54. Schlamp, M.C., Peng, X.G., and Alivisatos, A.P. (1997) *J. Appl. Phys.*, **82**, 5837.
55. Colvin, V.L., Schlamp, M.C., and Alivisatos, A.P. (1994) *Nature*, **370**, 354.
56. Sun, B. and Sirringhaus, H. (2005) *Nano Lett.*, **5**, 2408.
57. Talapin, D.V., Lee, J., Kovalenko, M.V., and Shevchenko, E.V. (2009) *Chem. Rev.*, Articel published online doi: 10.1021/cr900137k.
58. Sun, B. and Greenham, N.C. (2006) *Phys. Chem. Chem. Phys.*, **8**, 3557.

59. Sun, B.Q., Snaith, H.J., Dhoot, A.S., Westenhoff, S., and Greenham, N.C. (2005) *J. Appl. Phys.*, **97**, 6.
60. Gur, I., Fromer, N.A., Chen, C.P., Kanaras, A.G., and Alivisatos, A.P. (2007) *Nano Lett.*, **7**, 409.
61. Klem, E.J.D., Gregory, C.W., Cunningham, G.B., Hall, S., Temple, D.S., and Lewis, J.S. (2012) *Appl. Phys. Lett.*, **100**, 173109.
62. Wu, Y., Wadia, C., Ma, W., Sadtler, B., and Alivisatos, A.P. (2008) *Nano Lett.*, **8**, 2551.
63. Szendrei, K., Gomulya, W., Yarema, M., Heiss, W., and Loi, M.A. (2010) *Appl. Phys. Lett.*, **97**, 203501.
64. Jeong, K.S., Tang, J., Liu, H., Kim, J., Schaefer, A.W., Kemp, K., Levina, L., Wang, X., Hoogland, S., Debnath, R., Brzozowski, L., Sargent, E.H., and Asbury, J.B. (2011) *ACS Nano*, **6**, 89.
65. Kramer, I.J. and Sargent, E.H. (2011) *ACS Nano*, **5**, 8506.
66. Etgar, L., Moehl, T., Gabriel, S., Hickey, S.G., Eychmüller, A., and Grätzel, M. (2012) *ACS Nano*, **6**, 3092.
67. Ip, A.H., Thon, S.M., Hoogland, S., Voznyy, O., Zhitomirsky, D., Debnath, R., Levina, L., Rollny, L.R., Carey, G.H., Fischer, A., Kemp, K.W., Kramer, I.J., Ning, Z., Labelle, A.J., Chou, K.W., Amassian, A., and Sargent, E.H. (2012) *Nat. Nanotechnol.*, **7**, 577.
68. Sargent, E.H. (2012) *Nat. Photonics*, **6**, 133.
69. Maraghechi, P., Labelle, A.J., Kirmani, A.R., Lan, X., Adachi, M.M., Thon, S.M., Hoogland, S., Lee, A., Ning, Z., Fischer, A., Amassian, A., and Sargent, E.H. (2013) *ACS Nano*, **7**, 6111.
70. MISFITS' ARCHITECTURE The Rise of Solar Cell Efficiency. *http://misfitsarchitecture.com/tag/quantum-dot/.*
71. Massachusetts Institute of Technology *http://nanocluster.mit.edu/research.php* (accessed 25 July 2014).
72. OCRF *http://ocrf.kaust.edu.sa/Spotlight%20Stories/Pages/Nanoscale-Semiconductors-For-Next-Generation-Solar-Cells.aspx* (accessed 25 July 2014).
73. Park, H., Chang, S., Jean, J., Cheng, J.J., Araujo, P.T., Wang, M., Bawendi, M.G., Dresselhaus, M.S., Bulovic, V., Kong, J., and Gradecak, S. (2012) *Nano Lett.*, **13**, 233.

18
Scalable and Flexible Bioelectronics and Its Applications to Medicine

Salvatore Iannotta, Pasquale D'Angelo, Agostino Romeo, and Giuseppe Tarabella

18.1
Biosensing and Bioelectronics: A Fast Growing Field and a Challenging Research Area

Biosensing is one of the fastest growing fields motivated by the great promises impacting on health, well-being, food, energy, and society. This field has received a strong boost by the innovation introduced by nanoscience and technologies, but it faces a series of challenges that call for increased research and basic technological activities. The basic essential functionalities of any biosensing element or system are two: on one end, the proper selective affinity to the biorelevant components that need to be detected, traced, or monitored; on the other, the transduction, that is the physical mechanism of transforming in a "readable" and "storable" signal of each event. From this point of view, one of the most successful approaches is optical transduction, which takes advantage of photonics and optoelectronics. Optical methods can achieve fantastic records in term of sensitivity and selectivity, in particular by exploiting laser and spectroscopic methods. On the other end, direct electronic transduction is a very promising and potentially disruptive technology because it can be less cumbersome and inexpensive and forms a direct interface to all the great achievements of electronics-based technology, computing, and so on. The great potential of this approach is being strongly reinforced by the ability to work with flexible electronics, possibly fabricated on large areas. In order to sense biocomponents, such as biomarkers, biomolecules, proteins, and DNA, or even to achieve high specificity toward a target analyte, biological recognition elements should be immobilized on a surface or should be addressable somehow to produce an electrical signal induced by their interaction in a transducing element. In most cases, sensing requires that biocompatible and/or active materials (e.g., tissue, cell receptors, enzymes, antibodies, aptamers, and petites) as well as artificial biomimetic structures (e.g., synthetic receptors, imprinted polymers, and biomolecules) are used to obtain a surface both sensitive and selective to the target components.

One of the most relevant trends in current research deals with the development of strategies oriented toward human health and well-being, addressing some basic questions such as early diagnoses of diseases and novel (possibly personalized)

Large Area and Flexible Electronics, First Edition. Edited by Mario Caironi and Yong-Young Noh.

therapies up to the point of envisioning the challenging perspective of theragnostics, that is, developing devices, systems, and approaches where at the same time the detection and monitoring of a disease via biomolecular markers could produce automatically the needed release of drugs locally where needed or where the effect is maximized [1]. To this end, great support is given by the strong interdisciplinary character of materials science and related innovative technology. They are increasingly providing the needed combination of low cost, easy-to-use tools for diagnostics and point-of-care medicine, and treatment combined with viable industrial production (but, in broader terms, also many other mass devices and fields of interest). In turn, there is a worldwide increase in funding opportunities for research and innovation in bio-based technologies, as is clearly evident in the EU – Horizon 2020 calls as well as the U.S. NIH programs. Available literature provides a concrete marker of this collective effort. In this respect, the trend in published items per year in the field of bioelectronics indicates that, during the last year, the number of publications has doubled with respect to the previous 5 years [2]. In addition, the number of published papers dealing with biomedicine during the last 5 years underlines even better how topics focused on biology-based technologies are attracting more and more researchers' interest. We believe that the most promising developments in bioelectronics and biomedicine have been achieved by the many studies dealing with novel, fascinating approaches aimed at understanding the interaction of cells and biomolecules with electronic devices, allowing the study and monitoring in real time of biology-driven phenomena [3, 4].

The converging efforts of a large multidisciplinary community are making it more and more viable to promote devices and systems for biomedical applications. In such a context, the development of flexible, and possibly implantable, electronics is becoming increasingly reliable with solutions for bioelectric sensing, physiological monitoring, bioimaging, prosthetics, and drug delivery, while paving the way toward theragnostics. The great evolution of high-performance electronics in the last decades has mostly benefitted from the robust properties of hard crystalline materials. There is, therefore, a big mismatch with the mechanical and geometrical properties, namely flexibility and conformability/stretchability typical of bio- and living systems. Indeed, classic single-crystalline inorganic materials used for applications ranging from basic measurement of electrophysiological signals to sensing of biomolecules or bio-relevant systems, to delivery of advanced therapies, and to the establishment of human–machine interfaces are fundamentally rigid and planar, making integration with biological tissues poorly intimate and conformal. This rigidity leads to incomplete and ineffective contact of the devices with soft, convoluted biological tissues. Furthermore, *in vivo* studies have shown that this large mechanical mismatch can result in micromotion (relative movement between electrodes and tissues), which can contribute to the aggravation of tissue inflammation and/or loss of contact with the specific neurons or cells or nerves. Therefore, the ideal biointerfaced electronics should be flexible enough to be conformable with tissues and reduce micromotion-induced inflammation reactions.

On the other end, the branches of modern interdisciplinary science, focused on (multi)functional materials and devices, give rise to even faster developments, where flexible, large-area electronics become more and more technologically viable. This, in turn, is giving a very important boost to the above-mentioned applications because cost reduction and reliability represent the needed key feedbacks. In this framework, amazing technological breakthroughs have been achieved by developing new materials or by introducing improved technological solutions. In particular, silicon- and organic-based devices represent two interesting and complementary classes of technologies for flexible and biointegrated devices. Because of their suitability for efficient interfacing of biosystems, yet providing the necessary capabilities of flexibility and/or conformability, they will be treated in greater detail in this chapter.

Silicon and inorganic semiconductors are the key elements for the development of robust, reliable, and affordable circuits, possibly made of continuous films, suitable for high integration density. Silicon-based flexible and semitransparent circuits that still retain silicon's excellent electronic properties and show the potential to expand their use for both digital and analog devices would be desirable for the implementation of bioinspired applications and for future developments. This is because silicon allows, simultaneously, the considerable performance required by the controlling electronic tools, the functional properties of biomedical or biosensing devices (at least to a large extent by the proper functionalization processes), and the low-cost manufacturing processes, avoiding vacuum-based techniques and high processing temperatures. Currently, major achievements have been obtained in the field of silicon-based flexible electronics. Building-integrated photovoltaics (BIPVs) based on thin films made of inorganic semiconducting materials and compounds (amorphous silicon (a-Si) [5], cadmium telluride (CdTe), and copper indium gallium selenide (CIGS) [6]) have given rise to large-area, flexible, efficient solar cells that can realize the concept of a "smart city." Electronic paper displays are now more than simple laboratory prototypes [7]. Focusing more on the subject of the present chapter, low-dimensional silicon nanostructures have found use in a wide variety of bioapplications, such as ultrasensitive biomolecular detection [8] or fluorescent cellular imaging [9], thanks to their electromechanical properties, their high surface-to-volume ratio, their improved multifunctionality, and, especially, their compatibility with silicon-based technology. Flexible applications based on thin-film technology range from conventional electronic components and related assemblies, such as thin-film transistors (TFTs) [10] and logic gates [11], to bioapplications such as microelectromechanical systems (MEMS)s [12, 13] and sensor arrays [14]. Silicon-based technologies represents the fundamental basis for the very valuable developments in the field of microelectrode arrays (MEAs), which are contributing very strongly to the study, understanding, and applications of neurosciences [15].

As already mentioned, a key issue in interfacing semiconducting devices with living beings is the reliability in the operating conditions, in particular whenever the electronic devices must work in direct contact with biological liquids and tissues. Importantly, the development of strategies aimed at interfacing conventional semiconducting devices with living beings is strongly required as a preliminary step toward the implantation within the human body of electronic devices with important retention capability of their electronic properties upon bending [16]. In this respect, the silicon–cell junction developed by Fromherz [17] constitutes the first valid example of a direct coupling between living cells and electronic devices, essentially exploiting the properties of the silicon oxide layer that give a good (capacitive) electrical coupling and a good playground for the biofunctionalization processes.

Beyond the Si-based technology approach, which is by far the most reliable and mature available technology, it should be recognized that organic electronics-based devices have been increasingly successful in interfacing with biosystems. Organic electronic materials, based on π-conjugated organic small molecules or polymers, combine the advantageous characteristics of conventional polymers, such as low weight, easy processability, flexibility, and high conformability, with the functional physical properties of conventional semiconductors, such as absorption and emission of light and tunable conductivity.

In the last decades, organic semiconductors (OSCs) have been exploited in several fields, including organic light-emitting diodes (OLEDs) [18], organic photovoltaics (OPVs) [19], and organic thin-film transistors (OTFTs) [20]. Their low processing temperature is compatible with flexible plastic substrates, and their low elastic modulus allows bending. Deposition techniques involving, for instance, solution processing, spin coating, or roll-to-roll technology, are very simple and cost-effective techniques, and have paved the way to printable production schemes that are well suited to large-area and flexible applications. Another decisive advantage is the wide range of tunability of their properties by chemical/physical tailoring. This is of primary importance for sensors, which require specific chemical and physical properties in order to attach covalently to the recognition elements. The ability to detect a wide set of bioanalytes is intimately connected with the strong and controlled reactivity of the organic transducing materials in the presence of these species. Such reactivity arises from the porosity and permeability of organics, which result from the weak van der Waals interactions governing their solid-state structure. This reactivity, connected to their dual conducting character (ionic and electronic), is capable of inducing both structural and morphological changes. In other words, these changes can be transduced as changes of physical (e.g., doping or dedoping, electrochromism, or light emission) and mechanical properties (swelling), so that low-cost sensors and actuators, respectively, can be reasonably implemented on an industrial scale. In this respect, IDTechEx projections for the growth of the global market for printed sensors in 2020 indicate an increase by more than \$1 billion [21].

Even though organic devices will probably never compete with inorganic ones in terms of performance, robustness, and stability, their use is conceivable in applications that can withstand lower charge-carrier mobility. Thus, because of their characteristics, organic electronic materials are meant to be complementary to traditional silicon electronics and are to be considered for extending the functional abilities (in particular biofunctionalities) of more traditional electronics.

Within the fast growing and exciting field of carbon-based electronic materials, graphene is increasingly becoming a key player and being extensively investigated in bioelectronics and biomedical applications. Many examples have been published where graphene is playing a key role in biological sensors, drug delivery, tissue engineering, and others [22]. The bidimensional structure of graphene, its electronic surface sensitivity, and its optoelectronic properties are considered ideal for it to be integrated into biological systems for bioelectronic devices with flexible and/or stretchable properties. Flexible graphene-based devices could find interesting applications in many different areas, for example, in healthcare, food safety testing, or environmental monitoring. A good indication of the great expectation on graphene comes from the funding devoted to graphene research (the so-called Graphene Flagship) of the EU community, allocating up to €1 billion in 2012 [23], even though biomedical applications do not play a particularly relevant part yet. As is well known, the first stable graphene single layer was realized by Novoselov 10 years ago [24], and since then much work has been done, as proven by the number of papers published on graphene: many comprehensive reviews are present in the literature, which the readers can profitably refer to [22]. Recently, the possibility to integrate graphene-based TFTs on large-area and flexible plastics or stretchable rubber has been demonstrated [25–27].

Among the several organic-based devices developed so far, OTFTs are attracting a great deal of interest for sensing because of their simple electrical readout, high sensitivity, inherent signal amplification, miniaturization, and facile incorporation into arrays and circuits. To date, OTFTs have been intensively investigated for applications in flexible active-matrix displays [28–31], radio-frequency identification (RFID) tags [32, 33], and as sensors of mechanical deformation and pressure, humidity, organic vapors, pH, and ion concentrations, and a variety of biologically relevant analytes [34, 35].

OTFTs are three-terminal electronic devices consisting of a thin organic semiconducting layer, an insulating (dielectric) layer, and three electrodes – the source, drain, and gate. The active OSC layer can be made of oligomers, polymers, or organic nanostructured materials that are deposited as films (few tens of nanometers thick) by solution casting, spin coating, sublimation, Supersonic Molecular Beam Epitaxy (SuMBE), or printing techniques. Source and drain metallic electrodes, in contact with the OSC, are used to apply a source–drain voltage (V_{ds}) and measure the source–drain current (I_{ds}) that flows through the OSC, whereas a third electrode, the gate, is used to modulate the magnitude of I_{ds}. A transistor-based sensor usually achieves high sensitivity by exploiting the combination of sensing with the amplification intrinsically provided by the transistor, where, for example, a small change of the effective gate voltage induced by an analyte may lead to a pronounced variation of the channel current.

OTFTs can be classified into two primary categories: organic field-effect transistors (OFETs), based on OSCs, and organic electrochemical transistors (OECTs), based mainly on electrically conducting polymers. In the former, the working mechanism is based on a field-effect charge modulation at the interface between the OSC and the dielectric gate, while in the latter an electrochemical doping/de-doping takes place throughout the whole depth of the conducting polymer.

The most common architecture used in OTFT-based biosensors relies on the presence of an electrolyte in which the biological analytes or systems (proteins, antibodies, enzymes, DNA, cells, etc.) are dissolved [36]. This category includes both Electrolyte-Gated Organic Field Effect Transistors (EGOFETs) and OECTs, which will be described in detail in the next sections.

EGOFETs represent ideal candidates for a new biosensor generation, being characterized by bio- and mechanical compatibility, chemical stability, specific biorecognition, signal transduction efficiency, and topography [37]. In fact, these features are ideal to achieve optimal integration with a biological environment and to develop broad-based biosensors implementable on flexible technology. In addition, such devices are inexpensive, portable, and highly selective. They show ultrasensitive detection levels, besides having label-free detection capabilities, and can be easily integrated into a circuit. Hence, EGOFETs are the perfect candidates for smaller, faster, and cheaper devices designed to replace currently available time-consuming laboratory analyses, as well as to develop point-of-care diagnostic tools.

As already mentioned, this chapter reports and discusses the main features and recent achievements of flexible bioinspired devices and biosensors for biomedical applications, while identifying the major perspectives and the factors that still limit their widespread use. First of all, in Section 18.2, we will describe flexible electronics based on silicon and related bioinspired devices. Then, two classes of EGOFETs suitable for biosensing applications will be described, in particular EGOFETs in Section 18.3 and OECTs in Section 18.4.

18.2 Inorganic and Silicon-Based Flexible Electronics for Biosensing Devices

The peremptory impact of silicon on our daily lives is tangible in most of the operations we execute every day. In fact, it is well known that Si-based technologies are at the very basis of electronics and, broadly speaking, of the technological progress of our era. Silicon is presently viewed as the most common conventional reference semiconductor, even though, from the point of view of this chapter, a wide range of other semiconducting materials and alloys are used in electronic applications, in particular when multifunctional operations are needed. The target range of Si-based electronics covers in particular all those applications requiring mass production of high-speed semiconductor devices and related circuit assemblies whenever the biofunctionalities are not too stringent. Indeed, approaches aimed at exploring and studying alternative materials have given rise to novel and very promising perspectives, such as those of flexible organic electronics, particularly suitable for applications requiring specific complementary applications where the tailored functions and responses dominate over the electronic performance. Conversely, the limitations of organics, such as relatively poor robustness and electrical characteristics, in general do not allow embracing a generalized wide field of action, in particular when the operating conditions require endurance and a high level of reproducibility. Their application is more suitable to disposable devices and systems, in biocompatible/biofunctional environments, and when multifunctional properties should be simultaneously tailored, in particular, when bending, flexibility, and softness are key issues. Such features, in particular when several of them are required simultaneously, are by far much better guaranteed by compatible plastic substrates and by using π-conjugated OSCs.

Material processing aimed at manufacturing thin films can meet the requirements for combining the peculiar features of both classes of semiconducting materials. Beyond the combination of organics and inorganics in forming hybrid materials showing stability and conduction properties of inorganics and mechanical features of organics, solution processability appears to be the common way to be pursued by the conventional and alternative semiconducting materials industries.

In the light of the above considerations, an overview on the materials and protocols for flexible device manufacturing will be given in Section 18.2.1, prior to discussing the features of hybrid interfaces comprising inorganic semiconducting devices and living being and their perspective in bioinspired applications (Section 18.2.2).

18.2.1
Inorganic Semiconductors for Flexible Electronics: From Hybrids and Inorganic Semiconducting Composites to Silicon

The basic questions of producing highly performing flexible electronic materials, fabricated and processed on large areas, are thoroughly discussed in other chapters of this book. The growth/preparation of the materials, possibly in a thin-film form, is a key question that becomes particularly difficult when large-scale industrial production is required. A wet chemical approach is often the most suitable. On the other hand, covalently bonded atoms and molecules give rise, in most cases, to poorly soluble materials and compounds because of the strong force holding their structure together. For instance, the poor solubility in water, which is a polar solvent, can be attributed to the nonpolar character of covalent compounds, even though "like dissolves like" is not a general rule. It is more appropriate to consider that, beyond the interatomic and intermolecular forces sustaining the solid-state structure, networked solids such as nonionic crystalline materials can be viewed as an extended macromolecule, so their solvation is quite improbable. This feature limits largely the possibility of obtaining high-quality films from solutions. Nevertheless, many approaches offering higher mobility materials are available to produce films using inorganic semiconductors. Bath techniques, spin coating and printing, and processes based on relatively low temperatures, have proven suitable for preparing a wide class of materials, fully or partially based on inorganics, with good electrical performance. Several materials and compounds have been prepared, characterized, and finally implemented into device structures. In most cases, these materials belong to three main classes: layered organic–inorganic (OI) hybrids, metal chalcogenides, and metal oxides.

OI hybrids have gained a growing interest especially in the field of energy harvesting [38], even though the literature on sensing devices reports about novel approaches for biological and chemical detection carried out via functional hybrid materials and innovative device structures [39]. Metal chalcogenides and oxides in the form of quantum dots are used to tag biomolecules in ultrasensitive biological detection [40] upon surface functionalization aimed at conferring biocompatibility. Nanostructures can be used to encapsulate and release different drugs and biomolecules [41], while their surface can be functionalized also for drug targeting or biolabeling [42]. Thin films made of metal oxides have been used for manufacturing ion-sensitive field-effect transistors (ISFETs) based on Indium-Gallium-Zinc Oxide (IGZO), to be used as a pH-sensitive biosensor [43], while sputtered microporous Zinc Oxide (ZnO) has been used as the active element for a cholesterol biosensor [44].

OI hybrids combine the desirable properties from both their components. In particular, the organic component drives the self-assembly of the inorganic counterpart on a substrate starting from a solution precursor, while the inorganic component assists the charge transport to yield high mobility, in some cases comparable to that of silicon [45]. Among OI hybrids, the single-layer ($n \sim 1$) <100>-oriented layered perovskites are the simplest. These compounds contain single MX42 layers, where M is a divalent metal (e.g., Fe^{2+}, Cu^{2+}, and Mn^{2+}) and X is a halide (e.g., Cl^-, Br^-, I^-). The hybrid approach allows the development of layered compounds in which the overall properties can be engineered by acting on both the organic layer and the inorganic component, with the aim of enhancing their electronic properties and flexibility, respectively. Currently, perovskites are widely used in high-performance solar cell technology [46].

The class of chalcogenides has been studied for 50 years, and the first TFT, demonstrated by Weimer in 1962 [47], was based on CdS polycrystalline film. Even though metal chalcogenides have had only a minor role compared to silicon-based electronics, extensive studies of spin-coated nanoparticle (NP)-based films and nanowires (NWs) have focused attention on these materials for large-area electronics. The covalent nature of metal chalcogenides guarantees high n-type and p-type mobility, exceeding $100\,cm^2\,V^{-1}\,s^{-1}$ [48], but limits at the same time their solubility. Many efforts have focused on finding approaches alternative to vacuum-based techniques, such as chemical vapor deposition (CVD). Methods based on the decomposition of soluble precursors have enabled the deposition of metal chalcogenide thin films exhibiting a charge-carrier mobility exceeding $10\,cm^2\,V^{-1}\,s^{-1}$ [49]. Alternative solution-based deposition processes are galvanic deposition, chemical bath deposition, and its modified version, the successive ionic layer adsorption and reaction (SILAR) [50]. SILAR is generally used for semiconductor and ceramic NP film manufacturing [51], while galvanic deposition is used to deposit chalcogenides [52] and semiconducting oxides [53]. Galvanic deposition is a bath-based technique, wherein the driving force for the thin-film deposition is established by the half-cell potential between a suitable anode and the cathode (typically constituting the substrate). SILAR, indeed, involves the alternate immersion of a substrate into two precursor solution baths, cationic and anionic, in order to build up the film in a layer-by-layer approach. The main drawback of the bath technique is the slow growth rate, which makes its large-scale industrial production unthinkable. Finally, solution-processed metal oxides are grabbing the spotlight at OSCs because of their better electrical (mobility) and structural (stability) properties. Among the inorganic semiconductors, In- and Sn-doped ZnO-based thin films are the best candidates for flexible electronics, because of the low post-deposition processing temperature compatible with flexible plastic substrates. In oxide semiconductors, the amorphous phases are capable of exhibiting electron mobilities comparable to those of their crystalline counterparts. This allows avoiding the high-temperature post-deposition annealing aimed at enhancing the crystalline phase, which is typically required by other inorganic semiconductors. Several reports indicate methods for efficient low-temperature material preparation processes based on a precursor solution, such as NP dispersion or the sol–gel technique. Among these precursors, β-diketonates [54], metal halides [55], and metal carboxylates are frequently used for the purpose [56]. Solution-processed oxide semiconductors produced in this way have shown mobility values depending on the annealing procedure. A mobility higher than $10\,cm^2\,V^{-1}\,s^{-1}$ has been assessed in the case of In_2O_3 (if thermally annealed at 200 °C) [57]. A lower mobility ($1.3\,cm^2\,V^{-1}\,s^{-1}$) has been observed for ZnO processed at 140 °C using microwave-assisted annealing [58]. Even though these solutions-processable materials lead to devices characterized by promising performances, silicon always remains the reference material for electronics.

From the discussion so far, it is clear that there is a broad variety of solution-processable materials able to replace silicon in electronic application, but it remains a fact that solution processability of silicon would represent the optimum for integrating the electronic properties of this key semiconductor with the large-area, flexible supporting templates for electronic circuits (plastic, paper, or backplanes). Single-crystal silicon in the form of thin films counter their excellent electronic properties with robust rigidity. Film thickness in the case of polycrystalline silicon films could be a key factor in terms of flexibility. However, the integration of polycrystalline films onto flexible substrates implies an enhancement of the already copious grain boundaries upon bending of the flexible support. Grain boundaries are known to affect the stability of the electronic properties of thin films. In particular, the breakdown of long-range order induced by these defects in polycrystalline films implies a decrease of electrical (and thermal) conductivity. Strategies aimed at enlarging grain boundaries are mainly based on heat treatments, which are incompatible with the use of plastic substrates [59]. On the other hand, amorphous silicon (a-Si) lends itself to the manufacturing of flexible thin films, but the lack of long-range crystalline order results in very low charge carrier mobility (currently, overcome by state-of-the-art semiconducting solution-processable organics). Again, thin-film processing via heat treatments targeted at the enhancement of conductivity can damage the plastic substrate. An intuitive approach for the low-temperature deposition process of silicon on flexible substrates is based on the manufacture of microstructured silicon (μs-Si), which can be deposited using dry transfer or solution-based techniques.

The μs-Si manufacturing procedure is generally based on the anisotropic wet etching of some exposed regions of the SiO_2 wafer. The etching allows defining some micro/nano silicon structures that can be isolated from the substrate by using a further etching step, typically a Hydrofluoric Acid (HF) etching. The resulting micro/nanoscale objects made of single-crystal silicon can be deposited by the spin-coating technique, but also using a printing technique, in order to produce high-performance TFTs with a mobility of about $180\,cm^2\,V^{-1}\,s^{-1}$ [60]. Solution-processable silicon for thin-film deposition via liquid precursors and post-deposition annealing allows implementing high-performance TFTs applications, too. This method appears to be more versatile and suitable for large-scale production. Versatility is ascribable to the possibility of converting a liquid precursor such as the hydrogenated polysilanes, $-(SiH_2)_n-$, in a-Si and polycrystalline Si films using a thermal annealing after spin coating. Shimoda and coworkers [61] have reported on TFTs based on polycrystalline Si films showing a mobility of $108\,cm^2\,V^{-1}\,s^{-1}$ obtained upon a further heating process combined with a XeCl excimer laser irradiation of a-Si thin films. Annealing, then, serves both to convert films made of silane-based precursors and to restore the functional properties of solution-deposited Si films from the NP dispersion. Laser annealing [62] for large-area applications is expensive compared to other methods such as the microwave annealing [63]. Moreover, techniques requiring a post-annealing procedure are still unusable to convert silicon on plastic substrates, so some modification of the liquid precursor would be necessary to facilitate the proper integration at low temperatures of semiconducting films with plastic supports. Alternatively, n-type doped silicon inks can be used to implement thin-film technology within an industrial-scale process. So far, inks based on silicon NPs have been extensively used to fine-tune the architecture of solar cells [64] and to enhance the UV adsorption [65].

The manufacturing step of flexible inorganic-based electronic devices, device arrays, integrated circuits (ICs), and printed circuits boards (PCBs) relies on two main strategies. Integration can be carried out effectively at the material level or at the device component level. In the former case, devices are fabricated directly onto flexible substrates by means of techniques completely compatible with plastic substrates, using for this purpose techniques based on solution-processable materials such as dip/spin coating, already largely used in organic electronics, the industrial printing procedure, and, generally, low temperature-based techniques. The integration at the material level of foldable silicon complementary metal–oxide–semiconductor (Si-CMOS) ICs calls for the device integration step and successive methods for connecting the electrical components. For instance, Kim *et al.* [66] have reported the integration of aligned arrays of silicon nanoribbons onto ultrathin and elastomeric substrates with "wavy" structural layouts. They describe a procedure based on a transfer printing process by means of a poly(dimethylsiloxane) (PDMS) stamp in order to deliver arrays of n-doped and p-doped Si nanoribbons on a silicon substrate covered with a polymethylmethacrylate (PMMA)/polyimide (PI) double layer. Free-standing sheets and foldable substrates with a wavy layout are obtained by separating the PI layer from the rigid support by dissolving the PMMA layers in acetone.

The integration at the component level has been extensively studied following different approaches. Components have been glued using conductive epoxies [67], and electric contacts and interconnections have been patterned within an industrial-like approach (roll to roll) based on optimized inks made of metals and metal oxides [68]. Among these conventional approaches, self-assembly has been proposed as an alternative to the serial robotic assembly via pick-and-place method for the building of complex systems made of mechanical and electrical components. In fact, even though robotic assembly allows integrating 26 000 components per hour, the process becomes extremely slow and expensive if the component sizes are smaller than 300 μm [69]. The method uses the suspension of microcomponents in a liquid medium in order to actualize two-dimensional (2D) integration onto the desired flexible templates by controlling the movement of components via fluid flow and gravity forces. The control of both recognition and binding capability is a key factor for 2D electrical device integration. This control can be also identified with the ability to mimic 3D natural processes, since Nature exploits self-assembly in most of its forms. Whitesides' group reported the mimicking of the crystallization process for the implementation of symmetric electrical network in order to form interconnections between electronic devices [70]. The same group has also reported the biomimetic self-assembly of electronic devices in an asymmetric structural configuration, like that of microprocessors. The process mimics the folding of a polypeptide chain into a globular protein for the generation of an asymmetric multidomain electronic device, starting from linear chains of structures useful to implement microelectronic devices. Several pioneering works have focused on the optimal choice of the driving force for correct and efficient integration, so that surface energy [71], electrostatic [72] and electromagnetic forces [73], gravity [74], fluidic [75], or capillary forces [70] have been used for the heterogeneous assembly of electronic elements onto plastic substrates. The work by Stauth and Parviz [76] promotes the large-area integration of silicon-based microcomponents by a single-step self-assembly. The single-step process obviates the need for post-deposition processes.

The biomimetic character of electronic device self-assembly onto suitable substrates makes this technique interesting for bioinspired applications, whereas the production method of electronic devices and foldable substrates adapts itself well to the biological complexity of living systems. Hence, self-assembly paves the way toward a driven integration of flexible electronic systems with a specific functionality into living beings, without recourse to invasive graft techniques for implants, minimizing at the same time the rejection probability by the hosting organism.

Printing techniques have, finally, been demonstrated to be suitable for the development of an all-printed technology, above all in the case of NW-based devices. As a result, arrays made of semiconducting NW-based devices have been fabricated onto flexible substrates. Contact printing, in particular, enables the transfer of NWs from the growth donor substrate to the desired support, so that NW systems are chemically bound (as a result of van der Waals interactions between the donor surface and the host substrate) and aligned (using shear force) in a controllable way. Yerushalmi *et al.* [77] have demonstrated the growth of Si NWs directly onto cylindrical substrates acting as rolls in an Roll-to-Roll (R2R) process. The technique has been used by Fan *et al.* [78] to deposit CdTe, but also for aligning Ge NWs onto Si wafers functionalized by few monolayers of siloxane, or by poly-L-lysine, in order to enhance the chemical interaction between the substrate and the NWs.

18.2.2 Bioapplications: From Cell–Silicon Junctions Toward Neuroprosthesis and Neuromedicine

The aim of interfacing and integrating electronic devices and circuits into living beings represents the basic requirement for the development of hybrid neuro-electronic systems capable of carrying out different operations, with broad perspectives of application in several branches such as physiology, molecular biology, medicine, information technology, and, most importantly, bionics. The most critical technological questions concern the fabrication of a proper and functional electrical interface between semiconductor devices and living cells. There are a variety of strategies that essentially consist in efficient, robust, and reproducible electrical coupling to exchange signals between cells and electronic devices, and vice versa, so that the devices can record the cells' activity and control or mimic cellular processes, respectively. Even though the efforts developed by a large number of labs have produced working solutions, they all suffer from numerous limitations, which makes it difficult to apply them in a generalized way. In particular, the main difficulties arise from prosthetics, where reliable long-term electronic interfaces are strongly needed to produce fully functional systems and interfaces operated directly by signals and feedbacks directly generated by the human nervous system. The major difficulty arises from the very hostile environment represented by the human body and its reaction to an external material or body. In fact, the contact with living tissues, organs, and, in particular, blood represents one of the most hostile situations in which the survival of electrical functionalities of material and devices is still to be proven on the timescale desired by such applications. In his book, Saddow [79] has discussed in detail such problems by comparing different electronic materials from this point of view and has proposed SiC as a potentially viable technology thanks to its biofunctional properties and its robustness. On a broader perspective, some important limitations originate from the mismatching in terms of transport mechanisms taking place at the device/living being interface, because of the different nature of charge transport in semiconductor devices (where the carriers are electrons and holes) and in cells (where the key players are ions and ion channels in the membrane that regulate, monitor, and control all the basic cellular functions).

Even though such basic questions still need to be addressed by research exploring novel paradigms, the state of the art is very promising, and the available solutions, even though only partially, allow a wide spectrum of applications. From this point of view, the pioneering and extraordinary work carried out by Fromherz deserves a special mention because it has produced extensive results that represent real milestones in dealing with exchange signaling with cell junctions. His approach is fully based on Si technologies. The innovative character of Fromherz's approach emerges quite clearly even by a quick look at his papers, mainly focused on the hybrid cell–semiconductor interface. In particular, the interfaces that he has proposed provide precious tools for noninvasive monitoring and analysis of cells' activity and for studying and understanding cellular mechanisms and signaling. In addition, the coupling between inorganic solids and living beings relies on the extensive study of individual components of the junction. Therefore, the bioelectronic interfacing through the direct contact of cell membranes and semiconductor chips is strongly favoring the development of neuroelectronic devices, ultimately aimed at implementing neurocomputation and neuroprosthetics.

Fromherz's research activity consists of three stages based on two different approaches: the monitoring of cellular processes, and the capacitive stimulation of cells operated by the host chips. The starting point of Fromherz's work can be traced back to the development of a basic cell–field-effect transistor (FET) junction [17]. Cells have been cultivated on transistors, so that it has been possible to record the spontaneous action potentials of single cells, but also the stimulated action potential, upon stimulating the serotonin receptors in order to regulate the synaptic transmission [17]. Further developments have regarded the capacitive stimulation of cells by means of the supporting chips [80]. The second stage deals with the development of more complex systems made of neuronal cells and chips, so that it has become viable to monitor basic processes in neuronal network development starting from isolated cells [81]. The enhanced control of the interface properties becomes specific in the third period of Fromherz's activity. This control has been consolidated through several progressive finding such as follows:

- The understanding of cells adhesion features [82], which is at the basis of the signals exchange.
- The study of the device's working conditions for an efficient interaction at the cell/chip junction.
- The control of cells' network formation [83] and the results in terms of cells' activity monitoring, allowing the noninvasive analysis of large brain portions by CMOS arrays [84].

It is within this context that the first step toward the manufacturing of interfaces comprising chips and brain has been accomplished.

The proof-of-concept neuron/FET junction was demonstrated in a pioneering work published in 1991 [17]. In this work, a Retzius cell, specifically a neuron of the leech *Hirudo medicinalis* present in the marginal zone of its cerebral cortex, was interfaced with the open gate of a p-type FET. The neuron, implanted by an Ag/AgCl microelectrode, was mounted on the thin gate oxide surface of a p-type channel obtained by a boron-based doping procedure of the n-type silicon plate. Prior to mounting the neuronal cell, the gate oxide was coated with poly-L-lysine and placed in an electrolyte. The accumulation of holes was promoted at the gate/dielectric interface (inversion layer) by applying a positive bias between the bulk of the n-type silicon and the source electrode. In this scheme, a positive change in the electrical membrane potential (action potential) affects the electronic band structure of the semiconductor, causing a change in the density of holes accumulated at the interface, so that this cell–silicon interaction provokes modulation of the p-type current generated within the inversion layer by a positive source-to-drain bias. The device provides a capacitive coupling between the cell and the silicon substrate mediated by a metal-free cell–oxide interaction (neuron-to-silicon junction). In fact, this configuration prevents a direct contact between the source and drain contacts and the neuronal cell, avoiding thus electrochemical reactions between the same cell and the metal electrode. Importantly, this work shows that, in order to solve the problem connected with the impossibility of a direct coupling between ionic and electronic currents, the communication between cells and electronic channels can be implemented only by interjecting an insulating barrier, that is, the oxide, at the hybrid junction. The insulating layer executes two important roles. First, the oxide prevents the passage of electrons from the chip to the cell. Second, bare silicon shows different degrees of cytotoxicity because it is not stable in an aqueous ambient; the formation of silica and silicate nullifies the biocompatibility of the semiconductor, as they provoke harmful effects on the cells. The experimental results revealed that the lowering of the source-to-drain (I_{ds}) current is related to the increase in membrane potential (V_M) upon stimulation of the cell due to a current injected by the implanted microelectrode (Figure 18.1a). The system was able to record a single action potential as well as a sequence of action potentials (Figure 18.1b), also without stimulating the neuronal cell (Figure 18.1c). In the latter case, the shape of current modulation corresponded in some cases to the first derivative of the action potential (Figure 18.1d). The ratio between the effective gate voltage and the action potential is about 0.25.

By invoking a capacitive coupling combined with a variable electrical resistance of the cleft between the membrane and the gate, it has been possible to rationalize the experimental results using an equivalent circuit accounting for the membrane and gate capacitances (C_M and C_G) and the resistance of the cleft (R). Within this scheme (Figure 18.1e), in the case of high R and tight capacitive coupling, the potential V_J generated at the junction by an action potential V_M is directly proportional to V_M by a factor $C_M/C_M + C_{G.}$ Indeed, for low cleft resistance, V_J is proportional to the derivative of the action potential V_M, with the amplitude reduced by a factor RC_M.

The innovation of Fromherz's approach lies in the first approximation on the noninvasive neuron-to-silicon coupling. A broader view of the features connected with the communication between cells and semiconductor devices has suggested the way to implement the complementary operation [80]. In this respect, the silicon-to-neuron junction has been exploited to operate an extracellular stimulation of the Retzius cell from the leech by means of a p-type-silicon-based FET. The layout of the experiment is quite similar to that described for the recording of signals coming from the neuronal cell, so that the neuron is isolated from the p-type channel by means of a 10-nm-thick silicon

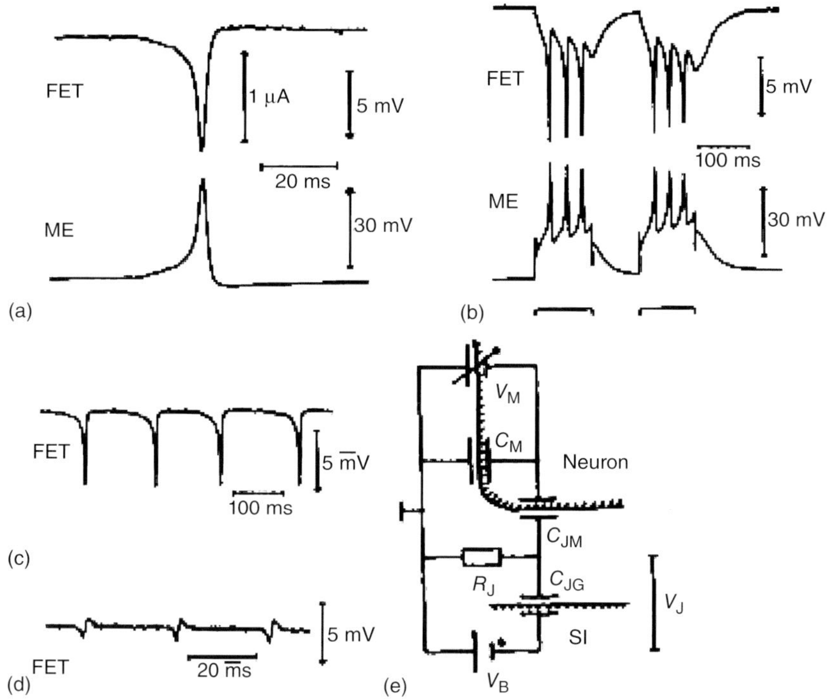

Figure 18.1 (a) I_{ds} due to a membrane potential (V_M) increase upon stimulation of the cell with an implanted microelectrode. (b,c) Sequence of action potentials upon stimulation (b) with and (c) without a microelectrode. (d) The shape of current modulation corresponds in some cases to the first derivative of the action potential. (e) Equivalent circuit accounting for the membrane and gate capacitances C_M and C_G and the resistance of the cleft, R [17]. (Reproduced with permission. Copyright © 1991, AAAS.)

oxide layer. The application of a bias voltage to the n-type bulk silicon results in a blocked p–n junction. Under this biasing condition, three different voltage biases – continuous, pulsed, and repetitive – were applied to the p-type FET channel in order to operate an extracellular stimulation. After that, the related responses were recorded by an extracellular microelectrode in contact with the cell membrane. Again, the tight capacitive coupling allowed the achievement of extracellular stimulation. The electrical excitation of the neuron upon application of a continuous voltage bias with the amplitude of 4.8, 4.9, and 5 V revealed that 4.8 V constituted a subthreshold stimulus for a Retzius cell characterized by a membrane capacitance of 1.3–1.6 nF and a cleft resistance of 30–40 MΩ. Action potentials were revealed for higher voltage steps. The delay of the cell response, attributed to the hyperpolarized Retzius cells under conditions of weak transient depolarization, was found to be inversely proportional to the height of the voltage step (Figure 18.2a). The response to 10-ms pulses of 3, 3.5, and 4 V applied to a cell with specifications (membrane capacitance of 600 pF) different from those of the cell of Figure 18.2a showed that pulses with amplitudes of 3.5 and 4 V elicited

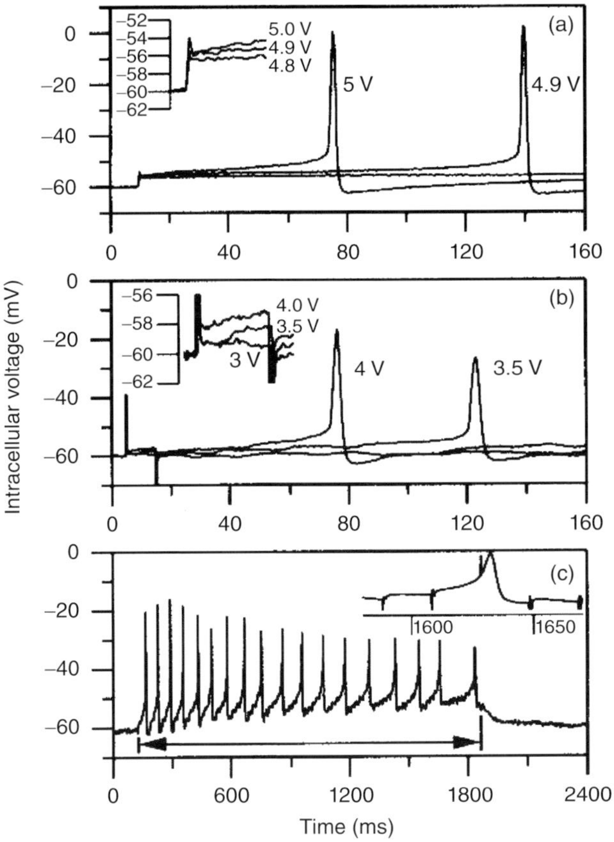

Figure 18.2 (a) Action potential of Retzius cells upon extracellular stimulation (inset) showing a subthreshold stimulus for a cell with 1.3–1.6 nF and 20 MΩ and a nominal cell/oxide surface contact of 50 μm. (b) Action potential of Retzius cells with a nominal cell/oxide surface contact of 40 μm, 600 pF, and 30–40 MΩ. (c) Response to repetitive stimulations applied at a frequency of 50 Hz for a period of 1.8 s [80]. (Reproduced with permission. Copyright © 1995, American Physical Society.)

in this case action potentials (Figure 18.2b). The application of repetitive pulses with duration of 10 μs resulted in a train of action potentials (Figure 18.2c).

The above-described findings represent the starting point for a wide range of experiments addressed to the fabrication of complex neural networks directly interfaced to electronic ICs. The activation of neurons by means of extracellular currents is a milestone toward implantable neuroprosthetic devices. Fromherz's work ranges from strategies suitable for elucidating the opportune working conditions of the junction for an efficient bidirectional cell–chip interaction [85, 86], through studies aimed at analyzing the physiology of cell networks [87, 88], to studies aimed at exploring the intimate nature of the hybrid junction [88].

Some examples of studies dedicated to the features of the junction have demonstrated that, using an AC stimulation, it is possible to reveal the features of the bottom side of attached cells [89]. In this respect, a distance of 20 nm between the membrane and the surface has been demonstrated to be sufficient for affecting the transmembrane voltage in order to provoke a rearrangement or a gating of certain ion channels. Extracellular recording has been found to depend on the inhomogeneity of ionic conductances in a cell membrane [90]. The adhesion of cells has been indicated as the key parameter for the efficient interaction between cells and chips. The variation of the membrane resistance, which is different in different junctions, has been correlated with the enhancement of the adhesion strength [91]. The electrical resistance of the cleft between the cell and the junction, whose importance is related to the control of the local extracellular voltage, could differ from that in the bulk extracellular electrolyte because of the interaction of the cell membrane with the substrate. Its evaluation by means of the junction has paved the way toward a finer analysis of the junction features, showing that an optimization of the junction with respect to adhesion is required for an efficient capacitive gating of ion channels, useful for the interfacing of the cell population. Scaling effects in terms of cell size (from leech to *Limnea Stagnalis* and rat neurons) and contact area, showing that also small capacitive currents can elicit neuronal activity upon providing a tight contact between neurons and silicon oxide [92], but also large-area applications based on sensor arrays [89] or exploiting the CMOS technology [81, 93], have been pursued for the implementation of neurobionic devices [94] and neuromedicine [95]. The adaptation of an industrial standard of CMOS technology is fundamental for implementing prosthesis and brain activity monitoring. A CMOS-compatible process, aimed at the realization of sensor arrays used in noninvasive, long-term recording of neural activity, has been implemented using chemically stable and biocompatible oxides with high dielectric constants (namely TiO_2 and ZrO_2) in order to efficiently couple the nerve cell signals to the CMOS circuitry. Action potentials of firing neurons which are capacitively coupled to the gates of Metal-Oxide Field Effect Transistor (MOSFETs) composing the CMOS have been detected [96]. Furthermore, a sensor array based on CMOS technology, allowing high-resolution imaging of extracellular signals from neural cells, has been shown [84]. The array provides 128×128 capacitive sensors with a pitch of $7.8\,\mu m \times 7.8\,\mu m$ on an area of $1\,mm^2$, allowing the detection of signals ranging from $100\,\mu V$ to $5\,mV$ peak to peak, with a full-frame sample rate per pixel of $2\,kS\,s^{-1}$ and a maximum internal bandwidth of 32 MHz. The application of CMOS technology exploiting capacitively coupled recording sites opens the way to the imaging of neuronal activity with high spatial and temporal resolution.

Most of the above-described experiments consider single neuronal cells interfaced with a chip. Importantly, in 2001, it was demonstrated how isolated cells cultivated on the chip could join together to form neuronal networks [87]. In 2002, a chip was interfaced with two neurons from the pond snail *Lymnaea stagnalis*, which were electrically joined through their grown neurites. The neuronal cells established a connection, as it happens in the nervous system, and formed a synapsis with the silicon electrode covered with a porous layer of poly-L-lysine, forming thus a hybrid network in which neurons responded to the stimuli applied by the electrodes, and electrodes were able to detect signals from neurons [97]. Later, it was demonstrated that chemical synapses showed a short-term memory that could be directly connected to a semiconductor chip [98].

The collection of these thorough results has entailed exhausting progress in terms of the fundamental steps toward the functional interfacing of chips and brain for the care of degenerative diseases [99]. Hence, the third stage of Fromherz's activity has culminated in the interfacing of electronic devices and retinal neurons, offering thus a tool for the partial restoration of visual function by means of reliable neuroprosthetic applications [94]. This study has demonstrated the effective selective electrical stimulation of different classes of retinal neurons (retinal ganglion cell, RGC; bipolar cells, BCs; photoreceptors, PRs), carried out with a multicapacitor array used for the stimulation with monophasic anodal or cathodal current pulses of low amplitude. The stimulation has been performed *ex vivo* on rabbit retinas interfaced in epiretinal (Figure 18.3a) and subretinal (Figure 18.3b) configurations to the multicapacitor array. The stimulated RGC activity with ganglion cells and PRs interfaced with the capacitor array was recorded in the form of action potentials using a tungsten electrode. The stimulation was carried out by applying a falling voltage ramp (amplitude 0.9 V, duration 1 ms, Figure 18.3c, i), so that the extracellular voltage, shown in Figure 18.3c, ii, was recorded for one stimulus and 50 stimuli, giving rise to stimulus-locked responses. The results showed that cathodal currents elicited negative voltage changes in the retina of about 15 mV upon application of a stimulation cathodal current set above the threshold for the activity of rabbit retinal. The trace of the signal (Figure 18.3c, iii) attributed to the ganglion cell was obtained by the subtraction of the average signal from each of the single simulation traces. A similar result has been shown for anodal currents (Figure 18.3d).

The application of neuron–chip junctions has been successfully demonstrated also in the context of research aimed at elucidating the effect of drugs on neurons directly cultivated on the chip, providing thus an innovative tool in pharmaceutical research [95]. The proof of feasibility of noninvasive receptor–cell–transistor (RCT) sensors was provided in 2007. The ion currents of a recombinant receptor were coupled directly to a microelectronic device to provide a cell–transistor junction which was able to record an extracellular voltage upon the opening of ion channels by means of an agonist stimulating the serotonin receptors. Overexpression of receptors can be a fingerprint of anomalies. In this context, the experiment showed how it was possible to record the activation of the ionotropic serotonin 5-HT3A, overexpressed in HEK293 cells, in terms of ion currents flowing through the activated channels.

In the light of Fromherz's studies, at present a large number of works deal with established strategies toward versatile bioelectronic devices exploiting unique combinations of mechanical and electronic properties. In particular, monitoring of living tissues' activity and manufacturing of efficient prosthesis have aroused the interest of researchers. The importance of the described approach is clearly evidenced by the extensive literature available on monitoring of cells activity via device arrays. The major questions related to flexibility and conformability

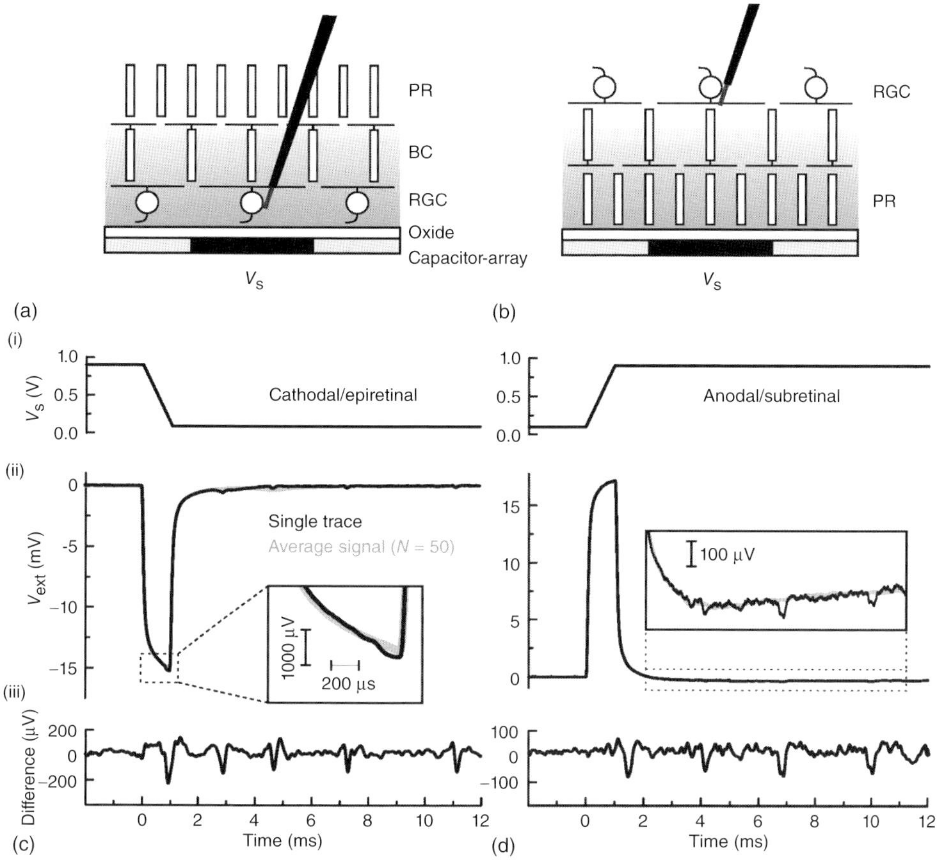

Figure 18.3 Selective electrical stimulation with a multicapacitor array of different classes of retinal neurons (retinal ganglion cell, RGC; bipolar cells, BC; photoreceptors, PR). Response of *ex vivo* rabbit retinas interfaced in (a) epiretinal and (b) subretinal configurations. (c) Ganglion cell activity following a monophasic cathodal current in epiretinal configuration. (c, i) Stimulating falling voltage ramp (amplitude 0.9 V, duration 1 ms) and (c, ii) extracellular voltage recorded for one stimulus and 50 stimuli (gray line in the inset). (c, iii) signal trace attributed to the ganglion cell obtained by subtraction of the average signal from each of the single simulation traces. (d) Identification of ganglion cell activity following a monophasic anodal current in subretinal configuration shows a similar behavior as in (c) [94]. (Reproduced with permission. Copyright © 2012, American Physiological Society.)

to the soft biotissue are now finding new explorative solutions by the massive adoption of nanomaterials and nanosystems based on Si and C. For instance, modifications to the cell–silicon junctions have been recently demonstrated using Si-NWFETs (nanowire field-effect transistors) and related flexible arrays, in order to implement the registration with the perspective of applicability down to the subcellular level. Cardiomyocyte cells cultivated on PDMS sheets have been interfaced with standard Si-NWFET arrays, so that it has been possible to record conductance signals from cardiomyocytes exhibiting excellent signal-to-noise ratios (SNRs) (>5) and tunable sensitivity from 31 to 7 nS [100]. Si-NWs, but also carbon nanotubes (CNTs) [101], show higher sensitivity than planar FETs when interfaced to electroactive cells. This is because FETs in general have a higher SNR and a reduced contact area for guaranteeing the efficient detection of cellular and, above all, subcellular signal recording [102]. The central role of nanostructures in terms of an efficient recording of signals emerges from the above considerations. In fact, the increase of the interface between cells and nanostructured semiconductors induces an enhancement of interfacial coupling and also increases both cellular adhesion and activity at the same time [103]. For these reasons, nanostructures represent the ideal candidates for building interfaces between living beings and electronic devices, above all with the perspective of interfacing flexible, large areas with tissues and organs. In this respect, the versatility of Si-NWFET arrays has been demonstrated by interfacing not only isolated cells but also entire organs such as spontaneously beating embryonic chicken hearts, providing a multiplexed recording from the arrays yielding signal propagation times across the myocardium with high spatial resolution [104]. The original experimental setup is shown in Figure 18.4a,b. Isolated embryonic chicken hearts beating at a frequency of 1–3 Hz were positioned (red arrow) on an NWFET placed in a temperature-regulated cell. The electrical response to the beating hearts was monitored by simultaneously recording signals from the NWFET (Figure 18.4c, red trace) and from a glass pipette inserted into the heart used as reference (Figure 18.4c, black trace). The comparison between the two recorded traces reveals a temporal correlation between the peaks shown in the two distinct traces, while the NW signals, showing excellent SNR, reveals an initial fast phase peak (associated with transient ion-channel currents) with a full width at half maximum (FWHM) of 6.8 ± 0.7 ms followed by a slower phase (FWHM 31.0 ± 0.9 ms, indicated by the black arrow).

A number of studies have focused on effectively interfacing organs and electronic devices made of nanostructures assembled into extended multiplexed amplifier arrays. Viventi *et al.* [105] have reported on foldable multiunit gold electrodes connected to amplifiers, each one made of seven n-type metal–oxide–semiconductor (nMOS) FETs based on single-crystal silicon nanoribbons transfer-printed on a silicon oxide substrate for the monitoring and stimulation of a beating heart. This work describes a mechanically flexible multilayer circuit fabricated on a 25-μm-thick plastic sheet of PI with a built-in array of 288 gold electrodes exploiting only 36 wires for external electrical connection and a total 2016 nMOSs. Flexibility was remarkably pushed for this sensor, as indicated by the maximum strains in Si and SiO_2 of 0.001% and 0.0001%, respectively (far below the critical values provoking electrical failures) at the bending radius of about 5 cm. These electrodes have been tested for monitoring the cardiac signaling in a porcine model, providing recordings of normal and stimulated heartbeats upon more than thousands of bending cycles, allowing thus the construction of an isochronal map of the heart's activation at high resolution, with an SNR as high as 34 dB. A similar approach has been used for the *in vivo* recording and stimulation

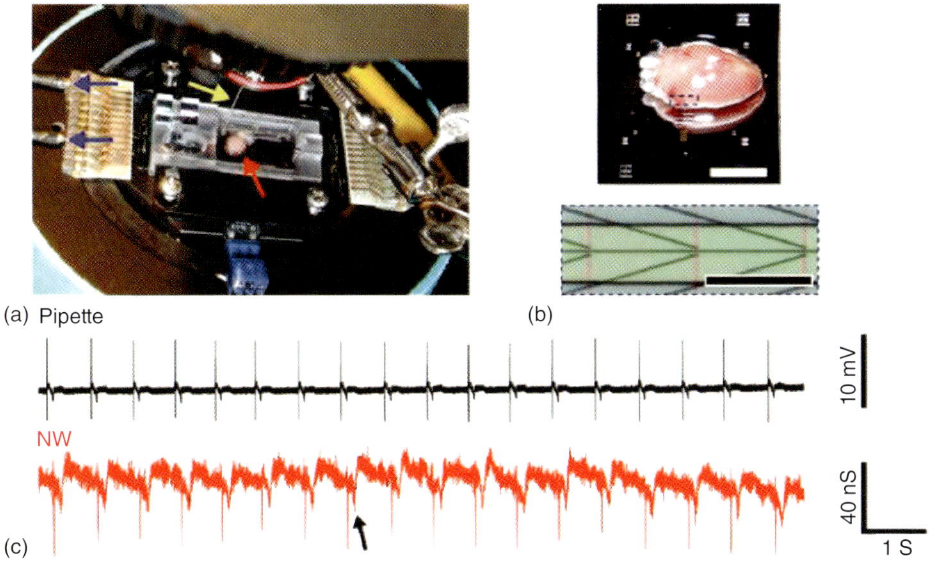

Figure 18.4 (a) Embryonic chicken hearts (red arrow) on an NWFET with a Ag/AgCl (yellow arrow) and source and drain (blue arrow) interconnections. (b) Magnification of the chicken heart (upper panel) and zoom of the dotted region in the upper panel showing three NWFETs (red lines denote the orientation of the NWs, while the scale bar is 150 μm). (c) Simultaneous recordings from a glass pipette (black) and an NWFET (red) [104]. (Reproduced with permission. Copyright © 2009, American Chemical Society.)

of cat brain activity [106]. A very important result is that high-resolution foldable arrays enable high signal quality over extended periods, preventing harmful effects associated with penetrating electrode arrays. The proposed high-density flexible device is basically a 360-channel high-density active electrode array connected to units made of single-crystal silicon nanomembrane transistors (720 in total), showing a fast high-resolution response with a mobility of about $350\,cm^2\,V^{-1}\,s^{-1}$. The transistors were fabricated using the transfer printing technology. The array, potentially fully implantable and nontoxic, has been shown to be capable of sampling a wide region of the brain (10 mm × 9 mm) with promisingly high spatial (500 μm spacing) and temporal ($>10\,kS\,s^{-1}$) resolutions. The spatial resolution is comparable to that of voltage-sensitive dyes, with an enhanced temporal resolution, allowing the recording from nonplanar and non-optically-accessible areas. Figure 18.5 shows the high-density, flexible array in contact with the visual cortex of a cat, while the inset shows the same electrode array folded and inserted into the interhemispheric fissure. In order to show the potentialities of this technology, the authors have reported, *inter alia*, the use of the array for monitoring electrographic seizures in a feline model induced by administering some picrotoxin directly into the brain. The micro-electrocorticogram (μECoG) signal recorded upon a single electrographic seizure has shown a large amplitude of 6.6 mV, a low noise (expressed as the root mean square voltage) of 45 μV, and a high SNR of 34 dB.

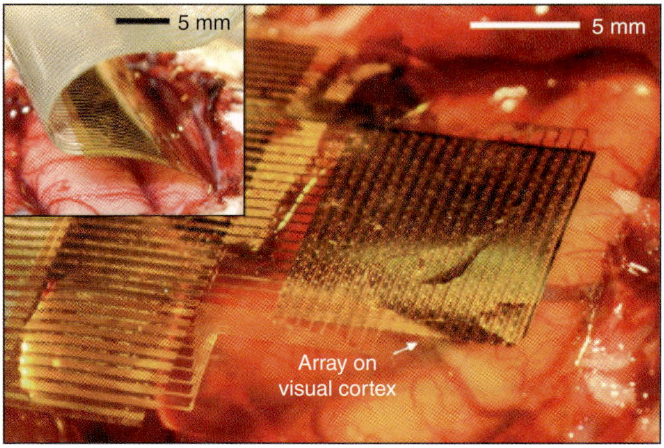

Figure 18.5 Electrode array in contact with the visual cortex (inset: the same array inserted into the interhemispheric fissure upon bending) [106]. (Reproduced with permission. Copyright © 2011, Nature Publishing Group.)

Progress in cell–device integration, on the other hand, has favored a growing interest on prosthesis, producing relevant results on sensitive artificial skin [107], three-dimensional (3D) synthetic biomaterials used as structural and bioactive scaffolds [108], and, above all, ocular prosthesis [109, 110], giving rise to improved and efficient protocols in regenerative medicine. 3D synthetic biomaterials allow extending the localized real-time monitoring of cellular activities and their physicochemical changes in the context of studies on cell/tissue interaction aimed at developing hybrid 3D electronics/tissue materials for synthetic biology. In this framework, Tian *et al.* [108] have shown robust electronic properties well suited for the 3D culture of neurons, cardiomyocytes, and smooth muscle cells by using 3D biocompatible materials based on macroporous, flexible, and free-standing NW nanoelectronic scaffolds (nano-ESs). They were prepared by self-organization of coplanar reticular networks and their coupling with natural biomaterials such as cardiomyocytes. The authors described an integrated system made of discrete electronic and biological building blocks. In particular, they followed three biomimetic and bottom-up steps: the fabrication of FET by using silicon nanowires; the self-assembly, by self- or manual organization, of 3D NW FET matrices forming free-standing nanoelectric scaffolds in order to mimic extracellular matrix structures; and, finally, the cultivation of cells within the nano-ES, so that a synthetic tissue was formed through biological processes. As a result, the as-prepared flexible, macroporous 3D sensor could explore novel routes for efficient and fully biocompatible merging of nanoelectronics with biological systems. This scheme paves the way toward a futuristic very innovative paradigm for prosthesis.

Finally, it is worth mentioning that reports dealing with ocular prosthesis consolidate the advances made in the field of regenerative medicine. These results are based on the pioneering works such as those of Cannata *et al.* [109]. They have reported the development of a 64-element 35-MHz composite ultrasonic array for ocular imaging with a resolution exceeding 50 μm axially and 100 μm laterally. Strategies toward the effective design of well-operating prosthesis have also been reported. In this perspective, Coulombe *et al.* [110] have reported the design of a flexible system for the microstimulation of the visual cortex. They followed a modular approach for the design of the implant. Their approach envisages the placement of sensitive electronic circuits near the sites of stimulation and the remaining control circuitry away from these sites. This approach allows effective miniaturization of the implant, distributing the electrodes over a wide area for enhancing the field of view, and contributes to the reduction in the wiring to the electrodes. By the way, the wiring is one of the major issues in prosthetics, requiring a much larger effort to be solved. In summary, their design is based on small stimulation modules (SMs) connected to penetrating MEAs and an interface module (IM) mounted away from the stimulation sites. On the basis of their design, a communication submodule (CSM) providing the IM and an SM has been fabricated on a standard 0.18-μm CMOS process. The resulting bioelectronic device supports several stimulation modes and is able to record pulsed stimuli with a rate exceeding 500 000 pulses s^{-1} [110].

18.3 EGOFETs for Flexible Biosensing

As already mentioned, OFETs show relatively low carrier mobilities and stability compared to silicon FETs. Nevertheless, they are characterized by better performance in terms of flexibility, biocompatibility, large area, and solution processability and, hence, promise also to be cost competitive so that, for example, they are well suited for disposable sensors and biodevices.

Conformability and flexibility of OSCs represent the main criteria while fabricating flexible biosensing technologies. Along with exploiting these characteristics, efforts toward flexible biosensing have also focused on developing processes that are compatible with flexible plastic substrate materials such as poly(ethylene terephthalate) (PET), poly(imide) (Kapton), poly(ether sulfone), cellulose, and silk fibroin.

A major unavoidable requirement for practical biosensing devices and systems is that they should operate in a wet, and sometimes chemically complex, environment with which the semiconducting layer is in direct contact. Furthermore, in order to avoid water electrolysis and chemical damage of the biosystems studied in the sample electrolyte (proteins, enzymes, cells, etc.), low voltages (<1.2 V) are strictly necessary. Low-voltage operation is also particularly important from the point of view of power consumption. This is critical for organic circuits, especially when low-voltages or battery power supplies are required, as it occurs, for example, in portable or wearable electronics and disposable sensors. In this respect, EGOFETs have recently attracted great attention as low-voltage operating devices capable of interfacing with liquid environments.

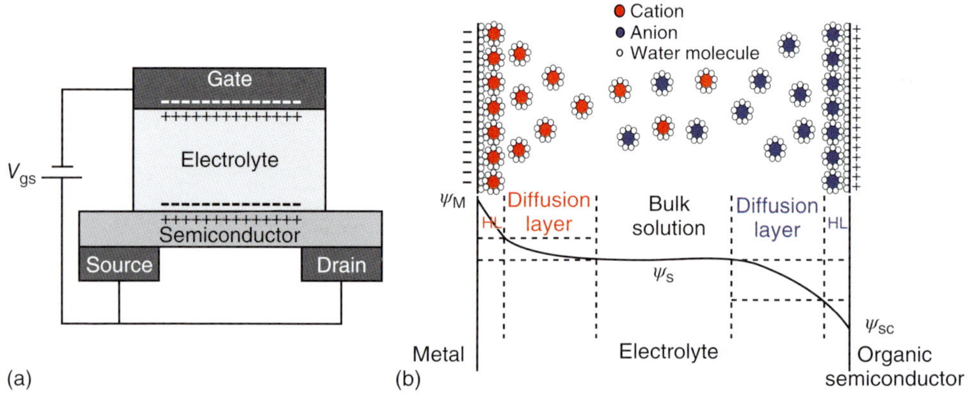

Figure 18.6 (a) Schematic of an EGOFET under bias (top-gate/bottom-contact configuration), where the conducting channel in the OSC is highlighted. (b) Illustration of the ion distribution within the Helmholtz layer (HL), the diffuse layer, and the bulk electrolyte (above), and spatial distribution of the potential between the gate and the semiconductor in an electrolyte-gated transistor (below) [36]. (Reproduced with permission. Copyright © 2012, Springer.)

In this section, we report on some interesting aspects of EGOFETs, particularly their architecture and working principle, materials and manufacturing processes, and some relevant examples of biosensing.

18.3.1 EGOFET: Architecture, Working Principle, and Materials

As mentioned in the introduction, EGOFETs are based on organic semiconducting materials, as for standard organic FETs, with the major difference that, instead of a solid dielectric, the gate is separated from the semiconductor by an electronically insulating, ionically conducting electrolyte (Figure 18.6, see also Chapter 8 for further details). EGOFETs, as OFETs in general, mostly operate in the so-called accumulation mode, where the I_{ds} current flowing through the semiconducting layer in the off state ($V_{gs} = 0$) is much lower than that flowing in the on state ($V_{gs} \neq 0$).

The current flow inside the semiconducting channel is well reproduced by the MOSFET's analytical expressions

$$I_{ds} = \frac{W}{L} C_i \mu (V_{gs} - V_T) V_{ds} \quad V_{ds} \ll (V_{gs} - V_T) \tag{18.1}$$

$$I_{ds}^{sat} = \frac{W}{2L} C_i \mu (V_{gs} - V_T)^2 \quad V_{ds} > (V_{gs} - V_T) \tag{18.2}$$

where I_{ds} is the channel current, V_g is the gate voltage, W and L are the channel width and length, respectively, μ is the field-effect mobility, C_i is the capacitance of the gate insulator, and V_t is the threshold voltage of the device, which is related to the doping level of the channel. Equation (18.1) refers to the linear region of the current–voltage characteristic curve, while Eq. (18.2) refers to the saturation regime.

A bias is applied to the gate electrode with a polarity opposite to that of the mobile charge carriers induced in the OSC at the semiconductor/dielectric interface; for p-type OSCs, for example, a negative bias is applied to the gate electrode. This induces a charge drift into the electrolyte, with anions in the electrolyte being repelled from the gate and a layer of anionic charge being created at the opposite electrolyte/semiconductor interface. The resulting large local electric field at the electrolyte/semiconductor interface causes holes to accumulate on the OSC surface to balance the negative charge of the anionic layer. Hence, an electrostatic capacitive coupling of the gate bias across the dielectric layer to the semiconductor is obtained, with the formation of two electric double layers (EDLs, also known as *Debye–Helmholtz layers*) at the gate/electrolyte interface and at the electrolyte/semiconductor interface, respectively. This accumulation of electric charges in the semiconductor at the interface with the electrolyte induces the formation of a conducting channel.

The working mechanism described so far, entirely driven by the accumulation of charges at the electrolyte/semiconductor interface, is of field-effect type. However, if ions penetrate the semiconductor, thus inducing a bulk electrochemical doping, the EDL model is not valid anymore and an electrochemical mechanism is rather involved. Coexistence of electrochemical switching and field-effect modulation of the organic channel results in transistors that are typically slow and exhibit a large hysteresis. Several studies have focused on clarifying the exact working mechanism of EGOFETs, although collective results point to an electrostatic (field-effect) doping mechanism when appropriate materials are chosen [111, 112]. Preventing electrochemical doping is typically realized by using large molecular or polymeric ions that cannot penetrate into the semiconductor channel (see following section).

The EDL at the electrolyte/semiconductor interface can have a very fast time response as a result of a charge separation of only a few angstroms within the Helmholtz layers formed in a few tens of microseconds [113]. Furthermore, and more importantly, the EDL can show extremely high capacitance, as may be estimated by the Helmholtz equation $C \sim k\varepsilon_0/\lambda$, where ε_0 is the vacuum permittivity, k is the effective dielectric constant of the EDL, and λ is the Debye screening length or the thickness of the double layer. With λ of the order of 1 nm, and $k \sim 10$, capacitances from a few to a few tens of microfarad per square centimeter are typically attainable with electrolytes [114, 115]. As a consequence, such huge capacitances induce a very large charge-carrier density ($>10^{14}$ cm^{-2}) [116] in the organic semiconducting channel, even for low (sub-volt) applied voltages [117].

The gate voltage required to switch the transistor from its "off" state to its "on" state, that is, to establish the conducting channel, is called the *threshold voltage*. The threshold voltage, the charge carrier mobility (μ), and the charge current modulation ratio between the "on" and "off" states (I_{on}/I_{off}) are the key parameters of a transistor.

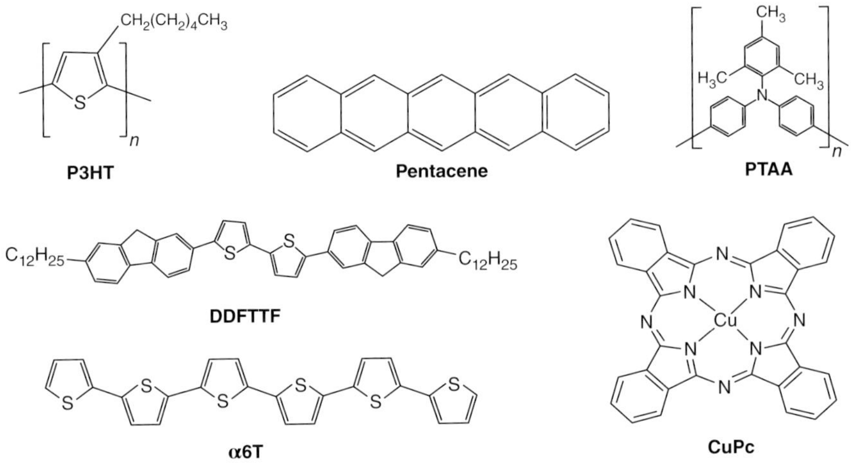

Figure 18.7 Chemical structure of some OSCs commonly used in OFETs.

Solution-gated graphene transistors (SGGTs) work on a similar concept and have been proposed in the wide and blossoming field of graphene-based devices as particularly well suited for chemical and biological sensing. In an SGGT, the gate and the graphene channel are separated by an electrolyte. By applying a gate voltage, the conductance of the graphene channel can be modulated as a result of change of carrier densities in the channel [22].

As in the case of organic molecular systems, all the essential parts (source, drain, and gate) and the graphene channel can be fabricated in a flexible architecture, which presents a very promising perspective for bioelectronics [118, 119].

Most relevant materials involved in EGOFETs are OSCs (for the active channel) and electrolytes (for the gate dielectric). The OSC used as active layer in EGOFETs can be made of oligomers or polymers that are deposited as thin films. Typical materials include conjugated polymers, such as poly-(3-hexylthiophene) (P3HT), poly(4,4′didecylbithiophene-*co*-2,5-thieno[2,3-*b*]thiophene) (PDTT), polytriarylamine (PTAA), alkyl and alkoxy-substituted polymers such as poly-(3,3″-dipentoxy-2,2″:5′,2″-terthiophene) (poly-DPOT), and 3,3″-didodecyl-2,2″:5′,2″-terthiophene (Poly-DDT), but also oligomers such as pentacene (and its soluble derivatives), α-sexithiophene (α6T), 5,5-bis-(7-dodecyl-9H-fluoren-2-yl)-2,2-bithiophene (DDFTTF), copper phthalocyanine (CuPc), and many other small-molecule OSCs (Figure 18.7) [120].

For EGOFETs in general, several electrolytes have been successfully investigated as the medium to electronically separate the gate from the transistor channel, yet providing low-voltage operation (<1 V). They have attracted much attention lately because they generate very high electric fields at the electrolyte/OSC interface even at very low voltages. Shown in Figure 18.8 are water and liquid electrolytes [121, 122], polymer electrolytes [123–125], ionic liquids [126], ionic gels [117, 127], and, finally, polyelectrolytes [121]. However, one drawback with electrolytes

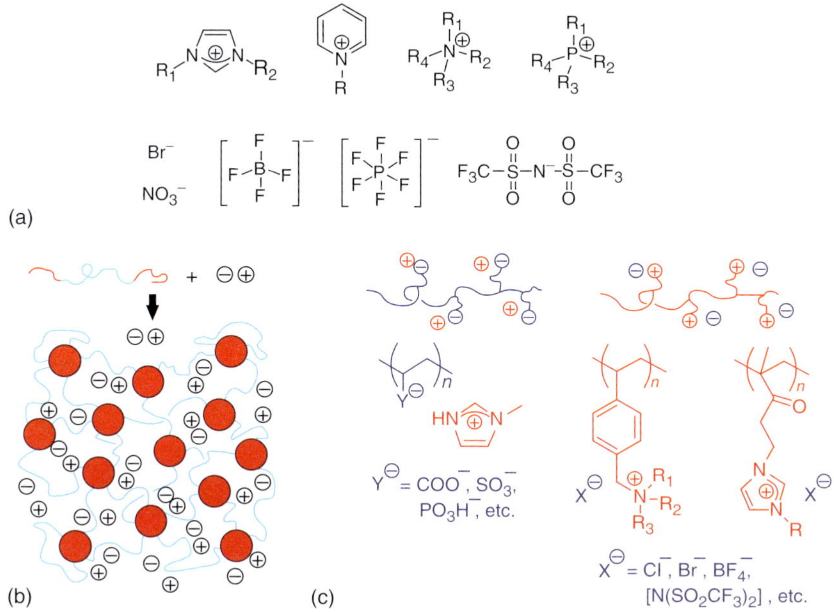

Figure 18.8 Examples of solid electrolytes. (a) Representative component ions for ionic liquids. (b) ABA triblock copolymer in an ionic liquid to form an ion gel. (c) Chemical structures of several polyelectrolytes [37]. (Reproduced with permission. Copyright © 2013, Wiley-VCH.)

is that electrostatic and electrochemical doping can occur simultaneously, especially when semiconducting polymers with flexible and open structures are used. Such electrochemical doping would result in a higher switching time and possible degradation of the semiconductor. Use of ion gels (ionic conducting liquid immobilized inside a polymer matrix) and polyelectrolytes (polymers with a backbone with substituent groups containing ionic functionalities) has been found to prevent this situation. In both cases, the ionic groups are virtually immobile, and cannot penetrate the semiconductor bulk.

Solubility in solvents of OSCs opens up cost-effective fabrication methods for organic electronic devices. Thermal evaporation at room temperature or solution-processable techniques, such as spin-coating, inkjet printing, and screen printing, are a few examples. These techniques can largely reduce the complexity of the manufacturing process and cost [128] compared to the fabrication of most inorganic semiconductor devices, which requires high-temperature processes and high-purity crystalline substrates, along with expensive clean room facilities.

From the point of view of carbon-based materials, serious efforts are on in developing new strategies for processing graphene: many synthesis methods have been explored and proposed to make graphene thin films: liquid-phase exfoliation, reduction Graphene-Oxide (GO), epitaxial growth on SiC, molecular assembly, and CVD [24, 129, 130], and all these methods are summarized and discussed in many reviews [131–134].

18.3.2 Biochemical Sensing

Biosensors conceived for clinical applications can be broadly classified according to their operational complexity and their compliance with homecare treatments. Complex equipment, operated by specialized personnel exclusively in clinical or research laboratories, constitute the first class. The approach based on biosensors follows a very different strategy: that is, they should be portable, easy-to-use, and disposable devices aimed at point-of-care diagnostics, and usable without specific training and not necessarily in specialized clinical structures. We confine our discussion here exclusively to the latter case, with particular attention to label-free sensing, that is, techniques based on EGOFETs which do not require an optical or electroactive probe to be conjugated to the biological analyte.

In order to confer biosensing selectivity to the EGOFET-based device, the main route is biofunctionalization by means of one or more biorecognition elements, that is, endowed with a specific reactivity toward the biological systems or markers to be detected (enzymes, proteins, DNA, etc.). The biofunctionalization can take place by grafting the biorecognition element at either of the two available Debye–Helmholtz interfaces: gate/electrolyte and electrolyte/OSC (Table 18.1) [142]. The overall system should be designed in such a way that any binding between the probe and target molecules would induce changes in the interfacial charge accumulation, and hence will be transduced in a specific modification of the charge distribution and/or carrier mobility in the OSC channel. This, in turn, is then converted to a specific electrical output, directly associable with the biorecognition event.

In the following, we introduce and discuss some relevant examples of EGOFETs applied to biochemical sensing, selecting a few cases for each of the most used biofunctionalization strategy.

As far as the biofunctionalization of the gate/electrolyte interface is concerned, an interesting example is the one described by Casalini *et al.* [122]. In their work,

Table 18.1 Summary of EGOFET-based biochemical sensors.

Bio-functionalization site	Strategy	Bio-recognition element	Analyte	References
Gate/electrolyte interface	—	Boronic acid	Dopamine	[122]
Electrolyte/OSC interface	–OH, –COOH,	PNA	DNA	[135]
	$-NH_2$ groups		DNA	[136]
		APTES + penicillinase	Penicillin	[137]
		BSA	BSA-antibody	[138]
		PL bilayer + biotin	Streptavidin	[139]
	Au-NPs	Thiolated aptamers	Hg^+	[140]
			Thrombin	[141]

PL, Phospholipids.

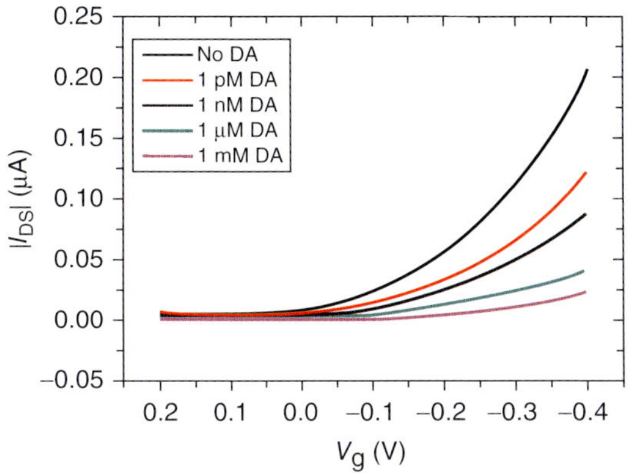

Figure 18.9 I_{ds} versus V_g transfer characteristics of a water-gated OFET in which the gold gate electrode was functionalized with a self-assembled monolayer of a boronic acid to sense dopamine down to the picomolar range [122]. (Reproduced with permission. Copyright © 2013, Elsevier.)

they introduced a water-gated organic field-effect transistor (WGOFET) sensor based on P3HT and specialized for detecting the dopamine neurotransmitter. They functionalized the gold gate electrode with a thiol-based self-assembled monolayer of boronic acid, which has a high chemical affinity toward molecules with geminal diol groups as occurs in dopamine. Upon binding with dopamine, a negatively charged boronic ester is formed, the charge of which is compensated by the amine group of the dopamine. As this electrical dipole is formed, the potential drop at the gate/electrolyte interface increases, thus causing a decrease in the current, because a higher gate voltage has to be applied in order to induce the same carrier density in the OSC channel. Detection of the analyte in the picomolar range was shown to be possible with such a device (Figure 18.9). This is due to the small capacitance changes induced by the dopamine–boronic acid complex which can dominate the in-series large capacitances of the electrolyte.

The other approach of biofunctionalizing the electrolyte/OSC interface is typically based on the activation of the OSC surface by attaching hydroxyl, carboxyl, or amino groups. These chemical moieties represent the chemical anchors to which it is easy to attach the biorecognition element that is designed, for example, for detecting DNA, for immunochemistry (involving antibody–antigen interactions), or for selective sensing of enzymes. A nice example of this approach has been reported by Khan *et al.* [135], who described the *in situ* label-free detection of short-chain DNA targets (Figure 18.10a). They used plasma-enhanced chemical vapor deposition (PE-CVD) to create a –COOH-rich layer (maleic anhydride (MA) polymer) on the OSC surface (DDFTTF). Then, they covalently attached to this layer peptide nucleic acid (PNA) strands, which are synthetic polymers similar to DNA but containing repeating *N*-(2-aminoethyl)-glycine units linked by peptide bonds [143]. They play the role of molecular probes, showing great selectivity, affinity, and stability compared to their DNA counterparts.

Whenever a DNA immobilization occurs, via hybridization reactions between the surface-bound PNA and the complementary target DNA strands, a change of the channel current occurs in the device due to the increase of the surface charge density on the channel. Indeed, in order to compensate the negative charges of the DNA backbone, more holes are injected into the channel, which is achieved by applying a more negative potential to the gate. This, in turn, induces a shift of the threshold potential toward negative values. The authors demonstrated the ability of the device to sense concentrations of the target DNA complement as low as 1 nM.

A deeper understanding of the role of the electrolyte ionic strength on the Debye length, which can screen negative DNA charges and hinder transduction, was achieved by Kergoat *et al.* [136]. In this case, a carboxylic acid derivative of P3HT was used to covalently immobilize amino-modified DNA probes via a coupling reaction between the amino groups of the DNA strands and the carboxylic acid moieties of the semiconductor surface. Analogous to the work previously described [135], clear changes in the electrical characteristics of the device were observed upon DNA immobilization and hybridization (Figure 18.10b). A very interesting feature was that the measured current changes were more pronounced when pure water was used as electrolyte instead of a buffer solution. This was explained considering that charged molecules are screened by dissolved counterions at rather high ionic concentrations. In particular, in the case of a DNA sensor, negatively charged DNA molecules could be surrounded by cations (solvation shell) because of electrostatic interactions. At the Debye length, the number of net charges approaches that of negative charges on the DNA backbone. This screening effect implies that the electrostatic potential arising from the DNA decays exponentially toward zero with distance. Because the Debye length is inversely proportional to the square root of the ionic force of the solution, reducing the ionic concentration can increase the Debye length (Figure 18.10c), thereby increasing the sensitivity of the WGOFET.

Figure 18.10 Responses of EGOFET-based biosensors where the biofunctionalization has been performed at the electrolyte/OSC interface. (a) Response of an EGOFET exposed to different DNA sequences, complementary targets (T2-MM0), one base mismatch (T1-MM1), and two base mismatches (T3-MM2) [135]. (Reproduced with permission. Copyright © 2010, Wiley-VCH). (b) Transfer curves for a P3PT-COOH OFET gated with water after immobilization of probe DNA (black line) and hybridization with complementary targets (red line). (c) Variation of the Debye length as a function of the Phosphate Buffered Saline (PBS) dilution [36]. (Reproduced with permission. Copyright © 2012, Springer.) (d) Time lapse of I_{ds} during a penicillin titration experiment. The green line shows the response of a device functionalized with APTES and penicillinase, while the gray line depicts the I_{ds} response of an untreated α6T OFET. The pH of the electrolyte is shown in red [137]. (Reproduced with permission. Copyright © 2012, Wiley-VCH.) (e) Normalized I_{ds} response to protein adsorption at two different pH values [138]. (Reproduced with permission. Copyright © 2011, Wiley-VCH.) (f) Normalized I_{ds} response of the EGOFET to streptavidin detection. The inset shows a schematics of the OSC surface functionalization, which explains the I_{ds} current increase [139]. (Reproduced with permission. Copyright © 2012, Wiley-VCH.)

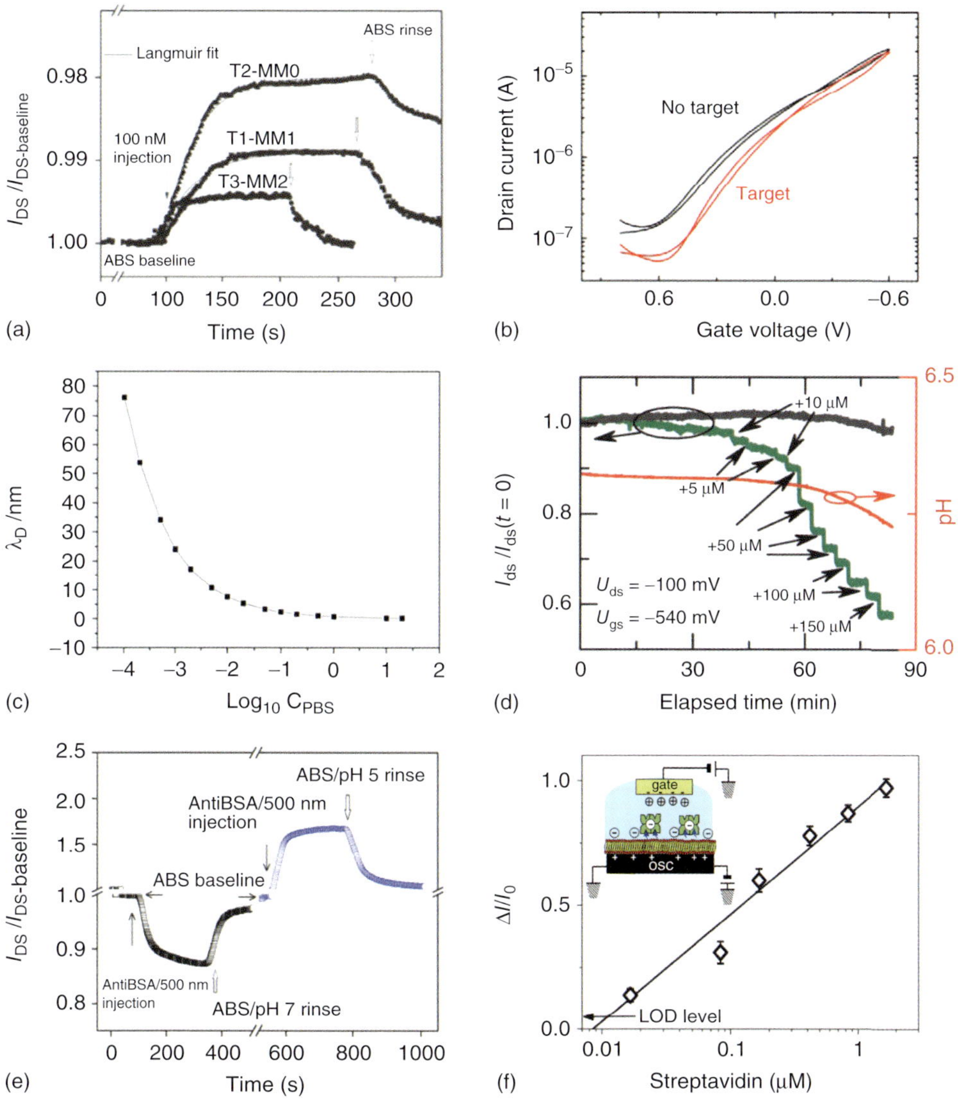

Langmuir fit
ABS rinse
T2-MM0
100 nM injection
T1-MM1
T3-MM2
ABS baseline
I_{DS} / $I_{DS\text{-baseline}}$
Time (s)
(a)
No target
Target
Drain current (A)
Gate voltage (V)
(b)
λ_D /nm
Log_{10} C_{PBS}
(c)
+10 μM
+5 μM
+50 μM
+100 μM
+150 μM
U_{ds} = −100 mV
U_{gs} = −540 mV
I_{ds} / $I_{ds}(t = 0)$
pH
Elapsed time (min)
(d)
ABS/pH 5 rinse
AntiBSA/500 nm injection
ABS baseline
AntiBSA/500 nm injection
ABS/pH 7 rinse
I_{DS} / $I_{DS\text{-baseline}}$
Time (s)
(e)
gate
osc
LOD level
$\Delta I/I_0$
Streptavidin (μM)
(f)

Also, selective enzyme sensing has been successfully achieved by specific surface-functionalization of OSCs in EGOFETs. Specific reactions of enzymes with selected analytes can locally induce sizable pH changes near the sensor surface. Since interfacial dipoles are highly responsive to pH changes, the channel current of the device gets modulated. This concept has been exploited by Buth *et al.* [137], who demonstrated an enzyme-based EGOFET sensor for penicillin. They first created hydroxyl groups by UV oxidation on the α6T surface; then they functionalized the surface with (3-aminopropyl)triethoxysilane (APTES); finally, they anchored penicillinase enzyme to the amino group of the APTES molecule. The sensing mechanism is based on the specific reaction of the enzyme with penicillin, which generates a proton, thereby inducing a local change of the pH in the electrolyte, which in turn is transduced by the transistor. With such a device, sensitivity down to micromolar changes in concentration of penicillin was demonstrated (Figure 18.10d).

Khan *et al.* [138] have also shown a label-free protein sensor based on pentacene. They passivated and chemically functionalized the pentacene active layer by PE-CVD of perfluor-1,3-dimethyl-cyclohexane (PFDMCH) and MA layers, respectively. Over the additional carboxyl groups contained in the thin perfluorinated passivation layer, they then grafted bovine serum albumin (BSA) by covalently bonding it to the sensor surface. In this way, they achieved selective and real-time detection of BSA antibody with a detection limit of 500 nM (Figure 18.10e) by measurement of the transistor drain current.

Another route has been proposed for biofunctionalizing OSC surfaces previously treated to form hydroxyl, carboxyl, or amino groups. It is the case of layers of amphiphilic molecules, such as phospholipids, as described by Magliulo *et al.* [139, 144]. They demonstrated the possibility to anchor to the OSC surface a biotinylated phospholipid bilayer containing the recognition element, specifically biotin. Phospholipid vesicles containing $-NH_2$ terminal groups were anchored to the OSC surface by a reaction with the previously generated –COOH groups. Eventually, the anchored vesicles spontaneously fuse, leading to the formation of a phospholipid bilayer covalently bound to the OSC surface. The biotinylated phospholipid bilayer was used to selectively react with streptavidin, and was able to detect concentrations of this protein as low as 10 nM (Figure 18.10f). In this case, the sensing mechanism is based on the extra gating produced by the streptavidin recognition. In fact, this protein is negatively charged at the operating pH. So, an overall electrochemical potential increase was observed, as confirmed by the V_t shift toward positive gate potentials, which, consequently, led to an increase of the I_{ds} current. Moreover, the streptavidin bound to the membrane still has biotin-binding sites available, so that it could be used for further anchoring and recognition of other biotinylated molecules such as antibodies.

A strategy based on the thiol/gold chemistry as an alternative to creating anchoring groups on the OSC surface has been described by Hammock *et al.* [140, 141]. They spun a layer of block-copolymer micelles containing $HAuCl_4$ in the pyridine core on the surface of the OSC, and subsequently exposed the layer to oxygen plasma to remove the block copolymer and simultaneously reduce

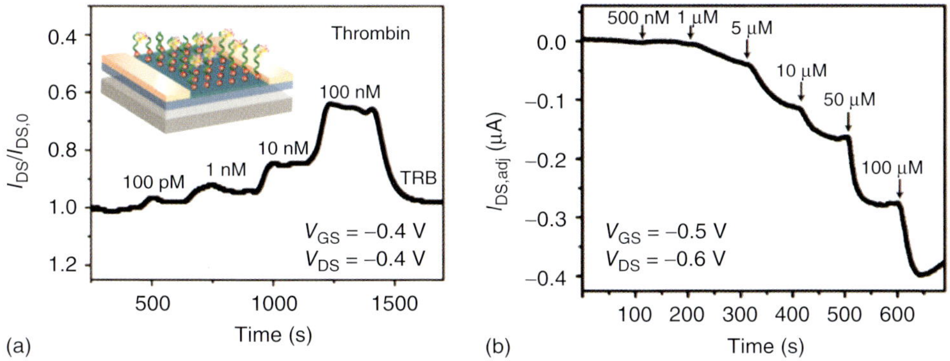

Figure 18.11 (a) Normalized I_{ds} current response upon thrombin detection by an OFET functionalized with Au-NP binding aptamers [141]. (Reproduced with permission. Copyright © 2013 American Chemical Society.) (b) I_{ds} current response upon titration of Hg^{2+} at increasing concentrations [140]. (Reproduced with permission. Copyright © 2012, American Chemical Society.)

the auric ions to metallic gold. This led to an array of uniformly spaced gold NPs immobilized on the OSC, which could be functionalized by using thiolated molecules. In the original papers, thiolated DNA oligomers were used for selective and sensitive detection of thrombin and mercury ions (Figure 18.11a,b, respectively). In the case of mercury ions, their binding to the DNA oligomers created an interfacial dipole, which, in turn, modulated the transistor current. Notably, the researchers demonstrated that for thrombin detection both the sign and magnitude of charge drastically affected the detection profile and the detection limit of the sensor. They found that, in general, increasing the overall net charge of the protein analyte resulted in increased sensitivity, which could be controlled by adjusting the pH of the buffer. The devices reported were sensitive to Hg^{2+} and thrombin concentrations of the order of 1 μM and 100 pM, respectively.

A comprehensive roadmap of graphene could be found in the paper by Novoselov *et al.* [131], in particular also for the applications of graphene in flexible electronics and biology [145].

18.3.3
Interfacing with Neural Tissue

The demand for precise and sensitive artificial systems capable of interfacing with neural tissue is increasing. Several examples of precise machine-to-brain interfacing that have been already established for therapeutic applications include cochlear implants, multielectrode arrays for neural firing analysis, deep-brain stimulation, and so on. Advanced biomedical devices have also been developed for drug screening purposes, which are able to stimulate and record in real time neural cells' bioelectrical activity. Neuro-prosthetics is another intriguing and crucial field, where devices are aimed to control, replace, or regenerate loss of neurological functions due to an injury or a disease. In many cases, noninvasive recording of neuronal activity is required, which relies on the detection of trans-membrane ionic currents that lead to small local potential changes at the sensor surface in contact with the neuron. Typical potential changes range between a few tens and hundreds of millivolts and occur in a bandwidth of up to a few kilohertz. Finally, efforts have been made also to study the interaction between neuronal cells and the substrate, in order to shed light on the chemical, mechanical, and topographic factors that contribute to the development of the neuronal cell morphology.

Among the several types of materials used so far for the applications described above, organic materials and OTFT devices offer significant advantages over traditional silicon-based devices, such as improved long-term biocompatibility, mechanical flexibility, and an adaptable form factor. Moreover, organic materials have previously been shown to be highly biocompatible, allowing the growth and survival of primary and secondary neural cells on pentacene [146, 147] and P3HT thin films [148, 149]. Examples where organic electronic materials have been used in actuators and electrodes to stimulate and regulate biological functions include drug-delivery systems, surfaces for regulating cell adhesion, and neuron-stimulating electrodes. Thus, OSCs, and in particular OTFTs, open new possibilities for neuronal transducers: operating in direct contact with water, they show large capacitive coupling to the liquid potential and efficient ion–electron conversion. Moreover, softness and flexibility are almost unique to thin films of organic electronic materials, and this represents a key feature for their exploitation in neuronal transducers. By designing implants whose softness corresponds to the surrounding tissue and which can adapt to the local topography, one can both optimize the electronic coupling to cells and minimize the mechanical stress induced by the cell–surface interaction, which can induce immune reactions and inflammatory processes on cells. Miniaturization of devices, and the consequent reduction of invasiveness of the implants, is another important advantage for organic-based devices. However, long-term biocompatibility of organic electronic materials in *in vivo* experiments has not been fully clarified yet and needs deeper investigation.

Apart from the above-mentioned features of organic-based devices, EGOFETs are well suited for interfacing to neural cells and networks because of the close relationship between the potential changes in the electrolyte (ΔV) and the resulting changes of I_{ds} current. According to Eq. (18.1), these two quantities are related to each other through the transconductance g of the transistor (defined as the ratio $\partial I_{ds}/\partial V_{gs}$), which depends on the capacitance of the EDL and the carrier mobility inside the OSC. Therefore, owing to the specific working mechanism of the device (which leads to extremely high values of capacitance) and to the material properties of the OSCs, OEGOFETs show remarkable values of g, thus maximizing the device response to the cellular events.

Some interesting applications of EGOFETs to neural tissues have been recently described, where these devices were applied to extracellular recording and stimulation of neuronal activity. Cramer *et al.* [147] used an ultrathin pentacene-based

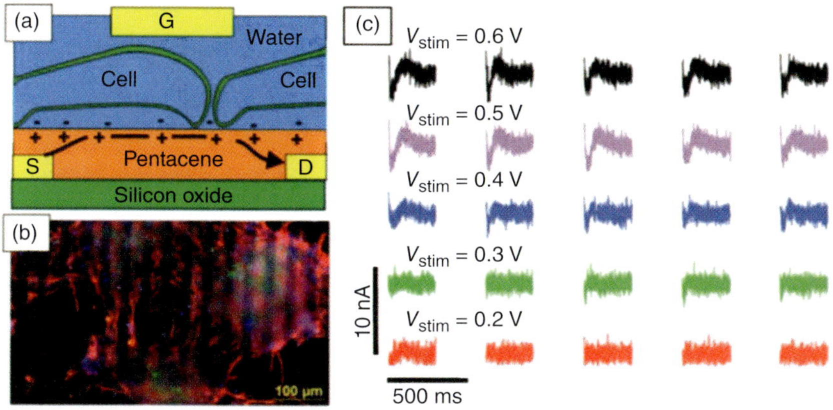

Figure 18.12 (a) Schematic of the device with cells adhering to the OSC surface. (b) Fluorescent microscopy showing the interdigitated network of differentiated neurons. (c) I_{ds} current versus time recorded from a differentiated neuronal network after electrical stimulation with various intensities [147]. (Reproduced with permission. Copyright © 2013, Royal Society of Chemistry.)

WGOFET to first induce differentiation of neural stem cells directly grown on the OSC surface into a dense neuronal network made of interconnected neurons (Figure 18.12a,b). In order to allow cell differentiation, the device was operated under cell culture conditions for up to at least 9 days, showing sufficient stability and sensitivity. Following this, the collective population of mature neurons was electrically stimulated with various intensities, and the I_{ds} current of the device was simultaneously monitored over time. Changes in the device response were recorded only for sufficiently strong stimulation (Figure 18.12c), and were traced back to changes of the extracellular potential adjacent to the transistor surface. Interestingly, the device turned out to be sensitive only to electrical signals elicited by neurons but not by undifferentiated stem cells. This work demonstrated that ultrathin film organic transistor could be a promising platform for neuronal activity transduction and stimulation as well as for stem cell-based therapies.

Benfenati *et al.* [150] reported on a transparent device that allows bidirectional stimulation and recording of primary neurons. The researchers demonstrated the possibility to detect intracellular potential changes of single neurons by means of patch-clamp measurements and fluorescence spectroscopy, providing higher SNR than MEAs.

18.3.4
Opportunities and Challenges

In this section, we provide an overview on electrical transduction with EGOFET-based devices applied to DNA sensing, enzymes, and pH detection and monitoring, as well as for recording electrical activity of neuronal networks and cells. Sensitivity to ions, compatibility with water, inherent low operating voltages coupled with current amplification, and easy and low-cost fabrication are all features of EGOFETs, ideally matching the requirements for biosensing applications. Hence, EGOFETs are good candidates for flexible, label-free, selective, *in situ*, and fast biosensing, which could lead to cheap manufactured devices with advantageous energy requirements.

Although EGOFETs are interesting and versatile devices, some issues need to be tackled to make their applications widespread, which require further research efforts to optimize their performances. For instance, the use of electrolyte gate insulators requires that all materials in contact with the electrolyte be electrochemically stable and corrosion resistant. Therefore only noble metals (Au, Pd, or Pt) or carbon-based conductors can be used for the transistor electrodes.

Another important issue is the relatively low switching speed of EGOFETs compared to Si-based OFETs, mainly due to the low ion mobility in solid polymer electrolytes. Furthermore, the device operating frequencies are intrinsically limited by the relatively poor OSC mobility. Problems come also from the source–gate and drain–gate leakage currents which cannot not be easily removed and are correlated to the structure and morphology of molecular materials. All these questions strongly motivate research activities aiming at improving the key material properties of the electrolyte dielectric layer, the OSC, and the contact interfaces. Quite relevant is also the position of the gate contact causing reliability problems, so that the need for a reference electrode has not been definitely ruled out.

18.4 OECTs for Biosensing and Biomonitoring

18.4.1 OECT Architecture and Working Principle

OECTs offer great advantages in biosensing because they can operate in liquid environments at low voltages (that is <1 V) [151]. The use of liquid electrolytes allows the ideal interfacing of electronics with biology. In an isolated device, the liquid interface is realized by confining the electrolyte reservoir by means of a micro-well generally made of an elastomeric polymer, PDMS, in contact with a conductive polymer. By far, the most common conducting polymer matrix used for OECTs is PEDOT:PSS (poly(3,4-ethylenedioxythiophene):polystyrene sulfonate), where the conducting polymer PEDOT is degenerately p-doped by the sulfonate anions contained in PSS. OECTs based on PEDOT:PSS work in the depletion mode, meaning that, upon application of a drain–source voltage (V_{ds}), holes drift within the transistor channel generating a drain–source current I_{ds} (referred to as the *on state of the transistor*). When a positive gate–source voltage (V_{gs}) is applied, cations from the electrolyte penetrate the PEDOT:PSS channel and compensate the doping, according to the following equation:

$$\mathrm{PEDOT^+ : PSS^- + M^+ + e^- \leftrightarrow PEDOT^0 + M^+ : PSS^-} \qquad (18.3)$$

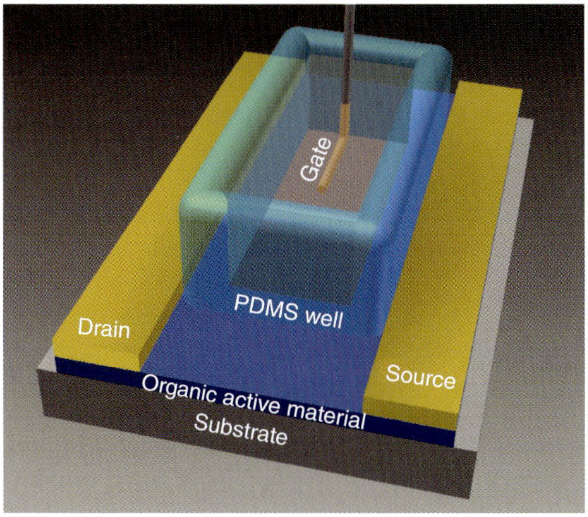

Figure 18.13 OECT device layout. The dark blue layer represents the active organic semiconductor. The electrolyte is confined on the organic layer through a PDMS well with the gate electrode immersed in it. The overlap defines the channel of the OECT. Metallic electrodes (source and drain, yellow) are placed at the edge of the organic stripe [151]. (Reproduced with permission. Copyright © 2012, Royal Society of Chemistry.)

This de-doping effect induces a decrease of the polymer conductivity, which results in a decrease of I_{ds} (referred to as the *off state of the transistor*) (Figure 18.13). The overlapping of the PEDOT:PSS stripe with the electrolyte defines the channel of the transistor, wherein the ionic interchange can occur [152]. Appropriate choice of the couple gate electrode–electrolyte allows control of the voltage drop at the gate/electrolyte and polymer/electrolyte interfaces, and hence the use of the transistor in a sensing mode of operation. Importantly, the OECT circuit schematic consists of both an electronic part and an ionic part. The electronic current in the channel is determined by the density of carriers, while the ionic flow is caused by ions contained in the electrolyte [153]. In the last years, great progress has been achieved in OECT-based bioapplications, especially with cells cultured, grown, and measured directly inside the OECT architecture.

Depending on the specific aim the OECT is used for, it can be exploited both as an *ion-to-electron amplifying* converter or as an *electrochemical sensor* [154]. When used as simple transducer, the principle of operation of the OECT is based on the de-doping induced into the channel by the cations coming from the electrolyte. Specifically, upon application of a positive source–gate bias (V_{gs}), cations drifting toward the organic polymer enter it and de-dope it. In this way, a decrease in the source–drain current (I_{ds}) takes place. This regime of operation is referred to sometimes as the *capacitive mode* or the *non-faradaic mode*. Under the electrochemical sensing mode, a redox reaction occurs at the gate electrode as a result of a charge transfer between analytes in the electrolyte and the gate electrode (this operation regime is referred to as the *faradaic mode*). Numerical analyses [154] have shown that, in the transduction mode, a gate electrode much larger than the channel gives a higher I_{ds} modulation and, as a consequence, an optimized value of the OECT transconductance g. Recently, it has been shown that OECT can yield a transconductance in the mS range, outperforming transistors based on traditional semiconductors and maintaining it constant up to 1 kHz of frequency [155].

18.4.2 The Applications of OECT as a Biological Sensor

As enzymatic sensors, OECTs have already been extensively investigated, and readers can refer to complete reviews that exhaustively cover this area of application [151, 120]. In the following, we report the newest and advanced applications of OECT in the biosensing and biological area. In particular we will discuss the use of OECT acting as a biosensor for drug carriers or for monitoring the dynamics of biomolecules. In addition, very recent applications of OECT coupled with cell cultures for monitoring the health and damage of barrier tissues and ion flux from cells will also be discussed.

18.4.2.1 Drug Nanocarriers for Drug Delivery

Micelles are of primary interest in drug delivery because of their potential use as drug carriers [156, 157]. An OECT sensor capable of detecting the onset of micelle formation has been demonstrated by using the cationic surfactant hexadecyl trimethyl ammonium bromide (CTAB) as electrolyte [158]. Above the critical micellar concentration (cmc), the surfactant forms positively charged spherical micelles detectable via the OECT. The device response undergoes a large increase above the cmc, revealing that positively charged CTAB micelles de-dope the PEDOT:PSS underneath more efficiently than the dissociated ions (Figure 18.14).

The ability to sensitively detect the assembly of this type of vesicles paves the way to real-time monitoring of drug loading and for realizing the mechanism and dynamics which are of paramount importance in pharmaceutical applications and nanomedicine and which presently require cumbersome procedures and sophisticated instrumentation, thereby strongly limiting the field.

In this framework, liposomes are an extensively investigated class of drug carriers [159], which have been studied by Tarabella *et al.* using an OECT platform [160]. Two types of liposomes have been analyzed: the first type with a bare surface made of lipids of lecithin, and the other with the surface functionalized by chitosan (referred to as *liposome-based nanoparticles*). The authors demonstrated that OECTs are ideal tools for sensing and real-time monitoring of lipids structures dissolved in solutions, and investigated in flow-through mode a dedicated microfluidic system (Figure 18.15). The detection limit of OECTs for the liposome-based NP has shown a wide dynamic range, down to 10^{-7} mg ml^{-1}. Such a dynamic range meets the requirement for typical drug-loading and drug-delivery conditions. The integration of the OECT with microfluidics leads to the possibility to discriminate different kinds of liposomes upon successive injections, providing

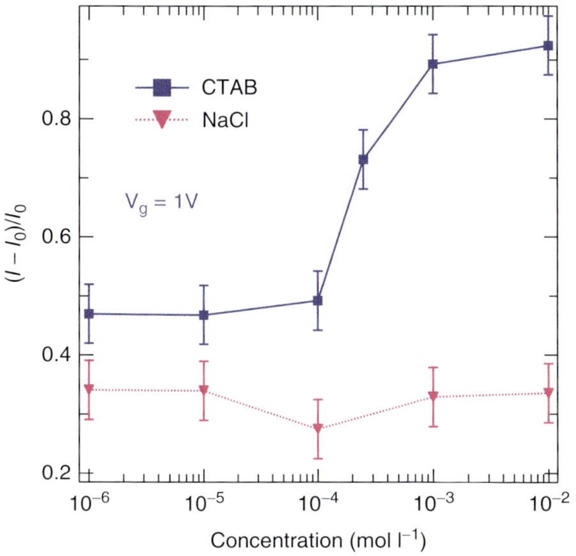

Figure 18.14 OECT current modulation versus [CTAB] and [NaCl] ($V_g = 1$ V, $V_d = -0.4$ V). At about 10^{-4} M in [CTAB], an increase of the OECT response is observable, which corresponds to the onset of the CTAB critical micellar concentration [158]. (Reproduced with permission. Copyright © 2012, Royal Society of Chemistry.)

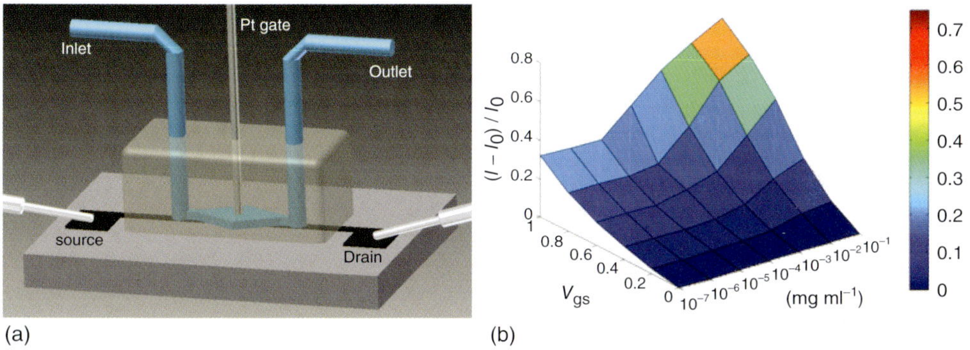

Figure 18.15 (a) OECT integrated with microfluidics. (b) 3D plot of the liposome-based nanoparticles sensor: OECT response plotted versus V_g and liposome concentration (expressed as milligram per milliliter) [161]. (Reproduced with permission. Copyright © 2013, Elsevier.)

a potential viable solution for assays and quality control in the pharmaceutical industry.

18.4.2.2 **Dopamine and Eumelanin Sensing**

OECT sensing mechanism is particularly well suited to monitoring the activity of catecholamine molecules because of the ability of catecholamine to transfer electrons by reacting with the surface of a platinum electrode (used as gate electrode),

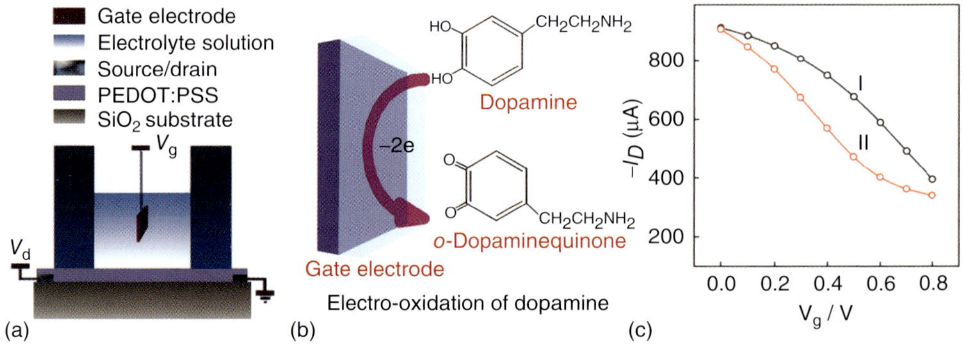

Figure 18.16 (a) Schematic layout of the OECT-based dopamine sensor. (b) Sensing mechanism of the dopamine detection. (c) Gate voltage shift upon addiction of dopamine (II) in the PBS batch (I) [162]. (Reproduced with permission. Copyright © 2011, Elsevier.)

so that the OECT can work in this case as an electrochemical sensor under the faradaic operating mode. Two examples are already present in the literature: the dopamine sensor developed in 2010 by Tang *et al.* [162], and the recent eumelanin sensor, developed by Tarabella *et al.* [163]. The latter study, in particular, shows that it is possible to extend the capability of the OECT sensor toward its application as a *tool* for investigating the redox properties of biomolecules, such as biopolymer pigments.

- *Dopamine sensor.* Dopamine is a neurotransmitter that plays an important role in neurophysiological functions. Some known illnesses, such as Parkinson's disease, are correlated with abnormal levels of dopamine. A highly sensitive OECT-based dopamine sensor was demonstrated for the first time in 2011 (Figure 18.16a) [162]. The sensing mechanism proposed for dopamine detection is shown in Figure 18.16b: dopamine undergoes an electroxidation at the platinum gate surface and an electron is transferred to the gate circuit of the OECT, generating a faradaic current. Then the potential drop at the gate/electrolyte becomes lower and the effective gate voltage (V_g^{eff}) acting on the ionic circuit increases, leading to a horizontal shift of the transfer curve (Figure 18.16c, I_{ds} vs V_g). Three different gate electrodes have been tested to figure out the best sensing performance (graphite, Au, and Pt electrodes) for an OECT transducer element made of PEDOT:PSS. The authors found that OECT sensor working with a Pt gate electrode showed the highest sensitivity, with a detection limit down to 5 nM.
- *Eumelanin sensor.* Eumelanins are the characteristic black, insoluble, heterogeneous biopolymers of human skin, hair, and eyes, produced in epidermal melanocytes by the tyrosinase-catalyzed oxidation of tyrosine [164]. Dysfunctions of eumelanin levels have been associated with different disorders, from melanoma (the most diffuse skin cancer) to neurodegenerative disorders, and the direct determination of eumelanin redox states and levels in a pathological setting would be a highly desirable goal [165].

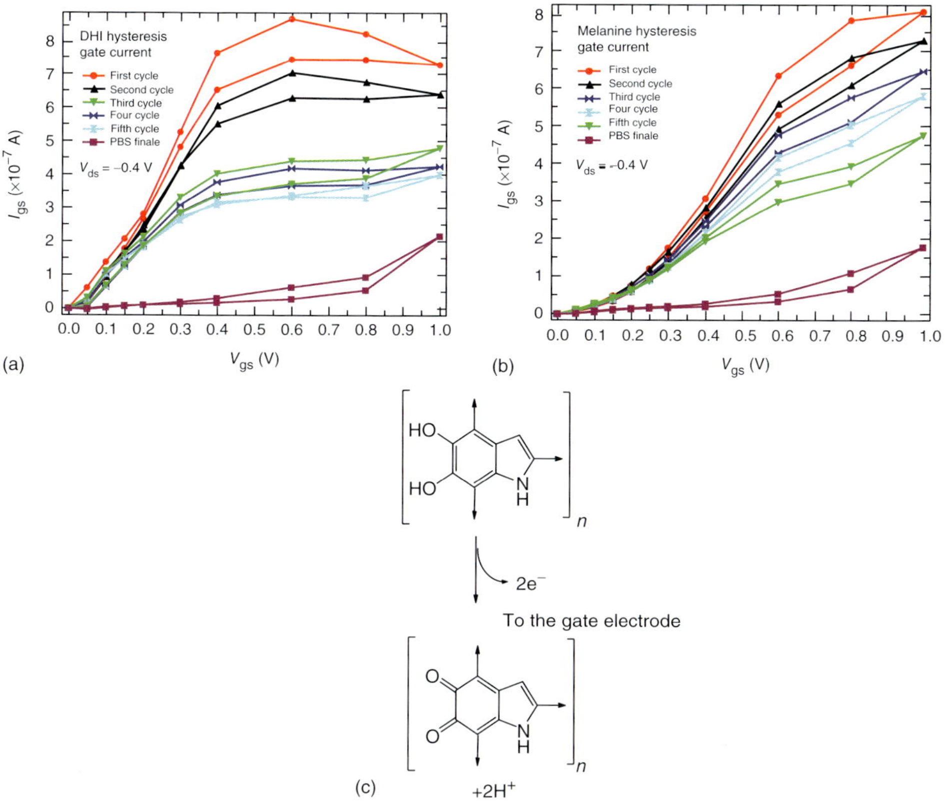

Figure 18.17 Hysteresis loop cycles of the gate current acquired for (a) DHI and (b) eumelanin pigment. $V_{ds} = -0.4$ V. (c) Proposed electron transfer mechanism at the Pt gate electrode of the catechol/quinone components [163]. (Reproduced with permission. Copyright © 2013, Royal Society of Chemistry.)

With the aim to directly investigate eumelanin's electrical response inside a biomimetic aqueous suspension state, an OECT-based eumelanin sensor has been developed that could effectively discriminate the monomer (5,6-dihydroxyindole, DHI) from the polymer (eumelanin) with a nominal concentration down to 10^{-6} M. The analysis of the gate currents demonstrated that eumelanin–OECT is a nice tool for investigating the redox properties of the analyzed molecules. In fact, the gate current directly measures the electron transfer from solid eumelanin to the Pt electrode surface, which is sustained by the redox-active catechol/quinone components of the polymer. The transfer curve of the gate current (I_{gs} vs V_g) gives hence an estimate of the threshold voltage at which the electron transfer begins to take place. The analysis of the gate current hysteresis loops observed for the DHI and eumelanin (Figure 18.17a,b) suggested an evolution of the polymer from a far-from-equilibrium redox state toward a more stable electronic arrangement promoted by redox exchange with the gate electrode (Figure 18.17c).

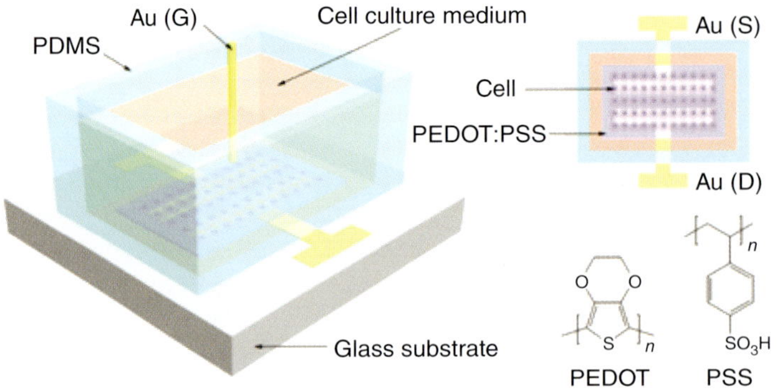

Figure 18.18 Schematic of the cell biosensor based on OECT technology. The pink color represents the cell culture medium used as an electrolyte [167]. (Reproduced with permission. Copyright © 2010, Wiley.)

18.4.2.3 **Sensing Cell and Bacterial Activity**

One of the first applications of OECT on cell studies was concerned with the control of the density gradient of epithelial cells grown along the channel of the OECT [166]. In this work, authors showed the possibility of controlling *in situ* the adhesion and proliferation of Madin–Darby canine kidney (MDCK) cells grown directly on the transistor channel. They used the PEDOT doped with tosylate ((poly(3,4-ethylenedioxythiophene):tosylate, PEDOT:TOS) and successfully observed the dependence of the cell density on the source–gate voltage applied (V_{gs}).

A cell-based biosensor in which PEDOT:PSS was the transducer element was successively demonstrated by Yan *et al.* [167]. In their study, normal and cancer cells were cultivated on the channel of the OECT and left to adhere. The transfer characteristics were measured after the adhesion and then upon the detachment of cells realized by a standard treatment with trypsin. The transfer characteristic showed a shift of the gate voltage of ~150 mV (Figure 18.18), demonstrating the sensitivity of the device to the presence of cells on the channel. Despite the fact that the sensor was capable of registering only gross cell detachment, this work demonstrated that OECT based on PEDOT:PSS was sensitive to the surface charge and morphology of the attached cells.

Another emerging and very promising application concerns barrier tissues that play many important roles in an organism's functions and as biological interfaces. They can act as selective filters or as elements maintaining highly regulated fluid compartments. Current technologies for measuring the permeability of barrier tissues are expensive and time consuming. A novel OECT design was conceived by Jimison *et al.* [168] with the aim of developing a low-cost and handy device able to dynamically address the health of a tissue layer, which may be reversibly damaged. A transwell (Twell) membrane support was incorporated into the OECT through a dedicated PDMS lodge in order to monitor in real-time experiments the barrier tissue integrity (Figure 18.19a). Caco-2 cells were grown and left to adhere on the

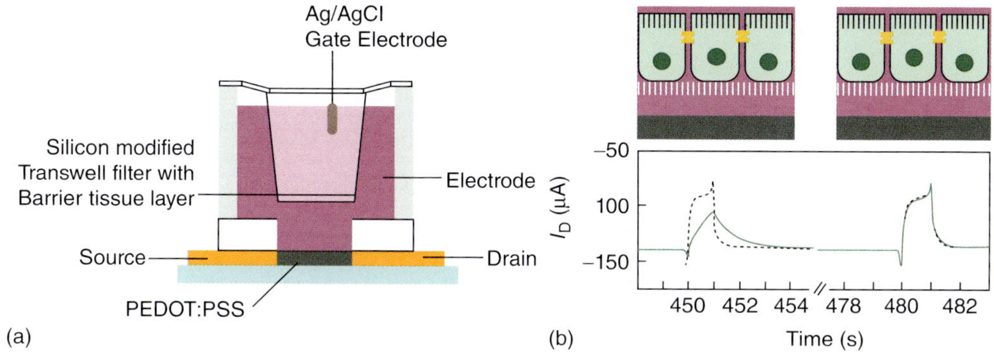

Figure 18.19 Integration of barrier tissue with an OECT. (a) Device architecture. (b) Monitoring the barrier tissue integrity with an OECT. Polarized Caco-2 cells with (top left) and without (top right) tight junctions (showed in yellow). Bottom panel: OECT I_{ds} transient response with cells before (left) and after (right) the addition of 100 mM H_2O_2, (solid lines). I_{ds} response without cells is represented by a dashed line [168]. (Reproduced with permission. Copyright © 2012, Wiley.)

bottom of the transwell for 21 days until a tight junction (TJ) was formed. Then the disruption of the TJ was induced by the injection of hydrogen peroxide (H_2O_2) or ethanol. The introduction of a Twell support imparts an added resistance in the OECT ionic circuit. Analysis of the kinetics curve, that is, the electrical signals measured with OECT, demonstrated the capability of monitoring the barrier tissue integrity by the OECT/Twell system. In particular, the time constant of the drain current $I_{ds}(t)$ has been demonstrated to change when the barrier tissue is damaged (Figure 18.19b). Finally, the OECT response compared to that of permeability assay shows the same sensitivity of this conventional test in the case of EtOH and higher sensitivity with respect to H_2O_2.

The same group of researchers has also shown in another work the possibility to monitor the disruption of the epithelial barrier tissue produced by the infection of the pathogen *Salmonella typhimurium* [169]. The monitoring process is based, in this case, on the rupture of TJs induced by the infection, which, in turn, causes an increase of the ionic flux from the apical to the base side of the cell layer (Figure 18.20).

Yao *et al.* [170] have proposed an innovative alternative system for monitoring the ionic flux of cells cultivated directly on the channels (again made of PEDOT:PSS) of an OECTs-based microarray. Calu-3 cells, a well-known model cell line capable of forming TJs, were cultivated and left to adhere for 7 days on the OECT array (Figure 18.21a). The sensing mechanism proposed in this work is based on the opening of an apical anion channel, the cystic fibrosis transmembrane conductance regulator (CFTR), responsible of transepithelial currents. Dysfunction of CFTR is related, for instance, to cystic fibrosis. The opening of CFTR, through the addiction of forskolin, causes a flux of chloride anions (Cl^-) inside the cells, from the basolateral to the apical side of the cell. A negative bias layer is therefore formed, generating on its turn a sodium cation

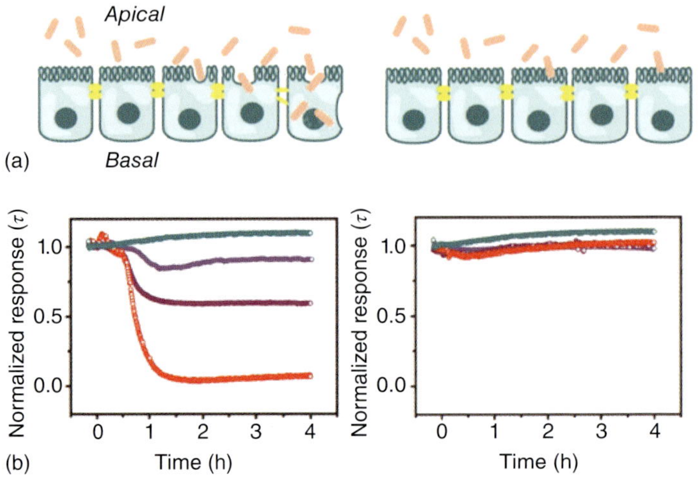

Figure 18.20 Kinetics of polarized epithelial monolayer infected with *Salmonella typhimurium*. (a) Cartoon illustrating infection with Wild-Type (WT) (left) and Non-Invasive (NI) *S. typhimurium* (right). (b) Mean normalized response (τ) of the OECT in the presence of WT (left) and NI *S. typhimurium* (right) at different MOI over 4 h, bacteria were added at $t = 0$. Noninfected represents OECT + cells with no added bacteria. Noninfected cells are in cyan, Multiplicity of Infection (MOI):10 in blue, MOI:100 in purple, and MOI:1000 in red [169]. (Reproduced with permission. Copyright © 2014, Wiley.)

flow from the medium culture at the basolateral side in the space between the cells and the PEDOT:PSS channels, toward the apical side. Because of the ion-sensitive properties of OECT, the authors noticed that the transistor was capable of detecting the change in concentration of Na^+. Control experiments performed with and without cells, and without the addiction of forskolin, seemed to confirm the proposed sensing mechanism (Figure 18.21b). This work is interesting because it represents the first proof of cells' physiological ion current monitoring by an OECT. An OECT-based bacterial sensor was proposed by He *et al.* [172] for the detection of enterohemorrhagic *Escherichia coli*. The sensing of *E. coli* was achieved by functionalizing the PEDOT:PSS channel of the OECT. The anti-*E. coli* antibody was immobilized on the surface of the PEDOT:PSS. Upon the capture of the bacteria by the antibody, an electrostatic interaction occurring at the polymer surface caused a change in the gate voltage shift. The achieved detection limit was down to 10^3 cfu ml^{-1} in bacterial concentration.

18.4.2.4 DNA

The first label-free DNA sensor based on OECT technology was reported by Krishnamoorthy *et al.* [173]. The channel of the OECT was made of PEDOT, which was realized by electropolymerization from EDOT (ethylenedioxythiophene) in the presence of the single-stranded deoxyribonucleic acid (ssDNA) probe. The sensing mechanism is based on the coupling of the DNA probe with its complementary DNA, inducing a change in the PEDOT conductivity recordable

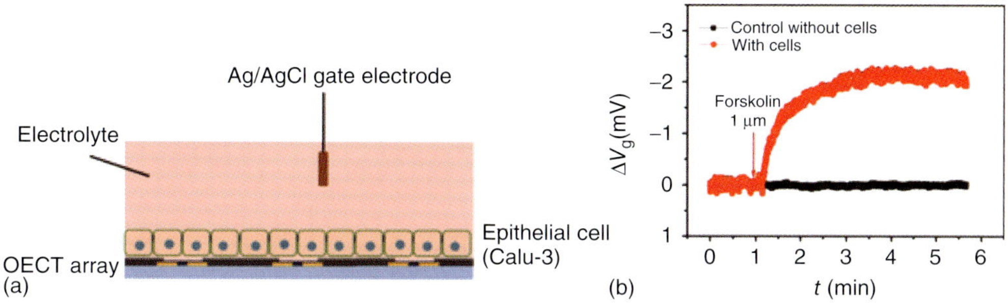

Figure 18.21 (a) Epithelial Calu-3 cells integrated with an OECT array. (b) OECT response with (red) and without (black) Calu-3 cells upon the addition of 1 μM CFTR agonist forskolin [171]. (Reproduced with permission. Copyright © 2013, Wiley.)

as a change in the source–drain current. A detection limit down to 80 ng ml^{-1} has been achieved, and the proposed mechanism has been confirmed by control experiments performed with and without noncomplementary ssDNA.

Recently, an OECT-based label-free DNA sensor was fabricated on a flexible substrate made of PET and integrated in a lab-on-chip system [174]. The system has been shown to be stable under bending and is capable of detecting complementary DNA targets down to 1 nM. The sensing mechanism is based on the immobilization of ssDNA probes on the Au gate electrode surface. The interaction between the electrode and the DNA molecules leads to a change in the work function of the electrode, which is hence reflected on the OECT response. Furthermore, the authors demonstrated the possibility of improving the sensing limit down to 1 pM by applying continuous voltage pulses on the gate electrode during DNA hybridization.

18.4.2.5 Biosensing Toward e-Textile Applications

The great potential of OECTs based on the PEDOT:PSS polymer has been also demonstrated by integrating the OECT device on a flexible cotton fiber substrate. Mattana *et al.* [175] first proposed an innovative deposition processes to obtain conductive yarns from insulating cotton fibers, which in turn were used as the base to build an OFET and an OECT device. Subsequently, Tarabella *et al.* [176] realized an wearable e-textile OECT – fully integrated on a cotton fiber – for the measurement of the saline concentration in real human sweat, with a concentration in the range 10^{-1} – 10^{-4} M; the device can work at very low voltages, as low as 0.2 V, thus implying a great potential in implementing this kind of OECT sensor in practical and physiological conditions.

18.4.3 Organic Electronic Ion Pumps (OEIPs)

OECT acting as a sensitive ion-to-electron transducer represents the first example of an organic device particularly well suited to interface electronics with

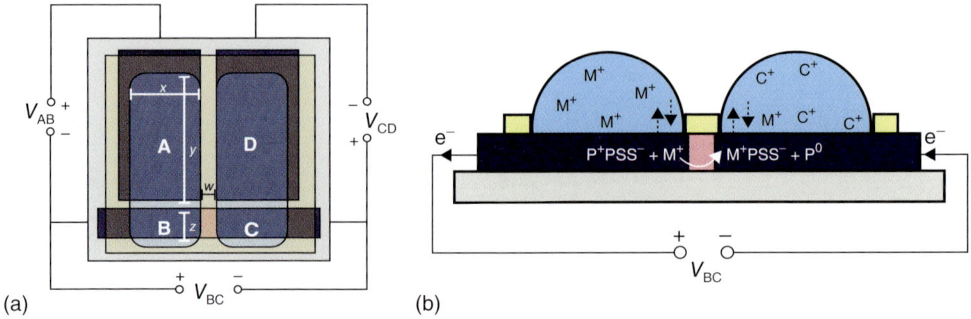

Figure 18.22 (a) Schematic top view of the OEIP layout with related driving biases and (b) schematic cross section of the electrodes and related ionic exchange between electrolyte reservoirs mediated by the organic layer underneath [178]. (Reproduced with permission. Copyright © 2007, McMillan Publisher Ltd.)

biology [151]. Another example of a device interfacing ionics and electronics is the organic electronic ion pump (OEIP). This device is able to deliver charged biomolecules with high and controllable spatiotemporal resolution [177]. The OEIP consists of two areas of PEDOT:PSS joined by an over-oxidized transport channel (Figure 18.22), the over-oxidation leading to the PEDOT:PSS being unable to conduct electrons, making only ionic diffusion possible [178]. By applying a bias between the electrodes at the edge of the two polymeric areas, a cationic migration is established between the oxidized and reduced areas, governed by the two electrochemical half-reactions:

$$\text{PEDOT}^0 + \text{Ion}^+ : \text{PSS}^- \rightarrow \text{PEDOT}^+ : \text{PSS}^- + \text{Ion}^+ + \text{e}^- \tag{18.4}$$

$$\text{PEDOT}^+ : \text{PSS}^- + \text{Ion}^+ + \text{e}^- \rightarrow \text{PEDOT}^0 + \text{Ion}^+ : \text{PSS}^- \tag{18.5}$$

Cations from the source of OEIP are pushed toward the target reservoir through the over-oxidized channel. It is worth noting that an OEIP resembles the transport of charged molecules in a standard gel-electrophoresis system. The main difference with the latter system is represented by the fact that in OEIP the cations can migrate inside a polymeric matrix without the presence of a liquid. A prototypical OEIP was able to transport laterally and electronically cations such as K^+, Na^+, and Ca^{2+} ions (and positively charged biomolecules) in a controlled way [178]. Later, OEIPs were applied for transporting also other kinds of molecules, in particular small-molecule neurotransmitters such as acetylcholine, glutamate, and aspartate [179].

In conclusion, the above results show some of the most recent applications of OECT in the biosensing field, but other novel and very advanced fields of research are growing quickly. For instance, the literature shows that it is possible to build logic circuits based on ionic transistors in which the logic states are generated by ionic current flows, the ions coming from electrolytes in the aqueous phase [180].

The combination of OECT and OEIP is of extraordinary potential because a fully novel field called *iontronics* [181] has been proposed that could profitably combine sensing and delivery of ions interacting directly with major biomedical functions including neuronal activities up to innovative prosthetic applications. In this framework, the efforts toward flexible, stretchable, and biofunctional electronic devices are being very much explored and under way.

In this context, one the most relevant and advanced application of OECT is in the area of neurosciences, where OECTs, realized on a flexible substrate, are used as electrodes interfacing directly with the brain for recording brain signals. Recording electrical signals generated by the brain is of paramount importance to gain information on the organization of the brain, including the discovery of place and grid cells and mirror neurons, and as a great tool for developing therapeutically important approaches. As already mentioned, recording brain signals is currently carried out with array of metal electrodes (MEA) made of silicon- or Ir-based probes detecting the average electrical local filed of a given brain volume. Khodagholy *et al.* [182] demonstrated the ability of OECT as an efficient *in vivo* electrode for recording brain activity. They fabricated arrays of OECTs on a flexible substrate to be conformable to the nonflat area of the cortex and used them to carry out ECoG on the somatosensory cortex of rats (OECT-based ECoGs). The OECT response was compared with the corresponding response obtained with surface electrodes made of PEDOT:PSS or Ir-penetrating electrodes (Figure 18.23a,b). The temporal coincidence of the peaks in the data indicated that the transistor recorded the same information as the electrodes. OECT showed superior performance in terms of SNR thanks to the amplification effect of the transistor. The calculated SNR for OECTs was 44 dB, whereas the PEDOT:PSS surface electrode yielded an SNR of 24.2 dB. The OECT performance is superior to that of the silicon probe already discussed previously, about 34 dB. The specific case studied refers to the firing of neurons under an epileptic attack, where the localization of a strong firing activity is of great importance and could guide surgical or other therapeutic approaches.

18.5
Conclusions and Outlook

The great progress and the fantastic acceleration achieved in flexible electronics, together with the ability to go into large-area production, are having a very strong impact in a field with high social relevance, such as bioelectronics. Independent of the substrate used, the different strategies require a strong interdisciplinary approach, where the convergence of know-how and technology makes the difference. Inorganic-, hybrid-, and organic-based electronics are all contributing quite significantly to the fast evolution of the field, and seem to be quite complementary depending strongly on the type of application. The nano-revolution is playing a major role in all cases where molecular and nanosystems are rapidly demonstrating their disruptive potential, with solutions that exploit new paradigms in sensing as well as in transduction efficiency and mechanisms. Combining high-performing electronic properties with flexibility and stretchability properties is

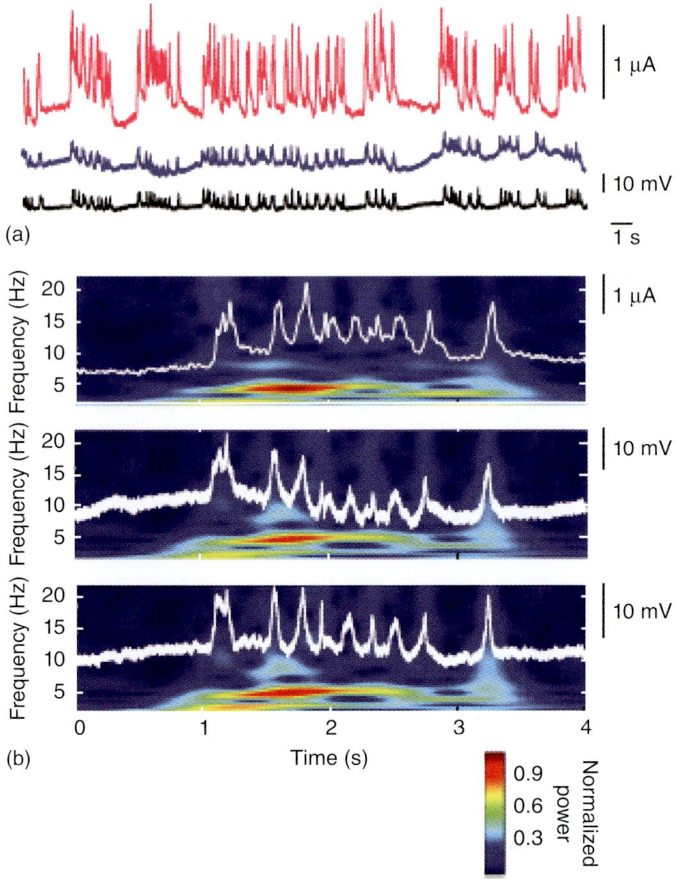

Figure 18.23 Validation in Genetic Absence Epilepsy in Rats from Strasbourg (GAERS) rats. (a) Recordings from an OECT (pink), a PEDOT:PSS surface electrode (blue), and an Ir-penetrating electrode (black). The transistor was biased with $V_d = 0.4\,V$ and $V_g = 0.3\,V$; the scale of 10 mV is for both surface and penetrating electrodes. Note the superior SNR of the OECT as compared with the surface electrode. (b) Time–frequency analysis of epileptiform activity during a short period, recorded by an OECT (top), a PEDOT:PSS surface electrode (middle), and an Ir-penetrating electrode (bottom) [182]. (Reproduced with permission. Copyright © 2007, McMillan Publisher Ltd.)

becoming more and more a reality with a huge number of applications among which the biomedical ones could be considered one of the most interesting and socially impacting. Of course, for many reasons we are still just at the beginning of the possible developments. To this end, one of the major questions, which is far from being fully resolved, is the ability to have reliable solutions in prosthetics, where the needed long-lasting performance in terms of active electronic biofunctionality is far from demonstrated. In this same framework, others key issues, which were not addressed in this chapter but are quite critical, are the problems connected to wiring and power supplies. The aim is to reduce cable connections to a real minimum for obvious reasons, knowing that the approach based on MEAs requires a large density of interconnections so that the need of other smart strategies is quite evident. From the energetic point of view, the progress in energy harvesting materials and systems could possibly bring in solutions that would not require external power supplies, even though ready-to-use solutions are not yet available.

We believe that bioelectronics and biosensing together are fields that challenge researchers and technologists from diverse fields. The first challenge is for material scientists, since the ultimate materials joining all the functionalities needed are not available and difficult to conceive so that a variety of specific, different solutions should be explored and tested. Then there are technological challenges, because flexible and innovative architectures are strongly required to smartly interface our standard electronic armamentarium with biomedical requirements. Then medical doctors have to come up with novel strategies for incorporating electronic systems in living bodies. The overall feeling is that this is a field of great interest and excitement that is attracting a rapidly increasing number of scientists.

References

1. Pene, F., Courtine, E., Cariou, A., and Mira, J.-P. (2009) *Crit. Care Med.*, **37**, S50–S58.
2. Owens, R.M. and Malliaras, G.G. (2010) *MRS Bull.*, **35**, 449.
3. Willner, I. and Katz, E. (2005) *Bioelectronics: From Theory to Applications*, Wiley-VCH Verlag GmbH, Weinheim.
4. Pethig, R.R. and Smith, S. (2012) *Introductory Bioelectronics: For Engineers and Physical Scientists*, John Wiley & Sons, Ltd, Chichester.
5. Söderström, T., Haug, F.-J., Terrazzoni-Daudrix, V., and Ballif, C. (2008) *J. Appl. Phys.*, **103**, 114509.
6. Rampino, S., Armani, N., Bissoli, F., Bronzoni, M., Calestani, D., Calicchio, M., Delmonte, N., Gilioli, E., Gombia, E., Mosca, R., Nasi, L., Pattini, F., Zappettini, A., and Mazzer, M. (2012) *Appl. Phys. Lett.*, **101**, 132107.
7. Heikenfeld, J., Drzaic, P., Yeo, J.-S., and Koch, T. (2011) *J. Soc. Inf. Disp.*, **19**, 129.
8. Fang, C., Agarwal, A., Widjaja, E., Garland, M.V., Wong, S.M., Linn, L., Khalid, N.M., Salim, S.M., and Balasubramanian, N. (2009) *Chem. Mater.*, **21**, 3542–3548.
9. He, Y., Kang, Z.-H., Li, Q.-S., Tsang, C.H.A., Fan, C.-H., and Lee, S.-T. (2009) *Angew. Chem.*, **121**, 134–138.
10. Das, S., Gulotty, R., Sumant, A.V., and Roelofs, A. (2014) *Nano Lett.*, **14**, 2861–2866.
11. Moon, T., Jung, J.-C., Han, Y., Jeon, Y., Koo, S.-M., and Kim, S. (2012) *IEEE Trans. Electron Devices*, **59**, 3288–3291.
12. Wang, W. and Soper, S.A. (2006) *Bio-MEMS: Technologies and Applications*, CRC Press, Boca Raton, FL.
13. Choudhary, V. and Iniewsky, K. (2013) *MEMS: Fundamental Technology and Applications*, CRC Press, Boca Raton, FL.
14. Yaman, M., Khudiyev, T., Ozgur, E., Kanik, M., Aktas, O., Ozgur, E.O., Deniz, H., Korkut, E., and Bayindir, M. (2011) *Nat. Mater.*, **10**, 494–501.
15. Ghane-motlagh, B. and Sawan, M. (2013) *Mater. Sci. Appl.*, **4**, 483–495.
16. Kim, D.-H., Ghaffari, R., Lu, N., and Rogers, J.a. (2012) *Annu. Rev. Biomed. Eng.*, **14**, 113–28.
17. Fromherz, P., Offenhausser, A., Vetter, T., and Weis, J. (1991) *Science*, **252**, 1290–1293.

18. Friend, R.H., Gymer, R.W., Holmes, A.B., Burroughes, J.H., Marks, R.N., Taliani, C., Bradley, D.D.C., Dos Santos, D.A., Bredas, J.L., Logdlund, M., and Salaneck, W.R. (1999) *Nature*, **397**, 121–128.
19. Nunzi, J.-M. (2002) *C.R. Phys.*, **3**, 523–542.
20. Klauk, H. (2010) *Chem. Soc. Rev.*, **39**, 2643–2666.
21. IDTechEx (2014) Printed and Flexible Sensors 2014-2024: Technologies, Players, Forecasts.
22. Yan, F., Zhang, M., and Li, J. (2013) *Adv. Healthcare Mater.*, **3** (3), 313–331.
23. European Union (2013) Graphene Flagship, *http://graphene-flagship.eu/* (accessed 18 July 2014).
24. Novoselov, K.S., Geim, A.K., Morozov, S.V., Jiang, D., Zhang, Y., Dubonos, S.V., Grigorieva, I.V., and Firsov, A.A. (2004) *Science*, **306**, 666.
25. Torrisi, F., Hasan, T., Wu, W., Sun, Z., Lombardo, A., Kulmala, T.S., Hsieh, G.-W., Jung, S., Bonaccorso, F., Paul, P.J., Chu, D., and Ferrari, A.C. (2012) *ACS Nano*, **6**, 2992–3006.
26. He, Q., Wu, S., Gao, S., Cao, X., Yin, Z., Li, H., Chen, P., and Zhang, H. (2011) *ACS Nano*, **5**, 5038–5044.
27. Lee, S.-K., Kim, B.J., Jang, H., Yoon, S.C., Lee, C., Hong, B.H., Rogers, J.A., Cho, J.H., and Ahn, J.-H. (2011) *Nano Lett.*, **11**, 4642–4646.
28. Sirringhaus, H., Tessler, N., and Friend, R. (1998) *Science*, **280**, 1741–1744.
29. Gelinck, G.H., Huitema, H.E.A., van Veenendaal, E., Cantatore, E., Schrijnemakers, L., van der Putten, J.B.P.H., Geuns, T.C.T., Beenhakkers, M., Giesbers, J.B., Huisman, B.-H., Meijer, E.J., Benito, E.M., Touwslager, F.J., Marsman, A.W., van Rens, B.J.E., and de Leeuw, D.M. (2004) *Nat. Mater.*, **3**, 106–110.
30. Zhou, L., Wanga, A., Wu, S.-C., Sun, J., Park, S., and Jackson, T.N. (2006) *Appl. Phys. Lett.*, **88**, 083502.
31. Andersson, P., Forchheimer, R., Tehrani, P., and Berggren, M. (2007) *Adv. Funct. Mater.*, **17**, 3074–3082.
32. Baude, P.F., Ender, D.A., Haase, M.A., Kelley, T.W., Muyres, D.V., and Theiss, S.D. (2003) *Appl. Phys. Lett.*, **82**, 3964.
33. Steudel, S., Myny, K., Arkhipov, V., Deibel, C., De Vusser, S., Genoe, J., and Heremans, P. (2005) *Nat. Mater.*, **4**, 597–600.
34. Mabeck, J.T. and Malliaras, G.G. (2006) *Anal. Bioanal. Chem.*, **384**, 343–353.
35. Lin, P. and Yan, F. (2012) *Adv. Mater. (Deerfield Beach)*, **24**, 34–51.
36. Kergoat, L., Piro, B., Berggren, M., Horowitz, G., and Pham, M.-C. (2012) *Anal. Bioanal. Chem.*, **402**, 1813–1826.
37. Kim, S.H., Hong, K., Xie, W., Lee, K.H., Zhang, S., Lodge, T.P., and Frisbie, C.D. (2013) *Adv. Mater.*, **25**, 1822–1846.
38. Chen, C.-H., Liu, K.-Y., Sudhakar, S., Lim, T.-S., Fann, W., Hsu, C.-P., and Luh, T.-Y. (2005) *J. Phys. Chem. B*, **109**, 17887–17891.
39. Mehrabani, S., Maker, A.J., and Armani, A.M. (2014) *Sensors (Basel)*, **14**, 5890–5928.
40. Jiang, W., Singhal, A., Fischer, H., Mardyani, S., and Chan, W.C.W. (2007) in *BioMEMS and Biomedical Nanotechnology, Vol. 3 Therapeutic Micro/Nanotechnology* (eds M. Ferrari, T. Desai, and S. Bhatia), Springer, Boston, MA, pp. 137–156.
41. Kwon, S.G. and Hyeon, T. (2008) *Acc. Chem. Res.*, **41**, 1696–1709.
42. Galian, R.E. and de la Guardia, M. (2009) *TrAC, Trends Anal. Chem.*, **28**, 279–291.
43. Lin, J.-C., Huang, B.-R., and Yang, Y.-K. (2013) *Sens. Actuators B*, **184**, 27–32.
44. Singh, S.P., Arya, S.K., Pandey, P., Malhotra, B.D., Saha, S., Sreenivas, K., and Gupta, V. (2007) *Appl. Phys. Lett.*, **91**, 063901.
45. Stoumpos, C.C., Malliakas, C.D., and Kanatzidis, M.G. (2013) *Inorg. Chem.*, **52**, 9019–9038.
46. Liu, M., Johnston, M.B., and Snaith, H.J. (2013) *Nature*, **501**, 395–398.
47. Weimer, P.K. (1983) *Proc. IRE*, **50**, 1462–1469.
48. Yoon, Y., Ganapathi, K., and Salahuddin, S. (2011) *Nano Lett.*, **11**, 3768–3773.
49. Mitzi, D.B., Kosbar, L.L., Murray, C.E., Copel, M., and Afzali, A. (2004) *Nature*, **428**, 299–303.

50. Sankapal, B.R., Mane, R.S., and Lokhande, C.D. (2000) *Mater. Res. Bull.*, **35**, 2027–2035.
51. Hodes, G. (2007) *Phys. Chem. Chem. Phys.*, **9**, 2181–2196.
52. McCandless, B.E., Mondal, A., and Birkmire, R.W. (1995) *Sol. Energy Mater. Sol. Cells*, **36**, 369–379.
53. Liu, G., Wang, L., and Xue, D. (2010) *Mater. Lett.*, **64**, 2475–2478.
54. Tiitta, B.M. and Niinisto, L. (1997) *Chem. Vap. Deposition*, **3**, 167–182.
55. Niederberger, M. (2007) *Acc. Chem. Res.*, **40**, 793–800.
56. Jansen, M. and Guenther, E. (1995) *Chem. Mater.*, **7**, 2110–2114.
57. Banger, K.K., Yamashita, Y., Mori, K., Peterson, R.L., Leedham, T., Rickard, J., and Sirringhaus, H. (2011) *Nat. Mater.*, **10**, 45–50.
58. Jun, T., Song, K., Jeong, Y., Woo, K., Kim, D., Bae, C., and Moon, J. (2011) *J. Mater. Chem.*, **21**, 1102.
59. Sun, B.Y. and Rogers, J.A. (2007) *Adv. Mater.*, **19**, 1897–1916.
60. Menard, E., Lee, K.J., Khang, D., Nuzzo, R.G., and Rogers, J.A. (2004) *Appl. Phys. Lett.*, **84**, 26–29.
61. Shimoda, T., Matsuki, Y., Furusawa, M., Aoki, T., Yudasaka, I., Tanaka, H., Iwasawa, H., Wang, D., Miyasaka, M., and Takeuchi, Y. (2006) *Nature*, **440**, 783–786.
62. Bet, S. and Kar, A. (2006) *Mater. Sci. Eng., B*, **130**, 228–236.
63. Drahi, E., Blayac, S., Saunier, S., Valdivieso, F., Bartholin, M., and Grosseau, P. (2011) *Energy Procedia*, **10**, 144–148.
64. Antoniadis, H. (2009) Silicon Ink High Efficiency Solar Cells. 34th IEEE Photovoltaic Specialist Conference, pp. 650–654.
65. Stupca, M., Alsalhi, M., Al Saud, T., Almuhanna, A., and Nayfeh, M.H. (2007) *Appl. Phys. Lett.*, **91**, 063107.
66. Kim, D.-H., Ahn, J.-H., Choi, W.M., Kim, H.-S., Kim, T.-H., Song, J., Huang, Y.Y., Liu, Z., Lu, C., and Rogers, J.A. (2008) *Science*, **320**, 507–511.
67. Siegel, B.A.C., Phillips, S.T., Dickey, M.D., Lu, N., Suo, Z., and Whitesides, G.M. (2010) *Adv. Funct. Mater.*, **20**, 28–35.
68. Perelaer, J., Smith, P.J., Mager, D., Soltman, D., Volkman, S.K., Subramanian, V., Korvink, G., and Schubert, U.S. (2010) *J. Mater. Chem.*, **20**, 8446–8453.
69. Tummala, R.R. (2001) *Fundamentals of Microsystems Packaging*, McGraw-Hill, NewYork.
70. Gracias, D.H., Tien, J., Breen, T.L., Hsu, C., and Whitesides, G.M. (2000) *Science*, **289**, 1170–1172.
71. Dendukuri, D., Hatton, T.A., and Doyle, P.S. (2006) *Langmuir*, **23**, 4669–4674.
72. Boote, J.J., Critchley, K., and Evans, S.D. (2006) *J. Exp. Nanosci.*, **1**, 125–142.
73. Tanase, M., Silevitch, D.M., Hultgren, A., Bauer, L.A., Searson, P.C., Meyer, G.J., and Reich, D.H. (2002) *J. Appl. Phys.*, **91**, 8549.
74. Snyder, E.J., Chideme, J., and Craig, G.S.W. (2002) *Jpn. J. Appl. Phys.*, **41**, 4366.
75. Salalha, W. and Zussman, E. (2005) *Phys. Fluids*, **17**, 063301.
76. Stauth, S.A. and Parviz, B.A. (2006) *Proc. Natl. Acad. Sci. U.S.A.*, **103**, 13922–13927.
77. Yerushalmi, R., Jacobson, Z.A., Ho, J.C., Fan, Z., and Javey, A. (2007) *Appl. Phys. Lett.*, **91**, 203104.
78. Fan, Z., Ho, J.C., Jacobson, Z.A., Razavi, H., and Javey, A. (2008) *Proc. Natl. Acad. Sci. U.S.A.*, **105**, 11066–11070.
79. Saddow, S. (2012) *Silicon Carbide Biotechnology: A Biocompatible Semiconductor for Advanced Biomedical Devices and Applications*, Elsevier.
80. Fromherz, P. and Stett, A. (1995) *Phys. Rev. Lett.*, **75**, 1670–1674.
81. Eversmann, B., Jenkner, M., Hofmann, F., Paulus, C., Brederlow, R., Holzapfl, B., Fromherz, P., Merz, M., Brenner, M., Schreiter, M., Gabl, R., Plehnert, K., Steinhauser, M., Eckstein, G., Schmitt-Landsiedel, D., and Thewes, R. (2003) *IEEE J. Solid-State Circuits*, **38**, 2306–2317.
82. Weis, R., Müller, B., and Fromherz, P. (1996) *Phys. Rev. Lett.*, **76**, 327–330.
83. Fromherz, P. (2002) *ChemPhysChem*, **3**, 276–284.
84. Eversmann, B., Jenkner, M., Hofmann, F., Paulus, C., Brederlow, R., Holzapfl,

B., Fromherz, P., Merz, M., Brenner, M., Schreiter, M., Gabl, R., Plehnert, K., Steinhauser, M., Eckstein, G., Schmitt-Landsiedel, D., and Thewes, R. (2003) *IEEE J. Solid-State Circuits*, **38**, 2306–2317.

85. Weis, R. and Fromherz, P. (1997) *Phys. Rev. E*, **55**, 877.
86. Gleixner, R. and Fromherz, P. (2006) *Biophys. J.*, **90**, 2600–2611.
87. Zeck, G. and Fromherz, P. (2001) *Proc. Natl. Acad. Sci. U.S.A.*, **98**, 10457–10462.
88. Schoen, I. and Fromherz, P. (2007) *Biophys. J.*, **92**, 1096–1111.
89. Weis, R., Müller, B., and Fromherz, P. (1996) *Phys. Rev. Lett.*, **76**, 327–330.
90. Fromherz, P. (1999) *Eur. Biophys. J.*, **28**, 254–258.
91. Weis, R. and Fromherz, P. (1997) *Phys. Rev. E.*, **55**, 877.
92. Hutzler, M. and Fromherz, P. (2004) *Eur. J. Neurosci.*, **19**, 2231–2238. doi: 10.1111/j.1460-9568.2004.03311.x
93. Hofmann, F., Eversmann, B., Jenkner, M., Frey, A., Menz, M., Birkenmaier, T., Fromherz, P., Schreiter, M., Gab, R., Plehnert, K., Steinhauser, M., Eckstein, G., and Thews, R. ESSDERC 2003, Proceeding of the 33rd European Solid State Device Research Conference, pp. 167–170.
94. Eickenscheidt, M., Jenkner, M., Thewes, R., Fromherz, P., and Zeck, G. (2012) *J. Neurophysiol.*, **107**, 2742–2755.
95. Peitz, I., Voelker, M., and Fromherz, P. (2007) *Angew. Chem. Int. Ed.*, **46**, 5787–5790.
96. Hofmann, F., Eversmann, B., Jenkner, M., Frey, A., Merz, M., Birkenmaier, T., Fromherz, P., Schreiter, M., Gabl, R., Plehnert, K., Steinhauser, M., Eckstein, G., and Thewes, R. (2003) *Electrical Performance of Electrical Packaging (IEEE Cat. No. 03TH8710)*, IEEE, pp. 167–170.
97. Bonifazi, P. and Fromherz, P. (2002) *Adv. Mater.*, **14**, 1190–1193.
98. Kaul, R.A., Syed, N.I., and Fromherz, P. (2004) *Phys. Rev. Lett.*, **92**, 1–4.
99. Fromherz, P. (2008) *Solid-State Electron.*, **52**, 1364–1373.
100. Cohen-Karni, T., Timko, B.P., Weiss, L.E., and Lieber, C.M. (2009), *Proceedings of the National Academy of Science of the United States of America*, **106** (18), 7309–7313.
101. Cellot, G., Cilia, E., Cipollone, S., Rancic, V., Sucapane, A., Giordani, S., Gambazzi, L., Markram, H., Grandolfo, M., Scaini, D., Gelain, F., Casalis, L., Prato, M., Giugliano, M., and Ballerini, L. (2009) *Nat. Nanotechnol.*, **4**, 126–133.
102. Banks, D.J., Balachandran, W., Richards, P.R., and Ewins, D. (2002) *Physiol. Meas.*, **23**, 437.
103. Sniadecki, N., Desai, R.A., Ruiz, S.A., and Chen, C.S. (2006) *Ann. Biomed. Eng.*, **34**, 59–74.
104. Timko, B.P., Cohen-karni, T., Yu, G., Qing, Q., Tian, B., and Lieber, C.M. (2009) *Nano Lett.*, **9** (2), 914–918.
105. Viventi, J., Kim, D., Moss, J.D., Kim, Y., Blanco, J.A., Annetta, N., Hicks, A., Xiao, J., Huang, Y., Callans, D.J., Rogers, J.A., and Litt, B. (2010) *Sci. Transl. Med.*, **22**, 24ra22. doi: 10.1126/scitranslmed.3000738
106. Viventi, J., Kim, D., Vigeland, L., Frechette, E.S., Blanco, J.A., Kim, Y., Avrin, A.E., Tiruvadi, V.R., Hwang, S., Vanleer, A.C., Wulsin, D.F., Davis, K., Gelber, C.E., Palmer, L., Van Der Spiegel, J., Wu, J., Xiao, J., Huang, Y., Contreras, D., Rogers, J.A., and Litt, B. (2011) *Nat. Neurosci.*, **14**, 1599–1605.
107. Lumelsky, V.J., Shur, M.S., and Wagner, S. (2001) *IEEE Trans. Sens.*, **1**, 41–51.
108. Tian, B., Liu, J., Dvir, T., Jin, L., Tsui, J.H., Qing, Q., Suo, Z., Langer, R., Kohane, D.S., and Lieber, C.M. (2012) *Nat. Mater.*, **11**, 986–994.
109. Cannata, J.M., Williams, J.A., Zhou, Q., Ritter, T.A., and Shung, K.K. (2006) *IEEE Trans. Ultrason. Ferroelectr. Freq. Control*, **53**, 224–236.
110. Coulombe, J., Sawan, M., and Gervais, J. (2008) *IEEE Trans. Biomed. Circuits Syst.*, **1**, 258–269.
111. Panzer, M.J. and Frisbie, C.D. (2007) *J. Am. Chem. Soc.*, **129**, 6599–6607.
112. Herlogsson, L., Cölle, M., Tierney, S., Crispin, X., and Berggren, M. (2010)

Adv. Mater. (Deerfield Beach), **22**, 72–76.
113. Cho, J.H., Lee, J., He, Y., Kim, B.S., Lodge, T.P., and Frisbie, C.D. (2008) *Adv. Mater.*, **20**, 686–690.
114. Mitra, S., Shukla, A., and Sampath, S. (2001) *J. Power Sources*, **101**, 213–218.
115. Cramer, T., Kyndiah, A., Murgia, M., Leonardi, F., Casalini, S., and Biscarini, F. (2012) *Appl. Phys. Lett.*, **100**, 143302.
116. Panzer, M.J. and Frisbie, C.D. (2006) *Adv. Funct. Mater.*, **16**, 1051–1056.
117. Cho, J.H., Lee, J., Xia, Y., Kim, B., He, Y., Renn, M.J., Lodge, T.P., and Frisbie, C.D. (2008) *Nat. Mater.*, **7**, 900–906.
118. Chen, F., Qing, Q., Xia, J.L., and Tao, N.J. (2010) *Chem. Asian J.*, **5**, 2144.
119. Stine, R., Mulvaney, S.P., Robinson, J.T., Tamanaha, C.R., and Sheehan, P.E. (2013) *Anal. Chem.*, **85**, 509–521.
120. Liao, C. and Yan, F. (2013) *Polym. Rev.*, **53**, 352–406.
121. Kergoat, L., Herlogsson, L., Braga, D., Piro, B., Pham, M.-C., Crispin, X., Berggren, M., and Horowitz, G. (2010) *Adv. Mater. (Deerfield Beach)*, **22**, 2565–2569.
122. Casalini, S., Leonardi, F., Cramer, T., and Biscarini, F. (2013) *Org. Electron.*, **14**, 156–163.
123. Panzer, M.J. and Frisbie, C.D. (2005) *J. Am. Chem. Soc.*, **127**, 6960–6961.
124. Said, E., Crispin, X., Herlogsson, L., Elhag, S., Robinson, N.D., and Berggren, M. (2006) *Appl. Phys. Lett.*, **89**, 143507.
125. Dumitru, L.M., Manoli, K., Magliulo, M., Sabbatini, L., Palazzo, G., and Torsi, L. (2013) *ACS Appl. Mater. Interfaces*, **5**, 10819–10823.
126. Ono, S., Miwa, K., Seki, S., and Takeya, J. (2009) *Appl. Phys. Lett.*, **94**, 063301.
127. Lee, J., Panzer, M.J., He, Y., Lodge, T.P., and Frisbie, C.D. (2007) *J. Am. Chem. Soc.*, **129**, 4532–4533.
128. Sokolov, A.N., Roberts, M.E., and Bao, Z. (2009) *Mater. Today*, **12**, 12–20.
129. Ohta, T., Bostwick, A., Seyller, T., Horn, K., and Rotenberg, E. (2008) *Science*, **313**, 951.
130. Li, X., Cai, W., An, J., Kim, S., Nah, J., Yang, D., Piner, R., Velamakanni, A., Jung, I., Tutuc, E., Banerjee, S.K., Colombo, L., and Ruoff, R.S. (2012) *Science*, **324**, 1312.
131. Novoselov, K.S., Fal'ko, V.I., Colombo, L., Gellert, P.R., Schwab, M.G., and Kim, K. (2012) *Nature*, **490**, 192–200.
132. Wan, X., Huang, Y., and Chen, Y. (2012) *Acc. Chem. Res.*, **45**, 598–607.
133. Jo, S.B., Park, J., Lee, W.H., Cho, K., and Hong, B.H. (2012) *Solid State Commun.*, **152**, 1350–1358.
134. Pang, S., Yang, S., Feng, X., and Müllen, K. (2012) *Adv. Mater. (Deerfield Beach)*, **24**, 1566–1570.
135. Khan, H.U., Roberts, M.E., Johnson, O., Förch, R., Knoll, W., and Bao, Z. (2010) *Adv. Mater. (Deerfield Beach)*, **22**, 4452–4456.
136. Kergoat, L., Piro, B., Berggren, M., Pham, M.-C., Yassar, A., and Horowitz, G. (2012) *Org. Electron.*, **13**, 1–6.
137. Buth, F., Donner, A., Sachsenhauser, M., Stutzmann, M., and Garrido, J.A. (2012) *Adv. Mater. (Deerfield Beach)*, **24**, 4511–4517.
138. Khan, H.U., Jang, J., Kim, J.-J., and Knoll, W. (2011) *J. Am. Chem. Soc.*, **133**, 2170–2176.
139. Magliulo, M., Mallardi, A., Mulla, M.Y., Cotrone, S., Pistillo, B.R., Favia, P., Vikholm-Lundin, I., Palazzo, G., and Torsi, L. (2013) *Adv. Mater. (Deerfield Beach)*, **25**, 2090–2094.
140. Hammock, M.L., Sokolov, A.N., Stoltenberg, R.M., Naab, B.D., and Bao, Z. (2012) *ACS Nano*, **6**, 3100–3108.
141. Hammock, M.L., Knopfmacher, O., Naab, B.D., Tok, J.B.-H., and Bao, Z. (2013) *ACS Nano*, **7**, 3970–3980.
142. Torsi, L., Magliulo, M., Manoli, K., and Palazzo, G. (2013) *Chem. Soc. Rev.*, **42**, 8612–8628.
143. Nielsen, P., Egholm, M., Berg, R., and Buchardt, O. (1991) *Science*, **254**, 1497–1500.
144. Cotrone, S., Ambrico, M., Toss, H., Angione, M.D., Magliulo, M., Mallardi, A., Berggren, M., Palazzo, G., Horowitz, G., Ligonzo, T., and Torsi, L. (2012) *Org. Electron.*, **13**, 638–644.
145. Kostarelos, K. and Novoselov, K.S. (2014) *Science (New York)*, **344**, 261–263.
146. Bystrenova, E., Jelitai, M., Tonazzini, I., Lazar, A.N., Huth, M., Stoliar, P.,

Dionigi, C., Cacace, M.G., Nickel, B., Madarasz, E., and Biscarini, F. (2008) *Adv. Funct. Mater.*, **18**, 1751–1756.
147. Cramer, T., Chelli, B., Murgia, M., Barbalinardo, M., Bystrenova, E., de Leeuw, D.M., and Biscarini, F. (2013) *Phys. Chem. Chem. Phys.*, **15**, 3897–3905.
148. Ghezzi, D., Antognazza, M.R., Dal Maschio, M., Lanzarini, E., Benfenati, F., and Lanzani, G. (2011) *Nat. Commun.*, **2**, 166.
149. Scarpa, G., Idzko, A., Yadav, A., Martin, E., and Thalhammer, S. (2010) *IEEE Trans. Nanotechnol.*, **9**, 527–532.
150. Benfenati, V., Toffanin, S., Bonetti, S., Turatti, G., Pistone, A., Chiappalone, M., Sagnella, A., Stefani, A., Generali, G., Ruani, G., Saguatti, D., Zamboni, R., and Muccini, M. (2013) *Nat. Mater.*, **12**, 672–680.
151. Tarabella, G., Mahvash Mohammadi, F., Coppedè, N., Barbero, F., Iannotta, S., Santato, C., and Cicoira, F. (2013) *Chem. Sci.*, **4**, 1395.
152. DeFranco, J.A., Schmidt, B.S., Lipson, M., and Malliaras, G.G. (2006) *Org. Electron.*, **7**, 22–28.
153. Bernards, D.A. and Malliaras, G.G. (2007) *Adv. Funct. Mater.*, **17**, 3538–3544.
154. Yaghmazadeh, O., Cicoira, F., Bernards, D., Yang, S.Y., Bonnassieux, Y., and Malliaras, G.G. (2011) *J. Polym. Sci., Part B: Polym. Phys.*, **49**, 34–39.
155. Khodagholy, D., Rivnay, J., Sessolo, M., Gurfinkel, M., Leleux, P., Jimison, L.H., Stavrinidou, E., Herve, T., Sanaur, S., Owens, R.M., and Malliaras, G.G. (2013) *Nat. Commun.*, **4**, 2133.
156. Leger, J.M. (2008) *Adv. Mater.*, **20**, 837–841.
157. Svennersten, K., Larsson, K.C., Berggren, M., and Richter-Dahlfors, A. (1810) *Biochim. Biophys. Acta*, **2011**, 276–285.
158. Tarabella, G., Nanda, G., Villani, M., Coppedè, N., Mosca, R., Malliaras, G.G., Santato, C., Iannotta, S., and Cicoira, F. (2012) *Chem. Sci.*, **3**, 3432.
159. Torchilin, V.P. (2005) *Nat. Rev. Drug Discovery*, **4**, 145–160.
160. Tarabella, G., Balducci, A.G., Coppedè, N., Marasso, S., D'Angelo, P., Barbieri, S., Cocuzza, M., Colombo, P., Sonvico, F., Mosca, R., and Iannotta, S. (2013) *Biochim. Biophys. Acta, Gen. Subj.*, **1830**, 4374–4380.
161. Tarabella, G., Balducci, A.G., Coppedè, N., Marasso, S., D'Angelo, P., Barbieri, S., Cocuzza, M., Colombo, P., Sonvico, F., Mosca, R., and Iannotta, S. (2013) *Biochim. Biophys. Acta*, **1830**, 4374–4380.
162. Tang, H., Lin, P., Chan, H.L.W., and Yan, F. (2011) *Biosens. Bioelectron.*, **26**, 4559–4563.
163. Tarabella, G., Pezzella, A., Romeo, A., D'Angelo, P., Coppede, N., Calicchio, M., d'Ischia, M., Mosca, R., and Iannotta, S. (2013) *J. Mater. Chem. B*, **1**, 3843–3849.
164. Land, E.J., Ramsden, C.A., and Riley, P.A. (2003) *Acc. Chem. Res.*, **36**, 300–308.
165. d'Ischia, M., Napolitano, A., Pezzella, A., Meredith, P., and Sarna, T. (2009) *Angew. Chem. Int. Ed.*, **48**, 3914–3921.
166. Svennersten, K., Bolin, M.H., Jager, E.W.H., Berggren, M., and Richter-Dahlfors, A. (2009) *Biomaterials*, **30**, 6257–6264.
167. Lin, P., Yan, F., Yu, J., Chan, H.L.W., and Yang, M. (2010) *Adv. Mater. (Deerfield Beach)*, **22**, 3655–3660.
168. Jimison, L.H., Tria, S.A., Khodagholy, D., Gurfinkel, M., Lanzarini, E., Hama, A., Malliaras, G.G., and Owens, R.M. (2012) *Adv. Mater. (Deerfield Beach)*, **24**, 5919–5923.
169. Tria, S.A., Ramuz, M., Huerta, M., Leleux, P., Rivnay, J., Jimison, L.H., Hama, A., Malliaras, G.G., and Owens, R.M. (2014) *Adv. Healthcare Mater.*, **3** (7), 1053–1060.
170. Yao, C., Xie, C., Lin, P., Yan, F., Huang, P., and Hsing, I. (2013) *Adv. Mater.*, **25** (45), 6575–6580.
171. Yao, C., Xie, C., Lin, P., Yan, F., Huang, P., and Hsing, I.-M. (2013) *Adv. Mater. (Deerfield Beach)*, **25**, 6575–6580.
172. He, R.-X., Zhang, M., Tan, F., Leung, P.H.M., Zhao, X.-Z., Chan, H.L.W., Yang, M., and Yan, F. (2012) *J. Mater. Chem.*, **22**, 22072.
173. Krishnamoorthy, K., Gokhale, R.S., Contractor, A.Q., and Kumar, A. (2004) *Chem. Commun.*, 820–821.

174. Lin, P., Luo, X., Hsing, I.-M., and Yan, F. (2011) *Adv. Mater. (Deerfield Beach)*, **23**, 4035–4040.
175. Mattana, G., Cosseddu, P., Fraboni, B., Malliaras, G.G., Hinestroza, J.P., and Bonfiglio, A. (2011) *Org. Electron.*, **12**, 2033–2039.
176. Tarabella, G., Villani, M., Calestani, D., Mosca, R., Iannotta, S., Zappettini, A., and Coppede, N. (2012) *J. Mater. Chem.*, **22**, 23830–23834.
177. Richter-Dahlfors, A., Simon, D.T., Kurup, S., Larsson, K.C., Hori, R., Tybrandt, K., Goiny, M., Jager, E.H., Berggren, M., and Canlon, B. (2009) *Nat. Mater.*, **8**, 742–746.
178. Isaksson, J., Kjäll, P., Nilsson, D., Robinson, N.D., Berggren, M., and Richter-Dahlfors, A. (2007) *Nat. Mater.*, **6**, 673–679.
179. Richter-Dahlfors, A., Tybrandt, K., Larsson, K.C., Kurup, S., Simon, D.T., Kjall, P., Isaksson, J., Sandberg, M., Jager, E.W.H., and Berggren, M. (2009) *Adv. Mater.*, **21**, 4442–4446. doi: 10.1002/Adma.200900187
180. Tybrandt, K., Forchheimer, R., and Berggren, M. (2012) *Nat. Commun.*, **3**, 871.
181. Leger, J., Berggren, M., and Carter, S. (2011) in *Iontronics. Ionic Carriers in Organic Electronic Materials and Devices* (eds J. Leger, M. Berggren, and S. Carter), Taylor & Francis Group.
182. Khodagholy, D., Doublet, T., Quilichini, P., Gurfinkel, M., Leleux, P., Ghestem, A., Ismailova, E., Hervé, T., Sanaur, S., Bernard, C., and Malliaras, G.G. (2013) *Nat. Commun.*, **4**, 1575.

Index

Large Area and Flexible Electronics, First Edition. Edited by Mario Caironi and Yong-Young Noh.

p

q

r

t